Geometry

Ray C. Jurgensen
Richard G. Brown
John W. Jurgensen

Teacher Consultants
Jean A. Giarrusso
Byron E. Gunsallus, Jr.
James R. Keeney
David Molina
Patricia Onodera Nicholson

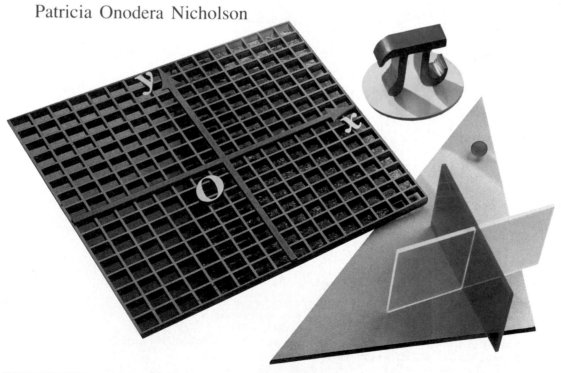

Houghton Mifflin Company • Boston
Atlanta   Dallas   Geneva, Ill.   Palo Alto   Princeton   Toronto

THE AUTHORS

Ray C. Jurgensen, formerly Chairman of the Mathematics Department and holder of the Eppley Chair of Mathematics, Culver Academies, Culver, Indiana.

Richard G. Brown, Mathematics Teacher, Phillips Exeter Academy, Exeter, New Hampshire.

John W. Jurgensen, teaches mathematics at the University of Houston-Downtown and is a mathematician for the National Aeronautics and Space Administration (NASA) at the Johnson Space Center.

TEACHER CONSULTANTS

Jean A. Giarrusso, Mathematics Teacher, Spanish River High School, Boca Raton, Florida

Byron E. Gunsallus, Jr., Mathematics Supervisor, Harrisburg High School, Harrisburg, Pennsylvania

James R. Keeney, Mathematics Teacher, Hillcrest High School, Country Club Hills, Illinois

David Molina, Doctoral Candidate in Mathematics Education, The University of Texas at Austin, Austin, Texas

Patricia Onodera Nicholson, Mathematics Teacher, Glen A. Wilson High School, Hacienda Heights, California

Printed in U.S.A.

ISBN: 0-395-46146-4

HIJ-RM-9876543

Contents

1 POINTS, LINES, PLANES, AND ANGLES

Some Basic Figures

Definitions and Postulates

2 DEDUCTIVE REASONING

Using Deductive Reasoning

Theorems about Angles and Perpendicular Lines

3 PARALLEL LINES AND PLANES

4 CONGRUENT TRIANGLES

5 QUADRILATERALS

Parallelograms

Special Quadrilaterals

Technology   Explorations 176, 189, 195   Computer Key-In 183

Special Topics   Biographical Note/*Benjamin Banneker* 171
Challenge 194   Application/*Rhombuses* 196

Reviews and Tests   Mixed Review Exercises 189   Self-Tests 182, 195
Chapter Summary 197   Chapter Review 197   Chapter Test 199
Cumulative Review 200

6 INEQUALITIES IN GEOMETRY

Inequalities and Indirect Proof

Inequalities in Triangles

Technology   Explorations 225   Computer Key-In 226

Special Topics   Challenges 207, 217
Career/*Cartographer* 213   Application/*Finding the Shortest Path* 224
Non-Euclidean Geometries 233

Reviews and Tests   Mixed Review Exercises 212   Self-Tests 218, 233
Chapter Summary 235   Chapter Review 235   Chapter Test 236
Algebra Review/*Fractions* 237   Preparing for College Entrance Exams 238
Cumulative Review 239

7 SIMILAR POLYGONS

Ratio, Proportion, and Similarity

Working with Similar Triangles

8 RIGHT TRIANGLES

Right Triangles

Trigonometry

9 CIRCLES

10 CONSTRUCTIONS AND LOCI

11 AREAS OF PLANE FIGURES

Areas of Polygons

Circles, Similar Figures, and Geometric Probability

12 AREAS AND VOLUMES OF SOLIDS

Important Solids

Similar Solids

13 COORDINATE GEOMETRY

14 TRANSFORMATIONS

Using Technology with This Course

There are two types of optional computer material in this book: Explorations and Computer Key-Ins. The Explorations sections are intended for use with computer software that draws and measures geometric figures, such as the *Geometric Supposer*. These sections provide exploratory exercises that lead students to discover geometric properties and develop geometric intuition.

The Computer Key-Ins do not require any supplementary computer software. These features teach some programming in BASIC and usually include a program that students can run to explore a topic covered in the chapter. Some writing of programs may be required in some of these features.

Calculator Key-In features and certain exercise sets also suggest appropriate use of scientific calculators with this course.

Symbols

$\lvert x \rvert$	absolute value of x (p. 12)
adj. ⓢ	adjacent angles (p. 19)
alt. int. ⓢ	alternate interior angles (p. 74)
$\angle$, ⓢ	angle(s) (pp. 17, 19)
a	apothem (p. 441)
$\approx$	is approximately equal to (p. 306)
$\overset{\frown}{BC}$	arc with endpoints B and C (p. 339)
A	area (p. 423)
B	area of base (p. 476)
b	length of base; y-intercept (p. 424; p. 548)
$\odot O$	circle with center O (p. 329)
C	circumference (p. 446)
comp. ⓢ	complementary angles (p. 50)
$S \circ T$	composite of S and T (p. 599)
$\cong$	congruent, is congruent to (p. 13)
$\leftrightarrow$	corresponds to (p. 117)
corr. ⓢ	corresponding angles (p. 74)
cos	cosine (p. 312)
$\circ$	degrees (p. 17)
diag.	diagonal (p. 187)
d	diameter; distance; length of diagonal (p. 446; p. 524; p. 430)
$D_{O,k}$	dilation with center O and scale factor k (p. 592)
e	edge length (p. 478)
$=$	equal(s); equality (pp. 13, 37)
ext. $\angle$	exterior angle (p. 103)
$>$, $\geq$	greater than; greater than or equal to (p. 16)
H_o	half turn about point O (p. 589)
h	height; length of altitude (p. 424; p. 435)
hyp.	hypotenuse (p. 141)
T^{-1}	inverse of transformation T (p. 605)
I	identity transformation (p. 605)
int. $\angle$	interior angle (p. 103)
L.A.	lateral area (p. 476)
JL	length of $\overline{JL}$, distance between points J and L (p. 11)
$<$, $\leq$	less than; less than or equal to (p. 16)
$\overset{\leftrightarrow}{AB}$	line containing points A and B (p. 5)
$S : A \rightarrow A'$	S maps point A to point A'. (p. 571)
$m\angle A$	measure of $\angle A$ (p. 17)
$\not\cong$	not congruent (p. 215)
$\neq$	not equal (p. 37)
$\not>$	not greater than (p. 220)
$\nparallel$	not parallel (p. 216)
opp. ⓢ	opposite angles (p. 187)
(x, y)	ordered pair (p. 113)
$\parallel$	parallel, is parallel to (p. 73)
$\square$	parallelogram (p. 167)
p	perimeter (p. 426)
$\perp$	perpendicular, is perpendicular to (p. 56)
π	pi (p. 446)
n-gon	polygon with n sides (p. 101)
quad.	quadrilateral (p. 168)
r	radius (p. 446)
$\dfrac{a}{b}$, $a:b$	ratio of a to b (pp. 241, 242)
$\overset{\rightarrow}{AB}$	ray with endpoint A, passing through point B (p. 11)
R_j	reflection in line j (p. 577)
rt. $\angle$	right angle (p. 19)
rt. $\triangle$	right triangle (p. 290)
$\mathscr{R}_{O,90}$	rotation about point O through $90°$ (p. 588)
s-s. int. ⓢ	same-side interior angles (p. 74)
$\overline{AB}$	segment with endpoints A and B (p. 11)
s	length of a side of a regular polygon (p. 423)
$\sim$	similar, is similar to (p. 249)
sin	sine (p. 312)
l	slant height (p. 482)
m	slope (p. 529)
$\sqrt{x}$	positive square root of x (p. 280)
supp. ⓢ	supplementary angles (p. 50)
T.A.	total area (p. 476)
tan	tangent (p. 305)
trap.	trapezoid (p. 198)
$\triangle$, ⚇	triangle(s) (pp. 93, 118)
$\overset{\rightarrow}{AB}$	vector from A to B (p. 539)
vert. ⓢ	vertical angles (p. 51)
V	volume (p. 476)

Reading Your Geometry Book

Reading mathematics is different from reading the newspaper or reading a novel because mathematics has its own vocabulary and symbols. You will find many new words in this book. Some of them are unique to mathematics, for example, *hypotenuse*, *isosceles*, and *secant*. Other words are used in everyday speech but have a different meaning in geometry, for example, *plane*, *line*, and *construction*. It is important to understand each new vocabulary word because ideas in geometry are built from the vocabulary and ideas that come before. If you don't remember the meaning of a word, you can look it up in the Glossary or Index.

Symbols

In order to read geometry, you must also know how to read its symbols. If you have trouble reading a statement expressed in symbols, first make sure you understand all the symbols. Use the list of symbols on page xi if you need to refresh your memory. Then, reread the statement. Sometimes you may find it helpful to write the statement out in words or just to say the words aloud.

Reading a Lesson

When you read a geometry lesson, first skim the lesson for main ideas. This book is organized to help you find the main ideas easily. Look for the title of the chapter and lesson. Each section of this book lists the objectives for the next two or three lessons. Look for displayed material in boxes or color, including postulates, theorems, and summaries. Look for any vocabulary words in **boldface** or *italic*. And look for any new symbols. Skimming the lesson should allow you to build a general framework of the information you are about to read.

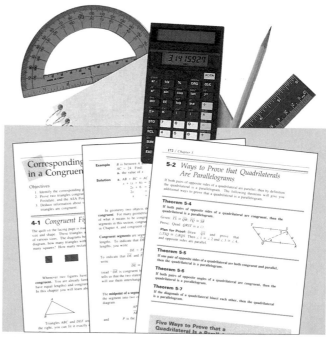

Now read the section. Think about each block of information as you read. Try to place it in the framework you built while skimming. There are many worked-out examples that can help you in doing the exercises. Think about how you would solve each example before reading its solution. Keep paper and pencil handy for doing the examples, taking notes, or writing down questions. If there is anything you don't understand after reading and rereading, make a note to ask your teacher or a classmate later.

After reading the lesson, but before attempting any of the exercises, ask yourself, "What did I just read?" Say the main words and ideas aloud to yourself or write down the ideas in your own words.

Diagrams

It is important to be able to read a diagram and to draw a diagram from given information. Look carefully for all the information given by a diagram, but be sure you don't read more into a diagram than is actually there. For example, don't assume that two segments are the same length just because they *look* that way; if the segments are the same length, they will be marked to show it. Other suggestions about reading and drawing diagrams are given on pages 19, 61, and 140.

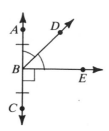

Exercises, Tests, and Reviews

Each lesson is followed by Classroom and Written Exercises. Self-Tests are mid-chapter progress tests. At the end of each chapter are other checks of your understanding and mastery: the Chapter Review and the Chapter Test. There are lesson numbers in the margin of the Chapter Review to indicate which lesson a group of exercises covers. Each chapter also has a Cumulative Review of the material covered through that chapter. And, at the end of the book, there are multiple-choice examinations for each chapter.

Other features you will find helpful are the Mixed Review Exercises and the Algebra Reviews, which review material you will need in the next lesson or chapter. Even-numbered chapters have multiple-choice tests, called Preparing for College Entrance Exams, with questions similar to those in some college entrance tests. There are also Chapter Summaries (at the end of each chapter), the list of symbols (page xi), the Glossary, the Index, and lists of postulates, theorems, and constructions (at the back of the book). Answers for all the Mixed Review Exercises, Self-Tests, and Preparing for College Entrance Exams, and for selected Written Exercises, Chapter Review Exercises, and Cumulative Review Exercises are also at the back of the book.

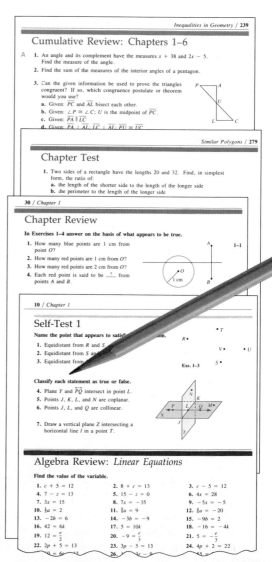

Table of Measures

Time

60 seconds (s) = 1 minute (min)
60 minutes = 1 hour (h)
24 hours = 1 day (d)
7 days = 1 week

$\left.\begin{array}{r}365 \text{ days}\\52 \text{ weeks (approx.)}\\12 \text{ months}\end{array}\right\}$ = 1 year

10 years = 1 decade
100 years = 1 century

Metric Units

Length

10 millimeters (mm) = 1 centimeter (cm)

$\left.\begin{array}{r}100 \text{ centimeters}\\1000 \text{ millimeters}\end{array}\right\}$ = 1 meter (m)

1000 meters = 1 kilometer (km)

Area

100 square millimeters (mm^2) = 1 square centimeter (cm^2)
10,000 square centimeters = 1 square meter (m^2)
10,000 square meters = 1 hectare (ha)

Volume

1000 cubic millimeters (mm^3) = 1 cubic centimeter (cm^3)
1,000,000 cubic centimeters = 1 cubic meter (m^3)

Liquid Capacity

1000 milliliters (mL) = 1 liter (L)
1000 cubic centimeters = 1 liter
1000 liters = 1 kiloliter (kL)

Mass

1000 milligrams (mg) = 1 gram (g)
1000 grams = 1 kilogram (kg)
1000 kilograms = 1 metric ton (t)

Temperature:
Degrees Celsius (°C)

0°C = freezing point of water
37°C = normal body temperature
100°C = boiling point of water

Notice that the same prefixes are used in many metric units of measure.

Examples   milli = thousandth: millimeter, milliliter, milligram
centi = hundredth: centimeter
kilo = thousand: kilometer, kiloliter, kilogram

United States Customary Units

Length

$$12 \text{ inches (in.)} = 1 \text{ foot (ft)}$$

$$\left.\begin{array}{r} 36 \text{ inches} \\ 3 \text{ feet} \end{array}\right\} = 1 \text{ yard (yd)}$$

$$\left.\begin{array}{r} 5280 \text{ feet} \\ 1760 \text{ yards} \end{array}\right\} = 1 \text{ mile (mi)}$$

Area

$$144 \text{ square inches (in.}^2) = 1 \text{ square foot (ft}^2)$$

$$9 \text{ square feet} = 1 \text{ square yard (yd}^2)$$

$$\left.\begin{array}{r} 43{,}560 \text{ square feet} \\ 4840 \text{ square yards} \end{array}\right\} = 1 \text{ acre (A)}$$

Volume

$$1728 \text{ cubic inches (in.}^3) = 1 \text{ cubic foot (ft}^3)$$

$$27 \text{ cubic feet} = 1 \text{ cubic yard (yd}^3)$$

Liquid Capacity

$$8 \text{ fluid ounces (fl oz)} = 1 \text{ cup (c)}$$

$$2 \text{ cups} = 1 \text{ pint (pt)}$$

$$2 \text{ pints} = 1 \text{ quart (qt)}$$

$$4 \text{ quarts} = 1 \text{ gallon (gal)}$$

Weight

$$16 \text{ ounces (oz)} = 1 \text{ pound (lb)}$$

$$2000 \text{ pounds} = 1 \text{ ton (t)}$$

Temperature:
Degrees Fahrenheit (°F)

$$32°F = \text{freezing point of water}$$

$$98.6°F = \text{normal body temperature}$$

$$212°F = \text{boiling point of water}$$

Compound units of metric units or U.S. customary units may be formed by multiplication or division.

Examples

square centimeters	cm^2
cubic yards	yd^3
kilometers per hour	km/h
feet per minute	ft/min

1 POINTS, LINES, PLANES, AND ANGLES

As ancient people studied the heavens, they saw and named many patterns of points, lines, and angles formed by the stars. Although modern astronomers use sophisticated observatories and equipment, they still base their calculations on geometric principles that have been known for many centuries.

Some Basic Figures

Objectives

1. Use the term *equidistant*.
2. Use the undefined terms *point*, *line*, and *plane*.
3. Draw representations of points, lines, and planes.
4. Use the terms *collinear*, *coplanar*, and *intersection*.

1-1 *A Game and Some Geometry*

Suppose that you and Pat are partners in a game in which you must locate various clues to win. You are told to pick up your next clue at a point that

1. is as far from the fountain as from the oak tree

   and

2. is 10 m (meters) from the flag pole.

You locate X, which satisfies both requirements, but grumble because there simply isn't any clue to be found at X.

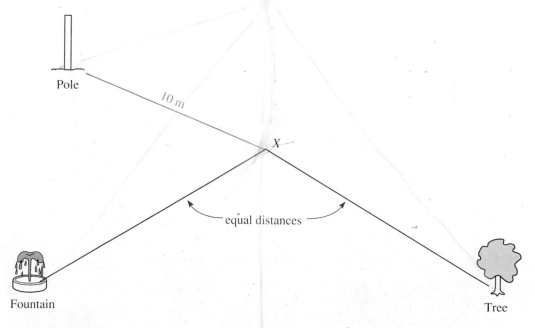

Then Pat realizes that there may be a different location that satisfies both requirements. (Before reading on, see if you can find another point that meets requirements 1 and 2.)

Suppose that you concentrate on points that satisfy requirement 1 while Pat works on points that meet requirement 2. In the diagram below, 1 cm represents 2 m, so the blue arc shows points that are 10 m from the pole. Each red point is equally distant, or *equidistant*, from *F* and *T*. Point *Y*, as well as *X*, meets both requirements. You and Pat find your clue at *Y* and proceed with the game.

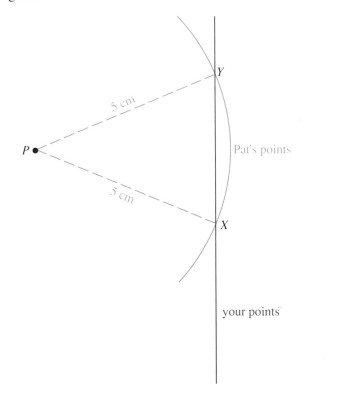

The game discussed above involves *points* and *distances*. When you approach the game systematically, you use *lines* and *circles*. Understanding the properties of geometric figures like these is an important part of geometry. The rest of this chapter will deal with the most basic figures of geometry.

Classroom Exercises

For Exercises 1–8 refer to the diagram above.

1. Suppose that the diagram showed, in blue, *all* the points that are 5 cm from *P*. What geometric figure would the points form?

2. In a more complete diagram, would there be a red point 15 cm from both *F* and *T*? How many such points?

3. It appears as if points *P*, *X*, and *T* might lie on a straight line. Use a ruler or the edge of a sheet of paper to see if they do.

4. It looks as if P might be equidistant from F and X. Is it?

5. Suppose Pat spoke of a line l joining F and T while you thought of a line n joining F and T. Is it better to say that l and n are two different lines, or to say that we have one line with two different names?

6. Point X is equidistant from F and T. Furthermore, point Y is equidistant from F and T. Does that mean that X and Y are equally distant from F?

7. Suppose you were asked to find a point 5 cm from P, 5 cm from F, and 5 cm from T. Is there such a point?

8. Do you believe there is any point that is equidistant from P, F, and T?

Written Exercises

A   **1.** Copy and complete the table. Refer to the diagrams on pages 1 and 2.

Distance between	Diagram distance	Ground distance
X and P	<u>5</u> cm	<u>10</u> m
X and F	<u>7</u> cm	<u>?</u> m
X and T	<u>?</u> cm	<u>?</u> m
Y and F	<u>?</u> cm	<u>19</u> m
F and T	<u>12</u> cm	<u>?</u> m

For Exercises 2–4 use a centimeter ruler. If you don't have a centimeter ruler, you may use the centimeter ruler shown below as a guide. Either open your compass to the appropriate distance or mark the appropriate distance on the edge of a sheet of paper.

Centimeter Ruler

2. Copy the points F, T, and P from the diagram on page 2. If you lay your paper over the page, you can see through the paper well enough to get the points.
   a. Draw a line to indicate all points equidistant from F and T.
   b. Draw a circle to indicate points 6 cm from P. If you don't have a compass, draw as well as you can freehand.
   c. How many points are equidistant from F and T, and are also 6 cm from P?

3. Repeat Exercise 2, but use 2 cm instead of 6 cm.

4. There is a distance you could use in parts (b) and (c) of Exercise 2 that would lead to the answer *one point* in part (c). Estimate that distance.

The spoon in the photograph appears to be broken because light rays bend as they go from air to water. As in the photograph, your eyes may mislead you in some of the exercises that follow, but you are asked to make estimates. You may want to check your estimates by measuring.

5. Which is greater, the distance from R to S or the distance from T to U?

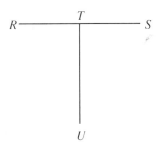

6. Which is greater, the distance from A to B or the distance from A to C?

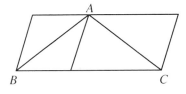

B   **7.** How does the area of the outer square compare with the area of the inner square?

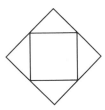

8. Compare the areas of the red and blue regions. (Area of circle $= \pi r^2$.)

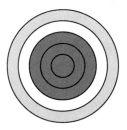

9. In the diagram a, b, c, and d are lengths. Which is greater, the product ab or the product cd?

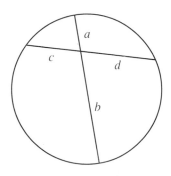

10. A path between opposite vertices of the square is made up of hundreds of horizontal and vertical segments. (The diagram shows a simplified version.) What is the best approximation to the length of the path—24, 34, 44, or more than 44?

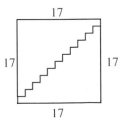

1-2 *Points, Lines, and Planes*

When you look at a color television picture, how many different colors do you see? Actually, the picture is made up of just three colors—red, green, and blue. Most color television screens are covered with more than 300,000 colored dots, as shown in the enlarged diagram below. Each dot glows when it is struck by an electron beam. Since the dots are so small, and so close together, your eye sees a whole image rather than individual dots.

Each dot on a television screen suggests the simplest figure studied in geometry—a *point*. Although a point doesn't have any size, it is often represented by a dot that does have some size. You usually name points by capital letters. Points *A* and *B* are pictured at the right.

All geometric figures consist of points. One familiar geometric figure is a *line*, which extends in two directions without ending. Although a picture of a line has some thickness, the line itself has no thickness.

Often a line is referred to by a single lower-case letter, such as *line l*. If you know that a line contains the points *A* and *B*, you can also call it *line AB* (denoted $\overleftrightarrow{AB}$) or *line BA* ($\overleftrightarrow{BA}$).

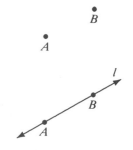

A geometric *plane* is suggested by a floor, wall, or table top. Unlike a table top, a plane extends without ending and has no thickness. Although a plane has no edges, we usually picture a plane by drawing a four-sided figure as shown below. We often label a plane with a capital letter.

Plane *M*   Plane *N*

In geometry, the terms *point*, *line*, and *plane* are accepted as intuitive ideas and are not defined. These *undefined terms* are then used in the definitions of other terms, such as those at the top of the next page.

Space is the set of all points. **Collinear points** are points all in one line.

Collinear points   Noncollinear points

Coplanar points are points all in one plane.

Coplanar points

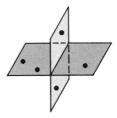

Noncoplanar points

    Some expressions commonly used to describe relationships between points, lines, and planes follow. In these expressions, *intersects* means "meets" or "cuts." The **intersection** of two figures is the set of points that are in both figures. Dashes in the diagrams indicate parts hidden from view in figures in space.

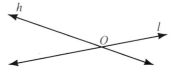

A is in *l*, or *A* is on *l*.
l contains *A*.
l passes through *A*.

l and *h* intersect in *O*.
l and *h* intersect at *O*.
O is the intersection of *l* and *h*.

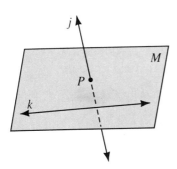

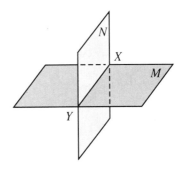

k and *P* are in *M*.
M contains *k* and *P*.
j intersects *M* at *P*.
P is the intersection of *j* and *M*.

M and *N* intersect in $\overleftrightarrow{XY}$.
$\overleftrightarrow{XY}$ is the intersection of *M* and *N*.
$\overleftrightarrow{XY}$ is in *M* and *N*.
M and *N* contain $\overleftrightarrow{XY}$.

    In this book, whenever we refer, for example, to "two points" or "three lines," we will mean *different* points or lines (or other geometric figures).

Classroom Exercises

Classify each statement as true or false.

1. $\overleftrightarrow{XY}$ intersects plane M at point O.
2. Plane M intersects $\overleftrightarrow{XY}$ in more than one point.
3. T, O, and R are collinear.
4. X, O, and Y are collinear.
5. R, O, S, and W are coplanar.
6. R, S, T, and X are coplanar.
7. R, X, O, and Y are coplanar.

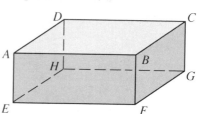

Exs. 1–7

8. Does a plane have edges?
9. Can a given point be in two lines? in ten lines?
10. Can a given line be in two planes? in ten planes?

Name a fourth point that is in the same plane as the given points.

11. A, B, C 12. E, F, H 13. D, C, H
14. A, D, E 15. B, E, F 16. B, G, C

The plane that contains the top of the box can be called plane ABCD.

17. Are there any points in $\overleftrightarrow{CG}$ besides C and G?
18. Are there more than four points in plane $ABCD$?
19. Name the intersection of planes $ABFE$ and $BCGF$.
20. Name two planes that do not intersect.

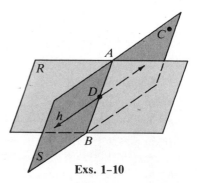

Exs. 11–20

Written Exercises

Classify each statement as true or false.

A
1. $\overleftrightarrow{AB}$ is in plane R.   2. S contains $\overleftrightarrow{AB}$.
3. R and S contain D.   4. D is on line h.
5. h is in S.   6. h is in R.
7. Plane R intersects plane S in $\overleftrightarrow{AB}$.
8. Point C is in R and S.
9. A, B, and C are collinear.
10. A, B, C, and D are coplanar.

Exs. 1–10

11. Make a sketch showing four coplanar points such that three, but not four, of them are collinear.

12. Make a sketch showing four points that are not coplanar.

A plane can be named by three or more noncollinear points it contains. In Chapter 12 you will study *pyramids* like the one shown at the right below.

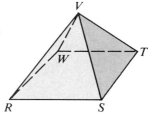

13. Name five planes that contain sides of the pyramid shown.

14. Of the five planes containing sides of the pyramid, are there any that do not intersect?

15. Name three lines that intersect at point *R*.

16. Name two planes that intersect in $\overleftrightarrow{ST}$.

17. Name three planes that intersect at point *S*.

18. Name a line and a plane that intersect in a point.

Exs. 13–18

Follow the steps shown to draw the figure named.

19. a rectangular solid or box

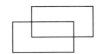

Step 1   *Step 2*   *Step 3*

20. a barn

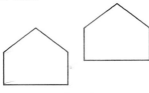   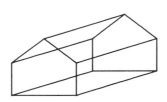   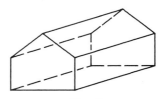

Step 1   *Step 2*   *Step 3*

Note: After drawing more figures in space, you will probably be able to go directly from Step 1 to Step 3.

21. Name two planes that intersect in $\overleftrightarrow{FG}$.

22. Name three lines that intersect at point *E*.

23. Name three planes that intersect at point *B*.

24. **a.** Are points *A*, *D*, and *C* collinear?
    b. Are points *A*, *D*, and *C* coplanar?

25. **a.** Are points *R*, *S*, *G*, and *F* coplanar?
    b. Are points *R*, *S*, *G*, and *C* coplanar?

26. **a.** Name two planes that do not intersect.
    b. Name two other planes that do not intersect.

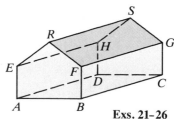

Exs. 21–26

You can think of the ceiling and floor of a room as parts of *horizontal planes*. The walls are parts of *vertical planes*. Vertical planes are represented by figures like those shown in which two sides are vertical. A horizontal plane is represented by a figure like that shown, with two sides horizontal and no sides vertical.

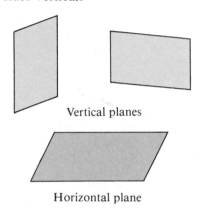

Vertical planes

Horizontal plane

B **27.** Can two horizontal planes intersect?

28. a. Can two vertical planes intersect?

   b. Suppose a line is known to be in a vertical plane. Does the line have to be a vertical line?

Sketch and label the figures described. Use dashes for hidden parts.

29. Vertical line l intersects a horizontal plane M at point O.

30. Horizontal plane P contains two lines k and n that intersect at point A.

31. Horizontal plane Q and vertical plane N intersect.

32. Vertical planes X and Y intersect in $\overleftrightarrow{AB}$.

33. Point P is not in plane N. Three lines through point P intersect N in points A, B, and C.

C **34.** Three vertical planes intersect in a line.

35. A vertical plane intersects two horizontal planes in lines l and n.

36. Three planes intersect in a point.

Challenge

If the area of the red square is 1 square unit, what is the area of the blue square? Give a convincing argument.

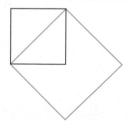

Self-Test 1

Name the point that appears to satisfy the description.

1. Equidistant from R and S
2. Equidistant from S and U
3. Equidistant from U and T

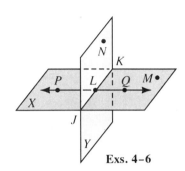

Exs. 1–3

Classify each statement as true or false.

4. Plane Y and $\overleftrightarrow{PQ}$ intersect in point L.
5. Points J, K, L, and N are coplanar.
6. Points J, L, and Q are collinear.

7. Draw a vertical plane Z intersecting a horizontal line l in a point T.

Exs. 4–6

Algebra Review: *Linear Equations*

Find the value of the variable.

1. $c + 5 = 12$
2. $8 + c = 13$
3. $c - 5 = 12$
4. $7 - z = 13$
5. $15 - z = 0$
6. $4x = 28$
7. $3x = 15$
8. $7x = -35$
9. $-5x = -5$
10. $\frac{1}{3}a = 2$
11. $\frac{3}{4}a = 9$
12. $\frac{4}{5}a = -20$
13. $-2b = 6$
14. $-3b = -9$
15. $-9b = 2$
16. $42 = 6k$
17. $5 = 10k$
18. $-16 = -4k$
19. $12 = \frac{e}{2}$
20. $-9 = \frac{e}{3}$
21. $5 = -\frac{e}{3}$
22. $2p + 5 = 13$
23. $3p - 5 = 13$
24. $4p + 2 = 22$
25. $60 = 6t + 12$
26. $12 = 3r - 9$
27. $55 = 7s - 8$
28. $8x + 2x = 90$
29. $8x - 2x = 90$
30. $x + 9x = 5$
31. $(2g - 15) + g = 9$
32. $3u + (u - 2) = 10$
33. $(w - 20) + 5w = 28$
34. $3x = 2x - 17$
35. $5y = 3y + 26$
36. $7z = 180 - 2z$
37. $12 + 3b = 2 + 5b$
38. $4c + 23 = 9c - 7$
39. $7h + (90 - h) = 210$
40. $5x + (180 - x) = 300$
41. $(4f + 5) + (5f + 40) = 180$
42. $(3g - 4) + (4g + 10) = 90$
43. $2(4d + 4) = d + 1$
44. $2(d + 5) = 3(d - 2)$
45. $180 - x = 3(90 - x)$
46. $3(180 - y) = 2(90 - y)$

Definitions and Postulates

Objectives

1. Use symbols for lines, segments, rays, and distances; find distances.
2. Name angles and find their measures.
3. State and use the Segment Addition Postulate and the Angle Addition Postulate.
4. Recognize what you can conclude from a diagram.
5. Use postulates and theorems relating points, lines, and planes.

1-3 *Segments, Rays, and Distance*

In the diagram, point B is *between* points A and C. Note that B must lie on $\overleftrightarrow{AC}$.

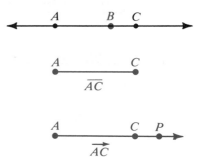

Segment AC, denoted $\overline{AC}$, consists of points A and C and all points that are between A and C. Points A and C are called the *endpoints* of $\overline{AC}$.

Ray AC, denoted $\overrightarrow{AC}$, consists of $\overline{AC}$ and all other points P such that C is between A and P. The *endpoint* of $\overrightarrow{AC}$ is A, the point named first.

$\overrightarrow{SR}$ and $\overrightarrow{ST}$ are called **opposite rays** if S is between R and T.

The hands of the clock shown suggest opposite rays.

On a *number line* every point is paired with a number and every number is paired with a point. In the diagram, point J is paired with -3, the *coordinate* of J.

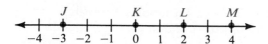

The **length** of $\overline{MJ}$, denoted by MJ, is the distance between point M and point J. You can find the length of a segment on a number line by subtracting the coordinates of its endpoints:

$$MJ = 4 - (-3) = 7$$

Notice that since a length must be a positive number, you subtract the lesser coordinate from the greater one. Actually, the distance between two points is the absolute value of the difference of their coordinates. When you use absolute value, the order in which you subtract coordinates doesn't matter.

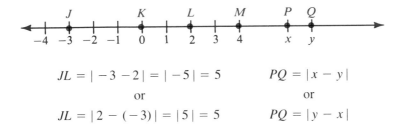

$$JL = |-3 - 2| = |-5| = 5 \qquad PQ = |x - y|$$

or   or

$$JL = |2 - (-3)| = |5| = 5 \qquad PQ = |y - x|$$

There are many different ways to pair the points on a line with numbers. For example, the red coordinates shown below would give distances in centimeters. The blue coordinates would give distances in inches.

Once you have chosen a unit of measure, the distance between any two points will be the same no matter where you place the coordinate 0. For example, the black coordinates below show another way of assigning coordinates to points on the line so that distances will be measured in inches.

Using number lines involves the following basic assumptions. Statements such as these that are accepted without proof are called **postulates** or **axioms**. Notice that the Ruler Postulate below allows you to measure distances using centimeters or inches or any other convenient unit. But once a unit of measure has been chosen for a particular problem, you must use that unit throughout the problem.

Postulate 1 *Ruler Postulate*

1. **The points on a line can be paired with the real numbers in such a way that any two points can have coordinates 0 and 1.**

2. **Once a coordinate system has been chosen in this way, the distance between any two points equals the absolute value of the difference of their coordinates.**

Postulate 2 *Segment Addition Postulate*

If *B* is between *A* and *C*, then

$$AB + BC = AC.$$

Example B is between A and C, with $AB = x$, $BC = x + 6$, and
$AC = 24$. Find:
a. the value of x **b.** BC

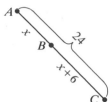

Solution   **a.** $AB + BC = AC$ **b.** $BC = x + 6$
$x + (x + 6) = 24$ $= 9 + 6$
$2x + 6 = 24$ $= 15$
$2x \quad\; = 18$
$x \quad\;\; = 9$

In geometry two objects that have the same size and shape are called
congruent. For many geometric figures we can give a more precise definition
of what it means to be congruent. For example, we will define congruent
segments in this section, congruent angles in the next section, congruent triangles
in Chapter 4, and congruent circles and arcs in Chapter 9.

Congruent segments are segments that have equal
lengths. To indicate that $\overline{DE}$ and $\overline{FG}$ have equal
lengths, you write

$$DE = FG.$$

To indicate that $\overline{DE}$ and $\overline{FG}$ are congruent, you
write

$$\overline{DE} \cong \overline{FG}$$

(read "$\overline{DE}$ is congruent to $\overline{FG}$"). The definition
tells us that the two statements are equivalent. We
will use them interchangeably.

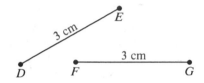

The **midpoint of a segment** is the point that divides
the segment into two congruent segments. In the
diagram

$$AP = PB,$$
$$\overline{AP} \cong \overline{PB},$$

and P is the midpoint of $\overline{AB}$.

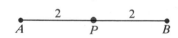

A **bisector of a segment** is a line, segment, ray,
or plane that intersects the segment at its midpoint.
Line l is a bisector of $\overline{AB}$. $\overline{PQ}$ and plane X also
bisect $\overline{AB}$.

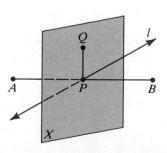

LET ME BREATHE!!!

Classroom Exercises

1. Does the symbol represent a line, segment, ray, or length?
   a. $\overline{PQ}$ **b.** $\overrightarrow{PQ}$ **c.** $\overleftrightarrow{PQ}$ **d.** PQ

2. How many endpoints does a segment have? a ray? a line?

3. Is $\overline{AB}$ the same as $\overline{BA}$?

4. Is $\overrightarrow{AB}$ the same as $\overrightarrow{BA}$?

5. Is $\overleftrightarrow{AB}$ the same as $\overleftrightarrow{BA}$?

6. Is AB the same as BA?

Exs. 3–6

7. What is the coordinate of P? of R?

8. Name the point with coordinate 2.

9. Find each distance: **a.** RS **b.** RQ **c.** PT

10. Name three segments congruent to $\overline{PQ}$.

11. Name the ray opposite to $\overrightarrow{SP}$.

12. Name the midpoint of $\overline{PT}$.

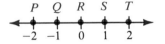

13. **a.** What number is halfway between 1 and 2? __
    b. What is the coordinate of the midpoint of $\overline{ST}$?

Exs. 7–14

14. **a.** Could you list all the numbers between 1 and 2?
    b. Is there a point on the number line for every number between 1 and 2?
    c. Is there any limit to the number of points between S and T?

State whether the figures *appear* to be congruent (that is, appear to have the same size and shape).

15.   16.   17.

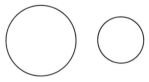

18.   19.   20.

21. Draw two points P and Q on a sheet of paper. Fold the paper so that fold line f contains both P and Q. Unfold the paper. Now fold so that P falls on Q. Call the second fold g. Lay the paper flat and label the intersection of f and g as point X. How are points P, Q, and X related? Explain.

22. If $AB = BC$, must point B be the midpoint of $\overline{AC}$? Explain.

The given numbers are the coordinates of two points on a number line. State the distance between the points.

23. -2 and 6   24. -2 and -6 25. 2 and -6 26. 7 and -1

Written Exercises

The numbers given are the coordinates of two points on a number line. State the distance between the points.

A   **1.** -6 and 9 **2.** -3 and -17 **3.** -1.2 and -5.7 **4.** -2.5 and 4.6

In the diagram, $\overline{HL}$ and $\overleftrightarrow{KT}$ intersect at the midpoint of $\overline{HL}$. Classify each statement as true or false.

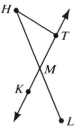

5. $\overline{LM} \cong \overline{MH}$ **6.** KM must equal MT.

7. $\overline{MT}$ bisects $\overline{LH}$.   **8.** $\overleftrightarrow{KT}$ is a bisector of $\overline{LH}$.

9. $\overrightarrow{MT}$ and $\overrightarrow{TM}$ are opposite rays.   **10.** $\overrightarrow{MT}$ and $\overrightarrow{MK}$ are opposite rays.

11. $\overleftrightarrow{LH}$ is the same as $\overline{HL}$.   **12.** $\overrightarrow{KT}$ is the same as $\overrightarrow{KM}$.

13. $\overleftrightarrow{KT}$ is the same as $\overleftrightarrow{KM}$.   **14.** $\overline{KT}$ is the same as $\overline{KM}$.

15. $HM + ML = HL$ **16.** $TM + MH = TH$

17. T is between H and M.   **18.** M is between K and T.

Exs. 5–18

Name each of the following.

19. The point on $\overrightarrow{DA}$ whose distance from D is 2

20. The point on $\overrightarrow{DG}$ whose distance from D is 2

21. Two points whose distance from E is 2

22. The ray opposite to $\overrightarrow{BE}$

23. The midpoint of $\overline{BF}$

24. The coordinate of the midpoint of $\overline{BD}$

25. The coordinate of the midpoint of $\overline{AE}$

26. A segment congruent to $\overline{AF}$

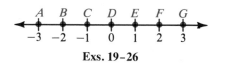

Exs. 19–26

In Exercises 27–30 draw $\overline{CD}$ and $\overline{RS}$ so that the conditions are satisfied.

27. $\overline{CD}$ and $\overline{RS}$ intersect, but neither segment bisects the other.

28. $\overline{CD}$ and $\overline{RS}$ bisect each other.

29. $\overline{CD}$ bisects $\overline{RS}$, but $\overline{RS}$ does not bisect $\overline{CD}$.

30. $\overline{CD}$ and $\overline{RS}$ do not intersect, but $\overrightarrow{CD}$ and $\overrightarrow{RS}$ do intersect.

B   **31.** In the diagram, $\overline{PR} \cong \overline{RT}$, S is the midpoint of $\overline{RT}$, $QR = 4$, and $ST = 5$. Complete.
     a. $RS = $ ___?___   **b.** $RT = $ ___?___
     c. $PR = $ ___?___   **d.** $PQ = $ ___?___

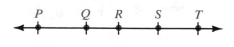

32. In the diagram, X is the midpoint of $\overline{VZ}$, $VW = 5$, and $VY = 20$. Find the coordinates of W, X, and Y.

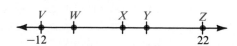

E **is the midpoint of** $\overline{DF}$. **Find the value of** *x*.

33. $DE = 5x + 3$, $EF = 33$

34. $DE = 45$, $EF = 5x - 10$

35. $DE = 3x$, $EF = x + 6$

36. $DE = 2x - 3$, $EF = 5x - 24$

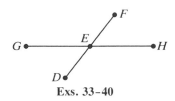

Exs. 33–40

Find the value of *y*.

37. $GE = y$, $EH = y - 1$, $GH = 11$

38. $GE = 3y$, $GH = 7y - 4$, $EH = 24$

Find the value of *z*. **Then find** *GE* **and** *EH* **and state whether** *E* **is the midpoint of** $\overline{GH}$.

39. $GE = z + 2$, $GH = 20$, $EH = 2z - 6$

40. $GH = z + 6$, $EH = 2z - 4$, $GE = z$

Name the graph of the given equation or inequality.

Example   **a.** $x \ge 2$ **b.** $4 \le x \le 6$

Solution   **a.** $\overrightarrow{NT}$ **b.** $\overline{TY}$

Exs. 41–45

41. $-2 \le x \le 2$ **42.** $x \le 0$ **43.** $|x| \le 4$ **44.** $|x| \ge 0$ **45.** $|x| = 0$

In Exercises 46 and 47 draw a diagram to illustrate your answer.

46. a. On $\overrightarrow{AB}$, how many points are there whose distance from point *A* is 3 cm?

   b. On $\overleftrightarrow{AB}$, how many points are there whose distance from point *A* is 3 cm?

C 47. On $\overrightarrow{AB}$, how many points are there whose distance from point *B* is 3 cm?

48. The Ruler Postulate suggests that there are many ways to assign coordinates to a line. The Fahrenheit and Celsius temperature scales on a thermometer indicate two such ways of assigning coordinates. A Fahrenheit temperature of 32° corresponds to a Celsius temperature of 0°. The formula, or rule, for converting a Fahrenheit temperature *F* into a Celsius temperature *C* is

$$C = \frac{5}{9}(F - 32).$$

a. What Celsius temperatures correspond to Fahrenheit temperatures of 212° and 98.6°?

b. Solve the equation above for *F* to obtain a rule for converting Celsius temperatures to Fahrenheit temperatures.

c. What Fahrenheit temperatures correspond to Celsius temperatures of −40° and 2000°?

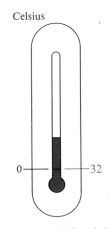

1-4 *Angles*

An **angle** ($\angle$) is the figure formed by two rays that have the same endpoint. The two rays are called the **sides** of the angle, and their common endpoint is the **vertex** of the angle.

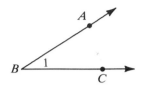

The sides of the angle shown are $\overrightarrow{BA}$ and $\overrightarrow{BC}$. The vertex is point B. The angle can be called $\angle B$, $\angle ABC$, $\angle CBA$, or $\angle 1$. If three letters are used to name an angle, the middle letter must name the vertex.

When you talk about this $\angle B$, everyone knows what angle you mean. But if you tried to talk about $\angle E$ in the diagram at the right, people wouldn't know which angle you meant. There are three angles with vertex E. To name any particular one of them you need to use either three letters or a number.

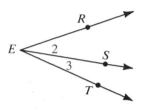

$\angle 2$ could also be called $\angle RES$ or $\angle SER$.
$\angle 3$ could also be called $\angle SET$ or $\angle TES$.
$\angle RET$ could also be called $\angle TER$.

You can use a protractor like the one shown below to find the *measure in degrees* of an angle. Although angles are sometimes measured in other units, this book will always use degree measure. Using the outer (red) scale of the protractor, you can see that $\angle XOY$ is a 40° angle. You can indicate that the (degree) measure of $\angle XOY$ is 40 by writing $m \angle XOY = 40$.

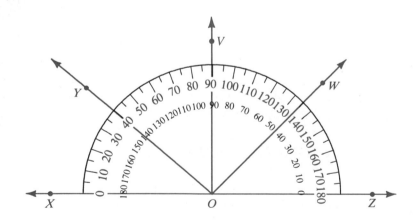

Using the inner scale of the protractor, you find that:

$m \angle YOZ = 140$ $m \angle WOZ = 45$ $m \angle YOW = 140 - 45 = 95$

Angles are classified according to their measures.

Acute angle: Measure between 0 and 90   **Examples:** $\angle XOY$ and $\angle VOW$
Right angle: Measure 90   **Examples:** $\angle XOV$ and $\angle VOZ$
Obtuse angle: Measure between 90 and 180   **Examples:** $\angle XOW$ and $\angle YOW$
Straight angle: Measure 180   **Example:** $\angle XOZ$

The two angle postulates below are very much like the Ruler Postulate and the Segment Addition Postulate on page 12.

Postulate 3   *Protractor Postulate*

On $\overleftrightarrow{AB}$ in a given plane, choose any point O between A and B. Consider $\overrightarrow{OA}$ and $\overrightarrow{OB}$ and all the rays that can be drawn from O on one side of $\overleftrightarrow{AB}$. These rays can be paired with the real numbers from 0 to 180 in such a way that:

a. $\overrightarrow{OA}$ is paired with 0, and $\overrightarrow{OB}$ with 180.

b. If $\overrightarrow{OP}$ is paired with x, and $\overrightarrow{OQ}$ with y, then $m \angle POQ = |x - y|$.

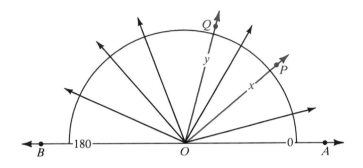

Postulate 4   *Angle Addition Postulate*

If point B lies in the interior of $\angle AOC$, then

$$m \angle AOB + m \angle BOC = m \angle AOC.$$

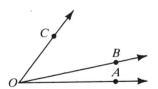

If $\angle AOC$ is a straight angle and B is any point not on $\overleftrightarrow{AC}$, then

$$m \angle AOB + m \angle BOC = 180.$$

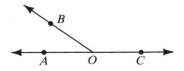

Congruent angles are angles that have equal measures. Since $\angle R$ and $\angle S$ both have measure 40, you can write

$$m \angle R = m \angle S \text{ or } \angle R \cong \angle S.$$

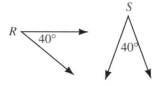

The definition of congruent angles tells us that these two statements are equivalent. We will use them interchangeably.

Adjacent angles (adj. ⓢ) are two angles in a plane that have a common vertex and a common side but no common interior points.

$\angle 1$ and $\angle 2$ are adjacent angles.　　$\angle 3$ and $\angle 4$ are not adjacent angles.

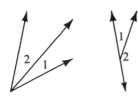 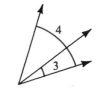

The **bisector of an angle** is the ray that divides the angle into two congruent adjacent angles. In the diagram,

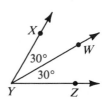

$$m \angle XYW = m \angle WYZ,$$

$$\angle XYW \cong \angle WYZ,$$

and　　$\overrightarrow{YW}$ bisects $\angle XYZ.$

　　　There are certain things that you can conclude from a diagram and others that you can't. The following are things you can conclude from the diagram shown below.

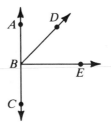

All points shown are coplanar.
$\overleftrightarrow{AB}$, $\overrightarrow{BD}$, and $\overrightarrow{BE}$ intersect at B.
A, B, and C are collinear.
B is between A and C.
$\angle ABC$ is a straight angle.
D is in the interior of $\angle ABE$.
$\angle ABD$ and $\angle DBE$ are adjacent angles.

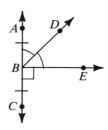

　　　The diagram above does *not* tell you that $\overline{AB} \cong \overline{BC}$, that $\angle ABD \cong \angle DBE$, or that $\angle CBE$ is a right angle. These three new pieces of information can be indicated in a diagram by using marks as shown at the right. Note that a small square is used to indicate a right angle (rt. $\angle$).

Classroom Exercises

Name the vertex and the sides of the given angle.

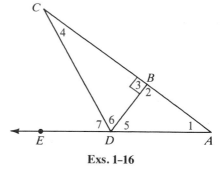

1. $\angle 4$ **2.** $\angle 1$ **3.** $\angle 6$

4. Name all angles adjacent to $\angle 6$.

5. Name three angles that have B as the vertex.

6. How many angles have D as the vertex?

State whether the angle appears to be acute, right, obtuse, or straight. Then estimate its measure.

Exs. 1–16

7. $\angle 1$ **8.** $\angle 2$ **9.** $\angle EDB$

10. $\angle CDB$ **11.** $\angle ADC$ **12.** $\angle ADE$

Complete.

13. $m\angle 7 + m\angle 6 = m\angle\underline{\ ?\ }$ **14.** $m\angle 6 + m\angle 5 = m\angle\underline{\ ?\ }$

15. $m\angle 2 + m\angle 3 = \underline{\ ?\ }$ **16.** If $\overrightarrow{DB}$ bisects $\angle CDA$, then $\angle\underline{\ ?\ } \cong \angle\underline{\ ?\ }$.

State the measure of each angle.

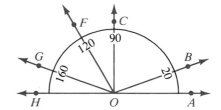

17. $\angle BOC$ **18.** $\angle GOH$

19. $\angle FOG$ **20.** $\angle COF$

21. $\angle GOB$ **22.** $\angle HOA$

23. Name four angles that are adjacent to $\angle FOG$.

24. What ray bisects which two angles?

25. Name a pair of congruent:

   a. acute angles   **b.** right angles   **c.** obtuse angles

Exs. 17–25

26. Study a corner of your classroom where two walls and the ceiling meet. How many right angles can you see at the corner?

27. Draw an angle, $\angle AOB$, on a sheet of paper. Fold the paper so that $\overrightarrow{OA}$ falls on $\overrightarrow{OB}$. Lay the paper flat and call the fold line $\overleftrightarrow{OK}$. How is $\overrightarrow{OK}$ related to $\angle AOB$? Explain.

Given the diagram, state whether you can reach the conclusion shown.

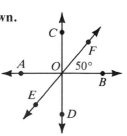

28. $m\angle FOB = 50$ **29.** $m\angle AOC = 90$

30. $m\angle DOC = 180$ **31.** $AO = OB$

32. $\angle AOC \cong \angle BOC$ **33.** $m\angle AOF = 130$

34. Points E, O, and F are collinear.

35. Point C is in the interior of $\angle AOF$.

36. $\angle AOE$ and $\angle AOD$ are adjacent angles.

Exs. 28–38

37. $\angle AOB$ is a straight angle.   **38.** $\overrightarrow{OA}$ and $\overrightarrow{OB}$ are opposite rays.

Written Exercises

A **1.** Name the vertex and the sides of ∠5.

2. Name all angles adjacent to ∠ADE.

State another name for the angle.

3. ∠1   **4.** ∠3   **5.** ∠5

6. ∠ALD   **7.** ∠AST   **8.** ∠LES

State whether the angle appears to be acute, right, obtuse, or straight.

9. ∠2   **10.** ∠LAS   **11.** ∠ATL

12. ∠S   **13.** ∠LTS   **14.** ∠EDT

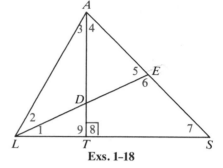

Exs. 1–18

Complete.

15. $m\angle 3 + m\angle 4 = m\angle$ __?__

16. $m\angle ALS - m\angle 2 = m\angle$ __?__

17. If $m\angle 1 = m\angle 2$, then __?__ bisects __?__.

18. $m\angle LDA + m\angle ADE =$ __?__

Without measuring, sketch each angle. Then use a protractor to check your accuracy.

19. 90° angle   **20.** 45° angle   **21.** 150° angle   **22.** 10° angle

Draw a line, $\overleftrightarrow{AB}$. Choose a point O between A and B. Use a protractor to investigate the following questions.

23. In the plane represented by your paper, how many lines can you draw through O that will form a 30° angle with $\overrightarrow{OB}$?

24. In the plane represented by your paper, how many lines can you draw through O that will form a 90° angle with $\overrightarrow{OB}$?

B **25.** Using a ruler, draw a large triangle. Then use a protractor to find the approximate measure of each angle and compute the sum of the three measures. Repeat this exercise for a triangle with a different shape. Did you get the same result?

26. Find $m\angle 2$, $m\angle 3$, and $m\angle 4$ when the measure of ∠1 is:
a. 90   **b.** 93

27. Express $m\angle 2$, $m\angle 3$, and $m\angle 4$ in terms of t when $m\angle 1 = t$.

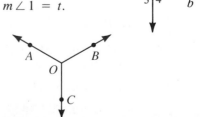

28. A careless person wrote, using the figure shown,

$$m\angle AOB + m\angle BOC = m\angle AOC.$$

What part of the Angle Addition Postulate did that person overlook?

$\overrightarrow{AL}$ **bisects** $\angle KAT$. **Find the value of** x.

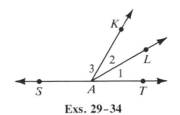

29. $m \angle 3 = 6x$, $m \angle KAT = 90 - x$

30. $m \angle 1 = 7x + 3$, $m \angle 2 = 6x + 7$

31. $m \angle 1 = 5x - 12$, $m \angle 2 = 3x + 6$

32. $m \angle 1 = x$, $m \angle 3 = 4x$

33. $m \angle 1 = 2x - 8$, $m \angle 3 = 116$

34. $m \angle 2 = x + 12$, $m \angle 3 = 6x - 20$

Exs. 29–34

C 35. a. Complete.

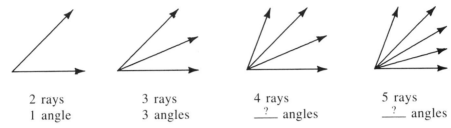

2 rays
1 angle

3 rays
3 angles

4 rays
__?__ angles

5 rays
__?__ angles

b. Study the pattern in the four cases shown, and predict the number of angles formed by six noncollinear rays that have the same endpoint.

c. Which of the expressions below gives the number of angles formed by n noncollinear rays that have the same endpoint?

$$n - 1 \qquad\qquad 2n - 3 \qquad\qquad n^2 - 3 \qquad\qquad \frac{n(n - 1)}{2}$$

36. $\overrightarrow{OC}$ bisects $\angle AOB$, $\overrightarrow{OD}$ bisects $\angle AOC$, $\overrightarrow{OE}$ bisects $\angle AOD$, $\overrightarrow{OF}$ bisects $\angle AOE$, and $\overrightarrow{OG}$ bisects $\angle FOC$.
a. If $m \angle BOF = 120$, then $m \angle DOE = \underline{}$.
b. If $m \angle COG = 35$, then $m \angle EOG = \underline{}$.

1-5 *Postulates and Theorems Relating Points, Lines, and Planes*

Recall that we have accepted, without proof, the following four basic assumptions.

The Ruler Postulate   The Segment Addition Postulate
The Protractor Postulate   The Angle Addition Postulate

These postulates deal with segments, lengths, angles, and measures. The following five basic assumptions deal with the way points, lines, and planes are related.

Postulate 5

A line contains at least two points; a plane contains at least three points not all in one line; space contains at least four points not all in one plane.

Postulate 6

Through any two points there is exactly one line.

Postulate 7

Through any three points there is at least one plane, and through any three noncollinear points there is exactly one plane.

Postulate 8

If two points are in a plane, then the line that contains the points is in that plane.

Postulate 9

If two planes intersect, then their intersection is a line.

Important statements that are *proved* are called **theorems.** In Classroom Exercise 1 you will see how Theorem 1-1 follows from the postulates. In Written Exercise 20 you will complete an argument that justifies Theorem 1-2. You will learn about writing proofs in the next chapter.

Theorem 1-1

If two lines intersect, then they intersect in exactly one point.

Theorem 1-2

Through a line and a point not in the line there is exactly one plane.

Theorem 1-3

If two lines intersect, then exactly one plane contains the lines.

The phrase "exactly one" appears several times in the postulates and theorems of this section. The phrase "one and only one" has the same meaning. For example, here is another correct form of Theorem 1-1:

If two lines intersect, then they intersect in one and only one point.

The theorem states that a point of intersection *exists* (there is *at least one* point of intersection) and the point of intersection is *unique* (*no more than one* such point exists).

Classroom Exercises

1. Theorem 1-1 states that two lines intersect in exactly one point. The diagram suggests what would happen if you tried to show two ''lines'' drawn through two points. State the postulate that makes this situation impossible.

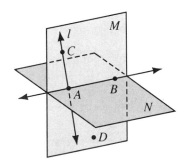

2. State Postulate 6 using the phrase *one and only one*.

3. Reword the following statement as two statements, one describing existence and the other describing uniqueness:

    A segment has exactly one midpoint.

Postulate 6 is sometimes stated as ''Two points *determine* a line.''

4. Restate Theorem 1-2 using the word *determine*.

5. Do two intersecting lines determine a plane?

6. Do three points determine a line?

7. Do three points determine a plane?

State a postulate, or part of a postulate, that justifies your answer to each exercise.

8. Name two points that determine line *l*.

9. Name three points that determine plane *M*.

10. Name the intersection of planes *M* and *N*.

11. Does $\overleftrightarrow{AD}$ lie in plane *M*?

12. Does plane *N* contain any points not on $\overleftrightarrow{AB}$?

Surveyors and photographers use a *tripod* for support.

13. Why does a three-legged support work better than one with four legs?

14. Explain why a four-legged table may rock even if the floor is level.

15. A carpenter checks to see if a board is warped by laying a straightedge across the board in several directions. State the postulate that is related to this procedure.

16. Think of the intersection of the ceiling and the front wall of your classroom as line *l*. Let the point in the center of the floor be point *C*.
    a. Is there a plane that contains line *l* and point *C*?
    b. State the theorem that applies.

Written Exercises

A **1.** State Theorem 1-2 using the phrase *one and only one.*

2. Reword Theorem 1-3 as two statements, one describing existence and the other describing uniqueness.

3. Planes *M* and *N* are known to intersect.
   a. What kind of figure is the intersection of *M* and *N*?
   b. State the postulate that supports your answer to part (a).

4. Points *A* and *B* are known to lie in a plane.
   a. What can you say about $\overleftrightarrow{AB}$?
   b. State the postulate that supports your answer to part (a).

In Exercises 5-11 you will have to visualize certain lines and planes not shown in the diagram of the box. When you name a plane, name it by using four points, no three of which are collinear.

5. Write the postulate that assures you that $\overleftrightarrow{AC}$ exists.

6. Name a plane that contains $\overleftrightarrow{AC}$.

7. Name a plane that contains $\overleftrightarrow{AC}$ but that is not shown in the diagram.

8. Name the intersection of plane *DCFE* and plane *ABCD*.

9. Name four lines shown in the diagram that don't intersect plane *EFGH*.

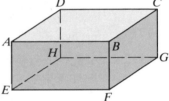

Exs. 5–12

10. Name two lines that are not shown in the diagram and that don't intersect plane *EFGH*.

11. Name three planes that don't intersect $\overleftrightarrow{EF}$ and don't contain $\overleftrightarrow{EF}$.

12. If you measure $\angle EFG$ with a protractor you get more than 90°. But you know that $\angle EFG$ represents a right angle in a box. Using this as an example, complete the table.

	$\angle EFG$	$\angle AEF$	$\angle DCB$	$\angle FBC$
In the diagram	obtuse	?	?	?
In the box	right	?	?	?

State whether it is possible for the figure described to exist. Write *yes* or *no*.

B **13.** Two points both lie in each of two lines.

14. Three points all lie in each of two planes.

15. Three noncollinear points all lie in each of two planes.

16. Two points lie in a plane *X*, two other points lie in a different plane *Y*, and the four points are coplanar but not collinear.

17. Points R, S, and T are noncollinear points.
  a. State the postulate that guarantees the existence of a plane X that contains R, S, and T.
  b. Draw a diagram showing plane X containing the noncollinear points R, S, and T.
  c. Suppose that P is any point of $\overleftrightarrow{RS}$ other than R and S. Does point P lie in plane X? Explain.
  d. State the postulate that guarantees that $\overleftrightarrow{TP}$ exists.
  e. State the postulate that guarantees that $\overleftrightarrow{TP}$ is in Plane X.

18. Points A, B, C, and D are four noncoplanar points.

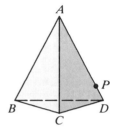

  a. State the postulate that guarantees the existence of planes ABC, ABD, ACD, and BCD.
  b. Explain how the Ruler Postulate guarantees the existence of a point P between A and D.
  c. State the postulate that guarantees the existence of plane BCP.
  d. Explain why there are an infinite number of planes through $\overline{BC}$.

C **19.** State how many segments can be drawn between the points in each figure. No three points are collinear.

  a.   **b.**   **c.**   **d.**

  3 points   4 points   5 points   6 points
  __?__ segments   __?__ segments   ___ ? segments   __?__ segments

  e. Without making a drawing, predict how many segments can be drawn between seven points, no three of which are collinear.
  f. How many segments can be drawn between n points, no three of which are collinear?

20. Parts (a) through (d) justify Theorem 1-2: Through a line and a point not in the line there is exactly one plane.

  a. If P is a point not in line k, what postulate permits us to state that there are two points R and S in line k?

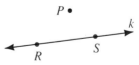

  b. Then there is at least one plane X that contains points P, R, and S. Why?
  c. What postulate guarantees that plane X contains line k? Now we know that there is a plane X that contains both point P and line k.
  d. There can't be another plane that contains point P and line k, because then *two* planes would contain noncollinear points P, R, and S. What postulate does this contradict?

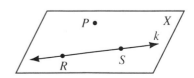

Application *Locating Points*

Suppose you lived in an area with streets laid out on a grid. If you lived in a house located at point *P* in the diagram at the right below, you could tell someone where you lived by saying:

From the crossing at the center of town, go three blocks east and two blocks north.

A friend of yours living at *Q* might say she lives two blocks west and three blocks north of the town center.

Mathematicians make such descriptions shorter by using a grid system and *coordinates*. They use (3, 2) for your house at point *P*, and (−2, 3) for your friend's house at *Q*. Point *O* at the center of town is (0, 0). Points *R* and *S* are (3, 0) and (−1, −2).

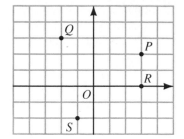

This grid system is not always the easiest way to describe a position. If you were a pilot and saw another airplane while flying, it would be difficult to give its position in this system. However, you might say the other plane is 4 km away at 11 o'clock, with 12 o'clock being straight ahead.

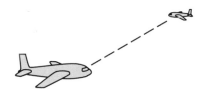

Mathematicians sometimes find it convenient to describe a point by a distance and an angle. Rotation in a clockwise direction is represented by a negative angle. Counterclockwise rotation is represented by a positive angle. A complete rotation, all the way around once, is 360° (or −360°). The labeled points in the diagram at the right are described as shown below.

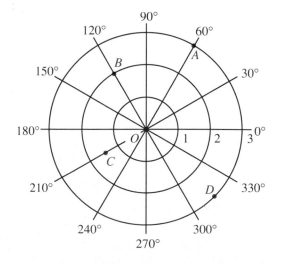

A	(3, 60°)
B	(2, 120°)
C	(1.5, 210°) or (1.5, −150°)
D	(3, 315°) or (3, −45°)

Sometimes you may want to change from one system to the other. For example, if you were at the town center and walked two blocks east and three blocks north, what would your position be in the distance-angle system? Use a centimeter ruler and draw the triangle suggested by your path. If you measure the triangle, you will get about (3.6, 56°).

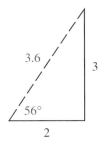

Exercises

1. Copy the grid system shown on the previous page onto a piece of graph paper. Then locate the following points.
   a. A point *T* five blocks due west of the center of town
   b. A point *U* five blocks east and two blocks south of the center of town
   c. A point *V* two blocks west and one block north of your house, which is located at point *P*

2. Give the letter that names each point.
   a. (2, 30°)
   b. (2.5, 120°)
   c. (1, −90°)

3. Give the distance and angle for each point.
   a. *C*
   b. *A*
   c. *T*

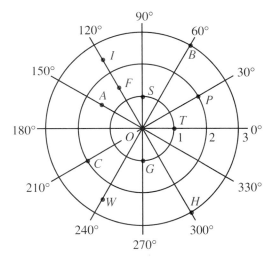

4. Give another way of naming each point.
   a. (1, −120°) **b.** (2, 300°) **c.** (2.5, −180°)

5. A point is given in the grid system. What would it be called in the distance-angle system? (*Hint*: See the discussion at the top of the page. Use a protractor and a centimeter ruler to help you answer the question.)
   a. (3, 4) **b.** (−2, 5)
   c. (4, 0) **d.** (8, −6)

6. A point is given in the distance-angle system. What would it be called, approximately, in the grid system? (*Hint*: Use a protractor and a centimeter ruler to draw the triangle suggested by the angle and distance. Measure the sides of the triangle.)
   a. (2, 50°) **b.** (1.5, −70°)
   c. (3, 90°) **d.** (1, 120°)

Self-Test 2

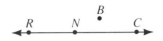

1. Write three names for the line pictured.
2. Name the ray that is opposite to $\overrightarrow{NC}$.
3. Is it correct to say that point B lies between points N and C?
4. When $RN = 7$, $NC = 3x + 5$, and $RC = 18$, what is the value of x?

Complete.

5. $m \angle 1 + m \angle 2 = m \angle \underline{\ ?\ }$
6. If $\angle 1 \cong \angle 2$, then $\underline{\ ?\ }$ is the bisector of $\angle \underline{\ ?\ }$.
7. $m \angle HOK = \underline{\ ?\ }$, and $\angle HOK$ is called a(n) $\underline{\ ?\ }$ angle.

8. Which of the four things stated *can't* you conclude from the diagram?
   a. A, B, and C are collinear. b. $\angle DBC$ is a right angle.
   c. B is the midpoint of $\overline{AC}$. d. E is in the interior of $\angle DBA$.

Apply postulates and theorems to complete the statements.

9. Through any two points $\underline{\ ?\ }$.   10. If points A and B are in plane Z, $\underline{\ ?\ }$.

11. If two planes intersect, then $\underline{\ ?\ }$.
12. If there is a line j and a point P not in the line, then $\underline{\ ?\ }$.

Chapter Summary

1. The concepts of *point*, *line*, and *plane* are basic to geometry. These undefined terms are used in the definitions of other terms.
2. $\overleftrightarrow{AB}$ represents a line, $\overline{AB}$ a segment, and $\overrightarrow{AB}$ a ray. AB represents the length of $\overline{AB}$; AB is a positive number.
3. Two rays with the same endpoint form an angle.
4. Congruent segments have equal lengths. Congruent angles have equal measures.
5. Angles are classified as acute, right, obtuse, or straight, according to their measures.
6. Diagrams enable you to reach certain conclusions. However, judgments about segment length and angle measure must not be made on the basis of appearances alone.
7. Statements that are accepted without proof are called postulates. Statements that are proved are called theorems.
8. Postulates and theorems in this chapter deal with distances, angle measures, points, lines, and planes.

Chapter Review

In Exercises 1–4 answer on the basis of what appears to be true.

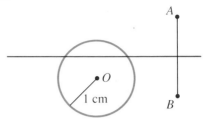

1. How many blue points are 1 cm from point *O*?

2. How many red points are 1 cm from *O*?

3. How many red points are 2 cm from *O*?

4. Each red point is said to be __?__ from points *A* and *B*.

1–1

Sketch and label the figures described.

5. Points *A*, *B*, *C*, and *D* are coplanar, but *A*, *B*, and *C* are the only three of those points that are collinear.

6. Line *l* intersects plane *X* in point *P*.

7. Plane *M* contains intersecting lines *j* and *k*.

8. Planes *X* and *Y* intersect in $\overleftrightarrow{AB}$.

1–2

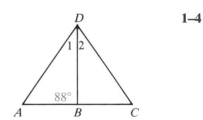

9. Name a point on $\overrightarrow{ST}$ that is not on $\overline{ST}$.

10. Complete: *RS* = __?__ and *ST* = __?__

11. Complete: $\overline{RS}$ and $\overline{ST}$ are called __?__ segments.

12. If *U* is the midpoint of $\overline{TV}$, find the value of *x*.

1–3

13. Name three angles that have vertex *D*. Which angles with vertex *D* are adjacent angles?

14. **a.** $m \angle CBD$ = __?__
    b. Name the postulate that justifies your answer in part (a).

15. What kind of angle is ∠*CBD*?

16. $\overrightarrow{DB}$ bisects ∠*ADC*, $m \angle 1 = 5x - 3$, and $m \angle 2 = x + 25$. Find the value of *x*.

1–4

Classify each statement as true or false.

17. It is possible to locate three points in such a position that an unlimited number of planes contain all three points.

18. It is possible for two intersecting lines to be noncoplanar.

19. Through any three points there is at least one line.

20. If points *A* and *B* lie in plane *P*, then so does any point of $\overrightarrow{AB}$.

1–5

Chapter Test

State how many points meet the requirements. For each answer write *none*, *one*, or *an unlimited number*.

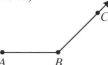

1. Equidistant from points A and B
2. On $\overrightarrow{BC}$ and equidistant from points A and B

Given the diagram, tell whether you can reach the conclusion shown.

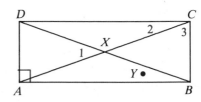

3. $\angle AXC$ is a straight angle.
4. Point Y lies in the interior of $\angle 3$.
5. $\angle ADC$ is a right angle.
6. X is the midpoint of $\overline{AC}$.
7. Point Y lies between points A and B.
8. Name three collinear points.
9. Name the intersection of $\overrightarrow{CX}$ and $\overleftrightarrow{AB}$.
10. Which postulate justifies the statement $AX + XC = AC$?
11. If $\overline{AC}$ bisects $\overline{BD}$, name two congruent segments.
12. Name the vertex and sides of $\angle 1$.
13. Name a right angle.
14. If $m \angle 1 = 46$, find $m \angle DXC$ and $m \angle CXB$.
15. If $m \angle DAX = 70$, find the measure of $\angle XAB$.

Exs. 3–15

Exercises 16–20 refer to a number line that is not pictured here. Point A has coordinate 2 and point B has coordinate 5.

16. What is the length of $\overline{AB}$?
17. What is the coordinate of the midpoint of $\overline{AB}$?
18. If A is the midpoint of $\overline{PB}$, what is the coordinate of P?
19. What is the coordinate of a point that is on $\overrightarrow{AB}$ and is 4 units from B?
20. What is the coordinate of a point that is 4 units from B, but is not on $\overrightarrow{AB}$?

21. Is it possible for a line and a point to be noncoplanar?
22. Is it possible for the intersection of two planes to consist of a segment?
23. Is a postulate an important proved statement, or is it a basic assumption?
24. Complete the statement of the postulate: If two points are in a plane, then __?__.

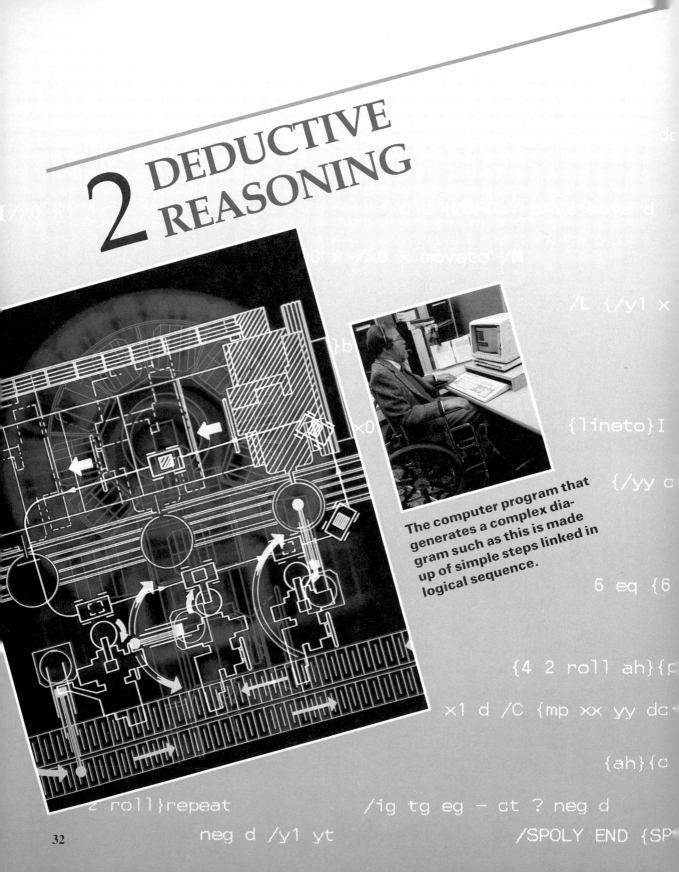

2 DEDUCTIVE REASONING

The computer program that generates a complex diagram such as this is made up of simple steps linked in logical sequence.

Using Deductive Reasoning

Objectives

1. Recognize the hypothesis and the conclusion of an if-then statement.
2. State the converse of an if-then statement.
3. Use a counterexample to disprove an if-then statement.
4. Understand the meaning of *if and only if.*
5. Use properties from algebra and properties of congruence in proofs.
6. Use the Midpoint Theorem and the Angle Bisector Theorem.
7. Know the kinds of reasons that can be used in proofs.

2-1 *If-Then Statements; Converses*

Your friend says, "If it rains after school, then I will give you a ride home."

A geometry student reads, "If B is between A and C, then $AB + BC = AC$."

These are examples of **if-then statements,** which are also called **conditional statements** or simply **conditionals.**

To represent an if-then statement symbolically, let p represent the **hypothesis,** shown in red, and let q represent the **conclusion,** shown in blue. Then we have the basic form of an if-then statement shown below:

If p, then q.

↑ ↑

p: **hypothesis** q: **conclusion**

The **converse** of a conditional is formed by interchanging the hypothesis and the conclusion.

Statement: If p, then q. Converse: If q, then p.

A statement and its converse say different things. In fact, some true statements have false converses.

Statement: If Ed lives in Texas, then he lives south of Canada.
False Converse: If Ed lives south of Canada, then he lives in Texas.

An if-then statement is false if an example can be found for which the hypothesis is true and the conclusion is false. Such an example is called a **counterexample.** It takes only one counterexample to disprove a statement. We know the converse above is false because we can find a counterexample: Ed could live in Kansas City, which *is* south of Canada and *is not* in Texas.

Some true statements have true converses.

Statement: If $4x = 20$, then $x = 5$.
True Converse: If $x = 5$, then $4x = 20$.

Conditional statements are not always written with the "if" clause first. Here are some examples. All these conditionals mean the same thing.

General Form	Example
If p, then q.	If $x^2 = 25$, then $x < 10$.
p implies q.	$x^2 = 25$ implies $x < 10$.
p only if q.	$x^2 = 25$ only if $x < 10$.
q if p.	$x < 10$ if $x^2 = 25$.

If a conditional and its converse are both true they can be combined into a single statement by using the words "if and only if." A statement that contains the words "if and only if" is called a **biconditional.** Its basic form is shown below.

$$p \text{ if and only if } q.$$

Every definition can be written as a biconditional as the statements below illustrate.

Definition: Congruent segments are segments that have equal lengths.

Biconditional: Segments are congruent if and only if their lengths are equal.

Classroom Exercises

State the hypothesis and the conclusion of each conditional.

1. If $2x - 1 = 5$, then $x = 3$.   **2.** If she's smart, then I'm a genius.

3. $8y = 40$ implies $y = 5$.   **4.** $RS = \frac{1}{2}RT$ if S is the midpoint of $\overline{RT}$.

5. $\angle 1 \cong \angle 2$ if $m\angle 1 = m\angle 2$.   **6.** $\angle 1 \cong \angle 2$ only if $m\angle 1 = m\angle 2$.

7. Combine the conditionals in Exercises 5 and 6 into a single biconditional.

Provide a counterexample to show that each statement is false. You may use words or draw a diagram.

8. If $\overline{AB} \cong \overline{BC}$, then B is the midpoint of $\overline{AC}$.

9. If a line lies in a vertical plane, then the line is vertical.

10. If a number is divisible by 4, then it is divisible by 6.

11. If $x^2 = 49$, then $x = 7$.

State the converse of each conditional. Is the converse true or false?

12. If today is Friday, then tomorrow is Saturday.   **13.** If $x > 0$, then $x^2 > 0$.

14. If a number is divisible by 6, then it is divisible by 3.   **15.** If $6x = 18$, then $x = 3$.

16. Give an example of a false conditional whose converse is true.

Written Exercises

Write the hypothesis and the conclusion of each conditional.

A **1.** If $3x - 7 = 32$, then $x = 13$.   **2.** I can't sleep if I'm not tired.

  3. I'll try if you will.   **4.** If $m \angle 1 = 90$, then $\angle 1$ is a right angle.

  5. $a + b = a$ implies $b = 0$.   **6.** $x = -5$ only if $x^2 = 25$.

Rewrite each pair of conditionals as a biconditional.

  7. If B is between A and C, then $AB + BC = AC$.
   If $AB + BC = AC$, then B is between A and C.

  8. If $m \angle AOC = 180$, then $\angle AOC$ is a straight angle.
   If $\angle AOC$ is a straight angle, then $m \angle AOC = 180$.

Write each biconditional as two conditionals that are converses of each other.

  9. Points are collinear if and only if they all lie in one line.

  10. Points lie in one plane if and only if they are coplanar.

Provide a counterexample to show that each statement is false. You may use words or a diagram.

  11. If $ab < 0$, then $a < 0$.   **12.** If $n^2 = 5n$, then $n = 5$.

  13. If point G is on $\overrightarrow{AB}$, then G is on $\overrightarrow{BA}$.   **14.** If $xy > 5y$, then $x > 5$.

  15. If a four-sided figure has four right angles, then it has four congruent sides.

  16. If a four-sided figure has four congruent sides, then it has four right angles.

Tell whether each statement is true or false. Then write the converse and tell whether it is true or false.

  17. If $x = -6$, then $|x| = 6$.   **18.** If $x^2 = 4$, then $x = -2$.

  19. If $b > 4$, then $5b > 20$.   **20.** If $m \angle T = 40$, then $\angle T$ is not obtuse.

  21. If Pam lives in Chicago, then she lives in Illinois.

  22. If $\angle A \cong \angle B$, then $m \angle A = m \angle B$.

B **23.** $a^2 > 9$ if $a > 3$.   **24.** $x = 1$ only if $x^2 = x$.

  25. $n > 5$ only if $n > 7$.   **26.** $ab = 0$ implies that $a = 0$ or $b = 0$.

  27. If points D, E, and F are collinear, then $DE + EF = DF$.

  28. P is the midpoint of $\overline{GH}$ implies that $GH = 2PG$.

  29. Write a definition of congruent angles as a biconditional.

  30. Write a definition of a right angle as a biconditional.

C **31.** What can you conclude if the following sentences are all true?
   (1) If p, then q.   (2) p (3) If q, then not r.   (4) s or r.

Geologist

Geologists study rock formations like those at Checkerboard Mountain in Zion National Park. Rock formations often occur in *strata*, or layers, beneath the surface of the Earth. Earthquakes occur at *faults*, breaks in the strata. In search of a fault, how would you determine the position of a stratum of rock buried deep beneath the surface of the Earth?

A geologist might start by picking three noncollinear points, A, B, and C, on the surface and drilling holes to find the depths of points A', B', and C' on the stratum. These three points determine the plane of the surface of the stratum.

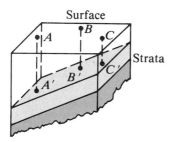

Geologists may work for industry, searching for oil or minerals. They may work in research centers, developing ways to predict earthquakes.

Today, geologists are trying to locate sources of geothermal energy, energy generated by the Earth's internal heat. A career in geology usually requires knowledge of mathematics, physics, and chemistry, as well as a degree in geology.

Mixed Review Exercises

Complete. You may find that drawing a diagram will help you.

1. If M is the midpoint of $\overline{AB}$, then __?__ $\cong$ __?__.

2. If $\overrightarrow{BX}$ is the bisector of $\angle ABC$, then __?__ $\cong$ __?__.

3. If point B lies in the interior of $\angle AOC$, then
   $m\angle$ __?__ $+ m\angle$ __?__ $= m\angle$ __?__.

4. If $\angle POQ$ is a straight angle and R is any point not on $\overleftrightarrow{PQ}$, then
   $m\angle$ __?__ $+ m\angle$ __?__ $=$ __?__.

2-2 *Properties from Algebra*

Since the length of a segment is a real number and the measure of an angle is a real number, the facts about real numbers and equality that you learned in algebra can be used in your study of geometry. The properties of equality that will be used most often are listed below.

Properties of Equality

Addition Property	If $a = b$ and $c = d$, then $a + c = b + d$.
Subtraction Property	If $a = b$ and $c = d$, then $a - c = b - d$.
Multiplication Property	If $a = b$, then $ca = cb$.
Division Property	If $a = b$ and $c \neq 0$, then $\dfrac{a}{c} = \dfrac{b}{c}$.
Substitution Property	If $a = b$, then either a or b may be substituted for the other in any equation (or inequality).
Reflexive Property	$a = a$
Symmetric Property	If $a = b$, then $b = a$.
Transitive Property	If $a = b$ and $b = c$, then $a = c$.

Recall that $DE = FG$ and $\overline{DE} \cong \overline{FG}$ can be used interchangeably, as can $m\angle D = m\angle E$ and $\angle D \cong \angle E$. Thus the following properties of congruence follow directly from the related properties of equality.

Properties of Congruence

Reflexive Property	$\overline{DE} \cong \overline{DE}$ $\qquad \angle D \cong \angle D$
Symmetric Property	If $\overline{DE} \cong \overline{FG}$, then $\overline{FG} \cong \overline{DE}$.
	If $\angle D \cong \angle E$, then $\angle E \cong \angle D$.
Transitive Property	If $\overline{DE} \cong \overline{FG}$ and $\overline{FG} \cong \overline{JK}$, then $\overline{DE} \cong \overline{JK}$.
	If $\angle D \cong \angle E$ and $\angle E \cong \angle F$, then $\angle D \cong \angle F$.

The properties of equality and other properties from algebra, such as the **Distributive Property,**

$$a(b + c) = ab + ac,$$

can be used to justify your steps when you solve an equation.

Example 1   Solve $3x = 6 - \frac{1}{2}x$ and justify each step.

Solution

Steps	*Reasons*
1. $3x = 6 - \frac{1}{2}x$	1. Given equation
2. $6x = 12 - x$	2. Multiplication Property of Equality
3. $7x = 12$	3. Addition Property of Equality
4. $x = \frac{12}{7}$	4. Division Property of Equality

Example 1 shows a proof of the statement "If $3x = 6 - \frac{1}{2}x$, then x *must* equal $\frac{12}{7}$." In other words, when given the information that $3x = 6 - \frac{1}{2}x$ we can use the properties of algebra to conclude, or *deduce*, that $x = \frac{12}{7}$.

Many proofs in geometry follow this same pattern. We use certain given information along with the properties of algebra and accepted statements, such as the Segment Addition Postulate and Angle Addition Postulate, to show that other statements *must* be true. Often a geometric proof is written in two-column form, with statements on the left and a reason for each statement on the right.

In the following examples, congruent segments are marked alike and congruent angles are marked alike. For example, in the diagram below, the marks show that $\overline{RS} \cong \overline{PS}$ and $\overline{ST} \cong \overline{SQ}$. In the diagram for Example 3 the marks show that $\angle AOC \cong \angle BOD$.

Example 2

Given: $\overline{RT}$ and $\overline{PQ}$ intersecting at S so that
$RS = PS$ and $ST = SQ$.

Prove: $RT = PQ$

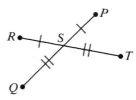

Proof:

Statements	Reasons
1. $RS = PS$; $ST = SQ$	1. Given
2. $RS + ST = PS + SQ$	2. Addition Prop. of $=$
3. $RS + ST = RT$; $PS + SQ = PQ$	3. Segment Addition Postulate
4. $RT = PQ$	4. Substitution Prop.

In Steps 1 and 3 of Example 2, notice how statements can be written in pairs when justified by the same reason.

Example 3

Given: $m \angle AOC = m \angle BOD$

Prove: $m \angle 1 = m \angle 3$

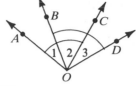

Proof:

Statements	Reasons
1. $m \angle AOC = m \angle BOD$	1. Given
2. $m \angle AOC = m \angle 1 + m \angle 2$; $\quad m \angle BOD = m \angle 2 + m \angle 3$	2. Angle Addition Postulate
3. $m \angle 1 + m \angle 2 = m \angle 2 + m \angle 3$	3. Substitution Prop.
4. $\qquad m \angle 2 = m \angle 2$	4. Reflexive Prop.
5. $m \angle 1 \qquad = \qquad m \angle 3$	5. Subtraction Prop. of $=$

Notice that the reason given for Step 4 is "Reflexive Property" rather than "Reflexive Property of Equality." Since the reflexive, symmetric, and transitive properties of equality are so closely related to the corresponding properties of congruence, we will simply use "Reflexive Property" to justify either

$$m \angle BOC = m \angle BOC \quad \text{or} \quad \angle BOC \cong \angle BOC.$$

Suppose, in a proof, you have made the statement that

$$m \angle R = m \angle S$$

and also the statement that

$$m \angle S = m \angle T.$$

You can then deduce that $m \angle R = m \angle T$ and use as your reason either "Transitive Property" or "Substitution Property." Similarly, if you know that

$$(1) \; m \angle R = m \angle S$$
$$(2) \; m \angle S = m \angle T$$
$$(3) \; m \angle T = m \angle V$$

you can go on to write $\qquad (4) \; m \angle R = m \angle V$

and use either "Transitive Property" or "Substitution Property" as your reason. Actually, you use the Transitive Property twice or else make a double substitution.

There are times when the Substitution Property is the simplest one to use. If you know that

$$(1) \; m \angle 4 + m \angle 2 + m \angle 5 = 180$$
$$(2) \; m \angle 4 = m \angle 1; \; m \angle 5 = m \angle 3$$

you can make a double substitution and get

$$(3) \; m \angle 1 + m \angle 2 + m \angle 3 = 180.$$

Note that you can't use the Transitive Property here.

Classroom Exercises

Justify each statement with a property from algebra or a property of congruence.

1. $\angle P \cong \angle P$

2. If $\overline{AB} \cong \overline{CD}$ and $\overline{CD} \cong \overline{EF}$, then $\overline{AB} \cong \overline{EF}$.

3. If $RS = TW$, then $TW = RS$.

4. If $x + 5 = 16$, then $x = 11$.

5. If $5y = -20$, then $y = -4$.

6. If $\frac{z}{5} = 10$, then $z = 50$.

7. $2(a + b) = 2a + 2b$

8. If $2z - 5 = -3$, then $2z = 2$.

9. If $2x + y = 70$ and $y = 3x$, then $2x + 3x = 70$.

10. If $AB = CD$, $CD = EF$, and $EF = 23$, then $AB = 23$.

Complete each proof by supplying missing reasons and statements.

11. Given: $m\angle 1 = m\angle 3$;
$\qquad\quad m\angle 2 = m\angle 4$

$\quad$ Prove: $m\angle ABC = m\angle DEF$

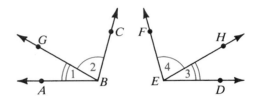

Proof:

Statements	Reasons
1. $m\angle 1 = m\angle 3$; $\quad m\angle 2 = m\angle 4$	1. _?_
2. $m\angle 1 + m\angle 2 = m\angle 3 + m\angle 4$	2. _?_
3. $m\angle 1 + m\angle 2 = m\angle ABC$; $\quad m\angle 3 + m\angle 4 = m\angle DEF$	3. _?_
4. $m\angle ABC = m\angle DEF$	4. _?_

12. Given: $ST = RN$; $IT = RU$

$\quad$ Prove: $SI = UN$

Proof:

Statements	Reasons
1. $ST = RN$	1. _?_
2. $\underline{\ ?\ } = SI + IT$; $\quad \underline{\ ?\ } = RU + UN$	2. _?_
3. $SI + IT = RU + UN$	3. _?_
4. $\quad IT = RU$	4. _?_
5. _?_	5. _?_

Written Exercises

Justify each step.

A

1. $4x - 5 = -2$
$4x \quad = 3$
$x \quad = \dfrac{3}{4}$

2. $\dfrac{3a}{2} = \dfrac{6}{5}$
$3a = \dfrac{12}{5}$
$a = \dfrac{4}{5}$

3. $\dfrac{z + 7}{3} = -11$
$z + 7 = -33$
$z \quad = -40$

4. $15y + 7 = 12 - 20y$
$35y + 7 = 12$
$35y \quad = 5$
$y \quad = \dfrac{1}{7}$

5. $\dfrac{2}{3}b = 8 - 2b$
$2b = 3(8 - 2b)$
$2b = 24 - 6b$
$8b = 24$
$b = 3$

6. $x - 2 = \dfrac{2x + 8}{5}$
$5(x - 2) = 2x + 8$
$5x - 10 = 2x + 8$
$3x - 10 = 8$
$3x \quad = 18$
$x \quad = 6$

Copy everything shown and supply missing statements and reasons.

7. Given: $\angle AOD$ as shown
Prove: $m\angle AOD = m\angle 1 + m\angle 2 + m\angle 3$

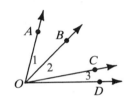

Proof:

Statements	Reasons
1. $m\angle AOD = m\angle AOC + m\angle 3$	1. _?_
2. $m\angle AOC = m\angle 1 + m\angle 2$	2. _?_
3. _?_	3. _?_

8. Given: $FL = AT$
Prove: $FA = LT$

Proof:

Statements	Reasons
1. _?_	1. Given
2. $LA = LA$	2. _?_
3. $FL + LA = AT + LA$	3. _?_
4. $FL + LA = FA$; $LA + AT = LT$	4. _?_
5. _?_	5. Substitution Prop.

9. Given: $DW = ON$
　　　Prove: $DO = WN$

Proof:

Statements	Reasons
1. $DW = ON$	1. _?_
2. $DW = DO + OW$;	2. _?_
$ON = $ _?_ $+$ _?_	
3. _?_	3. Substitution Prop.
4. $OW = OW$	4. _?_
5. _?_	5. _?_

10. Given: $m \angle 4 + m \angle 6 = 180$
　　　Prove: $m \angle 5 = m \angle 6$

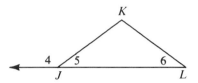

Proof:

Statements	Reasons
1. $m \angle 4 + m \angle 6 = 180$	1. _?_
2. $m \angle 4 + m \angle 5 = 180$	2. _?_
3. $m \angle 4 + m \angle 5 = m \angle 4 + m \angle 6$	3. _?_
4. $m \angle 4 \qquad = m \angle 4$	4. _?_
5. _?_	5. _?_

Copy everything shown and write a two-column proof.

B **11.** Given: $m \angle 1 = m \angle 2$;
　　　　　$m \angle 3 = m \angle 4$
　　　　Prove: $m \angle SRT = m \angle STR$

12. Given: $RP = TQ$;
　　　　　$PS = QS$
　　　　Prove: $RS = TS$

13. Given: $RQ = TP$;
　　　　　$ZQ = ZP$
　　　　Prove: $RZ = TZ$

14. Given: $m \angle SRT = m \angle STR$;
　　　　　$m \angle 3 = m \angle 4$
　　　　Prove: $m \angle 1 = m \angle 2$

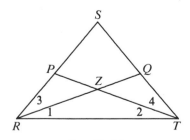

Exs. 11–14

C **15.** Consider the following statements:

Reflexive Property: Robot A is as rusty as itself.
Symmetric Property: If Robot A is as rusty as Robot B,
then Robot B is as rusty as Robot A.
Transitive Property: If Robot A is as rusty as Robot B and
Robot B is as rusty as Robot C, then
Robot A is as rusty as Robot C.

A *relation* such as "is as rusty as" that is reflexive, symmetric,
and transitive is an *equivalence relation*. Which of the following are equivalence relations?

a. is rustier than   **b.** has the same length as
c. is opposite (for rays)   **d.** is coplanar with (for lines)

2-3 *Proving Theorems*

Chapter 1 included three *theorems*, statements that are proved. The theorems
were deduced from *postulates*, statements that are accepted without proof. We
will prove additional theorems throughout the book. When writing proofs, we
will treat properties from algebra as postulates.

Suppose you are told that Y is the midpoint of $\overline{XZ}$ and that $XZ = 12$. You
probably realize that $XY = 6$. Your conclusion about one particular situation
suggests the general statement shown below as Theorem 2-1. The theorem
uses the definition of a midpoint to prove additional properties of a midpoint
that are not explicitly included in the definition. In this case, the theorem
states something obvious. Later theorems may not be so obvious. In fact,
some of them may surprise you.

Theorem 2-1   *Midpoint Theorem*

If M is the midpoint of $\overline{AB}$, then $AM = \frac{1}{2}AB$ and $MB = \frac{1}{2}AB$.

Given: M is the midpoint of $\overline{AB}$.
Prove: $AM = \frac{1}{2}AB$; $MB = \frac{1}{2}AB$

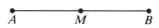

Proof:

Statements	Reasons
1. M is the midpoint of $\overline{AB}$.	1. Given
2. $\overline{AM} \cong \overline{MB}$, or $AM = MB$	2. Definition of midpoint
3. $AM + MB = AB$	3. Segment Addition Postulate
4. $AM + AM = AB$, or $2AM = AB$	4. Substitution Prop. (Steps 2 and 3)
5. $AM = \frac{1}{2}AB$	5. Division Prop. of $=$
6. $MB = \frac{1}{2}AB$	6. Substitution Prop. (Steps 2 and 5)

Example 1 Given: M is the midpoint of $\overline{AB}$;

N is the midpoint of $\overline{CD}$;

$AB = CD$

What can you deduce?

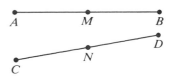

Solution Because M and N are midpoints, you know that $AM = MB$ and $CN = ND$.
From the Midpoint Theorem, you know that $AM = \frac{1}{2}AB$ and $CN = \frac{1}{2}CD$.
Since $AB = CD$, you know that $\frac{1}{2}AB = \frac{1}{2}CD$.
By substitution, you get $AM = CN$.
Thus you can deduce that AM, MB, CN, and ND are all equal.

The next theorem is similar to the Midpoint Theorem. It proves properties of the angle bisector that are not given in the definition. The proof is left as Classroom Exercise 10.

Theorem 2-2   *Angle Bisector Theorem*

If $\overrightarrow{BX}$ is the bisector of $\angle ABC$, then

$$m \angle ABX = \tfrac{1}{2}m \angle ABC \text{ and } m \angle XBC = \tfrac{1}{2}m \angle ABC.$$

Given: $\overrightarrow{BX}$ is the bisector of $\angle ABC$.

Prove: $m \angle ABX = \frac{1}{2}m \angle ABC$; $m \angle XBC = \frac{1}{2}m \angle ABC$

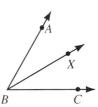

In addition to postulates and definitions, theorems may be used to justify steps in a proof. Notice the use of the Angle Bisector Theorem in Example 2.

Example 2

Given: $\overrightarrow{EG}$ is the bisector of $\angle DEF$;

$\overrightarrow{SW}$ is the bisector of $\angle RST$;

$m \angle DEG = m \angle RSW$

Prove: $m \angle DEF = m \angle RST$

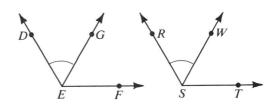

Proof:

Statements	Reasons
1. $\overrightarrow{EG}$ is the bisector of $\angle DEF$; $\overrightarrow{SW}$ is the bisector of $\angle RST$.	1. Given
2. $m \angle DEG = \frac{1}{2}m \angle DEF$; $m \angle RSW = \frac{1}{2}m \angle RST$	2. Angle Bisector Theorem
3. $m \angle DEG = m \angle RSW$	3. Given
4. $\frac{1}{2}m \angle DEF = \frac{1}{2}m \angle RST$	4. Substitution Prop. (Steps 2 and 3)
5. $m \angle DEF = m \angle RST$	5. Multiplication Prop. of =

The two-column proofs you have seen in this section and the previous one are examples of **deductive reasoning.** We have proved statements by reasoning from postulates, definitions, theorems, and given information. The kinds of reasons you can use to justify statements in a proof are listed below.

Reasons Used in Proofs

Given information

Definitions

Postulates (These include properties from algebra.)

Theorems that have already been proved

Classroom Exercises

What postulate, definition, or theorem justifies the statement about the diagram?

1. $m \angle AEB + m \angle BEC = m \angle AEC$

2. $AE + EF = AF$

3. $m \angle AEB + m \angle BEF = 180$

4. If E is the midpoint of $\overline{AF}$, then $\overline{AE} \cong \overline{EF}$.

5. If E is the midpoint of $\overline{AF}$, then $AE = \frac{1}{2}AF$.

6. If E is the midpoint of $\overline{AF}$, then $\overrightarrow{EC}$ bisects $\overline{AF}$.

7. If $\overrightarrow{EB}$ bisects $\overline{AF}$, then E is the midpoint of $\overline{AF}$.

8. If $\overrightarrow{EB}$ is the bisector of $\angle AEC$, then $m \angle AEB = \frac{1}{2}m \angle AEC$.

9. If $\angle BEC \cong \angle CEF$, then $\overrightarrow{EC}$ is the bisector of $\angle BEF$.

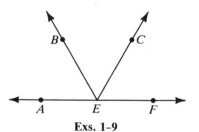

Exs. 1–9

10. Complete the proof of Theorem 2-2.

Given: $\overrightarrow{BX}$ is the bisector of $\angle ABC$.

Prove: $m \angle ABX = \frac{1}{2}m \angle ABC$; $m \angle XBC = \frac{1}{2}m \angle ABC$

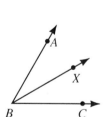

Proof:

Statements	Reasons
1. $\overrightarrow{BX}$ is the bisector of $\angle ABC$.	1. __?__
2. $\angle ABX \cong$ __?__, or $m \angle ABX =$ __?__	2. __?__
3. $m \angle ABX + m \angle XBC = m \angle ABC$	3. __?__
4. $m \angle ABX + m \angle ABX = m \angle ABC$, or $2m \angle ABX = m \angle ABC$	4. __?__
5. $m \angle ABX = \frac{1}{2}m \angle ABC$	5. __?__
6. $m \angle XBC = \frac{1}{2}m \angle ABC$	6. Substitution Prop. (Steps __?__ and __?__)

Written Exercises

Name the definition, postulate, or theorem that justifies the statement about the diagram.

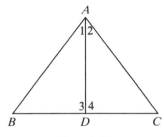

A 1. If D is the midpoint of $\overline{BC}$, then $\overline{BD} \cong \overline{DC}$.

  2. If $\angle 1 \cong \angle 2$, then $\overrightarrow{AD}$ is the bisector of $\angle BAC$.

  3. If $\overrightarrow{AD}$ bisects $\angle BAC$, then $\angle 1 \cong \angle 2$.

  4. $m\angle 3 + m\angle 4 = 180$

  5. If $\overline{BD} \cong \overline{DC}$, then D is the midpoint of $\overline{BC}$.

  6. If D is the midpoint of $\overline{BC}$, then $BD = \frac{1}{2}BC$.

  7. $m\angle 1 + m\angle 2 = m\angle BAC$

  8. $BD + DC = BC$

Exs. 1–8

Write the number that is paired with the bisector of $\angle CDE$.

9.
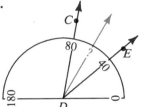

10.
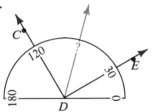

11.
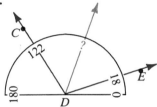

12. **a.** Draw a diagram similar to the one shown.
  b. Use a protractor to draw the bisectors of $\angle LMP$ and $\angle PMN$.
  c. What is the measure of the angle formed by these bisectors?
  d. Explain how you could have known the answer to part (c) without measuring.

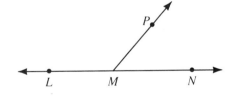

B 13. The coordinates of points L and X are 16 and 40, respectively. N is the midpoint of $\overline{LX}$, and Y is the midpoint of $\overline{LN}$. Sketch a diagram and find:
  a. LN **b.** the coordinate of N **c.** LY **d.** the coordinate of Y

  14. $\overrightarrow{SW}$ bisects $\angle RST$ and $m\angle RST = 72$. $\overrightarrow{SZ}$ bisects $\angle RSW$, and $\overrightarrow{SR}$ bisects $\angle NSW$. Sketch a diagram and find $m\angle RSZ$ and $m\angle NSZ$.

  15. **a.** Suppose M and N are the midpoints of $\overline{LK}$ and $\overline{GH}$, respectively. What segments are congruent?
  b. What additional information about the figure would enable you to deduce that $LM = NH$?

  16. **a.** Suppose $\overrightarrow{SV}$ bisects $\angle RST$ and $\overrightarrow{RU}$ bisects $\angle SRT$. What angles are congruent?
  b. What additional information would enable you to deduce that $m\angle VSU = m\angle URV$?

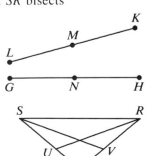

What can you deduce from the given information?

17. Given: $AE = DE$;
     $CE = BE$

18. Given: $\overline{AC}$ bisects $\overline{DB}$;
     $\overline{DB}$ bisects $\overline{AC}$;
     $CE = BE$

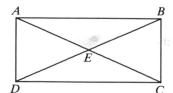

19. Copy and complete the following proof of the statement: If points A and B have coordinates a and b, with $b > a$, and the midpoint M of $\overline{AB}$ has coordinate x, then $x = \dfrac{a + b}{2}$.

    Given: Points A and B have coordinates a and b;
    $b > a$; midpoint M of $\overline{AB}$ has coordinate x.

    Prove: $x = \dfrac{a + b}{2}$

Proof:

Statements	Reasons
1. A, M, and B have coordinates a, x, and b respectively; $b > a$	1. __?__
2. $AM = x - a$; $MB = b - x$	2. __?__
3. M is the midpoint of $\overline{AB}$.	3. __?__
4. $\overline{AM} \cong \overline{MB}$, or $AM = MB$	4. __?__
5. $x - a = b - x$	5. __?__
6. $2x = $ __?__	6. __?__
7. $x = \dfrac{a + b}{2}$	7. __?__

C 20. Fold down a corner of a rectangular sheet of paper. Then fold the next corner so that the edges touch as in the figure. Measure the angle formed by the fold lines. Repeat with another sheet of paper, folding the corner at a different angle. Explain why the angles formed are congruent.

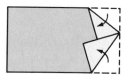

21. M is the midpoint of $\overline{AB}$, Q is the midpoint of $\overline{AM}$, and T is the midpoint of $\overline{QM}$. If the coordinates of A and B are a and b, find the coordinates of Q and T in terms of a and b.

22. Point T is the midpoint of $\overline{RS}$, W is the midpoint of $\overline{RT}$, and Z is the midpoint of $\overline{WS}$. If the length of $\overline{TZ}$ is x, find the following lengths in terms of x. (*Hint*: Sketch a diagram and let $y = WT$.)
    a. RW **b.** ZS **c.** RS **d.** WZ

◆ Computer Key-In

A bee starts at point P_0, flies to point P_1, and lands. The bee then returns half of the way to P_0, landing at P_2. From P_2, the bee returns half of the way to P_1, landing at P_3, and so forth. Can you predict the bee's location after 10 trips?

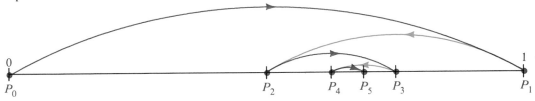

Assuming that P_0 and P_1 have coordinates 0 and 1, respectively, the BASIC program below will compute and print the bee's location at the end of trips 2 through 10. P_n represents the position of the bee after n trips. Since P_n is the midpoint of the bee's previous two positions, P_{n-1} and P_{n-2}, line 50 calculates $P(N)$ by using the statement proved in Exercise 19, page 47.

```
10  DIM P(50)
20  LET P(0) = 0
30  LET P(1) = 1
40  FOR N = 2 TO 10
50  LET P(N) = (1/2) * (P(N - 2) + P(N - 1))
60  PRINT N, P(N)
70  NEXT N
80  END
```

Exercises

1. Enter the program on your computer and RUN it. Do you notice any patterns or trends in the coordinates? Change line 40 so that the computer will print the coordinates up to P_{40}. What simple fraction is approximated by P_{40}?

2. In line 50, $P(n)$ could instead be computed from the *series*
$$1 - \tfrac{1}{2} + \tfrac{1}{4} - \tfrac{1}{8} + \cdots + (-\tfrac{1}{2})^{n-1}$$
where each term of the series reflects the bee's return half of the way from P_{n-1} to P_{n-2}. Replace line 50 with the line below and RUN the new program.

```
50  LET P(N) = P(N - 1) + (-1/2) ↑ (N - 1)
```

Check that both programs produce the same results. (Some slight variations will be expected, due to rounding off.)

3. Suppose that on each trip the bee returned one third of the way to the previous point instead of half of the way. How would the series in Exercise 2 be modified? How would line 50 of Exercise 2 be modified? RUN a modified program for 30 trips and determine what point the bee seems to be approaching.

Self-Test 1

Use the conditional: If $\overleftrightarrow{AB}$ and $\overleftrightarrow{CD}$ intersect, then $\overrightarrow{AB}$ and $\overrightarrow{CD}$ intersect.

1. Write the hypothesis and the conclusion of the conditional.

2. Write the converse of the conditional. Is the converse true or false?

3. Rewrite the following pair of conditionals as a biconditional:
   $\overline{AB} \cong \overline{CD}$ if $AB = CD$; $\overline{AB} \cong \overline{CD}$ only if $AB = CD$.

4. Provide a counterexample to disprove the statement:
   If $m \angle A$ is less than 100, then $\angle A$ is an acute angle.

5. Given: $m \angle A + m \angle B = 180$; $m \angle C = m \angle B$
   What property of equality justifies the statement $m \angle A + m \angle C = 180$?

6. Point M is the midpoint of $\overline{RT}$. $RM = x$ and $RT = 4x - 6$. Find the value of x.

7. The measure of $\angle ABC$ is 108. $\overrightarrow{BD}$ is the bisector of $\angle ABC$, and $\overrightarrow{BE}$ is the bisector of $\angle ABD$. Find the measure of $\angle EBC$.

8. You can use given information and theorems as reasons in proofs. Name two other kinds of reasons you can use.

| **Biographical Note** | *Julia Morgan* |

Julia Morgan (1872–1959), the first successful woman architect in the United States, was born in San Francisco. Though best known for her design of San Simeon, the castle-like former home of William Randolph Hearst pictured at the left, she designed numerous public buildings and private homes. Even today, to own "a Julia Morgan house" carries considerable prestige.

To become an architect, Morgan needed great determination as well as a brilliant mind. Since the University of California did not have an architecture curriculum at that time, she prepared for graduate work in Paris by studying civil engineering. In Paris the École des Beaux-Arts, which had just begun to admit foreigners, was particularly reluctant to admit a foreign woman. She persisted, however, and became the school's first woman graduate.

Theorems about Angles and Perpendicular Lines

Objectives

1. Apply the definitions of complementary and supplementary angles.
2. State and use the theorem about vertical angles.
3. Apply the definition and theorems about perpendicular lines.
4. State and apply the theorems about angles supplementary to, or complementary to, congruent angles.
5. Plan proofs and then write them in two-column form.

2-4 *Special Pairs of Angles*

Complementary angles (comp. $\angle s$) are two angles whose measures have the sum 90. Each angle is called a *complement* of the other.

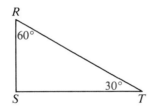

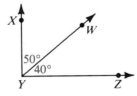

$\angle R$ and $\angle T$ are complementary.

$\angle XYW$ is a complement of $\angle WYZ$.

Supplementary angles (supp. $\angle s$) are two angles whose measures have the sum 180. Each angle is called a *supplement* of the other.

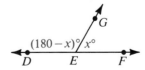

$\angle A$ and $\angle B$ are supplementary.

$\angle DEG$ is a supplement of $\angle GEF$.

Example 1 A supplement of an angle is three times as large as a complement of the angle. Find the measure of the angle.

Solution Let x = the measure of the angle.
Then $180 - x$ = the measure of its supplement,
and $90 - x$ = the measure of its complement.
$180 - x = 3(90 - x)$
$180 - x = 270 - 3x$
$2x = 90$
$x = 45$ The measure of the angle is 45.

Vertical angles (vert. &) are two angles such that the sides of one angle are opposite rays to the sides of the other angle. When two lines intersect, they form two pairs of vertical angles.

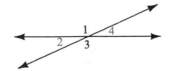

∠1 and ∠3 are vert. &.   ∠2 and ∠4 are vert. &.

Theorem 2-3

Vertical angles are congruent.

Given: ∠1 and ∠2 are vertical angles.

Prove: ∠1 ≅ 2

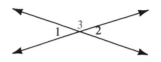

Proof:

Statements	Reasons
1. $m\angle 1 + m\angle 3 = 180$; $m\angle 2 + m\angle 3 = 180$	1. Angle Addition Postulate
2. $m\angle 1 + m\angle 3 = m\angle 2 + m\angle 3$	2. Substitution Prop.
3. $m\angle 3 = $ $m\angle 3$	3. Reflexive Prop.
4. $m\angle 1$ $= m\angle 2$, or $\angle 1 \cong \angle 2$	4. Subtraction Prop. of =

Example 2 In the diagram, ∠4 ≅ ∠5.
Name two other angles congruent to ∠5.

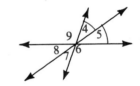

Solution   ∠8 ≅ ∠5 since vertical angles are congruent.
Since ∠7 ≅ ∠4 and ∠4 ≅ ∠5, ∠7 ≅ ∠5
by the Transitive Property.

Classroom Exercises

Find the measures of a complement and a supplement of ∠A.

1. $m\angle A = 10$ **2.** $m\angle A = 75$ **3.** $m\angle A = 89$ **4.** $m\angle A = y$

5. Name two right angles.

6. Name two adjacent complementary angles.

7. Name two complementary angles that are not adjacent.

8. a. Name a supplement of ∠MLQ.
   b. Name another pair of supplementary angles.

9. In the diagram, $m\angle AXB = 90$. Name:
   a. two congruent supplementary angles
   b. two supplementary angles that are not congruent
   c. two complementary angles
   d. a straight angle

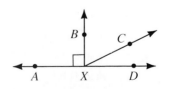

Complete.

10. $\angle AOB \cong$ ___?___   11. $\angle AOE \cong$ ___?___

12. $\angle FOB \cong$ ___?___   13. $\angle COA \cong$ ___?___

14. $m \angle FOE =$ ___?___   15. $m \angle COD =$ ___?___

16. $m \angle DOB =$ ___?___   17. $m \angle AOB =$ ___?___

18. $m \angle COE =$ ___?___   19. $m \angle FOB =$ ___?___

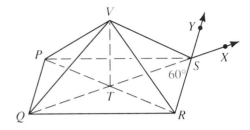

20. The four angles of figure $PQRS$ are right angles. $\angle VTR$ is a right angle. $m \angle QSR = 60$. Find the measures.
    a. $m \angle VTP$
    b. $m \angle XSY$
    c. $m \angle RSX$
    c. $m \angle PSY$

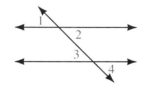

21. Given: $\angle 2 \cong \angle 3$
    a. What can you deduce?
    b. Explain how you would prove your conclusion.

Written Exercises

Find the measures of a complement and a supplement of $\angle K$.

A   1. $m \angle K = 20$ 2. $m \angle K = 72\frac{1}{2}$ 3. $m \angle K = x$ 4. $m \angle K = 2y$

5. Two complementary angles are congruent. Find their measures.

6. Two supplementary angles are congruent. Find their measures.

In the diagram, $\angle AFB$ is a right angle. Name the figures described.

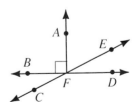

7. Another right angle

8. Two complementary angles

9. Two congruent supplementary angles

10. Two noncongruent supplementary angles

11. Two acute vertical angles

12. Two obtuse vertical angles

In the diagram, $\overrightarrow{OT}$ bisects $\angle SOU$, $m \angle UOV = 35$, and $m \angle YOW = 120$. Find the measure of each angle.

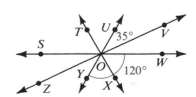

13. $m \angle ZOY$

14. $m \angle ZOW$

15. $m \angle VOW$

16. $m \angle SOU$

17. $m \angle TOU$

18. $m \angle ZOT$

Find the value of *x*.

19.

20.

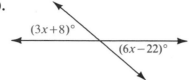

21.

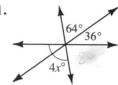

22. ∠1 and ∠2 are supplements.
∠3 and ∠4 are supplements.

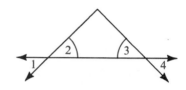

a. If $m\angle 1 = m\angle 3 = 27$, find $m\angle 2$ and $m\angle 4$.
b. If $m\angle 1 = m\angle 3 = x$, find $m\angle 2$ and $m\angle 4$ in terms of *x*.
c. If two angles are congruent, must their supplements be congruent?

23. Copy everything shown. Complete the proof.

Given: $\angle 2 \cong \angle 3$
Prove: $\angle 1 \cong \angle 4$

Proof:

Statements	Reasons
1. $\angle 1 \cong \angle 2$	1. _?_
2. $\angle 2 \cong \angle 3$	2. _?_
3. $\angle 3 \cong \angle 4$	3. _?_
4. _?_	4. Transitive Property (used twice)

If ∠*A* and ∠*B* are supplementary, find the value of *x*, *m* ∠*A*, and *m* ∠*B*.

B **24.** $m\angle A = 2x,\ m\angle B = x - 15$ **25.** $m\angle A = x + 16,\ m\angle B = 2x - 16$

If ∠*C* and ∠*D* are complementary, find the value of *y*, *m* ∠*C*, and *m* ∠*D*.

26. $m\angle C = 3y + 5,\ m\angle D = 2y$ **27.** $m\angle C = y - 8,\ m\angle D = 3y + 2$

Use the given information to write an equation and solve the problem.

28. Find the measure of an angle that is twice as large as its supplement.

29. Find the measure of an angle that is half as large as its complement.

30. The measure of a supplement of an angle is 12 more than twice the measure of the angle. Find the measures of the angle and its supplement.

31. A supplement of an angle is six times as large as a complement of the angle. Find the measures of the angle, its supplement, and its complement.

Find the values of *x* and *y* for each diagram.

32.

$x°$ $(3x-8)°$
$(2y-17)°$

33.

$50°$ $(3x-y)°$
$x°$
$(2x-16)°$

C **34.** Can the measure of a complement of an angle ever equal exactly half the measure of a supplement of the angle? Explain.

35. You are told that the measure of an acute angle is equal to the difference between the measure of a supplement of the angle and twice the measure of a complement of the angle. What can you deduce about the angle? Explain.

Application   *Orienteering*

The sport of orienteering involves finding your way from control point to control point in a wilderness area, using a map and protractor-type compass. Similar methods can be used by hikers, hunters, boaters, and backpackers.

One thing you want to be able to do is locate your position on the map. This can be done by taking sightings of specific objects. For example, suppose you can see a lookout tower (on Number Four Mountain at ● on the map shown below).

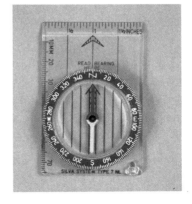

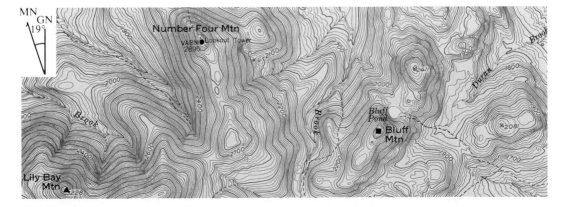

You sight across your compass and discover the tower is 33° east of magnetic north (MN). On your map you draw a line through the tower at a 33° angle to magnetic north. Be sure to use magnetic north rather than true north, for they may differ by as much as 20°. Hiking maps and nautical charts usually give both. All compass readings here are given in terms of magnetic north.

You are somewhere on the line you have drawn. If there is a feature near you (a trail, stream or pond), then your position is where the line crosses the feature on the map. Otherwise, you will need to take a second sighting, on the peak of Lily Bay Mountain (at ▲ on the map). It is 50° west of north. Draw a line on your map through the peak at a 50° angle with magnetic north. You are close to the point where the lines cross.

Since a third landmark is visible, the summit of Bluff Mountain (■ on the map), you can check your position with a third sighting. The three lines might cross at a single point. However, there is usually some error in sighting and drawing the angles, so instead of meeting exactly at a point, the three lines drawn often form a triangle. If the triangle is small, it gives you a good idea of your true position.

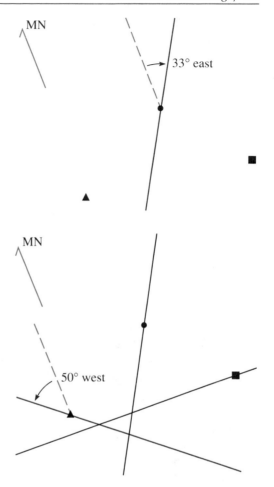

Exercises

1. Another orienteering party sights on Lily Bay Mountain and the lookout tower and finds the following angles: mountain, 58° west of north; tower, 40° east of north. Are they north or south of you?

2. If you head due east from Lily Bay Mountain (90° east of magnetic north), will you pass Bluff Mountain on your right or on your left?

3. Lillian and Ray both sight Lily Bay Mountain at 70° west of north, but Lillian sees the lookout tower at 40° east of north, while Ray sees it at 20° east of north. Which person is closer to Bluff Mountain?

4. Sailors use this method of finding their position when they are navigating near shore, sighting on lighthouses, smokestacks, and other landmarks shown on their charts. They call the small triangle formed by the three sighting lines a ''cocked hat,'' and usually mark their position at the corner closest to the nearest hazard. Why is this a sensible rule?

2-5 *Perpendicular Lines*

In the town shown, roads that run east-west are called streets, while those that run north-south are called avenues. Each of the streets is *perpendicular* to each of the avenues.

Perpendicular lines are two lines that intersect to form right angles (90° angles). Because lines that form one right angle always form four right angles (see Exercise 26, page 21), you can conclude that two lines are perpendicular, by definition, once you know that any one of the angles they form is a right angle. The definition of perpendicular lines can be used in the two ways shown below.

1. If $\overleftrightarrow{JK}$ is perpendicular to $\overleftrightarrow{MN}$ (written $\overleftrightarrow{JK} \perp \overleftrightarrow{MN}$), then each of the numbered angles is a right angle (a 90° angle).

2. If any one of the numbered angles is a right angle (a 90° angle), then $\overleftrightarrow{JK} \perp \overleftrightarrow{MN}$.

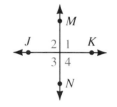

The word *perpendicular* is also used for intersecting rays and segments. For example, if $\overleftrightarrow{JK} \perp \overleftrightarrow{MN}$ in the diagram, then $\overline{JK} \perp \overline{MN}$ and the sides of $\angle 2$ are perpendicular.

The definition of perpendicular lines is closely related to the following theorems. Notice that Theorem 2-4 and Theorem 2-5 are *converses* of each other. For the proofs of the theorems, see the exercises.

Theorem 2-4
If two lines are perpendicular, then they form congruent adjacent angles.

Theorem 2-5
If two lines form congruent adjacent angles, then the lines are perpendicular.

Theorem 2-6
If the exterior sides of two adjacent acute angles are perpendicular, then the angles are complementary.

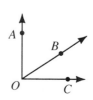

Given: $\overrightarrow{OA} \perp \overrightarrow{OC}$

Prove: $\angle AOB$ and $\angle BOC$ are comp. $\angle$ s.

Classroom Exercises

1. Complete the proof of Theorem 2-4: If two lines are perpendicular, then they form congruent adjacent angles.

Given: $l \perp n$
Prove: $\angle 1$, $\angle 2$, $\angle 3$, and $\angle 4$ are congruent angles.

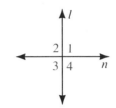

Proof:

Statements	Reasons
1. $l \perp n$	1. _?_
2. $\angle 1$, $\angle 2$, $\angle 3$, $\angle 4$ are 90° ⓢ.	2. Definition of _?_
3. $\angle 1$, $\angle 2$, $\angle 3$, $\angle 4$ are ≅ ⓢ.	3. Definition of _?_

2. In the diagram, $\overleftrightarrow{AB} \perp \overleftrightarrow{CD}$ and $\overleftrightarrow{EF} \perp \overleftrightarrow{GH}$. Name eight right angles.

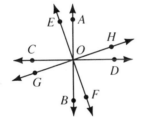

3. In the diagram, $\overrightarrow{OZ} \perp \overleftrightarrow{PQ}$, $\overrightarrow{OZ} \perp \overleftrightarrow{XY}$, and $\overleftrightarrow{PQ} \perp \overleftrightarrow{XY}$. Name eight right angles.

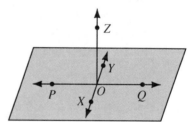

In the diagram, $\overrightarrow{BE} \perp \overleftrightarrow{AC}$ and $\overrightarrow{BD} \perp \overrightarrow{BF}$. Find the measures of the following angles.

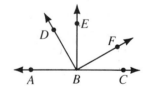

	$m \angle CBF$	$m \angle EBF$	$m \angle DBE$	$m \angle DBA$	$m \angle DBC$
4.	40	?	?	?	?
5.	x	?	?	?	?

Name the definition or state the theorem that justifies the statement about the diagram.

6. If $\angle 6$ is a right angle, then $\overleftrightarrow{RS} \perp \overleftrightarrow{TV}$.
7. If $\overleftrightarrow{RS} \perp \overleftrightarrow{TV}$, then $\angle 5$, $\angle 6$, $\angle 7$, and $\angle 8$ are right angles.
8. If $\overleftrightarrow{RS} \perp \overleftrightarrow{TV}$, then $\angle 8 \cong \angle 7$.
9. If $\overleftrightarrow{RS} \perp \overleftrightarrow{TV}$, then $m \angle 6 = 90$.
10. If $\angle 5 \cong \angle 6$, then $\overleftrightarrow{RS} \perp \overleftrightarrow{TV}$.
11. If $m \angle 5 = 90$, then $\overleftrightarrow{RS} \perp \overleftrightarrow{TV}$.

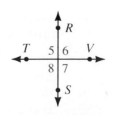

Written Exercises

A   **1.** In the diagram, $\overrightarrow{UL} \perp \overleftrightarrow{MJ}$ and $m \angle JUK = x$. Express in terms of x the measures of the angles named.

    a. $\angle LUK$ **b.** $\angle MUK$

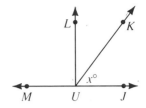

2. Copy and complete the proof of Theorem 2-5: If two lines form congruent adjacent angles, then the lines are perpendicular.

    Given: $\angle 1 \cong \angle 2$
    Prove: $l \perp n$

Proof:

Statements	Reasons
1. $\angle 1 \cong \angle 2$, or $m \angle 1 = m \angle 2$	1. ___?___
2. $m \angle 1 + m \angle 2 = 180$	2. ___?___
3. $m \angle 2 + m \angle 2 = 180$, or $2m \angle 2 = 180$	3. ___?___
4. $m \angle 2 = 90$	4. ___?___
5. ___?___	5. Def. of $\perp$ lines

Name the definition or state the theorem that justifies the statement about the diagram.

3. If $\angle EBC$ is a right angle, then $\overrightarrow{BE} \perp \overleftrightarrow{AC}$.

4. If $\overleftrightarrow{AC} \perp \overrightarrow{BE}$, then $\angle ABE$ is a right angle.

5. If $\overrightarrow{BE} \perp \overleftrightarrow{AC}$, then $\angle ABD$ and $\angle DBE$ are complementary.

6. If $\angle ABD$ and $\angle DBE$ are complementary angles, then $m \angle ABD + m \angle DBE = 90$.

7. If $\overrightarrow{BE} \perp \overleftrightarrow{AC}$, then $m \angle ABE = 90$.

8. If $\angle ABE \cong \angle EBC$, then $\overleftrightarrow{AC} \perp \overrightarrow{BE}$.

Exs. 3–12

In the diagram, $\overrightarrow{BE} \perp \overleftrightarrow{AC}$ and $\overrightarrow{BD} \perp \overrightarrow{BF}$. Find the value of x.

9. $m \angle ABD = 2x - 15$, $m \angle DBE = x$

10. $m \angle DBE = 3x$, $m \angle EBF = 4x - 1$

11. $m \angle ABD = 3x - 12$, $m \angle DBE = 2x + 2$, $m \angle EBF = 2x + 8$

12. $m \angle ABD = 6x$, $m \angle DBE = 3x + 9$, $m \angle EBF = 4x + 18$,
    $m \angle FBC = 4x$

13. Copy and complete the proof of Theorem 2-6: If the exterior sides of two adjacent acute angles are perpendicular, then the angles are complementary.

Given: $\overrightarrow{OA} \perp \overrightarrow{OC}$
Prove: $\angle AOB$ and $\angle BOC$ are comp. $\angle$s.

Proof:

Statements	Reasons
1. $\overrightarrow{OA} \perp \overrightarrow{OC}$	1. __?__
2. $m \angle AOC = 90$	2. Def. of $\perp$ lines
3. $m \angle AOB + m \angle BOC = m \angle AOC$	3. __?__
4. __?__	4. Substitution Prop.
5. __?__	5. Def. of comp. $\angle$s

In the figure $\overleftrightarrow{BF} \perp \overleftrightarrow{AE}$, $m \angle BOC = x$, and $m \angle GOH = y$. Express the measure of the angle in terms of x, y, or both.

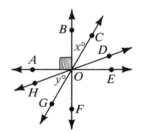

B 14. $\angle COA$

15. $\angle COH$

16. $\angle HOF$

17. $\angle DOE$

Can you conclude from the information given for each exercise that $\overrightarrow{XY} \perp \overrightarrow{XZ}$?

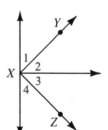

18. $m \angle 1 = 46$ and $m \angle 4 = 44$

19. $\angle 1$ and $\angle 3$ are complementary.

20. $\angle 2 \cong \angle 3$

21. $m \angle 1 = m \angle 4$

22. $\angle 1$ and $\angle 3$ are congruent and complementary.

23. $m \angle 1 = m \angle 2$ and $m \angle 3 = m \angle 4$

24. $\angle 1 \cong \angle 3$ and $\angle 2 \cong \angle 4$

25. $\angle 1 \cong \angle 4$ and $\angle 2 \cong \angle 3$

What can you conclude from the information given?

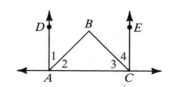

26. Given: $\overrightarrow{AB}$ bisects $\angle DAC$;
    $\overrightarrow{CB}$ bisects $\angle ECA$;
    $m \angle 2 = 45$;
    $m \angle 3 = 45$

27. Given: $\overrightarrow{AD} \perp \overleftrightarrow{AC}$; $\overrightarrow{CE} \perp \overleftrightarrow{AC}$; $m \angle 1 = m \angle 4$

28. Copy everything shown and write a two-column proof.

Given: $\overleftrightarrow{AO} \perp \overleftrightarrow{CO}$

Prove: $\angle 1$ and $\angle 3$ are comp. $\angle s$.

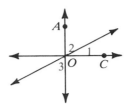

C   29. First find two lines (other than $\overleftrightarrow{YD}$ and $\overleftrightarrow{YF}$) that are perpendicular. Then write a two-column proof that the lines are perpendicular.

Given: $\overleftrightarrow{YD} \perp \overleftrightarrow{YF}$;
$\qquad m \angle 7 = m \angle 5$;
$\qquad m \angle 8 = m \angle 6$

Prove: _____?_____ $\perp$ _____?_____

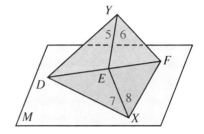

Mixed Review Exercises

Write something you can conclude from the given information.

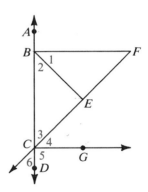

1. Given: $m \angle 1 = m \angle 4$; $m \angle 2 = m \angle 3$

2. Given: $AB = CD$

3. Given: $m \angle 6 = m \angle 4$

4. Given: $\overrightarrow{FB} \perp \overrightarrow{AD}$; $\overrightarrow{BE}$ bisects $\angle FBC$.

5. Given: $BE = EF$; E is the midpoint of $\overline{FC}$.

6. Given: $\angle 1$ and $\angle 2$ are complements.

7. Given: $\angle 4$ and $\angle 6$ are complements.

2-6 *Planning a Proof*

As you have seen in the last few sections, a proof of a theorem consists of five parts:

1. *Statement* of the theorem

2. A *diagram* that illustrates the given information

3. A list, in terms of the figure, of what is *given*

4. A list, in terms of the figure, of what you are to *prove*

5. A series of *statements and reasons* that lead from the given information to the statement that is to be proved

In many of the proofs in this book, the diagram and the statements of what is given and what is to be proved will be supplied for you. Sometimes you will be asked to provide them.

When you draw a diagram, try to make it reasonably accurate, avoiding special cases that might mislead. For example, when a theorem refers to an angle, don't draw a *right* angle.

Before you write the steps in a two-column proof you will need to plan your proof. Sometimes you will read the statement of a theorem and see immediately how to prove it. Other times you may need to try several approaches before you find a plan that works.

If you don't see a method of proof immediately, try reasoning back from what you would like to prove. Think: "This conclusion will be true if _?_ is true. This, in turn, will be true if _?_ is true . . . ." Sometimes this procedure leads back to a given statement. If so, you have found a method of proof.

Studying the proofs of previous theorems may suggest methods to try. For example, the proof of the theorem that vertical angles are congruent suggests the proof of the following theorem.

Theorem 2-7

If two angles are supplements of congruent angles (or of the same angle), then the two angles are congruent.

Given: $\angle 1$ and $\angle 2$ are supplementary;
$\qquad \angle 3$ and $\angle 4$ are supplementary;
$\qquad \angle 2 \cong \angle 4$

Prove: $\angle 1 \cong \angle 3$

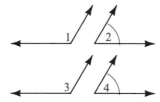

Proof:

Statements	Reasons
1. $\angle 1$ and $\angle 2$ are supplementary; $\angle 3$ and $\angle 4$ are supplementary.	1. Given
2. $m\angle 1 + m\angle 2 = 180$; $m\angle 3 + m\angle 4 = 180$	2. Def. of supp. $\angle s$
3. $m\angle 1 + m\angle 2 = m\angle 3 + m\angle 4$	3. Substitution Prop.
4. $\angle 2 \cong \angle 4$, or $m\angle 2 = m\angle 4$	4. Given
5. $m\angle 1 = m\angle 3$, or $\angle 1 \cong \angle 3$	5. Subtraction Prop. of $=$

The proof of the following theorem is left as Exercise 18.

Theorem 2-8

If two angles are complements of congruent angles (or of the same angle), then the two angles are congruent.

There is often more than one way to prove a particular statement, and the amount of detail one includes in a proof may differ from person to person. You should show enough steps so the reader can follow your argument and see why the theorem you are proving is true. As you gain more experience in writing proofs, you and your teacher may agree on what steps may be combined or omitted.

Classroom Exercises

a. **In each exercise use the information given to conclude that two angles are congruent.**

b. **Name or state the definition or theorem that justifies your conclusion.**

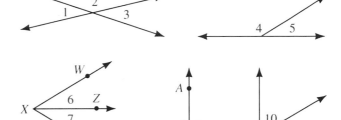

1. $\angle 6$ is comp. to $\angle 10$;
   $\angle 7$ is comp. to $\angle 10$.

2. $m \angle 5 = 31$; $m \angle 7 = 31$

3. $\overline{AB} \perp \overline{CD}$

4. $\overrightarrow{XZ}$ bisects $\angle WXY$.

5. $\angle 4$ is supp. to $\angle 6$;
   $\angle 2$ is supp. to $\angle 7$;
   $\angle 6 \cong \angle 7$

6. Given only the diagrams, and no additional information

Describe your plan for proving the following. You don't need to give all the details.

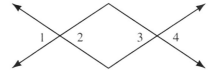

7. Given: $\angle 2 \cong \angle 3$
   Prove: $\angle 1 \cong \angle 4$

8. Given: $\angle 3$ is supp. to $\angle 1$;
   $\angle 4$ is supp. to $\angle 2$.
   Prove: $\angle 3 \cong \angle 4$

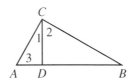

9. Given: $\overline{AC} \perp \overline{BC}$;
   $\angle 3$ is comp. to $\angle 1$.
   Prove: $\angle 3 \cong \angle 2$

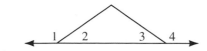

10. Given: $m \angle 1 = m \angle 4$
    Prove: $m \angle 2 = m \angle 3$

Written Exercises

Write the name or statement of the definition, postulate, property, or theorem that justifies the statement about the diagram.

A 1. $AD + DB = AB$

2. $m \angle 1 + m \angle 2 = m \angle CDB$

3. $\angle 2 \cong \angle 6$

4. If D is the midpoint of $\overline{AB}$, then $AD = \frac{1}{2}AB$.

5. If $\overrightarrow{DF}$ bisects $\angle CDB$, then $\angle 1 \cong \angle 2$.

6. $m \angle ADF + m \angle FDB = 180$

7. If $\overline{CD} \perp \overline{AB}$, then $m \angle CDB = 90$.

8. If $\angle 4 \cong \angle 3$, then $\overrightarrow{DG}$ bisects $\angle BDE$.

9. If $m \angle 3 + m \angle 4 = 90$, then $\angle 3$ and $\angle 4$ are complements.

10. If $\angle ADF$ and $\angle 4$ are supplements, then $m \angle ADF + m \angle 4 = 180$.

11. If $\overline{AB} \perp \overline{CE}$, then $\angle ADC \cong \angle ADE$.

12. If $\angle 4$ is complementary to $\angle 5$ and $\angle 6$ is complementary to $\angle 5$, then $\angle 4 \cong \angle 6$.

13. If $\angle FDG$ is a right angle, then $\overrightarrow{DF} \perp \overrightarrow{DG}$.

14. If $\angle FDG \cong \angle GDH$, then $\overrightarrow{DG} \perp \overleftrightarrow{HF}$.

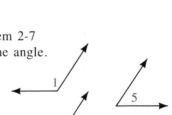

Exs. 1–14

15. Copy everything shown and complete the proof of Theorem 2-7 for the case where two angles are supplements of the same angle.

Given: $\angle 1$ and $\angle 5$ are supplementary; $\angle 3$ and $\angle 5$ are supplementary.

Prove: $\angle 1 \cong \angle 3$

Proof:

Statements	Reasons
1. $\angle 1$ and $\angle 5$ are supplementary; $\angle 3$ and __?__.	1. __?__
2. $m \angle 1 + m \angle 5 = 180$; $m \angle 3 + m \angle 5 = 180$	2. __?__
3. $m \angle 1 + m \angle 5 = m \angle 3 + m \angle 5$	3. __?__
4. $\qquad m \angle 5 = \qquad m \angle 5$	4. Reflexive Prop.
5. $m \angle 1 = m \angle 3$, or $\angle 1 \cong \angle 3$	5. __?__

16. **a.** Are there any angles in the diagram that must be congruent to $\angle 4$? Explain.

b. If $\angle 4$ and $\angle 5$ are supplementary, name all angles shown that must be congruent to $\angle 4$.

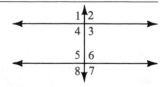

17. a. Copy everything shown and complete the proof.

Given: $\overline{PQ} \perp \overline{QR}$;
$\overline{PS} \perp \overline{SR}$;
$\angle 1 \cong \angle 4$

Prove: $\angle 2 \cong \angle 5$

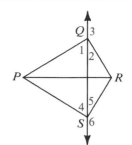

Proof:

Statements	Reasons
1. $\overline{PQ} \perp \overline{QR}$; $\overline{PS} \perp \overline{SR}$	1. _?_
2. $\angle 2$ is comp. to $\angle 1$; $\angle 5$ is comp. to $\angle 4$.	2. _?_
3. $\angle 1 \cong \angle 4$	3. _?_
4. $\angle 2 \cong \angle 5$	4. _?_

b. After proving that $\angle 2 \cong \angle 5$ in part (a), tell how you could go on to prove that $\angle 3 \cong \angle 6$.

B **18.** Prove Theorem 2-8: If two angles are complements of congruent angles, then the two angles are congruent. *Note*: You will need to draw your own diagram and state what is given and what you are to prove in terms of your diagram. (*Hint*: See the proof of Theorem 2-7 on page 61.)

Copy everything shown and write a two-column proof.

19. Given: $\angle 2 \cong \angle 3$
Prove: $\angle 1 \cong \angle 4$

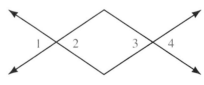

20. Given: $\angle 3$ is supp. to $\angle 1$;
$\angle 4$ is supp. to $\angle 2$.
Prove: $\angle 3 \cong \angle 4$

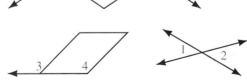

21. Given: $\overline{AC} \perp \overline{BC}$;
$\angle 3$ is comp. to $\angle 1$.
Prove: $\angle 3 \cong \angle 2$

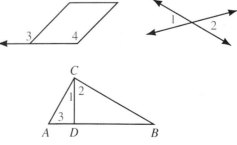

22. Given: $m\angle 1 = m\angle 2$;
$m\angle 3 = m\angle 4$
Prove: $\overrightarrow{YS} \perp \overleftrightarrow{XZ}$

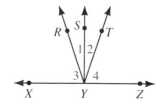

23. Draw any ∠*AOB* and its bisector $\overrightarrow{OE}$. Now draw the rays opposite to $\overrightarrow{OA}$, $\overrightarrow{OB}$, and $\overrightarrow{OE}$. What can you conclude about the part of the diagram shown in red? Prove your conclusion.

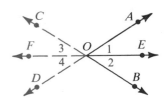

C 24. Make a diagram showing ∠*PQR* bisected by $\overrightarrow{QX}$. Choose a point *Y* on the ray opposite to $\overrightarrow{QX}$.
Prove: ∠*PQY* ≅ ∠*RQY*

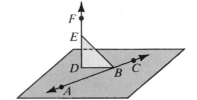

25. Given: *m*∠*DBA* = 45;
    m∠*DEB* = 45
    Prove: ∠*DBC* ≅ ∠*FEB*

Self-Test 2

1. It is known that ∠*HOK* has a supplement, but can't have a complement. Name one possible measure for ∠*HOK*.

2. *m*∠1 = 3*x* − 5 and *m*∠2 = *x* + 25
   a. *x* = __?__   **b.** *m*∠1 = __?__ (numerical value)

For Exercises 3 and 4 you are given that $\overrightarrow{OB}$ ⊥ *l* and $\overrightarrow{OA}$ ⊥ $\overrightarrow{OC}$.

3. If *m*∠3 = 37, complete:
   m∠4 = __?__   *m*∠5 = __?__   *m*∠6 = __?__
4. If *m*∠3 = *t*, express the measures of the other numbered angles in terms of *t*.

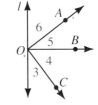

In the diagram, $\overline{DE}$ ⊥ *n*. State the theorem or name the definition that justifies the statement about the diagram.

5. ∠8 is a 90° angle.
6. ∠7 ≅ ∠10
7. ∠9 and ∠10 are complementary.

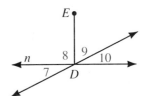

8. Give a plan for the following proof.
   Given: ∠1 is supp. to ∠3;
          ∠2 is supp. to ∠3.
   Prove: *j* ⊥ *k*

9. Write a proof for Exercise 8 in two-column form.

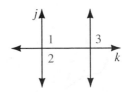

| **Extra** | *Möbius Bands* |

Take a long narrow strip of paper. Give the strip a half-twist. Tape the ends together. The result is a *Möbius band*.

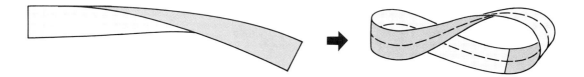

Exercises

1. Make a Möbius band. Color one side of the Möbius band. How much of the band is left uncolored? The original strip of paper had two sides. How many sides does a Möbius band have?

2. Cut the Möbius band lengthwise down the middle. (Start at a point midway between the edges and cut around the band.) What is the result? Cut the band a second time down the middle. Write a sentence or two describing what happens.

3. Give a full twist to a long, narrow strip of paper. Tape the ends together. How many sides does this band have? Cut this band lengthwise down the middle. Write a brief description of what is formed.

4. Make a Möbius band. Let the band be 3 cm wide. Make a lengthwise cut, staying 1 cm from the right-hand edge. Describe the result.

5. Take two long narrow strips of paper. Fasten them together so they are perpendicular and form a plus sign. Twist one strip so it is a Möbius band and fasten its ends together. Don't twist the other strip at all, just fasten its ends. Cut the Möbius band down the middle lengthwise. Then cut the other band down the middle. Describe the final result.

Chapter Summary

1. *If p, then q* is a conditional statement. *p* is the hypothesis and *q* is the conclusion. *If q, then p* is the converse. The statement *p if and only if q* is a biconditional that means both the conditional and its converse are true.

2. Properties of algebra (see page 37) can be used to reach conclusions in geometry. Properties of congruence are related to some of the properties of equality.

3. Deductive reasoning is a process of proving conclusions. Given information, definitions, postulates, and previously proved theorems are the four kinds of reasons that can be used to justify statements in a proof.

4. When $m \angle A + m \angle B = 90$, $\angle A$ and $\angle B$ are complementary. When $m \angle C + m \angle D = 180$, $\angle C$ and $\angle D$ are supplementary. Complements (or supplements) of the same angle or of congruent angles are congruent.

5. Vertical angles are congruent.

6. Perpendicular lines are two lines that form right angles (90° angles). If two lines are perpendicular, then they form congruent adjacent angles. If two lines form congruent adjacent angles, then the lines are perpendicular.

7. If the exterior sides of two adjacent acute angles are perpendicular, then the angles are complementary.

8. The proof of a theorem consists of five parts, which are listed on page 60.

Chapter Review

Use the conditional: If $m \angle 1 = 120$, then $\angle 1$ is obtuse.

1. Write the hypothesis and the conclusion of the conditional. **2-1**

2. Write the converse of the conditional.

3. Provide a counterexample to disprove the converse.

4. Write a definition of a straight angle as a biconditional.

Justify each statement with a property from algebra or a property of congruence.

5. If $m \angle A + m \angle B + m \angle C = 180$ and $m \angle C = 50$, then   **2-2**
$m \angle A + m \angle B + 50 = 180$.

6. If $m \angle A + m \angle B + 50 = 180$, then $m \angle A + m \angle B = 130$.

7. If $6x = 18$, then $x = 3$.

8. If $\overline{AB} \cong \overline{CD}$ and $\overline{CD} \cong \overline{EF}$, then $\overline{AB} \cong \overline{EF}$.

Name the definition, postulate, or theorem that justifies the statement.

9. If $\overline{RS} \cong \overline{ST}$, then S is the midpoint of $\overline{RT}$. **2-3**

10. If $\overrightarrow{SW}$ bisects $\angle VST$, then $\angle VSW \cong \angle WST$.

11. If $\overrightarrow{SW}$ bisects $\angle VST$, then $m \angle WST = \frac{1}{2} m \angle VST$.

)

12. If $\angle BOC$ is a right angle and $m \angle COD = 58$, **2-4**
then $m \angle DOE = \underline{\ ?\ }$, $m \angle BOA = \underline{\ ?\ }$,
and $m \angle AOC = \underline{\ ?\ }$.

13. Name a supplement of $\angle AOE$.

14. A supplement of a given angle is four times as large as a complement of the angle. Find the measure of the given angle.

Name the definition or state the theorem that justifies the statement about the diagram.

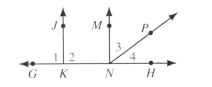

15. If $\overrightarrow{KJ} \perp \overleftrightarrow{GH}$, then $\angle 1$ is a right angle.   2–5

16. If $\angle 2$ is a 90° angle, then $\overrightarrow{KJ} \perp \overleftrightarrow{GH}$.

17. If $\overrightarrow{NM} \perp \overleftrightarrow{GH}$, then $\angle MNK \cong \angle MNH$.

18. If $\overrightarrow{NM} \perp \overleftrightarrow{GH}$, then $\angle 3$ and $\angle 4$ are complementary.

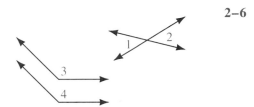

19. Write a plan for a proof.   2–6
    Given: $\angle 3$ is a supplement of $\angle 1$;
              $\angle 4$ is a supplement of $\angle 2$.
    Prove: $\angle 3 \cong \angle 4$

20. Write a proof in two-column form for Exercise 19.

Chapter Test

1. Use the conditional: Two angles are congruent if they are vertical angles.
   a. Write the hypothesis.   **b.** Write the converse.

2. Provide a counterexample to disprove the statement:
   If $x^2 > 4$, then $x > 2$.

3. Write the biconditional as two conditionals that are converses of each other:
   Angles are congruent if and only if their measures are equal.

4. Supply reasons to justify the steps:

	Steps	*Reasons*
1.	$y = 12$	1. Given
2.	$5x = 2x + y$	2. Given
3.	$5x = 2x + 12$	3. _?_
4.	$3x = 12$	4. _?_
5.	$x = 4$	5. _?_

5. $\overrightarrow{OB}$ is the bisector of $\angle AOC$ and $\overrightarrow{OC}$ is the bisector of $\angle BOD$.
   $m \angle AOC = 60$. Find $m \angle COD$.

6. S is the midpoint of $\overline{RT}$ and W is the midpoint of $\overline{ST}$. If $RT = 32$, find ST, WT, and RW.

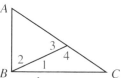

7. In the diagram, $\overline{AB} \perp \overline{BC}$. Name:
   a. two supplementary angles
   b. two complementary angles

8. Given: $\angle 5$ is supplementary to $\angle 4$.
   a. What can you conclude about $\angle 5$ and $\angle 3$?
   b. State the theorem that justifies your conclusion.   **Exs. 7–9**

9. Suppose $m \angle 3 = 3x + 5$ and $m \angle 4 = 6x + 13$. Find the value of x.

10. State the theorem that justifies the statement $\angle 6 \cong \angle 7$.

11. Suppose you have already stated that $\angle 6 \cong \angle 7$ and $\angle 7 \cong \angle 8$. What property of congruence justifies the conclusion that $\angle 6 \cong \angle 8$?

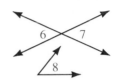

12. Write a proof in two-column form.

Given: $\overrightarrow{DC} \perp \overleftrightarrow{BD};\ \angle 1 \cong \angle 2$

Prove: $\overrightarrow{BA} \perp \overleftrightarrow{BD}$

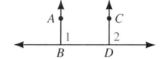

Algebra Review: *Systems of Equations*

Solve each system of equations by the substitution method.

Example 1 (1) $y = 5 - 2x$
   (2) $5x - 6y = 21$

Solution   Substitute $5 - 2x$ for y in (2): $5x - 6(5 - 2x) = 21$
$17x - 30 = 21;\ x = 3$

Substitute 3 for x in (1): $y = 5 - 2(3) = -1$

The solution is $x = 3,\ y = -1$.

1. $y = 3x$
$\quad 5x + y = 24$

2. $y = 2x + 5$
$\quad 3x - y = 4$

3. $x = 8 + 3y$
$\quad 2x - 5y = 8$

4. $3x + 2y = 71$
$\quad y = 4 + 2x$

5. $4x - 5y = 92$
$\quad x = 7y$

6. $y = 3x + 8$
$\quad x = y$

7. $8x + 3y = 26$
$\quad 2x = y - 4$

8. $x - 7y = 13$
$\quad 3x - 5y = 23$

9. $3x + y = 19$
$\quad 2x - 5y = -10$

Solve each system by the method of addition or subtraction.

Example 2 (1) $3x - y = 13$
   (2) $4x + y = 22$

Solution   Add (1) and (2):
$7x = 35;\ x = 5$

Substitute 5 for x in (2):
$4(5) + y = 22;\ y = 2$

The solution is $x = 5,\ y = 2$.

Example 3 (1) $6x + 15y = 90$
   (2) $6x - 14y = 32$

Solution   Subtract (2) from (1):
$29y = 58;\ y = 2$

Substitute 2 for y in (1):
$6x + 15(2) = 90;\ x = 10$

The solution is $x = 10,\ y = 2$.

10. $5x - y = 20$
$\quad 3x + y = 12$

11. $x + 3y = 7$
$\quad x + 2y = 4$

12. $3x - 2y = 11$
$\quad 3x - y = 7$

13. $7x + y = 29$
$\quad 5x + y = 21$

14. $8x - y = 17$
$\quad 6x + y = 11$

15. $9x - 2y = 50$
$\quad 6x - 2y = 32$

16. $7y = 2x + 35$
$\quad 3y = 2x + 15$

17. $2y = 3x - 1$
$\quad 2y = x + 21$

18. $19 = 5x + 2y$
$\quad 1 = 3x - 4y$

Preparing for College Entrance Exams

Strategy for Success
When you are taking a college entrance exam, be sure to read the directions, the questions, and the answer choices very carefully. In the test booklet, you may want to underline important words such as *not*, *exactly*, *false*, *never*, and *except*, and to cross out answer choices that are clearly incorrect.

Indicate the best answer by writing the appropriate letter.

1. On a number line, point M has coordinate -3 and point R has coordinate 6. Point Z is on $\overrightarrow{RM}$ and $RZ = 4$. Find the coordinate of Z.
   (A) -7 **(B)** 1 **(C)** 2 **(D)** 10 **(E)** cannot be determined

2. $\angle 1$ and $\angle 2$ are complementary. $m \angle 1 = 5x + 15$ and $m \angle 2 = 10x$. The measure of $\angle 1$ is:
   (A) 5 **(B)** 11 **(C)** 40 **(D)** 70 **(E)** 30

3. Vertical angles are never:
   (A) complementary **(B)** supplementary **(C)** right angles
   (D) adjacent **(E)** congruent

4. A reason that cannot be used to justify a statement in a proof is:
   (A) a postulate **(B)** a definition **(C)** given information
   (D) yesterday's theorem **(E)** tomorrow's theorem

5. Which of the following must be true?
   (I)   If two lines form congruent adjacent angles, then the lines are perpendicular.
   (II) If two lines are perpendicular, then they form congruent adjacent angles.
   (III) If the exterior sides of two adjacent obtuse angles are perpendicular, then the angles are complementary.

   (A) I only **(B)** II only **(C)** III only
   (D) I and II only **(E)** I, II, and III

6. $\angle 1$ and $\angle 2$ are congruent angles. $m \angle 1 = 10x - 20$ and $m \angle 2 = 8x + 2$. $\angle 1$ is a(n) __?__ angle.
   (A) acute **(B)** right **(C)** obtuse **(D)** straight
   (E) answer cannot be determined

7. If you know that $m \angle A = m \angle B$ and $m \angle B = m \angle C$, then what reason can you give for the statement that $m \angle A = m \angle C$?
   (I) Reflexive Property (II) Transitive Property (III) Substitution Property
   (A) I only **(B)** II only **(C)** III only
   (D) either I or II **(E)** either II or III

8. Which of the following is *not* the converse of the statement: If b, then c.
   (A) If c, then b. **(B)** b if c. **(C)** c if and only if b.
   (D) c only if b. **(E)** c implies b.

Cumulative Review: Chapters 1 and 2

Name or state the postulate, property, definition, or theorem that justifies the statement.

A   **1.** If $8x = 16$, then $x = 2$.

    2. If $\angle K \cong \angle L$ and $\angle L \cong \angle M$, then $\angle K \cong \angle M$.

    3. If $\angle AOB$ is a right angle, then $\overleftrightarrow{OA} \perp \overleftrightarrow{OB}$.

    4. If $a + 7 = b$ and $b = 4$, then $a + 7 = 4$.

    5. If $a + 7 = 4$, then $a = -3$.

    6. There is a line through F and H.

    7. The intersection of plane $CDEH$ and plane $FGHE$ is $\overleftrightarrow{EH}$.

    8. If W is the midpoint of $\overline{XV}$, then $XW = \frac{1}{2}XV$.

    9. $MW + WN = MN$

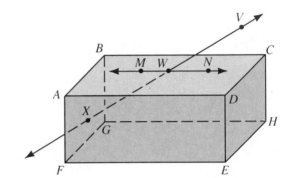

Classify each statement as true or false.

  10. $\overrightarrow{WV}$ contains point X.

  11. $\overline{MN}$ lies in plane $ABCD$.

  12. $\overleftrightarrow{WV}$ intersects plane $ABGF$.

  13. F, E, H, and C are coplanar.

  14. A, B, and V are coplanar.

Exs. 6–14

Classify each statement as true or false. If it is false, provide a counterexample.

  15. Through any three points, there is exactly one plane.

  16. Perpendicular lines form congruent adjacent angles.

  17. If points A and B are in plane M, then $\overline{AB}$ is in plane M.

  18. Complementary angles must be adjacent.

B   **19.** If $m \angle A = 45$, then the complement of $\angle A$ is one third of its supplement.

    20. If $m \angle RUN = m \angle SUN$, then $\overrightarrow{UN}$ is the bisector of $\angle RUS$.

In the diagram, $\overrightarrow{OB}$ bisects $\angle AOC$ and $\overleftrightarrow{EC} \perp \overrightarrow{OD}$. Find the value of x.

  21. $m \angle 5 = 2x$, $m \angle 3 = x$

  22. $m \angle 1 = 2x$, $m \angle 2 = 6x + 2$

  23. $m \angle 2 = 6x + 9$, $m \angle 5 = 2x + 49$

  24. $m \angle 2 = 3x$, $m \angle 3 = 2x - 4$

  25. $m \angle 1 = x - 8$, $m \angle 2 = 2x + 5$, $m \angle 4 = 3x - 26$

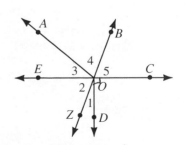

3 PARALLEL LINES AND PLANES

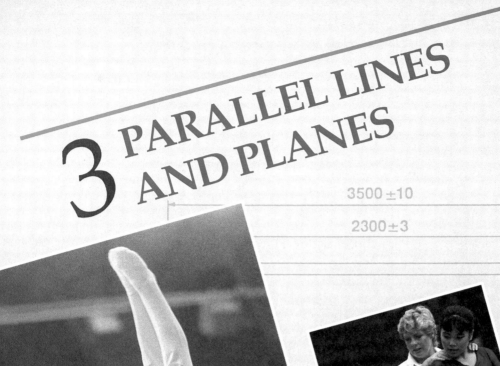

3500 ±10

2300 ±3

650 MIN.

1750 ±5

Parallel lines are suggested by many pieces of apparatus in a gymnasium and also by some of the positions taken by a performer's body.

Height Adjustment Assembly

Adjusting Pin

Reducing Tube

When Lines and Planes Are Parallel

Objectives

1. Distinguish between intersecting lines, parallel lines, and skew lines.
2. State and apply the theorem about the intersection of two parallel planes by a third plane.
3. Identify the angles formed when two lines are cut by a transversal.
4. State and apply the postulates and theorems about parallel lines.
5. State and apply the theorems about a parallel and a perpendicular to a given line through a point outside the line.

3-1 *Definitions*

Two lines that do not intersect are either *parallel* or *skew*.

Parallel lines (∥ lines) are coplanar lines that do not intersect.

Skew lines are noncoplanar lines. Therefore, they are neither parallel nor intersecting.

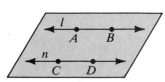
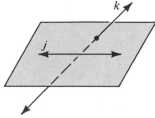

l and *n* are parallel lines.
l is parallel to *n* (*l* ∥ *n*).

j and *k* are skew lines.

Segments and rays contained in parallel lines are also called parallel. For example, in the figure at the left above, $\overline{AB} \parallel \overline{CD}$ and $\overrightarrow{AB} \parallel \overrightarrow{CD}$.

In the diagram at the right, $\overline{PQ}$ and $\overline{RS}$ do not intersect, but they are parts of lines, $\overleftrightarrow{PQ}$ and $\overleftrightarrow{RS}$, that do intersect. Thus $\overline{PQ}$ is *not* parallel to $\overline{RS}$.

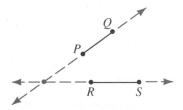

The box pictured below may help you understand the following definitions. Think of the top of the box as part of plane *X* and the bottom of the box as part of plane *Y*.

Parallel planes (∥ planes) do not intersect.
Plane *X* is parallel to plane *Y* (*X* ∥ *Y*).

A line and a plane are parallel if they do not intersect.
For example, $\overleftrightarrow{EF} \parallel Y$ and $\overleftrightarrow{FG} \parallel Y$.
Also, $\overleftrightarrow{AB} \parallel X$ and $\overleftrightarrow{BC} \parallel X$.

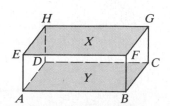

Our first theorem about parallel lines and planes is given below. Notice the importance of definitions in the proof.

Theorem 3-1

If two parallel planes are cut by a third plane, then the lines of intersection are parallel.

Given: Plane $X \parallel$ plane Y;
    plane Z intersects X in line l;
    plane Z intersects Y in line n.

Prove: $l \parallel n$

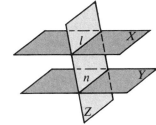

Proof:

Statements	Reasons
1. l is in Z; n is in Z.	1. Given
2. l and n are coplanar.	2. Def. of coplanar
3. l is in X; n is in Y; $X \parallel Y$.	3. Given
4. l and n do not intersect.	4. Parallel planes do not intersect. (Def. of $\parallel$ planes)
5. $l \parallel n$	5. Def. of $\parallel$ lines (Steps 2 and 4)

The following terms, which are needed for future theorems about parallel lines, apply only to coplanar lines.

A **transversal** is a line that intersects two or more coplanar lines in different points. In the next diagram, t is a transversal of h and k. The angles formed have special names.

Interior angles: angles 3, 4, 5, 6

Exterior angles: angles 1, 2, 7, 8

Alternate interior angles (alt. int. $\angle$s) are two nonadjacent interior angles on opposite sides of the transversal.

 $\angle 3$ and $\angle 6$ $\angle 4$ and $\angle 5$

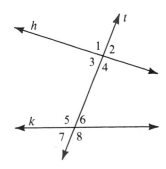

Same-side interior angles (s-s. int. $\angle$s) are two interior angles on the same side of the transversal.

 $\angle 3$ and $\angle 5$ $\angle 4$ and $\angle 6$

Corresponding angles (corr. $\angle$s) are two angles in corresponding positions relative to the two lines.

 $\angle 1$ and $\angle 5$ $\angle 2$ and $\angle 6$ $\angle 3$ and $\angle 7$ $\angle 4$ and $\angle 8$

Classroom Exercises

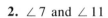

1. The blue line is a transversal.
   a. Name four pairs of corresponding angles.
   b. Name two pairs of alternate interior angles.
   c. Name two pairs of same-side interior angles.
   d. Name two pairs of angles that could be called *alternate exterior angles*.
   e. Name two pairs of angles that could be called *same-side exterior angles*.

Classify each pair of angles as alternate interior angles, same-side interior angles, corresponding angles, or none of these.

2. ∠7 and ∠11
3. ∠14 and ∠16
4. ∠4 and ∠10
5. ∠3 and ∠6
6. ∠6 and ∠11
7. ∠2 and ∠10
8. ∠2 and ∠3
9. ∠7 and ∠12

10. Classify each pair of lines as intersecting, parallel, or skew.
    a. $\overleftrightarrow{AB}$ and $\overleftrightarrow{EJ}$
    b. $\overleftrightarrow{AB}$ and $\overleftrightarrow{FK}$
    c. $\overleftrightarrow{AB}$ and $\overleftrightarrow{ID}$
    d. $\overleftrightarrow{EF}$ and $\overleftrightarrow{IH}$
    e. $\overleftrightarrow{EF}$ and $\overleftrightarrow{NM}$
    f. $\overleftrightarrow{CN}$ and $\overleftrightarrow{FG}$

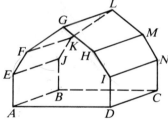
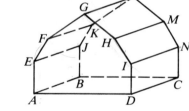

11. Name six lines parallel to $\overleftrightarrow{GL}$.
12. Name several lines skew to $\overleftrightarrow{GL}$.
13. Name five lines parallel to plane *ABCD*.
14. Name two coplanar segments that do not intersect and yet are not parallel.

Complete each statement with the word *always*, *sometimes*, or *never*.

15. Two skew lines are __?__ parallel.
16. Two parallel lines are __?__ coplanar.
17. A line in the plane of the ceiling and a line in the plane of the floor are __?__ parallel.
18. Two lines in the plane of the floor are __?__ skew.
19. A line in the plane of a wall and a line in the plane of the floor are
    a. __?__ parallel.   b. __?__ intersecting.   c. __?__ skew.

Written Exercises

Classify each pair of angles as alternate interior angles, same-side interior angles, or corresponding angles.

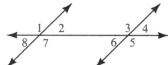

A **1.** ∠2 and ∠6   **2.** ∠8 and ∠6

  3. ∠2 and ∠3   **4.** ∠3 and ∠7

  5. ∠5 and ∠7   **6.** ∠3 and ∠1

Name the two lines and the transversal that form each pair of angles.

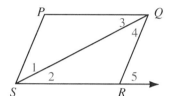

  7. ∠2 and ∠3

  8. ∠1 and ∠4

  9. ∠P and ∠PSR

  10. ∠5 and ∠PSR

  11. ∠5 and ∠PQR

Classify each pair of angles as alternate interior, same-side interior, or corresponding angles.

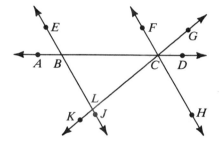

  12. ∠EBA and ∠FCB

  13. ∠DCH and ∠CBJ

  14. ∠FCB and ∠CBL

  15. ∠FCL and ∠BLC

  16. ∠HCB and ∠CBJ

  17. ∠GCH and ∠GLJ

In Exercises 18–20 use two lines of notebook paper as parallel lines and draw any transversal. Use a protractor to measure.

  18. Measure one pair of corresponding angles. Repeat the experiment with another transversal. What appears to be true?

  19. Measure one pair of alternate interior angles. Repeat the experiment with another transversal. What appears to be true?

  20. Measure one pair of same-side interior angles. Repeat the experiment with another transversal. What appears to be true?

B **21.** Draw a large diagram showing three transversals intersecting two nonparallel lines *l* and *n*. Number three pairs of same-side interior angles *on the same sides of the transversals*, as shown in the diagram.

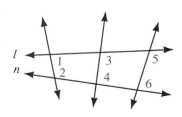

  a. Find $m\angle 1 + m\angle 2$.

  b. Find $m\angle 3 + m\angle 4$.

  c. Predict the value of $m\angle 5 + m\angle 6$. Then check your prediction by measuring.

  d. What do you conclude?

22. Draw a diagram of a six-sided box by following the steps below.

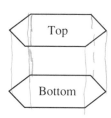

Top

Bottom

Step 1

Draw a six-sided top. Then draw an exact copy of the top directly below it.

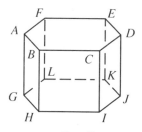

Step 2

Draw vertical edges. Make invisible edges dashed.

Exercises 23–29 refer to the diagram in Step 2 of Exercise 22.

23. Name five lines that appear to be parallel to $\overline{AG}$.

24. Name three lines that appear to be parallel to $\overline{AB}$.

25. Name four lines that appear to be skew to $\overline{AB}$.

26. Name two planes parallel to $\overleftrightarrow{AF}$.

27. Name four planes parallel to $\overleftrightarrow{FL}$.

28. How many pairs of parallel planes are shown?

29. Suppose the top and bottom of the box lie in parallel planes. Explain how Theorem 3-1 can be used to prove $\overline{CD} \parallel \overline{IJ}$.

Complete each statement with the word *always*, *sometimes*, or *never*.

30. When there is a transversal of two lines, the three lines are __?__ coplanar.

31. Three lines intersecting in one point are __?__ coplanar.

32. Two lines that are not coplanar __?__ intersect.

33. Two lines parallel to a third line are __?__ parallel to each other.

34. Two lines skew to a third line are __?__ skew to each other.

35. Two lines perpendicular to a third line are __?__ perpendicular to each other.

36. Two planes parallel to the same line are __?__ parallel to each other.

37. Two planes parallel to the same plane are __?__ parallel to each other.

38. Lines in two parallel planes are __?__ parallel to each other.

39. Two lines parallel to the same plane are __?__ parallel to each other.

Draw each figure described.

C **40.** Lines a and b are skew, lines b and c are skew, and $a \parallel c$.

41. Lines d and e are skew, lines e and f are skew, and $d \perp f$.

42. Line $l \parallel$ plane X, plane $X \parallel$ plane Y, and l is not parallel to Y.

Explorations

These exploratory exercises can be done using a computer with a program that draws and measures geometric figures.

Draw two parallel segments and a transversal and label the points of intersection. Measure all eight angles formed. Repeat several times. Do you notice any patterns? What kinds of angles are congruent? What kinds of angles are supplementary?

3-2 *Properties of Parallel Lines*

By experimenting with parallel lines, transversals, and a protractor in Exercise 18, page 76, you probably discovered that corresponding angles are congruent. There is not enough information in our previous postulates and theorems to deduce this property as a theorem. We will accept it as a postulate.

Postulate 10

If two parallel lines are cut by a transversal, then corresponding angles are congruent.

From this postulate we can easily prove the next three theorems.

Theorem 3-2

If two parallel lines are cut by a transversal, then alternate interior angles are congruent.

Given: $k \parallel n$; transversal t cuts k and n.

Prove: $\angle 1 \cong \angle 2$

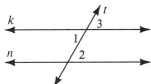

Proof:

Statements	Reasons
1. $k \parallel n$	1. Given
2. $\angle 1 \cong \angle 3$	2. Vert. $\angle$ are $\cong$.
3. $\angle 3 \cong \angle 2$	3. If two parallel lines are cut by a transversal, then corr. $\angle$ are $\cong$.
4. $\angle 1 \cong \angle 2$	4. Transitive Property

Theorem 3-3

If two parallel lines are cut by a transversal, then same-side interior angles are supplementary.

Given: $k \parallel n$; transversal t cuts k and n.

Prove: $\angle 1$ is supplementary to $\angle 4$.

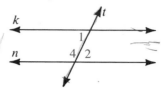

The proof is left as Exercise 22.

Theorem 3-4

If a transversal is perpendicular to one of two parallel lines, then it is perpendicular to the other one also.

Given: Transversal t cuts l and n;
$\quad\quad t \perp l$; $l \parallel n$

Prove: $t \perp n$

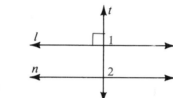

The proof is left as Exercise 13.

For the rest of this book, arrowheads will no longer be used in diagrams to suggest that a line extends in both directions without ending. Instead, pairs of arrowheads (and double arrowheads when necessary) will be used to indicate parallel lines, as shown in the following examples.

Example 1   Find the measure of $\angle PQR$.

Solution   The diagram shows that
$$\overleftrightarrow{QR} \perp \overleftrightarrow{RS} \text{ and } \overleftrightarrow{QP} \parallel \overleftrightarrow{RS}.$$
Then by Theorem 3-4, $\overleftrightarrow{QR} \perp \overleftrightarrow{QP}$ and $m \angle PQR = 90$.

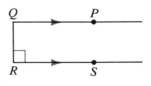

Example 2   Find the values of x, y, and z.

Solution   Since $a \parallel b$, $2x = 40$. (Why?)
Thus, $x = 20$.
Since $c \parallel d$, $y = 40$. (Why?)
Since $a \parallel b$, $y + z = 180$. (Why?)
            $40 + z = 180$
                 $z = 140$

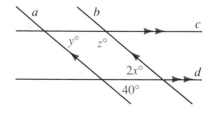

Classroom Exercises

1. What do the arrowheads in the diagram tell you?

State the postulate or theorem that justifies each statement.

2. $\angle 1 \cong \angle 5$ **3.** $\angle 3 \cong \angle 6$

4. $m \angle 4 + m \angle 6 = 180$ **5.** $m \angle 4 = m \angle 8$

6. $m \angle 4 = m \angle 5$ **7.** $\angle 6 \cong \angle 7$

8. $k \perp p$ **9.** $\angle 3$ is supplementary to $\angle 5$.

Exs. 1–13

10. If $m \angle 1 = 130$, what are the measures of the other numbered angles?

11. If $m \angle 1 = x$, what are the measures of the other numbered angles?

12. If $m \angle 4 = 2m \angle 3$, find $m \angle 6$.

13. If $m \angle 5 = m \angle 6 + 20$, find $m \angle 1$.

14. Alan tried to prove Postulate 10 as shown below. However, he did *not* have a valid proof. Explain why not.

If two parallel lines are cut by a transversal, then corresponding angles are congruent.

Given: $k \parallel l$; transversal t cuts k and l.

Prove: $\angle 1 \cong \angle 2$

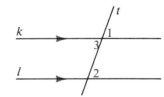

Proof:

Statements	Reasons
1. $k \parallel l$	1. Given
2. $\angle 3 \cong \angle 2$	2. If two parallel lines are cut by a transversal, then alt. int. $\angle$s are $\cong$.
3. $\angle 1 \cong \angle 3$	3. Vert. $\angle$s are $\cong$.
4. $\angle 1 \cong \angle 2$	4. Transitive Prop.

Written Exercises

A   **1.** If $a \parallel b$, name all angles that must be congruent to $\angle 1$.

    2. If $c \parallel d$, name all angles that must be congruent to $\angle 1$.

Assume that $a \parallel b$ and $c \parallel d$.

    3. Name all angles congruent to $\angle 2$.

    4. Name all angles supplementary to $\angle 2$.

    5. If $m \angle 13 = 110$, then $m \angle 15 = \underline{}$ and $m \angle 3 = \underline{}$.

    6. If $m \angle 7 = x$, then $m \angle 12 = \underline{}$ and $m \angle 6 = \underline{}$.

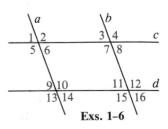

Exs. 1–6

Find the values of x and y.

7.

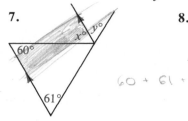

60 + 61 + x = y

8.

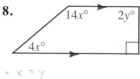

9.

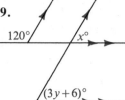

10.

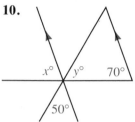

11.

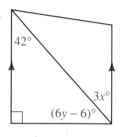

12.

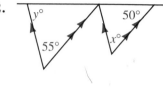

13. Copy and complete the proof of Theorem 3-4.

  Given: Transversal t cuts l and n;
    $t \perp l$; $l \parallel n$

  Prove: $t \perp n$

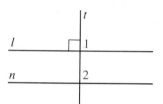

Proof:

Statements	Reasons
1. $t \perp l$	1. ___?___
2. $m \angle 1 = 90$	2. ___?___
3. ___?___	3. Given
4. $\angle 2 \cong \angle 1$ or $m \angle 2 = m \angle 1$	4. ___?___
5. ___?___	5. Substitution Property
6. $t \perp n$	6. ___?___

Find the values of x, y, and z.

B **14.**

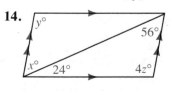

15.

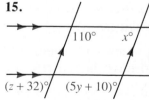

16.

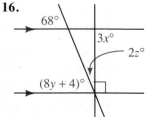

17. Given: $\overline{AB} \parallel \overline{CD}$; $m \angle D = 116$;
$\quad\quad\;\; \overrightarrow{AK}$ bisects $\angle DAB$.

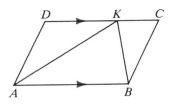

   a. Find the measures of $\angle DAB$, $\angle KAB$, and $\angle DKA$.

   b. Is there enough information for you to conclude that $\angle D$ and $\angle C$ are supplementary, or is more information needed?

Find the values of x and y.

18.

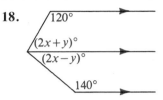

19.

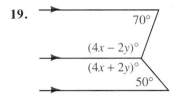

Write proofs in two-column form.

20. Given: $k \parallel l$
   Prove: $\angle 2 \cong \angle 7$

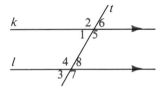

21. Given: $k \parallel l$
   Prove: $\angle 1$ is supplementary to $\angle 7$.

22. Copy what is shown for Theorem 3-3 on page 79. Then write a proof in two-column form.

23. Draw a four-sided figure $ABCD$ with $\overline{AB} \parallel \overline{DC}$ and $\overline{AD} \parallel \overline{BC}$.
   a. Prove that $\angle A \cong \angle C$.
   b. Is $\angle B \cong \angle D$?

C **24.** Given: $\overline{AS} \parallel \overline{BT}$;
        $m \angle 4 = m \angle 5$
   Prove: $\overrightarrow{SA}$ bisects $\angle BSR$.

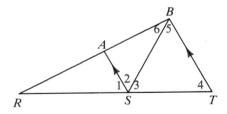

25. Given: $\overline{AS} \parallel \overline{BT}$;
        $m \angle 4 = m \angle 5$;
        $\overrightarrow{SB}$ bisects $\angle AST$.
   Find the measure of $\angle 1$.

Mixed Review Exercises

For each statement (a) tell whether the statement is true or false, (b) write the converse, and (c) tell whether the converse is true or false.

  1. If two lines are perpendicular, then they form congruent adjacent angles.

  2. If two lines are parallel, then they are not skew.

  3. Two angles are supplementary if the sum of their measures is 180.

  4. Two planes are parallel only if they do not intersect.

3-3 *Proving Lines Parallel*

In the preceding section you saw that when two lines are parallel, you can conclude that certain angles are congruent or supplementary. In this section the situation is reversed. From two angles being congruent or supplementary you will conclude that certain lines forming the angles are parallel. The key to doing this is Postulate 11 below. Postulate 10 is repeated so you can compare the wording of the postulates. Notice that these two postulates are converses of each other.

Postulate 10

If two parallel lines are cut by a transversal, then corresponding angles are congruent.

Postulate 11

If two lines are cut by a transversal and corresponding angles are congruent, then the lines are parallel.

The next three theorems can be deduced from Postulate 11.

Theorem 3-5

If two lines are cut by a transversal and alternate interior angles are congruent, then the lines are parallel.

Given: Transversal t cuts lines k and n;
$\qquad \angle 1 \cong \angle 2$
Prove: $k \parallel n$

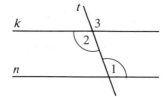

Proof:

Statements	Reasons
1. $\angle 1 \cong \angle 2$	1. Given
2. $\angle 2 \cong \angle 3$	2. Vert. $\angle$s are $\cong$.
3. $\angle 1 \cong \angle 3$	3. Transitive Property
4. $k \parallel n$	4. If two lines are cut by a transversal and corr. $\angle$s are $\cong$, then the lines are $\parallel$.

You may have recognized that Theorem 3-5 is the converse of Theorem 3-2, "If two parallel lines are cut by a transversal, then alternate interior angles are congruent." The next theorem is the converse of Theorem 3-3, "If two parallel lines are cut by a transversal, then same-side interior angles are supplementary."

Theorem 3-6

If two lines are cut by a transversal and same-side interior angles are supplementary, then the lines are parallel.

Given: Transversal t cuts lines k and n;
  $\angle 1$ is supplementary to $\angle 2$.
Prove: $k \parallel n$

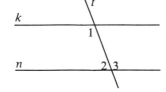

The proof is left as Exercise 22.

Theorem 3-7

In a plane two lines perpendicular to the same line are parallel.

Given: $k \perp t$; $n \perp t$
Prove: $k \parallel n$

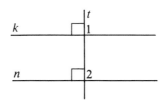

The proof is left as Exercise 23.

Example 1   Which segments are parallel?

Solution   (1) $\overline{HI}$ and $\overline{TN}$ are parallel since corresponding angles have the same measure:
  $m\angle HIL = 23 + 61 = 84$
  $m\angle TNI = 22 + 62 = 84$
  (2) $\overline{WI}$ and $\overline{AN}$ are *not* parallel since $61 \neq 62$.

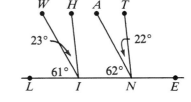

Example 2   Find the values of x and y that make $\overline{AC} \parallel \overline{DF}$ and $\overline{AE} \parallel \overline{BF}$.

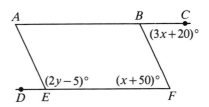

Solution   If $m\angle CBF = m\angle BFE$,
  then $\overline{AC} \parallel \overline{DF}$. (Why?)
  $$3x + 20 = x + 50$$
  $$2x = 30$$
  $$x = 15$$

If $\angle AEF$ and $\angle F$ are supplementary,
  then $\overline{AE} \parallel \overline{BF}$. (Why?)
  $$(2y - 5) + (x + 50) = 180$$
  $$(2y - 5) + (15 + 50) = 180$$
  $$2y = 120$$
  $$y = 60$$

The following theorems can be proved using previous postulates and theorems. We state the theorems without proof, however, for you to use in future work.

Theorem 3-8

Through a point outside a line, there is exactly one line parallel to the given line.

Theorem 3-9

Through a point outside a line, there is exactly one line perpendicular to the given line.

Given this:

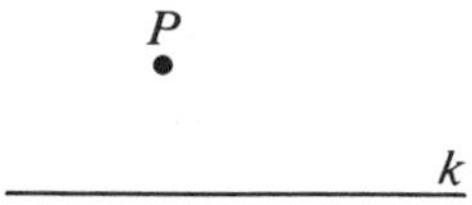

P

k

Theorem 3-8 says that line n exists and is unique.

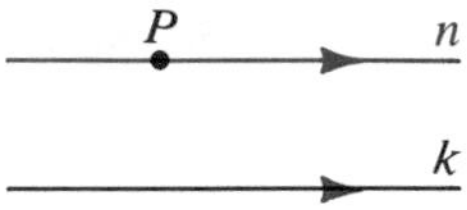

P n

k

Given this:

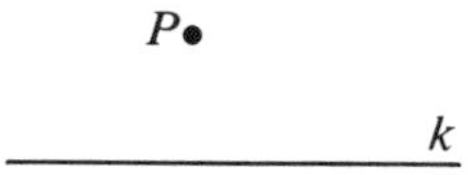

P

k

Theorem 3-9 says that line h exists and is unique.

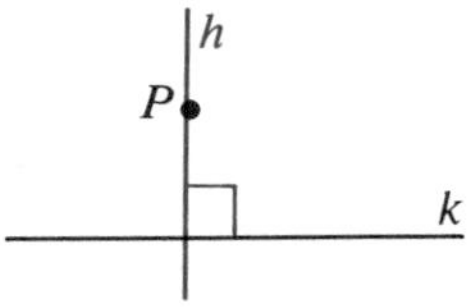

h

P

k

Theorem 3-10

Two lines parallel to a third line are parallel to each other.

Given: $k \parallel l$; $k \parallel n$
Prove: $l \parallel n$

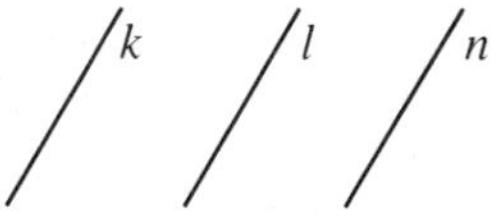

k l n

In Classroom Exercise 20 you will explain why this theorem is true when all three lines are coplanar. The theorem also holds true for lines in space.

Ways to Prove Two Lines Parallel

1. Show that a pair of corresponding angles are congruent.
2. Show that a pair of alternate interior angles are congruent.
3. Show that a pair of same-side interior angles are supplementary.
4. In a plane show that both lines are perpendicular to a third line.
5. Show that both lines are parallel to a third line.

Classroom Exercises

State which segments (if any) are parallel. State the postulate or theorem that justifies your answer.

1.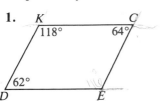

2.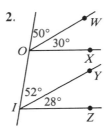

3.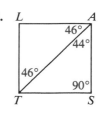

4.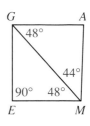

In each exercise some information is given. Use this information to name the segments that must be parallel. If there are no such segments, say so.

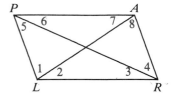

5. $m \angle 1 = m \angle 8$

6. $\angle 2 \cong \angle 7$

7. $\angle 5 \cong \angle 3$

8. $m \angle 5 = m \angle 4$

9. $m \angle 5 + m \angle 6 = m \angle 3 + m \angle 4$

10. $m \angle APL + m \angle PAR = 180$

11. $m \angle 1 + m \angle 2 + m \angle 5 + m \angle 6 = 180$

12. Reword Theorem 3-8 as two statements, one describing existence and the other describing uniqueness.

13. Reword Theorem 3-9 as two statements, one describing existence and the other describing uniqueness.

14. How many lines can be drawn through P parallel to $\overleftrightarrow{QR}$?

15. How many lines can be drawn through Q parallel to $\overleftrightarrow{PR}$?

16. How many lines can be drawn through P perpendicular to $\overleftrightarrow{QR}$?

17. In the plane of P, Q, and R, how many lines can be drawn through R perpendicular to $\overleftrightarrow{QR}$? What postulate or theorem justifies your answer?

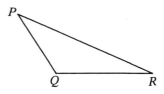

18. In space, how many lines can be drawn through R perpendicular to $\overleftrightarrow{QR}$?

19. True or false?
    a. Two lines perpendicular to a third line must be parallel.
    b. In a plane two lines perpendicular to a third line must be parallel.
    c. In a plane two lines parallel to a third line must be parallel.
    d. Any two lines parallel to a third line must be parallel.

20. Use the diagram to explain why Theorem 3-10 is true for coplanar lines. That is, if $k \parallel l$ and $k \parallel n$, why does it follow that $l \parallel n$?

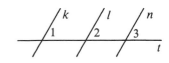

Written Exercises

In each exercise some information is given. Use this information to name the
segments that must be parallel. If there are no such segments, write *none*.

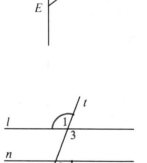

A 1. $\angle 2 \cong \angle 9$ 2. $\angle 6 \cong \angle 7$

3. $m\angle 1 = m\angle 8 = 90$ 4. $\angle 5 \cong \angle 9$

5. $m\angle 2 = m\angle 5$ 6. $\angle 3 \cong \angle 11$

7. $m\angle 1 = m\angle 4 = 90$ 8. $m\angle 10 = m\angle 11$

9. $m\angle 8 + m\angle 5 + m\angle 6 = 180$

10. $\overline{FC} \perp \overline{AE}$ and $\overline{FC} \perp \overline{BD}$

11. $m\angle 5 + m\angle 6 = m\angle 9 + m\angle 10$

12. $\angle 7$ and $\angle EFB$ are supplementary.

13. $\angle 2$ and $\angle 3$ are complementary and $m\angle 1 = 90$.

14. $m\angle 2 + m\angle 3 = m\angle 4$

15. $m\angle 7 = m\angle 3 = m\angle 10$

16. $m\angle 4 = m\angle 8 = m\angle 1$

17. Write the reasons to complete the proof: If two lines
    are cut by a transversal and alternate exterior angles
    are congruent, then the lines are parallel.

   Given: Transversal t cuts lines l and n;
          $\angle 2 \cong \angle 1$

   Prove: $l \parallel n$

Proof:

Statements	Reasons
1. $\angle 2 \cong \angle 1$	1. _?_
2. $\angle 1 \cong \angle 3$	2. _?_
3. $\angle 2 \cong \angle 3$	3. _?_
4. $l \parallel n$	4. _?_

Find the values of *x* and *y* that make the red lines parallel *and* the blue lines
parallel.

B 18.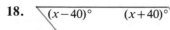

19.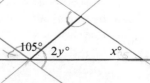

20. Given: $\angle 1 \cong \angle 2$; $\angle 4 \cong \angle 5$
    What can you prove about $\overline{PQ}$ and $\overline{RS}$? Be prepared
    to give your reasons in class, if asked.

21. Given: $\angle 3 \cong \angle 6$
    What can you prove about other angles? Be prepared
    to give your reasons in class, if asked.

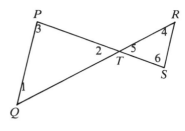

22. Copy what is shown for Theorem 3-6 on page 84. Then write a proof in
    two-column form.

23. Copy what is shown for Theorem 3-7 on page 84. Then write a proof in
    two-column form.

24. Given: $\overline{BE}$ bisects $\angle DBA$; $\angle 3 \cong \angle 1$
    Prove: $\overline{CD} \parallel \overline{BE}$

25. Given: $\overline{BE} \perp \overline{DA}$; $\overline{CD} \perp \overline{DA}$
    Prove: $\angle 1 \cong \angle 2$

26. Given: $\angle C \cong \angle 3$; $\overline{BE} \perp \overline{DA}$
    Prove: $\overline{CD} \perp \overline{DA}$

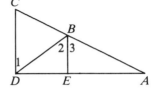

**Find the measure of $\angle RST$. (*Hint*: Draw a line through S parallel to $\overrightarrow{RX}$
and $\overrightarrow{TY}$.)**

27.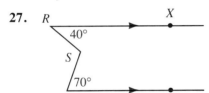

28.

29. Find the values of x and y that make the lines shown in
    red parallel.

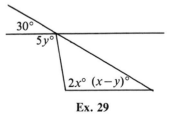

Ex. 29

C 30. Draw two parallel lines cut by a transversal. Then draw
    the bisectors of two corresponding angles. What appears
    to be true about the bisectors? Prove that your conclusion
    is true.

31. Find the value of x that makes the lines shown in red
    parallel.

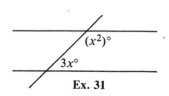

Ex. 31

Self-Test 1

Complete each statement with the word *always*, *sometimes*, or *never*.

1. Two lines that do not intersect are ___?___ parallel.

2. Two skew lines ___?___ intersect.

3. Two lines parallel to a third line are ___?___ parallel.

4. If a line is parallel to plane *X* and also to plane *Y*, then plane *X* and plane *Y* are ___?___ parallel.

5. Plane *X* is parallel to plane *Y*. If plane *Z* intersects *X* in line *l* and *Y* in line *n*, then *l* is ___?___ parallel to *n*.

6. Name two pairs of congruent alternate interior angles.

7. Name two pairs of congruent corresponding angles.

8. Name a pair of supplementary same-side interior angles.

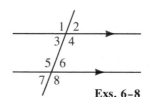

Exs. 6–8

9. Complete: If $\overline{AE} \parallel \overline{BD}$, then $\angle 1 \cong$ ___?___ and $\angle 9 \cong$ ___?___.

10. If $\overline{ED} \parallel \overline{AC}$, name all pairs of angles that must be congruent.

11. If $\overline{ED} \parallel \overline{AC}$ and $\overline{EB} \parallel \overline{DC}$, name all angles that must be congruent to $\angle 5$.

12. Complete: If $\overline{ED} \parallel \overline{AC}$, $\overline{EB} \parallel \overline{DC}$, and $m \angle 2 = 65$, then $m \angle 8 =$ ___?___ and $m \angle EDC =$ ___?___.

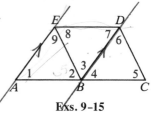

Exs. 9–15

Use the given information to name the segments (if any) that must be parallel.

13. $\angle 3 \cong \angle 6$ **14.** $\angle 9 \cong \angle 6$ **15.** $m \angle 7 + m \angle AED = 180$

16. Complete: Through a point outside a line, ___?___ line(s) can be drawn parallel to the given line, and ___?___ line(s) can be drawn perpendicular to the given line.

Explorations

These exploratory exercises can be done using a computer with a program that draws and measures geometric figures.

Draw a triangle *ABC*. At each vertex extend one side, as shown in the diagram. Measure all six angles formed. Repeat on several triangles. What do you notice?

What is the sum of the measures of the angles inside the triangle? of the angles outside the triangle?

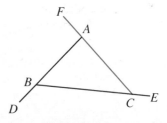

Application *Technical Drawing*

Can the shape of a three-dimensional object be determined from a single two-dimensional image? For example, if you photographed the barn shown on page 75 from a point directly above, then your photograph might look something like the sketch shown at the right. You cannot tell from this one photograph how the roof slopes or anything about the sides of the barn. You would have a much better idea of the shape of the barn if you could also see it from the front and from one side.

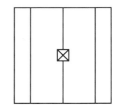

The three views of the barn are the parts of an *orthographic projection*, a set of projections of an object into three planes perpendicular to one another. They show the actual shape of the building much more clearly than any single picture.

To make an orthographic projection of an object, draw a top view, a front view, and a side view. Arrange the three views in an "L" shaped pattern as illustrated in the figure on the left below. Some corresponding vertices have been connected with red lines.

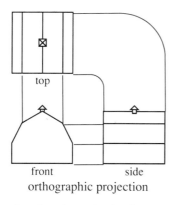

orthographic projection

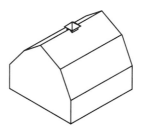

isometric drawing

A related method of representing a three-dimensional object by a two-dimensional image is an *isometric drawing*. In this type of representation the object is viewed at an angle that allows simultaneous vision of the top, front, and one side. Unlike most drawings, however, an isometric drawing does not show perspective. Rather, congruent sides are drawn congruent. Because we are accustomed to seeing objects in perspective, an isometric representation can appear distorted to us. The following figures illustrate the difference between a perspective drawing (left) and an isometric drawing (right).

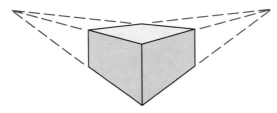

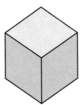

To make an orthographic projection into an isometric drawing, we fold the top, front, and side together. The base of the box becomes angled, but vertical lines remain vertical, parallel lines remain parallel, and congruence is preserved.

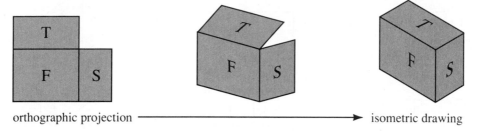

orthographic projection ⟶ isometric drawing

To illustrate, we shall make an isometric drawing of the solid whose orthographic projection is shown below. Visible edges and intersections are shown as solid lines. Hidden edges are shown as dashed lines.

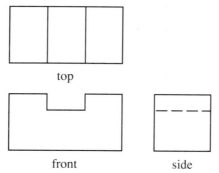

top

front   side

Begin by drawing three rays with a common endpoint such that one ray is vertical and the other two rays are 30° off of horizontal. Mark off the lengths of the front, side, and height of the solid. By drawing congruent segments and by showing parallel edges as parallel in the figure, you can finish the isometric drawing. The figures that follow suggest the procedure.

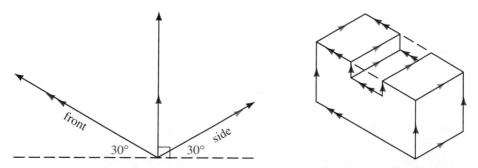

Industry requires millions of drawings similar to these each year. The drafters who make these drawings professionally combine the knowledge of parallel lines with the skills of using a compass, a protractor, and a ruler. More recently, drafters make drawings like these using a computer and a special printer.

Exercises

Match the orthographic projections with their isometric drawings. If there is no isometric drawing, then make one.

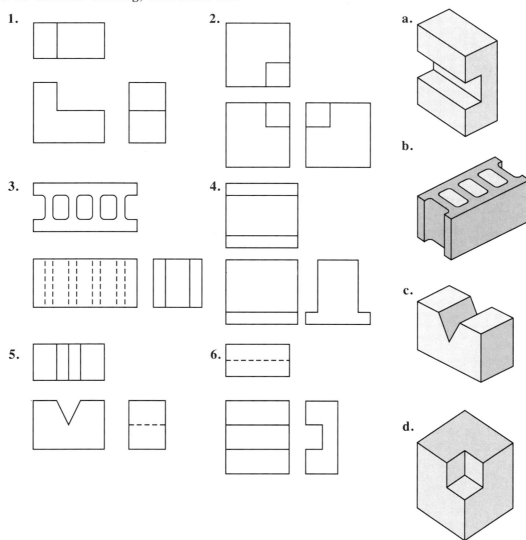

Trace each figure. Then make an orthographic projection of the figure.

7.

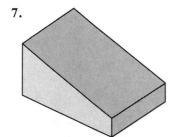

8.

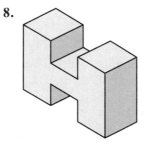

Applying Parallel Lines to Polygons

Objectives

1. Classify triangles according to sides and to angles.
2. State and apply the theorem and the corollaries about the sum of the measures of the angles of a triangle.
3. State and apply the theorem about the measure of an exterior angle of a triangle.
4. Recognize and name convex polygons and regular polygons.
5. Find the measures of interior angles and exterior angles of convex polygons.
6. Understand and use inductive reasoning.

3-4 *Angles of a Triangle*

A **triangle** is the figure formed by three segments joining three noncollinear points. Each of the three points is a **vertex** of the triangle. (The plural of *vertex* is *vertices*.) The segments are the **sides** of the triangle.

Triangle *ABC* (△*ABC*) is shown.

Vertices of △*ABC*: points *A*, *B*, *C*

Sides of △*ABC*: $\overline{AB}$, $\overline{BC}$, $\overline{CA}$

Angles of △*ABC*: ∠*A*, ∠*B*, ∠*C*

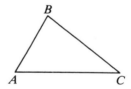

A triangle is sometimes classified by the number of congruent sides it has.

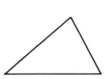

Scalene triangle
No sides congruent

Isosceles triangle
At least two sides congruent

Equilateral triangle
All sides congruent

Triangles can also be classified by their angles.

Acute △
Three acute ⩘

Obtuse △
One obtuse ∠

Right △
One right ∠

Equiangular △
All ⩘ congruent

An **auxiliary line** is a line (or ray or segment) added to a diagram to help in a proof. An auxiliary line is used in the proof of the next theorem, one of the best-known theorems of geometry. The auxiliary line is shown as a dashed line in the diagram.

Theorem 3-11

The sum of the measures of the angles of a triangle is 180.

Given: $\triangle ABC$

Prove: $m \angle 1 + m \angle 2 + m \angle 3 = 180$

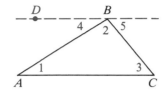

Proof:

Statements	Reasons
1. Through B draw $\overleftrightarrow{BD}$ parallel to $\overleftrightarrow{AC}$.	1. Through a point outside a line, there is exactly one line $\parallel$ to the given line.
2. $m \angle DBC + m \angle 5 = 180$; $m \angle DBC = m \angle 4 + m \angle 2$	2. Angle Addition Postulate
3. $m \angle 4 + m \angle 2 + m \angle 5 = 180$	3. Substitution Property
4. $\angle 4 \cong \angle 1$, or $m \angle 4 = m \angle 1$; $\angle 5 \cong \angle 3$, or $m \angle 5 = m \angle 3$	4. If two parallel lines are cut by a transversal, then alt. int. $\angle$s are $\cong$.
5. $m \angle 1 + m \angle 2 + m \angle 3 = 180$	5. Substitution Property

A statement that can be proved easily by applying a theorem is often called a **corollary** of the theorem. Corollaries, like theorems, can be used as reasons in proofs. Each of the four statements that are shown below is a corollary of Theorem 3-11.

Corollary 1
If two angles of one triangle are congruent to two angles of another triangle, then the third angles are congruent.

Corollary 2
Each angle of an equiangular triangle has measure 60.

Corollary 3
In a triangle, there can be at most one right angle or obtuse angle.

Corollary 4
The acute angles of a right triangle are complementary.

In the classroom exercises you will explain how these corollaries follow from Theorem 3-11.

Example 1   Is $\angle P \cong \angle V$?

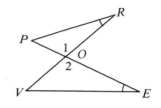

Solution $\angle R \cong \angle E$ (Given in diagram)

$\angle 1 \cong \angle 2$ (Vertical angles are congruent.)

Thus two angles of $\triangle PRO$ are congruent to two angles of $\triangle VEO$, and therefore $\angle P \cong \angle V$ by Corollary 1.

When one side of a triangle is extended, an *exterior angle* is formed as shown in the diagrams below. Because an exterior angle of a triangle is always a supplement of the adjacent interior angle of the triangle, its measure is related in a special way to the measure of the other two angles of the triangle, called the *remote interior angles*.

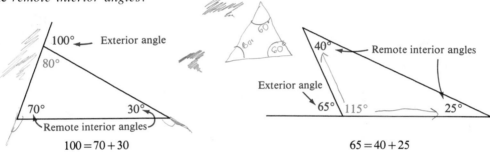

$$100 = 70 + 30$$

$$65 = 40 + 25$$

Theorem 3-12

The measure of an exterior angle of a triangle equals the sum of the measures of the two remote interior angles.

The proof of Theorem 3-12 is left as Classroom Exercise 15.

Example 2   In $\triangle ABC$, $m\angle A = 120$ and an exterior angle at C is five times as large as $\angle B$. Find $m\angle B$.

Solution   Let $m\angle B = x$.
Draw a diagram that shows the given information.
Then apply Theorem 3-12.

$$5x = 120 + x$$
$$4x = 120$$
$$x = 30$$
$$m\angle B = 30$$

Classroom Exercises

Complete each statement with the word *always*, *sometimes*, or *never*.

1. If a triangle is isosceles, then it is __?__ equilateral.
2. If a triangle is equilateral, then it is __?__ isosceles.
3. If a triangle is scalene, then it is __?__ isosceles.
4. If a triangle is obtuse, then it is __?__ isosceles.

Explain how each corollary of Theorem 3-11 follows from the theorem.

5. Corollary 1   6. Corollary 2   7. Corollary 3   8. Corollary 4

Find the value of *x*.

9.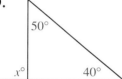

10.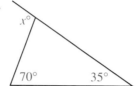

11.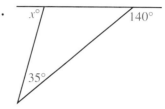

What is wrong with each of the following instructions?

12. Draw the bisector of ∠*J* to the midpoint of $\overline{PE}$.
13. Draw the line from *P* perpendicular to $\overline{JE}$ at its midpoint.
14. Draw the line through *P* and *X* parallel to $\overleftrightarrow{JE}$.

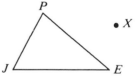

15. In the diagram you know that
    (1) $m\angle 1 + m\angle 2 + m\angle 3 = 180$
    (2) $m\angle 3 + m\angle 4 = 180$
    Explain how these equations allow you
    to prove Theorem 3-12.

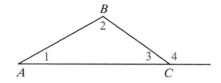

16. Fold a corner of a sheet of paper and then cut along the fold to get a right triangle. Let the right angle be ∠*C*. Fold each of the other two vertices so that they coincide with point *C*. What result of this section does this illustrate?

17. Cut out any large △*XYZ*. (If the triangle has a longest side, let that side be $\overline{YZ}$.) Fold so that *X* lies on the fold line and *Y* falls on $\overrightarrow{YZ}$. Let *P* be the intersection of $\overline{YZ}$ and the fold line. Unfold. Now fold the paper so that *Y* coincides with *P*. Fold it twice more so that both *X* and *Z* coincide with *P*. What result of this section does this illustrate?

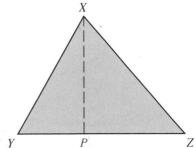

Written Exercises

Draw a triangle that satisfies the conditions stated. If no triangle can satisfy the conditions, write *not possible*.

A **1. a.** An acute isosceles triangle
    b. A right isosceles triangle
    c. An obtuse isosceles triangle

2. a. An acute scalene triangle
    b. A right scalene triangle
    c. An obtuse scalene triangle

3. A triangle with two acute exterior angles

4. A triangle with two obtuse exterior angles

Complete.

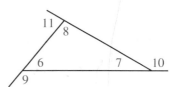

5. $m \angle 6 + m \angle 7 + m \angle 8 = \underline{\ ?\ }$.

6. If $m \angle 6 = 52$ and $m \angle 11 = 82$, then $m \angle 7 = \underline{\ ?\ }$.

7. If $m \angle 6 = 55$ and $m \angle 10 = 150$, then $m \angle 8 = \underline{\ ?\ }$.

8. If $m \angle 6 = x$, $m \angle 7 = x - 20$, and $m \angle 11 = 80$, then $x = \underline{\ ?\ }$.

9. If $m \angle 8 = 4x$, $m \angle 7 = 30$, and $m \angle 9 = 6x - 20$, then $x = \underline{\ ?\ }$.

10. $m \angle 9 + m \angle 10 + m \angle 11 = \underline{\ ?\ }$.

Exs. 5–10

Find the values of *x* and *y*.

11.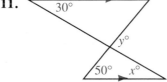

12.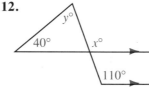

13.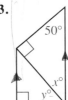

B **14.**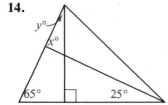

15.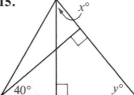

16.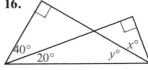

17. The lengths of the sides of a triangle are $4n$, $2n + 10$, and $7n - 15$. Is there a value of n that makes the triangle equilateral? Explain.

18. The lengths of the sides of a triangle are $3t$, $5t - 12$, and $t + 20$.
    a. Find the value(s) of t that make the triangle isosceles.
    b. Does any value of t make the triangle equilateral? Explain.

19. The largest two angles of a triangle are two and three times as large as the smallest angle. Find all three measures.

20. The measure of one angle of a triangle is 28 more than the measure of the smallest angle of the triangle. The measure of the third angle is twice the measure of the smallest angle. Find all three measures.

21. In $\triangle ABC$, $m \angle A = 60$ and $m \angle B < 60$. What can you say about $m \angle C$?

22. In $\triangle RST$, $m \angle R = 90$ and $m \angle S > 20$. What can you say about $m \angle T$?

23. Given: $\overline{AB} \perp \overline{BC}$; $\overline{BD} \perp \overline{AC}$
    a. If $m \angle C = 22$, find $m \angle ABD$.
    b. If $m \angle C = 23$, find $m \angle ABD$.
    c. Explain why $m \angle ABD$ always equals $m \angle C$.

24. The bisectors of $\angle EFG$ and $\angle EGF$ meet at I.
    a. If $m \angle EFG = 40$, find $m \angle FIG$.
    b. If $m \angle EFG = 50$, find $m \angle FIG$.
    c. Generalize your results in (a) and (b).

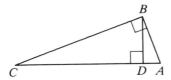

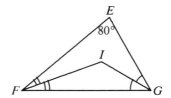

25. Given: $\angle ABD \cong \angle AED$
    Prove: $\angle C \cong \angle F$

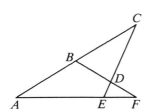

26. Find the measures of $\angle 1$ and $\angle 2$.

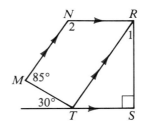

27. Prove Theorem 3-11 by using the diagram below. (Begin by stating what is given and what is to be proved. Draw the auxiliary ray shown.)

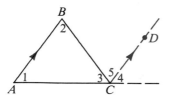

28. Given: $\overrightarrow{GK}$ bisects $\angle JGI$;
        $m \angle H = m \angle I$
    Prove: $\overline{GK} \parallel \overline{HI}$

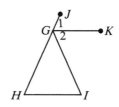

Find the values of *x* and *y*.

29.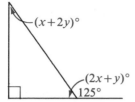

30.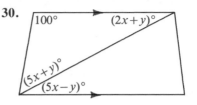

31. Given: $\overline{AB} \perp \overline{BF}$; $\overline{HD} \perp \overline{BF}$;
   $\overline{GF} \perp \overline{BF}$; $\angle A \cong \angle G$

   Which numbered angles must be congruent?

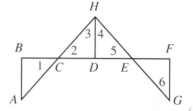

C **32.** Given: $\overrightarrow{PR}$ bisects $\angle SPQ$;
   $\overline{PS} \perp \overline{SQ}$; $\overline{RQ} \perp \overline{PQ}$

   Which numbered angles must be congruent?

33. **a.** Draw two parallel lines and a transversal.
   b. Use a protractor to draw bisectors of two same-side interior angles.
   c. Measure the angles formed by the bisectors. What do you notice?
   d. Prove your answer to part (c).

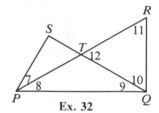

Ex. 32

34. A pair of same-side interior angles are *trisected* (divided into three congruent angles) by the red lines in the diagram. Find out what you can about the angles of *ABCD*.

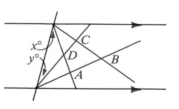

Explorations

These exploratory exercises can be done using a computer with a program that draws and measures geometric figures.

Decide if the following statements are true or false. If you think the statement is true, give a convincing argument to support your belief. If you think the statement is false, make a sketch and give all the measurements of the triangle that you find as your counterexample. For each false statement, also discover if there are types of triangles for which the statement is true.

1. The measure of an exterior angle is greater than the measure of any interior angle of a triangle.

2. An exterior angle is always an obtuse angle.

3. An exterior angle and some interior angle are supplementary.

4. The sum of the measures of an exterior angle and the remote interior angles is 180.

Carpenter

Carpenters work in all parts of the construction industry. A self-employed carpenter may work on relatively small-scale projects—for example, remodeling rooms or making other

alterations in existing houses or even building new single-family houses. As an employee of a large building con-tractor, a carpenter may be part of the work force building apartment or office complexes, stores, factories, and other major projects. Some carpenters are employed solely to provide maintenance to a large structure, where they do repairs and upkeep and make any alterations in the structure that are required.

Carpenters with adequate experience and expertise may become specialists in some skill of their own choice, for example, framing, interior finishing, or cabinet making. A carpenter who learns all aspects of the building industry thoroughly may decide to go into business as a general contractor, responsible for all work on an entire project.

Although some carpenters learn the trade through four-year apprenticeships, most learn on the job. These workers begin as laborers or as carpenters' helpers. While they work in these jobs they gradually acquire the skills necessary to become carpenters themselves. Carpenters must be able to measure accurately and to apply their knowledge of arithmetic, geometry, and informal algebra. They also benefit from being able to read and understand plans, blueprints, and charts.

3-5 *Angles of a Polygon*

The word **polygon** means "many angles." Look at the figures at the left below and note that each polygon is formed by coplanar segments (called *sides*) such that:

(1) Each segment intersects exactly two other segments, one at each endpoint.

(2) No two segments with a common endpoint are collinear.

Polygons   Not Polygons

Can you explain why each of the figures at the right above is *not* a polygon?

A **convex polygon** is a polygon such that no line containing a side of the polygon contains a point in the interior of the polygon. The outline of the state flag of Arizona, shown at the left below, is a convex polygon. At the right below is the state flag of Ohio, whose outline is a nonconvex polygon.

 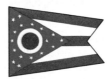

When we refer to a polygon in this book we will mean a convex polygon.

Polygons are classified according to the number of sides they have. Listed below are some of the special names for polygons you will see in this book.

Number of Sides	Name
3	triangle
4	quadrilateral
5	pentagon
6	hexagon
8	octagon
10	decagon
n	n-gon

A triangle is the simplest polygon. The terms that we applied to triangles (such as *vertex* and *exterior angle*) also apply to other polygons.

When referring to a polygon, we list its consecutive vertices in order. Pentagon *ABCDE* and pentagon *BAEDC* are two of the many correct names for the polygon shown at the right.

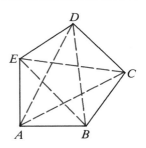

A segment joining two nonconsecutive vertices is a **diagonal** of the polygon. The diagonals of the pentagon at the right are indicated by dashes.

To find the sum of the measures of the angles of a polygon draw all the diagonals from just *one* vertex of the polygon to divide the polygon into triangles.

     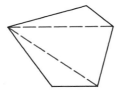   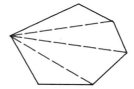

4 sides, 2 triangles   5 sides, 3 triangles   6 sides, 4 triangles
Angle sum = 2(180)   Angle sum = 3(180)   Angle sum = 4(180)

Note that the number of triangles formed in each polygon is two less than the number of sides. This result suggests the following theorem.

Theorem 3-13

The sum of the measures of the angles of a convex polygon with *n* sides is $(n - 2)180$.

Since the sum of the measures of the *interior* angles of a polygon depends on the number of sides, *n*, of the polygon, you would think that the same is true for the sum of the exterior angles. This is *not* true, as Theorem 3-14 reveals. The experiment suggested in Exercise 7 should help convince you of the truth of Theorem 3-14.

Theorem 3-14

The sum of the measures of the exterior angles of any convex polygon, one angle at each vertex, is 360.

Example 1   A polygon has 32 sides. Find (a) the sum of the measures of the interior angles and (b) the sum of the measures of the exterior angles, one angle at each vertex.

Solution   (a) Interior angle sum = $(32 - 2)180 = 5400$ (Theorem 3-13)
(b) Exterior angle sum = 360   (Theorem 3-14)

Polygons can be equiangular or equilateral. If a polygon is both equiangular and equilateral, it is called a **regular polygon.**

Hexagon that is
neither equiangular
nor equilateral

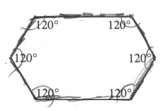

Equiangular hexagon

Equilateral
hexagon

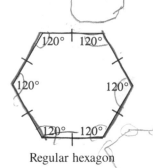

Regular hexagon

Example 2   A regular polygon has 12 sides. Find the measure of each interior angle.

Solution 1   Interior angle sum = $(12 - 2)180 = 1800$
Each of the 12 congruent interior angles has measure $1800 \div 12$, or 150.

Solution 2   Each exterior angle has measure $360 \div 12$, or 30.
Each interior angle has measure $180 - 30$, or 150.

Classroom Exercises

Is the figure a convex polygon, a nonconvex polygon, or neither?

1.

2.

3.

4.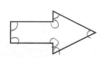

5.

6.

7. Imagine stretching a rubber band around each of the figures in Exercises 1–6. What is the relationship between the rubber band and the figure when the figure is a convex polygon?

8. A polygon has 102 sides. What is the interior angle sum? the exterior angle sum?

9. Complete the table for regular polygons.

Number of sides	6	10	20	?	?	?	?
Measure of each ext. $\angle$	?	?	?	10	20	?	?
Measure of each int. $\angle$	?	?	?	?	?	179	90

Written Exercises

For each polygon, find (a) the interior angle sum and (b) the exterior angle sum.

A   **1.** Quadrilateral   **2.** Pentagon   **3.** Hexagon

     4. Octagon   **5.** Decagon   **6.** *n*-gon

7. Draw a pentagon with one exterior angle at each vertex. Cut out the exterior angles and arrange them so that they all have a common vertex, as shown at the far right. What is the sum of the measures of the exterior angles? Repeat the experiment with a hexagon. Do your results support Theorem 3-14?

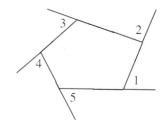

8. Complete the table for regular polygons.

Number of sides	9	15	30	?	?	?	?
Measure of each ext. ∠	?	?	?	6	8	?	?
Measure of each int. ∠	?	?	?	?	?	165	178

9. A baseball diamond's home plate has three right angles. The other two angles are congruent. Find their measure.

10. Four of the angles of a pentagon have measures 40, 80, 115, and 165. Find the measure of the fifth angle.

11. The face of a honeycomb consists of interlocking regular hexagons. What is the measure of each angle of these hexagons?

Sketch the polygon described. If no such polygon exists, write *not possible*.

12. A quadrilateral that is equiangular but not equilateral

13. A quadrilateral that is equilateral but not equiangular

14. A regular pentagon, one of whose angles has measure 120

15. A regular polygon, one of whose angles has measure 130

B **16.** The sum of the measures of the interior angles of a polygon is five times the sum of the measures of its exterior angles, one angle at each vertex. How many sides does the polygon have?

17. The measure of each interior angle of a regular polygon is eleven times that of an exterior angle. How many sides does the polygon have?

18. a. What is the measure of each interior angle of a regular pentagon?
   b. Can you tile a floor with tiles shaped like regular pentagons? (Ignore the difficulty in tiling along the edges of the room.)

19. Make a sketch showing how to tile a floor using both squares and regular octagons.

20. The cover of a soccerball consists of interlocking regular pentagons and regular hexagons, as shown at the right. The second diagram shows that regular pentagons and hexagons cannot be interlocked in this pattern to tile a floor. Why not?

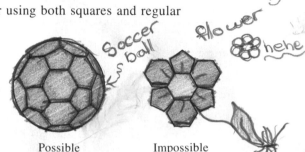

Possible   Impossible

21. In quadrilateral *ABCD*, $m \angle A = x$, $m \angle B = 2x$, $m \angle C = 3x$, and $m \angle D = 4x$. Find the value of x and then state which pair of sides of *ABCD* must be parallel.

22. In pentagon *PQRST*, $m \angle P = 60$ and $m \angle Q = 130$. $\angle S$ and $\angle T$ are each three times as large as $\angle R$.
   a. Find the measures of $\angle R$, $\angle S$, and $\angle T$.
   b. Which pair of sides of *PQRST* must be parallel?

23. *ABCDEFGHIJ* is a regular decagon. If sides $\overline{AB}$ and $\overline{CD}$ are extended to meet at *K*, find the measure of $\angle K$.

24. $\overline{BC}$ is one side of a regular *n*-gon. The sides next to $\overline{BC}$ are extended to meet at *W*. Find the measure of $\angle W$ in terms of *n*.

25. The sum of the measures of the interior angles of a polygon is known to be between 2100 and 2200. How many sides does the polygon have?

C **26.** The sum of the measures of the interior angles of a polygon with *n* sides is *S*. Without using *n* in your answer, express in terms of *S* the sum of the measures of the angles of a polygon with:
   a. $n + 1$ sides   **b.** $2n$ sides

27. The formula $S = (n - 2)180$ can apply to nonconvex polygons if you allow the measure of an interior angle to be more than 180.
   a. Illustrate this with a diagram that shows interior angles with measures greater than 180.
   b. Does the reasoning leading up to Theorem 3-13 apply to your figure?

28. Given: The measure of each interior angle of a regular *n*-gon is *x* times that of an exterior angle.
   a. Express *x* in terms of *n*.
   b. For what values of *n* will *x* be an integer?

3-6 *Inductive Reasoning*

Throughout these first three chapters, we have been using deductive reasoning. Now we'll consider **inductive reasoning,** a kind of reasoning that is widely used in science and in everyday life.

Example 1 After picking marigolds for the first time, Connie began to sneeze. She also began sneezing the next four times she was near marigolds. Based on this past experience, Connie reasons inductively that she is allergic to marigolds.

Example 2 Every time Pitch has thrown a high curve ball to Slugger, Slugger has gotten a hit. Pitch concludes from this experience that it is not a good idea to pitch high curve balls to Slugger.

In coming to this conclusion, Pitch has used inductive reasoning. It may be that Slugger just happened to be lucky those times, but Pitch is too bright to feed another high curve to Slugger.

From these examples you can see how inductive reasoning differs from deductive reasoning.

Deductive Reasoning	**Inductive Reasoning**
Conclusion based on accepted statements (definitions, postulates, previous theorems, corollaries, and given information)	Conclusion based on several past observations
Conclusion *must* be true if hypotheses are true.	Conclusion is *probably* true, but not necessarily true.

Often in mathematics you can reason inductively by observing a pattern.

Example 3 Look for a pattern and predict the next number in each sequence.
a. 3, 6, 12, 24, __?__   **b.** 11, 15, 19, 23, __?__   **c.** 5, 6, 8, 11, 15, __?__

Solution **a.** Each number is twice the preceding number. The next number will be 2 × 24, or 48.

b. Each number is 4 more than the preceding number. The next number will be 23 + 4, or 27.

c. Look at the differences between the numbers.

Numbers   5   6   8   11   15   ?

Differences   1   2   3   4   ?

The next difference will be 5, and thus the next number will be 15 + 5, or 20.

Classroom Exercises

Tell whether the reasoning process is deductive or inductive.

1. Ramon noticed that spaghetti had been on the school menu for the past five Wednesdays. Ramon decides that the school always serves spaghetti on Wednesday.

2. Ky did his assignment, adding the lengths of the sides of triangles to find the perimeters. Noticing the results for several equilateral triangles, he guesses that the perimeter of every equilateral triangle is three times the length of a side.

3. By using the definitions of equilateral triangle (a triangle with three congruent sides) and of perimeter (the sum of the lengths of the sides of a figure), Katie concludes that the perimeter of every equilateral triangle is three times the length of a side.

4. Linda observes that $(-1)^2 = +1$, $(-1)^4 = +1$, and $(-1)^6 = +1$. She concludes that every even power of (-1) is equal to $+1$.

5. John knows that multiplying a number by -1 merely changes the sign of the number. He reasons that multiplying a number by an even power of -1 will change the sign of the number an even number of times. He concludes that this is equivalent to multiplying a number by $+1$, so that every even power of -1 is equal to $+1$.

6. Look at the discussion leading up to the statement of Theorem 3-13 on page 102. Is the thinking inductive or deductive?

Written Exercises

Look for a pattern and predict the next two numbers in each sequence.

A
1. 1, 4, 16, 64, . . .
2. 18, 15, 12, 9, . . .
3. $1, \frac{1}{3}, \frac{1}{9}, \frac{1}{27}, \ldots$

4. 1, 4, 9, 16, . . .
5. 2, 3, 5, 8, 12, . . .
6. 10, 12, 16, 22, 30, . . .

7. 40, 39, 36, 31, 24, . . .
8. $8, -4, 2, -1, \frac{1}{2}, \ldots$
9. 2, 20, 10, 100, 50, . . .

Accept the two statements as given information. State a conclusion based on *deductive* reasoning. If no conclusion can be reached, write *none*.

10. Chan is older than Pedro.
    Pedro is older than Sarah.

11. Valerie is older than Greg.
    Dan is older than Greg.

12. Polygon G has more than 6 sides.
    Polygon G has fewer than 8 sides.

13. Polygon G has more than 6 sides.
    Polygon K has more than 6 sides.

14. There are three sisters. Two of them are athletes and two of them like tacos. Can you be sure that both of the athletes like tacos? Do you reason deductively or inductively to conclude the following? *At least one of the athletic sisters likes tacos.*

For each exercise, write the equation you think should come next. Check your prediction with a calculator.

15. $\quad 1 \times 9 + 2 = 11$
$\quad\quad 12 \times 9 + 3 = 111$
$\quad\quad 123 \times 9 + 4 = 1111$

16. $\quad 9 \times 9 + 7 = 88$
$\quad\quad 98 \times 9 + 6 = 888$
$\quad\quad 987 \times 9 + 5 = 8888$

17. $\quad 9^2 = 81$
$\quad\quad 99^2 = 9801$
$\quad\quad 999^2 = 998001$

Draw several diagrams to help you decide whether each statement is true or false. If it is false, show a counterexample. If it is true, draw and label a diagram you could use in a proof. List, in terms of the diagram, what is given and what is to be proved. Do *not* write a proof.

B **18.** If a triangle has two congruent sides, then the angles opposite those sides are congruent.

19. If a triangle has two congruent angles, then the sides opposite those angles are congruent.

20. If two triangles have equal perimeters, then they have congruent sides.

21. All diagonals of a regular pentagon are congruent.

22. If both pairs of opposite sides of a quadrilateral are parallel, then the diagonals bisect each other.

23. If the diagonals of a quadrilateral are congruent and also perpendicular, then the quadrilateral is a regular quadrilateral.

24. The diagonals of an equilateral quadrilateral are congruent.

25. The diagonals of an equilateral quadrilateral are perpendicular.

26. a. Study the diagrams below. Then guess the number of regions for the fourth diagram. Check your answer by counting.

2 points   3 points   4 points   5 points
2 regions   4 regions   8 regions   __?__ regions

b. Using 6 points on a circle as shown, guess the number of regions within the circle. Carefully check your answer by counting.
Important note: This exercise shows that a pattern predicted on the basis of a few cases may be incorrect. To be sure of a conclusion, use a deductive proof.

27. a. Draw several quadrilaterals whose opposite sides are parallel. With a protractor measure both pairs of opposite angles of each figure. On the basis of the diagrams and measurements, what do you guess is true for all such quadrilaterals? (*Note:* See Exercise 23, page 82.)

b. State and prove the converse of your conclusion about opposite angles in part (a).

c. Write a biconditional about pairs of opposite angles of a quadrilateral.

C **28. a.** Substitute each of the integers from 1 to 9 for n in the expression
$n^2 + n + 11$.
    b. Using inductive reasoning, guess what kind of number you will get
when you substitute any positive integer for n in the expression
$n^2 + n + 11$.
    c. Test your guess by substituting 10 and 11 for n.

29. Complete the table for convex polygons.

Number of sides	3	4	5	6	7	8	n
Number of diagonals	0	2	?	?	?	?	?

30. Find the sum of the measures of the angles formed at the tips of each star.
    a. five-pointed star   **b.** six-pointed star

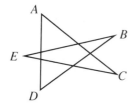 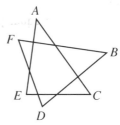

    c. Using inductive reasoning, suggest a formula for the sum of the angle
measures at the tips of an n-pointed star.
    d. Using deductive reasoning, justify your formula.

♦ Calculator Key-In

**Complete the right side of the first three equations in each exercise. Then
use inductive reasoning to predict what the fourth equation would be if the
pattern were continued. Check your prediction with your calculator.**

1. $1 \times 1 = $ __?__
     $11 \times 11 = $ __?__
     $111 \times 111 = $ __?__
     __?__ $\times$ __?__ $= $ __?__

2. $6 \times 7 = $ __?__
     $66 \times 67 = $ __?__
     $666 \times 667 = $ __?__
     __?__ $\times$ __?__ $= $ __?__

3. $8 \times 8 = $ __?__
     $98 \times 98 = $ __?__
     $998 \times 998 = $ __?__
     __?__ $\times$ __?__ $= $ __?__

4. $7 \times 9 = $ __?__
     $77 \times 99 = $ __?__
     $777 \times 999 = $ __?__
     __?__ $\times$ __?__ $= $ __?__

Self-Test 2

Complete.

1. If the measure of each angle of a triangle is less than 90, the triangle is called __?__.

2. If a triangle has no congruent sides, it is called __?__.

3. Each angle of an equiangular triangle has measure __?__.

4. In the diagram, m∠1 = __?__ and m∠2 = __?__.

5. If the measures of the acute angles of a right triangle are $2x + 4$ and $3x - 9$, then $x =$ __?__.

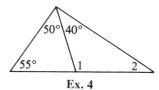

Ex. 4

6. Find the values of y and z.

7. The lengths of the sides of a triangle are $2x + 5$, $3x + 10$, and $x + 12$. Find all values of x that make the triangle isosceles.

8. An octagon has __?__ sides.

9. A regular polygon is both __?__ and __?__.

10. In a regular decagon, the sum of the measures of the exterior angles is __?__ and the measure of each interior angle is __?__.

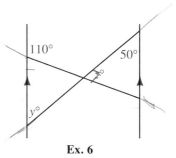

Ex. 6

11. If the measure of each angle of a polygon is 174, then the measure of each exterior angle is __?__ and the polygon has __?__ sides.

Use inductive reasoning to predict the next number in each sequence.

12. $2, -4, 8, -16, \ldots$

13. $7, 12, 17, 22, 27, \ldots$

14. $1, 4, 9, 16, 25, \ldots$

15. $1, 4, 2, 8, 4, 16, 8, 32, \ldots$

Chapter Summary

1. Lines that do not intersect are either parallel or skew.

2. When two parallel lines are cut by a transversal:
   a. corresponding angles are congruent;
   b. alternate interior angles are congruent;
   c. same-side interior angles are supplementary;
   d. if the transversal is perpendicular to one of the two parallel lines, it is also perpendicular to the other one.

3. The chart on page 85 lists five ways to prove lines parallel.

4. Through a point outside a line, there is exactly one line parallel to, and exactly one line perpendicular to, the given line.

5. Two lines parallel to a third line are parallel to each other.

6. Triangles are classified (page 93) by the lengths of their sides and by the measures of their angles. In any $\triangle ABC$, $m \angle A + m \angle B + m \angle C = 180$.

7. The measure of an exterior angle of a triangle equals the sum of the measures of the two remote interior angles.

8. The sum of the measures of the angles of a convex polygon with n sides is $(n - 2)180$. The sum of the measures of the exterior angles, one angle at each vertex, is 360.

9. Polygons that are both equiangular and equilateral are regular polygons.

10. Inductive reasoning is the process of observing individual cases and then reaching a general conclusion suggested by them. The conclusion is probably, but not necessarily, true.

Chapter Review

1. $\angle 5$ and $\angle$ __?__ are same-side interior angles.   3–1

2. $\angle 5$ and $\angle 1$ are __?__ angles.

3. $\angle 5$ and $\angle 3$ are __?__ angles.

4. Line j, not shown, does not intersect line r. Must lines r and j be parallel?

Exs. 1–7

In the diagram above, $r \parallel s$.

5. If $m \angle 1 = 105$, then $m \angle 5 = $ __?__ and $m \angle 7 = $ __?__.   3–2

6. Solve for x: $m \angle 2 = 70$ and $m \angle 8 = 6x - 2$

7. Solve for y: $m \angle 3 = 8y - 40$ and $m \angle 8 = 2y + 20$

8. Lines a, b, and c are coplanar, $a \parallel b$, and $a \perp c$. What can you conclude? Explain.

9. Which line is parallel to $\overleftrightarrow{AB}$? Why?   3–3

10. Name a pair of parallel lines other than the pair in Exercise 9. Why must they be parallel?

11. Name five ways to prove two lines parallel.

Exs. 9, 10

12. If x and $2x - 15$ represent the measures of the acute angles of a right triangle, find the value of x.   3–4

13. $m \angle 6 + m \angle 7 + m \angle 8 = $ __?__

14. If $m \angle 1 = 30$ and $m \angle 4 = 130$, then $m \angle 2 = $ __?__.

15. If $\angle 4 \cong \angle 5$ and $\angle 1 \cong \angle 7$, name two other pairs of congruent angles and give a reason for each answer.

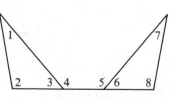

Exs. 13–15

16. **a.** Sketch a hexagon that is equiangular but not equilateral. 3–5
    b. What is its interior angle sum?
    c. What is its exterior angle sum?

17. A regular polygon has 18 sides. Find the measure of each interior angle.

18. A regular polygon has 24 sides. Find the measure of each exterior angle.

19. Each interior angle of a regular polygon has measure 150. How many sides does the polygon have?

Use inductive reasoning to predict the next two numbers in each sequence.

20. 15, 30, 45, 60, . . .   21. 100, -10, 1, $-\dfrac{1}{10}$, . . .   3–6

Chapter Test

Complete each statement with the word *always*, *sometimes*, or *never*.

1. Two lines that have no points in common are __?__ parallel.

2. If a line is perpendicular to one of two parallel lines, then it is __?__ perpendicular to the other one.

3. If two lines are cut by a transversal and same-side interior angles are complementary, then the lines are __?__ parallel.

4. An obtuse triangle is __?__ a right triangle.

5. In $\triangle ABC$, if $\overline{AB} \perp \overline{BC}$, then $\overline{AC}$ is __?__ perpendicular to $\overline{BC}$.

6. As the number of sides of a regular polygon increases, the measure of each exterior angle __?__ decreases.

Find the value of x.

7. $m \angle 1 = 3x - 20$, $m \angle 2 = x$
8. $m \angle 2 = 2x + 12$, $m \angle 3 = 4(x - 7)$

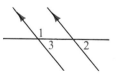

Find the measures of the numbered angles.

9. *XYZ* is regular.   10.   11. *ABCDE* is regular.

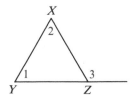   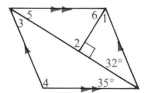   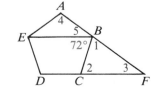

12. In the diagram for Exercise 11, explain why $\overline{EB}$ and $\overline{DF}$ must be parallel.

13. Given: $\overleftrightarrow{AB} \parallel \overleftrightarrow{CD}$; $\overrightarrow{BF}$ bisects $\angle ABE$;
    $\overrightarrow{DG}$ bisects $\angle CDB$.
    Prove: $\overleftrightarrow{BF} \parallel \overleftrightarrow{DG}$

14. Predict the next two numbers in the
    sequence 7, 9, 11, 13, ...

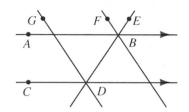

Algebra Review: *The Coordinate Plane*

1. What is the x-coordinate of point P?

2. What is the y-coordinate of point P?

3. What are the coordinates of the origin, point O?

4. Name the graph of the ordered pair $(0, -2)$.

Name the coordinates of each point.

5. M 6. N 7. K 8. R 9. S

10. T 11. U 12. V 13. W 14. Q

15. Name all the points shown that lie on the x-axis.

16. Name all the points shown that lie on the y-axis.

17. What is the x-coordinate of every point that lies
    on a vertical line through P?

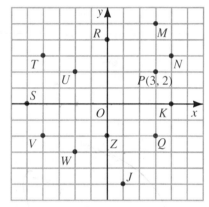

Exs. 1–22

18. Which of the following points lie on a horizontal line through W?
       a. $(-2, 1)$ **b.** $(2, 3)$ **c.** $(1, -3)$ **d.** $(-2, 0)$ **e.** $(0, -3)$ **f.** $(2, 0)$

**Name all the points shown that lie in the quadrant indicated. (A point on
an axis is not in any quadrant.)**

19. Quadrant I   20. Quadrant II   21. Quadrant III   22. Quadrant IV

Plot each point on graph paper.

23. O $(0, 0)$ 24. A $(2, 1)$ 25. B $(3, 4)$ 26. C $(5, 0)$

27. D $(0, 3)$ 28. E $(-3, 1)$ 29. F $(-2, -1)$ 30. G $(1, -2)$

31. H $(0, -4)$ 32. I $(-4, 0)$ 33. J $(4, -2)$ 34. K $(-4, -3)$

**Find the coordinates of the midpoint of $\overline{AB}$. (You may want to draw a
diagram.)**

35. A $(0, 1)$, B $(4, 1)$ 36. A $(2, 0)$, B $(2, 10)$ 37. A $(0, 1)$, B $(0, 5)$

38. A $(-3, 4)$, B $(-3, -4)$ 39. A $(-5, -2)$, B $(-3, -2)$ 40. A $(4, -1)$, B $(-2, -1)$

Cumulative Review: Chapters 1–3

Complete each statement with the word *always*, *sometimes*, or *never*.

A 1. If $\overleftrightarrow{AB}$ intersects $\overline{CD}$, then $\overline{AB}$ __?__ intersects $\overline{CD}$.

2. If two planes intersect, their intersection is __?__ a line.

3. If $a \perp c$ and $b \perp c$, then a and b are __?__ parallel.

4. If two parallel planes are cut by a third plane, then the lines of intersection are __?__ coplanar.

5. A scalene triangle __?__ has an acute angle.

Draw a diagram that satisfies the conditions stated. If the conditions cannot be satisfied, write *not possible*.

6. $\overline{AB}$ and $\overline{XY}$ intersect and A is the midpoint of $\overline{XY}$.

7. A triangle is isosceles but not equilateral.

8. Three points all lie in both plane M and plane N.

9. Two lines intersect to form adjacent angles that are not supplementary.

10. Points A and B on a number line have coordinates -3.5 and 8.5. Find the coordinate of the midpoint of $\overline{AB}$.

11. $\overrightarrow{QX}$ bisects $\angle PQR$, $m \angle PQX = 5x + 13$, and $m \angle XQR = 9x - 39$. Find (a) the value of x and (b) $m \angle PQR$.

12. The measure of a supplement of an angle is 35 more than twice the complement of the angle. Find the measures of the angle, its supplement, and its complement.

13. The measures of two angles of a triangle are five and six times as large as the measure of the smallest angle. Find all three measures.

In the diagram $\overleftrightarrow{AB}$ bisects $\angle DHF$, $\overleftrightarrow{AB} \perp \overline{GH}$, $\overleftrightarrow{AB} \parallel \overleftrightarrow{CD}$, and $m \angle AHD = 60$. Find the measure of each angle.

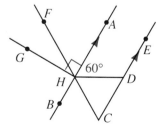

14. $\angle FHD$	**15.** $\angle AHG$	**16.** $\angle FHG$
17. $\angle GHB$	**18.** $\angle BHC$	**19.** $\angle DHC$
20. $\angle HDE$	**21.** $\angle HDC$	**22.** $\angle HCD$

Tell whether each statement is true or false. Then write the converse and tell whether it is true or false.

23. If two lines do not intersect, then they are parallel.

24. If two lines intersect to form right angles, then the lines are perpendicular.

25. An angle is acute only if it is not obtuse.

26. A triangle is isosceles if it is equilateral.

Name or state the postulate, definition, or theorem that justifies each statement about the diagram.

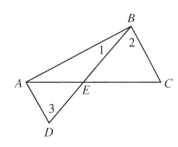

27. $\angle AED \cong \angle BEC$

28. $AE + EC = AC$

29. $m\angle 1 + m\angle 2 = m\angle ABC$

30. If $\angle 2 \cong \angle 3$, then $\overline{AD} \parallel \overline{BC}$.

31. $m\angle AEB = m\angle 2 + m\angle C$

32. If $\overline{DA} \perp \overline{AB}$, then $m\angle DAB = 90$.

33. $m\angle 1 + m\angle 3 + m\angle DAB = 180$

34. If $\angle ABC$ is a right angle, then $\overline{AB} \perp \overline{BC}$.

Complete.

35. The endpoint of $\overrightarrow{XY}$ is point __?__.

36. If the sum of the measures of two angles is 180, then the angles are __?__.

37. If the measure of each interior angle of a regular polygon is 108, then the polygon is a(n) __?__.

38. If M is the midpoint of $\overline{AB}$ and $AM = 12$, then $AB = $ __?__.

39. If two parallel lines are cut by a transversal, then alternate interior angles are __?__.

40. The process of forming a conclusion based on past observations or patterns is called __?__ reasoning.

41. When a statement and its converse are both true, they can be combined into one statement called a __?__.

42. In a decagon the sum of the measures of the exterior angles is __?__.

43. In an octagon the sum of the measures of the interior angles is __?__.

44. Every triangle has at least two __?__ angles.

Write a two column proof.

B 45. Given: $\overline{WX} \perp \overline{XY}$;
$\quad\quad\quad$ $\angle 1$ is comp. to $\angle 3$.
$\quad\quad$ Prove: $\angle 2 \cong \angle 3$

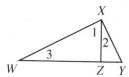

46. Given: $\overline{RU} \parallel \overline{ST}$; $\angle R \cong \angle T$
$\quad\quad$ Prove: $\overline{RS} \parallel \overline{UT}$

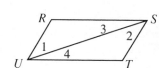

4 CONGRUENT TRIANGLES

The simplest plane figure—a triangle—was used to create the striking pattern of this traditional American quilt.

Corresponding Parts in a Congruence

Objectives

1. Identify the corresponding parts of congruent figures.
2. Prove two triangles congruent by using the SSS Postulate, the SAS Postulate, and the ASA Postulate.
3. Deduce information about segments and angles after proving that two triangles are congruent.

4-1 *Congruent Figures*

The quilt on the facing page is made up of many triangles that are all the same size and shape. These triangles are arranged to form squares and rectangles of various sizes. The diagrams below feature the pattern in the quilt. In each diagram, how many triangles with the same size and shape do you see? How many squares? How many rectangles?

Whenever two figures have the same size and shape, they are called **congruent.** You are already familiar with congruent segments (segments that have equal lengths) and congruent angles (angles that have equal measures). In this chapter you will learn about congruent triangles.

Triangles *ABC* and *DEF* are congruent. If you mentally slide △*ABC* to the right, you can fit it exactly over △*DEF* by matching up the vertices like this:

$$A \longleftrightarrow D \qquad B \longleftrightarrow E \qquad C \longleftrightarrow F$$

The sides and angles will then match up like this:

Corresponding angles	Corresponding sides
$\angle A \longleftrightarrow \angle D$	$\overline{AB} \longleftrightarrow \overline{DE}$
$\angle B \longleftrightarrow \angle E$	$\overline{BC} \longleftrightarrow \overline{EF}$
$\angle C \longleftrightarrow \angle F$	$\overline{AC} \longleftrightarrow \overline{DF}$

Do you see that the following statements are true?

(1) Since congruent triangles have the same shape, their corresponding angles are congruent.

(2) Since congruent triangles have the same size, their corresponding sides are congruent.

We have the following definition for *congruent triangles*.

Two triangles are **congruent** if and only if their vertices can be matched up so that the *corresponding parts* (angles and sides) of the triangles are congruent.

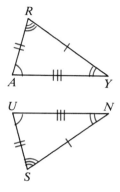

The congruent parts of the triangles shown are marked alike. Imagine sliding $\triangle SUN$ up until $\overline{UN}$ falls on $\overline{AY}$ and then flipping $\triangle SUN$ over so that point S falls on point R. The vertices are matched like this:

$$S \leftrightarrow R \qquad U \leftrightarrow A \qquad N \leftrightarrow Y$$

$\triangle SUN$ fits over $\triangle RAY$. The corresponding parts are congruent, and the triangles are congruent.

When referring to congruent triangles, we name their corresponding vertices in the same order. For the triangles shown,

$$\triangle SUN \text{ is congruent to } \triangle RAY.$$
$$\triangle SUN \cong \triangle RAY$$

The following statements about these triangles are also correct, since corresponding vertices of the triangles are named in the same order.

$$\triangle NUS \cong \triangle YAR \qquad\qquad \triangle SNU \cong \triangle RYA$$

Suppose you are given that $\triangle XYZ \cong \triangle ABC$. From the definition of congruent triangles you know, for example, that

$$\overline{XY} \cong \overline{AB} \qquad \text{and} \qquad \angle X \cong \angle A.$$

When the definition of congruent triangles is used to justify either of these statements, the wording commonly used is

Corresponding parts of congruent triangles are congruent,

which is often written:

Corr. parts of $\cong$ $\triangle$ are $\cong$.

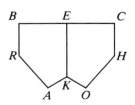

BRAKE $\cong$ CHOKE

Two *polygons* are **congruent** if and only if their vertices can be matched up so that their corresponding parts are congruent. Just as for triangles, there are many ways to list the congruence between the two pentagons at the right so that corresponding vertices are written in the same order.

Notice that side $\overline{KE}$ of pentagon *BRAKE* corresponds to side $\overline{KE}$ of pentagon *CHOKE*. $\overline{KE}$ is called a *common side* of the two pentagons.

Classroom Exercises

Suppose you know that $\triangle FIN \cong \triangle WEB$.

1. Name the three pairs of corresponding sides.
2. Name the three pairs of corresponding angles.
3. Is it correct to say $\triangle NIF \cong \triangle BEW$?
4. Is it correct to say $\triangle INF \cong \triangle EWB$?

The two triangles shown are congruent. Complete.

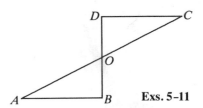

5. $\triangle ABO \cong$ __?__
6. $\angle A \cong$ __?__
7. $\overline{AO} \cong$ __?__
8. $BO =$ __?__

9. Can you deduce that O is the midpoint of any segment? Explain.
10. Explain how you can deduce that $\overline{DC} \parallel \overline{AB}$.
11. Suppose you know that $\overline{DB} \perp \overline{DC}$. Explain how you can deduce that $\overline{DB} \perp \overline{BA}$.

Exs. 5–11

The pentagons shown are congruent. Complete.

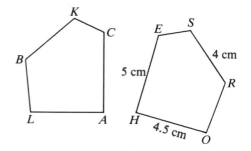

12. B corresponds to __?__.
13. $BLACK \cong$ __?__
14. __?__ $= m\angle E$
15. $KB =$ __?__ cm
16. If $\overline{CA} \perp \overline{LA}$, name two right angles in the figures.

17. The five leaves shown are all congruent, but one differs from the others. Which one is different and how?

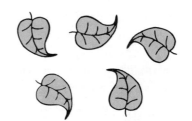

18. **a.** Name the coordinates of points A, B, and C.
    b. Name the coordinates of a point D such that $\triangle ABC \cong \triangle ABD$.
19. Name the coordinates of a point G such that $\triangle ABC \cong \triangle EFG$. Is there another location for G such that $\triangle ABC \cong \triangle EFG$?
20. Name the coordinates of two possible points H such that $\triangle ABC \cong \triangle FEH$.

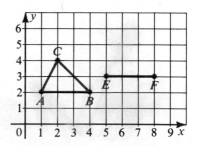

Written Exercises

Suppose $\triangle BIG \cong \triangle CAT$. **Complete.**

A
1. $\angle G \cong$ ___?___
2. ___?___ $= m \angle A$
3. $BI =$ ___?___
4. ___?___ $\cong \overline{AT}$
5. $\triangle IGB \cong$ ___?___
6. ___?___ $\cong \triangle CTA$

7. If $\triangle DEF \cong \triangle RST$, $m \angle D = 100$, and $m \angle F = 40$, name four congruent angles.

8. Is the statement ''Corresponding parts of congruent triangles are congruent'' based on a definition, postulate, or theorem?

9. Suppose $\triangle LXR \cong \triangle FNE$. List six congruences that can be justified by the following reason: Corr. parts of $\cong \triangle$ are $\cong$.

10. The two triangles shown are congruent. Complete.
    a. $\triangle STO \cong$ ___?___
    b. $\angle S \cong$ ___?___ because ___?___.
    c. $\overline{SO} \cong$ ___?___ because ___?___.
       Then point O is the midpoint of ___?___.
    d. $\angle T \cong$ ___?___ because ___?___.
       Then $\overline{ST} \parallel \overline{RK}$ because ___?___.

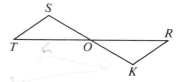

11. The two triangles shown are congruent. Complete.
    a. $\triangle PAL \cong$ ___?___
    b. $\overline{PA} \cong$ ___?___
    c. $\angle 1 \cong$ ___?___ because ___?___.
       Then $\overline{PA} \parallel$ ___?___ because ___?___.
    d. $\angle 2 \cong$ ___?___ because ___?___.
       Then ___?___ $\parallel$ ___?___ because ___?___.

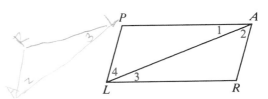

Plot the given points on graph paper. Draw $\triangle FAT$. **Locate point** C **so that** $\triangle FAT \cong \triangle CAT$.

12. $F(1, 2)$ $A(4, 7)$ $T(4, 2)$
13. $F(7, 5)$ $A(-2, 2)$ $T(5, 2)$

Plot the given points on graph paper. Draw $\triangle ABC$ **and** $\triangle DEF$. **Copy and complete the statement** $\triangle ABC \cong$ ___?___.

B
14. $A(-1, 2)$ $B(4, 2)$ $C(2, 4)$
    $D(5, -1)$ $E(7, 1)$ $F(10, -1)$

15. $A(-7, -3)$ $B(-2, -3)$ $C(-2, 0)$
    $D(0, 1)$ $E(5, 1)$ $F(0, -2)$

16. $A(-3, 1)$ $B(2, 1)$ $C(2, 3)$
    $D(4, 3)$ $E(6, 3)$ $F(6, 8)$

17. $A(1, 1)$ $B(8, 1)$ $C(4, 3)$
    $D(3, -7)$ $E(5, -3)$ $F(3, 0)$

Plot the given points on graph paper. Draw $\triangle ABC$ **and** $\overline{DE}$. **Find two locations of point** F **such that** $\triangle ABC \cong \triangle DEF$.

18. $A(1, 2)$ $B(4, 2)$ $C(2, 4)$ $D(6, 4)$ $E(6, 7)$
19. $A(-1, 0)$ $B(-5, 4)$ $C(-6, 1)$ $D(1, 0)$ $E(5, 4)$

$\overline{OR}$ **is a common side of two congruent quadrilaterals.**

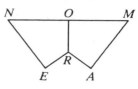

20. Complete: quad. *NERO* ≅ quad. __?__

21. In your own words explain why each of the
following statements must be true.
   a. *O* is the midpoint of $\overline{NM}$.
   b. ∠*NOR* ≅ ∠*MOR*
   c. $\overline{RO} \perp \overline{NM}$

Exs. 20, 21

22. Accurately draw each triangle described. Predict whether your triangle
will be congruent to your classmates'.
   a. In △*RST*, *RS* = 4 cm, *m*∠*S* = 45, and *ST* = 6 cm.
   b. In △*UVW*, *m*∠*U* = 30, *UV* = 5 cm, and *m*∠*V* = 100.
   c. In △*DEF*, *m*∠*D* = 30, *m*∠*E* = 68, and *m*∠*F* = 82.
   d. In △*XYZ*, *XY* = 3 cm, *YZ* = 5 cm, and *XZ* = 6 cm. (Try for a
reasonably accurate drawing. You may find it helpful to cut a thin
strip of paper for each side, then form the triangle.)

23. Does congruence of triangles have the reflexive property? the symmetric
property? the transitive property?

C **24.** Suppose you are given a scalene triangle and a point *P* on some line *l*.
How many triangles are there with one vertex at *P*, another vertex on *l*,
and each triangle congruent to the given triangle?

Challenge

Twelve toothpicks are arranged as shown to form a regular hexagon.
a. Copy the figure and show how six more toothpicks of the same size
could be used to divide it into three congruent regions.
b. Keeping two of the toothpicks from part (a) in the same place and
moving four, use the six toothpicks to divide the figure into two
congruent regions.

Mixed Review Exercises

Write proofs in two-column form.

1. Given: $\overline{AD} \perp \overline{BC}$; $\overline{BA} \perp \overline{AC}$
   Prove: ∠1 ≅ ∠2

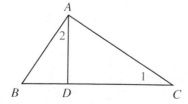

2. Given: $\overline{FC}$ and $\overline{SH}$ bisect each
       other at *A*; *FC* = *SH*
   Prove: *SA* = *AC*

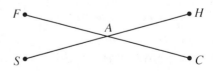

4-2 *Some Ways to Prove Triangles Congruent*

If two triangles are congruent, the six parts of one triangle are congruent to the six corresponding parts of the other triangle. If you are not sure whether two triangles are congruent, however, it is not necessary to compare all six parts. As you saw in Written Exercise 22 of the preceding section, sometimes three pairs of congruent corresponding parts will guarantee that two triangles are congruent. The following postulates give you three ways to show that two triangles are congruent by comparing only three pairs of corresponding parts.

Postulate 12   *SSS Postulate*

If three sides of one triangle are congruent to three sides of another triangle, then the triangles are congruent.

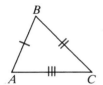 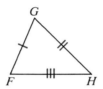

By the SSS Postulate, $\triangle ABC \cong \triangle FGH$ and $\triangle POE \cong \triangle TRY$.

Sometimes it is helpful to describe the parts of a triangle in terms of their relative positions.

$\overline{AB}$ is *opposite* $\angle C$.

$\overline{AB}$ is *included* between $\angle A$ and $\angle B$.

$\angle A$ is *opposite* $\overline{BC}$.

$\angle A$ is *included* between $\overline{AB}$ and $\overline{AC}$.

Postulate 13   *SAS Postulate*

If two sides and the included angle of one triangle are congruent to two sides and the included angle of another triangle, then the triangles are congruent.

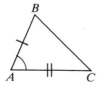 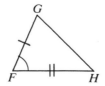 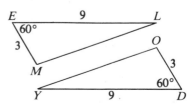

By the SAS Postulate, $\triangle ABC \cong \triangle FGH$ and $\triangle MEL \cong \triangle ODY$.

Postulate 14 *ASA Postulate*

If two angles and the included side of one triangle are congruent to two angles and the included side of another triangle, then the triangles are congruent.

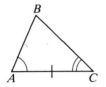 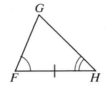 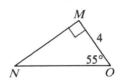 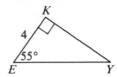

By the ASA Postulate, $\triangle ABC \cong \triangle FGH$ and $\triangle MON \cong \triangle KEY$.

Example   Supply the missing statements and reasons in the following proof.

Given: E is the midpoint of $\overline{MJ}$;
$\overline{TE} \perp \overline{MJ}$

Prove: $\triangle MET \cong \triangle JET$

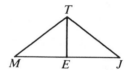

Proof:

Statements	Reasons
1. E is the midpoint of $\overline{MJ}$.	1. Given
2. ___?___ $\cong$ ___?___	2. Def. of midpoint
3. $\overline{TE} \perp \overline{MJ}$	3. ___?___
4. $\angle MET \cong \angle JET$	4. ___?___
5. $\overline{TE} \cong$ ___?___	5. ___?___
6. $\triangle MET \cong \triangle JET$	6. ___?___

Solution   Statement 2 $\overline{ME} \cong \overline{JE}$
Reason 3   Given
Reason 4   If two lines are $\perp$, then they form $\cong$ adj. $\angle$s.
Statement 5 $\overline{TE}$
Reason 5   Reflexive Prop.
Reason 6   SAS Postulate

Classroom Exercises

Does the SAS Postulate justify that the two triangles are congruent?

1.   **2.** 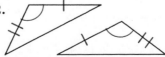   **3.**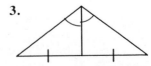

Can the two triangles be proved congruent? If so, what postulate can be used?

4. 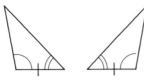 5. 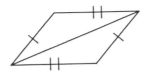 6.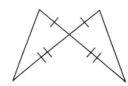

7. 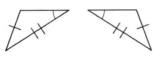 8.  9.

10. Explain how you would prove the following.
    Given: $\overline{HY} \cong \overline{LY}$;
    $\quad\quad\overline{WH} \parallel \overline{LF}$
    Prove: $\triangle WHY \cong \triangle FLY$

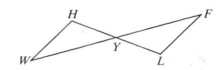

11. **a.** List two pairs of congruent correspond-
    ing sides and one pair of congruent
    corresponding angles in $\triangle YTR$ and
    $\triangle XTR$.
    b. Notice that, in each triangle, you listed
    two sides and a *nonincluded* angle. Do
    you think that SSA is enough to guarantee
    that two triangles are congruent?

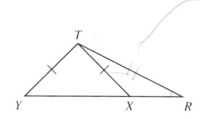

Written Exercises

Decide whether you can deduce by the SSS, SAS, or ASA Postulate that another triangle is congruent to $\triangle ABC$. If so, write the congruence and name the postulate used. If not, write *no congruence can be deduced*.

A 1. 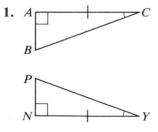 2. 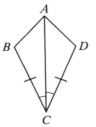 3.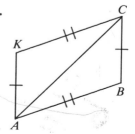

4.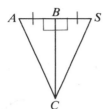

5.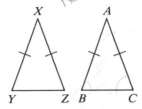

6.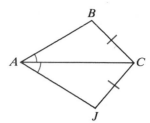

7.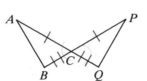

8.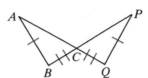

9.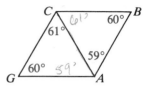

10.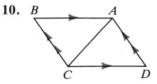

11.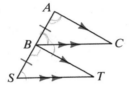

12.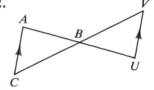

13.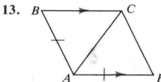

14.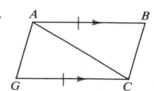

15.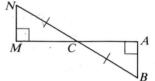

16. Supply the missing reasons.
Given: $\overline{AB} \parallel \overline{DC}$; $\overline{AB} \cong \overline{DC}$
Prove: $\triangle ABC \cong \triangle CDA$

Proof:

Statements	Reasons
1. $\overline{AB} \cong \overline{DC}$	1. ?
2. $\overline{AC} \cong \overline{AC}$	2. ?
3. $\overline{AB} \parallel \overline{DC}$	3. ?
4. $\angle BAC \cong \angle DCA$	4. ?
5. $\triangle ABC \cong \triangle CDA$	5. ?

17. Supply the missing statements and reasons.

Given: $\overline{RS} \perp \overline{ST}$; $\overline{TU} \perp \overline{ST}$;

V is the midpoint of $\overline{ST}$.

Prove: $\triangle RSV \cong \triangle UTV$

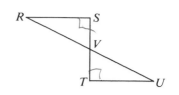

Proof:

Statements	Reasons
1. $\overline{RS} \perp \overline{ST}$; $\overline{TU} \perp \overline{ST}$	1. __?__
2. $m \angle S = 90$; $m \angle$ __?__ $= 90$	2. __?__
3. $\angle S \cong \angle T$	3. __?__
4. V is the midpoint of $\overline{ST}$.	4. __?__
5. $\overline{SV} \cong$ __?__	5. __?__
6. $\angle RVS \cong \angle$ __?__	6. __?__
7. $\triangle$ __?__ $\cong \triangle$ __?__	7. __?__

Write proofs in two-column form.

B **18.** Given: $\overline{TM} \cong \overline{PR}$; $\overline{TM} \parallel \overline{RP}$

Prove: $\triangle TEM \cong \triangle PER$

19. Given: E is the midpoint of $\overline{TP}$;

E is the midpoint of $\overline{MR}$.

Prove: $\triangle TEM \cong \triangle PER$

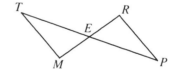

20. Given: Plane M bisects $\overline{AB}$; $\overline{PA} \cong \overline{PB}$

Prove: $\triangle POA \cong \triangle POB$

21. Given: Plane M bisects $\overline{AB}$; $\overline{PO} \perp \overline{AB}$

Prove: $\triangle POA \cong \triangle POB$

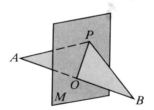

Draw and label a diagram. List, in terms of the diagram, what is given and what is to be proved. Then write a two-column proof.

22. In an isosceles triangle, if the angle between the congruent sides is bisected, then two congruent triangles are formed.

23. In an isosceles triangle, if a segment is drawn from the vertex of the angle between the congruent sides to the midpoint of the opposite side, then congruent triangles are formed.

24. If a line perpendicular to $\overline{AB}$ passes through the midpoint of $\overline{AB}$, and segments are drawn from any other point on that line to A and B, then two congruent triangles are formed.

25. If pentagon $ABCDE$ is equilateral and has right angles at B and E, then diagonals $\overline{AC}$ and $\overline{AD}$ form congruent triangles.

Copy each three-dimensional figure and with colored pencils outline the triangles listed. What postulate proves that these triangles are congruent?

C 26.

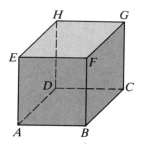

27.

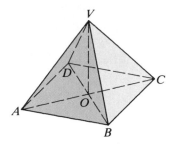

Given: Cube whose faces are
      congruent squares
Show: △*ABF*, △*BCG*

Given: Pyramid with square base;
      $VA = VB = VC = VD$
Show: △*VAB*, △*VBC*

4-3 *Using Congruent Triangles*

Our goal in the preceding section was to prove that two triangles are congruent. Our goal in this section is to deduce information about segments or angles once we have shown that they are corresponding parts of congruent triangles.

Example 1

Given: $\overline{AB}$ and $\overline{CD}$ bisect each other at M.
Prove: $\overline{AD} \parallel \overline{BC}$

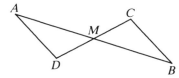

Plan for Proof: You can prove $\overline{AD} \parallel \overline{BC}$ if you can show that alternate interior angles $\angle A$ and $\angle B$ are congruent. You will know that $\angle A$ and $\angle B$ are congruent if they are corresponding parts of congruent triangles. The diagram suggests that you try to prove △*AMD* ≅ △*BMC*.

Proof:

Statements	Reasons
1. $\overline{AB}$ and $\overline{CD}$ bisect each other at M.	1. Given
2. M is the midpoint of $\overline{AB}$ and of $\overline{CD}$.	2. Def. of a bisector of a segment
3. $\overline{AM} \cong \overline{MB}$; $\overline{DM} \cong \overline{MC}$	3. Def. of midpoint
4. $\angle AMD \cong \angle BMC$	4. Vertical ⩘ are ≅.
5. △*AMD* ≅ △*BMC*	5. SAS Postulate
6. $\angle A \cong \angle B$	6. Corr. parts of ≅ ⩘ are ≅.
7. $\overline{AD} \parallel \overline{BC}$	7. If two lines are cut by a transversal and alt. int. ⩘ are ≅, then the lines are ∥.

Some proofs require the idea of a line perpendicular to a plane. **A line and a plane are perpendicular** if and only if they intersect and the line is perpendicular to all lines in the plane that pass through the point of intersection. Suppose you are given $\overleftrightarrow{PO} \perp$ plane X. Then you know that $\overleftrightarrow{PO} \perp \overleftrightarrow{OA}$, $\overleftrightarrow{PO} \perp \overleftrightarrow{OB}$, $\overleftrightarrow{PO} \perp \overleftrightarrow{OC}$, $\overleftrightarrow{PO} \perp \overline{OC}$, and so on. The ice-fishing equipment shown below suggests a line perpendicular to a plane.

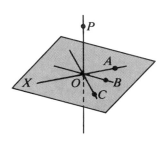 

Example 2

Given: $\overline{PO} \perp$ plane X;
       $\overline{AO} \cong \overline{BO}$
Prove: $\overline{PA} \cong \overline{PB}$

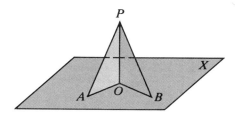

Plan for Proof: You can prove $\overline{PA} \cong \overline{PB}$ if you can show that these segments are corresponding parts of congruent triangles. The diagram suggests that you try to prove $\triangle POA \cong \triangle POB$.

Proof:

Statements	Reasons
1. $\overline{PO} \perp$ plane X	1. Given
2. $\overline{PO} \perp \overline{OA}$; $\overline{PO} \perp \overline{OB}$	2. Def. of a line perpendicular to a plane
3. $m\angle POA = 90$; $m\angle POB = 90$	3. Def. of $\perp$ lines
4. $\angle POA \cong \angle POB$	4. Def. of $\cong$ $\angle\!\!\angle$
5. $\overline{AO} \cong \overline{BO}$	5. Given
6. $\overline{PO} \cong \overline{PO}$	6. Reflexive Prop.
7. $\triangle POA \cong \triangle POB$	7. SAS Postulate
8. $\overline{PA} \cong \overline{PB}$	8. Corr. parts of $\cong$ $\triangle$ are $\cong$.

A Way to Prove Two Segments or Two Angles Congruent

1. Identify two triangles in which the two segments or angles are corresponding parts.
2. Prove that the triangles are congruent.
3. State that the two parts are congruent, using the reason
   Corr. parts of ≅ △ are ≅.

Classroom Exercises

Describe your plan for proving the following.

1. Given: $\overleftrightarrow{PR}$ bisects $\angle QPS$; $\overline{PQ} \cong \overline{PS}$
   Prove: $\angle Q \cong \angle S$

2. Given: $\overleftrightarrow{PR}$ bisects $\angle QPS$ and $\angle QRS$
   Prove: $\overline{RQ} \cong \overline{RS}$

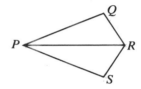

3. Given: $\overline{WX} \cong \overline{YZ}$; $\overline{ZW} \cong \overline{XY}$
   Prove: $\overline{WX} \parallel \overline{ZY}$

4. Given: $\overline{ZW} \parallel \overline{YX}$; $\overline{ZW} \cong \overline{XY}$
   Prove: $\overline{ZY} \parallel \overline{WX}$

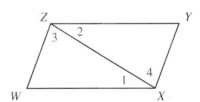

5. Given: $\overline{CD} \perp \overline{AB}$;
   D is the midpoint of $\overline{AB}$.
   Prove: $\overline{CA} \cong \overline{CB}$

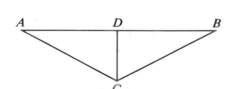

6. Given: M is the midpoint of $\overline{AB}$;
   plane $X \perp \overline{AB}$ at M.
   What can you deduce about $\overline{AP}$ and $\overline{BP}$?
   Describe a plan for proving that your conclusion is correct.

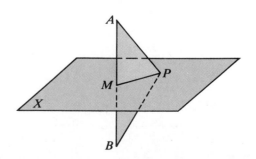

Written Exercises

Copy and complete the proof.

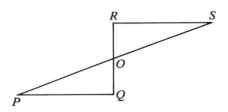

A **1.** Given: $\angle P \cong \angle S$;

   O is the midpoint of $\overline{PS}$.

   Prove: O is the midpoint of $\overline{RQ}$.

Proof:

Statements	Reasons
1. $\angle P \cong \angle S$	1. ___?___
2. O is the midpoint of $\overline{PS}$.	2. ___?___
3. $\overline{PO} \cong \overline{SO}$	3. ___?___
4. $\angle POQ \cong \angle SOR$	4. ___?___
5. $\triangle POQ \cong \triangle SOR$	5. ___?___
6. $\overline{QO} \cong \overline{RO}$	6. ___?___
7. O is the midpoint of $\overline{RQ}$.	7. ___?___

The statements in Exercise 2 might be used as statements in a proof but they are given out of order. Find an appropriate order for the statements. (There may be more than one correct order.)

2. Given: $\overline{AM} \cong \overline{BM}$; $\overline{TM} \perp \overline{AB}$

   Prove: $\overline{AT} \cong \overline{BT}$

   (a) $\overline{AM} \cong \overline{BM}$

   (b) $\triangle AMT \cong \triangle BMT$

   (c) $\angle 1 \cong \angle 2$

   (d) $\overline{AT} \cong \overline{BT}$

   (e) $\overline{TM} \perp \overline{AB}$

   (f) $\overline{TM} \cong \overline{TM}$

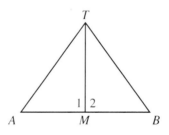

Write proofs in two-column form.

3. Given: $\overline{WO} \cong \overline{ZO}$; $\overline{XO} \cong \overline{YO}$

   Prove: $\angle W \cong \angle Z$

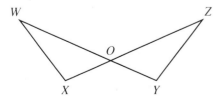

4. Given: M is the midpoint of $\overline{AB}$;

   $\angle 1 \cong \angle 2$; $\angle 3 \cong \angle 4$

   Prove: $\overline{AC} \cong \overline{BD}$

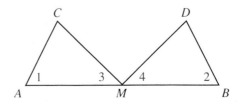

5. Prove the following statement: If both pairs of opposite sides of a quadrilateral are parallel, then they are also congruent.
   Given: $\overline{SK} \parallel \overline{NR}$; $\overline{SN} \parallel \overline{KR}$
   Prove: $\overline{SK} \cong \overline{NR}$; $\overline{SN} \cong \overline{KR}$

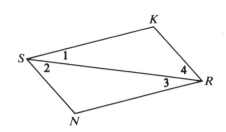

6. Prove the converse of the statement in Exercise 5: If both pairs of opposite sides of a quadrilateral are congruent, then they are also parallel.
   Given: $\overline{SK} \cong \overline{NR}$; $\overline{SN} \cong \overline{KR}$
   Prove: $\overline{SK} \parallel \overline{NR}$; $\overline{SN} \parallel \overline{KR}$

Write proofs in two-column form.

7. Given: $\overline{AD} \parallel \overline{ME}$; $\overline{MD} \parallel \overline{BE}$;
   M is the midpoint of $\overline{AB}$.
   Prove: $\overline{MD} \cong \overline{BE}$

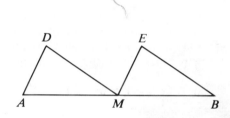

B 8. Given: M is the midpoint of $\overline{AB}$;
   $\overline{AD} \cong \overline{ME}$; $\overline{AD} \parallel \overline{ME}$
   Prove: $\overline{MD} \parallel \overline{BE}$

In Exercises 9 and 10 you are given more information than you need. For each exercise state one piece of given information that you do not need for the proof. Then give a two-column proof that does not use that piece of information.

9. Given: $\overline{PQ} \cong \overline{PS}$; $\overline{QR} \cong \overline{SR}$;
   $\angle 1 \cong \angle 2$
   Prove: $\angle 3 \cong \angle 4$

10. Given: $\overline{LM} \cong \overline{LN}$; $\overline{KM} \cong \overline{KN}$;
    $\overrightarrow{KO}$ bisects $\angle MKN$.
    Prove: $\overrightarrow{LO}$ bisects $\angle MLN$.

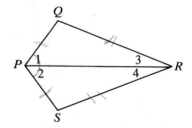

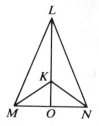

11. Given: $\overline{WX} \perp \overline{YZ}$; $\angle 1 \cong \angle 2$; $\overline{UX} \cong \overline{VX}$
    Which one(s) of the following statements *must* be true?
    (1) $\overline{XW} \perp \overline{UV}$　　(2) $\overline{UV} \parallel \overline{YZ}$　　(3) $\overline{VX} \perp \overline{UX}$

12. Given: $\overline{WX} \perp \overline{UV}$; $\overline{WX} \perp \overline{YZ}$; $\overline{WU} \cong \overline{WV}$
    Prove whatever you can about angles 1, 2, 3, and 4.

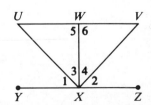

13. Given: $\overline{RS} \perp$ plane Y;
$\quad\quad\quad \angle TRS \cong \angle VRS$
$\quad$ Prove: $\triangle RTV$ is isosceles.

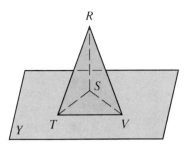

14. Given: $\overline{PA} \perp$ plane X; $\overline{QB} \perp$ plane X;
$\quad\quad\quad O$ is the midpoint of $\overline{AB}$.
$\quad$ Prove: O is the midpoint of $\overline{PQ}$.

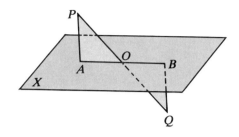

15. A young tree on level ground is supported at P by three wires of equal length. The wires are staked to the ground at points A, B, and C, which are equally distant from the base of the tree, T. Explain in a paragraph how you can prove that the angles the wires make with the ground are all congruent.

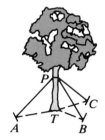

C **16.** Napoleon, on a river bank, wanted to know the width of the stream. A young soldier faced directly across the stream and adjusted the visor of his cap until the tip of the visor was in line with his eye and the opposite bank. Next he did an about-face and noted the spot on the ground now in line with his eye and visor-tip. He paced off the distance to this spot, made his report, and earned a promotion. What postulate is this method based on? Draw a diagram to help you explain.

Self-Test 1

Given: $\triangle KOP \cong \triangle MAT$

1. What can you conclude about $\angle P$? Why?
2. Name three pairs of corresponding sides.

Decide whether the two triangles must be congruent. If so, write the congruence and name the postulate used. If not, write *no congruence can be deduced*.

3.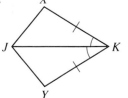

4.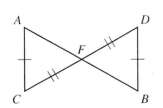

5.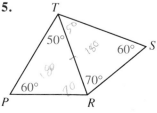

Write proofs in two-column form.

6. Given: $\angle 1 \cong \angle 2$; $\angle 3 \cong \angle 4$
Prove: $\triangle ADB \cong \triangle CBD$

7. Given: $\overline{CD} \cong \overline{AB}$; $\overline{CB} \cong \overline{AD}$
Prove: $\angle 1 \cong \angle 2$

8. Given: $\overline{AD} \parallel \overline{BC}$; $\overline{AD} \cong \overline{CB}$
Prove: $\overline{DC} \parallel \overline{AB}$

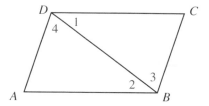

| **Application** | *Bracing With Triangles* |

The two famous landmarks pictured above have much in common. They were completed within a few years of each other, the Eiffel Tower in 1889 and the Statue of Liberty in 1886. The French engineer Gustave Eiffel designed both the tower's sweeping form and the complex structure that supports Liberty's copper skin. And both designs gain strength from the rigidity of the triangular shape.

The strength of triangular bracing is related to the SSS Postulate, which tells us that a triangle with given sides can have only one shape. A rectangle formed by four bars joined at their ends can flatten into a parallelogram, but the structural triangle cannot be deformed except by bending or stretching the bars.

The Eiffel Tower's frame is tied together by a web of triangles. A portion of the statue's armature is shown in the photograph at the right. The inner tower of wide members is strengthened by double diagonal bracing. A framework of lighter members, also joined in triangular patterns, surrounds this core.

Structural engineers use geometry in designing bridges, towers, and large-span roofs. See what you can find out about Eiffel's bridges and about the work of some of the other great modern builders.

Explorations

These exploratory exercises can be done using a computer with a program that draws and measures geometric figures.

Draw several isosceles triangles. For each triangle, measure all sides and angles. What do you notice?
What is the relationship between the congruent sides and some of the angles?

Draw several triangles with two congruent angles. Measure all sides. What do you notice?
What is the relationship between the congruent angles and some of the sides?

Some Theorems Based on Congruent Triangles

Objectives

1. Apply the theorems and corollaries about isosceles triangles.
2. Use the AAS Theorem to prove two triangles congruent.
3. Use the HL Theorem to prove two right triangles congruent.
4. Prove that two overlapping triangles are congruent.

4-4 *The Isosceles Triangle Theorems*

The photograph shows the Transamerica Pyramid in San Francisco. Each of its four faces is an isosceles triangle, with two congruent sides. These congruent sides are called **legs** and the third side is called the **base.** The angles at the base are called *base angles* and the angle opposite the base is called the *vertex angle* of the isosceles triangle.

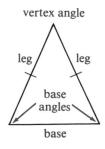

vertex angle

leg   leg

base angles

base

You can use the steps described below to form an isosceles triangle. Refer to the diagrams shown.

(1) Fold a sheet of paper in half.
(2) Cut off a double-thickness corner piece along the dashed line.
(3) Open the corner piece and lay it flat. You will have a triangle, which is labeled △*PRS* in the diagram. The fold line is labeled $\overline{PQ}$.
(4) Since $\overline{PR}$ and $\overline{PS}$ were formed by the same cut line, you can conclude that they are congruent segments and that △*PRS* is isosceles.

Since △*PRQ* fits exactly over △*PSQ* when you fold along $\overline{PQ}$, you can also conclude the following about isosceles △*PRS*:

$$\angle PRS \cong \angle PSR$$
$$\overline{PQ} \text{ bisects } \angle RPS.$$
$$\overline{PQ} \text{ bisects } \overline{RS}.$$
$$\overline{PQ} \perp \overline{RS} \text{ at } Q.$$
$$\triangle PQR \cong \triangle PQS$$

These observations suggest some of the following results.

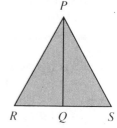

Theorem 4-1   *The Isosceles Triangle Theorem*

If two sides of a triangle are congruent, then the angles opposite those sides are congruent.

Given: $\overline{AB} \cong \overline{AC}$

Prove: $\angle B \cong \angle C$

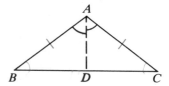

Plan for Proof: You can show that $\angle B$ and $\angle C$ are corresponding parts of congruent triangles if you draw an auxiliary line that will give you such triangles. For example, draw the bisector of $\angle A$.

Theorem 4-1 is often stated as follows: Base angles of an isosceles triangle are congruent. The following corollaries of Theorem 4-1 will be discussed as classroom exercises.

Corollary 1
An equilateral triangle is also equiangular.

Corollary 2
An equilateral triangle has three 60° angles.

Corollary 3
The bisector of the vertex angle of an isosceles triangle is perpendicular to the base at its midpoint.

Theorem 4-2

If two angles of a triangle are congruent, then the sides opposite those angles are congruent.

Given: $\angle B \cong \angle C$
Prove: $\overline{AB} \cong \overline{AC}$

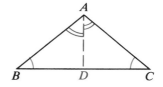

Plan for Proof: You can show that $\overline{AB}$ and $\overline{AC}$ are corresponding parts of congruent triangles. Draw the bisector of $\angle A$ as your auxiliary line, show that $\angle ADB \cong \angle ADC$, and use ASA.

Corollary

An equiangular triangle is also equilateral.

Notice that Theorem 4-2 is the converse of Theorem 4-1, and the corollary of Theorem 4-2 is the converse of Corollary 1 of Theorem 4-1.

Classroom Exercises

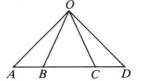

1. If $\triangle AOD$ is isosceles, with $\overline{OA} \cong \overline{OD}$, then $\angle \underline{\ ?\ } \cong \angle \underline{\ ?\ }$.
2. If $\triangle BOC$ is isosceles, with $\overline{OB} \cong \overline{OC}$, then $\angle \underline{\ ?\ } \cong \angle \underline{\ ?\ }$.
3. If $\triangle AOD$ is an isosceles right triangle with right $\angle AOD$, then the measure of $\angle A$ is $\underline{\ ?\ }$.

4. Given the triangles at the right, which of the following can you conclude are true?

   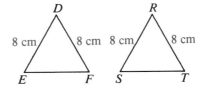

   a. $\angle D \cong \angle R$ **b.** $\overline{DE} \cong \overline{DF}$
   c. $\overline{DF} \cong \overline{RT}$ **d.** $\angle E \cong \angle F$
   e. $\angle E \cong \angle S$ **f.** $\angle S \cong \angle T$

Given the two congruent angles, name two segments that must be congruent.

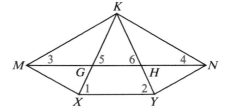

5. $\angle 1 \cong \angle 2$

6. $\angle 3 \cong \angle 4$

7. $\angle 5 \cong \angle 6$

8. Is the statement "$\overline{MK} \cong \overline{NK}$ if and only if $\angle 3 \cong \angle 4$" true or false?

9. Explain how Corollary 1 follows from Theorem 4-1.
10. Explain how Corollary 2 follows from Corollary 1.
11. Explain how Corollary 3 follows from Theorem 4-1.
12. Explain how the Corollary follows from Theorem 4-2.

Written Exercises

Find the value of *x*.

A **1.**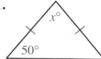

2.

3.

4.

5.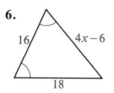

6.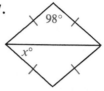

7.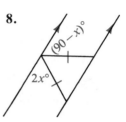

8.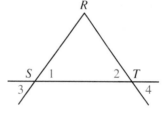

For each exercise place the statements in an appropriate order for a proof. (There may be more than one correct order.)

9. Given: $\overline{RS} \cong \overline{RT}$
   Prove: $\angle 3 \cong \angle 4$
   (a) $\angle 3 \cong \angle 4$
   (b) $\angle 3 \cong \angle 1;\ \angle 2 \cong \angle 4$
   (c) $\overline{RS} \cong \overline{RT}$
   (d) $\angle 1 \cong \angle 2$

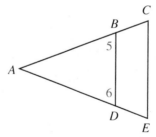

10. Given: $\overline{BD} \parallel \overline{CE};\ \angle 5 \cong \angle 6$
   Prove: $\overline{AC} \cong \overline{AE}$
   (a) $\overline{BD} \parallel \overline{CE}$
   (b) $\overline{AC} \cong \overline{AE}$
   (c) $\angle 5 \cong \angle C;\ \angle 6 \cong \angle E$
   (d) $\angle 5 \cong \angle 6$
   (e) $\angle C \cong \angle E$

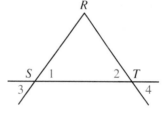

Write proofs in two-column form.

11. Theorem 4-1

12. Theorem 4-2

13. Given: M is the midpoint of $\overline{JK}$;
   $\angle 1 \cong \angle 2$
   Prove: $\overline{JG} \cong \overline{MK}$

14. Given: $\overline{XY} \cong \overline{XZ}$
   Prove: $\angle 3 \cong \angle 5$

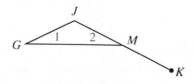

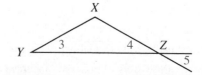

B **15.** Given: $\overline{PQ} \cong \overline{PR}$; $\overline{TR} \cong \overline{TS}$
Which one(s) of the following *must* be true?
(1) $\overline{ST} \parallel \overline{QP}$ (2) $\overline{ST} \cong \overline{QP}$ (3) $\angle T \cong \angle P$

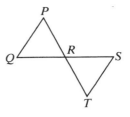

16. Given: $\angle S \cong \angle T$; $\overline{ST} \parallel \overline{QP}$
Which one(s) of the following *must* be true?
(1) $\angle P \cong \angle Q$ (2) $PR = QR$
(3) R is the midpoint of $\overline{PT}$.

Write proofs in two-column form.

17. Given: $\overline{XY} \cong \overline{XZ}$; $\overline{OY} \cong \overline{OZ}$
Prove: $m \angle 1 = m \angle 4$

18. Given: $\overline{XY} \cong \overline{XZ}$;
     $\overrightarrow{YO}$ bisects $\angle XYZ$;
     $\overrightarrow{ZO}$ bisects $\angle XZY$.
Prove: $\overline{YO} \cong \overline{ZO}$

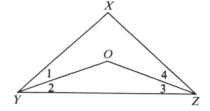

19. Given: $\overline{AB} \cong \overline{AC}$; $\overline{AL}$ and $\overline{AM}$ trisect $\angle BAC$.
     (This means $\angle 1 \cong \angle 2 \cong \angle 3$.)
Prove: $\overline{AL} \cong \overline{AM}$

20. Given: $\angle 4 \cong \angle 7$; $\angle 1 \cong \angle 3$
Prove: $\triangle ABC$ is isosceles.

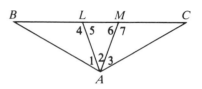

21. Given: $\overline{OP} \cong \overline{OQ}$; $\angle 3 \cong \angle 4$
Prove: $\angle 5 \cong \angle 6$

22. Given: $\overline{PO} \cong \overline{QO}$; $\overline{RO} \cong \overline{SO}$
  a. If you are also given that $m \angle 1 = 40$, find the measures of $\angle 2$, $\angle 7$, $\angle 5$, and $\angle 6$. Then decide whether $\overline{PQ}$ must be parallel to $\overline{SR}$.
  b. Repeat part (a), but use $m \angle 1 = k$.

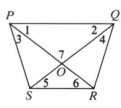

23. Complete.
  a. If $m \angle 1 = 20$, then $m \angle 3 = \underline{\ ?\ }$, $m \angle 4 = \underline{\ ?\ }$, and $m \angle 5 = \underline{\ ?\ }$.
  b. If $m \angle 1 = x$, then $m \angle 3 = \underline{\ ?\ }$, $m \angle 4 = \underline{\ ?\ }$, and $m \angle 5 = \underline{\ ?\ }$.

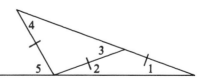

24. a. If $m \angle 1 = 35$, find $m \angle ABC$.
    b. If $m \angle 1 = k$, find $m \angle ABC$.

25. a. If $m \angle 1 = 23$, find $m \angle 7$.
    b. If $m \angle 1 = k$, find $m \angle 7$.

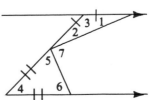

26. Draw an isosceles △*ABC* whose vertex angle, ∠*A*, has measure 80.

    a. Draw $\overrightarrow{AX}$, the bisector of an exterior angle at *A*. Is $\overrightarrow{AX} \parallel \overline{BC}$? Explain.

    b. Would your answer change if the measure of ∠*A* changed?

Find the values of *x* and *y*.

27. In equiangular △*ABC*, *AB* = 4*x* − *y*, *BC* = 2*x* + 3*y*, and *AC* = 7.

28. In equilateral △*DEF*, *m*∠*D* = *x* + *y* and *m*∠*E* = 2*x* − *y*.

29. In △*JKL*, $\overline{JK} \cong \overline{KL}$, *m*∠*J* = 2*x* − *y*, *m*∠*K* = 2*x* + 2*y*, and *m*∠*L* = *x* + 2*y*.

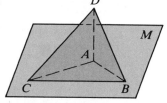

30. Given: △*ABC* in plane *M*; *D* not in plane *M*;
        ∠*ACB* ≅ ∠*ABC*; ∠*DCB* ≅ ∠*DBC*
Name a pair of congruent triangles.
Prove that your answer is correct.

31. Given: $\overline{JL} \perp$ plane *Z*;
        △*KMN* is isosceles, with $\overline{KM} \cong \overline{KN}$.

    a. Prove that two other triangles are isosceles.

    b. Must these two isosceles triangles be congruent?

        Explain.

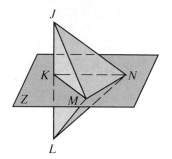

32. Draw an isosceles triangle and then join the midpoints of its sides to form another triangle. What can you deduce about this second triangle? Explain.

C   **33.** *ABCDE* is a regular pentagon and *DEFG* is a square. Find the measures of ∠*EAF*, ∠*AFD*, and ∠*DAF*.

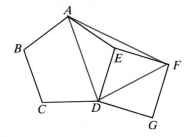

34. Given: △*ABC* is equilateral;
        ∠*CAD* ≅ ∠*ABE* ≅ ∠*BCF*
Prove something interesting about △*DEF*.

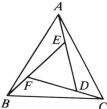

Challenge

The figure shown at the right can be dissected into three congruent pieces, as shown by the dashed lines. Can you dissect the figure into **(a)** two congruent pieces? **(b)** four congruent pieces?

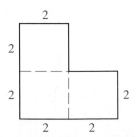

4-5 *Other Methods of Proving Triangles Congruent*

The SSS, SAS, and ASA Postulates give us three methods of proving triangles congruent. In this section we will develop two other methods.

Theorem 4-3   *AAS Theorem*

If two angles and a non-included side of one triangle are congruent to the corresponding parts of another triangle, then the triangles are congruent.

Given: $\triangle ABC$ and $\triangle DEF$; $\angle B \cong \angle E$;
$\qquad \angle C \cong \angle F$; $\overline{AC} \cong \overline{DF}$

Prove: $\triangle ABC \cong \triangle DEF$

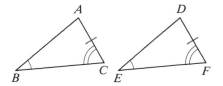

Plan for Proof: You can prove the triangles congruent if you can apply one of the SSS, SAS, or ASA Postulates. You can use the ASA Postulate if you first show that $\angle A \cong \angle D$. To do that, use the fact that the other two angles of $\triangle ABC$ are congruent to the other two angles of $\triangle DEF$.

Do you see overlapping triangles in the photograph? Sometimes you want to prove that certain overlapping triangles are congruent. For example, suppose you have the following problem:

Given: $\overline{GJ} \cong \overline{GK}$;
$\qquad \angle H \cong \angle I$

Prove: $\triangle GHJ \cong \triangle GIK$

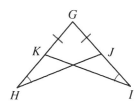

You may find it helps you visualize the congruence if you redraw the two triangles, as shown below. Now you can see that since $\angle G$ is common to both triangles, the triangles must be congruent by the AAS Theorem.

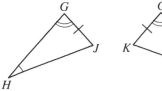

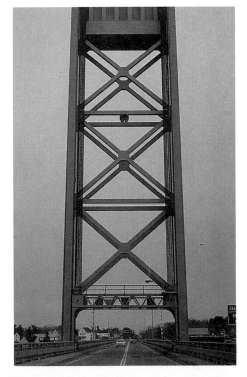

Our final method of proving triangles congruent applies only to right triangles. In a right triangle the side opposite the right angle is called the **hypotenuse** (hyp.). The other two sides are called **legs**.

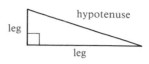

A proof in two-column form for the next theorem would be too long and involved. The proof shown below is written instead in *paragraph form*, which emphasizes the *key steps* in the proof. You will learn to write paragraph proofs in the next section.

Theorem 4-4 *HL Theorem*

If the hypotenuse and a leg of one right triangle are congruent to the corresponding parts of another right triangle, then the triangles are congruent.

Given: $\triangle ABC$ and $\triangle DEF$;
 $\angle C$ and $\angle F$ are right $\angle$s;
 $\overline{AB} \cong \overline{DE}$ (hypotenuses);
 $\overline{BC} \cong \overline{EF}$ (legs)
Prove: $\triangle ABC \cong \triangle DEF$

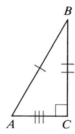
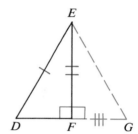

Proof:

By the Ruler Postulate there is a point G on the ray opposite to $\overrightarrow{FD}$ such that $\overline{FG} \cong \overline{CA}$. Draw $\overline{GE}$. Because $\angle DFE$ is a right angle, $\angle GFE$ is also a right angle. $\triangle ABC \cong \triangle GEF$ by the SAS Postulate. Then $\overline{AB} \cong \overline{GE}$. Since $\overline{DE} \cong \overline{AB}$, we have $\overline{DE} \cong \overline{GE}$. In isosceles $\triangle DEG$, $\angle G \cong \angle D$. Since $\triangle ABC \cong \triangle GEF$, $\angle A \cong \angle G$. Then $\angle A \cong \angle D$. Finally, $\triangle ABC \cong \triangle DEF$ by the AAS Theorem.

Recall from Exercise 22 on page 121 and Exercise 11 on page 124 that AAA and SSA correspondences do not guarantee congruent triangles. We can now summarize the methods available for proving triangles congruent.

Summary of Ways to Prove Two Triangles Congruent

All triangles:	SSS	SAS	ASA	AAS
Right triangles:	HL			

Which of these methods are postulates and which are theorems?

Classroom Exercises

State which congruence method(s) can be used to prove the triangles congruent. If no method applies, say *none*.

1.

2.

3.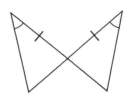

4.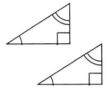

5.

6.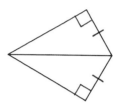

7.

8.

9.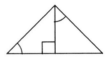

For each diagram, name a pair of overlapping triangles. Tell whether the triangles are congruent by the SSS, SAS, ASA, AAS, or HL method.

10. Given: $\overline{AB} \cong \overline{DC}$;
$\overline{AC} \cong \overline{DB}$

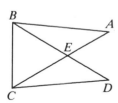

11. Given: $\angle 2 \cong \angle 3$;
$\angle 1 \cong \angle 4$

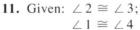
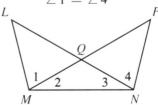

12. Given: $\overline{WU} \cong \overline{ZV}$;
$WX = YZ$;
$\angle U$ and $\angle V$ are rt. $\triangle$.

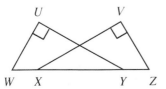

13. Given: $\angle ABC \cong \angle ACB$;
$\overline{AE} \perp \overline{EC}$;
$\overline{AD} \perp \overline{DB}$

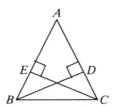

14. To prove that right triangles are congruent, some geometry books also use the methods stated below. For each method, draw two right triangles that appear to be congruent. Mark the given information on your triangles. Use your marks to determine which of our methods (SSS, SAS, ASA, AAS, or HL) could be used instead of each method listed.
   a. **Leg-Leg Method** (LL) If two legs of one right triangle are congruent to the two legs of another right triangle, then the triangles are congruent.
   b. **Hypotenuse-Acute Angle Method** (HA) If the hypotenuse and an acute angle of one right triangle are congruent to the hypotenuse and an acute angle of another right triangle, then the triangles are congruent.
   c. **Leg-Acute Angle Method** (LA) If a leg and an acute angle of one right triangle are congruent to the corresponding parts in another right triangle, then the triangles are congruent.

Written Exercises

A 1. Supply the missing statements and reasons.

Given: $\angle W$ and $\angle Y$ are rt. $\angle$s;
$\overline{WX} \cong \overline{YX}$

Prove: $\overline{WZ} \cong \overline{YZ}$

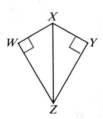

Proof:

Statements	Reasons
1. $\angle W$ and $\angle Y$ are rt. $\angle$s.	1. _?_
2. $\triangle XWZ$ and $\triangle XYZ$ are rt. $\triangle$s.	2. _?_
3. $\overline{WX} \cong \overline{YX}$	3. _?_
4. _?_	4. Reflexive Prop.
5. $\triangle XWZ \cong$ _?_	5. _?_
6. _?_	6. _?_

2. Place the statements in an appropriate order for a proof.

Given: $\overline{KL} \perp \overline{LA}$; $\overline{KJ} \perp \overline{JA}$;
$\overrightarrow{AK}$ bisects $\angle LAJ$.

Prove: $\overline{LK} \cong \overline{JK}$

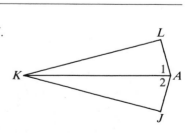

(a) $\overline{KL} \perp \overline{LA}$; $\overline{KJ} \perp \overline{JA}$
(b) $\angle 1 \cong \angle 2$
(c) $\overrightarrow{AK}$ bisects $\angle LAJ$.
(d) $m \angle L = 90$; $m \angle J = 90$

(e) $\overline{LK} \cong \overline{JK}$
(f) $\angle L \cong \angle J$
(g) $\triangle LKA \cong \triangle JKA$
(h) $\overline{KA} \cong \overline{KA}$

Write proofs in two-column form.

3. Given: $\overline{EF} \perp \overline{EG}$; $\overline{HG} \perp \overline{EG}$;
$\overline{EH} \cong \overline{GF}$
Prove: $\angle H \cong \angle F$

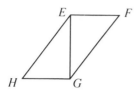

4. Given: $\overline{RT} \cong \overline{AS}$;
$\overline{RS} \cong \overline{AT}$
Prove: $\angle TSA \cong \angle STR$

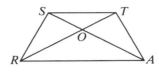

Use the information given in each exercise to name the method (SSS, SAS, ASA, AAS, or HL) you could use to prove $\triangle AOB \cong \triangle AOC$. You need not write the proofs.

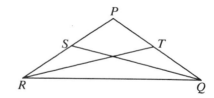

5. Given: $\overline{AO} \perp$ plane M; $\overline{BO} \cong \overline{CO}$
6. Given: $\overline{AO} \perp$ plane M; $\angle B \cong \angle C$
7. Given: $\overline{AO} \perp$ plane M; $\overline{AB} \cong \overline{AC}$

B   **8.** Given: $\overline{AB} \cong \overline{AC}$; $\overline{OB} \cong \overline{OC}$
   a. Is it possible to prove that $\angle AOB \cong \angle AOC$?
   b. Is it possible to prove that $\angle AOB$ and $\angle AOC$ are right angles?

9. In many proofs you may find that different methods can be used. You may not know in advance which method will be better. There are *two* possible pairs of overlapping triangles that could be used in this proof. To compare the two methods, write a two-column proof for each plan.
Given: $\overline{PR} \cong \overline{PQ}$; $\overline{SR} \cong \overline{TQ}$
Prove: $\overline{QS} \cong \overline{RT}$
   a. Plan for Proof: Show that $\triangle RQS \cong \triangle QRT$ by SAS.
   b. Plan for Proof: Show that $\triangle PQS \cong \triangle PRT$ by SAS.

10. a. Draw an isosceles $\triangle RST$ with $\overline{RT} \cong \overline{ST}$. Let M be the midpoint of $\overline{ST}$ and N be the midpoint of $\overline{RT}$. Draw $\overline{RM}$ and $\overline{SN}$ and label their common point O. Now draw $\overline{NM}$.
   b. Name four *pairs* of congruent triangles.

Tell which pairs of congruent parts and what method (SSS, SAS, ASA, AAS, or HL) you would use to prove the triangles are congruent.

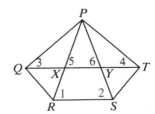

11. Given: $\angle 1 \cong \angle 2$; $\angle 3 \cong \angle 4$; $\overline{QR} \cong \overline{TS}$
$\triangle QPR \cong \triangle TPS$ by what method?

12. Given: $\angle 3 \cong \angle 4$; $\angle 5 \cong \angle 6$
$\triangle PQX \cong \triangle PTY$ by what method?

13. Given: $\angle 3 \cong \angle 4$; $\angle 5 \cong \angle 6$
$\triangle QPY \cong \triangle TPX$ by what method?

Write proofs in two-column form.

14. Given: $\angle R \cong \angle T$; $\overline{RS} \parallel \overline{QT}$
Prove: $\overline{RS} \cong \overline{TQ}$
(*Hint*: What auxiliary line can you draw
to form congruent triangles?)

15. Given: $\angle 1 \cong \angle 2 \cong \angle 3$;
$\overline{EN} \cong \overline{DG}$
Prove: $\angle 4 \cong \angle 5$

For Exercises 16–19 draw and label a diagram. List, in terms of the diagram, what is given and what is to be proved. Then write a two-column proof.

16. In two congruent triangles, if segments are drawn from two corresponding vertices perpendicular to the opposite sides, then those segments are congruent.

17. If segments are drawn from the endpoints of the base of an isosceles triangle perpendicular to the opposite legs, then those segments are congruent.

18. If $\angle A$ and $\angle B$ are the base angles of isosceles $\triangle ABC$, and the bisector of $\angle A$ meets $\overline{BC}$ at X and the bisector of $\angle B$ meets $\overline{AC}$ at Y, then $\overline{AX} \cong \overline{BY}$.

19. If segments are drawn from the midpoints of the legs of an isosceles triangle perpendicular to the base, then those segments are congruent.

20. Write a detailed plan for proof.
Given: $\overline{FL} \cong \overline{AK}$;
$\overline{SF} \cong \overline{SK}$;
M is the midpoint of $\overline{SF}$;
N is the midpoint of $\overline{SK}$.
Prove: $\overline{AM} \cong \overline{LN}$

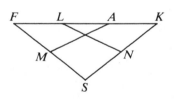

Write proofs in two-column form. Use the facts that the sides of a square are all congruent and that the angles of a square are all right angles.

C **21.** The diagram shows three squares and an equilateral triangle.
Prove: $\overline{AE} \cong \overline{FC} \cong \overline{ND}$

22. Use the results of Exercise 21 to prove that $\triangle FAN$ is equilateral.

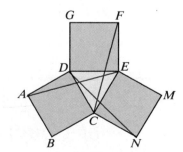

Self-Test 2

Find the value of x.

1.

2.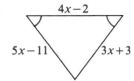

3.

4. Given: $\overline{AB} \cong \overline{AC}$; $\overline{BN} \perp \overline{AC}$; $\overline{CM} \perp \overline{AB}$

Explain how you could prove that $\triangle ABN \cong \triangle ACM$.

5. Given: $\overline{MB} \cong \overline{NC}$; $\overline{BN} \perp \overline{AC}$; $\overline{CM} \perp \overline{AB}$

Prove: $\overline{CM} \cong \overline{BN}$

More about Proof in Geometry

Objectives

1. Prove two triangles congruent by first proving two other triangles congruent.
2. Apply the definitions of the median and the altitude of a triangle and the perpendicular bisector of a segment.
3. State and apply the theorem about a point on the perpendicular bisector of a segment, and the converse.
4. State and apply the theorem about a point on the bisector of an angle, and the converse.

4-6 *Using More than One Pair of Congruent Triangles*

Sometimes two triangles that you want to prove congruent have common parts with two *other* triangles that you can easily prove congruent. You may then be able to use corresponding parts of these other triangles to prove the original triangles congruent.

Example

Given: $\angle 1 \cong \angle 2$; $\angle 5 \cong \angle 6$

Prove: $\overline{AC} \perp \overline{BD}$

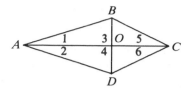

Plan for Proof: It may be helpful here to *reason backward* from what you want to prove. You can show $\overline{AC} \perp \overline{BD}$ if you can show that $\angle 3 \cong \angle 4$. You can prove $\angle 3 \cong \angle 4$ if you can prove that the angles are corresponding parts of congruent triangles. To prove $\triangle ABO \cong \triangle ADO$, you need $\overline{AB} \cong \overline{AD}$. You can prove this congruence by proving that $\triangle ABC \cong \triangle ADC$. You should prove this congruence first.

Proof:

Statements	Reasons
1. $\angle 1 \cong \angle 2$; $\angle 5 \cong \angle 6$	1. Given
2. $\overline{AC} \cong \overline{AC}$	2. Reflexive Property
3. $\triangle ABC \cong \triangle ADC$	3. ASA Postulate
4. $\overline{AB} \cong \overline{AD}$	4. Corr. parts of $\cong \triangle$ are $\cong$.
5. $\overline{AO} \cong \overline{AO}$	5. Reflexive Property
6. $\triangle ABO \cong \triangle ADO$	6. SAS Postulate (Steps 1, 4, and 5)
7. $\angle 3 \cong \angle 4$	7. Corr. parts of $\cong \triangle$ are $\cong$.
8. $\overline{AC} \perp \overline{BD}$	8. If two lines form $\cong$ adj. $\angle$s, then the lines are $\perp$.

If you were to outline this two-column proof, you might pick out the following *key steps*.

Key steps of proof:

1. $\triangle ABC \cong \triangle ADC$ (ASA Postulate)
2. $\overline{AB} \cong \overline{AD}$ (Corr. parts of $\cong \triangle$ are $\cong$.)
3. $\triangle ABO \cong \triangle ADO$ (SAS Postulate)
4. $\angle 3 \cong \angle 4$ (Corr. parts of $\cong \triangle$ are $\cong$.)
5. $\overline{AC} \perp \overline{BD}$ (If two lines form $\cong$ adj. $\angle$s, then the lines are $\perp$.)

In mathematics a proof is often given in paragraph form rather than in two-column form. A *paragraph proof* usually focuses on the key ideas and omits details that the writer thinks will be clear to the reader. The following paragraph proof might be given for the example above.

Paragraph proof:

$\triangle ABC \cong \triangle ADC$ by the ASA Postulate. Therefore, corresponding parts $\overline{AB}$ and $\overline{AD}$ are congruent. $\overline{AB}$ and $\overline{AD}$ are also corresponding parts of $\triangle ABO$ and $\triangle ADO$, which can now be proved congruent by the SAS Postulate. So corresponding parts $\angle 3$ and $\angle 4$ are congruent, and $\overline{AC} \perp \overline{BD}$.

Classroom Exercises

In Exercises 1–3 you are given a diagram that is marked with given information.
Give the reason for each key step of the proof.

1. Prove: $\overline{AS} \cong \overline{DT}$
   Key steps of proof:
   a. $\triangle ABC \cong \triangle DEF$
   b. $\angle C \cong \angle F$
   c. $\triangle ACS \cong \triangle DFT$
   d. $\overline{AS} \cong \overline{DT}$

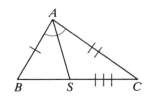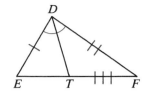

2. Prove: $\overline{AX} \cong \overline{AY}$
   Key steps of proof:
   a. $\triangle PAL \cong \triangle KAN$
   b. $\angle L \cong \angle N$
   c. $\triangle LAX \cong \triangle NAY$
   d. $\overline{AX} \cong \overline{AY}$

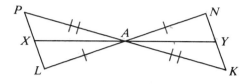

3. Prove: $\angle 3 \cong \angle 4$
   Key steps of proof:
   a. $\triangle LOB \cong \triangle JOB$
   b. $\angle 1 \cong \angle 2$
   c. $\triangle LBA \cong \triangle JBA$
   d. $\angle 3 \cong \angle 4$

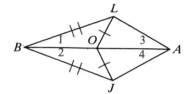

4. Suggest a plan for proving that $\angle D \cong \angle F$.

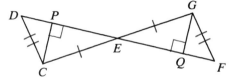

Written Exercises

In Exercises 1–6 you are given a diagram that is marked with given information.
Give the reason for each key step of the proof.

A 1. Prove: $\overline{NE} \cong \overline{OS}$
   Key steps of proof:
   a. $\triangle RNX \cong \triangle LOY$
   b. $\angle X \cong \angle Y$
   c. $\triangle NEX \cong \triangle OSY$
   d. $\overline{NE} \cong \overline{OS}$

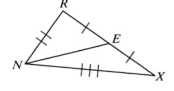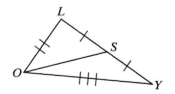

2. Prove: $\overline{BE} \cong \overline{DF}$
   Key steps of proof:
   a. $\triangle ABC \cong \triangle CDA$
   b. $\angle 1 \cong \angle 2$
   c. $\triangle ABE \cong \triangle CDF$
   d. $\overline{BE} \cong \overline{DF}$

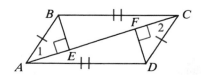

3. Prove: $\angle G \cong \angle T$
Key steps of proof:
a. $\triangle RAJ \cong \triangle NAK$
b. $\overline{RJ} \cong \overline{NK}$
c. $\triangle GRJ \cong \triangle TNK$
d. $\angle G \cong \angle T$

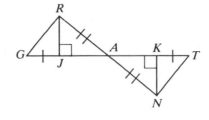

4. Prove: $\overline{AL} \cong \overline{CM}$
Key steps of proof:
a. $\triangle ABD \cong \triangle CDB$
b. $\overline{AD} \cong \overline{CB};\ \angle 1 \cong \angle 2$
c. $\triangle ADL \cong \triangle CBM$
d. $\overline{AL} \cong \overline{CM}$

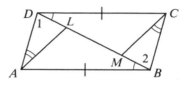

5. Prove: $\overline{DX} \cong \overline{EX}$
Key steps of proof:
a. $\triangle POD \cong \triangle POE$
b. $\overline{PD} \cong \overline{PE}$
c. $\triangle PDX \cong \triangle PEX$
d. $\overline{DX} \cong \overline{EX}$

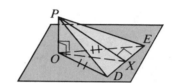

6. Prove: $\angle CBA \cong \angle DBA$
Key steps of proof:
a. $\overline{OC} \cong \overline{OD}$
b. $\triangle CAO \cong \triangle DAO$
c. $\angle CAO \cong \angle DAO$
d. $\triangle CAB \cong \triangle DAB$
e. $\angle CBA \cong \angle DBA$

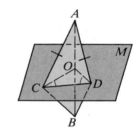

B **7.** Given: $\overline{LF} \cong \overline{KF};\ \overline{LA} \cong \overline{KA}$
Prove: $\overline{LJ} \cong \overline{KJ}$

a. List the key steps of a proof.
b. Write a proof in two-column form.

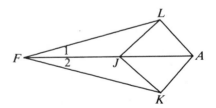

8. Given: $\overline{PR}$ bisects $\angle SPT$ and $\angle SRT$.
Prove: $\overline{PR}$ bisects $\angle SQT$.

a. List the key steps of a proof.
b. Write a proof in paragraph form.

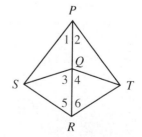

Write proofs in the form specified by your teacher (two-column form, paragraph form, or a list of key steps).

9. Given: $\triangle RST \cong \triangle XYZ$;

$\overrightarrow{SK}$ bisects $\angle RST$;

$\overrightarrow{YL}$ bisects $\angle XYZ$.

Prove: $\overline{SK} \cong \overline{YL}$

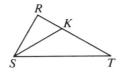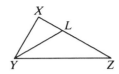

10. Given: Congruent parts as marked in the diagram.

Prove: $\angle B \cong \angle F$

(*Hint*: First draw two auxiliary lines.)

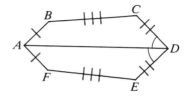

11. Given: $\overline{DE} \cong \overline{FG}$; $\overline{GD} \cong \overline{EF}$;

$\angle HDE$ and $\angle KFG$ are rt. $\angle$.

Prove: $\overline{DH} \cong \overline{FK}$

12. Given: $\overline{PQ} \perp \overline{QR}$;

$\overline{PS} \perp \overline{SR}$;

$\overline{PQ} \cong \overline{PS}$

Prove: O is the midpoint of $\overline{QS}$.

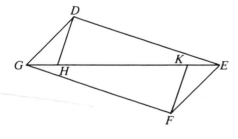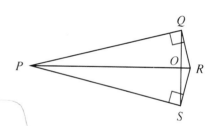

13. Draw two line segments, $\overline{KL}$ and $\overline{MN}$, that bisect each other at O. Mark a point P on $\overrightarrow{KN}$ and let Q be the point where $\overrightarrow{PO}$ intersects $\overline{ML}$. Prove that O is the midpoint of $\overline{PQ}$. (First state what is given and what is to be proved.)

14. This figure is like the one that Euclid used to prove that the base angles of an isosceles triangle are congruent (our Theorem 4-1). Write a paragraph proof following the key steps shown below.

Given: $\overline{AB} \cong \overline{AC}$;

$\overline{AB}$ and $\overline{AC}$ are extended so $\overline{BD} \cong \overline{CE}$.

Prove: $\angle ABC \cong \angle ACB$

Key steps of proof:
1. $\triangle DAC \cong \triangle EAB$
2. $\triangle DBC \cong \triangle ECB$
3. $\angle DBC \cong \angle ECB$
4. $\angle ABC \cong \angle ACB$

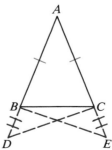

C 15. Given: $\overline{AM} \cong \overline{MB}$; $\overline{AD} \cong \overline{BC}$;
$\angle MDC \cong \angle MCD$
Prove: $\overline{AC} \cong \overline{BD}$

16. Given: $\angle 1 \cong \angle 2$;
$\angle 3 \cong \angle 4$;
$\angle 5 \cong \angle 6$
Prove: $\overline{BC} \cong \overline{ED}$

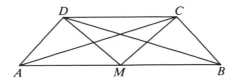

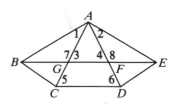

17. *A*, *B*, *C*, and *D* are noncoplanar. $\triangle ABC$, $\triangle ACD$, and $\triangle ABD$ are equilateral. *X* and *Y* are midpoints of $\overline{AC}$ and $\overline{AD}$. *Z* is a point on $\overline{AB}$. What kind of triangle is $\triangle XYZ$? Explain.

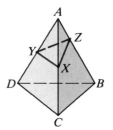

Mixed Review Exercises

1. Write the Isosceles Triangle Theorem (Theorem 4-1) and its converse (Theorem 4-2) as a single biconditional statement.

Complete each statement with the word *always*, *sometimes*, or *never*.

2. Two isosceles triangles with congruent bases are __?__ congruent.
3. Two isosceles triangles with congruent vertex angles are __?__ congruent.
4. Two equilateral triangles with congruent bases are __?__ congruent.

Draw a diagram for each of the following.

5. **a.** *M* is between *A* and *B*.
   b. *M* is the midpoint of $\overline{AB}$.

6. **a.** $\overleftrightarrow{XY}$ bisects $\overline{CD}$.
   b. $\overrightarrow{XY}$ bisects $\angle CXD$.

7. **a.** acute scalene $\triangle JKL$
   b. obtuse scalene $\triangle JKL$

8. **a.** acute isosceles $\triangle XYZ$
   b. obtuse isosceles $\triangle XYZ$

9. **a.** right scalene $\triangle RST$
   b. right isosceles $\triangle RST$

10. **a.** equilateral $\triangle EFG$
    b. equiangular $\triangle EFG$

11. Write a proof in two-column form.
    Given: $\overline{BE} \cong \overline{CD}$; $\overline{BD} \cong \overline{CE}$
    Prove: $\triangle ABC$ is isosceles.

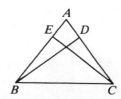

4-7 Medians, Altitudes, and Perpendicular Bisectors

A **median** of a triangle is a segment from a vertex to the midpoint of the opposite side. The three medians of △ABC are shown below in red.

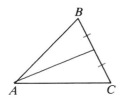 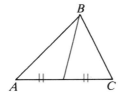 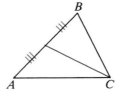

An **altitude** of a triangle is the perpendicular segment from a vertex to the line that contains the opposite side. In an acute triangle, the three altitudes are all inside the triangle.

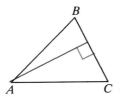 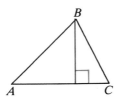 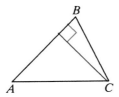

In a right triangle, two of the altitudes are parts of the triangle. They are the legs of the right triangle. The third altitude is inside the triangle.

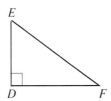 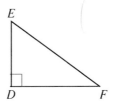 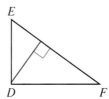

In an obtuse triangle, two of the altitudes are outside the triangle. For obtuse △KLN, $\overline{LH}$ is the altitude from L, and $\overline{NI}$ is the altitude from N.

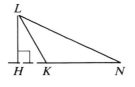 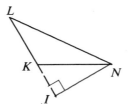 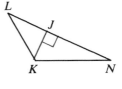

A **perpendicular bisector** of a segment is a line (or ray or segment) that is perpendicular to the segment at its midpoint. In the figure at the right, line l is a perpendicular bisector of $\overline{JK}$.

In a given plane, there is exactly one line perpendicular to a segment at its midpoint. We speak of *the* perpendicular bisector of a segment in such a case.

Proofs of the following theorems are left as Exercises 14 and 15.

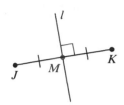

Theorem 4-5

If a point lies on the perpendicular bisector of a segment, then the point is equidistant from the endpoints of the segment.

Given: Line l is the perpendicular bisector of $\overline{BC}$; A is on l.

Prove: $AB = AC$

Theorem 4-6

If a point is equidistant from the endpoints of a segment, then the point lies on the perpendicular bisector of the segment.

Given: $AB = AC$

Prove: A is on the perpendicular bisector of $\overline{BC}$.

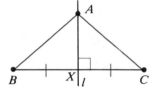

Plan for Proof: The perpendicular bisector of $\overline{BC}$ must contain the midpoint of $\overline{BC}$ and be perpendicular to $\overline{BC}$. Draw an auxiliary line containing A that has one of these properties and prove that it has the other property as well. For example, first draw a segment from A to the midpoint X of $\overline{BC}$. You can show that $AX \perp \overline{BC}$ if you can show that $\angle 1 \cong \angle 2$. Since these angles are corresponding parts of two triangles, first show that $\triangle AXB \cong \triangle AXC$.

In the proof of Theorem 4-6 other auxiliary lines could have been chosen instead. For example, we can draw the altitude to $\overline{BC}$ from A, meeting $\overline{BC}$ at a point Y as shown in the diagram at the right. Here, since $\overline{AY} \perp \overline{BC}$ we need to prove that $\overline{YB} \cong \overline{YC}$. Either method can be used to prove Theorem 4-6.

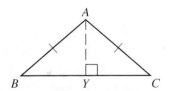

Example   Suppose you know that line *l* is the perpendicular bisector of $\overline{RS}$. What can you deduce if you also know that

a. *P* lies on *l*?
b. there is a point *Q* such that *QR* = 7 and *QS* = 7?

Solution   **a.** *PR* = *PS* (Theorem 4-5)
b. *Q* lies on *l*. (Theorem 4-6)

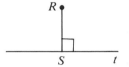

The **distance from a point to a line** (or plane) is defined to be the length of the perpendicular segment from the point to the line (or plane). Since $\overline{RS} \perp t$, *RS* is the distance from *R* to line *t*.

In Exercises 16 and 17 you will prove the following theorems, which are similar to Theorems 4-5 and 4-6.

Theorem 4-7

If a point lies on the bisector of an angle, then the point is equidistant from the sides of the angle.

Given: $\overrightarrow{BZ}$ bisects $\angle ABC$; *P* lies on $\overrightarrow{BZ}$;
$\overline{PX} \perp \overrightarrow{BA}$; $\overline{PY} \perp \overrightarrow{BC}$

Prove: *PX* = *PY*

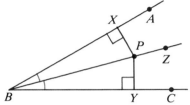

Theorem 4-8

If a point is equidistant from the sides of an angle, then the point lies on the bisector of the angle.

Given: $\overline{PX} \perp \overrightarrow{BA}$; $\overline{PY} \perp \overrightarrow{BC}$;
$PX = PY$

Prove: $\overrightarrow{BP}$ bisects $\angle ABC$.

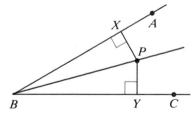

Theorem 4-5 and its converse, Theorem 4-6, can be combined into a single biconditional statement. The same is true for Theorems 4-7 and 4-8.

A point is on the perpendicular bisector of a segment if and only if it is equidistant from the endpoints of the segment.

A point is on the bisector of an angle if and only if it is equidistant from the sides of the angle.

Classroom Exercises

Complete.

1. If K is the midpoint of $\overline{ST}$, then $\overline{RK}$ is called a(n) __?__ of $\triangle RST$.
2. If $\overline{RK} \perp \overline{ST}$, then $\overline{RK}$ is called a(n) __?__ of $\triangle RST$.
3. If K is the midpoint of $\overline{ST}$ and $\overline{RK} \perp \overline{ST}$, then $\overline{RK}$ is called a(n) __?__ of $\overline{ST}$.
4. If $\overline{RK}$ is both an altitude and a median of $\triangle RST$, then:
    a. $\triangle RSK \cong \triangle RTK$ by __?__.   **b.** $\triangle RST$ is a(n) __?__ triangle.
5. If R is on the perpendicular bisector of $\overline{ST}$, then R is equidistant from __?__ and __?__. Thus __?__ = __?__.

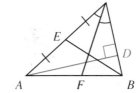

6. Refer to $\triangle ABC$ and name each of the following.
    a. a median of $\triangle ABC$
    b. an altitude of $\triangle ABC$
    c. a bisector of an angle of $\triangle ABC$

7. Draw $\overline{XY}$. Label its midpoint Q.
    a. Select a point P equidistant from X and Y. Draw $\overline{PX}$, $\overline{PY}$, and $\overline{PQ}$.
    b. What postulate justifies the statement $\triangle PQX \cong \triangle PQY$?
    c. What reason justifies the statement $\angle PQX \cong \angle PQY$?
    d. What reason justifies the statement $\overleftrightarrow{PQ} \perp \overline{XY}$?
    e. What name for $\overleftrightarrow{PQ}$ best describes the relationship between $\overleftrightarrow{PQ}$ and $\overline{XY}$?

8. Given: $\triangle DEF$ is isosceles with $DF = EF$;
    $\overline{FX}$ bisects $\angle DFE$.
    a. Would the median drawn from F to $\overline{DE}$ be the same segment as $\overline{FX}$?
    b. Would the altitude drawn from F to $\overline{DE}$ be the same segment as $\overline{FX}$?

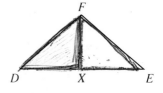

9. What kind of triangle has three angle bisectors that are also altitudes and medians?
10. Given: $\overrightarrow{NO}$ bisects $\angle N$.
    What can you conclude from each of the following additional statements?
    a. P lies on $\overrightarrow{NO}$.
    b. The distance from a point Q to each side of $\angle N$ is 13.

11. Plane M is the *perpendicular bisecting plane* of $\overline{AB}$ at O (that is, M is the plane that is perpendicular to $\overline{AB}$ at its midpoint, O). Points C and D also lie in plane M. List three pairs of congruent triangles and tell which congruence method can be used to prove each pair congruent.

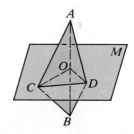

Written Exercises

A **1. a.** Draw a large scalene triangle *ABC*. Carefully draw the bisector of ∠*A*, the altitude from *A*, and the median from *A*. These three should all be different.

b. Draw a large isosceles triangle *ABC* with vertex angle *A*. Carefully draw the bisector of ∠*A*, the altitude from *A*, and the median from *A*. Are these three different?

2. Draw a large obtuse triangle. Then draw its three altitudes in color.

3. Draw a right triangle. Then draw its three altitudes in color.

4. Draw a large acute scalene triangle. Then draw the perpendicular bisectors of its three sides.

5. Draw a large scalene right triangle. Then draw the perpendicular bisectors of its three sides and tell whether they appear to meet in a point. If so, where is this point?

6. Cut out any large triangle. Fold the two sides of one angle of the triangle together to form the angle bisector. Use the same method to form the bisectors of the other two angles. What do you notice?

Complete each statement.

7. If *X* is on the bisector of ∠*SKN*, then *X* is equidistant from ___?___ and ___?___.

8. If *X* is on the bisector of ∠*SNK*, then *X* is equidistant from ___?___ and ___?___.

9. If *X* is equidistant from $\overline{SK}$ and $\overline{SN}$, then *X* lies on the ___?___.

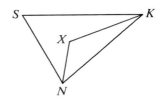

10. If *O* is on the perpendicular bisector of $\overline{LA}$, then *O* is equidistant from ___?___ and ___?___.

11. If *O* is on the perpendicular bisector of $\overline{AF}$, then *O* is equidistant from ___?___ and ___?___.

12. If *O* is equidistant from *L* and *F*, then *O* lies on the ___?___.

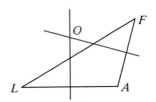

13. Given: *P* is on the perpendicular bisector of $\overline{AB}$;
       P is on the perpendicular bisector of $\overline{BC}$.
Prove: *PA* = *PC*

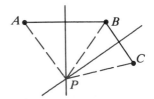

Use the diagrams on pages 153 and 154 to prove the following theorems.

B **14.** Theorem 4-5   **15.** Theorem 4-6
    16. Theorem 4-7   **17.** Theorem 4-8

18. Given: *S* is equidistant from *E* and *D*;
      V is equidistant from *E* and *D*.
   Prove: $\overleftrightarrow{SV}$ is the perpendicular bisector of $\overline{ED}$.

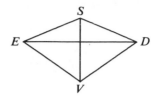

19. a. A town wants to build a beach house on the lake front equidistant from the recreation center and the school. Copy the diagram and show the point *B* where the beach house should be located.

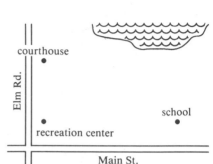

   b. The town also wants to build a boat-launching site that is equidistant from Elm Road and Main Street. Find the point *L* where it should be built.
   c. On your diagram, locate the spot *F* for a flagpole that is to be the same distance from the recreation center, the school, and the courthouse.

20. Given: $\triangle LMN \cong \triangle RST$;
      $\overline{LX}$ and $\overline{RY}$ are altitudes.
   Prove: $\overline{LX} \cong \overline{RY}$

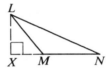 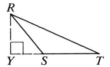

21. a. Given: $\overline{AB} \cong \overline{AC}$; $\overline{BD} \perp \overline{AC}$; $\overline{CE} \perp \overline{AB}$
   Prove: $\overline{BD} \cong \overline{CE}$

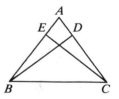

   b. The result you proved in part (a) can be stated as a theorem about certain altitudes. State this theorem in your own words.

22. Prove that the medians drawn to the legs of an isosceles triangle are congruent. Write the proof in two-column form.

For Exercises 23–27 write proofs in paragraph form. (*Hint:* **You can use theorems from this section to write fairly short proofs for Exercises 23 and 24.**)

23. Given: $\overleftrightarrow{SR}$ is the $\perp$ bisector of $\overline{QT}$;
      $\overleftrightarrow{QR}$ is the $\perp$ bisector of $\overline{SP}$.
   Prove: $PQ = TS$

24. Given: $\overrightarrow{DP}$ bisects $\angle ADE$;
      $\overrightarrow{EP}$ bisects $\angle DEC$.
   Prove: $\overrightarrow{BP}$ bisects $\angle ABC$.

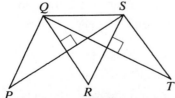 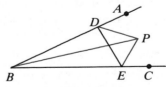

25. Given: Plane M is the perpendicular bisecting plane of $\overline{AB}$.
(That is, $\overline{AB} \perp$ plane M and O is the midpoint of $\overline{AB}$.)

Prove: **a.** $\overline{AD} \cong \overline{BD}$
b. $\overline{AC} \cong \overline{BC}$
c. $\angle CAD \cong \angle CBD$

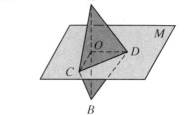

C **26.** Given: $m \angle RTS = 90$;
$\overleftrightarrow{MN}$ is the $\perp$ bisector of $\overline{TS}$.

Prove: $\overline{TM}$ is a median.

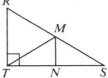

27. Given: $\overline{EH}$ and $\overline{FJ}$ are medians of scalene $\triangle EFG$; P is on $\overrightarrow{EH}$ such that $\overline{EH} \cong \overline{HP}$; Q is on $\overrightarrow{FJ}$ such that $\overline{FJ} \cong \overline{JQ}$.

Prove: **a.** $\overline{GQ} \cong \overline{GP}$
b. $\overline{GQ}$ and $\overline{GP}$ are both parallel to $\overline{EF}$.
c. P, G, and Q are collinear.

Write paragraph proofs. (In this book a star designates an exercise that is unusually difficult.)

★ **28.** Given: $\overline{AE} \parallel \overline{BD}$; $\overline{BC} \parallel \overline{AD}$;
$\overline{AE} \cong \overline{BC}$; $\overline{AD} \cong \overline{BD}$

Prove: **a.** $\overline{AC} \cong \overline{BE}$
b. $\overline{EC} \parallel \overline{AB}$

★ **29.** Given: $\overleftrightarrow{AM}$ is the $\perp$ bis. of $\overline{BC}$;
$\overline{AE} \perp \overline{BD}$; $\overline{AF} \perp \overline{DF}$;
$\angle 1 \cong \angle 2$

Prove: $\overline{BE} \cong \overline{CF}$

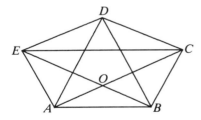

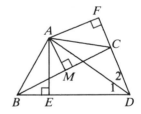

Explorations

These exploratory exercises can be done using a computer with a program that draws and measures geometric figures.

Decide if the following statements are true or false. If you think the statement is true, give a convincing argument to support your belief. If you think the statement is false, make a sketch and give all the measurements of the triangle that you find as your counterexample. For each false statement, also discover if there are types of triangles for which the statement is true.

1. An angle bisector bisects the side opposite the bisected angle.

2. A median bisects the angle at the vertex from which it is drawn.

3. The length of a median is equal to half of the length of the side it bisects.

Self-Test 3

1. Suppose you wish to prove $\triangle AFE \cong \triangle BFD$. If you have already proved $\triangle ABE \cong \triangle BAD$, what corresponding parts from this second pair of congruent triangles would you use to prove the first pair of triangles congruent?

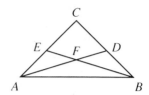

2. Given: $\triangle MPQ \cong \triangle PMN$;
   $\overline{MS} \cong \overline{PR}$

   Prove: $\triangle MSN \cong \triangle PRQ$

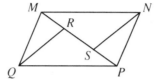

3. In $\triangle JKL$ name each of the following.

   a. an altitude **b.** a median

4. Note that $ZL = ZJ$. Can you deduce that $\overrightarrow{KZ}$ bisects $\angle LKJ$?

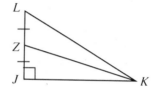

5. $\overrightarrow{UV}$ bisects $\angle WUX$. Write the theorem that justifies the statement that V is equidistant from $\overrightarrow{UW}$ and $\overrightarrow{UX}$.

6. In $\triangle ABC$, $AB = 7$ and $BC = 7$. Write the theorem that allows you to conclude that B is on the perpendicular bisector of $\overline{AC}$.

Chapter Summary

1. Congruent figures have the same size and shape. Two triangles are congruent if their corresponding sides and angles are congruent.

2. We have five ways to prove two triangles congruent:
   SSS   SAS   ASA   AAS   HL (rt. △)

3. A common way to prove that two segments or two angles are congruent is to show that they are corresponding parts of congruent triangles.

4. A line and plane are perpendicular if and only if they intersect and the line is perpendicular to all lines in the plane that pass through the point of intersection.

5. If two sides of a triangle are congruent, then the angles opposite those sides are congruent. An equilateral triangle is also equiangular, with three $60°$ angles.

6. If two angles of a triangle are congruent, then the sides opposite those angles are congruent. An equiangular triangle is also equilateral.

7. Sometimes you can prove one pair of triangles congruent and then use corresponding parts from those triangles to prove that another pair of triangles are congruent.

8. Proofs in geometry are commonly written in two-column form, as a list of key steps, or in paragraph form.

9. Every triangle has three medians and three altitudes.

10. The perpendicular bisector of a segment is the line that is perpendicular to the segment at its midpoint.

11. A point lies on the perpendicular bisector of a segment if and only if the point is equidistant from the endpoints of the segment.

12. A point lies on the bisector of an angle if and only if the point is equidistant from the sides of the angle.

Chapter Review

The two triangles shown are congruent.
Complete.

1. $\triangle STW \cong$ __?__

2. $\triangle PQR \cong$ __?__

3. $\angle R \cong$ __?__ ω

4. __?__ $= RP$

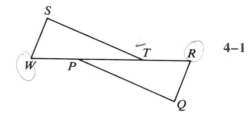

4–1

Can you deduce from the given information that $\triangle RXY \cong \triangle SXY$? If so, what postulate can you use?

5. Given: $\overline{RX} \cong \overline{SX}$; $\overline{RY} \cong \overline{SY}$

6. Given: $\overline{RY} \cong \overline{SY}$; $\angle R \cong \angle S$

7. Given: $\overline{XY}$ bisects $\angle RXS$ and $\angle RYS$.

8. Given: $\angle RXY \cong \angle SXY$; $\overline{RX} \cong \overline{SX}$

4–2

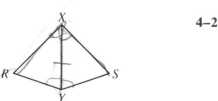

Write proofs in two-column form.

9. Given: $\overline{JM} \cong \overline{LM}$; $\overline{JK} \cong \overline{LK}$
   Prove: $\angle MJK \cong \angle MLK$

10. Given: $\angle JMK \cong \angle LMK$; $\overline{MK} \perp$ plane P
    Prove: $\overline{JK} \cong \overline{LK}$

4–3

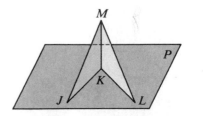

Complete.

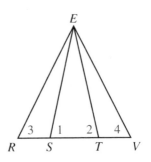

11. If $\angle 3 \cong \angle 4$, then which segments must be congruent?

12. If $\triangle REV$ is an equiangular triangle, then $\triangle REV$ is also a(n) ___?___ triangle.

13. If $\overline{ES} \cong \overline{ET}$, $m\angle 1 = 75$, and $m\angle 2 = 3x$, then $x =$ ___?___.

14. If $\angle 1 \cong \angle 2$, $ES = 3y + 5$, and $ET = 25 - y$, then $y =$ ___?___.

Write proofs in two-column form.

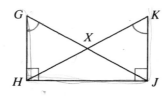

15. Given: $\overline{GH} \perp \overline{HJ}$; $\overline{KJ} \perp \overline{HJ}$;
      $\angle G \cong \angle K$
   Prove: $\triangle GHJ \cong \triangle KJH$

16. Given: $\overline{GH} \perp \overline{HJ}$; $\overline{KJ} \perp \overline{HJ}$;
      $\overline{GJ} \cong \overline{KH}$
   Prove: $\overline{GH} \cong \overline{KJ}$

17. Give the reason for each key step of the proof.

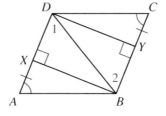

   Given: $\overline{AX} \cong \overline{CY}$; $\angle A \cong \angle C$;
        $\overline{BX} \perp \overline{AD}$; $\overline{DY} \perp \overline{BC}$
   Prove: $\overline{AD} \parallel \overline{BC}$

   1. $\triangle ABX \cong \triangle CDY$
   2. $\overline{BX} \cong \overline{DY}$
   3. $\triangle BDX \cong \triangle DBY$
   4. $\angle 1 \cong \angle 2$
   5. $\overline{AD} \parallel \overline{BC}$

18. Refer to $\triangle DEF$ and name each of the following:

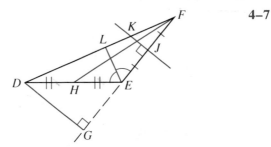

   a. an altitude
   b. a median
   c. the perpendicular bisector of a side of the triangle

19. Point G lies on the perpendicular bisector of $\overline{EF}$. Write the theorem that justifies the statement that $GE = GF$.

20. $\triangle ABC$ and $\triangle ABD$ are congruent right triangles with common hypotenuse $\overline{AB}$. Write the theorem that allows you to conclude that point B lies on the bisector of $\angle DAC$.

Chapter Test

Complete.

1. If $\triangle BAD \cong \triangle TOP$, then $\overline{DB} \cong \underline{\ ?\ }$ and $\triangle PTO \cong \underline{\ ?\ }$.

2. $\triangle EFG$ is isosceles, with $m \angle G = 94$. The legs are sides $\underline{\ ?\ }$ and $\underline{\ ?\ }$. $m \angle E = \underline{\ ?\ }$ (numerical answer).

3. You want to prove $\triangle ABC \cong \triangle XYZ$. You have shown $\overline{AB} \cong \overline{XY}$ and $\overline{AC} \cong \overline{XZ}$. To prove the triangles congruent by SAS you must show that $\underline{\ ?\ } \cong \underline{\ ?\ }$. To prove the triangles congruent by SSS you must show that $\underline{\ ?\ } \cong \underline{\ ?\ }$.

4. A method that can be used to prove right triangles congruent, but cannot be used with other types of triangles, is the $\underline{\ ?\ }$ method.

5. $\triangle CAP$ and $\triangle TAP$ are equilateral and coplanar. $\overline{AP}$ is a common side of the two triangles. $m \angle CAT = \underline{\ ?\ }$ (numerical answer).

6. A segment from a vertex of a triangle to the midpoint of the opposite side is called a(n) $\underline{\ ?\ }$ of the triangle.

7. A point lies on the bisector of an angle if and only if it is equidistant from $\underline{\ ?\ }$.

8. If in $\triangle ABC$ $m \angle A = 50$, $m \angle C = 80$, $AC = 7x + 8$, and $BC = 38 - 3x$, then $x = \underline{\ ?\ }$.

Can two triangles be proved congruent? If so, by which method, SSS, SAS, ASA, AAS, or HL?

9.
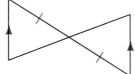

10.
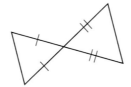

11.
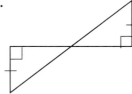

12.
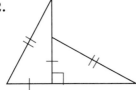

13.
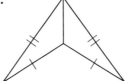

14.
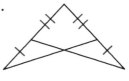

$\overline{WX}$ **and** $\overline{YZ}$ **are perpendicular bisectors of each other.**

15. W is equidistant from $\underline{\ ?\ }$ and $\underline{\ ?\ }$.

16. Z is equidistant from $\underline{\ ?\ }$ and $\underline{\ ?\ }$.

17. Name four isosceles triangles.

18. How many pairs of congruent triangles are shown in the diagram?

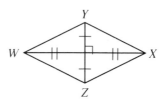

19. Given: $\angle 1 \cong \angle 2$; $\angle PQR \cong \angle SRQ$
Prove: $\overline{PR} \cong \overline{SQ}$

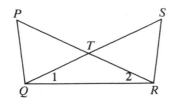

20. Given: $\angle 1 \cong \angle 2$; $\angle 3 \cong \angle 4$
Prove: $\triangle ZXY$ is isosceles.

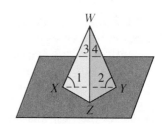

Algebra Review: *Quadratic Equations*

Solve each equation by factoring or by using the quadratic formula. The quadratic formula is:

$$\text{If } ax^2 + bx + c = 0, \text{ with } a \neq 0, \text{ then } x = \frac{-b \pm \sqrt{b^2 - 4ac}}{2a}.$$

Example　　$3x^2 + 14x + 8 = 0$

Solution 1　　*By factoring*

$$3x^2 + 14x + 8 = 0$$

$$(3x + 2)(x + 4) = 0$$

$$3x + 2 = 0 \quad \text{or} \quad x + 4 = 0$$

$$x = -\frac{2}{3} \quad \text{or} \quad x = -4$$

Solution 2　　*By quadratic formula*

$$3x^2 + 14x + 8 = 0 \qquad a = 3, \ b = 14, \ c = 8$$

$$x = \frac{-b \pm \sqrt{b^2 - 4ac}}{2a} = \frac{-14 \pm \sqrt{14^2 - 4(3)(8)}}{2(3)}$$

$$x = \frac{-14 \pm \sqrt{196 - 96}}{6} = \frac{-14 \pm 10}{6}$$

$$x = -\frac{2}{3} \quad \text{or} \quad x = -4$$

1. $x^2 + 5x - 6 = 0$

2. $n^2 - 6n + 8 = 0$

3. $y^2 - 7y - 18 = 0$

4. $x^2 + 8x = 0$

5. $y^2 = 13y$

6. $2z^2 + 7z = 0$

7. $n^2 - 144 = 25$

8. $50x^2 = 200$

9. $50x^2 = 2$

10. $49z^2 = 1$

11. $y^2 - 6y + 9 = 0$

12. $x^2 - 7x + 12 = 0$

13. $y^2 + 8y + 12 = 0$

14. $t^2 + 5t = 24$

15. $v^2 + 25 = 10v$

16. $x^2 = 3x + 4$

17. $t^2 - t = 20$

18. $y^2 = 20y - 36$

19. $3x^2 + 3x = 4$

20. $15 + 4y^2 = 17y$

21. $x^2 + 5x + 2 = 0$

22. $x^2 + 2x - 1 = 0$

23. $x^2 - 5x + 3 = 0$

24. $x^2 + 3x - 2 = 0$

25. $(y - 5)^2 = 16$

26. $z^2 = 4(2z - 3)$

27. $x(x + 5) = 14$

In Exercises 28–33 x represents the length of a segment. When a value of x doesn't make sense as a length, eliminate that value of x.

28. $x(x - 50) = 0$

29. $x^2 - 400 = 0$

30. $x^2 - 17x + 72 = 0$

31. $2x^2 + x - 3 = 0$

32. $2x^2 - 7x - 4 = 0$

33. $6x^2 = 5x + 6$

Preparing for College Entrance Exams

Strategy For Success
Some college entrance exam questions ask you to decide if several state-ments are true based on given information (see Exercises 3 and 8). In these exercises, check each statement separately and then choose the answer with the correct combination of true statements.

Indicate the best answer by writing the appropriate letter.

1. The measures of the angles of a triangle are $2x + 10$, $3x$, and $8x - 25$. The triangle is:
   (**A**) obtuse   (**B**) right   (**C**) acute   (**D**) equilateral   (**E**) isosceles

2. A regular polygon has an interior angle of measure 120. How many vertices does the polygon have?
   (**A**) 3   (**B**) 5   (**C**) 6   (**D**) 9   (**E**) 12

3. Plane M is parallel to plane N. Line l lies in M and line k lies in N. Which of the following statement(s) are possible?
   (I) Lines l and k are parallel.   (II) Lines l and k intersect.
   (III) Lines l and k are skew.
   (**A**) I only   (**B**) II only   (**C**) III only
   (**D**) I and III only   (**E**) I, II, and III

4. Given: $\overline{BE}$ bisects $\overline{AD}$. To prove that the triangles are congruent by the AAS method, you must show that:
   (**A**) $\angle A \cong \angle E$ (**B**) $\angle A \cong \angle D$ (**C**) $\angle B \cong \angle E$
   (**D**) $\angle B \cong \angle D$ (**E**) $\overline{AD}$ bisects $\overline{BE}$.

5. Given: $\triangle RGA$ and $\triangle PMC$ with $\overline{RG} \cong \overline{PM}$, $\overline{RA} \cong \overline{PC}$, and $\angle R \cong \angle P$. Which method could be used to prove that $\triangle RGA \cong \triangle PMC$?
   (**A**) SSS   (**B**) SAS   (**C**) HL   (**D**) ASA
   (**E**) There is not enough information for a proof.

6. Predict the next number in the sequence, 2, 6, 12, 20, 30, 42, __?__.
   (**A**) 52   (**B**) 54   (**C**) 56   (**D**) 58   (**E**) 60

7. In $\triangle JKL$, $\overline{KL} \cong \overline{JL}$, $m\angle K = 2x - 36$, and $m\angle L = x + 2$. Find $m\angle J$.
   (**A**) 56   (**B**) 52   (**C**) 53   (**D**) 55   (**E**) 64

8. In $\triangle RST$, $\overleftrightarrow{SU}$ is the perpendicular bisector of $\overline{RT}$ and U lies on $\overline{RT}$. Which statement(s) must be true?
   (I) $\triangle RST$ is equilateral.   (II) $\triangle RSU \cong \triangle TSU$
   (III) $\overrightarrow{SU}$ is the bisector of $\angle RST$.
   (**A**) I only   (**B**) II only   (**C**) III only
   (**D**) II and III only   (**E**) I, II, and III

9. Given: $\triangle SUN \cong \triangle TAN$. You can conclude that:
   (**A**) $\angle S \cong \angle A$ (**B**) $\overline{SN} \cong \overline{TN}$ (**C**) $\angle T \cong \angle U$
   (**D**) $\overline{SU} \cong \overline{TN}$ (**E**) $\overline{UN} \cong \overline{TA}$

Cumulative Review: Chapters 1–4

Complete each sentence with the most appropriate word, phrase, or value.

A   **1.** If S is between R and T, then $RS + ST = RT$ by the ___?___.

2. If two parallel planes are cut by a third plane, then the lines of intersection are ___?___.

3. $\overrightarrow{BD}$ bisects $\angle ABC$, $m\angle ABC = 5x - 4$, and $m\angle CBD = 2x + 10$. $\angle ABC$ is a(n) ___?___ angle.

4. If two intersecting lines form congruent adjacent angles, then the lines are ___?___.

5. If $\angle 1$ and $\angle 2$ are complements and $m\angle 1 = 74$, then $m\angle 2 = $ ___?___.

6. Given the conditional "If $x = 9$, then $3x = 27$," its converse is ___?___.

7. If the measure of each interior angle of a polygon is 144, then the polygon has ___?___ sides.

8. In quadrilateral $EFGH$, $\overline{EF} \parallel \overline{HG}$, $m\angle E = y + 10$, $m\angle F = 2y - 40$, and $m\angle H = 2y - 31$. $m\angle G = $ ___?___ (numerical answer)

9. If a diagonal of an equilateral quadrilateral is drawn, the two triangles formed can be proved congruent by the ___?___ method.

Find the measure of each numbered angle.

10.

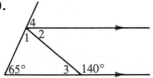

11.

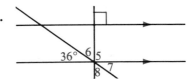

Could the given information be used to prove that two lines are parallel? If so, which lines?

12. $m\angle 8 + m\angle 9 = 180$

13. $\angle 1 \cong \angle 4$

14. $m\angle 2 = m\angle 6$

15. $\angle 8$ and $\angle 5$ are rt. $\angle$s.

B   **16.** Given: $\overline{MN} \cong \overline{MP}$; $\angle NMO \cong \angle PMO$
Prove: $\overleftrightarrow{MO}$ is the $\perp$ bisector of $\overline{NP}$.

17. Given: $\overline{MO} \perp \overline{NP}$; $\overline{NO} \cong \overline{PO}$
Prove: $\overline{MN} \cong \overline{MP}$

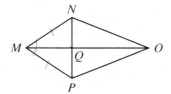

18. Write a paragraph proof: If $\overline{AX}$ is both a median and an altitude of $\triangle ABC$, then $\triangle ABC$ is isosceles.

5 QUADRILATERALS

All artists seek ways to express their own visions of life and the world. Here, Hans Hofmann has created a painting composed entirely of quadrilaterals.

Parallelograms

Objectives

1. Apply the definition of a parallelogram and the theorems about properties of a parallelogram.
2. Prove that certain quadrilaterals are parallelograms.
3. Apply theorems about parallel lines and the segment that joins the midpoints of two sides of a triangle.

5-1 *Properties of Parallelograms*

A **parallelogram** ($\square$) is a quadrilateral with both pairs of opposite sides parallel. The following theorems state some properties common to all parallelograms. Your proofs of these theorems (Written Exercises 13–15) will be based on what you have learned about parallel lines and congruent triangles.

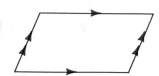

Theorem 5-1

Opposite sides of a parallelogram are congruent.

Given: $\square EFGH$

Prove: $\overline{EF} \cong \overline{HG}$; $\overline{FG} \cong \overline{EH}$

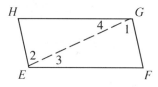

Plan for Proof: Draw $\overline{EG}$ to form triangles with corresponding sides $\overline{EF}$ and $\overline{HG}$, $\overline{FG}$ and $\overline{EH}$. Use the pairs of alternate interior angles $\angle 1$ and $\angle 2$, $\angle 3$ and $\angle 4$, to prove the triangles congruent by ASA.

Theorem 5-2

Opposite angles of a parallelogram are congruent.

Theorem 5-3

Diagonals of a parallelogram bisect each other.

Given: $\square QRST$ with diagonals $\overline{QS}$ and $\overline{TR}$

Prove: $\overline{QS}$ and $\overline{TR}$ bisect each other.

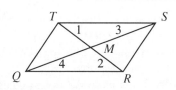

Plan for Proof: You can prove that $\overline{QM} \cong \overline{MS}$ and $\overline{RM} \cong \overline{MT}$ by showing that they are corresponding parts of congruent triangles. Since $\overline{QR} \cong \overline{TS}$ by Theorem 5-1, you can show that $\triangle QMR \cong \triangle SMT$ by ASA.

Classroom Exercises

1. Quad. *GRAM* is a parallelogram.
   a. Why is $\angle G$ supplementary to $\angle M$?
   b. Why is $\angle M$ supplementary to $\angle A$?
   c. Complete: Consecutive angles of a parallelogram are ___?___, while opposite angles are ___?___.

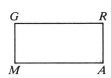

2. Suppose that $\angle M$ is a right angle. What can you deduce about angles G, R, and A?

In Exercises 3–5 quad. *ABCD* is a parallelogram. Find the values of *x*, *y*, and *z*.

3.

4.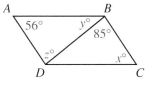

5.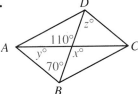

Must quad. *EFGH* be a parallelogram? Can it be a parallelogram? Explain.

6.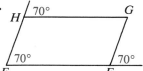

7.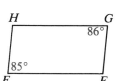

8.

Quad. *ABCD* is a parallelogram. Name the principal theorem or definition that justifies the statement.

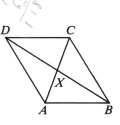

9. $\overline{AD} \parallel \overline{BC}$

10. $\angle ADX \cong \angle CBX$

11. $m \angle ABC = m \angle CDA$

12. $\overline{AD} \cong \overline{BC}$

13. $AX = \frac{1}{2}AC$

14. $DX = BX$

15. Draw a quadrilateral that isn't a parallelogram but does have two 60° angles opposite each other.

16. State each theorem in if-then form. (Begin ''If a quadrilateral is a ....'')
    a. Theorem 5-1   **b.** Theorem 5-2   **c.** Theorem 5-3

17. **a.** Draw any two segments, $\overline{AC}$ and $\overline{BD}$, that bisect each other at O. What appears to be true of quad. *ABCD*?
    b. This exercise investigates the converse of what theorem?

18. Draw two segments that are both parallel and congruent. Connect their endpoints to form a quadrilateral. What appears to be true of the quadrilateral?

Written Exercises

Exercises 1–4 refer to ▱*CREW*.

A **1.** If $OE = 4$ and $WE = 8$, name two segments congruent to $\overline{WE}$.

2. If $\overline{WR} \perp \overline{CE}$, name all angles congruent to $\angle RCE$.

3. If $\overline{WR} \perp \overline{CE}$, name all segments congruent to $\overline{WE}$.

4. If $RE = EW$, name all angles congruent to $\angle ERW$.

In Exercises 5–10 quad. *PQRS* is a parallelogram. Find the values of *a*, *b*, *x*, and *y*.

5.
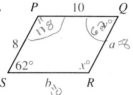

6.
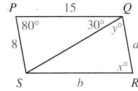

7.
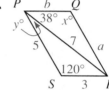

8.
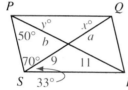

9.
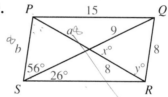

10.
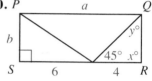

11. Find the perimeter of ▱*RISK* if $RI = 17$ and $IS = 13$.

12. The perimeter of ▱*STOP* is 54 cm, and $\overline{ST}$ is 1 cm longer than $\overline{SP}$. Find ST and SP.

13. Prove Theorem 5-1.

14. Prove Theorem 5-2. (Draw and label a diagram. List what is given and what is to be proved.)

15. Prove Theorem 5-3.

16. Given: *ABCX* is a ▱;
　　　　　 DXFE is a ▱.
　　Prove: $\angle B \cong \angle E$

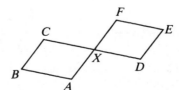

The coordinates of three vertices of $\square ABCD$ are given. Plot the points and find the coordinates of the fourth vertex.

17. $A(1, 0)$, $B(5, 0)$, $C(7, 2)$, $D(\underline{\quad?\quad}, \underline{\quad?\quad})$

18. $A(3, 2)$, $B(8, 2)$, $C(\underline{\quad?\quad}, \underline{\quad?\quad})$, $D(0, 5)$

Each figure in Exercises 19–24 is a parallelogram with its diagonals drawn. Find the values of x and y.

19.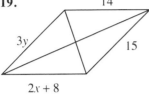

20.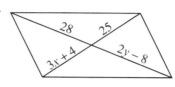

21.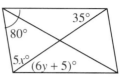

B 22.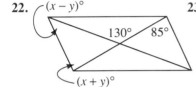

23.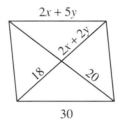

24.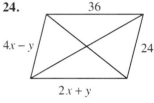

Quad. *DECK* is a parallelogram. Complete.

25. If $KT = 2x + y$, $DT = x + 2y$, $TE = 12$, and $TC = 9$, then $x = \underline{\quad?\quad}$ and $y = \underline{\quad?\quad}$.

26. If $DE = x + y$, $EC = 12$, $CK = 2x - y$, and $KD = 3x - 2y$, then $x = \underline{\quad?\quad}$, $y = \underline{\quad?\quad}$, and the perimeter of $\square DECK = \underline{\quad?\quad}$.

27. If $m \angle 1 = 3x$, $m \angle 2 = 4x$, and $m \angle 3 = x^2 - 70$, then $x = \underline{\quad?\quad}$ and $m \angle CED = \underline{\quad?\quad}$ (numerical answers).

28. If $m \angle 1 = 42$, $m \angle 2 = x^2$, and $m \angle CED = 13x$, then $m \angle 2 = \underline{\quad?\quad}$ or $m \angle 2 = \underline{\quad?\quad}$ (numerical answers).

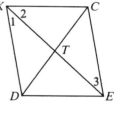

29. Given: $\square PQRS$; $\overline{PJ} \cong \overline{RK}$
    Prove: $\overline{SJ} \cong \overline{QK}$

30. Given: $\square JQKS$; $\overline{PJ} \cong \overline{RK}$
    Prove: $\angle P \cong \angle R$

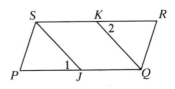

31. Given: $ABCD$ is a $\square$; $\overline{CD} \cong \overline{CE}$
    Prove: $\angle A \cong \angle E$

32. Given: $ABCD$ is a $\square$; $\angle A \cong \angle E$
    Prove: $\overline{AB} \cong \overline{CE}$

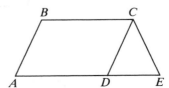

Find something interesting to prove. Then prove it. Answers may vary.

33. Given: □*ABCD*; ∠1 ≅ ∠2

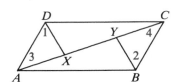

34. Given: □*EFIH*; □*EGJH*; ∠1 ≅ ∠2

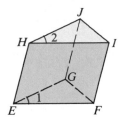

The coordinates of three vertices of a parallelogram are given. Find all the possibilities you can for the coordinates of the fourth vertex.

C **35.** (3, 4), (9, 4), (6, 8)

36. (−1, 0), (2, −2), (2, 2)

37. **a.** Given: Plane *P* ∥ plane *Q*; *j* ∥ *k*
   Prove: *AX* = *BY*
   b. State a theorem about parallel planes and lines that you proved in part (a).

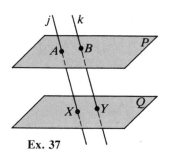

Ex. 37

38. Prove: If a segment whose endpoints lie on opposite sides of a parallelogram passes through the midpoint of a diagonal, that segment is bisected by the diagonal.

★ **39.** Write a paragraph proof: The sum of the lengths of the segments drawn from any point in the base of an isosceles triangle perpendicular to the legs is equal to the length of the altitude drawn to one leg.

Biographical Note   *Benjamin Banneker*

Benjamin Banneker (1731–1806) was a noted American scholar, largely self-taught, who became both a surveyor and an astronomer. As a surveyor, Banneker was a member of the commission that defined the boundary line and laid out the streets of the District of Columbia. As an astronomer, he accurately predicted a solar eclipse in 1789. From 1791 until his death he published almanacs containing information on astronomy, tide tables, and also such diverse subjects as insect life and medicinal products. Banneker's almanacs included ideas that were far ahead of their time, for example, the formation of a Department of the Interior and an organization like the United Nations.

5-2 *Ways to Prove that Quadrilaterals Are Parallelograms*

If both pairs of opposite sides of a quadrilateral are parallel, then by definition the quadrilateral is a parallelogram. The following theorems will give you additional ways to prove that a quadrilateral is a parallelogram.

Theorem 5-4

If both pairs of opposite sides of a quadrilateral are congruent, then the quadrilateral is a parallelogram.

Given: $\overline{TS} \cong \overline{QR}$; $\overline{TQ} \cong \overline{SR}$

Prove: Quad. $QRST$ is a $\square$.

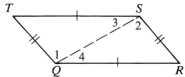

Plan for Proof: Draw $\overline{QS}$ and prove that $\triangle TSQ \cong \triangle RQS$. Then $\angle 1 \cong \angle 2$ and $\angle 3 \cong \angle 4$, and opposite sides are parallel.

Theorem 5-5

If one pair of opposite sides of a quadrilateral are both congruent and parallel, then the quadrilateral is a parallelogram.

Theorem 5-6

If both pairs of opposite angles of a quadrilateral are congruent, then the quadrilateral is a parallelogram.

Theorem 5-7

If the diagonals of a quadrilateral bisect each other, then the quadrilateral is a parallelogram.

Five Ways to Prove that a Quadrilateral Is a Parallelogram

1. Show that *both* pairs of opposite sides are parallel.
2. Show that *both* pairs of opposite sides are congruent.
3. Show that *one* pair of opposite sides are both congruent and parallel.
4. Show that both pairs of opposite angles are congruent.
5. Show that the diagonals bisect each other.

Classroom Exercises

Study the markings on each figure and decide whether *ABCD* *must* **be a parallelogram. If the answer is** *yes*, **state the definition or theorem that applies.**

1.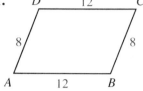

2.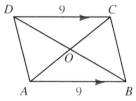

3.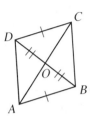

4.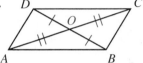

5.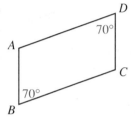

6.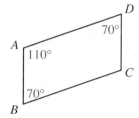

7.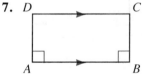

8.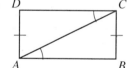

9.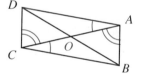

10. Draw a quadrilateral that has two pairs of congruent sides but that is *not* a parallelogram.

11. Draw a quadrilateral that is *not* a parallelogram but that has one pair of congruent sides and one pair of parallel sides.

12. *Parallel rulers*, used to draw parallel lines, are constructed so that *EF* = *HG* and *HE* = *GF*. Since there are hinges at points *E*, *F*, *G*, and *H*, you can vary the distance between $\overleftrightarrow{HG}$ and $\overleftrightarrow{EF}$. Explain why $\overleftrightarrow{HG}$ and $\overleftrightarrow{EF}$ are always parallel.

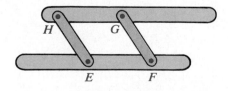

13. The pliers shown are made in such a way that the jaws are always parallel. Explain.

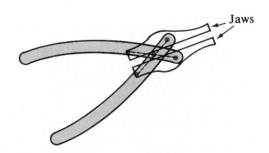

Jaws

Written Exercises

State the principal definition or theorem that enables you to deduce, from the information given, that quad. *SACK* is a parallelogram.

A   1. $\overline{SA} \parallel \overline{KC}$; $\overline{SK} \parallel \overline{AC}$

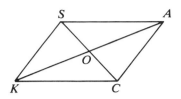

2. $\overline{SA} \cong \overline{KC}$; $\overline{SK} \cong \overline{AC}$

3. $\overline{SA} \cong \overline{KC}$; $\overline{SA} \parallel \overline{KC}$

4. $SO = \frac{1}{2}SC$; $KO = \frac{1}{2}KA$

5. $\angle SKC \cong \angle CAS$; $\angle KCA \cong \angle ASK$

6. Suppose you know that $\triangle SOK \cong \triangle COA$. Explain how you could prove that quad. *SACK* is a parallelogram.

7. The legs of this ironing board are built so that $BO = AO = RO = DO$. What theorem guarantees that the board is parallel to the floor $(\overline{AR} \parallel \overline{BD})$?

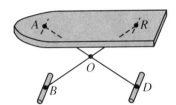

8. The quadrilaterals numbered 1, 2, 3, 4, and 5 are parallelograms. If you wanted to show that quadrilateral 6 is also a parallelogram, which of the five methods listed on page 172 would be easiest to use?

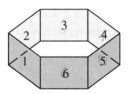

9. What theorem in this section is the converse of each theorem?
   a. Theorem 5-1   **b.** Theorem 5-2   **c.** Theorem 5-3

10. Give the reasons for each step in the following proof of Theorem 5-6.

Given: $m \angle A = m \angle C = x$;
$\quad\quad\quad m \angle B = m \angle D = y$

Prove: *ABCD* is a $\square$.

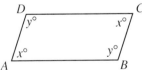

Proof:

Statements	Reasons
1. $m \angle A = m \angle C = x$; $\quad m \angle B = m \angle D = y$	1. ___?___
2. $2x + 2y = 360$	2. ___?___
3. $x + y = 180$	3. ___?___
4. $\overline{AB} \parallel \overline{DC}$ and $\overline{AD} \parallel \overline{BC}$	4. ___?___
5. *ABCD* is a $\square$.	5. ___?___

Draw and label a diagram. List what is given and what is to be proved. Then write a two-column proof of the theorem.

B   **11.** Theorem 5-4   **12.** Theorem 5-5   **13.** Theorem 5-7

For Exercises 14–18 write paragraph proofs.

14. Given: $\square ABCD$; M and N are the midpoints of $\overline{AB}$ and $\overline{DC}$.
Prove: $AMCN$ is a $\square$.

15. Given: $\square ABCD$; $\overline{AN}$ bisects $\angle DAB$; $\overline{CM}$ bisects $\angle BCD$.
Prove: $AMCN$ is a $\square$.

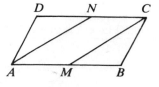

16. Given: $\square ABCD$; W, X, Y, Z are midpoints of $\overline{AO}$, $\overline{BO}$, $\overline{CO}$, and $\overline{DO}$.
Prove: $WXYZ$ is a $\square$.

17. Given: $\square ABCD$; $DE = BF$
Prove: $AFCE$ is a $\square$.

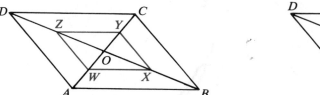

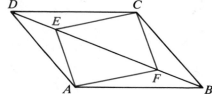

18. Given: $\square KGLJ$; $FK = HL$
Prove: $FGHJ$ is a $\square$.

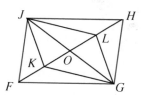

What values must x and y have to make the quadrilateral a parallelogram?

19.

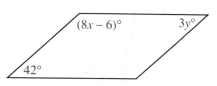

20.

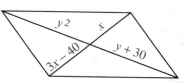

21.

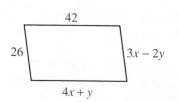

22.

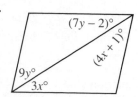

23. Given: $\square ABCD$;
$\overline{DE} \perp \overline{AC}$; $\overline{BF} \perp \overline{AC}$
Prove: $DEBF$ is a $\square$.

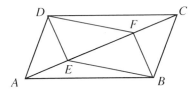

24. Given: Plane $X \parallel$ plane Y;
$\overline{LM} \cong \overline{ON}$
Prove: $LMNO$ is a $\square$.

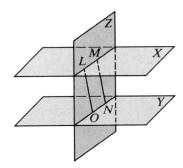

C **25.** Write a paragraph proof.

Given: $\square ABCD$; $\square BEDF$
Prove: $AECF$ is a $\square$.

(*Hint:* A short proof is possible if certain auxiliary segments are drawn.)

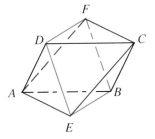

Explorations

These exploratory exercises can be done using a computer with a program that draws and measures geometric figures.

Draw any $\triangle ABC$. Label the midpoint of $\overline{AB}$ as D. Draw a segment through D parallel to $\overline{BC}$ that intersects $\overline{AC}$ at E. Measure AE and EC. What do you notice?

Draw any $\triangle ABC$. Label the midpoints of $\overline{AB}$ and $\overline{AC}$ as D and E, respectively. Draw $\overline{DE}$. Measure $\angle AED$ and $\angle ACB$. What do you notice? What is true of $\overline{DE}$ and $\overline{BC}$? Measure DE and BC. What do you notice?

Write an equation that relates DE and BC. Repeat the drawing and measurements until you are sure of your equation.

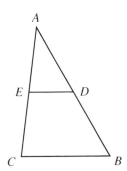

5-3 *Theorems Involving Parallel Lines*

In this section we will prove four useful theorems about parallel lines. The first theorem uses the definition of the distance from a point to a line. (See page 154.)

Theorem 5-8

If two lines are parallel, then all points on one line are equidistant from the other line.

Given: $l \parallel m$; A and B are any points on l;
$\overline{AC} \perp m$; $\overline{BD} \perp m$

Prove: $AC = BD$

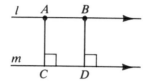

Proof:

Since $\overline{AB}$ and $\overline{CD}$ are contained in parallel lines, $\overline{AB} \parallel \overline{CD}$. Since $\overline{AC}$ and $\overline{BD}$ are coplanar and are both perpendicular to m, they are parallel. Thus $ABDC$ is a parallelogram, by the definition of a parallelogram. Since opposite sides $\overline{AC}$ and $\overline{BD}$ are congruent, $AC = BD$.

Theorem 5-9

If three parallel lines cut off congruent segments on one transversal, then they cut off congruent segments on every transversal.

Given: $\overleftrightarrow{AX} \parallel \overleftrightarrow{BY} \parallel \overleftrightarrow{CZ}$;
$\overline{AB} \cong \overline{BC}$

Prove: $\overline{XY} \cong \overline{YZ}$

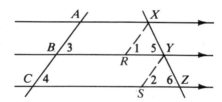

Proof:

Through X and Y draw lines parallel to $\overleftrightarrow{AC}$, intersecting $\overleftrightarrow{BY}$ at R and $\overleftrightarrow{CZ}$ at S, as shown. Then $AXRB$ and $BYSC$ are parallelograms, by the definition of a parallelogram. Since the opposite sides of a parallelogram are congruent, $\overline{XR} \cong \overline{AB}$ and $\overline{BC} \cong \overline{YS}$. It is given that $\overline{AB} \cong \overline{BC}$, so using the Transitive Property twice gives $\overline{XR} \cong \overline{YS}$. Parallel lines are cut by transversals to form the following pairs of congruent corresponding angles:

$$\angle 1 \cong \angle 3 \qquad \angle 3 \cong \angle 4 \qquad \angle 4 \cong \angle 2 \qquad \angle 5 \cong \angle 6$$

Then $\angle 1 \cong \angle 2$ (Transitive Property), and $\triangle XYR \cong \triangle YZS$ by AAS. Since $\overline{XY}$ and $\overline{YZ}$ are corresponding parts of these triangles, $\overline{XY} \cong \overline{YZ}$.

Theorem 5-10

A line that contains the midpoint of one side of a triangle and is parallel to another side passes through the midpoint of the third side.

Given: M is the midpoint of $\overline{AB}$;
$\overleftrightarrow{MN} \parallel \overline{BC}$

Prove: N is the midpoint of $\overline{AC}$.

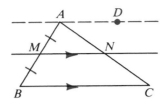

Proof:

Let $\overleftrightarrow{AD}$ be the line through A parallel to $\overleftrightarrow{MN}$. Then $\overleftrightarrow{AD}$, $\overleftrightarrow{MN}$, and $\overleftrightarrow{BC}$ are three parallel lines that cut off congruent segments on transversal $\overleftrightarrow{AB}$. By Theorem 5-9 they also cut off congruent segments on $\overleftrightarrow{AC}$. Thus $\overline{AN} \cong \overline{NC}$ and N is the midpoint of $\overline{AC}$.

The next theorem has two parts, the first of which is closely related to Theorem 5-10.

Theorem 5-11

The segment that joins the midpoints of two sides of a triangle

(1) is parallel to the third side;
(2) is half as long as the third side.

Given: M is the midpoint of $\overline{AB}$;
N is the midpoint of $\overline{AC}$.

Prove: (1) $\overline{MN} \parallel \overline{BC}$
(2) $MN = \frac{1}{2}BC$

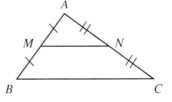

Proof of (1):

There is exactly one line through M parallel to $\overline{BC}$. By Theorem 5-10 that line passes through N, the midpoint of $\overline{AC}$. Thus $\overline{MN} \parallel \overline{BC}$.

Proof of (2):

Let L be the midpoint of $\overline{BC}$, and draw $\overline{NL}$. By part (1), $\overline{MN} \parallel \overline{BC}$ and also $\overline{NL} \parallel \overline{AB}$. Thus quad. $MNLB$ is a parallelogram. Since its opposite sides are congruent, $MN = BL$. Since L is the midpoint of $\overline{BC}$, $BL = \frac{1}{2}BC$. Therefore $MN = \frac{1}{2}BC$.

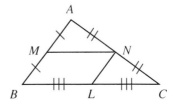

Example   *P, Q,* and *R* are midpoints of the sides of △*DEF*.
        a. What kind of figure is *DPQR*?
        b. What is the perimeter of *DPQR*?

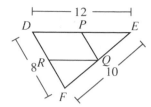

Solution   **a.** Since $\overline{RQ} \parallel \overline{DE}$ and $\overline{PQ} \parallel \overline{DF}$, quad. *DPQR*
        is a parallelogram.
      b. $RQ = \frac{1}{2}DE = DP = 6$ and $PQ = \frac{1}{2}DF = DR = 4$.
        Thus the perimeter of *DPQR* is $6 + 4 + 6 + 4$, or 20.

Classroom Exercises

1. You can use a sheet of lined notebook paper to divide a segment
   into a number of congruent parts. Here a piece of cardboard
   with edge $\overline{AB}$ is placed so that $\overline{AB}$ is separated into five congruent
   parts. Explain why this works.

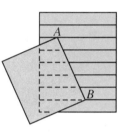

M, N, and T are the midpoints of the sides of △XYZ.

2. If $XZ = 10$, then $MN = \underline{\quad?\quad}$.

3. If $TN = 7$, then $XY = \underline{\quad?\quad}$.

4. If $ZN = 8$, then $TM = \underline{\quad?\quad}$.

5. If $XY = k$, then $TN = \underline{\quad?\quad}$.

6. Suppose $XY = 10$, $YZ = 14$, and $XZ = 8$.
   What are the lengths of the three sides of
     a. △*TNZ*?   **b.** △*MYN*?
     c. △*XMT*?   **d.** △*NTM*?

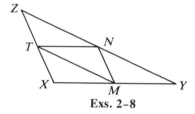

Exs. 2–8

7. State a theorem suggested by Exercise 6.

8. How many parallelograms are in the diagram?

9. What result of this section do the railings suggest?

Written Exercises

Points *A*, *B*, *E*, and *F* are the midpoints of $\overline{XC}$, $\overline{XD}$, $\overline{YC}$, and $\overline{YD}$. Complete.

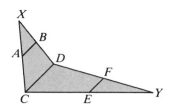

A   **1.** If $CD = 24$, then $AB = \underline{\quad?\quad}$ and $EF = \underline{\quad?\quad}$.

    2. If $AB = k$, then $CD = \underline{\quad?\quad}$ and $EF = \underline{\quad?\quad}$.

    3. If $AB = 5x - 8$ and $EF = 3x$, then $x = \underline{\quad?\quad}$.

    4. If $CD = 8x$ and $AB = 3x + 2$, then $x = \underline{\quad?\quad}$.

5. Given: *L*, *M*, and *N* are midpoints of the sides of $\triangle TKO$. Find the perimeter of each figure.

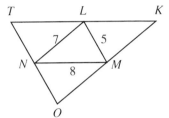

    a. $\triangle TKO$ **b.** $\triangle LMK$

    c. $\square TNML$ **d.** quad. *LNOK*

6. a. Name all triangles congruent to $\triangle TNL$.

    b. Suppose you are told that the area of $\triangle NLM$ is 17.32 cm². What is the area of $\triangle TKO$?

Name all the points shown that *must* be midpoints of the sides of the large triangle.

7.

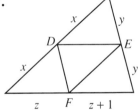

8.

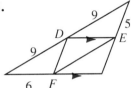

9.

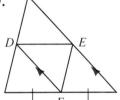

$\overleftrightarrow{AE}$, $\overleftrightarrow{BF}$, $\overleftrightarrow{CG}$, and $\overleftrightarrow{DH}$ are parallel, with $EF = FG = GH$. Complete.

10. If $AB = 5$, then $AD = \underline{\quad?\quad}$.

11. If $AC = 12$, then $CD = \underline{\quad?\quad}$.

12. If $AB = 5x$ and $BC = 2x + 12$, then $x = \underline{\quad?\quad}$.

13. If $AC = 22 - x$ and $BD = 3x - 22$, then $x = \underline{\quad?\quad}$.

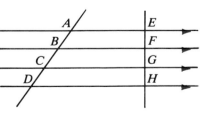

Exs. 10–15

B   **14.** If $AB = 15$, $BC = 2x - y$, and $CD = x + y$, then $x = \underline{\quad?\quad}$ and $y = \underline{\quad?\quad}$.

    15. If $AB = 12$, $BC = 2x + 3y$, and $BD = 8x$, then $x = \underline{\quad?\quad}$ and $y = \underline{\quad?\quad}$.

In Exercises 16–17 a segment joins the midpoints of two sides of a triangle. Find the values of *x* and *y*.

16.

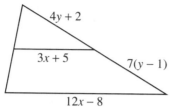

17.

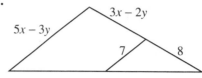

18. Given: A is the midpoint of $\overline{OX}$;
$\overline{AB} \parallel \overline{XY}$; $\overline{BC} \parallel \overline{YZ}$
Prove: $\overline{AC} \parallel \overline{XZ}$

19. Given: $\square ABCD$; $\overline{BE} \parallel \overline{MD}$;
M is the midpoint of $\overline{AB}$.
Prove: $DE = BC$

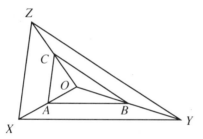

20. Given: $\overline{PQ}$, $\overline{RS}$, and $\overline{TU}$ are each
perpendicular to $\overleftrightarrow{UQ}$;
R is the midpoint of $\overline{PT}$.
Prove: R is equidistant from U and Q.

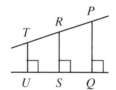

21. *EFGH* is a parallelogram whose diagonals intersect at *P*. *M* is the midpoint of $\overline{FG}$. Prove that $MP = \frac{1}{2}EF$.

22. A *skew quadrilateral SKEW* is shown. *M, N, O,* and *P* are the midpoints of $\overline{SK}$, $\overline{KE}$, $\overline{WE}$, and $\overline{SW}$. Explain why *PMNO* is a parallelogram.

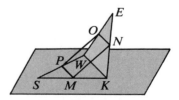

23. Draw $\triangle ABC$ and label the midpoints of $\overline{AB}$, $\overline{AC}$, and $\overline{BC}$ as *X*, *Y*, and *Z*, respectively. Let *P* be the midpoint of $\overline{BZ}$ and *Q* be the midpoint of $\overline{CZ}$. Prove that $PX = QY$.

24. Draw $\triangle ABC$ and let *D* be the midpoint of $\overline{AB}$. Let *E* be the midpoint of $\overline{CD}$. Let *F* be the intersection of $\overrightarrow{AE}$ and $\overline{BC}$. Draw $\overline{DG}$ parallel to $\overline{EF}$ meeting $\overline{BC}$ at *G*. Prove that $BG = GF = FC$.

C **25.** Given: Parallel planes P, Q, and R cutting transversals $\overleftrightarrow{AC}$ and $\overleftrightarrow{DF}$; $AB = BC$
Prove: $DE = EF$
(*Hint:* You can't assume that $\overleftrightarrow{AC}$ and $\overleftrightarrow{DF}$ are
coplanar. Draw $\overline{AF}$, cutting plane Q at X. Using
the plane of $\overline{AC}$ and $\overline{AF}$, apply Theorems 3-1 and
5-10. Then use the plane of $\overline{AF}$ and $\overline{FD}$.)

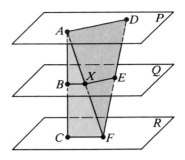

Self-Test 1

The diagonals of $\square ABCD$ intersect at Z. Tell
whether each statement *must be*, *may be*, or *cannot
be* true.

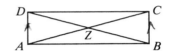

1. $\overline{AC} \cong \overline{BD}$ **2.** $\overline{DZ} \cong \overline{BZ}$
3. $\overline{AD} \parallel \overline{BC}$ **4.** $m \angle DAB = 85$ and $m \angle BCD = 95$

5. List five ways to prove that quad. $ABCD$ is a parallelogram.

6. a. State a theorem that allows you to
conclude that $3x - 7 = 11$.
b. Find the values of x and y.

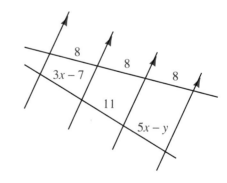

7. Given: $\square ABCD$;
M is the midpoint of $\overline{AB}$.
Prove: $MO = \frac{1}{2}AD$

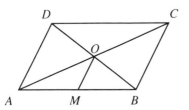

8. Given: $\square PQRS$;
$\overline{PX}$ bisects $\angle QPR$;
$\overline{RY}$ bisects $\angle SRP$.
Prove: $RYPX$ is a $\square$.

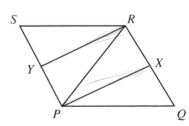

♦ Computer Key-In

The BASIC computer program below will calculate the lengths of sides and diagonals of a quadrilateral. (Line 200 uses the distance formula, which you will study in Chapter 13.) Do the exercises to see what you can discover before studying special quadrilaterals in the next two sections.

```
10   DIM X(4), Y(4), A$(4)
15   A$(1) = "A":A$(2) = "B":A$(3) = "C":A$(4) = "D"
20   FOR I = 1 TO 4
30   PRINT "VERTEX ";A$(I);" ";
40   INPUT X(I), Y(I)
50   NEXT I
60   PRINT "SIDES"
70   FOR I = 1 TO 4
80   LET J = I + 1 - 4 * INT(I/4)
90   GOSUB 200
95   NEXT I
100  PRINT "DIAGONALS"
110  LET I = 1:LET J = 3
120  GOSUB 200
130  LET I = 2:LET J = 4
140  GOSUB 200
150  END
200  LET D = SQR((X(I)-X(J))↑2 + (Y(I)-Y(J))↑2)
210  LET D = INT(100 * D + .5)/100
220  PRINT A$(I);A$(J);" = ";D
230  RETURN
```

Exercises

Plot the given points and draw quad. *ABCD* and its diagonals. Also RUN the program above, inputting the given coordinates for the vertices. Then tell which of the following statements are true for quad. *ABCD*.

I. Quad. *ABCD* is a parallelogram.

II. Both pairs of opposite sides are congruent.

III. All sides are congruent.

IV. The diagonals are congruent.

V. The diagonals are perpendicular.

VI. The sides form four right angles.

1. a. $A(1, 1)$, $B(3, 4)$, $C(10, 4)$, $D(8, 1)$
   b. $A(0, -3)$, $B(-4, -1)$, $C(-2, 1)$, $D(2, -1)$
2. a. $A(1, 2)$, $B(-3, -1)$, $C(-7, 2)$, $D(-3, 5)$
   b. $A(1, 1)$, $B(3, -3)$, $C(-1, -1)$, $D(-3, 3)$
3. a. $A(4, -1)$, $B(-2, -1)$, $C(-2, 2)$, $D(4, 2)$
   b. $A(-5, 0)$, $B(-3, 4)$, $C(5, 0)$, $D(3, -4)$
4. a. $A(2, 2)$, $B(2, -2)$, $C(-2, -2)$, $D(-2, 2)$
   b. $A(0, 3)$, $B(3, 0)$, $C(0, -3)$, $D(-3, 0)$
5. a. $A(-7, 0)$, $B(-4, 4)$, $C(-1, 4)$, $D(-1, 0)$
   b. $A(4, 6)$, $B(10, 3)$, $C(4, -3)$, $D(1, 3)$
6. a. $A(0, 0)$, $B(2, 4)$, $C(4, 0)$, $D(2, -2)$
   b. $A(6, 0)$, $B(3, -5)$, $C(-4, 0)$, $D(3, 5)$

Special Quadrilaterals

Objectives

1. Apply the definitions and identify the special properties of a rectangle, a rhombus, and a square.
2. Determine when a parallelogram is a rectangle, rhombus, or square.
3. Apply the definitions and identify the properties of a trapezoid and an isosceles trapezoid.

5-4 *Special Parallelograms*

In this section you will study the properties of special parallelograms: *rectangles*, *rhombuses*, and *squares*.

A **rectangle** is a quadrilateral with four right angles. Therefore, every rectangle is a parallelogram. (Why?)

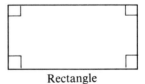

Rectangle

A **rhombus** is a quadrilateral with four congruent sides. Therefore, every rhombus is a parallelogram. (Why?)

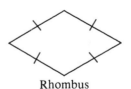

Rhombus

A **square** is a quadrilateral with four right angles and four congruent sides. Therefore, every square is a rectangle, a rhombus, and a parallelogram. (Why?)

Square

Since rectangles, rhombuses, and squares are parallelograms, they have all the properties of parallelograms. They also have the special properties given in the theorems on the next page. Proofs of these theorems are left as exercises.

Theorem 5-12

The diagonals of a rectangle are congruent.

Theorem 5-13

The diagonals of a rhombus are perpendicular.

Theorem 5-14

Each diagonal of a rhombus bisects two angles of the rhombus.

Example   Given: *ABCD* is a rhombus.
What can you conclude?

Solution   *ABCD* is a parallelogram, with all the properties of
a parallelogram. Also:
By Theorem 5-13, $\overline{AC} \perp \overline{BD}$.
By Theorem 5-14, $\overline{AC}$ bisects $\angle DAB$ and $\angle BCD$;
$\overline{BD}$ bisects $\angle ABC$ and $\angle ADC$.

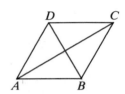

The properties of rectangles lead to the following interesting
conclusion about any right triangle.

Begin with rt. $\triangle XYZ$.
1. Draw lines to form rectangle *XZYK*. (How?)
2. Draw $\overline{ZK}$. $ZK = XY$ (Why?)
3. $\overline{ZK}$ and $\overline{XY}$ bisect each other. (Why?)
4. $MX = MY = MZ = MK$, by (2) and (3).
Since $MX = MY = MZ$, we have shown the following.

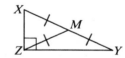

Theorem 5-15

**The midpoint of the hypotenuse of a right triangle is equidistant from the
three vertices.**

Proofs of the next two theorems will be discussed as Classroom Exercises.

Theorem 5-16

**If an angle of a parallelogram is a right angle, then the parallelogram is a
rectangle.**

Theorem 5-17

**If two consecutive sides of a parallelogram are congruent, then the parallelogram
is a rhombus.**

Classroom Exercises

1. Name each figure shown that appears to be:
   a. a parallelogram
   b. a rectangle
   c. a rhombus
   d. a square

2. Name each figure that is *both* a rectangle *and* a rhombus.

3. Name each figure that is a rectangle but not a square.

4. Name each figure that is a rhombus but not a square.

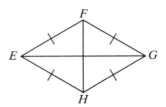

5. When you know that one angle of a parallelogram is a right angle, you can prove that the parallelogram is a rectangle. Draw a diagram and explain.

6. When you know that two consecutive sides of a parallelogram are congruent, you can prove that the parallelogram is a rhombus. Draw a diagram and explain.

7. Given: Rhombus *EFGH*
   a. *F*, being equidistant from *E* and *G*, must lie on the __?__ of $\overline{EG}$.
   b. *H*, being equidistant from *E* and *G*, must lie on the __?__ of $\overline{EG}$.
   c. From (a) and (b) you can deduce that $\overline{FH}$ is the __?__ of $\overline{EG}$.
   d. State the theorem of this section that you have just proved.

$\angle KAP$ **is a right angle, and** $\overline{AM}$ **is a median. Complete.**

8. If $MP = 6\frac{1}{2}$, then $MA = $ __?__.

9. If $MA = t$, then $KP = $ __?__.

10. If $m \angle K = 40$, then $m \angle KAM = $ __?__.

11. In the diagrams below, the red figures are formed by joining the midpoints of the sides of the quadrilaterals.
    a. What seems to be the common property of the red figures?
    b. Describe how you would prove your answer to part (a).

 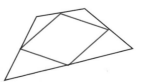 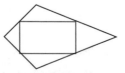

Written Exercises

Copy the chart. Then place check marks in the appropriate spaces.

		Property	Parallelogram	Rectangle	Rhombus	Square
A	**1.**	Opp. sides are ∥.				
	2.	Opp. sides are ≅.				
	3.	Opp. ⩘ are ≅.				
	4.	A diag. forms two ≅ ⧍.				
	5.	Diags. bisect each other.				
	6.	Diags. are ≅.				
	7.	Diags. are ⊥.				
	8.	A diag. bisects two ⩘.				
	9.	All ⩘ are rt. ⩘.				
	10.	All sides are ≅.				

Quad. *SLTM* is a rhombus.

11. If $m\angle 1 = 25$, find the measures of $\angle 2$, $\angle 3$, $\angle 4$, and $\angle 5$.

12. If $m\angle 1 = 3x + 8$ and $m\angle 2 = 11x - 24$, find the value of x.

13. If $m\angle 1 = 3x + 1$ and $m\angle 3 = 7x - 11$, find the value of x.

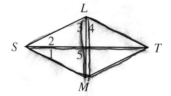

Quad. *FLAT* is a rectangle.

14. If $m\angle 1 = 18$, find the measures of $\angle 2$, $\angle 3$, and $\angle 4$.

15. If $FA = 27$, find LO.

16. If $TO = 4y + 7$ and $FA = 30$, find the value of y.

$\overline{GM}$ is a median of right $\triangle IRG$.

17. If $m\angle 1 = 32$, find the measures of $\angle 2$, $\angle 3$, and $\angle 4$.

18. If $m\angle 4 = 7x - 3$ and $m\angle 3 = 6(x + 1)$, find the value of x.

19. If $GM = 2y + 3$ and $RI = 12 - 8y$, find the value of y.

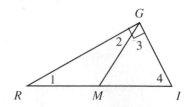

The coordinates of three vertices of a rectangle are given. Plot the points and find the coordinates of the fourth vertex. Is the rectangle a square?

20. *O*(0, 0), *P*(0, 5), *Q*(__?__, __?__), *R*(2, 0)   **21.** *A*(2, 1), *B*(4, 1), *C*(4, 5), *D*(__?__, __?__)

22. *O*(0, 0), *E*(4, 0), *F*(4, 3), *G*(__?__, __?__)   **23.** *H*(1, 3), *I*(4, 3), *J*(__?__, __?__), *K*(1, 6)

$\overline{RA}$ is an altitude of △*SAT*. *P* and *Q* are midpoints of $\overline{SA}$ and $\overline{TA}$. *SR* = 9, *RT* = 16, *QT* = 10, and *PR* = 7.5.

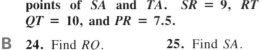

B **24.** Find *RQ*.   **25.** Find *SA*.

26. Find the perimeter of △*PQR*.

27. Find the perimeter of △*SAT*.

28. Given: ▱*ABZY*; $\overline{ZY} \cong \overline{BX}$;
              ∠1 ≅ ∠2
      Prove: *ABZY* is a rhombus.

29. Given: ▱*ABZY*; $\overline{AY} \cong \overline{BX}$
      Prove: ∠1 ≅ ∠2 and ∠1 ≅ ∠3

30. Given: Rectangle *QRST*;
              ▱*RKST*
      Prove: △*QSK* is isosceles.

31. Given: Rectangle *QRST*;
              ▱*RKST*; ▱*JQST*
      Prove: $\overline{JT} \cong \overline{KS}$

32. Prove Theorem 5-12.

33. Prove Theorem 5-14 for one diagonal of the rhombus. (Note that a proof for the other would be similar, step-by-step.)

34. Prove: If the diagonals of a parallelogram are perpendicular, then the parallelogram is a rhombus.

35. Prove: If the diagonals of a parallelogram are congruent, then the parallelogram is a rectangle.

36. a. The bisectors of the angles of ▱*ABCD* intersect to form quad. *WXYZ*. What special kind of quadrilateral is *WXYZ*?
   b. Prove your answer to part (a).

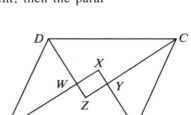

37. Draw a rectangle and bisect its angles. The bisectors intersect to form what special kind of quadrilateral?

The coordinates of three vertices of a rhombus are given, not necessarily in order. Plot the points and find the coordinates of the fourth vertex. Measure the sides to check your answer.

38. *O*(0, 0), *L*(5, 0), *D*(4, 3), *V*(__?__, __?__)   **39.** *O*(0, 0), *S*(0, 10), *E*(6, 18), *W*(__?__, __?__)

C **40. a.** Suppose that two sides of a quadrilateral are parallel and that one diagonal bisects an angle. Does that quadrilateral have to be special in other ways? If so, write a proof. If not, draw a convincing diagram.
   b. Repeat part (a) with these conditions: Suppose that two sides are parallel and that one diagonal bisects two angles of the quadrilateral.

41. Draw a regular pentagon *ABCDE*. Let *X* be the intersection of $\overline{AC}$ and $\overline{BD}$. What special kind of quadrilateral is *AXDE*? Write a paragraph proof.

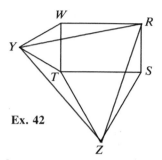

42. Given: Rectangle *RSTW*;
   equilateral △ *YWT* and *STZ*
What is true of △*RYZ*?
Write a paragraph proof.

Ex. 42

Explorations

These exploratory exercises can be done using a computer with a program that draws and measures geometric figures.

As you will learn in the next section, a *trapezoid* is a quadrilateral with exactly one pair of parallel sides.
Draw trapezoid *ABCD* with $\overline{BA} \parallel \overline{CD}$. Label the midpoints of $\overline{AD}$ and $\overline{BC}$ as *E* and *F* respectively, and draw $\overline{FE}$.
Measure ∠*BFE* and ∠*BCD*. What is true of $\overline{CD}$ and $\overline{FE}$? What postulate or theorem tells you this?
What is true of $\overline{FE}$ and $\overline{BA}$? Why?
Measure the lengths of $\overline{BA}$, $\overline{CD}$, and $\overline{FE}$. What do you notice?
Write an equation that relates *BA*, *CD*, and *FE*. Repeat the drawing and measurements until you are sure of your equation.

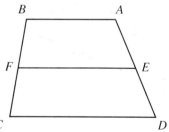

Mixed Review Exercises

Find the average of the given numbers. (The *average* is the sum of the numbers divided by the number of numbers.)

1. 17, 9 **2.** 15, 25 **3.** 18, 2, 13 **4.** 7, 8, 5, 15, 10
5. 7.9, 8.5 **6.** 4, −7 **7.** −3, 4, −7, 10 **8.** 1.7, 2.6, 9.1, 0.4

9. The numbers given are the coordinates of the endpoints of a segment on a number line. Find the coordinate of the midpoint by taking the average.
   a. 12, 34 **b.** −3, 7 **c.** 17, −9 **d.** −5, −7

5-5 *Trapezoids*

A quadrilateral with exactly one pair of parallel sides is called a **trapezoid**. The parallel sides are called the **bases**. The other sides are **legs**.

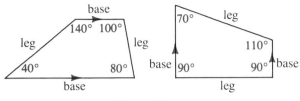

Trapezoids

A trapezoid with congruent legs is called an **isosceles trapezoid**. If you fold any isosceles trapezoid so that the legs coincide, you will find that both pairs of *base angles* are congruent.

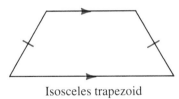

Isosceles trapezoid

A trapezoidal shape can be seen in the photograph. Is it isosceles?

Theorem 5-18

Base angles of an isosceles trapezoid are congruent.

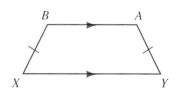

Given: Trapezoid $ABXY$ with $\overline{BX} \cong \overline{AY}$

Prove: $\angle X \cong \angle Y$; $\angle B \cong \angle A$

Plan for Proof: Note that the diagram does not contain any parallelograms or congruent triangles that might be used. To obtain such figures, you could draw auxiliary lines. You could use either diagram below to prove the theorem.

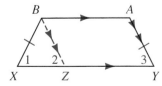

Draw $\overline{BZ} \parallel \overline{AY}$ so that $ABZY$ is a $\square$.

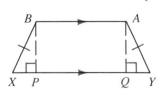

Draw $\overline{BP} \perp \overline{XY}$ and $\overline{AQ} \perp \overline{XY}$. Since $\overline{BA} \parallel \overline{XY}$, $BP = AQ$ by Theorem 5-8.

The **median** of a trapezoid is the segment that joins the midpoints of the legs. Note the difference between the median of a trapezoid and a median of a triangle.

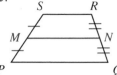

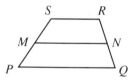

$\overline{MN}$ is *the* median of trapezoid *PQRS*. $\overline{AT}$ is *a* median of $\triangle ABC$.

Theorem 5-19

The median of a trapezoid

(1) is parallel to the bases;
(2) has a length equal to the average of the base lengths.

Given: Trapezoid *PQRS* with median $\overline{MN}$

Prove: (1) $\overline{MN} \parallel \overline{PQ}$ and $\overline{MN} \parallel \overline{SR}$ (2) $MN = \frac{1}{2}(PQ + SR)$

Plan for Proof: Again it is necessary to introduce auxiliary lines, and again there is more than one way to do this. Although any of the diagrams below could be used to prove the theorem, the proof below uses the second diagram.

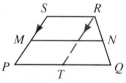

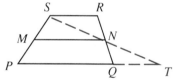

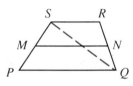

Draw $\overline{RT} \parallel \overline{SP}$.   Draw $\overrightarrow{SN}$ intersecting $\overrightarrow{PQ}$ at *T*.   Draw $\overline{SQ}$.

Proof:

Extend $\overline{SN}$ to intersect $\overrightarrow{PQ}$ at *T*. $\triangle TNQ \cong \triangle SNR$ by ASA. Then $SN = NT$ and *N* is the midpoint of $\overline{ST}$. Using Theorem 5-11 and $\triangle PST$, (1) $\overline{MN} \parallel \overline{PQ}$ (and also $\overline{MN} \parallel \overline{SR}$), and (2) $MN = \frac{1}{2}PT = \frac{1}{2}(PQ + QT) = \frac{1}{2}(PQ + SR)$, since $\overline{QT} \cong \overline{SR}$.

Example A trapezoid and its median are shown. Find the value of *x*.

Solution
$$10 = \frac{1}{2}[(2x - 4) + (x - 3)]$$
$$20 = (2x - 4) + (x - 3)$$
$$20 = 3x - 7$$
$$27 = 3x$$
$$9 = x$$

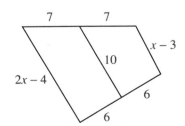

Classroom Exercises

X, Y, and *Z* are midpoints of the sides of isosceles △*ABC*.

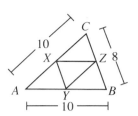

1. Explain why *XZBA* is a trapezoid.
2. Name two trapezoids other than *XZBA*.
3. Name an isosceles trapezoid and find its perimeter.

Find the length of the median of each trapezoid.

4.
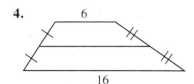

5.
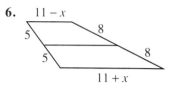

6.

Draw the trapezoid described. If such a trapezoid cannot be drawn, explain why not.

7. with two right angles
8. with both bases shorter than the legs
9. with congruent bases
10. with three acute angles

11. Draw a quadrilateral such that exactly three sides are congruent and two pairs of angles are congruent.

Written Exercises

Each diagram shows a trapezoid and its median. Find the value of *x*.

A **1.**
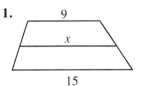

2.
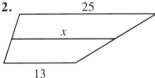

3.
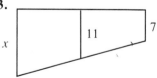

4.
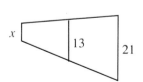

5.
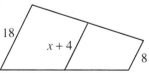

6.
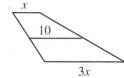

7.
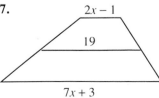

8.
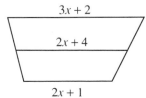

9.
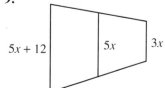

10. One angle of an isosceles trapezoid has measure 57. Find the measures of the other angles.

11. Two congruent angles of an isosceles trapezoid have measures $3x + 10$ and $5x - 10$. Find the value of x and then give the measures of all angles of the trapezoid.

In Exercises 12–20, $TA = AB = BC$ and $TD = DE = EF$.

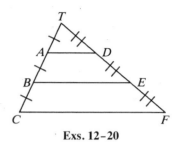

Exs. 12–20

12. Write an equation that relates AD and BE. (*Hint:* Think of $\triangle TBE$.)

13. Write an equation that relates AD, BE, and CF. (*Hint:* Think of trapezoid $CFDA$.)

14. If $AD = 7$, then $BE = \underline{\ ?\ }$ and $CF = \underline{\ ?\ }$.

15. If $BE = 26$, then $AD = \underline{\ ?\ }$ and $CF = \underline{\ ?\ }$.

16. If $AD = x$ and $BE = x + 6$, then $x = \underline{\ ?\ }$ and $CF = \underline{\ ?\ }$ (numerical answers).

B 17. If $AD = x + 3$, $BE = x + y$, and $CF = 36$, then $x = \underline{\ ?\ }$ and $y = \underline{\ ?\ }$.

18. If $AD = x + y$, $BE = 20$, and $CF = 4x - y$, then $CF = \underline{\ ?\ }$ (numerical answer).

19. Tony makes up a problem for the figure, setting $AD = 5$ and $CF = 17$. Katie says, "You can't do that." Explain.

20. Mike makes up a problem for the figure, setting $AD = 2x + 1$, $BE = 4x + 2$, and $CF = 6x + 3$ and asking for the value of x. Katie says, "Anybody can do that problem." Explain.

39

Draw a quadrilateral of the type named. Join, in order, the midpoints of the sides. What special kind of quadrilateral do you appear to get?

21. rhombus
22. rectangle
23. isosceles trapezoid
24. non-isosceles trapezoid
25. quadrilateral with no congruent sides

26. Carefully draw an isosceles trapezoid and measure its diagonals. What do you discover? Write a proof of your discovery.

27. Prove Theorem 5-18.

A *kite* is a quadrilateral that has two pairs of congruent sides, but opposite sides are not congruent.

28. Draw a convex kite. Discover, state, and prove whatever you can about the diagonals and angles of a kite.

29. **a.** Draw a convex kite. Join, in order, the midpoints of the sides. What special kind of quadrilateral do you appear to get?
    b. Repeat part (a), but draw a nonconvex kite.

ABCD is a trapezoid with median $\overline{MN}$.

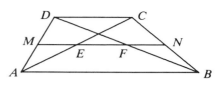

30. If *DC* = 6 and *AB* = 16, find *ME*, *FN*, and *EF*.

31. Prove that $EF = \frac{1}{2}(AB - DC)$.

32. If *DC* = 3*x*, *AB* = $2x^2$, and *EF* = 7, find the value of *x*.

C **33.** $\overline{VE}$ and $\overline{FG}$ are congruent. *J*, *K*, *L*, and *M* are the midpoints of $\overline{EF}$, $\overline{VF}$, $\overline{VG}$, and $\overline{EG}$. What name best describes *JKLM*? Explain.

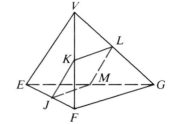

34. When the midpoints of the sides of quad. *ABCD* are joined, rectangle *PQRS* is formed.
   a. Draw other quadrilaterals *ABCD* with this property.
   b. What must be true of quad. *ABCD* if *PQRS* is to be a rectangle?

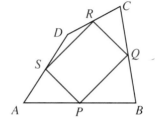

35. When the midpoints of the quad. *ABCD* are joined, rhombus *PQRS* is formed.
   a. Draw other quadrilaterals *ABCD* with this property.
   b. What must be true of quad. *ABCD* if *PQRS* is to be a rhombus?

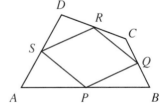

36. *P*, *Q*, *R*, and *S* are the midpoints of the sides of quad. *ABCD*. In this diagram $\overline{PR}$ and $\overline{SQ}$ have the same midpoint, point *O*. If you think this will be the case for *any* quad. *ABCD*, prove it. If not, tell what other information you need to know about quad. *ABCD* before you can conclude that $\overline{PR}$ and $\overline{SQ}$ have the same midpoint.

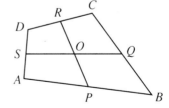

Challenge

The three-dimensional figure shown has six congruent edges. Draw four such figures. On your diagrams show how a plane can intersect the figure to form **(a)** a triangle with three congruent sides, **(b)** a triangle with sides not all congruent, **(c)** a rectangle, and **(d)** an isosceles trapezoid.

Explorations

These exploratory exercises can be done using a computer with a program that draws and measures geometric figures.

Draw any $\triangle ABC$. Label the midpoint of $\overline{AB}$ as D, of $\overline{AC}$ as E, and of $\overline{BC}$ as F.

Form a quadrilateral (*ABFE*, *BCED*, or *CADF*) by using two midpoints and two vertices.

What kind of quadrilateral is each of *ABFE*, *BCED*, and *CADF*? How do you know?

Form a quadrilateral (*ADFE*, *BFED*, or *CEDF*) by using three midpoints and a vertex.

What kind of quadrilateral is each of *ADFE*, *BFED*, and *CEDF*? How do you know?

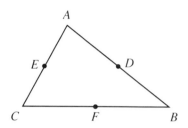

Self-Test 2

Quad. *WXYZ* must be a special figure to meet the conditions stated. Write the best name for that special quadrilateral.

1. $\overline{WX} \cong \overline{YZ}$ and $\overline{WX} \parallel \overline{YZ}$ 2. $\overline{WX} \parallel \overline{YZ}$ and $\overline{WX} \not\cong \overline{YZ}$

3. $\overline{WX} \cong \overline{YZ}$, $\overline{XY} \cong \overline{ZW}$, and diag. $\overline{WY} \cong$ diag. $\overline{XZ}$

4. Diagonals $\overline{WY}$ and $\overline{XZ}$ are congruent and are perpendicular bisectors of each other.

5. An isosceles trapezoid has sides of lengths 5, 8, 5, and 14. Find the length of the median.

6. *M* is the midpoint of hypotenuse $\overline{AB}$. Find *AM* and $m\angle ACM$.

7. Given: $\angle 1 \cong \angle 2 \cong \angle 3 \cong \angle 4$
   Prove: *EFGH* is a rhombus.

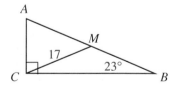

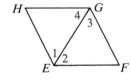

8. *PQRS* is a $\square$.
   a. If *X* is the midpoint of $\overline{PQ}$ and *Y* is the midpoint of $\overline{SR}$, what special kind of quadrilateral is *XQRY*?
   b. Prove your answer to part (a).
   c. Draw a line through *O* intersecting $\overline{PQ}$ at *J* and $\overline{SR}$ at *K*. If *J* and *K* are not midpoints, what special kind of quadrilateral is *JQRK*?

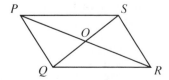

| Application | *Rhombuses* |

Many objects that need to change in size or shape are built in the shape of a rhombus. What makes this shape so useful is that if you keep the lengths of the sides the same, opposite sides remain parallel as you change the measures of the angles. A rhombus also has the property that as you change the measures of the angles, the vertices slide along the lines that contain the diagonals and the diagonals remain perpendicular. Two applications of this property are illustrated in the photographs below.

A rhombic shape can be used to support weight when the height of the object changes but the load must remain balanced. To use the jack shown above, you turn a crank. This brings the two hinges on the horizontal diagonal closer together and forces the hinges on the vertical diagonal farther apart, which lifts the car. Could a jack be in the shape of a parallelogram that is not a rhombus? Would a jack in the shape of a kite work?

The folding elevator gate shown at the right changes in width, but the vertical bars remain vertical. Some types of fireplace tongs can extend and retract in a similar way. Electric trains sometimes use rhombic arrangements to maintain contact with overhead wires even when the distance between the top of the coach and the wire changes. Can you think of other objects that use this property?

Chapter Summary

1. A parallelogram has these properties:
   a. Opposite sides are parallel.
   b. Opposite sides are congruent.
   c. Opposite angles are congruent.
   d. Diagonals bisect each other.

2. The chart on page 172 lists five ways to prove that a quadrilateral is a parallelogram.

3. If two lines are parallel, then all points on one line are equidistant from the other line.

4. If three parallel lines cut off congruent segments on one transversal, then they cut off congruent segments on every transversal.

5. A line that contains the midpoint of one side of a triangle and is parallel to another side bisects the third side.

6. The segment that joins the midpoints of two sides of a triangle is parallel to the third side and has a length equal to half the length of the third side.

7. The midpoint of the hypotenuse of a right triangle is equidistant from all three vertices.

8. Rectangles, rhombuses, and squares are parallelograms with additional properties. Trapezoids and kites are not parallelograms, but are special quadrilaterals with additional properties.

9. The median of a trapezoid is parallel to the bases and has a length equal to half the sum of the lengths of the bases.

Chapter Review

In parallelogram *EFGH*, *m* ∠ *EFG* = 70.

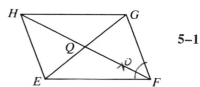

1. $m \angle HEF = \underline{\quad?\quad}$
2. If $m \angle EFH = 32$, then $m \angle EHF = \underline{\quad?\quad}$.
3. If $HQ = 14$, then $HF = \underline{\quad?\quad}$.
4. If $EH = 8x - 7$ and $FG = 5x + 11$, then $x = \underline{\quad?\quad}$.

5–1

In each exercise you could prove that quad. *SANG* is a parallelogram if one more fact, in addition to those stated, were given. State that fact.

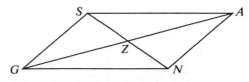

5. $GN = 9$; $NA = 5$; $SA = 9$
6. $\angle ASG \cong \angle GNA$
7. $\overline{SZ} \cong \overline{NZ}$
8. $\overline{SA} \parallel \overline{GN}$; $SA = 17$

5–2

State the principal theorem that justifies the statement about the diagram.

9. If $\overline{DE} \parallel \overline{BC}$, then D is the midpoint of $\overline{AB}$.

10. If D is the midpoint of $\overline{AB}$, then $\overline{DE} \parallel \overline{BC}$.

11. If D is the midpoint of $\overline{AB}$, then $DE = 6$.

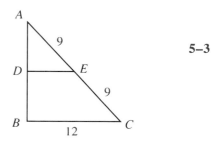

5–3

12. Given: $\square CDEF$; S and T are the midpoints of $\overline{EF}$ and $\overline{ED}$.

Prove: $\overline{SR} \cong \overline{FD}$

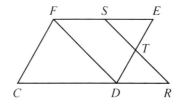

Give the most descriptive name for quad. *MNOP*.

13. $\overline{MN} \cong \overline{PO}$; $\overline{MN} \parallel \overline{PO}$

5–4

14. $\overline{MN} \parallel \overline{PO}$; $\overline{NO} \parallel \overline{MP}$; $\overline{MO} \perp \overline{NP}$

15. $\angle M \cong \angle N \cong \angle O \cong \angle P$

16. *MNOP* is a rectangle with $MN = NO$.

17. Given: *ABCD* is a rhombus;
$\qquad$ $DE = BF$
$\quad$ **Prove:** *AECF* is a rhombus.

Draw and label a diagram. List, in terms of the diagram, what is given and what is to be proved. Then write a proof.

18. $\overline{PX}$ and $\overline{QY}$ are altitudes of acute $\triangle PQR$, and Z is the midpoint of $\overline{PQ}$. Prove that $\triangle XYZ$ is isosceles.

$\overline{MN}$ **is the median of trapezoid *ZOID*.**

19. The bases of trap. *ZOID* are __?__ and __?__.

5–5

20. If $ZO = 8$ and $MN = 11$, then $DI = $ __?__.

21. If $ZO = 8$, then $TN = $ __?__.

22. If trap. *ZOID* is isosceles and $m \angle D = 80$, then $m \angle O = $ __?__.

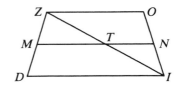

Chapter Test

Complete each statement with the word *always*, *sometimes*, or *never*.

1. A square is __?__ a rectangle.

2. A rectangle is __?__ a rhombus.

3. A rhombus is __?__ a square.

4. A rhombus is __?__ a parallelogram.

5. A trapezoid __?__ has three congruent sides.

6. The diagonals of a trapezoid __?__ bisect each other.

7. The diagonals of a rectangle are __?__ congruent.

8. The diagonals of a parallelogram __?__ bisect the angles.

Trapezoid *ABCD* has median $\overline{MN}$.

9. If $DC = 42$ and $MN = 35$, then $AB = $ __?__ .

10. If $FC = 9$, then $EN = $ __?__ .

11. If $AB = 5j + 7k$ and $DC = 9j - 3k$, then $MN = $ __?__ .

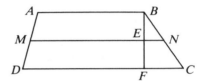

Can you deduce from the given information that quad. *ABCD* is a parallelogram? If so, what theorem can you use?

12. $\angle ADC \cong \angle CBA$ and $\angle BAD \cong \angle DCB$

13. $\overline{AD} \parallel \overline{BC}$ and $\overline{AD} \cong \overline{BC}$

14. $AT = CT$ and $DT = \frac{1}{2}DB$

15. $\overline{AB}, \overline{BC}, \overline{CD}$, and $\overline{DA}$ are all congruent.

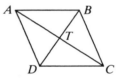

16. $\overline{RE}$ is an altitude of $\triangle RST$.
Find *MN*, *NE*, and *RT*.

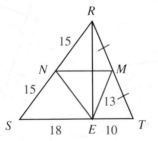

17. $l \parallel m \parallel n$
Find the values of *x*, *y*, and *z*.

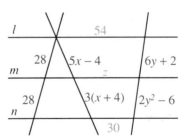

18. Given: $\square PQRS$; $PA = RB$
Prove: $AS = BQ$

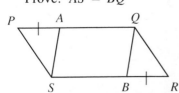

19. Given: $\overline{PR} \parallel \overline{VO}$; $\overline{RO} \parallel \overline{PV}$; $\overline{PR} \cong \overline{RO}$
Prove: $\angle 1$ and $\angle 2$ are complementary.

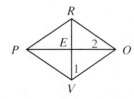

Cumulative Review: Chapters 1–5

A 1. Given two parallel lines n and k, how many planes contain n and k?

2. **a.** Is it possible for two lines to be neither intersecting nor parallel? If so, what are the lines called?
   b. Repeat part (a), replacing *lines* with *planes*.

3. Write the converse of the statement: If you are a member of the skiing club, then you enjoy winter weather.

4. On a number line, point A has a coordinate -5 and B has a coordinate 3. Find the coordinate of the midpoint of $\overline{AB}$.

5. Name the property that justifies the statement: If $\angle 1 \cong \angle 2$ and $\angle 2 \cong \angle 3$, then $\angle 1 \cong \angle 3$.

In Exercises 6–10, complete each statement about the diagram. Then state the definition, postulate, or theorem that justifies your answer.

6. $m \angle 1 + m \angle 2 + m \angle 3 = \underline{\quad ? \quad}$

7. $m \angle 1 + m \angle 4 = \underline{\quad ? \quad}$

8. $m \angle 1 + m \angle 2 = m \angle \underline{\quad ? \quad}$

9. If $\overleftrightarrow{EC} \parallel \overleftrightarrow{BD}$, then $\angle 7 \cong \underline{\quad ? \quad}$.

10. If $\angle 2 \cong \angle 3$, then $\overline{EC} \cong \underline{\quad ? \quad}$.

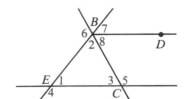

Complete each statement.

11. The median to the base of an isosceles triangle $\underline{\quad ? \quad}$ the vertex angle and is $\underline{\quad ? \quad}$ to the base.

12. **a.** If a point lies on the perpendicular bisector of $\overline{AB}$, then the point is equidistant from $\underline{\quad ? \quad}$.
    b. If a point lies on the bisector of $\angle RST$, then the point is equidistant from $\underline{\quad ? \quad}$.

13. Suppose $\triangle ART \cong \triangle DEB$.
    a. $\triangle EBD \cong \underline{\quad ? \quad}$ **b.** $\overline{AT} \cong \underline{\quad ? \quad}$ **c.** $m \angle R = \underline{\quad ? \quad}$

14. If a regular polygon has 40 sides, the measure of each interior angle is $\underline{\quad ? \quad}$.

15. When two parallel lines are cut by a transversal, a pair of corresponding angles have measures $2x + 50$ and $3x$. The measures of the angles are $\underline{\quad ? \quad}$ and $\underline{\quad ? \quad}$.

B 16. In $\triangle SUN$, $\angle S \cong \angle N$. Given that $SU = 2x + 7$, $UN = 4x - 1$, and $SN = 3x + 4$, find the numerical length of each side.

17. M and N are the midpoints of the legs of trapezoid $EFGH$. If bases $\overline{EF}$ and $\overline{HG}$ have lengths $2r + s$ and $4r - 3s$, express the length of $\overline{MN}$ in terms of r and s.

Given: *ABCD* is a parallelogram;
$\overline{AD} \cong \overline{AC}$; $\overline{AE} \cong \overline{EC}$
$\angle ADF \cong \angle CDF$; $m \angle DAC = 36$

Complete each statement about the diagram.

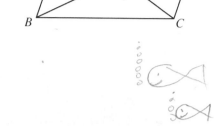

18. $\overline{AE} \cong \overline{EC}$, so $\overline{BE}$ is a(n) ___?___ of
$\triangle ABC$.

19. $\angle ADF \cong \angle CDF$, so $\overline{DF}$ is a(n) ___?___ of
$\angle ADC$.

20. $\triangle ADC$ is a(n) ___?___ triangle.

21. $m \angle DAC = 36$, so $m \angle ADC =$ ___?___ and
$m \angle ADF =$ ___?___.

22. $\triangle ADF$ is a(n) ___?___ triangle.

23. $\angle ADC \cong \angle$ ___?___ $\cong \angle$ ___?___ $\cong \angle$ ___?___ $\cong \angle$ ___?___.

In the diagram, $m \angle VOZ = 90$.
$\overline{OW}$ is an altitude of $\triangle VOZ$.
$\overline{OX}$ bisects $\angle VOZ$.
$\overline{OY}$ is a median of $\triangle VOZ$.
Find the measures of the four numbered angles.

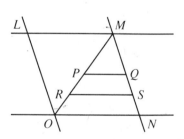

24. $m \angle Z = 30$ **25.** $m \angle Z = k$

**In Exercises 26–29, complete each statement about the diagram. Then state
the definition, postulate, or theorem that justifies your answer.**

26. If $LM = ON$ and $LO = MN$, then $LMNO$ is
a ___?___.

27. If $LMNO$ is a rhombus, then $\angle LOM \cong$ ___?___ $\cong$
___?___ $\cong$ ___?___.

28. If $MP = PO$ and $\overline{PQ} \parallel \overline{ON}$, then Q is the ___?___
of ___?___.

29. If $\overline{PQ} \parallel \overline{ON}$, $PR = RO$ and $QS = SN$, then
$RS = \frac{1}{2}($ ___?___ $+$ ___?___ $)$.

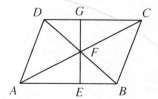

30. Given: $WP = ZP$; $PY = PX$
Prove: $\angle WXY \cong \angle ZYX$

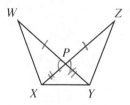

31. Given: $\overline{AD} \cong \overline{BC}$; $\overline{AD} \parallel \overline{BC}$
Prove: $\overline{EF} \cong \overline{FG}$

6 INEQUALITIES IN GEOMETRY

Crystallographers can easily identify sulfur crystals by the characteristic faces and angles produced by the symmetrical arrangement of the atoms.

Inequalities and Indirect Proof

Objectives

1. Apply properties of inequality to positive numbers, lengths of segments, and measures of angles.
2. State the contrapositive and inverse of an if-then statement.
3. Understand the relationship between logically equivalent statements.
4. Draw correct conclusions from given statements.
5. Write indirect proofs in paragraph form.

6-1 *Inequalities*

Our geometry up until now has emphasized congruent segments and angles, and the triangles and polygons they form. To deal with segments whose lengths are equal and angles whose measures are equal, you have used properties of equality taken from algebra. They are stated on page 37. In this chapter you will work with segments having unequal lengths and angles having unequal measures. You will use properties of inequality taken from algebra.

Example 1   Complete each conclusion by inserting one of the symbols $<$, $=$, or $>$.

a. Given: $AC > AB$; $AB > BC$
   Conclusion: $AC \underline{\quad?\quad} BC$

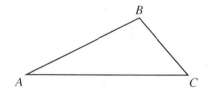

b. Given: $m \angle BAC + m \angle CAD = m \angle BAD$
   Conclusion: $m \angle BAD \underline{\quad?\quad} m \angle BAC$;
   $m \angle BAD \underline{\quad?\quad} m \angle CAD$

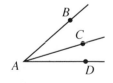

Solution   a. $AC > BC$ (Equivalently, $BC < AC$.)

b. $m \angle BAD > m \angle BAC$ (Equivalently, $m \angle BAC < m \angle BAD$.)
   $m \angle BAD > m \angle CAD$ (Equivalently, $m \angle CAD < m \angle BAD$.)

The properties of inequality you will use most often in geometry are stated on the following page. When you use any one of them in a proof, you can write as your reason *A Prop. of Ineq.* Can you see which properties were used in Example 1?

Properties of Inequality

If $a > b$ and $c \geq d$, then $a + c > b + d$.

If $a > b$ and $c > 0$, then $ac > bc$ and $\dfrac{a}{c} > \dfrac{b}{c}$.

If $a > b$ and $c < 0$, then $ac < bc$ and $\dfrac{a}{c} < \dfrac{b}{c}$.

If $a > b$ and $b > c$, then $a > c$.
If $a = b + c$ and $c > 0$, then $a > b$.

Example 2

Given: $AC > BC$; $CE > CD$

Prove: $AE > BD$

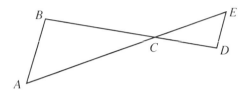

Proof:

Statements	Reasons
1. $AC > BC$; $CE > CD$	1. Given
2. $AC + CE > BC + CD$	2. A Prop. of Ineq.
3. $AC + CE = AE$; $BC + CD = BD$	3. Segment Addition Postulate
4. $AE > BD$	4. Substitution Prop.

Example 3

Given: $\angle 1$ is an exterior angle of $\triangle DEF$.

Prove: $m\angle 1 > m\angle D$;
$\qquad m\angle 1 > m\angle E$

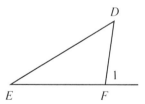

Proof:

Statements	Reasons
1. $m\angle 1 = m\angle D + m\angle E$	1. The measure of an ext. $\angle$ of a $\triangle$ equals the sum of the measures of the two remote int. $\angle\!s$.
2. $m\angle 1 > m\angle D$; $m\angle 1 > m\angle E$	2. A Prop. of Ineq.

Example 3 above proves the following theorem.

Theorem 6-1 *The Exterior Angle Inequality Theorem*
The measure of an exterior angle of a triangle is greater than the measure of either remote interior angle.

Classroom Exercises

Classify each conditional as true or false.

1. If $3a > 9$, then $a > 27$. **2.** If $4b > 20$, then $b > 5$.

3. If $x > 4$, then $x + 1 > 5$. **4.** If $x + 1 > 5$, then $x > 4$.

5. If $c - 5 > 45$, then $c > 48$. **6.** If $a + b = n$ and $c > b$, then $a + c > n$.

7. If $y > 18$, then $y > 20$. **8.** If $y > 20$, then $y > 18$.

9. If $a > 5$ and $5 > b$, then $a > b$. **10.** If $d > e$ and $f > e$, then $d > f$.

11. If $g > h$ and $j = h$, then $g > j$. **12.** If $p = q + 6$, then $p > q$.

13. If $c > d$ and $e = f$, then $c + e = d + f$.

14. If $g > h$ and $i > j$, then $g + h > i + j$.

15. If $k > l$ and $m > n$, then $k + m > l + n$.

16. If $a > b$, then $100 - a > 100 - b$.

Complete each statement by writing $<$, $=$, or $>$.

17. X   Y   Z

a. $XZ \underline{\quad ? \quad} XY + YZ$
b. $XZ \underline{\quad ? \quad} XY$
c. $XZ \underline{\quad ? \quad} YZ$

18.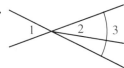

a. $m \angle 1 \underline{\quad ? \quad} m \angle 3$
b. $m \angle 2 \underline{\quad ? \quad} m \angle 3$
c. $m \angle 1 \underline{\quad ? \quad} m \angle 2$

19.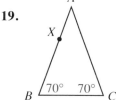

a. $AB \underline{\quad ? \quad} AC$
b. $AB \underline{\quad ? \quad} AX + XB$
c. $AB \underline{\quad ? \quad} XB$
d. $AC \underline{\quad ? \quad} XB$

20. Supply reasons to complete the proof.

Given: $m \angle 2 > m \angle 1$

Prove: $m \angle 2 > m \angle 4$

Proof:

Statements	Reasons
1. $m \angle 2 > m \angle 1$	1. $\underline{\ ?\ }$
2. $m \angle 1 > m \angle 3$	2. $\underline{\ ?\ }$
3. $m \angle 2 > m \angle 3$	3. $\underline{\ ?\ }$
4. $\angle 3 \cong \angle 4$, or $m \angle 3 = m \angle 4$	4. $\underline{\ ?\ }$
5. $m \angle 2 > m \angle 4$	5. $\underline{\ ?\ }$

Written Exercises

Some information about the diagram is given. Tell whether the other statements can be deduced from what is given. (Write *yes* or *no*.)

A **1.** Given: Point Y lies between points X and Z.

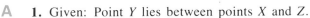

   a. $XY = \frac{1}{2}XZ$ **b.** $XZ = XY + YZ$
   c. $XZ > XY$ **d.** $YZ > XY$
   e. $XZ > YZ$ **f.** $XZ > 2XY$

2. Given: Point B lies in the interior of $\angle AOC$.

   a. $m\angle 1 = m\angle 2$ **b.** $m\angle AOC = m\angle 1 + m\angle 2$
   c. $m\angle AOC > m\angle 1$ **d.** $m\angle AOC > m\angle 2$
   e. $m\angle 1 > m\angle 2$ **f.** $m\angle AOC > 90$

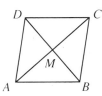

3. Given: $\square ABCD$; $AC > BD$

   a. $AB > AD$ **b.** $AM > MC$
   c. $DM = MB$ **d.** $AM > MB$

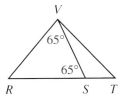

4. Given: $m\angle RVS = m\angle RSV = 65$

   a. $RT > RS$ **b.** $RT > RV$
   c. $RS > ST$ **d.** $VT < RS$

5. When some people are given that $j > k$ and $l > m$, they carelessly conclude that $j + k > l + m$. Find values for $j, k, l,$ and m that show this conclusion is false.

Write the reasons that justify the statements.

6.

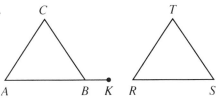

Given: $\triangle ABC \cong \triangle RST$
Prove: $AK > RS$

Statements of proof:

1. $\triangle ABC \cong \triangle RST$
2. $\overline{AB} \cong \overline{RS}$, or $AB = RS$
3. $AK = AB + BK$
4. $AK > AB$
5. $AK > RS$

7.

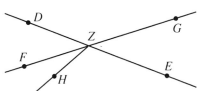

Given: $\overleftrightarrow{DE}$, $\overleftrightarrow{FG}$ and $\overrightarrow{ZH}$ contain point Z.
Prove: $m\angle DZH > m\angle GZE$

Statements of proof:

1. $\angle DZF \cong \angle GZE$,
   or $m\angle DZF = m\angle GZE$
2. $m\angle DZH = m\angle DZF + m\angle FZH$
3. $m\angle DZH > m\angle DZF$
4. $m\angle DZH > m\angle GZE$

Write proofs in two-column form.

B   **8.** Given: $KL > NL$; $LM > LP$
Prove: $KM > NP$

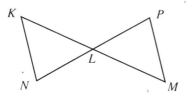

9. Given: $m \angle ROS > m \angle TOV$
Prove: $m \angle ROT > m \angle SOV$

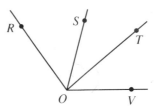

10. Given: $\overline{VY} \perp \overline{YZ}$
Prove: $\angle VXZ$ is an obtuse angle.

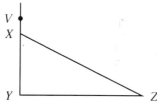

11. Given: The diagram
Prove: $m \angle 1 > m \angle 4$

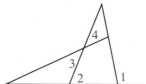

12. Given: $\overline{QR}$ and $\overline{ST}$ bisect each other.
Prove: $m \angle XRT > m \angle S$

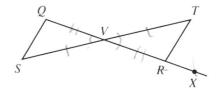

C   **13.** Given: Point K lies inside $\triangle ABC$.
Prove: $m \angle K > m \angle C$

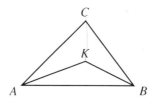

Challenge

A cube with sides n cm long is painted on all faces. It is then cut into cubes with sides 1 cm long. If $n = 4$, as the diagram at the right illustrates, how many of these smaller cubes will have paint on

a. 3 surfaces?   **b.** 2 surfaces?
c. 1 surface?   **d.** 0 surfaces?

Answer the questions for any positive integer n.

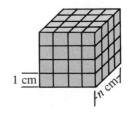

6-2 *Inverses and Contrapositives*

You have already studied the converse of an if-then statement. Now we consider two other related conditionals called the *inverse* and the *contrapositive*.

Statement:	If p, then q.
Inverse:	If not p, then not q.
Contrapositive:	If not q, then not p.

Example   Write (**a**) the inverse and (**b**) the contrapositive of the true conditional:
If two lines are not coplanar, then they do not intersect.

Solution   **a.** Inverse: If two lines are coplanar, then they intersect. (False)
b. Contrapositive: If two lines intersect, then they are coplanar. (True)

As you can see, the inverse of a true conditional is *not* necessarily true.

You can use a **Venn diagram** to represent a conditional. Since any point inside circle p is also inside circle q, this diagram represents "If p, then q." Similarly, if a point is *not* inside circle q, then it *can't* be inside circle p. Therefore, the same diagram also represents "If not q, then not p." Since the same diagram represents both a conditional and its contrapositive, these statements are either both true or both false. They are called **logically equivalent** statements.

Since a conditional and its contrapositive are logically equivalent, you may prove a conditional by proving its contrapositive. Sometimes this is easier, as you will see in Written Exercises 21 and 22.

The Venn diagram at the right represents both the converse "If q, then p" and the inverse "If not p, then not q." Therefore, the converse and the inverse of a conditional are also logically equivalent statements.

Summary of Related If-Then Statements

Given statement:	If p, then q.
Contrapositive:	If not q, then not p.
Converse:	If q, then p.
Inverse:	If not p, then not q.

A statement and its contrapositive are logically equivalent.
A statement is *not* logically equivalent to its converse or to its inverse.

Using a Venn diagram to illustrate a conditional statement can also help you determine whether an argument leads to a valid conclusion.

Suppose this conditional is true:

All runners are athletes.

(If a person is a runner, then that person is an athlete.)

What can you conclude from each additional statement?
1. Leroy is a runner.
2. Lucia is not an athlete.
3. Linda is an athlete.
4. Larry is not a runner.

The conditional is paired with the four different statements as shown below.

1. *Given:* If *p*, then *q*;   All runners are athletes.
           p   Leroy is a runner.
    Conclusion: *q*   Leroy is an athlete.

2. *Given:* If *p*, then *q*;   All runners are athletes.
           not *q*   Lucia is not an athlete.
    Conclusion: not *p*   Lucia is not a runner.

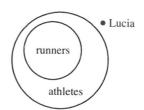

3. *Given:* If *p*, then *q*;   All runners are athletes.
           q   Linda is an athlete.
    No conclusion follows.   Linda might be a runner
                                or she might not be.

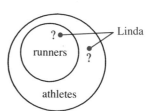

4. *Given:* If *p*, then *q*;   All runners are athletes.
           not *p*   Larry is not a runner.
    No conclusion follows.   Larry might be an athlete
                                or he might not be.

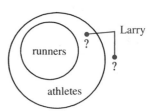

Classroom Exercises

1. State the contrapositive of each statement.
    a. If I can sing, then you can dance.
    b. If you can't play baseball, then I can't ride a horse.
    c. If $x = 4$, then $x^2 - 5 = 11$.
    d. If $y < 3$, then $y \neq 4$.
    e. If a polygon is a triangle, then the sum of the measures of its angles is 180.

2. State the inverse of each statement in Exercise 1.

3. A certain conditional is true. Must its converse be true? Must its inverse be true? Must its contrapositive be true?

4. A certain conditional is false. Must its converse be false? Must its inverse be false? Must its contrapositive be false?

Classify each conditional as true or false. Then state its inverse and contra-positive, and classify each of these as true or false.

5. If a triangle is equilateral, then it is equiangular.

6. If $\angle A$ is acute, then $m \angle A \neq 100$.

7. If a triangle is not isosceles, then it is not equilateral.

8. If two planes do not intersect, then they are parallel.

Express each statement in if-then form.

9. All squares are rhombuses.

10. No trapezoids are equiangular.

11. All marathoners have stamina.

12. Suppose "All marathoners have stamina" is a true conditional. What, if anything, can you conclude from each additional statement? If no conclusion is possible, say so.
    a. Nick is a marathoner.
    b. Heidi has stamina.
    c. Mimi does not have stamina.
    d. Arlo is not a marathoner.

Written Exercises

Write (a) the contrapositive and (b) the inverse of each statement.

A 1. If $n = 17$, then $4n = 68$.

2. If those are red and white, then this is blue.

3. If x is not even, then $x + 1$ is not odd.

4. If Abby is not here, then she is not well.

For each statement in Exercises 5–10 copy and complete a table like the one shown below.

If ___?___, then ___?___.   True/False

Statement	?	?
Contrapositive	?	?
Converse	?	?
Inverse	?	?

5. If I live in Los Angeles, then I live in California.
6. If $\angle 1$ and $\angle 2$ are vertical angles, then $m\angle 1 = m\angle 2$.
7. If $AM = MB$, then M is the midpoint of $\overline{AB}$.
8. If a triangle is scalene, then it has no congruent sides.

B 9. If $-2n < 6$, then $n > -3$.
10. If $x^2 > 1$, then $x > 1$.

Reword the given statement in if-then form and illustrate it with a Venn diagram. What can you conclude by using the given statement together with each additional statement? If no conclusion is possible, say so.

11. Given: All senators are at least 30 years old.
    a. Jose Avila is 48 years old.
    b. Rebecca Castelloe is a senator.
    c. Constance Brown is not a senator.
    d. Ling Chen is 29 years old.

12. Given: Math teachers assign hours of homework.
    a. Bridget Sullivan is a math teacher.
    b. August Campos assigns hours of homework.
    c. Andrew Byrnes assigns no homework at all.
    d. Jason Babler is not a math teacher.

What can you conclude by using the given statement together with each additional statement? If no conclusion is possible, say so.

13. Given: If it is not raining, then I am happy.
    a. I am not happy.
    b. It is not raining.
    c. I am overjoyed.
    d. It is raining.

14. Given: All my students love geometry.
    a. Stu is my student.
    b. Luis loves geometry.
    c. Stella is not my student.
    d. George does not love geometry.

15. Given: If two angles are vertical angles, then they are congruent.
    a. $\angle 1 \cong \angle 2$
    b. $m\angle ABC \neq m\angle DBF$
    c. $\angle 3$ and $\angle 4$ are adjacent angles.
    d. $\overline{RS}$ and $\overline{TU}$ intersect at V.

What can you conclude by using the given statement together with each additional statement? If no conclusion is possible, say so.

16. Given: The diagonals of a rhombus are perpendicular.
    a. *JKLM* is a rhombus.   **b.** In quad. *DIME*, $\overline{DM} \perp \overline{IE}$.
    c. *STUV* is not a rhombus.   **d.** In quad. *NOPQ*, $\overline{NP} \not\perp \overline{OQ}$.

17. Given: The diagonals of a rectangle are congruent.
    a. *PQRS* is a rectangle.   **b.** In quad. *ABCD*, $AC = BD$.
    c. *WXYZ* is not a rectangle.   **d.** In quad. *STAR*, $SA > TR$.

18. Given: Every square is a rhombus.
    a. *ABCD* is a rhombus.   **b.** In quad. *LAST*, $LA \neq LT$.
    c. *PQRS* is a square.   **d.** *GHIJ* is not a square.

C   **19.** What simpler name can be used for the converse of the inverse of a conditional?

     20. Write the contrapositive of the converse of the inverse of the conditional: If *r*, then *s*.

Prove each of the following statements by proving its contrapositive. Begin by writing what is given and what is to be proved.

21. If $m \angle A + m \angle B \neq 180$, then $m \angle D + m \angle C \neq 180$.

22. If n^2 is not a multiple of 3, then *n* is not a multiple of 3.

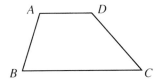

Mixed Review Exercises

Complete each statement with the word *always*, *sometimes*, or *never*.

1. Two lines that do not intersect are __?__ parallel.

2. Two lines parallel to the same plane __?__ intersect.

3. The diagonals of a parallelogram __?__ bisect each other.

4. An acute triangle is __?__ a right triangle.

5. Two lines parallel to a third line are __?__ parallel.

6. A square is __?__ a rectangle.

7. An altitude of a triangle is __?__ a median.

8. Find the measures of $\angle 1$, $\angle 2$, $\angle 3$, and $\angle 4$ in the figure shown.

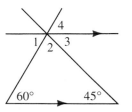

9. Find the value of *x*.

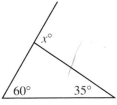

Career

Cartographer

When you think of a map, do you think of a piece of paper with colored areas and lines? Surprisingly, some maps now consist of thousands, or even millions, of numbers stored on computer tapes. Obviously *cartography*, or map-making, is changing.

Several technical advances have led to changes in mapping. Space satellites carrying scanners produce extremely detailed images of the entire world at regular intervals. Besides conventional photographs, these scanners also record images using infrared and other wavelengths beyond the range of visible light. After processing by computer, such images provide many more kinds of information than the traditional political boundaries and topographic features of conventional maps. For example, they can map soil types and land use, dis-

tinguishing among farm fields, forests, and urban areas. In fact, they can even differentiate a corn field from a soybean field, or a freshly plowed field from a field with a mature crop.

In the false-color map shown below of Oregon and Washington, vegetation appears as red, dry regions are blue, water is black, and snow on the Cascade Mountains and the Olympic Mountains is white.

Both new images and conventional map data are being digitized, that is, converted to numerical codes and stored on computer tape. As new images are received, changes in physical features are coded and recorded. Map users can thus be provided with maps that are constantly being revised and kept up to date.

6-3 *Indirect Proof*

Until now, the proofs you have written have been *direct* proofs. Sometimes it is difficult or even impossible to find a direct proof. In that case it may be possible to reason *indirectly*. Indirect reasoning is commonplace in everyday life. Suppose, for example, that after walking home, Sue enters the house carrying a dry umbrella. You can conclude that it is not raining outside. Why? Because if it were raining, then her umbrella would be wet. The umbrella is not wet. Therefore, it is not raining.

In an **indirect proof** you begin by assuming temporarily that the desired conclusion is not true. Then you reason logically until you reach a contradiction of the hypothesis or some other known fact. Because you've reached a contradiction, you know that the temporary assumption is impossible and therefore the desired conclusion must be true.

The procedure for writing an indirect proof is summarized below. Notice how these steps are applied in the examples of indirect proof that follow.

How to Write an Indirect Proof

1. Assume temporarily that the conclusion is not true.
2. Reason logically until you reach a contradiction of a known fact.
3. Point out that the temporary assumption must be false, and that the conclusion must then be true.

Example 1

Given: n is an integer and n^2 is even.

Prove: n is even.

Proof:

Assume temporarily that n is not even. Then n is odd, and

$$n^2 = n \times n$$
$$= \text{odd} \times \text{odd} = \text{odd}.$$

But this contradicts the given information that n^2 is even. Therefore the temporary assumption that n is not even must be false. It follows that n is even.

Example 2

Prove that the bases of a trapezoid have unequal lengths.

Given: Trap. *PQRS* with bases $\overline{PQ}$ and $\overline{SR}$

Prove: $PQ \neq SR$

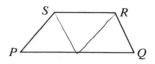

Proof:

Assume temporarily that $PQ = SR$. We know that $\overline{PQ} \parallel \overline{SR}$ by the definition of a trapezoid. Since quadrilateral *PQRS* has two sides that are both congruent and parallel, it must be a parallelogram, and $\overline{PS}$ must be parallel to $\overline{QR}$. But this contradicts the fact that, by definition, trapezoid *PQRS* can have only one pair of parallel sides. The temporary assumption that $PQ = SR$ must be false. It follows that $PQ \neq SR$.

Can you see how proving a statement by proving its contrapositive is related to *indirect proof*? If you want to prove the statement "If *p*, then *q*," you could prove the contrapositive "If not *q*, then not *p*." Or you could write an indirect proof—assume that *q* is false and show that this assumption implies that *p* is false.

Classroom Exercises

1. An indirect proof is to be used to prove the following:

   If $AB = AC$, then $\triangle ABD \cong \triangle ACD$.

   Which one of the following is the correct way to begin?
   a. Assume temporarily that $AB \neq AC$.
   b. Assume temporarily that $\triangle ABD \not\cong \triangle ACD$.

What is the first sentence of an indirect proof of the statement shown?

2. $\triangle ABC$ is equilateral.   **3.** Doug is a Canadian.

4. $a \geq b$ **5.** Kim isn't a violinist.

6. $m \angle X > m \angle Y$ **7.** $\overline{CX}$ isn't a median of $\triangle ABC$.

8. Planning to write an indirect proof that $\angle A$ is an obtuse angle, Becky began by saying "Assume temporarily that $\angle A$ is an acute angle." What has Becky overlooked?

9. Wishing to prove that *l* and *m* are skew lines, John began an indirect proof by supposing that *l* and *m* are intersecting lines. What possibility has John overlooked?

10. Arrange sentences (a)–(e) in an order that completes an indirect proof of the following statement: In a plane, two lines perpendicular to a third line are parallel to each other.

Given: Lines *a*, *b*, and *t* lie in a plane;
     $a \perp t$; $b \perp t$
Prove: $a \parallel b$

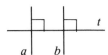

(a) Then *a* intersects *b* in some point *Z*.
(b) But this contradicts the theorem which says that there is exactly one line perpendicular to a given line through a point outside the line.
(c) It is false that *a* is not parallel to *b*, and it follows that $a \parallel b$.
(d) Assume temporarily that *a* is not parallel to *b*.
(e) Then there are two lines through *Z* and perpendicular to *t*.

Written Exercises

Suppose someone plans to write an indirect proof of each conditional. Write a correct first sentence of the indirect proof.

A
  1. If $m \angle A = 50$, then $m \angle B = 40$.
  2. If $\overline{DF} \not\cong \overline{RT}$, then $\overline{DE} \not\cong \overline{RS}$.
  3. If $a \neq b$, then $a - b \neq 0$.
  4. If $x^2 \neq y^2$, then $x \neq y$.
  5. If $\overline{EF} \not\cong \overline{GH}$, then $\overleftrightarrow{EF}$ and $\overleftrightarrow{GH}$ aren't parallel.

Write an indirect proof in paragraph form.

  6. Given: People wearing coats are shivering as they come to the door.
     Prove: It's cold outside.
  7. Given: $\triangle XYZ$; $m \angle X = 100$
     Prove: $\angle Y$ is not a rt. $\angle$.
  8. Given: *n* is an integer and n^2 is odd.
     Prove: *n* is odd.

  9. Given: Transversal *t* cuts lines *a* and *b*;
          $m \angle 1 \neq m \angle 2$
     Prove: $a \not\parallel b$

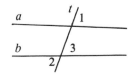

10. Given: $\overline{OJ} \cong \overline{OK}$; $\overline{JE} \not\cong \overline{KE}$
     Prove: $\overrightarrow{OE}$ doesn't bisect $\angle JOK$.

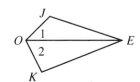

B **11.** Given: $\overleftrightarrow{AB} \nparallel \overleftrightarrow{CD}$
Prove: Planes P and Q intersect.

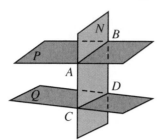

12. Given: $\triangle RVT$ and SVT are equilateral;
$\triangle RVS$ is not equilateral.
Prove: $\triangle RST$ is not equilateral.

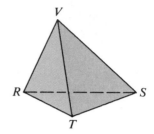

13. Given: quad. *EFGH* in which $m \angle EFG = 93$;
$m \angle FGH = 20$; $m \angle GHE = 147$; $m \angle HEF = 34$
Prove: *EFGH* is not a convex quadrilateral.

14. Given: $AT = BT = 5$; $CT = 4$
Prove: $\angle ACB$ is not a rt. $\angle$.

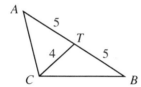

15. Given: Coplanar lines l, k, n;
n intersects l in P; $l \parallel k$
Prove: n intersects k.

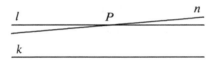

16. Prove that if two angles of a triangle are not congruent, then the sides opposite those angles are not congruent.

17. Prove that there is no regular polygon with an interior angle whose measure is 155.

18. Prove that the diagonals of a trapezoid do not bisect each other.

C **19.** Prove that if two lines are perpendicular to the same plane, then the lines do not intersect.

20. Given: Points R, S, T, and W; $\overleftrightarrow{RS}$ and $\overleftrightarrow{TW}$ are skew.
Prove: $\overleftrightarrow{RT}$ and $\overleftrightarrow{SW}$ are skew.

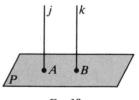

Ex. 19

Challenge

One of four children ate the last piece of lasagna. When questioned they responded as follows:

        Joan: I didn't eat it.   Ken: Leo ate it.
        Leo: Martha ate it.   Martha: Leo is lying.

If only one of the four children lied, who ate the last piece?

Self-Test 1

Classify each conditional as true or false.

1. If $j > k$, then $k < j$.
2. If $a > b$ and $b = c$, then $a > c$.
3. If $r > t$ and $s > t$, then $r > s$.
4. If $\angle BCD$ is an exterior angle of $\triangle ABC$, then
   $m\angle BCD > m\angle A + m\angle B$.

Use the conditional: If $\triangle ABC$ is acute, then $m\angle C \neq 90$.

5. Write the inverse of the statement. Is it true or false?
6. Write the contrapositive of the statement. Is it true or false?

7. Write the letter paired with the statement that is logically equivalent to
   "If Dan can't go, then Valerie can go."
   A. If Valerie can go, then Dan can't go.   B. If Dan can go, then Valerie can't go.
   C. If Valerie can't go, then Dan can go.   D. If Dan can go, then Valerie can go.

8. Given: All rhombuses are parallelograms.
   What can you conclude from each additional statement? If no conclusion
   is possible, write *no conclusion*.
   a. *ABCD* is not a parallelogram.   **b.** *QRST* is not a rhombus.
   c. *MNOP* is a parallelogram.   **d.** *GHIJ* is a rhombus.

9. Suppose you plan to write an indirect proof of the statement: If $AB = 7$,
   then $AC = 14$. Write a correct first sentence of the indirect proof.

10. Write the letters (a)–(d) in an order that completes an indirect proof of
    the statement: Through a point outside a line, there is at most one line
    perpendicular to the given line.

    Given: Point P not on line k
    Prove: There is at most one line through P
           perpendicular to k.

    (a) But this contradicts Corollary 3 of Theorem 3-11: In
    a triangle, there can be at most one right angle or
    obtuse angle.
    (b) Then $\angle PAB$ and $\angle PBA$ are right angles, and $\triangle PAB$ has two right
    angles.
    (c) Thus our temporary assumption is false, and there is at most one line
    through P perpendicular to k.
    (d) Assume temporarily that there are two lines through P perpendicular
    to k at A and B.

Inequalities in Triangles

Objectives

1. State and apply the inequality theorems and corollaries for one triangle.
2. State and apply the inequality theorems for two triangles.

6-4 *Inequalities for One Triangle*

From the information given in the diagram at the left below you can deduce that $\angle C \cong \angle B$.

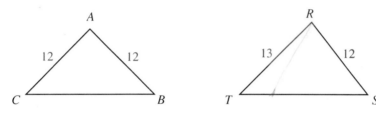

Using the information in the diagram at the right you could write an indirect proof showing that $m \angle S \neq m \angle T$. The following theorem enables you to reach an even stronger conclusion, the conclusion that $m \angle S > m \angle T$.

Theorem 6-2

If one side of a triangle is longer than a second side, then the angle opposite the first side is larger than the angle opposite the second side.

Given: $\triangle RST$; $RT > RS$
Prove: $m \angle RST > m \angle T$

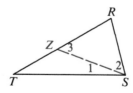

Proof:

By the Ruler Postulate there is a point Z on $\overline{RT}$ such that $RZ = RS$. Draw $\overline{SZ}$.
In isosceles $\triangle RZS$, $m \angle 3 = m \angle 2$.
Because $m \angle RST = m \angle 1 + m \angle 2$, you have $m \angle RST > m \angle 2$.
Substitution of $m \angle 3$ for $m \angle 2$ yields $m \angle RST > m \angle 3$.
Because $\angle 3$ is an ext. $\angle$ of $\triangle ZST$, you have $m \angle 3 > m \angle T$.
From $m \angle RST > m \angle 3$ and $m \angle 3 > m \angle T$, you get $m \angle RST > m \angle T$.

Theorem 6-3

If one angle of a triangle is larger than a second angle, then the side opposite the first angle is longer than the side opposite the second angle.

Given: $\triangle RST$; $m \angle S > m \angle T$

Prove: $RT > RS$

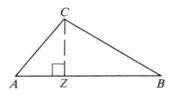

Proof:

Assume temporarily that $RT \not> RS$. Then either $RT = RS$ or $RT < RS$.

  Case 1: If $RT = RS$, then $m \angle S = m \angle T$.
  Case 2: If $RT < RS$, then $m \angle S < m \angle T$ by Theorem 6-2.

In either case there is a contradiction of the given fact that $m \angle S > m \angle T$. The assumption that $RT \not> RS$ must be false. It follows that $RT > RS$.

Corollary 1

The perpendicular segment from a point to a line is the shortest segment from the point to the line.

Corollary 2

The perpendicular segment from a point to a plane is the shortest segment from the point to the plane.

    See Classroom Exercises 18 and 19 for proofs of the corollaries.

Theorem 6-4   *The Triangle Inequality*

The sum of the lengths of any two sides of a triangle is greater than the length of the third side.

Given: $\triangle ABC$

Prove: (1) $AB + BC > AC$
      (2) $AB + AC > BC$
      (3) $AC + BC > AB$

Proof:

One of the sides, say $\overline{AB}$, is the longest side. (Or $\overline{AB}$ is at least as long as each of the other sides.) Then (1) and (2) are true. To prove (3), draw a line, $\overleftrightarrow{CZ}$, through C and perpendicular to $\overleftrightarrow{AB}$. (Through a point outside a line, there is exactly one line perpendicular to the given line.) By Corollary 1 of Theorem 6-3, $\overline{AZ}$ is the shortest segment from A to $\overleftrightarrow{CZ}$. Also, $\overline{BZ}$ is the shortest segment from B to $\overleftrightarrow{CZ}$. Therefore $\quad AC > AZ$ and $BC > ZB$.

$$AC + BC > AZ + ZB \text{ (Why?)}$$
$$AC + BC > AB \text{ (Why?)}$$

Example The lengths of two sides of a triangle are 3 and 5. The length of the third side must be greater than __?__, but less than __?__.

Solution Let x be the length of the third side.

$$x + 3 > 5 \qquad 3 + 5 > x \qquad x + 5 > 3$$
$$x > 2 \qquad\quad 8 > x \qquad\quad x > -2$$

The length of the third side must be greater than 2 but less than 8.

Note that the inequality $x + 5 > 3$ did not give us any useful information. Since $5 > 3$, the sum of 5 and *any* positive number is greater than 3.

Classroom Exercises

Name the largest angle and the smallest angle of the triangle.

1.

2.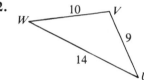

3.

Name the longest side and the shortest side of the triangle.

4.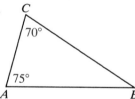

5.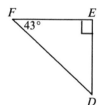

6.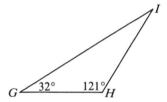

Is it possible for a triangle to have sides with the lengths indicated?

7. 10, 9, 8

8. 6, 6, 20

9. 7, 7, 14.1

10. 16, 11, 5

11. 0.6, 0.5, 1

12. 18, 18, 0.06

13. An isosceles triangle is to have a base that is 20 cm long. Draw a diagram to show the following.
  a. The legs can be very long.
  b. Although the legs must be more than 10 cm long, each length can be very close to 10 cm.

14. The base of an isosceles triangle has length 12. What can you say about the length of a leg?

15. Two sides of a parallelogram have lengths 10 and 12. What can you say about the lengths of the diagonals?

16. Two sides of a triangle have lengths 15 and 20. The length of the third side can be any number between __?__ and __?__.

17. Suppose you know only that the length of one side of a rectangle is 100. What can you say about the length of a diagonal?

18. Use the diagram below to explain how Corollary 1 follows from Theorem 6-3.

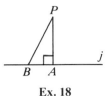

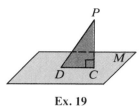

Ex. 18   Ex. 19

19. Use the diagram, in which $\overline{PC} \perp$ plane M, to explain how Corollary 2 follows from Theorem 6-3 or from Corollary 1.

20. Which is the largest angle of a right triangle? Which is the longest side of a right triangle? Explain.

Written Exercises

The lengths of two sides of a triangle are given. Write the numbers that best complete the statement: The length of the third side must be greater than __?__, but less than __?__.

A   **1.** 6, 9
2. 15, 13
3. 100, 100

4. $7n$, $10n$
5. a, b (where $a > b$)
6. k, $k + 5$

In Exercises 7–9 the diagrams are not drawn to scale. If each diagram were drawn to scale, which numbered angle would be the largest?

7.

8.

9.

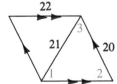

In Exercises 10–14 the diagrams are not drawn to scale. If each diagram were drawn to scale, which segment shown would be the longest?

10.

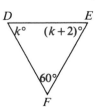

11.

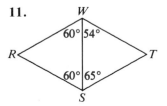

12.

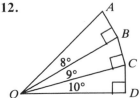

B **13.**

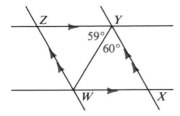

14.

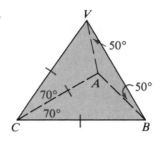

15. Use the lengths *a*, *b*, *c*, *d*, and *e* to complete:

___?___ > ___?___ > ___?___ > ___?___ > ___?___

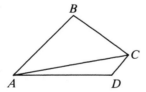

16. Use $m \angle 1$, $m \angle 2$, and $m \angle 3$ to complete:

___?___ > ___?___ > ___?___

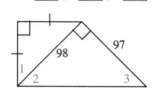

17. The diagram is not drawn to scale. Use $m \angle 1$, $m \angle 2$, $m \angle X$, $m \angle Y$, and $m \angle XZY$ to complete:

___?___ > ___?___ > ___?___ > ___?___ > ___?___

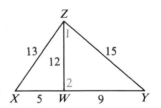

18. Given: Quad. *ABCD*
Prove: $AB + BC + CD + DA > 2(AC)$

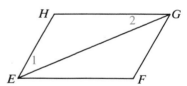

19. Given: $\square EFGH$; $EF > FG$
Prove: $m \angle 1 > m \angle 2$

C **20.** Discover, state, and prove a theorem that compares the perimeter of a quadrilateral with the sum of the lengths of the diagonals.

21. Prove that the sum of the lengths of the medians of a triangle is greater than half the perimeter.

22. If you replace "medians" with "altitudes" in Exercise 21, can you prove the resulting statement? Explain.

In Exercises 23 and 24, begin your proofs by drawing auxiliary lines.

23. Discover, state, and prove a theorem that compares the length of the longest side of a quadrilateral with the sum of the lengths of the other three sides.

24. Prove: If *P* is any point inside $\triangle XYZ$, then $ZX + ZY > PX + PY$.

Application   *Finding the Shortest Path*

The owners of pipeline *l* plan to construct a pumping station at a point *S* on line *l* in order to pipe oil to two major customers, located at *A* and *B*. To minimize the cost of constructing lines from *S* to *A* and *B*, they wish to locate *S* along *l* so that the distance *SA* + *SB* is as small as possible.

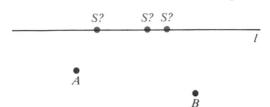

The construction engineer uses the following method to locate *S*:

1. Draw a line through *B* perpendicular to *l*, intersecting *l* at point *P*.

2. On this perpendicular, locate point *C* so that *PC* = *PB*.

3. Draw $\overline{AC}$.

4. Locate *S* at the intersection of $\overline{AC}$ and *l*.

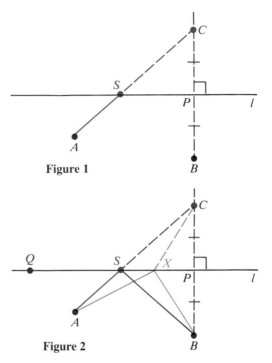

Figure 1

Figure 2 shows the path of the new pipelines through the pumping station located at *S*, and an alternative path going through a different point, *X*, on *l*. You can use Theorem 6-4 (the Triangle Inequality) to show that if *X* is any point on *l* other than *S*, then *AX* + *XB* > *AS* + *SB*. So any alternative path is longer than the path through *S*.

Figure 2

Exercises

Exercises 1 and 2 refer to Figure 2 on the preceding page.

1. Supply the reason for each key step of the proof that the method given for finding S yields the shortest total length for the pipelines serving A and B.

   1. l is the perpendicular bisector of $\overline{BC}$.
   2. $SC = SB$
   3. $AS + SC = AC$
   4. $AS + SB = AC$
   5. $XC = XB$
   6. $AX + XC > AC$
   7. $AX + XB > AS + SB$

2. This method for finding S is sometimes called a *solution by reflection*, since it involves *reflecting* point B in line l. (See Chapter 14 for more on reflections.) Show that $\overline{AS}$ and $\overline{SB}$, like reflected paths of light, make congruent angles with l. That is, prove that $\angle QSA \cong \angle PSB$. (*Hint*: Draw your own diagram, omitting the part of Figure 2 shown in blue.)

Explorations

These exploratory exercises can be done using a computer with a program that draws and measures geometric figures.

Draw several *pairs* of triangles, varying the size of just one side or just one angle. Make charts like the ones below to record your data. Record the lengths of the sides and measures of the angles you give, as well as the measurements you get. Do as many pairs as you need to help you recognize a pattern.

Enter the lengths of all three sides (SSS).

pair 1	
$AB =$	same $AB =$
$BC =$	same $BC =$
$AC =$	longer $AC =$
$m \angle ABC =$	$m \angle ABC =$

What happened to the angle opposite the side you made longer?

Enter the lengths of two sides and the included angle (SAS).

pair 1	
$AB =$	same $AB =$
$\angle BAC =$	larger $\angle BAC =$
$AC =$	same $AC =$
$BC =$	$BC =$

What happened to the side opposite the angle you made larger?

♦ Computer Key-In

If you break a stick into three pieces, what is the probability that you can join the pieces end-to-end to form a triangle?

It's easy to see that if the sum of the lengths of any two of the pieces is less than or equal to that of the third, a triangle can't be formed. This is the Triangle Inequality (Theorem 6-4).

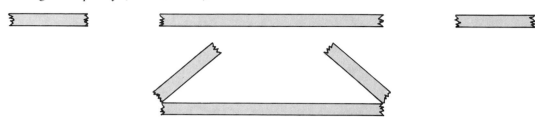

By an experiment, your class can estimate the probability that three pieces of broken stick will form a triangle. Suppose everyone in your class has a stick 1 unit long and breaks it into three pieces. If there are thirty people in your class and eight people are able to form a triangle with their pieces, we estimate that the probability of forming a triangle is about $\frac{8}{30}$, or $\frac{4}{15}$.

Of course, this experiment is not very practical. You can get much better results by having a computer simulate the breaking of many, many sticks, as in the program in BASIC on the next page.

Computer simulations are useful whenever large numbers of operations need to be done in a short period of time. In this problem, for example, an accurate probability depends on using a large number of sticks. Computer simulations have been used when a real experiment would be costly or dangerous; aeronautics companies use real-time flight simulators on the ground to train pilots. Simulations are also applied to investigate statistical data where many variables determine the outcome, as in the analysis and prediction of weather patterns. In the stick-triangle problem, using the computer has another advantage— a computer can break very small pieces that a human couldn't, so the probability figure will be theoretically more accurate, if less "realistic."

In lines 30 and 40 of the following program, you tell the computer how many sticks you want to break. Each stick is 1 unit long, and the computer breaks each stick by choosing two random numbers x and y between 0 and 1. These numbers divide the stick into three lengths r, s, and t.

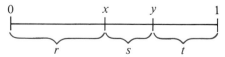

The computer then keeps count of the number of sticks (N) which form a triangle when broken.

Notice that RND is used in lines 70 and 80. Since usage of RND varies, check this with the manual for your computer and make any necessary changes. The computer print-outs shown in this text use capital letters. The x, y, r, s, and t used in the discussion above appear as X, Y, R, S, T.

```
10  PRINT "SIMULATION--BREAKING STICKS TO MAKE TRIANGLES"
20  PRINT
30  PRINT "HOW MANY STICKS DO YOU WANT TO BREAK";
40  INPUT D
50  LET N = 0
60  FOR I = 1 TO D
70  LET X = RND (1)
80  LET Y = RND (1)
90  IF X > = Y THEN 70
100 LET R = X
110 LET S = Y - R
120 LET T = 1 - R - S
130 IF R + S < = T THEN 170
140 IF S + T < = R THEN 170
150 IF R + T < = S THEN 170
160 LET N = N + 1
170 NEXT I
180 LET P = N/D
190 PRINT
200 PRINT "THE EXPERIMENTAL PROBABILITY THAT"
210 PRINT "A BROKEN STICK CAN FORM A TRIANGLE IS ";P
220 END
```

Line Number	Explanation
60–120	These lines simulate the breaking of each stick. When I = 10, for example, the computer is "breaking" the tenth stick.
130–150	Here the computer uses the Triangle Inequality to check whether the pieces of the broken stick can form a triangle. If not, the computer goes on to the next stick (line 170) and the value of N is not affected.
160	If the broken stick has survived the tests of steps 130–150, then the pieces can form a triangle and the value of N is increased by 1.
170	Lines 60–170 form a loop that is repeated D times. After I = D, the probability P is calculated and printed (lines 180–210).

Exercises

1. Pick any two numbers x and y between 0 and 1 with $x < y$. With paper and pencil, carry out the instructions in lines 100 through 150 of the program to see how the computer finds r, s, and t and tests to see whether the values can be the lengths of the sides of a triangle.

2. If you use a language other than BASIC, write a similar program for your computer.

3. Run the program several times for large values of D, say 100, 400, 800, and compare your results with those of some classmates. Does the probability that the pieces of a broken stick form a triangle appear to be less than or greater than $\frac{1}{2}$?

6-5 *Inequalities for Two Triangles*

Begin with two matched pairs of sticks joined loosely at B and E. Open them so that $m \angle B > m \angle E$ and you find that $AC > DF$. Conversely, if you open them so that $AC > DF$, you see that $m \angle B > m \angle E$. Two theorems are suggested by these examples. The first theorem is surprisingly difficult to prove. The second theorem has an indirect proof.

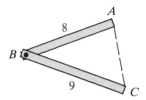
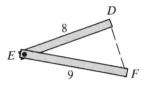

Theorem 6-5   *SAS Inequality Theorem*

If two sides of one triangle are congruent to two sides of another triangle, but the included angle of the first triangle is larger than the included angle of the second, then the third side of the first triangle is longer than the third side of the second triangle.

Given: $\overline{BA} \cong \overline{ED}$; $\overline{BC} \cong \overline{EF}$;
      $m \angle B > m \angle E$

Prove: $AC > DF$

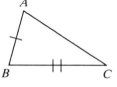
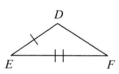

Proof:
Draw $\overrightarrow{BZ}$ so that $m \angle ZBC = m \angle E$. On $\overrightarrow{BZ}$ take point X so that $BX = ED$.
Then either X is on $\overline{AC}$ or X is not on $\overline{AC}$.
In both cases $\triangle XBC \cong \triangle DEF$ by SAS, and $XC = DF$.

  Case 1: X is on $\overline{AC}$.
$AC > XC$ (Seg. Add. Post. and a Prop. of Ineq.)
$AC > DF$ (Substitution Property, using the equation in red above)

  Case 2: X is not on $\overline{AC}$.
Draw the bisector of $\angle ABX$, intersecting $\overline{AC}$ at Y.
Draw $\overline{XY}$ and $\overline{XC}$.
$BA = ED = BX$
Since $\triangle ABY \cong \triangle XBY$ (SAS), $AY = XY$.
$XY + YC > XC$ (Why?)
$AY + YC > XC$ (Why?), or $AC > XC$
    $AC > DF$ (Substitution Property)

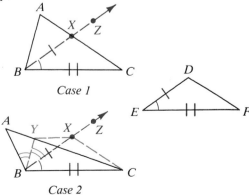

Theorem 6-6   *SSS Inequality Theorem*

If two sides of one triangle are congruent to two sides of another triangle, but the third side of the first triangle is longer than the third side of the second, then the included angle of the first triangle is larger than the included angle of the second.

Given: $\overline{BA} \cong \overline{ED}$; $\overline{BC} \cong \overline{EF}$;
$\quad\quad\;\; AC > DF$

Prove: $m \angle B > m \angle E$

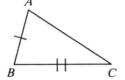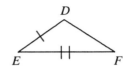

Proof:

Assume temporarily that $m \angle B \ngtr m \angle E$.
Then either $m \angle B = m \angle E$ or $m \angle B < m \angle E$.

Case 1: If $m \angle B = m \angle E$, then $\triangle ABC \cong \triangle DEF$ by the SAS Postulate, and $AC = DF$.

Case 2: If $m \angle B < m \angle E$, then $AC < DF$ by the SAS Inequality Theorem.

In both cases there is a contradiction of the given fact that $AC > DF$. What was temporarily assumed to be true, that $m \angle B \ngtr m \angle E$, must be false. It follows that $m \angle B > m \angle E$.

Example 1   Given: $\overline{RS} \cong \overline{RT}$; $m \angle 1 > m \angle 2$
What can you deduce?

Solution   In the two triangles you have $\overline{RV} \cong \overline{RV}$ as
well as $\overline{RS} \cong \overline{RT}$. Since $m \angle 1 > m \angle 2$,
you can apply the SAS Inequality Theorem
to get $SV > TV$.

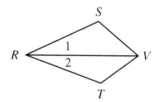

Example 2   Given: $\overline{EF} \cong \overline{EG}$; $DF > DG$
What can you deduce?

Solution $\overline{DE}$ and $\overline{EF}$ of $\triangle DEF$ are congruent to $\overline{DE}$
and $\overline{EG}$ of $\triangle DEG$. Since $DF > DG$, you
can apply the SSS Inequality Theorem to
get $m \angle DEF > m \angle DEG$.

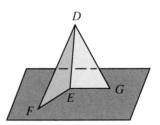

Classroom Exercises

What can you deduce? Name the theorem that supports your answer.

1.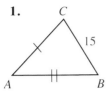

2.

3. $m \angle 1 > m \angle 2$

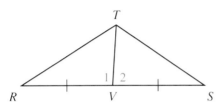

4. $\angle 1$ is a rt. $\angle$; $\angle 2$ is an obtuse $\angle$.

 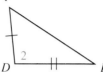

5.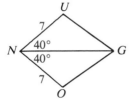

6.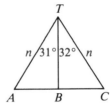

7.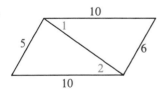

8.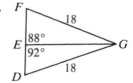

You will need a centimeter ruler and a protractor for Exercises 9 and 10.

9. a. Draw an isosceles triangle with legs 7 cm long and a vertex angle of 120°. Measure the length of the base of the triangle.

 b. If you keep the legs at 7 cm in length but halve the measure of the vertex angle to 60°, what happens to the length of the base? What kind of triangle is this new triangle? What is the length of the third side?

10. a. Draw a right triangle with legs of 6 cm and 8 cm. Measure the length of the hypotenuse.

 b. If you keep the 6 cm and 8 cm sides the same lengths but halve the measure of the included angle to 45°, what happens to the length of the third side? Test your answer by drawing the new triangle.

Written Exercises

What can you deduce? Name the theorem that supports your answer.

A   **1.** Given: $\overline{AM}$ is a median of $\triangle ABC$;
          $AB > AC$

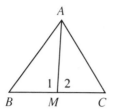

2. Given: $\square RSTV$;
        $m \angle TSR > m \angle VRS$

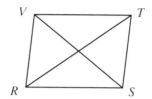

Complete the statements by writing $<$, $=$, or $>$.

3. $XY \underline{^?} XZ$;
   $XW \underline{^?} 12$

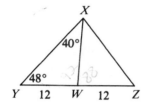

4. $AD \underline{^?} CE$

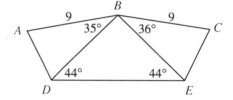

5. $m \angle 1 \underline{^?} m \angle 2$;
   $m \angle 3 \underline{^?} m \angle 4$

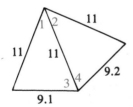

6. $m \angle 5 \underline{^?} m \angle 6$

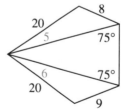

7. $a \underline{^?} b$

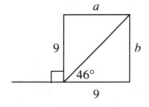

8. $c \underline{^?} 10$

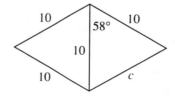

Complete the statements by writing <, =, or >.

B   **9.** $m \angle 1 \underline{\quad ? \quad} m \angle 2$

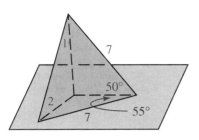

Write proofs in two-column form.

11. Given: $\overline{TU} \cong \overline{US} \cong \overline{SV}$
   Prove: $ST > SV$

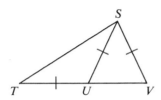

C   **13.** Given: $\overline{PA} \cong \overline{PC} \cong \overline{QC} \cong \overline{QB}$
   Prove: $m \angle PCA < m \angle QCB$

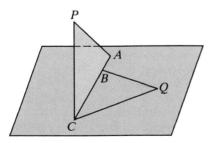

10. $SR = ST$; $VX = VT$
   $m \angle RSV \underline{\quad ? \quad} m \angle TSV$

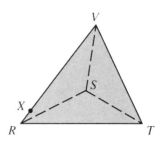

12. Given: Plane P bisects $\overline{XZ}$ at Y;
   $WZ > WX$
   Discover and prove something about
   the figure.

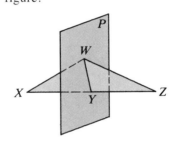

14. Given: $\overline{DE} \perp$ plane M; $EK > EJ$
   Prove: $DK > DJ$
   (*Hint:* On $\overline{EK}$, take Z so that $EZ = EJ$.)

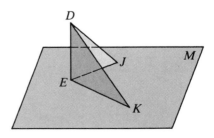

15. In the three-dimensional figure shown, all the edges *except*
$\overline{VC}$ are congruent. What can you say about the measures of
the largest angles of the twelve angles in the figure
   a. if $\overline{VC}$ is longer than the other edges?
   b. if $\overline{VC}$ is shorter than the other edges?

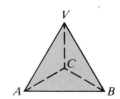

Self-Test 2

1. In $\triangle XYZ$, $m \angle X = 50$, $m \angle Y = 60$, and $m \angle Z = 70$. Name the longest side of the triangle.

2. In $\triangle DOM$, $\angle O$ is a right angle and $m \angle D > m \angle M$. Which side of $\triangle DOM$ is the shortest side?

Complete each statement by writing $<$, $=$, or $>$.

3. If $ER > EN$, then $m \angle R \underline{^?} m \angle N$.

4. If $\overline{AG} \cong \overline{ER}$, $\overline{AP} \cong \overline{EN}$, and $\angle A \cong \angle E$, then $GP \underline{^?} RN$.

5. If $\overline{GA} \cong \overline{RE}$, $\overline{GP} \cong \overline{RN}$, and $AP > EN$, then $m \angle G \underline{^?} m \angle R$.

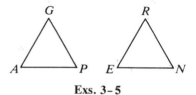

Exs. 3–5

6. The lengths of the sides of a triangle are 5, 6, and x. Then x must be greater than $\underline{^?}$ and less than $\underline{^?}$.

The longer diagonal of $\square QRST$ is $\overline{QS}$. Tell whether each statement *must be*, *may be*, or *cannot be* true.

7. $\angle R$ is an acute angle.   **8.** $QS > RS$ **9.** $RS > RT$

| **Extra** | *Non-Euclidean Geometries* |

When you develop a geometry, you have some choice as to which statements you are going to postulate and which you are going to prove. For example, consider these two statements:

(A) If two parallel lines are cut by a transversal, then corresponding angles are congruent.

(B) Through a point outside a line, there is exactly one line parallel to the given line.

In this book, statement (A) is Postulate 10 and statement (B) is Theorem 3-8. In some books, statement (B) is a postulate (commonly called Euclid's *Parallel Postulate*) and statement (A) is a theorem. In still other developments, both of these statements are proved on the basis of some third statement chosen as a postulate.

A geometry that provides for a unique parallel to a line through a point not on the line is called *Euclidean*, so this text is a Euclidean geometry book. In the nineteenth century, it was discovered that geometries exist in which the Parallel Postulate is *not true*. Such geometries are called *non-Euclidean*. The statements at the top of the next page show the key differences between Euclidean geometry and two types of non-Euclidean geometry.

Euclidean geometry	Through a point outside a line, there is *exactly one* line parallel to the given line.
Hyperbolic geometry	Through a point outside a line, there is *more than one* line parallel to the given line. (This geometry was discovered by Bolyai, Lobachevsky, and Gauss.)
Elliptic geometry	Through a point outside a line, there is *no* line parallel to the given line. (This geometry was discovered by Riemann and is used by ship and airplane navigators.)

To see a model of a no-parallel geometry, visualize the surface of a sphere. Think of a line as being a great circle of the sphere, that is, the intersection of the sphere and a plane that passes through the center of the sphere. On the sphere, through a point outside a line, there is no line parallel to the given line. All lines, as defined, intersect. In the figure, for example, X is a point not on the red great circle. A line has been drawn through X, namely the great circle shown in blue. You can see that the two lines intersect in *two* points, A and B.

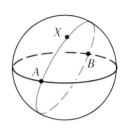

To see how statement (B) follows from our postulates, notice that Postulates 10 and 11 play a crucial role in the following proof. In fact, without such assumptions about parallels there couldn't be a proof. Before the discovery of non-Euclidean geometries people didn't know that this was the case and tried, without success, to find a proof that was independent of any assumption about parallels.

Given: Point P outside line k.

Prove: (1) There is a line through P parallel to k.

(2) There is only one line through P parallel to k.

Key steps of proof of (1):

1. Draw a line through P and some point Q on k. (Postulates 5 and 6)

2. Draw line l so that $\angle 2$ and $\angle 1$ are corresponding angles and $m \angle 2 = m \angle 1$. (Protractor Postulate)

3. $l \parallel k$, so there is a line through P parallel to k. (Postulate 11)

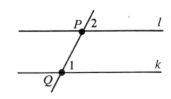

Indirect proof of (2):

Assume temporarily that there are at least two lines, x and y, through P parallel to k. Draw a line through P and some point R on k. $\angle 4 \cong \angle 3$ and $\angle 5 \cong \angle 3$ by Postulate 10, so $\angle 5 \cong \angle 4$. But since x and y are different lines we also have $m \angle 5 > m \angle 4$. This is impossible, so our assumption must be false, and it follows that there is only one line through P parallel to k.

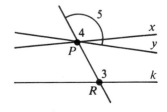

Chapter Summary

1. The properties of inequality most often used are stated on page 204.

2. The measure of an exterior angle of a triangle is greater than the measure of either remote interior angle. (The Exterior Angle Inequality Theorem)

3. The summary on page 208 gives the relationship between an if-then statement, its converse, its inverse, and its contrapositive. An if-then statement and its contrapositive are logically equivalent.

4. You begin an indirect proof by assuming temporarily that what you wish to prove true is *not* true. If this temporary assumption leads to a contradiction of a known fact, then your temporary assumption must be false and what you wish to prove true must be true.

5. In $\triangle RST$, if $RT > RS$, then $m \angle S > m \angle T$. Conversely, if $m \angle S > m \angle T$, then $RT > RS$.

6. The perpendicular segment from a point to a line (or plane) is the shortest segment from the point to the line (or plane).

7. The sum of the lengths of any two sides of a triangle is greater than the length of the third side. (The Triangle Inequality)

8. You can use the SAS Inequality and SSS Inequality Theorems to compare the lengths of sides and measures of angles in two triangles.

Chapter Review

Complete each statement by writing $<$, $=$, or $>$.

1. $m \angle 1 \underline{^?} m \angle 5$ **6–1**

2. $m \angle 1 \underline{^?} m \angle 2$

3. $m \angle 3 \underline{^?} m \angle 4$

4. $m \angle 5 \underline{^?} m \angle 2$

5. If $a > b$, $c < b$, and $d = c$, then $a \underline{^?} d$.

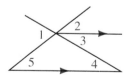

Given: All registered voters must be at least 18 years old.
What, if anything, can you conclude from each additional statement?

6. Eric is 19 years old.   7. Bonnie is not registered to vote.   **6–2**

8. Will is 15 years old.   9. Barbara is a registered voter.

10. Write the letters (a)–(d) in an order that completes an indirect proof of   6–3
    the statement: If $n^2 + 6 = 32$, then $n \neq 5$.
    (a) But this contradicts the fact that $n^2 + 6 = 32$.
    (b) Our temporary assumption must be false, and it follows that $n \neq 5$.
    (c) Assume temporarily that $n = 5$.
    (d) Then $n^2 + 6 = 31$.

11. In $\triangle TOP$, if $OT > OP$, then $m \angle P > \underline{}$.   6–4

12. In $\triangle RED$, if $m \angle D < m \angle E$, then $RD > \underline{}$.

13. Points X and Y are in plane M. If $\overline{PX} \perp$ plane M, then $PX \underline{} PY$.

14. Two sides of a triangle have lengths 6 and 8. The length of the third side
    must be greater than $\underline{}$ and less than $\underline{}$.

Complete each statement by writing $<$, $=$, or $>$.

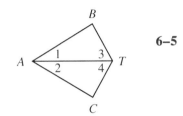

15. If $\overline{AB} \cong \overline{AC}$ and $m \angle 1 > m \angle 2$, then $BT \underline{} CT$.   6–5

16. If $\overline{TB} \cong \overline{TC}$ and $AB < AC$, then $m \angle 3 \underline{} m \angle 4$.

17. If $\angle 1 \cong \angle 2$ and $\angle 3 \cong \angle 4$, then $AB \underline{} AC$.

18. If $\overline{TB} \cong \overline{TC}$ and $m \angle 3 > m \angle 4$, then $AB \underline{} AC$.

Chapter Test

Complete each statement by writing $<$, $=$, or $>$.

1. If $x > y$ and $y = z$, then $x \underline{} z$.   2. If $a > b$, and $c < b$, then $c \underline{} a$.

3. If $s = t + 4$, then $s \underline{} t$.   4. If $e + 5 = f + 4$, then $e \underline{} f$.

5. Write (a) the inverse and (b) the contrapositive of
    "If point P is on $\overline{AB}$, then $AB > AP$."

6. Pair each statement below with the given statement above and tell what
    conclusion, *if any*, must follow.
    a. P is not on $\overline{AB}$.   **b.** P is on $\overline{AB}$.   **c.** $AB \leq AP$ **d.** $AB > AP$

7. If the lengths of the sides of a triangle are x, 15, and 21, then x must be
    greater than $\underline{}$ and less than $\underline{}$.

**In Exercises 8–10 the diagrams are not drawn to scale. If each diagram were
drawn accurately, which segment shown would be the shortest?**

8.

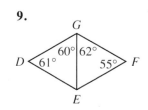

9.

10.

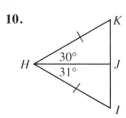

11. If $VE > VO$, then $m \angle \underline{\ ?\ } > m \angle \underline{\ ?\ }$.

12. If $m \angle UEO > m \angle UOE$, then $\underline{\ ?\ } > \underline{\ ?\ }$.

13. If $\overline{VE} \cong \overline{VO}$ and $m \angle UVE > m \angle UVO$, then $\underline{\ ?\ } > \underline{\ ?\ }$.

14. If $m \angle EVU = 60$, $\overline{OE} \cong \overline{OU}$, and $m \angle VOE > m \angle VOU$, then the largest angle of $\triangle UVE$ is $\angle \underline{\ ?\ }$.

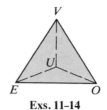

Exs. 11–14

15. Write an indirect proof.
Given: Trap. $ABCD$ with $\overline{AB} \parallel \overline{DC}$
Prove: $\angle C$ and $\angle D$ are not both right angles.

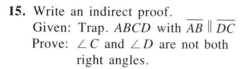

16. Given: $XS > YS$; $\overline{RX} \cong \overline{TY}$; S is the midpoint of $\overline{RT}$.
Prove: $m \angle R > m \angle T$

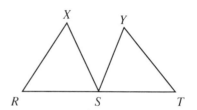

Algebra Review: *Fractions*

Simplify the following fractions.

Example   **a.** $\dfrac{8w}{2}$ **b.** $\dfrac{5t - 10}{15}$ **c.** $\dfrac{x + 6}{36 - x^2}$

Solution   **a.** $4w$ **b.** $\dfrac{5(t - 2)}{15}$ **c.** $\dfrac{x + 6}{(6 - x)(6 + x)}$

$= \dfrac{t - 2}{3}$ $= \dfrac{1}{6 - x}$

1. $\dfrac{14}{70}$ **2.** $\dfrac{75}{15}$ **3.** $\dfrac{18a}{36}$ **4.** $\dfrac{3x}{x}$

5. $\dfrac{x}{3x}$ **6.** $\dfrac{5bc}{10b^2}$ **7.** $\dfrac{-8y^3}{2y}$ **8.** $\dfrac{-18r^3t}{12rt}$

9. $\dfrac{3ab^2}{6bc}$ **10.** $\dfrac{6a + 12}{6}$ **11.** $\dfrac{9x - 6y}{3}$ **12.** $\dfrac{33ab - 22b}{11b}$

13. $\dfrac{x + 2}{3x + 6}$ **14.** $\dfrac{2c - 2d}{2c + 2d}$ **15.** $\dfrac{t^2 - 1}{t - 1}$ **16.** $\dfrac{5a + 5b}{a^2 - b^2}$

17. $\dfrac{b^2 - 25}{b^2 - 12b + 35}$ **18.** $\dfrac{a^2 + 8a + 16}{a^2 - 16}$ **19.** $\dfrac{3x^2 - 6x - 24}{3x^2 + 2x - 8}$

Preparing for College Entrance Exams

Strategy for Success
You may find it helpful to sketch figures or do calculations in your test booklet. Be careful not to make extra marks on your answer sheet.

Indicate the best answer by writing the appropriate letter.

1. The diagonals of quadrilateral *MNOP* intersect at *X*. Which statement guarantees that *MNOP* is a rectangle?
   (A) $MX = NX = OX = PX$ **(B)** $\angle PMN \cong \angle MNO \cong \angle NOP$
   (C) $MO = NP$ **(D)** $\overline{MO} \perp \overline{NP}$ **(E)** $\overline{MN} \perp \overline{MP}$

2. Which statement does *not* guarantee that quadrilateral *WXYZ* is a parallelogram?
   (A) $\overline{WX} \cong \overline{YZ}; \overline{XY} \parallel \overline{WZ}$ **(B)** $\angle W \cong \angle Y; \angle X \cong \angle Z$
   (C) $\overline{WX} \cong \overline{YZ}; \overline{XY} \cong \overline{WZ}$ **(D)** $\overline{XY} \parallel \overline{WZ}; \overline{WX} \parallel \overline{ZY}$
   (E) $\overline{XY} \cong \overline{WZ}; \overline{XY} \parallel \overline{WZ}$

3. In $\triangle ABC$, if $AB = BC$ and $AC > BC$, then:
   (A) $AB < AC - BC$ **(B)** $m \angle B > m \angle C$ **(C)** $m \angle B < m \angle A$
   (D) $m \angle B = 60$ **(E)** $m \angle B = m \angle A$

4. Which statement is not always true for every rhombus *ABCD*?
   (A) $AB = BC$ **(B)** $AC = BD$ **(C)** $\angle B \cong \angle D$
   (D) $\overline{AC} \perp \overline{BD}$ **(E)** $\angle ABD \cong \angle CBD$

5. Given: $m \angle 3 > m \angle 4$
   Compare: $x = m \angle 1 + m \angle 4$ and
   $\qquad\qquad y = m \angle 2 + m \angle 3$

   (A) $x > y$ **(B)** $y > x$ **(C)** $x = y$
   (D) No comparison possible with information given

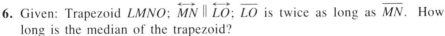

6. Given: Trapezoid *LMNO*; $\overleftrightarrow{MN} \parallel \overleftrightarrow{LO}$; $\overline{LO}$ is twice as long as $\overline{MN}$. How long is the median of the trapezoid?

   (A) $\frac{4}{3}LO$ **(B)** $\frac{3}{2}LO$ **(C)** $\frac{2}{3}MN$ **(D)** $\frac{3}{4}MN$ **(E)** $\frac{3}{2}MN$

7. Quad. *CAKE* is a rectangle. Find *CK*.
   (A) 2 **(B)** 3 **(C)** 4 **(D)** 6 **(E)** 8

8. Which of the following statement(s) are true?
   (I) If $a > b$, then $ax > bx$ for all numbers x.
   (II) If $ax > bx$ for some number x, then $a > b$.
   (III) If $a > b$, then for some number x, $ax < bx$.

   (A) I only **(B)** II only **(C)** III only **(D)** all of the above
   (E) none of the above

Cumulative Review: Chapters 1–6

A **1.** An angle and its complement have the measures $x + 38$ and $2x - 5$. Find the measure of the angle.

2. Find the sum of the measures of the interior angles of a pentagon.

3. Can the given information be used to prove the triangles congruent? If so, which congruence postulate or theorem would you use?

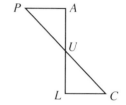

   a. Given: $\overline{PC}$ and $\overline{AL}$ bisect each other.
   b. Given: $\angle P \cong \angle C$; U is the midpoint of $\overline{PC}$.
   c. Given: $\overline{PA} \parallel \overline{LC}$
   d. Given: $\overline{PA} \perp \overline{AL}$; $\overline{LC} \perp \overline{AL}$; $\overline{PU} \cong \overline{UC}$

4. Tell whether the statement is *always*, *sometimes*, or *never* true for a parallelogram $ABCD$ with diagonals that intersect at P.

   a. $AB = BC$ **b.** $\overline{AC} \perp \overline{BD}$ **c.** $\angle A$ and $\angle B$ are complementary $\angle$.
   d. $\angle ADB \cong \angle CBD$ **e.** $\overline{AP} \cong \overline{PC}$ **f.** $\triangle ABC \cong \triangle CDA$

5. In $\triangle XYZ$, $m \angle X = 64$ and $m \angle Y = 54$. Name **(a)** the longest and **(b)** the shortest side of $\triangle XYZ$.

6. a. Which segment is longer: $\overline{RS}$ or $\overline{JK}$?
   b. Name the theorem that supports your answer.

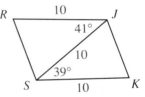

B **7.** The difference between the measures of two supplementary angles is 38. Find the measure of each angle.

8. The lengths of the sides of a triangle are z, $z + 3$, and $z + 6$. What can you conclude about the value of z?

9. Write an indirect proof of the following statement: If $PQRS$ is a quadrilateral, then $\angle Q$, $\angle R$, and $\angle S$ are not all $120°$.

10. Given: $m \angle B > m \angle A$;
         $m \angle E > m \angle D$
      Prove: $AD > BE$

11. Given: $\overline{DC} \parallel \overline{AB}$; $\overline{CE} \perp \overline{AB}$; $\overline{AF} \perp \overline{AB}$
      Prove: $AECF$ is a rectangle.

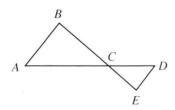

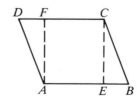

7 SIMILAR POLYGONS

In mapping a landform like this river valley, cartographers must apply the geometric principles of similarity and draw accurately to scale.

Ratio, Proportion, and Similarity

Objectives

1. Express a ratio in simplest form.
2. Solve for an unknown term in a given proportion.
3. Express a given proportion in an equivalent form.
4. State and apply the properties of similar polygons.

7-1 *Ratio and Proportion*

The **ratio** of one number to another is the quotient when the first number is divided by the second. This quotient is usually expressed in *simplest form*.

$$\text{The ratio of 8 to 12 is } \frac{8}{12}, \text{ or } \frac{2}{3}.$$

$$\text{If } y \neq 0, \text{ the ratio of } x \text{ to } y \text{ is } \frac{x}{y}.$$

Since we cannot divide by zero, a ratio $\frac{r}{s}$ is defined only if $s \neq 0$. When an expression such as $\frac{r}{s}$ appears in this book, you may assume that $s \neq 0$.

Example 1
  a. Find the ratio of OI to ZD.

  b. Find the ratio of the measure of the smallest angle of the trapezoid to that of the largest angle.

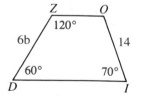

Solution
  a. $\dfrac{OI}{ZD} = \dfrac{14}{6b} = \dfrac{7}{3b}$

    The ratio of OI to ZD is 7 to $3b$.

  b. $\angle O$ has measure $180 - 70$, or 110. Thus $\angle D$ is the smallest angle and $\angle Z$ is the largest angle.

$$\frac{m \angle D}{m \angle Z} = \frac{60}{120} = \frac{1}{2}$$

    The ratio of the measure of the smallest angle of the trapezoid to that of the largest angle is 1 to 2.

Ratios can be used to compare two numbers. To find the ratio of the lengths of two segments, the segments must be measured in terms of the same unit.

Example 2 A poster is 1 m long and 52 cm wide. Find the ratio of the width to the length.

Solution *Method 1*

Use centimeters.

1 m = 100 cm

$$\frac{\text{width}}{\text{length}} = \frac{52}{100} = \frac{13}{25}$$

Method 2

Use meters.

52 cm = 0.52 m

$$\frac{\text{width}}{\text{length}} = \frac{0.52}{1} = \frac{52}{100} = \frac{13}{25}$$

Example 2 shows that the ratio of two quantities is not affected by the unit chosen.

Sometimes the ratio of a to b is written in the form $a:b$. This form can also be used to compare three or more numbers. The statement that three numbers are in the ratio $c:d:e$ (read "c to d to e") means:

(1) The ratio of the first two numbers is $c:d$.
(2) The ratio of the last two numbers is $d:e$.
(3) The ratio of the first and last numbers is $c:e$.

Example 3 The measures of the three angles of a triangle are in the ratio $2:2:5$. Find the measure of each angle.

Solution Let $2x$, $2x$, and $5x$ represent the measures.

$$2x + 2x + 5x = 180$$
$$9x = 180$$
$$x = 20$$

Then $2x = 40$ and $5x = 100$.

The measures of the angles are 40, 40, and 100.

A **proportion** is an equation stating that two ratios are equal. For example,

$$\frac{a}{b} = \frac{c}{d} \quad \text{and} \quad a:b = c:d$$

are equivalent forms of the same proportion. Either form can be read "a is to b as c is to d." The number a is called the first *term* of the proportion. The numbers b, c, and d are the second, third, and fourth terms, respectively.

When three or more ratios are equal, you can write an *extended proportion:*

$$\frac{a}{b} = \frac{c}{d} = \frac{e}{f}$$

Classroom Exercises

Express the ratio in simplest form.

1. $\dfrac{12}{20}$　　**2.** $\dfrac{3p}{5p}$　　**3.** $\dfrac{4n}{n^2}$　　**4.** $\dfrac{n^2}{4n}$

5. Is the ratio $a:b$ always, sometimes, or never equal to the ratio $b:a$? Explain.

6. An office copy machine can make a reduction to 90%, thus making the copy slightly smaller than the original. What is the ratio of the length of a line of text in the original to the length of a copy of that line?

7. Barbara is making oatmeal for breakfast. The instructions say to use 3 cups of water with 2 cups of oatmeal.
  a. What is the ratio of water to oatmeal?
  b. If Barbara uses 6 cups of water, how much oatmeal does she need?

Express the ratio in simplest form.

8. $DI:IS$　　　　　**9.** $ST:DI$　　　　　**10.** $IT:DT$

11. $DI:IT$　　　　**12.** $IT:DS$　　　　**13.** $IS:DI:IT$

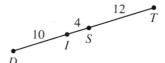

14. What is the ratio of 750 mL to 1.5 L?

15. Can you find the ratio of 2 L to 4 km? Explain.

16. The ratio of the lengths of two segments is $4:3$ when they are measured in centimeters. What is their ratio when they are measured in inches?

17. Three numbers aren't known, but the ratio of the numbers is $1:2:5$. Is it possible that the numbers are 1, 2, and 5? 10, 20, and 50? 3, 6, and 20? x, $2x$, and $5x$?

18. What is the second term of the proportion $\dfrac{a}{b} = \dfrac{x}{y}$?

Written Exercises

ABCD **is a parallelogram. Find the value of each ratio.**

A　**1.** $AB:BC$　　　　　　　**2.** $AB:CD$

　　3. $m\angle C:m\angle D$　　　　**4.** $m\angle B:m\angle C$

　　5. $AD:$ perimeter of $ABCD$

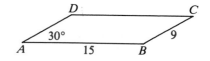

In Exercises 6–14, $x = 12$, $y = 10$, and $z = 24$. Write each ratio in simplest form.

6. x to y　　　　　　**7.** z to x　　　　　　**8.** $x + y$ to z

9. $\dfrac{x}{x + z}$　　　　　**10.** $\dfrac{x + y}{z + y}$　　　　**11.** $\dfrac{y + z}{x - y}$

12. $x:y:z$　　　　　**13.** $z:x:y$　　　　**14.** $x:(x + y):(y + z)$

Exercises 15–20 refer to a triangle. Express the ratio of the height to the base in simplest form.

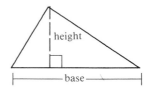

height

base

	15.	16.	17.	18.	19.	20.
height	5 km	1 m	0.6 km	1 m	8 cm	40 mm
base	45 km	0.6 m	0.8 km	85 cm	50 mm	0.2 m

Write the algebraic ratio in simplest form.

21. $\dfrac{3a}{4ab}$

22. $\dfrac{2cd}{5c^2}$

23. $\dfrac{3(x+4)}{a(x+4)}$

In Exercises 24–29 find the measure of each angle.

B **24.** The ratio of the measures of two complementary angles is $4:5$.

25. The ratio of the measures of two supplementary angles is $11:4$.

26. The measures of the angles of a triangle are in the ratio $3:4:5$.

27. The measures of the acute angles of a right triangle are in the ratio $5:7$.

28. The measures of the angles of an isosceles triangle are in the ratio $3:3:2$.

29. The measures of the angles of a hexagon are in the ratio $4:5:5:8:9:9$.

30. The perimeter of a triangle is 132 cm and the lengths of its sides are in the ratio $8:11:14$. Find the length of each side.

31. The measures of the consecutive angles of a quadrilateral are in the ratio $5:7:11:13$. Find the measure of each angle, draw a quadrilateral that satisfies the requirements, and explain why two sides must be parallel.

32. What is the ratio of the measure of an interior angle to the measure of an exterior angle in a regular hexagon? A regular decagon? A regular *n*-gon?

33. A team's best hitter has a lifetime batting average of .320. He has been at bat 325 times.
   a. How many hits has he made?
   b. The same player goes into a slump and doesn't get any hits at all in his next ten times at bat. What is his current batting average to the nearest thousandth?

C **34.** A basketball player has made 24 points out of 30 free throws. She hopes to make all her next free throws until her free-throw percentage is 85 or better. How many consecutive free throws will she have to make?

35. Points *B* and *C* lie on $\overline{AD}$. Find *AC* if $\dfrac{AB}{BD} = \dfrac{3}{4}$, $\dfrac{AC}{CD} = \dfrac{5}{6}$, and $BD = 66$.

36. Find the ratio of *x* to *y*: $\dfrac{4}{y} + \dfrac{3}{x} = 44$

$$\dfrac{12}{y} - \dfrac{2}{x} = 44$$

7-2 *Properties of Proportions*

The first and last terms of a proportion are called the *extremes*. The middle terms are the *means*. In the proportions below, the extremes are shown in red. The means are shown in black.

$$a:b = c:d \qquad 6:9 = 2:3 \qquad \frac{6}{9} = \frac{2}{3}$$

Notice that $6 \cdot 3 = 9 \cdot 2$. This illustrates a property of all proportions, called the *means-extremes* property of proportions:

> The product of the extremes equals the product of the means.

$$\frac{a}{b} = \frac{c}{d} \text{ is } \textit{equivalent} \text{ to } ad = bc.$$

The two equations are equivalent because we can change either of them into the other by multiplying (or dividing) each side by bd. Try this yourself.

It is often necessary to replace one proportion by an equivalent proportion. When you do so in a proof, you can use the reason "A property of proportions." The following properties will be justified in the exercises.

Properties of Proportions

1. $\dfrac{a}{b} = \dfrac{c}{d}$ is equivalent to:

    a. $ad = bc$ **b.** $\dfrac{a}{c} = \dfrac{b}{d}$ **c.** $\dfrac{b}{a} = \dfrac{d}{c}$ **d.** $\dfrac{a + b}{b} = \dfrac{c + d}{d}$

2. If $\dfrac{a}{b} = \dfrac{c}{d} = \dfrac{e}{f} = \cdots$, then $\dfrac{a + c + e + \cdots}{b + d + f + \cdots} = \dfrac{a}{b} = \cdots$.

Example   Use the proportion $\dfrac{x}{y} = \dfrac{5}{2}$ to complete each statement.

    a. $5y = \underline{\ ?\ }$ **b.** $\dfrac{x + y}{y} = \dfrac{?}{?}$

    c. $\dfrac{2}{5} = \dfrac{?}{?}$ **d.** $\dfrac{x}{5} = \dfrac{?}{?}$

Solution   **a.** $5y = 2x$ **b.** $\dfrac{x + y}{y} = \dfrac{7}{2}$

    c. $\dfrac{2}{5} = \dfrac{y}{x}$ **d.** $\dfrac{x}{5} = \dfrac{y}{2}$

Classroom Exercises

1. If $\frac{e}{f} = \frac{g}{h}$, which equation is correct?

   a. $ef = gh$ **b.** $eh = fg$ **c.** $eg = fh$

2. Which proportions are equivalent to $\frac{x}{12} = \frac{3}{4}$?

   a. $\frac{x}{3} = \frac{12}{4}$ **b.** $\frac{x}{4} = \frac{12}{3}$ **c.** $\frac{12}{x} = \frac{4}{3}$ **d.** $\frac{x + 12}{12} = \frac{7}{4}$

Complete the statement.

3. If $\frac{a}{b} = \frac{2}{3}$, then $3a = \underline{\ ?\ }$.

4. If $\frac{c}{d} = \frac{4}{7}$, then $\frac{d}{c} = \frac{?}{?}$.

5. If $\frac{e}{f} = \frac{5}{9}$, then $\frac{e}{5} = \frac{?}{?}$.

6. If $\frac{g}{h} = \frac{j}{8}$, then $\frac{j}{g} = \frac{?}{?}$.

7. If $\frac{k}{m} = \frac{2}{3}$, then $\frac{k + m}{m} = \frac{?}{?}$.

8. If $\frac{n}{p} = \frac{q}{r} = \frac{7}{9}$, then $\frac{n + q + 7}{p + r + 9} = \frac{?}{?}$.

9. **a.** Apply the means-extremes property of proportions to the proportion

   $\frac{e}{f} = \frac{g}{5}$ and you get $5e = \underline{\ ?\ }$.

   b. Apply the property to the proportion $\frac{5}{f} = \frac{g}{e}$ and you get $\underline{\ ?\ } = \underline{\ ?\ }$.

   c. Are the proportions $\frac{e}{f} = \frac{g}{5}$ and $\frac{5}{f} = \frac{g}{e}$ equivalent? Why?

10. Explain an easy way to show that the proportions $\frac{x}{7} = \frac{2}{3}$ and $\frac{x}{2} = \frac{3}{7}$ are not equivalent.

11. Apply the means-extremes property to $\frac{x}{10} = \frac{4}{5}$ and you get $5x = \underline{\ ?\ }$ and $x = \underline{\ ?\ }$.

12. If $\frac{4}{y} = \frac{7}{9}$, then $\underline{\ ?\ } = \underline{\ ?\ }$ and $y = \underline{\ ?\ }$.

What can you conclude from the given information?

13. $\frac{b}{a} = \frac{t}{x}$ and $\frac{a}{b} = \frac{x}{p}$

14. $\frac{2}{5} = \frac{y}{k}$ and $\frac{2}{z} = \frac{5}{k}$

15. Apply the means-extremes property to $\frac{a}{b} = \frac{c}{d}$ and also to $\frac{a}{c} = \frac{b}{d}$.
    (Note that you have justified Property 1(b) on page 245 by showing that each proportion is equivalent to the same equation.)

16. Explain why $\frac{a}{b} = \frac{c}{d}$ and $\frac{b}{a} = \frac{d}{c}$ are equivalent. (This justifies Property 1(c) on page 245.)

Written Exercises

Complete each statement.

A

1. If $\dfrac{x}{3} = \dfrac{2}{5}$, then $5x =$ ___?___ .

2. If $\dfrac{4}{x} = \dfrac{2}{7}$, then $2x =$ ___?___ .

3. If $a:3 = 7:4$, then $4a =$ ___?___ .

4. If $4:t = 8:9$, then $8t =$ ___?___ .

5. If $\dfrac{a}{4} = \dfrac{b}{7}$, then $\dfrac{a}{b} = \dfrac{?}{?}$.

6. If $\dfrac{x}{y} = \dfrac{3}{8}$, then $\dfrac{y}{x} = \dfrac{?}{?}$.

7. If $\dfrac{x}{2} = \dfrac{y}{3}$, then $\dfrac{x+2}{2} =$ ___?___ .

8. If $\dfrac{a}{b} = \dfrac{5-x}{x}$, then $\dfrac{a+b}{b} =$ ___?___ .

Find the value of x.

9. $\dfrac{x}{4} = \dfrac{3}{5}$

10. $\dfrac{4}{x} = \dfrac{2}{5}$

11. $\dfrac{2}{5} = \dfrac{3x}{7}$

12. $\dfrac{8}{x} = \dfrac{2}{5}$

13. $\dfrac{x+5}{4} = \dfrac{1}{2}$

14. $\dfrac{x+3}{2} = \dfrac{4}{3}$

15. $\dfrac{x+2}{x+3} = \dfrac{4}{5}$

16. $\dfrac{2x+1}{4x-1} = \dfrac{2}{3}$

17. $\dfrac{x+3}{2} = \dfrac{2x-1}{3}$

18. $\dfrac{x+4}{x-4} = \dfrac{6}{5}$

19. $\dfrac{7}{6x-4} = \dfrac{9}{4x+6}$

20. $\dfrac{3x+5}{3} = \dfrac{18x+5}{7}$

For the figure shown, it is given that $\dfrac{KR}{RT} = \dfrac{KS}{SU}$. Copy and complete the table.

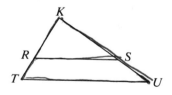

	KR	RT	KT	KS	SU	KU
21.	12	9	?	16	?	?
22.	8	?	10	12	?	?
23.	16	? 6	?	? 20	10	30
24.	?	2	?	9	?	12
25.	?	?	12	10	5	?
26.	12	4	?	?	?	20
27.	?	9	36	?	?	48
28.	?	?	30	28	?	42

B

(*Hint for Ex. 25:* Let $KR = x$, then $RT = 12 - x$.)

29. Show that the proportions $\dfrac{a+b}{b} = \dfrac{c+d}{d}$ and $\dfrac{a}{b} = \dfrac{c}{d}$ are equivalent.
(Note that this exercise justifies property 1(d) on page 245.)

30. Given the proportions $\dfrac{x+y}{y} = \dfrac{r}{s}$ and $\dfrac{x-y}{x+y} = \dfrac{s}{y}$, what can you conclude?

Show that the given proportions are equivalent.

31. $\dfrac{a - b}{a + b} = \dfrac{c - d}{c + d}$ and $\dfrac{a}{b} = \dfrac{c}{d}$

32. $\dfrac{a + c}{b + d} = \dfrac{a - c}{b - d}$ and $\dfrac{a}{b} = \dfrac{c}{d}$

Find the value of x.

33. $\dfrac{x}{x + 5} = \dfrac{x - 4}{x}$

34. $\dfrac{x - 2}{x} = \dfrac{x}{x + 3}$

35. $\dfrac{x + 1}{x - 2} = \dfrac{x + 5}{x - 6}$

C **36.** $\dfrac{x - 1}{x - 2} = \dfrac{x + 4}{x + 2}$

37. $\dfrac{x(x + 5)}{4x + 4} = \dfrac{9}{5}$

38. $\dfrac{x - 1}{x + 2} = \dfrac{10}{3x - 2}$

Find the values of x and y.

39. $\dfrac{y}{x - 9} = \dfrac{4}{7}$

  $\dfrac{x + y}{x - y} = \dfrac{5}{3}$

40. $\dfrac{x - 3}{4} = \dfrac{y + 2}{2}$

  $\dfrac{x + y - 1}{6} = \dfrac{x - y + 1}{5}$

41. Prove: If $\dfrac{a}{b} = \dfrac{c}{d} = \dfrac{e}{f}$, then $\dfrac{a + c + e}{b + d + f} = \dfrac{a}{b}$. (*Hint*: Let $\dfrac{a}{b} = r$. Then $a = br$, $c = dr$, and $e = \underline{\ ?\ }$.)

42. Explain how to extend the proof of Exercise 41 to justify Property 2 on page 245.

43. If $\dfrac{4a - 9b}{4a} = \dfrac{a - 2b}{b}$, find the numerical value of the ratio $a : b$.

 *Similar Polygons*

When you draw a diagram of a soccer field, you don't need an enormous piece of paper. You use a convenient sheet and draw *to scale*. That is, you show the right shape, but in a convenient size. Two figures, such as those below, that have the same shape are called *similar*.

Two polygons are **similar** if their vertices can be paired so that:

(1) Corresponding angles are congruent.

(2) Corresponding sides are in proportion. (Their lengths have the same ratio.)

When you refer to similar polygons, their corresponding vertices must be listed in the same order. If polygon *PQRST* is similar to polygon *VWXYZ*, you write polygon *PQRST* ~ polygon *VWXYZ*.

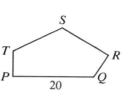 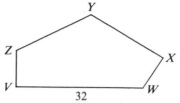

From the definition of similar polygons, we have:

(1) $\angle P \cong \angle V$ $\angle Q \cong \angle W$ $\angle R \cong \angle X$ $\angle S \cong \angle Y$ $\angle T \cong \angle Z$

(2) $\dfrac{PQ}{VW} = \dfrac{QR}{WX} = \dfrac{RS}{XY} = \dfrac{ST}{YZ} = \dfrac{TP}{ZV}$

Similarity has some of the same properties as equality and congruence (page 37). Similarity is reflexive, symmetric, and transitive.

If two polygons are similar, then the ratio of the lengths of two corresponding sides is called the **scale factor** of the similarity. The scale factor of pentagon *PQRST* to pentagon *VWXYZ* is $\dfrac{PQ}{VW} = \dfrac{20}{32} = \dfrac{5}{8}$.

The example that follows shows one convenient way to label corresponding vertices: *A* and *A'* (read *A* prime), *B* and *B'*, and so on.

Example Quad. *ABCD* ~ quad. *A'B'C'D'*. Find:

    a. their scale factor

    b. the values of *x*, *y*, and *z*

    c. the perimeters of the two quadrilaterals

    d. the ratio of the perimeters

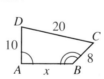 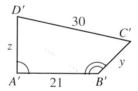

Solution **a.** The scale factor is $\dfrac{DC}{D'C'} = \dfrac{20}{30} = \dfrac{2}{3}$.

    b. $\dfrac{DC}{D'C'} = \dfrac{AB}{A'B'}$ $\dfrac{DC}{D'C'} = \dfrac{BC}{B'C'}$ $\dfrac{DC}{D'C'} = \dfrac{AD}{A'D'}$

         $\dfrac{2}{3} = \dfrac{x}{21}$ $\dfrac{2}{3} = \dfrac{8}{y}$ $\dfrac{2}{3} = \dfrac{10}{z}$

         $x = 14$ $y = 12$ $z = 15$

    c. The perimeter of quad. *ABCD* is $10 + 20 + 8 + 14 = 52$.

        The perimeter of quad. *A'B'C'D'* is $15 + 30 + 12 + 21 = 78$.

    d. The ratio of the perimeters is $\dfrac{52}{78}$, or $\dfrac{2}{3}$.

If you compare the ratio of the perimeters with the scale factor of the similarity, you discover they are the same. This property will be discussed further in Exercise 23 on page 251 and in Theorem 11-7.

Classroom Exercises

Are the quadrilaterals similar? If they aren't, tell why not.

1. *ABCD* and *EFGH*　　　　　　　　　**2.** *ABCD* and *JKLM*

3. *ABCD* and *NOPQ*　　　　　　　　　**4.** *JKLM* and *NOPQ*

5. If the corresponding angles of two polygons are congruent, must the polygons be similar?

6. If the corresponding sides of two polygons are in proportion, must the polygons be similar?

7. Two polygons are similar. Do they have to be congruent?

8. Two polygons are congruent. Do they have to be similar?

9. Are all regular pentagons similar?

10. Quad. *JUDY* ~ quad. *J'U'D'Y'*. Complete.
  a. $m \angle Y' = \underline{}$ and $m \angle D = \underline{}$.
  b. The scale factor of quad. *JUDY* to quad. *J'U'D'Y'* is $\underline{}$.
  c. Find *DU*, *Y'J'*, and *J'U'*.
  d. The ratio of the perimeters is $\underline{}$.
  e. Explain why it is not true that quad. *DUJY* ~ quad. *Y'J'U'D'*.

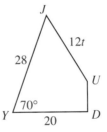
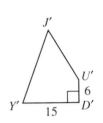

Written Exercises

Tell whether the two polygons are *always*, *sometimes*, or *never* similar.

1. Two equilateral triangles　　　　　　**2.** Two right triangles

3. Two isosceles triangles　　　　　　　**4.** Two scalene triangles

5. Two squares　　　　　　　　　　　　**6.** Two rectangles

7. Two rhombuses　　　　　　　　　　　**8.** Two isosceles trapezoids

9. Two regular hexagons　　　　　　　**10.** Two regular polygons

11. A right triangle and an acute triangle

12. An isosceles triangle and a scalene triangle

13. A right triangle and a scalene triangle

14. An equilateral triangle and an equiangular triangle

In Exercises 15-23 quad. *TUNE* ~ quad. *T'U'N'E'*.

15. What is the scale factor of quad. *TUNE* to quad. *T'U'N'E'*?

16. What special kind of quadrilateral must quad. *T'U'N'E'* be? Explain.

17. Find $m \angle T'$.   **18.** Find $m \angle E'$.

19. Find *UN*.   **20.** Find *T'U'*.

21. Find *TE*.   **22.** Find the ratio of the perimeters.

B **23.** What property of proportions on page 245 would you use to show that the ratio of the perimeters is equal to the ratio of the lengths of any two corresponding sides?

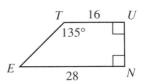
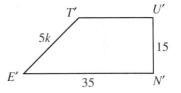

Two similar polygons are shown. Find the values of *x*, *y*, and *z*.

24.

25.

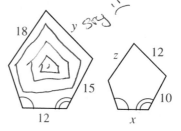

26.

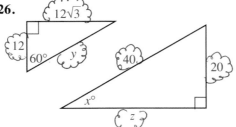

27.

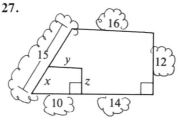

28. Draw two equilateral hexagons that are clearly not similar.

29. Draw two equiangular hexagons that are clearly not similar.

30. If $\triangle ABC \sim \triangle DEF$, express *AB* in terms of other lengths. (There are two possible answers.)

31. Explain how you can tell at once that quadrilateral *RSWX* is not similar to quadrilateral *RSYZ*.

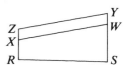

Plot the given points on graph paper. Draw quadrilateral *ABCD* and $\overline{A'B'}$. Locate points *C'* and *D'* so that *A'B'C'D'* is similar to *ABCD*.

32. $A(0, 0)$, $B(4, 0)$, $C(2, 4)$, $D(0, 2)$, $A'(-10, -2)$, $B'(-2, -2)$

33. $A(0, 0)$, $B(4, 0)$, $C(2, 4)$, $D(0, 2)$, $A'(7, 2)$, $B'(7, 0)$

34. The card shown was cut into four congruent pieces with each piece similar to the original. Find the value of x.

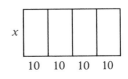

35. Quad. *WHAT* is a figure such that $WHAT \sim HATW$. Find the measure of each angle. What special kind of figure must the quadrilateral be?

C 36. What can you deduce from the diagram shown at the right? Explain.

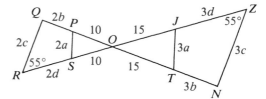

37. The large rectangle shown is a *golden rectangle*. This means that when a square is cut off, the rectangle that remains is similar to the original rectangle.
 a. How wide is the original rectangle?
 b. The ratio of length to width in a golden rectangle is called the *golden ratio*. Write the golden ratio in simplified radical form. Then use a calculator to find an approximation to the nearest hundredth.

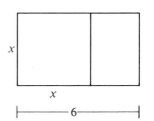

Self-Test 1

Express the ratio in simplest form.

1. $9:15$

2. 60 cm to 2 m

3. $\dfrac{4ab}{6b^2}$

Solve for x.

4. $\dfrac{x}{8} = \dfrac{9}{12}$

5. $\dfrac{x-2}{2} = \dfrac{x+6}{4}$

6. $\dfrac{x}{5-x} = \dfrac{12}{8}$

Tell whether the equation is equivalent to the proportion $\dfrac{a}{b} = \dfrac{5}{7}$.

7. $\dfrac{a}{7} = \dfrac{b}{5}$

8. $7a = 5b$

9. $\dfrac{a+b}{b} = \dfrac{12}{7}$

10. If $\triangle ABC \sim \triangle RST$, $m\angle A = 45$, and $m\angle C = 60$, then $m\angle R = \underline{\ ?\ }$, $m\angle T = \underline{\ ?\ }$, and $m\angle S = \underline{\ ?\ }$.

The quadrilaterals shown are similar.

11. The scale factor of the smaller quadrilateral to the larger quadrilateral is $\underline{\ ?\ }$.

12. $x = \underline{\ ?\ }$ **13.** $y = \underline{\ ?\ }$ **14.** $z = \underline{\ ?\ }$

15. The measures of the angles of a hexagon are in the ratio $5:5:5:6:7:8$. Find the measures.

◆ Calculator Key-In

Before the pediment on top of the Parthenon in Athens was destroyed, the front of the building fit almost exactly into a golden rectangle. In a **golden rectangle,** the length l and width w satisfy the equation $\dfrac{l}{w} = \dfrac{l + w}{l}$. The ratio $\dfrac{l}{w}$ is called the **golden ratio.**

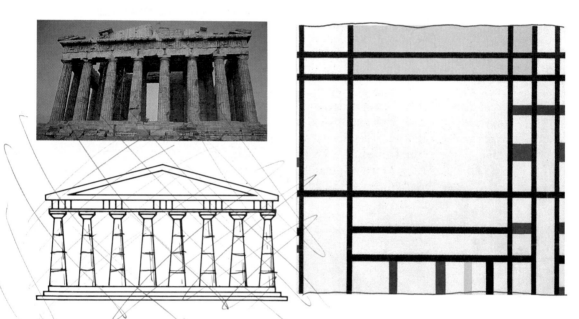

Over the centuries, artists and architects have found the golden rectangle to be especially pleasing to the eye. How many golden rectangles can you find in the painting by Piet Mondrian (1872–1944) that is shown?

Exercises

1. A regular pentagon is shown. It happens to be true that $\dfrac{AD}{AC}$, $\dfrac{AC}{AB}$, and $\dfrac{AB}{BC}$ all equal the golden ratio. Measure the appropriate lengths to the nearest millimeter and compute the ratios with a calculator.

2. From the equation $\dfrac{l}{w} = \dfrac{l + w}{l}$ it can be shown that the numerical value of $\dfrac{l}{w}$ is $\dfrac{1 + \sqrt{5}}{2}$. Express the value of $\dfrac{l}{w}$, the golden ratio, as a decimal.

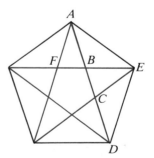

Explorations

These exploratory exercises can be done using a computer with a program that draws and measures geometric figures.

Draw any triangle and a median of the triangle. Measure and record the lengths of the sides and the median, the measures of the angles, and the perimeter of the triangle.

Change the scale of your triangle. Remeasure and record the lengths of the sides and the median, the measures of the angles, and the perimeter of the triangle.

Compare the measurements of the corresponding angles of the two triangles. What do you notice?

Divide the length of each side of the original triangle by the length of the corresponding side of the second triangle. What do you notice?

What do you know about the two triangles?

Divide the length of the median of the original triangle by the length of the median of the second triangle. What do you notice?

Divide the perimeter of the original triangle by the perimeter of the second triangle. What do you notice?

Working with Similar Triangles

Objectives

1. Use the AA Similarity Postulate, the SAS Similarity Theorem, and the SSS Similarity Theorem to prove triangles similar.
2. Use similar triangles to deduce information about segments or angles.
3. Apply the Triangle Proportionality Theorem and its corollary.
4. State and apply the Triangle Angle-Bisector Theorem.

7-4 *A Postulate for Similar Triangles*

You can always prove that two triangles are similar by showing that they satisfy the definition of similar polygons. However, there are simpler methods. The following experiment suggests the first of these methods: Two triangles are similar whenever two pairs of angles are congruent.

1. Draw any two segments $\overline{AB}$ and $\overline{A'B'}$.

2. Draw any angle at A and a congruent angle at A'. Draw any angle at B and a congruent angle at B'. Label points C and C' as shown. $\angle ACB \cong \angle A'C'B'$. (Why?)

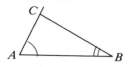

3. Measure each pair of corresponding sides and compute an approximate decimal value for the ratio of their lengths:

$$\frac{AB}{A'B'} \qquad \frac{BC}{B'C'} \qquad \frac{AC}{A'C'}$$

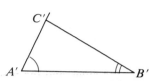

4. Are the ratios computed in Step 3 approximately the same?

If you worked carefully, your answer in Step 4 was *yes*. Corresponding angles of the two triangles are congruent and corresponding sides are in proportion. By the definition of similar polygons, $\triangle ABC \sim \triangle A'B'C'$.

Whenever you draw two triangles with two angles of one triangle congruent to two angles of the other, you will find that the third angles are also congruent and that corresponding sides are in proportion.

Postulate 15 *AA Similarity Postulate*

If two angles of one triangle are congruent to two angles of another triangle, then the triangles are similar.

Example

Given: $\angle H$ and $\angle F$ are rt. $\angle$s.
Prove: $HK \cdot GO = FG \cdot KO$

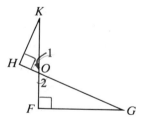

Plan for Proof: You can prove that $HK \cdot GO = FG \cdot KO$ if you show that $\dfrac{HK}{FG} = \dfrac{KO}{GO}$. You can get this proportion if you show that $\triangle HKO \sim \triangle FGO$.

Proof:

Statements	Reasons
1. $\angle 1 \cong \angle 2$	1. Vertical $\angle$s are $\cong$.
2. $\angle H$ and $\angle F$ are rt. $\angle$s.	2. Given
3. $m \angle H = 90 = m \angle F$	3. Def. of a rt. $\angle$
4. $\angle H \cong \angle F$	4. Def. of $\cong$ $\angle$s
5. $\triangle HKO \sim \triangle FGO$	5. AA Similarity Postulate
6. $\dfrac{HK}{FG} = \dfrac{KO}{GO}$	6. Corr. sides of $\sim$ $\triangle$ are in proportion.
7. $HK \cdot GO = FG \cdot KO$	7. A property of proportions

The example shows one way to prove that the product of the lengths of two segments is equal to the product of the lengths of two other segments. You prove that two triangles are similar, write a proportion, and then apply the means-extremes property of proportions.

Classroom Exercises

In Exercises 1–8 $\triangle ABC \sim \triangle DEF$. Tell whether each statement must be true.

1. $\triangle BAC \sim \triangle EFD$
2. If $m \angle D = 45$, then $m \angle A = 45$.
3. If $m \angle B = 70$, then $m \angle F = 70$.
4. $AB:DE = EF:BC$
5. $AC:DF = AB:DE$
6. If $\dfrac{DF}{AC} = \dfrac{8}{5}$, then $\dfrac{m \angle D}{m \angle A} = \dfrac{8}{5}$.
7. If $\dfrac{DF}{AC} = \dfrac{8}{5}$, then $\dfrac{EF}{BC} = \dfrac{8}{5}$.

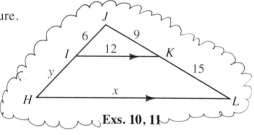

8. If the scale factor of $\triangle ABC$ to $\triangle DEF$ is 5 to 8, then the scale factor of $\triangle DEF$ to $\triangle ABC$ is 8 to 5.

9. One right triangle has an angle with measure 37. Another right triangle has an angle with measure 53. Are the two triangles similar? Explain.

10. Name all pairs of congruent angles in the figure.

11. Complete.
   a. $\triangle IKJ \sim \underline{\quad ? \quad}$
   b. $\dfrac{?}{x} = \dfrac{9}{24}$ and $x = \underline{\quad ? \quad}$
   c. $\dfrac{9}{24} = \dfrac{6}{?}$ and $y = \underline{\quad ? \quad}$

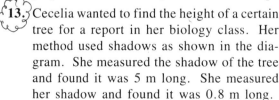

Exs. 10, 11

12. Suppose you want to show $AB \cdot YZ = CD \cdot WX$. What are some proportions that are equivalent to that equation?

13. Cecelia wanted to find the height of a certain tree for a report in her biology class. Her method used shadows as shown in the diagram. She measured the shadow of the tree and found it was 5 m long. She measured her shadow and found it was 0.8 m long.
   a. $\triangle \underline{\ ? \ } \sim \triangle \underline{\ ? \ }$
   b. Complete: $\dfrac{SC}{?} = \dfrac{CH}{?}$
   c. If Cecelia is 1.6 m tall, about how tall is the tree?

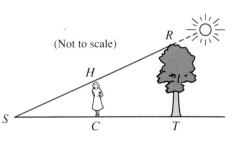

(Not to scale)

Written Exercises

Tell whether the triangles are similar or not similar. If you can't reach a conclusion, write *no conclusion is possible*.

A **1.**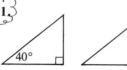

2.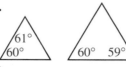

3.

4.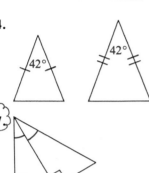

5.

no conclusion

6.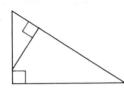

7.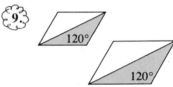

8.

Trapezoid given

9.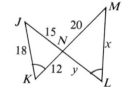

Parallelograms given

10. Complete.

   a. $\triangle JKN \sim$ ___?___

   b. $\dfrac{JK}{?} = \dfrac{JN}{?} = \dfrac{KN}{?}$

   c. $\dfrac{15}{?} = \dfrac{18}{?}$ and $\dfrac{15}{?} = \dfrac{12}{?}$

   d. $x =$ ___?___ and $y =$ ___?___

Find the values of x and y.

11.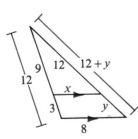

12.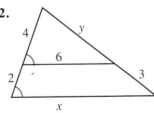

13.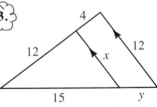

B **14. a.** Name two triangles that are similar to $\triangle ABC$.

   b. Find the values of x and y.

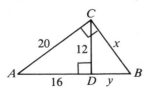

15. If $IV = 36$ m, $VE = 20$ m, and $EB = 15$ m, find the width, RI, of the river.

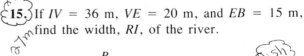

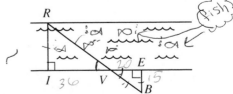

16. To estimate the height of a pole, a basketball player exactly 2 m tall stood so that the ends of his shadow and the shadow of the pole coincided. He found that $\overline{DE}$ and $\overline{DF}$ measured 1.6 m and 4.4 m, respectively. About how tall was the pole?

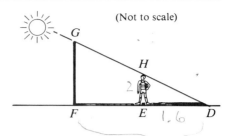

(Not to scale)

17. The diagram, *not* drawn to scale, shows a film being projected on a screen. $LF = 6$ cm and $LS = 24$ m. The screen image is 2.2 m tall. How tall is the film image?

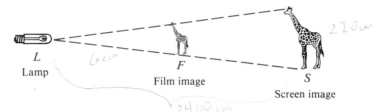

L
Lamp

F
Film image

S
Screen image

In Exercises 18 and 19 *ABCD* is a parallelogram. Find the values of *x* and *y*.

18.

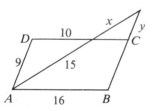

19.

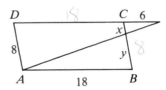

20. You can estimate the height of a flagpole by placing a mirror on level ground so that you see the top of the flagpole in it. The girl shown is 172 cm tall. Her eyes are about 12 cm from the top of her head. By measurement, *AM* is about 120 cm and *A'M* is about 4.5 m. From physics it is known that $\angle 1 \cong \angle 2$. Explain why the triangles are similar and find the approximate height of the pole.

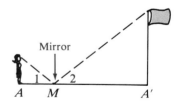

Mirror

21. Given: $\overline{EF} \parallel \overline{RS}$
Prove: **a.** $\triangle FXE \sim \triangle SXR$
    b. $\dfrac{FX}{SX} = \dfrac{EF}{RS}$

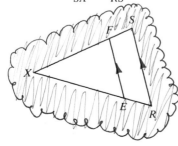

22. Given: $\angle 1 \cong \angle 2$
Prove: **a.** $\triangle JIG \sim \triangle JZY$
    b. $\dfrac{JG}{JY} = \dfrac{GI}{YZ}$

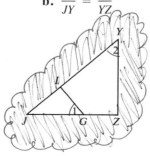

23. Given: $\angle B \cong \angle C$
    Prove: $NM \cdot CM = LM \cdot BM$

24. Given: $\overline{BN} \parallel \overline{LC}$
    Prove: $BN \cdot LM = CL \cdot NM$

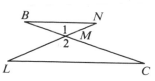

25. Given: $\triangle ABC \sim \triangle XYZ$;
    $\overline{AD}$ and $\overline{XW}$ are altitudes.
    Prove: $\dfrac{AD}{XW} = \dfrac{AB}{XY}$

26. Given: $\triangle PQR \sim \triangle GHI$;
    $\overrightarrow{PS}$ and $\overrightarrow{GJ}$ are angle bisectors.
    Prove: $\dfrac{PS}{GJ} = \dfrac{PQ}{GH}$

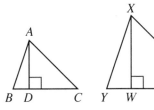

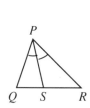

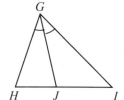

27. Given: $\overline{AH} \perp \overline{EH}$; $\overline{AD} \perp \overline{DG}$
    Prove: $AE \cdot DG = AG \cdot HE$

28. Given: $\overline{QT} \parallel \overline{RS}$
    Prove: $\dfrac{QU}{RV} = \dfrac{UT}{VS}$

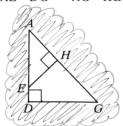

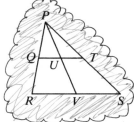

29. Given: $\angle 1 \cong \angle 2$
    Prove: $(AB)^2 = AD \cdot AC$

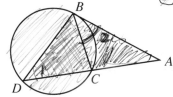

In the diagram for Exercises 30 and 31, the plane of
$\triangle A'B'C'$ is parallel to the plane of $\triangle ABC$.

30. $VA' = 15$ and $A'A = 20$
    a. If $VC' = 18$, then $VC = \underline{\ ?\ }$.
    b. If $VB = 49$, then $BB' = \underline{\ ?\ }$.
    c. If $A'B' = 24$, then $AB = \underline{\ ?\ }$.

31. If $VA' = 10$, $VA = 25$, $AB = 20$, $BC = 14$,
    and $AC = 16$, find the perimeter of $\triangle A'B'C'$.

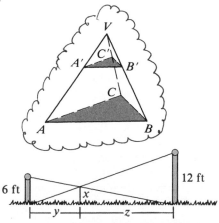

C **32.** Two vertical poles have heights 6 ft and 12 ft.
    A rope is stretched from the top of each pole
    to the bottom of the other. How far above the
    ground do the ropes cross? (*Hint*: The lengths
    y and z do not affect the answer.)

In Exercises 33 and 34 write a paragraph proof for anything you are asked to prove.

33. Given: Regular pentagon *ABCDE*

    a. Make a large copy of the diagram.

    b. Write the angle measures on your diagram.

    c. Prove that $\dfrac{DA}{DK} = \dfrac{DK}{AK}$.

★ 34. *ABCD* is a square.

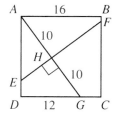

    a. Find the distance from *H* to each side of the square.

    b. Find *BF*, *FC*, *CG*, *DE*, *EA*, *EH*, and *HF*.

★ 35. Related to any doubly convex lens there is a focal distance *OF*. Physicists have determined experimentally that a vertical lens, a vertical object $\overline{JT}$ (with $\overline{JO}$ horizontal), a vertical image $\overline{IM}$, and a focus *F* are related as shown in the diagram. Once the relationship is known, geometry can be used to establish a lens law:

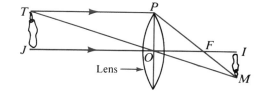

$$\frac{1}{\text{object distance}} + \frac{1}{\text{image distance}} = \frac{1}{\text{focal distance}}$$

    a. Prove that $\dfrac{1}{OJ} + \dfrac{1}{OI} = \dfrac{1}{OF}$.

    b. Show algebraically that $OF = \dfrac{OJ \cdot OI}{OJ + OI}$.

Challenges

1. Explain how to pass a plane through a cube so that the intersection is:
   a. an equilateral triangle
   b. a trapezoid
   c. a pentagon
   d. a hexagon

2. The six edges of the three-dimensional figure are congruent. Each of the four corners is cut off by a plane that passes through the midpoints of the three edges that intersect at that corner. For example, corner A is cut off by plane *MNT*. Describe the three-dimensional figure that remains.

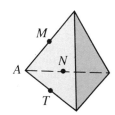

◆ Computer Key-In

The sequence 1, 1, 2, 3, 5, 8, 13, 21, . . . is called a *Fibonacci sequence* after its discoverer, Leonardo Fibonacci, a thirteenth century mathematician. The first two terms are 1 and 1. You then add two consecutive terms to get the next term.

$$\frac{1\text{st}}{\text{term}} + \frac{2\text{nd}}{\text{term}} = \frac{3\text{rd}}{\text{term}} \qquad \frac{2\text{nd}}{\text{term}} + \frac{3\text{rd}}{\text{term}} = \frac{4\text{th}}{\text{term}} \qquad \frac{3\text{rd}}{\text{term}} + \frac{4\text{th}}{\text{term}} = \frac{5\text{th}}{\text{term}}$$

The following computer program computes the first twenty-five terms of the Fibonacci sequence shown above and finds the ratio of any term to its preceding term. For example, we want to look at the ratios

$$\frac{1}{1} = 1, \ \frac{2}{1} = 2, \ \frac{3}{2} = 1.5, \ \frac{5}{3} \approx 1.66667, \ \text{and so on.}$$

```
10  PRINT "TERM NO.", "TERM", "RATIO"
20  LET A = 1
30  LET B = 1
40  PRINT "1",A,"-"              100    PRINT N,B,G
50  FOR N = 2 TO 25              110    LET C = B + A
60    LET D = B/A                120    LET A = B
70    LET E = 10000 * D          130    LET B = C
80    LET F = INT(E)             140  NEXT N
90    LET G = F/10000            150  END
```

Exercises

Type the program into your computer and use it in Exercises 1–4.

1. RUN the given computer program. As the terms become larger, what happens to the values of the ratios?

2. Suppose another sequence is formed by choosing starting numbers different from 1 and 1. For example, suppose the sequence is 3, 11, 14, 25, 39, . . . , where the pattern for creating the terms of the sequence is still the same. Change lines 20 and 30 to:

   ```
   20  LET A = 3
   30  LET B = 11
   ```

   RUN the modified program. What happens to the values of the ratios as the terms become larger and larger?

3. Modify the program again so that another pair of starting numbers is used and the first thirty terms are computed. RUN the program. What can you conclude from the results?

4. Compare this ratio to the golden ratio calculated in Exercise 2 on page 253. Do you see a connection?

Application   *Scale Drawings*

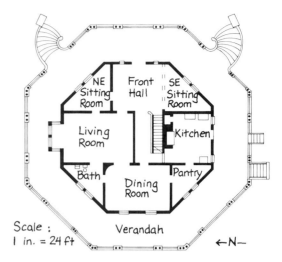

Scale :
1 in. = 24 ft   Verandah   ←N—

This "octagon house" was built in Irvington, New York, in 1860. The plan shows the rooms on the first floor. The scale on this *scale drawing* tells you that a length of 1 in. on the plan represents a true length of 24 ft.

$$\frac{\text{Plan length in inches}}{\text{True length in feet}} = \frac{1}{24}$$

The following examples show how you can use this formula to find actual dimensions of the house from the plan or to convert dimensions of full-sized objects to plan size.

The verandah measures $\frac{3}{8}$ in. wide on the plan. Find its true width, T.

$\frac{\frac{3}{8}}{T} = \frac{1}{24}$, so $1 \cdot T = \frac{3}{8} \cdot 24$

$T = 9$ The real verandah is 9 ft wide.

A sofa is 6 ft long. Find its plan length, P.

$\frac{P}{6} = \frac{1}{24}$, so $24 \cdot P = 6 \cdot 1$

$P = \frac{1}{4}$ The plan length is $\frac{1}{4}$ in.

Exercises

1. Find the true length and width of the dining room.

2. A rug measures 9 ft by $7\frac{1}{2}$ ft. What would its dimensions be on the floor plan? Would it fit in the northeast sitting room?

3. If a new floor plan is drawn with a scale of 1 in. = 10 ft, how many times longer is each line segment on the new plan than the corresponding segment on the plan shown?

4. Suppose that on the architect's drawings each side of the verandah (the outer octagon) measured 12 in. What was the scale of these drawings?

7-5 *Theorems for Similar Triangles*

You can prove two triangles similar by using the definition of similar polygons or by using the AA Postulate. Of course, in practice you would always use the AA Postulate instead of the definition. (Why?) Two additional methods are established in the theorems below. The proofs involve proportions, congruence, and similarity.

Theorem 7-1   *SAS Similarity Theorem*

If an angle of one triangle is congruent to an angle of another triangle and the sides including those angles are in proportion, then the triangles are similar.

Given: $\angle A \cong \angle D$;

$$\frac{AB}{DE} = \frac{AC}{DF}$$

Prove: $\triangle ABC \sim \triangle DEF$

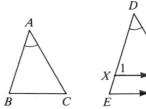

Plan for Proof: Take X on $\overline{DE}$ so that $DX = AB$. Draw a line through X parallel to $\overleftrightarrow{EF}$. Then $\triangle DEF \sim \triangle DXY$ and $\frac{DX}{DE} = \frac{DY}{DF}$. Since $\frac{AB}{DE} = \frac{AC}{DF}$ and $DX = AB$, you can deduce that $DY = AC$. Thus $\triangle ABC \cong \triangle DXY$ by SAS. Therefore, $\triangle ABC \sim \triangle DXY$, and $\triangle ABC \sim \triangle DEF$ by the Transitive Property of Similarity.

Theorem 7-2   *SSS Similarity Theorem*

If the sides of two triangles are in proportion, then the triangles are similar.

Given: $\dfrac{AB}{DE} = \dfrac{BC}{EF} = \dfrac{AC}{DF}$

Prove: $\triangle ABC \sim \triangle DEF$

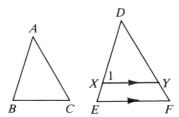

Plan for Proof: Take X on $\overline{DE}$ so that $DX = AB$. Draw a line through X parallel to $\overleftrightarrow{EF}$. Then $\triangle DEF \sim \triangle DXY$ and $\frac{DX}{DE} = \frac{XY}{EF} = \frac{DY}{DF}$. With the given proportion and $DX = AB$, you can deduce that $AC = DY$ and $BC = XY$. Thus $\triangle ABC \cong \triangle DXY$ by SSS. Therefore, $\triangle ABC \sim \triangle DXY$, and $\triangle ABC \sim \triangle DEF$ by the Transitive Property of Similarity.

To prove two polygons similar, you might need to compare the corresponding sides. As shown in the example below, a useful technique is to compare the longest sides, the shortest sides, and so on.

Example Can the information given in each part be used to prove $\triangle RST \sim \triangle WZT$? If so, how?

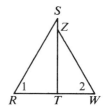

a. $RS = 18$, $ST = 15$, $RT = 10$, $WT = 6$, $ZT = 9$, $WZ = 10.8$

b. $\angle 1 \cong \angle 2$, $\dfrac{WZ}{RS} = \dfrac{TZ}{TS}$

c. $\overline{ST} \perp \overline{RW}$, $ST = 32$, $SZ = 8$, $RT = 20$, $WT = 15$

Solution a. Comparing the longest sides, $\dfrac{RS}{WZ} = \dfrac{18}{10.8} = \dfrac{5}{3}$.

Comparing the shortest sides, $\dfrac{RT}{WT} = \dfrac{10}{6} = \dfrac{5}{3}$.

Comparing the remaining sides, $\dfrac{ST}{ZT} = \dfrac{15}{9} = \dfrac{5}{3}$.

Thus, $\dfrac{RS}{WZ} = \dfrac{RT}{WT} = \dfrac{ST}{ZT}$.

$\triangle RST \sim \triangle WZT$ by the SSS Similarity Theorem.

b. Notice that $\angle 1$ and $\angle 2$ are not the angles included by the sides that are in proportion. Therefore, the triangles cannot be proved similar.

c. Comparing the shorter legs, $\dfrac{RT}{WT} = \dfrac{20}{15} = \dfrac{4}{3}$.

Comparing the other legs, $\dfrac{ST}{ZT} = \dfrac{32}{32 - 8} = \dfrac{32}{24} = \dfrac{4}{3}$.

$\dfrac{RT}{WT} = \dfrac{ST}{ZT}$ and $\angle RTS \cong \angle WTS$.

$\triangle RST \sim \triangle WZT$ by the SAS Similarity Theorem.

The perimeters of similar polygons are in the same ratio as the corresponding sides. By using similar triangles, you can prove that corresponding segments such as diagonals of similar polygons also have this ratio. (See Exercises 25 and 26, page 259 and Exercises 17 and 18, page 267.)

Classroom Exercises

Can the two triangles shown be proved similar? If so, state the similarity and tell which similarity postulate or theorem you would use.

1.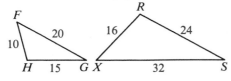

2.

3.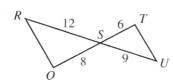

4.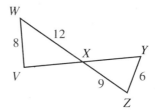

5.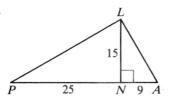

6.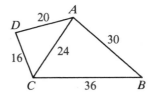

7. Suppose you want to prove that $\triangle RST \sim \triangle XYZ$ by the SSS Similarity Theorem. State the extended proportion you would need to prove first.

8. Suppose you want to prove that $\triangle RST \sim \triangle XYZ$ by the SAS Similarity Theorem. If you know that $\angle R \cong \angle X$, what else do you need to prove?

9. A *pantograph* is a tool for enlarging or reducing maps and drawings. Four bars are pinned together at A, B, C, and D so that $ABCD$ is a parallelogram and points P, D, and E lie on a line. Point P is fixed to the drawing board. To enlarge a figure, the artist inserts a stylus at D, a pen or pencil at E, and guides the stylus so that it traces the original. As D moves, the angles of the parallelogram change, but P, D, and E remain collinear. Suppose PA is 3 units and AB is 7 units.

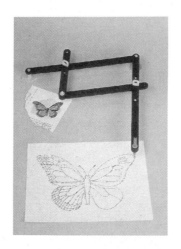
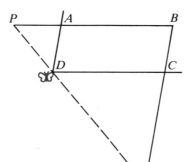
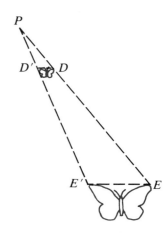

   a. Explain why $\triangle PBE \sim \triangle PAD$.
   b. What is the ratio of PB to PA?
   c. What is the ratio of PE to PD?
   d. What is the ratio of the butterfly's wingspan, $E'E$, in the enlargement to its wingspan, $D'D$, in the original?

Written Exercises

Name two similar triangles. What postulate or theorem justifies your answer?

A **1.**
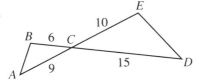

2.
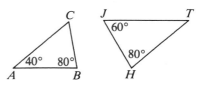

3.
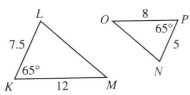

4.
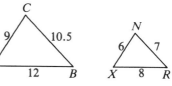

5.
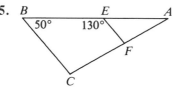

6.

One triangle has vertices A, B, and C. Another has vertices T, R, and I. Are the two triangles similar? If so, state the similarity and the scale factor.

	AB	BC	AC	TR	RI	TI
7.	6	8	10	9	12	15
8.	6	8	10	12	22	16
9.	6	8	10	20	25	15
10.	6	8	10	10	7.5	12.5

11. Given: $\dfrac{DE}{GH} = \dfrac{DF}{GI} = \dfrac{EF}{HI}$

Prove: $\angle E \cong \angle H$

12. Given: $\dfrac{DE}{GH} = \dfrac{EF}{HI}$; $\angle E \cong \angle H$

Prove: $\dfrac{EF}{HI} = \dfrac{DF}{GI}$

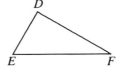

B **13.** Given: $\dfrac{VW}{VX} = \dfrac{VZ}{VY}$

Prove: $\overline{WZ} \parallel \overline{XY}$

14. Given: $\dfrac{VW}{VY} = \dfrac{VZ}{VX}$

Which one(s) of the following *must* be true?

(1) $\triangle VWZ \sim \triangle VXY$ (2) $\overline{WZ} \parallel \overline{XY}$ (3) $\angle 1 \cong \angle Y$

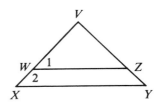

15. Given: $\dfrac{JL}{NL} = \dfrac{KL}{ML}$

    Prove: $\angle J \cong \angle N$

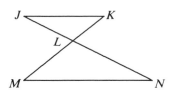

16. Given: $\dfrac{AB}{SR} = \dfrac{BC}{RA} = \dfrac{CA}{AS}$

    Prove: $\overline{BC} \parallel \overline{AR}$

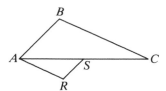

Draw and label a diagram. List, in terms of the diagram, what is given and what is to be proved. Then write a proof.

17. If two triangles are similar, then the lengths of corresponding medians are in the same ratio as the lengths of corresponding sides.

18. If two quadrilaterals are similar, then the lengths of corresponding diagonals are in the same ratio as the lengths of corresponding sides.

19. If the vertex angle of one isosceles triangle is congruent to the vertex angle of another isosceles triangle, then the triangles are similar.

20. The faces of a cube are congruent squares. The cube shown is cut by plane $ABCD$. $VA = VB$ and $VW = 4 \cdot VA$. Find, in terms of AB, the length of the median of trap. $ABCD$.

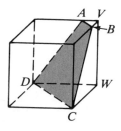

21. Given: $OR' = 2 \cdot OR$;
          $OS' = 2 \cdot OS$;
          $OT' = 2 \cdot OT$
  Prove: $\triangle RST \sim \triangle R'S'T'$

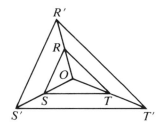

22. Prove Theorem 5-11 on page 178: The segment that joins the midpoints of two sides of a triangle is parallel to the third side and is half as long as the third side.

Given: M is the midpoint of $\overline{AB}$;
      N is the midpoint of $\overline{AC}$.
Prove: $\overline{MN} \parallel \overline{BC}$; $MN = \frac{1}{2}BC$

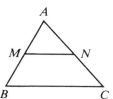

C 23. Given: $\square WXYZ$
    Prove: $\triangle ATB \sim \triangle A'TB'$
    $\left(\textit{Hint:} \text{ Show that } \dfrac{AT}{A'T} \text{ and } \dfrac{BT}{B'T} \text{ both equal } \dfrac{TW}{TY}.\right)$

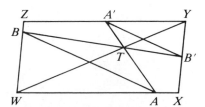

Explorations

These exploratory exercises can be done using a computer with a program that draws and measures geometric figures.

1. Draw any triangle ABC. Choose a point D on $\overline{AB}$. Draw a line through D parallel to $\overline{BC}$ and intersecting $\overline{AC}$ at point E.

   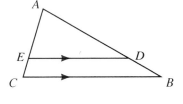

   What do you know about $\triangle ABC$ and $\triangle ADE$?

   What do you know about $\dfrac{AE}{AC}$ and $\dfrac{AD}{AB}$?

   Calculate $\dfrac{AE}{EC}$ and $\dfrac{AD}{DB}$. What do you notice?

   Calculate $\dfrac{EC}{AC}$ and $\dfrac{DB}{AB}$. What do you notice?

   Repeat on other triangles. What do you notice?

2. Draw any triangle ABC. Draw the bisector of $\angle A$ intersecting $\overline{CB}$ at point D.

   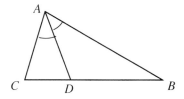

   Measure and record the four lengths: AB, AC, BD, DC.

   Calculate $\dfrac{BD}{DC}$ and $\dfrac{AB}{AC}$. What do you notice?

   Repeat on other triangles. What do you notice?

Mixed Review Exercises

Complete.

1. Given: $\overline{AF} \parallel \overline{BE} \parallel \overline{CD}$; $\overline{AB} \cong \overline{BC}$

   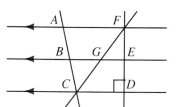

   a. $\overline{GF} \cong \underline{\ ?\ }$ and $\overline{ED} \cong \underline{\ ?\ }$
   b. If $AB = 9$, then $AC = \underline{\ ?\ }$
   c. If $FG = 3x + 2$ and $GC = 7x - 10$, then $x = \underline{\ ?\ }$.
   d. $m\angle FEG = \underline{\ ?\ }$

2. Given: W and U are the midpoints of $\overline{RV}$ and $\overline{VT}$; $\overline{SW} \parallel \overline{VT}$

   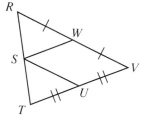

   a. S is the $\underline{\ ?\ }$ of $\overline{RT}$ and $\overline{SU} \parallel \underline{\ ?\ }$.
   b. If $TV = 12x + 4$ and $SW = 3x + 8$, then $x = \underline{\ ?\ }$.
   c. If $RW = 4y + 1$ and $SU = 9y - 19$, then $y = \underline{\ ?\ }$

7-6 *Proportional Lengths*

Points L and M lie on $\overline{AB}$ and $\overline{CD}$, respectively. If $\dfrac{AL}{LB} = \dfrac{CM}{MD}$, we say that $\overline{AB}$ and $\overline{CD}$ are **divided proportionally**.

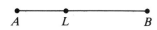 

Theorem 7-3 *Triangle Proportionality Theorem*

If a line parallel to one side of a triangle intersects the other two sides, then it divides those sides proportionally.

Given: $\triangle RST$; $\overleftrightarrow{PQ} \parallel \overline{RS}$

Prove: $\dfrac{RP}{PT} = \dfrac{SQ}{QT}$

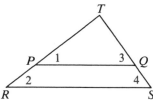

Proof:

Statements	Reasons
1. $\overleftrightarrow{PQ} \parallel \overline{RS}$	1. __?__
2. $\angle 1 \cong \angle 2$; $\angle 3 \cong \angle 4$	2. __?__
3. $\triangle RST \sim \triangle PQT$	3. __?__
4. $\dfrac{RT}{PT} = \dfrac{ST}{QT}$	4. Corr. sides of $\sim \triangle$ are in proportion.
5. $RT = RP + PT$; $ST = SQ + QT$	5. __?__
6. $\dfrac{RP + PT}{PT} = \dfrac{SQ + QT}{QT}$	6. __?__
7. $\dfrac{RP}{PT} = \dfrac{SQ}{QT}$	7. A property of proportions (Property 1(d), page 245.)

We will use the Triangle Proportionality Theorem to justify any proportion equivalent to $\dfrac{RP}{PT} = \dfrac{SQ}{QT}$. For the diagram at the right, some of the proportions that may be justified by the Triangle Proportionality Theorem include:

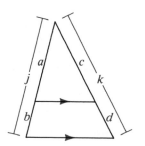

$$\frac{a}{j} = \frac{c}{k} \qquad \frac{a}{c} = \frac{j}{k} \qquad \frac{b}{j} = \frac{d}{k}$$

$$\frac{a}{b} = \frac{c}{d} \qquad \frac{a}{c} = \frac{b}{d} \qquad \frac{b}{d} = \frac{j}{k}$$

Example   Find the numerical value.

a. $\dfrac{TN}{NR}$ b. $\dfrac{TR}{NR}$ c. $\dfrac{RN}{RT}$

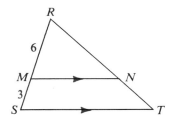

Solution   a. $\dfrac{TN}{NR} = \dfrac{SM}{MR} = \dfrac{3}{6} = \dfrac{1}{2}$

b. $\dfrac{TR}{NR} = \dfrac{SR}{MR} = \dfrac{9}{6} = \dfrac{3}{2}$

c. $\dfrac{RN}{RT} = \dfrac{RM}{RS} = \dfrac{6}{9} = \dfrac{2}{3}$

Compare the following corollary with Theorem 5-9 on page 177.

Corollary

If three parallel lines intersect two transversals, then they divide the transversals proportionally.

Given: $\overleftrightarrow{RX} \parallel \overleftrightarrow{SY} \parallel \overleftrightarrow{TZ}$

Prove: $\dfrac{RS}{ST} = \dfrac{XY}{YZ}$

Plan for Proof: Draw $\overline{TX}$, intersecting $\overleftrightarrow{SY}$ at N. Note that $\overleftrightarrow{SY}$ is parallel to one side of $\triangle RTX$, and also to one side of $\triangle TXZ$. You can apply the Triangle Proportionality Theorem to both of these triangles. Use those proportions to show $\dfrac{RS}{ST} = \dfrac{XY}{YZ}$.

Theorem 7-4   *Triangle Angle-Bisector Theorem*

If a ray bisects an angle of a triangle, then it divides the opposite side into segments proportional to the other two sides.

Given: $\triangle DEF$; $\overrightarrow{DG}$ bisects $\angle FDE$.

Prove: $\dfrac{GF}{GE} = \dfrac{DF}{DE}$

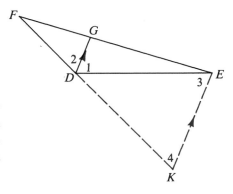

Plan for Proof: Draw a line through E parallel to $\overrightarrow{DG}$ and intersecting $\overrightarrow{FD}$ at K. Apply the Triangle Proportionality Theorem to $\triangle FKE$. $\triangle DEK$ is isosceles with $DK = DE$. Substitute this into your proportion to complete the proof.

Classroom Exercises

1. The two segments are divided proportionally. State several correct
   proportions.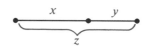

2. Complete the proportions stated informally below.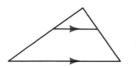

   $$\frac{\text{lower left}}{\text{whole left}} = \frac{\text{lower right}}{?} \qquad \frac{\text{upper left}}{\text{lower left}} = \frac{?}{?}$$

   $$\frac{\text{upper left}}{\text{whole left}} = \frac{\text{upper parallel}}{?} = \frac{?}{?}$$

State a proportion for each diagram.

3.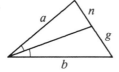

4.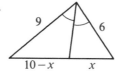

5.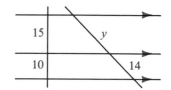

6. Suppose you want to find the length of the upper left segment
   in the diagram at the right. Three methods are suggested below.
   Complete each solution.

   a. 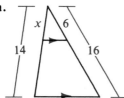   b.   c.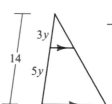

7. Explain why the expressions $3y$ and $5y$ can be used in Exercise 6(c).

8. The converse of the corollary of the Triangle Proportionality Theorem is:
   If three lines divide two transversals proportionally, then the lines are
   parallel. Is the converse true? (*Hint:* Can you draw a diagram with
   lengths like those shown below, but in which lines r, s, and t are not
   parallel?)

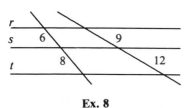

   Ex. 8   **Ex. 9**

9. Must lines a, b, and c shown above be parallel? Explain.

Written Exercises

A 1. Tell whether the proportion is correct.

a. $\dfrac{r}{s} = \dfrac{a}{b}$ b. $\dfrac{j}{a} = \dfrac{s}{r}$ c. $\dfrac{a}{b} = \dfrac{n}{t}$

d. $\dfrac{t}{k} = \dfrac{a}{j}$ e. $\dfrac{r}{s} = \dfrac{n}{k}$ f. $\dfrac{b}{j} = \dfrac{t}{k}$

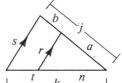

2. Tell whether the proportion is correct.

a. $\dfrac{d}{f} = \dfrac{g}{e}$ b. $\dfrac{f}{g} = \dfrac{e}{d}$

c. $\dfrac{g}{f} = \dfrac{e}{d}$ d. $\dfrac{d}{f} = \dfrac{e}{g}$

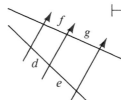

Find the value of x.

3.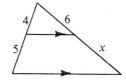

4.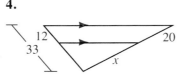

5.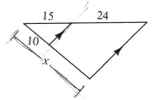

6.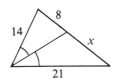

7.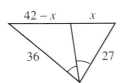

8.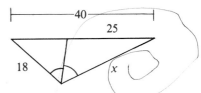

9.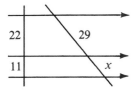

10.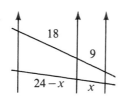

11.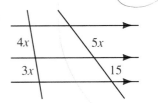

Copy the table and fill in as many spaces *as possible*. It may help to draw a new sketch for each exercise and label lengths as you find them.

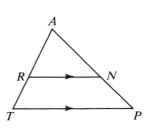

	AR	RT	AT	AN	NP	AP	RN	TP
B 12.	6	4	?	9	?	?	?	15
13.	?	?	?	?	6	16	?	?
14.	18	?	?	?	?	?	30	40
15.	12	?	20	?	?	30	15	?
16.	?	18	?	?	26	?	12	36
17.	?	8	16	6	?	?	?	?

18. Prove the corollary of the Triangle Proportionality Theorem.

19. Prove the Triangle Angle-Bisector Theorem.

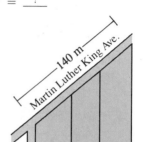

Complete.

20. $AD = 21$, $DC = 14$, $AC = 25$, $AB = $ __?__

21. $AC = 60$, $CD = 30$, $AD = 50$, $BC = $ __?__

22. $AB = 27$, $BC = x$, $CD = \frac{4}{3}x$, $AD = x$, $AC = $ __?__

23. $AB = 2x - 12$, $BC = x$, $CD = x + 5$, $AD = 2x - 4$, $AC = $ __?__

24. Three lots with parallel side boundaries extend from the avenue to the boulevard as shown. Find, to the nearest tenth of a meter, the frontages of the lots on Martin Luther King Avenue.

25. The lengths of the sides of $\triangle ABC$ are $BC = 12$, $CA = 13$, and $AB = 14$. If M is the midpoint of $\overline{CA}$, and P is the point where $\overline{CA}$ is cut by the bisector of $\angle B$, find MP.

26. Prove: If a line bisects both an angle of a triangle and the opposite side, then the triangle is isosceles.

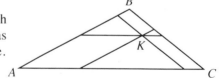

C **27.** Discover and prove a theorem about planes and transversals suggested by the corollary of the Triangle Proportionality Theorem.

28. Prove that there cannot be a triangle in which the trisectors of an angle also trisect the opposite side.

29. Can there exist a $\triangle ROS$ in which the trisectors of $\angle O$ intersect $\overline{RS}$ at D and E, with $RD = 1$, $DE = 2$, and $ES = 4$? Explain.

30. Angle E of $\triangle ZEN$ is obtuse. The bisector of $\angle E$ intersects $\overline{ZN}$ at X. J and K lie on $\overline{ZE}$ and $\overline{NE}$ with $ZJ = ZX$ and $NK = NX$. Discover and prove something about quadrilateral $ZNKJ$.

★ **31.** In $\triangle ABC$, $AB = 8$, $BC = 6$, and $AC = 12$. Each of the three segments drawn through point K has length x and is parallel to a side of the triangle. Find the value of x.

★ **32.** In $\triangle RST$, U lies on $\overline{TS}$ with $TU:US = 2:3$. M is the midpoint of $\overline{RU}$. $\overrightarrow{TM}$ intersects $\overline{RS}$ in V. Find the ratio $RV:RS$.

★ **33.** Prove *Ceva's Theorem*: If P is any point inside $\triangle ABC$, then $\dfrac{AX}{XB} \cdot \dfrac{BY}{YC} \cdot \dfrac{CZ}{ZA} = 1$.

(*Hint:* Draw lines parallel to $\overline{CX}$ through A and B. Apply the Triangle Proportionality Theorem to $\triangle ABM$. Show that $\triangle APN \sim \triangle MPB$, $\triangle BYM \sim \triangle CYP$, and $\triangle CZP \sim \triangle AZN$.)

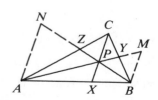

Self-Test 2

State the postulate or theorem you can use to prove that two triangles are similar.

1.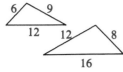

2.

3.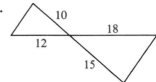

4. Complete.

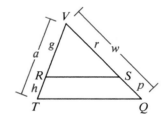

 a. $\triangle ABC \sim$ ____

 b. $\dfrac{AB}{?} = \dfrac{AC}{?} = \dfrac{BC}{?}$

 c. $\dfrac{15}{?} = \dfrac{21}{?}$, and $x =$ ____

 d. $\dfrac{15}{?} = \dfrac{?}{12}$, and $y =$ ____

In the figure, it is given that $\overline{RS} \parallel \overline{TQ}$. Complete each proportion.

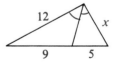

5. $\dfrac{g}{h} = \dfrac{?}{p}$

6. $\dfrac{a}{h} = \dfrac{w}{?}$

7. $\dfrac{r}{g} = \dfrac{p}{?}$

8. $\dfrac{h}{p} = \dfrac{?}{w}$

Find the value of x.

9.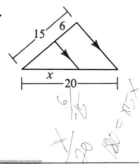

10.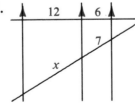

11.

Challenge

Given: $\overline{FD} \parallel \overline{AC}$; $\overline{BD} \parallel \overline{AE}$; $\overline{FB} \parallel \overline{EC}$

Show that B, D, and F are midpoints of $\overline{AC}$, $\overline{CE}$, and $\overline{EA}$.

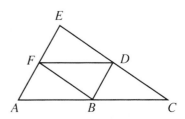

Extra

Topology

In the geometry we have been studying, our interest has been in congruent figures and similar figures, that is, figures with the same size and shape or at least the same shape. If we were studying the branch of geometry called *topology*, we would be interested in properties of figures that are even more basic than size and shape. For example, imagine taking a rubber band and stretching it into all kinds of figures.

  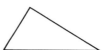

These figures have different sizes and shapes, but they still have something in common: Each one can be turned into any of the others by stretching and bending the rubber band. In topology figures are classified according to this kind of family resemblance. Figures that can be stretched, bent, or molded into the same shape without cutting or puncturing belong to the same family and are called *topologically equivalent*. Thus circles, squares, and triangles are equivalent. Likewise the straight line segment and wiggly curves below are equivalent.

Notice that to make one of the figures above out of the rubber band you would have to cut the band, so these two-ended curves are not equivalent to the closed curves in the first illustration.

Suppose that in the plane figures below, the lines are joined where they cross. Then these figures belong to a third family. They are equivalent to each other but not to any of the figures above.

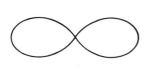 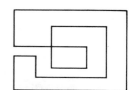

One of the goals of topology is to identify and describe the different families of equivalent figures. A person who studies topology (called a *topologist*) is interested in classifying solid figures as well as figures in a plane. For example, a topologist considers an orange, a teaspoon, and a brick equivalent to each other.

Orange

Teaspoon

Brick

In fact, a doughnut is topologically equivalent to a coffee cup. (See the diagrams below.) For this reason, a topologist has been humorously described as a mathematician who can't tell the difference between a doughnut and a coffee cup!

Think of the objects as made of modeling clay.

Push thumb into clay to make room for coffee.

Exercises

In each exercise tell which figure is *not* topologically equivalent to the rest. Exercises 1 and 2 show plane figures.

1. a.　　　b.　　　c.　　　d.

2. a.　　　b.　　　c.　　　d.

3. **a.** solid ball　　**b.** hollow ball　　**c.** crayon　　**d.** comb

4. **a.** saucer　　**b.** house key　　**c.** coffee cup　　**d.** wedding ring

5. **a.** hammer　　**b.** screwdriver　　**c.** thimble　　**d.** sewing needle

6. Group the block numbers shown into three groups such that the numbers in each group are topologically equivalent to each other.

7. Make a series of drawings showing that the items in each pair are topologically equivalent to each other.
   a. a drinking glass and a dollar bill　　**b.** a tack and a paper clip

Chapter Summary

1. The ratio of a to b is the quotient $\frac{a}{b}$ (b cannot be 0). The ratio $\frac{a}{b}$ can also be written $a:b$.

2. A proportion is an equation, such as $\frac{a}{b} = \frac{c}{d}$, stating that two ratios are equal.

3. The properties of proportions (see page 245) are used to change proportions into equivalent equations. For example, the product of the extremes equals the product of the means.

4. Similar figures have the same shape. Two polygons are similar if and only if corresponding angles are congruent and corresponding sides are in proportion.

5. Ways to prove two triangles similar:
   AA Similarity Postulate   SAS Similarity Theorem   SSS Similarity Theorem

6. Ways to show that segments are proportional:
   a. Corresponding sides of similar polygons are in proportion.
   b. If a line is parallel to one side of a triangle and intersects the other two sides, then it divides those sides proportionally.
   c. If three parallel lines intersect two transversals, they divide the transversals proportionally.
   d. If a ray bisects an angle of a triangle, then it divides the opposite side into segments proportional to the other two sides.

Chapter Review

Write the ratio in simplest form.

1. $15:25$ 2. $6:12:9$ 3. $\dfrac{16xy}{24x^2}$ 7–1

4. The measures of the angles of a triangle are in the ratio $4:4:7$. Find the three measures.

Is the equation equivalent to the proportion $\dfrac{30 - x}{x} = \dfrac{8}{7}$?

5. $7x = 8(30 - x)$ 6. $\dfrac{x}{30 - x} = \dfrac{7}{8}$ 7–2

7. $8x = 210 - 7x$ 8. $\dfrac{30}{x} = \dfrac{15}{7}$

9. If △ABC ~ △NJT, then ∠B ≅ __?__.

7–3

10. If quad. *DEFG* ~ quad. *PQRS*, then $\dfrac{FG}{RS} = \dfrac{GD}{?}$.

11. △ABC ~ △JET, and the scale factor of △ABC to △JET is $\dfrac{5}{3}$.

  a. If *BC* = 20, then *ET* = __?__.
  b. If the perimeter of △JET is 30, then the perimeter of △ABC is __?__.

12. The quadrilaterals are similar.
  Find the values of *x* and *y*.

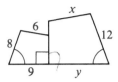

13. a. △RTS ~ __?__

7–4

  b. What postulate or theorem justifies the statement
    in part (a)?

14. $\dfrac{RT}{?} = \dfrac{TS}{?} = \dfrac{RS}{?}$

15. Suppose you wanted to prove

$$RS \cdot UV = RT \cdot UH.$$

  You would first use similar triangles to show that
  $\dfrac{RS}{?} = \dfrac{?}{?}.$

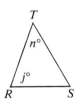
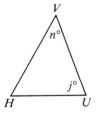

**Can the two triangles be proved similar? If so, state the similarity and the
postulate or theorem you would use. If not, write *no*.**

16. ∠A ≅ ∠D   **17.** ∠B ≅ ∠D

7–5

18. *CN* = 16, *ND* = 14,   **19.** *AN* = 7, *AB* = 13,
  BN = 7, *AN* = 8   *DN* = 14, *DC* = 26

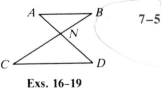

Exs. 16–19

20.

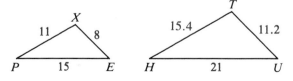

21. Which proportion is *incorrect*?

7–6

  (1) $\dfrac{OS}{ST} = \dfrac{OV}{VW}$ (2) $\dfrac{SV}{TW} = \dfrac{OS}{ST}$ (3) $\dfrac{OT}{OW} = \dfrac{OS}{OV}$

22. If *OS* = 8, *ST* = 12, and *OV* = 10, then *OW* = __?__.

23. If *OS* = 8, *ST* = 12, and *OW* = 24, then *VW* = __?__.

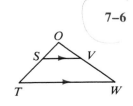

24. In △ABC, the bisector of ∠B meets $\overline{AC}$ at *K*. *AB* = 18, *BC* = 24, and
  AC = 28. Find *AK*.

Chapter Test

1. Two sides of a rectangle have the lengths 20 and 32. Find, in simplest form, the ratio of:
   a. the length of the shorter side to the length of the longer side
   b. the perimeter to the length of the longer side

2. If quad. *ABCD* ~ quad. *THUS*, then:
   a. $\angle U \cong$ ___?___
   b. $\dfrac{BC}{HU} = \dfrac{AD}{?}$

3. If $x:y:z = 4:6:9$ and $z = 45$, then $x =$ ___?___ and $y =$ ___?___.

4. If $\dfrac{8}{9} = \dfrac{x}{15}$, then $x =$ ___?___.   5. If $\dfrac{a}{b} = \dfrac{c}{10}$, then $\dfrac{a + b}{?} = \dfrac{?}{10}$.

6. What postulate or theorem justifies the statement
   $\triangle AVB \sim \triangle NVK$?

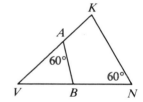

7. $\dfrac{AB}{NK} = \dfrac{VA}{?}$ 8. $\angle VBA \cong$ ___?___

9. The scale factor of $\triangle AVB$ to $\triangle NVK$ is $\dfrac{5}{8}$.
   If $VA = 2.5$ and $VB = 1.7$, then $VN =$ ___?___.

10. If $PR = 10$, $RS = 6$, and $PT = 15$, then $TU =$ ___?___.
11. If $PT = 32$, $PU = 48$, and $RS = 10$, then $PR =$ ___?___.
12. If $PR = 14$, $RS = 7$, and $RT = 26$, then $SU =$ ___?___.

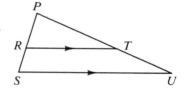

In $\triangle GEB$, the bisector of $\angle E$ meets $\overline{GB}$ at K.

13. If $GK = 5$, $KB = 8$, and $GE = 7$, then $EB =$ ___?___.
14. If $GE = 14$, $EB = 21$, and $GB = 30$, then $GK =$ ___?___.

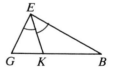

15. Given: $\overleftrightarrow{DE} \parallel \overleftrightarrow{FG} \parallel \overleftrightarrow{HJ}$
    Prove: $DF \cdot GJ = FH \cdot EG$

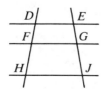

16. Given: $BX = 6$; $AX = 8$;
    $CX = 9$; $DX = 12$
    Prove: $\overline{AB} \parallel \overline{CD}$

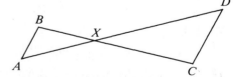

Algebra Review: *Radical Expressions*

The symbol $\sqrt{}$ always indicates the positive square root of a number. The radical $\sqrt{64}$ can be *simplified*.

Simplify.

Example 1   **a.** $\sqrt{56}$ **b.** $\sqrt{\dfrac{16}{3}}$ **c.** $(3\sqrt{7})^2$

Solution   **a.** $\sqrt{56} = \sqrt{4 \cdot 14} = \sqrt{4} \cdot \sqrt{14} = 2\sqrt{14}$

           b. $\sqrt{\dfrac{16}{3}} = \dfrac{\sqrt{16}}{\sqrt{3}} = \dfrac{4}{\sqrt{3}} \cdot \dfrac{\sqrt{3}}{\sqrt{3}} = \dfrac{4\sqrt{3}}{3}$

           c. $(3\sqrt{7})^2 = 3\sqrt{7} \cdot 3\sqrt{7} = 3 \cdot 3 \cdot \sqrt{7} \cdot \sqrt{7} = 9 \cdot 7 = 63$

1. $\sqrt{36}$ **2.** $\sqrt{81}$ **3.** $\sqrt{24}$ **4.** $\sqrt{98}$ **5.** $\sqrt{300}$

6. $\sqrt{\dfrac{1}{4}}$ **7.** $\dfrac{\sqrt{5}}{\sqrt{3}}$ **8.** $\sqrt{\dfrac{80}{25}}$ **9.** $\dfrac{2\sqrt{3}}{\sqrt{12}}$ **10.** $\sqrt{\dfrac{250}{48}}$

11. $\sqrt{13^2}$ **12.** $(\sqrt{17})^2$ **13.** $(2\sqrt{3})^2$ **14.** $(3\sqrt{8})^2$ **15.** $(9\sqrt{2})^2$

16. $5\sqrt{18}$ **17.** $4\sqrt{27}$ **18.** $6\sqrt{24}$ **19.** $5\sqrt{8}$ **20.** $9\sqrt{40}$

Solve for *x*. Assume *x* represents a positive number.

Example 2 $2^2 + x^2 = 4^2$ **Example 3** $x^2 + (3\sqrt{2})^2 = 9^2$

Solution $4 + x^2 = 16$ **Solution** $x^2 + 18 = 81$

                 $x^2 = 12$ $x^2 = 63$

                   $x = \sqrt{12}$ $x = \sqrt{63}$

                     $x = 2\sqrt{3}$ $x = 3\sqrt{7}$

21. $3^2 + 4^2 = x^2$ **22.** $x^2 + 4^2 = 5^2$ **23.** $5^2 + x^2 = 13^2$

24. $x^2 + 3^2 = 4^2$ **25.** $4^2 + 7^2 = x^2$ **26.** $x^2 + 5^2 = 10^2$

27. $1^2 + x^2 = 3^2$ **28.** $x^2 + 5^2 = (5\sqrt{2})^2$ **29.** $(x)^2 + (7\sqrt{3})^2 = (2x)^2$

Challenge

Given regular hexagon *ABCDEF*, with center *O* and sides of length 12. Let *G* be the midpoint of $\overline{BC}$. Let *H* be the midpoint of $\overline{DE}$. $\overline{AH}$ intersects $\overline{EB}$ at *J* and $\overline{FG}$ intersects $\overline{EB}$ at *K*.

Find *JK*.

(*Hint*: Draw auxiliary lines $\overline{HG}$ and $\overline{DA}$.)

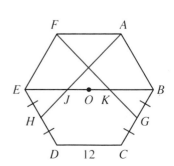

Cumulative Review: Chapters 1–7

True-False Exercises

Write T or F to indicate your answer.

A   **1.** If $AX = XB$, then X must be the midpoint of $\overline{AB}$.

     2. Definitions may be used to justify statements in a proof.

     3. If a line and a plane are parallel, then the line is parallel to every line in the plane.

     4. When two parallel lines are cut by a transversal, any two angles formed are either congruent or supplementary.

     5. If the sides of one triangle are congruent to the corresponding sides of another triangle, then the corresponding angles must also be congruent.

     6. Every isosceles trapezoid contains two pairs of congruent angles.

B   **7.** If a quadrilateral has two pairs of supplementary angles, then it must be a parallelogram.

     8. If the diagonals of a quadrilateral bisect each other and are congruent, then the quadrilateral must be a square.

     9. In $\triangle PQR$, $m\angle P = m\angle R = 50$. If T lies on $\overline{PR}$ and $m\angle PQT = 42$, then $PT < TR$.

    10. In quad. $WXYZ$, if $WX = XY = 25$, $YZ = 20$, $ZW = 16$, and $WY = 20$, then $\overline{WY}$ divides the quadrilateral into two similar triangles.

    11. Two equiangular hexagons are always similar.

Multiple-Choice Exercises

Indicate the best answer by writing the appropriate letter.

A   **1.** Which pair of angles must be congruent?
     a. $\angle 1$ and $\angle 4$ **b.** $\angle 2$ and $\angle 3$
     c. $\angle 2$ and $\angle 4$ **d.** $\angle 4$ and $\angle 5$
     e. $\angle 2$ and $\angle 8$

     2. If a, b, c, and d are coplanar lines such that $a \perp b$, $c \perp d$, and $b \parallel c$, then:
     a. $a \perp d$ **b.** $b \parallel d$ **c.** $a \parallel d$ **d.** $a \parallel c$ **e.** none of these

     3. If $\triangle ABC \cong \triangle NDH$, then it is also true that:
     a. $\angle B \cong \angle H$ **b.** $\angle A \cong \angle H$ **c.** $\overline{AB} \cong \overline{HD}$
     d. $\overline{CA} \cong \overline{HN}$ **e.** $\triangle CBA \cong \triangle DHN$

B   **4.** If $PQRS$ is a parallelogram, which of the following *must* be true?
     a. $PQ = QR$ **b.** $PQ = RS$ **c.** $PR = QS$ **d.** $\overline{PR} \perp \overline{QS}$ **e.** $\angle Q \cong \angle R$

     5. Which of the following can be the lengths of the sides of a triangle?
     a. 3, 7, 10   **b.** 3, 7, 11   **c.** 0.5, 7, 7   **d.** $\dfrac{1}{2}, \dfrac{1}{4}, \dfrac{1}{5}$ **e.** 1, 3, 5

Always-Sometimes-Never Exercises

Write A, S, or N to indicate your choice.

A 1. If a conditional is false, then its converse is __?__ false.

2. Two vertical angles are __?__ adjacent.

3. An angle __?__ has a complement.

4. Two parallel lines are __?__ coplanar.

5. Two perpendicular lines are __?__ both parallel to a third line.

6. A scalene triangle is __?__ equiangular.

7. A regular polygon is __?__ equilateral.

8. A rectangle is __?__ a rhombus.

9. If $\overline{RS} \cong \overline{MN}$, $\overline{ST} \cong \overline{NO}$, and $\angle R \cong \angle M$, then $\triangle RST$ and $\triangle MNO$ are __?__ congruent.

10. The HL method is __?__ appropriate for proving that two acute triangles are congruent.

11. If $AX = BX$, $AY = BY$, and points A, B, X, and Y are coplanar, then $\overline{AB}$ and $\overline{XY}$ are __?__ perpendicular.

B 12. The diagonals of a trapezoid are __?__ perpendicular.

13. If a line parallel to one side of a triangle intersects the other two sides, then the triangle formed is __?__ similar to the given triangle.

14. If $\triangle JKL \cong \triangle NET$ and $\overline{NE} \perp \overline{ET}$, then it is __?__ true that $LJ < TE$.

15. If $AB + BC > AC$, then A, B, and C are __?__ collinear points.

16. A triangle with sides of length $x - 1$, x, and x is __?__ an obtuse triangle.

Completion Exercises

Complete each statement in the best way.

A 1. If $\overrightarrow{YW}$ bisects $\angle XYZ$ and $m \angle WYX = 60$, then $m \angle XYZ = $ __?__.

2. The acute angles of a right triangle are __?__.

3. A supplement of an acute angle is a(n) __?__ angle.

4. Adjacent angles formed by __?__ lines are congruent.

5. The measure of each interior angle of a regular pentagon is __?__.

6. In $\triangle ABC$ and $\triangle DEF$, $\angle A \cong \angle D$ and $\angle B \cong \angle E$. $\triangle ABC$ and $\triangle DEF$ must be __?__.

B 7. When the midpoints of the sides of a rhombus are joined in order, the resulting quadrilateral is best described as a __?__.

8. If $\frac{r}{s} = \frac{t}{u}$, then $\frac{r + s}{t + u} = \frac{?}{?}$.

9. The ratio of the measures of the acute angles of a right triangle is $3 : 2$. The measure of the smaller acute angle is __?__.

Algebraic Exercises

In Exercises 1–9 find the value of *x*.

A　**1.** On a number line, R and S have coordinates -8 and x, and the midpoint of $\overline{RS}$ has coordinate -1.

2. Two vertical angles have measures $x^2 + 18x$ and $x^2 + 54$.

3. The measures of the angles of a quadrilateral are x, $x + 4$, $x + 8$, and $x + 12$.

4. The lengths of the legs of an isosceles triangle are $7x - 13$ and $2x + 17$.

5. Consecutive angles of a parallelogram have measures $6x$ and $2x + 20$.

6. A trapezoid has bases of length x and $x + 8$ and a median of length 15.

7. $\dfrac{3x - 1}{4x + 2} = \dfrac{2}{3}$　　　　**8.** $\dfrac{5}{8} = \dfrac{x - 1}{6}$　　　　**9.** $\dfrac{x}{x + 4} = \dfrac{x + 3}{x + 9}$

B　**10.** The measure of a supplement of an angle is 8 more than three times the measure of a complement. Find the measure of the angle.

11. In a regular polygon, the ratio of the measure of an exterior angle to the measure of an interior angle is $2:13$. How many sides does the polygon have?

12. The sides of a parallelogram have lengths 12 cm and 15 cm. Find the lengths of the sides of a similar parallelogram with perimeter 90 cm.

13. A triangle with perimeter 64 cm has sides with lengths in the ratio $4:5:7$. Find the length of each side.

14. In $\triangle XYZ$, $XY = YZ$. Find the measure of $\angle Z$ if $m\angle X : m\angle Y = 5:2$.

15. In the diagram, $\overline{AB} \parallel \overline{DC}$ and $\overline{AD} \parallel \overline{GC}$. Find the values of x and y.

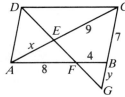

Proof Exercises

A　**1.** Given: $\overline{SU} \cong \overline{SV}$; $\angle 1 \cong \angle 2$
Prove: $\overline{UQ} \cong \overline{VQ}$

2. Given: $\overrightarrow{QS}$ bisects $\angle RQT$; $\angle R \cong \angle T$
Prove: $\overrightarrow{SQ}$ bisects $\angle RST$.

B　**3.** Given: $\triangle QRU \cong \triangle QTV$; $\overline{US} \cong \overline{VS}$
Prove: $\triangle QRS \cong \triangle QTS$

4. Given: $\overline{QS}$ bisects $\angle UQV$ and $\angle USV$; $\angle R \cong \angle T$
Prove: $\overline{RQ} \cong \overline{TQ}$

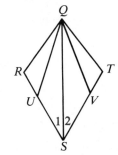

5. Given: $\overline{EF} \parallel \overline{JK}$; $\overline{JK} \parallel \overline{HI}$
Prove: $\triangle EFG \sim \triangle IHG$

6. Given: $\dfrac{JG}{HG} = \dfrac{KG}{IG}$, $\angle 1 \cong \angle 2$
Prove: $\overline{EF} \parallel \overline{HI}$

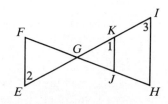

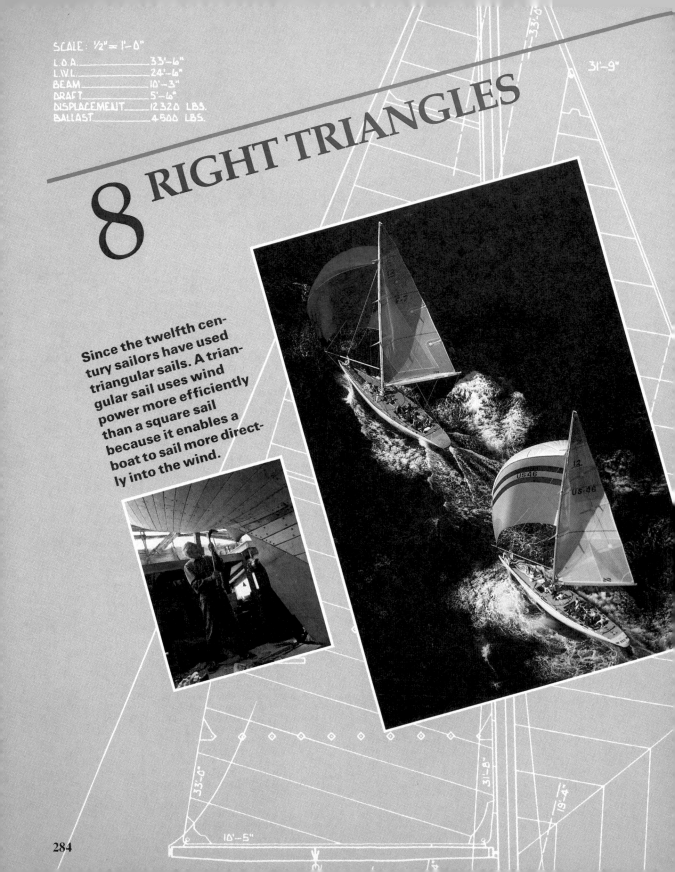

SCALE: ½"= 1'-0"

L.O.A. .... 33'-6"
L.W.L. .... 24'-6"
BEAM .... 10'-3"
DRAFT .... 5'-6"
DISPLACEMENT .... 12,320 LBS.
BALLAST .... 4,500 LBS.

8 RIGHT TRIANGLES

Since the twelfth century sailors have used triangular sails. A triangular sail uses wind power more efficiently than a square sail because it enables a boat to sail more directly into the wind.

Right Triangles

Objectives

1. Determine the geometric mean between two numbers.
2. State and apply the relationships that exist when the altitude is drawn to the hypotenuse of a right triangle.
3. State and apply the Pythagorean Theorem.
4. State and apply the converse of the Pythagorean Theorem and related theorems about obtuse and acute triangles.
5. Determine the lengths of two sides of a 45°-45°-90° or a 30°-60°-90° triangle when the length of the third side is known.

8-1 *Similarity in Right Triangles*

Recall that in the proportion $\dfrac{a}{x} = \dfrac{y}{b}$, the terms shown in red are called the *means*. If a, b, and x are positive numbers and $\dfrac{a}{x} = \dfrac{x}{b}$, then x is called the **geometric mean** between a and b. If you solve this proportion for x, you will find that $x = \sqrt{ab}$, a positive number. (The other solution, $x = -\sqrt{ab}$, is discarded because x is defined to be positive.)

Example 1 Find the geometric mean between 5 and 11.

Solution 1 Solve the proportion $\dfrac{5}{x} = \dfrac{x}{11}$: $x^2 = 5 \cdot 11$; $x = \sqrt{55}$.

Solution 2 Use the equation $x = \sqrt{ab} = \sqrt{5 \cdot 11} = \sqrt{55}$.

Theorem 8-1

If the altitude is drawn to the hypotenuse of a right triangle, then the two triangles formed are similar to the original triangle and to each other.

Given: $\triangle ABC$ with rt. $\angle ACB$;
        altitude $\overline{CN}$

Prove: $\triangle ACB \sim \triangle ANC \sim \triangle CNB$

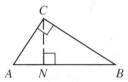

Plan for Proof: Begin by redrawing the three triangles you want to prove similar. Mark off congruent angles and apply the AA Similarity Postulate.

The proof of Theorem 8-1 is left as Exercise 40. The altitude to the hypotenuse divides the hypotenuse into two segments. Corollaries 1 and 2 of Theorem 8-1 deal with geometric means and the lengths of these segments.

For simplicity in stating these corollaries, the words *segment*, *side*, *leg*, and *hypotenuse* are used to refer to the *length* of a segment rather than the segment itself. We will use this convention throughout the book when the context makes this meaning clear.

Corollary 1

When the altitude is drawn to the hypotenuse of a right triangle, the length of the altitude is the geometric mean between the segments of the hypotenuse.

Given: $\triangle ABC$ with rt. $\angle ACB$; altitude $\overline{CN}$

Prove: $\dfrac{AN}{CN} = \dfrac{CN}{BN}$

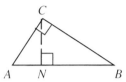

Proof:

By Theorem 8-1, $\triangle ANC \sim \triangle CNB$. Because corresponding sides of similar triangles are in proportion, $\dfrac{AN}{CN} = \dfrac{CN}{BN}$.

Corollary 2

When the altitude is drawn to the hypotenuse of a right triangle, each leg is the geometric mean between the hypotenuse and the segment of the hypotenuse that is adjacent to that leg.

Given: $\triangle ABC$ with rt. $\angle ACB$; altitude $\overline{CN}$

Prove: (1) $\dfrac{AB}{AC} = \dfrac{AC}{AN}$ and (2) $\dfrac{AB}{BC} = \dfrac{BC}{BN}$

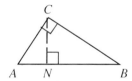

Proof of (1):

By Theorem 8-1, $\triangle ACB \sim \triangle ANC$. Because corresponding sides of similar triangles are in proportion, $\dfrac{AB}{AC} = \dfrac{AC}{AN}$. The proof of (2) is very similar.

Example 2   Use the diagram to find the values of h, a, and b.

Solution   First determine what parts of the "big" triangle are labeled h, a, and b:

h is the altitude to the hypotenuse, a is a leg, and b is a leg.

By Corollary 1, $\dfrac{3}{h} = \dfrac{h}{7}$ and $h = \sqrt{21}$.

By Corollary 2, $\dfrac{10}{a} = \dfrac{a}{3}$ and $a = \sqrt{30}$.

By Corollary 2, $\dfrac{10}{b} = \dfrac{b}{7}$ and $b = \sqrt{70}$.

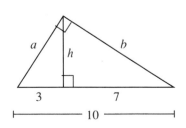

Working with geometric means may involve working with radicals. Radicals should always be written in **simplest form.** This means writing them so that

1. No perfect square factor other than 1 is under the radical sign.
2. No fraction is under the radical sign.
3. No fraction has a radical in its denominator.

Example 3 Simplify: **a.** $5\sqrt{18}$ **b.** $\sqrt{\dfrac{3}{2}}$ **c.** $\dfrac{15}{\sqrt{5}}$

Solution **a.** Since $18 = 9 \cdot 2$, there is a perfect square factor, 9, under the radical sign.
$$5\sqrt{18} = 5 \cdot \sqrt{9 \cdot 2} = 5 \cdot \sqrt{9} \cdot \sqrt{2} = 5 \cdot 3 \cdot \sqrt{2} = 15\sqrt{2}$$

b. There is a fraction, $\dfrac{3}{2}$, under the radical sign.
$$\sqrt{\dfrac{3}{2}} = \sqrt{\dfrac{3}{2} \cdot \dfrac{2}{2}} = \sqrt{\dfrac{6}{4}} = \dfrac{\sqrt{6}}{\sqrt{4}} = \dfrac{\sqrt{6}}{2}$$

c. There is a radical in the denominator of the fraction.
$$\dfrac{15}{\sqrt{5}} = \dfrac{15}{\sqrt{5}} \cdot \dfrac{\sqrt{5}}{\sqrt{5}} = \dfrac{15\sqrt{5}}{5} = 3\sqrt{5}$$

Example 4 Find the values of w, x, y, and z.

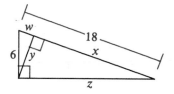

Solution

$$\dfrac{18}{6} = \dfrac{6}{w} \text{(Cor. 2)}$$
$$18w = 36$$
$$w = 2$$
Then $x = 18 - 2 = 16$.

$$\dfrac{16}{y} = \dfrac{y}{2} \text{(Cor. 1)}$$
$$y^2 = 16 \cdot 2$$
$$y = \sqrt{16 \cdot 2}$$
$$y = \sqrt{16} \cdot \sqrt{2}$$
$$y = 4\sqrt{2}$$

$$\dfrac{18}{z} = \dfrac{z}{16} \text{(Cor. 2)}$$
$$z^2 = 16 \cdot 18$$
$$z = \sqrt{16 \cdot 18}$$
$$z = \sqrt{16 \cdot 9 \cdot 2}$$
$$z = 4 \cdot 3 \cdot \sqrt{2} = 12\sqrt{2}$$

Classroom Exercises

Use the diagram to complete each statement.

1. If $m \angle R = 30$, then $m \angle RSU = \underline{\ ?\ }$,
   $m \angle TSU = \underline{\ ?\ }$, and $m \angle T = \underline{\ ?\ }$.
2. If $m \angle R = k$, then $m \angle RSU = \underline{\ ?\ }$,
   $m \angle TSU = \underline{\ ?\ }$, and $m \angle T = \underline{\ ?\ }$.
3. $\triangle RST \sim \triangle \underline{\ ?\ } \sim \triangle \underline{\ ?\ }$
4. $\triangle RSU \sim \triangle \underline{\ ?\ } \sim \triangle \underline{\ ?\ }$

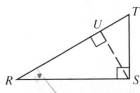

Exs. 1–4

Simplify.

5. $\sqrt{50}$ **6.** $3\sqrt{8}$ **7.** $\sqrt{225}$ **8.** $7\sqrt{63}$ **9.** $\sqrt{288}$

10. $\sqrt{\dfrac{3}{4}}$ **11.** $\sqrt{\dfrac{1}{5}}$ **12.** $\dfrac{\sqrt{5}}{\sqrt{2}}$ **13.** $\sqrt{\dfrac{5}{2}}$ **14.** $\dfrac{3}{4}\sqrt{\dfrac{28}{3}}$

15. Give the geometric mean between:
    a. 2 and 3   **b.** 2 and 6   **c.** 4 and 25

Study the diagram. Then complete each statement.

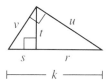

16. a. t is the geometric mean between __?__ and __?__.
     b. u is the geometric mean between __?__ and __?__.
     c. v is the geometric mean between __?__ and __?__.

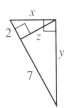

17. a. z is the geometric mean between __?__ and __?__.
       Thus $z =$ __?__.
     b. x is the geometric mean between __?__ and __?__.
       Thus $x =$ __?__.
     c. y is the geometric mean between __?__ and __?__.
       Thus $y =$ __?__.

Written Exercises

Simplify.

A   **1.** $\sqrt{12}$ **2.** $\sqrt{72}$ **3.** $\sqrt{45}$ **4.** $\sqrt{75}$ **5.** $\sqrt{800}$

     6. $\sqrt{54}$ **7.** $9\sqrt{40}$ **8.** $4\sqrt{28}$ **9.** $\sqrt{30} \cdot \sqrt{6}$ **10.** $\sqrt{5} \cdot \sqrt{35}$

     11. $\sqrt{\dfrac{3}{7}}$ **12.** $\sqrt{\dfrac{9}{5}}$ **13.** $\dfrac{18}{\sqrt{3}}$ **14.** $\dfrac{24}{3\sqrt{2}}$ **15.** $\dfrac{\sqrt{15}}{3\sqrt{45}}$

Find the geometric mean between the two numbers.

16. 2 and 18   **17.** 3 and 27   **18.** 49 and 25

19. 1 and 1000   **20.** 16 and 24   **21.** 22 and 55

Exercises 22–30 refer to the figure at the right.

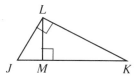

22. If $LM = 4$ and $MK = 8$, find JM.
23. If $LM = 6$ and $JM = 4$, find MK.
24. If $JM = 3$ and $MK = 6$, find LM.
25. If $JM = 4$ and $JK = 9$, find LK.
26. If $JM = 3$ and $MK = 9$, find LJ.

B   **27.** If $JM = 3$ and $JL = 6$, find MK.   **28.** If $JL = 9$ and $JM = 6$, find MK.
     29. If $LK = 3\sqrt{6}$ and $MK = 6$, find JM.   **30.** If $LK = 7$ and $MK = 6$, find JM.

Find the values of *x*, *y*, and *z*.

31.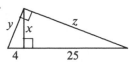

32.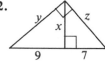

33.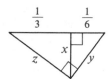

34.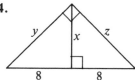

35.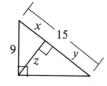

36.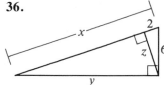

37.

38.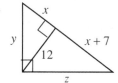

39.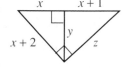

40. Prove Theorem 8-1.

41. a. Refer to the figure at the right, and use Corollary 2
to complete:
$$a^2 = \underline{\quad?\quad} \text{ and } b^2 = \underline{\quad?\quad}$$

b. Add the equations in part (a), factor the sum on the
right, and show that $a^2 + b^2 = c^2$.

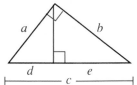

C **42.** Prove: In a right triangle, the product of the hypotenuse and the length of
the altitude drawn to the hypotenuse equals the product of the two legs.

43. Given: *PQRS* is a rectangle;
PS is the geometric mean
between *ST* and *TR*.
Prove: $\angle PTQ$ is a right angle.

44. Given: *PQRS* is a rectangle;
$\angle A$ is a right angle.
Prove: $BS \cdot RC = PS \cdot QR = (PS)^2$

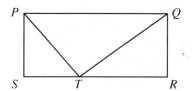

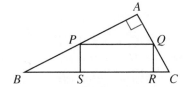

45. The *arithmetic mean* between two numbers r and s is defined to be $\dfrac{r+s}{2}$.

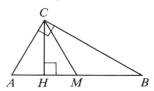

a. $\overline{CM}$ is the median and $\overline{CH}$ is the altitude to the hypotenuse of right $\triangle ABC$. Show that CM is the arithmetic mean between AH and BH, and that CH is the geometric mean between AH and BH. Then use the diagram to show that the arithmetic mean is greater than the geometric mean.

b. Show algebraically that the arithmetic mean between two different numbers r and s is greater than the geometric mean. (*Hint:* The geometric mean is $\sqrt{rs}$. Work backward from $\dfrac{r+s}{2} > \sqrt{rs}$ to $(r-s)^2 > 0$ and then reverse the steps.)

8-2 *The Pythagorean Theorem*

One of the best known and most useful theorems in all of mathematics is the *Pythagorean Theorem*. It is believed that Pythagoras, a Greek mathematician and philosopher, proved this theorem about twenty-five hundred years ago. Many different proofs exist, including one by President Garfield (Exercise 32, page 438) and the proof suggested by the Challenge on page 294.

Theorem 8-2 *Pythagorean Theorem*

In a right triangle, the square of the hypotenuse is equal to the sum of the squares of the legs.

Given: $\triangle ABC$; $\angle ACB$ is a rt. $\angle$.
Prove: $c^2 = a^2 + b^2$

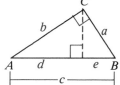

Proof:

Statements	Reasons
1. Draw a perpendicular from C to $\overline{AB}$.	1. Through a point outside a line, there is exactly one line __?__.
2. $\dfrac{c}{a} = \dfrac{a}{e}$; $\dfrac{c}{b} = \dfrac{b}{d}$	2. When the altitude is drawn to the hypotenuse of a rt. $\triangle$, each leg is the geometric mean between __?__.
3. $ce = a^2$; $cd = b^2$	3. A property of proportions
4. $ce + cd = a^2 + b^2$	4. Addition Property of $=$
5. $c(e + d) = a^2 + b^2$	5. Distributive Property
6. $c^2 = a^2 + b^2$	6. Substitution Property

Example Find the value of *x*. Remember that the length of a segment must be a positive number.

a.

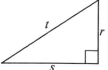

b.

Solution

a. $x^2 = 7^2 + 3^2$
$x^2 = 49 + 9$
$x^2 = 58$
$x = \sqrt{58}$

b. $x^2 + (x + 2)^2 = 10^2$
$x^2 + x^2 + 4x + 4 = 100$
$2x^2 + 4x - 96 = 0$
$x^2 + 2x - 48 = 0$
$(x + 8)(x - 6) = 0$
$\cancel{x = -8}, x = 6$

Classroom Exercises

1. The early Greeks thought of the Pythagorean Theorem in this form: *The
area of the square on the hypotenuse of a right triangle equals the sum
of the areas of the squares on the legs.* Draw a diagram to illustrate that
interpretation.

2. Which equations are correct for the right triangle shown?
 a. $r^2 = s^2 + t^2$ **b.** $s^2 = r^2 + t^2$ **c.** $s^2 + r^2 = t^2$
 d. $s^2 = t^2 - r^2$ **e.** $t = r + s$ **f.** $t^2 = (r + s)^2$

Complete each simplification.

3. $(\sqrt{3})^2 = \sqrt{3} \cdot \underline{\ ?\ } = \underline{\ ?\ }$ **4.** $(3\sqrt{11})^2 = \underline{\ ?\ } \cdot \underline{\ ?\ } = 9 \cdot \underline{\ ?\ } = \underline{\ ?\ }$

Simplify each expression.

5. $(\sqrt{5})^2$ **6.** $(2\sqrt{7})^2$ **7.** $(7\sqrt{2})^2$ **8.** $(2n)^2$

9. $\left(\dfrac{3}{\sqrt{5}}\right)^2$ **10.** $\left(\dfrac{\sqrt{2}}{2}\right)^2$ **11.** $\left(\dfrac{n}{\sqrt{3}}\right)^2$ **12.** $\left(\dfrac{2}{3}\sqrt{6}\right)^2$

**State an equation you could use to find the value of *x*. Then find the value
of *x* in simplest radical form.**

13.   **14.**   **15.**

  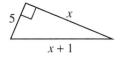

16.   **17.**   **18.**

  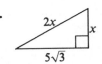

Written Exercises

Find the value of x.

A **1.**

2.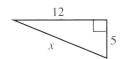

3.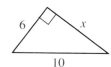

4.

5.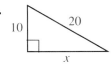

6.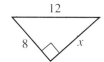

7.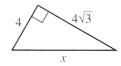

8.

9.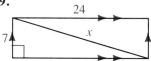

10.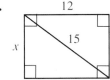

11.

12.

13. A rectangle has length 2.4 and width 1.8. Find the length of a diagonal.

14. A rectangle has a diagonal of 2 and length of $\sqrt{3}$. Find its width.

15. Find the length of a diagonal of a square with perimeter 16.

16. Find the length of a side of a square with a diagonal of length 12.

17. The diagonals of a rhombus have lengths 16 and 30. Find the perimeter of the rhombus.

18. The perimeter of a rhombus is 40 cm, and one diagonal is 12 cm long. How long is the other diagonal?

Find the value of x.

B **19.**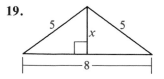

20.

21.

Find the value of *x*.

22.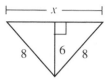

23.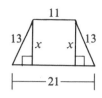

24.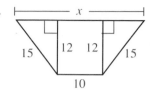

25.

26.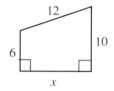

27.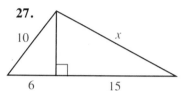

28.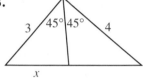

(*Hint*: Use the Angle-Bisector Theorem, p. 270.)

29.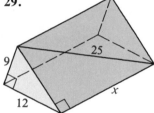

30.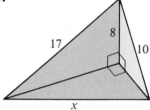

31. A right triangle has legs of 6 and 8. Find the lengths of:
  a. the median to the hypotenuse   **b.** the altitude to the hypotenuse.

32. A rectangle is 2 cm longer than it is wide. The diagonal of the rectangle is 10 cm long. Find the perimeter of the rectangle.

In Exercises 33–36 the dimensions of a rectangular box are given. Sketch the box and find the length of a diagonal of the box.

Example   Dimensions 6, 4, 3

Solution

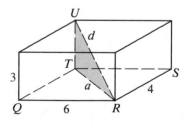

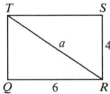

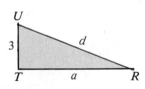

$$a^2 = 6^2 + 4^2$$
$$a^2 = 36 + 16$$
$$a^2 = 52$$

$$d^2 = a^2 + 3^2$$
$$d^2 = 52 + 9$$
$$d^2 = 61$$
$$d = \sqrt{61}$$

33. 12, 4, 3   **34.** 5, 5, 2   **35.** *e, e, e*   **36.** *l, w, h*

Find the value of h.

C **37.**

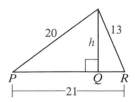

(*Hint*: Let $PQ = x$; $QR = 21 - x$.)

38.

(*Hint*: Let $TU = x$; $SU = x + 11$.)

39. O is the *center* of square $ABCD$ (the point of intersection of the diagonals) and $\overline{VO}$ is perpendicular to the plane of the square. Find OE, the distance from O to the plane of $\triangle VBC$.

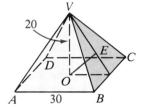

Mixed Review Exercises

Given: $\triangle ABC$. Complete.

1. If $m \angle A > m \angle B$, then $BC > \underline{\ \ ?\ \ }$.
2. If $AB > BC$, then $m \angle C \underline{\ \ ?\ \ } m \angle \underline{\ \ ?\ \ }$.
3. $AB + BC \underline{\ \ ?\ \ } AC$
4. If $\angle C$ is a right angle, then $\underline{\ \ ?\ \ }$ is the longest side.
5. If $AB = AC$, then $\angle \underline{\ \ ?\ \ } \cong \angle \underline{\ \ ?\ \ }$.
6. If $\angle A \cong \angle C$, then $BC = \underline{\ \ ?\ \ }$.
7. If $\angle C$ is a right angle and X is the midpoint of the hypotenuse, then $AX = \underline{\ \ ?\ \ } = \underline{\ \ ?\ \ }$.

Challenge

Start with a right triangle. Build a square on each side. Locate the center of the square drawn on the longer leg. Through the center, draw a parallel to the hypotenuse and a perpendicular to the hypotenuse.

Cut out the pieces numbered 1–5. Can you arrange the five pieces to cover exactly the square built on the hypotenuse? (This suggests another proof of the Pythagorean Theorem.)

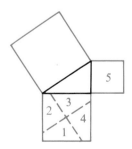

8-3 *The Converse of the Pythagorean Theorem*

We have seen that the converse of a theorem is not necessarily true. However, the converse of the Pythagorean Theorem *is* true. It is stated below as Theorem 8-3.

Theorem 8-3

If the square of one side of a triangle is equal to the sum of the squares of the other two sides, then the triangle is a right triangle.

Given: $\triangle ABC$ with $c^2 = a^2 + b^2$

Prove: $\triangle ABC$ is a right triangle.

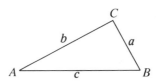

Key steps of proof:

1. Draw rt. $\triangle EFG$ with legs a and b.
2. $d^2 = a^2 + b^2$ (Pythagorean Theorem)
3. $c^2 = a^2 + b^2$ (Given)
4. $c = d$ (Substitution)
5. $\triangle ABC \cong \triangle EFG$ (SSS Postulate)
6. $\angle C$ is a rt. $\angle$. (Corr. parts of $\cong$ $\triangle$ are $\cong$.)
7. $\triangle ABC$ is a rt. $\triangle$. (Def. of a rt. $\triangle$)

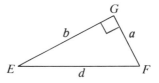

A triangle with sides 3 units, 4 units, and 5 units long is called a 3-4-5 triangle. The numbers 3, 4, and 5 satisfy the equation $a^2 + b^2 = c^2$, so we can apply Theorem 8-3 to conclude that a 3-4-5 triangle is a right triangle. The side lengths shown in the table all satisfy the equation $a^2 + b^2 = c^2$, so the triangles formed are right triangles.

Some Common Right Triangle Lengths

3, 4, 5	5, 12, 13	8, 15, 17	7, 24, 25
6, 8, 10	10, 24, 26		
9, 12, 15			
12, 16, 20			
15, 20, 25			

Theorem 8-3 is restated on the next page, along with Theorems 8-4 and 8-5. If you know the lengths of the sides of a triangle, you can use these theorems to determine whether the triangle is right, acute, or obtuse. In each theorem, *c is the length of the longest side* of $\triangle ABC$. Exercises 20 and 19 ask you to state Theorems 8-4 and 8-5 more formally and then prove them.

Theorem 8-3

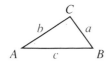

If $c^2 = a^2 + b^2$,
then $m \angle C = 90$,
and $\triangle ABC$ is right.

Theorem 8-4

If $c^2 < a^2 + b^2$,
then $m \angle C < 90$,
and $\triangle ABC$ is acute.

Theorem 8-5

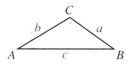

If $c^2 > a^2 + b^2$,
then $m \angle C > 90$,
and $\triangle ABC$ is obtuse.

Example A triangle has sides of the given lengths. Is it acute, right, or obtuse?
  a. 9, 40, 41   **b.** 6, 7, 8   **c.** 7, 8, 11

Solution **a.** $41^2 \underline{\ ?\ } 9^2 + 40^2$ **b.** $8^2 \underline{\ ?\ } 6^2 + 7^2$ **c.** $11^2 \underline{\ ?\ } 7^2 + 8^2$

  $1681 \underline{\ ?\ } 81 + 1600$ $64 \underline{\ ?\ } 36 + 49$ $121 \underline{\ ?\ } 49 + 64$

  $1681 = 1681$ $64 < 85$ $121 > 113$

  The triangle is right.   The triangle is acute.   The triangle is obtuse.

Classroom Exercises

If a triangle is formed with sides having the lengths given, is it acute, right, or obtuse? If a triangle can't be formed, say *not possible.*

1. 6, 8, 10   **2.** 4, 6, 8   **3.** 1, 4, 6

4. 8, 10, 12   **5.** $\sqrt{7}, \sqrt{7}, \sqrt{14}$ **6.** 4, $4\sqrt{3}$, 8

7. Specify all values of x that make the statement true.
  a. $\angle 1$ is a right angle.   **b.** $\angle 1$ is an acute angle.
  c. $\angle 1$ is an obtuse angle.   **d.** The triangle is isosceles.
  e. No triangle is possible.

Exercises 8–10 refer to the figures below.

8. Explain why x must equal 5.

9. Explain why $\angle D$ must be a right angle.

10. Explain why $\angle P$ must be a right angle.

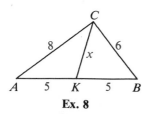

Ex. 8

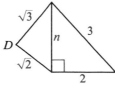

Ex. 9

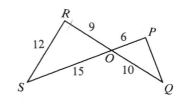

Ex. 10

Written Exercises

Tell whether a triangle with sides of the given lengths is acute, right, or obtuse.

A **1.** 11, 11, 15   **2.** 9, 9, 13   **3.** 8, $8\sqrt{3}$, 16

4. 6, 6, $6\sqrt{2}$ **5.** 8, 14, 17   **6.** 0.6, 0.8, 1

7. a. 0.5, 1.2, 1.3   **b.** $5n$, $12n$, $13n$ where $n > 0$

8. a. 33, 44, 55   **b.** $3n$, $4n$, $5n$ where $n > 0$

9. Given: $\angle UTS$ is a rt. $\angle$.
Show that $\triangle RST$ must be a rt. $\triangle$.

10. Given: $\overline{AC} \perp$ plane P
Show that $\triangle BCD$ must be obtuse.

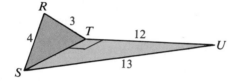

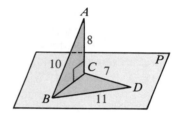

Use the information to decide if $\triangle ABC$ is acute, right, or obtuse.

B **11.** $AC = 13$, $BC = 15$, $CD = 12$

12. $AC = 10$, $BC = 17$, $CD = 8$

13. $AC = 13$, $BC = \sqrt{34}$, $CD = 3$

14. $AD = 2$, $DB = 8$, $CD = 4$

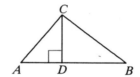

15. The sides of a triangle have lengths x, $x + 4$, and 20. Specify those values of x for which the triangle is acute with longest side 20.

16. Sketch $\square EFGH$ with $EF = 13$, $EG = 24$, and $FH = 10$. What special kind of parallelogram is $EFGH$? Explain.

17. Sketch $\square RSTU$, with diagonals intersecting at M. $RS = 9$, $ST = 20$, and $RM = 11$. Which segment is longer, $\overline{SM}$ or $\overline{RM}$? Explain.

18. If x and y are positive numbers with $x > y$, show that a triangle with sides of lengths $2xy$, $x^2 - y^2$, and $x^2 + y^2$ is always a right triangle.

19. a. Complete this statement of Theorem 8-5:
If the square of the longest side of a triangle ___?___.

b. Prove Theorem 8-5.
Given: $\triangle RST$; $l^2 > j^2 + k^2$
Prove: $\triangle RST$ is an obtuse triangle.
(*Hint:* Start by drawing right $\triangle UVW$ with legs j and k.
Compare lengths l and n.)

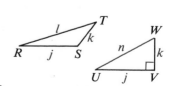

20. **a.** Complete this statement of Theorem 8-4:
    If the square of the longest side of a triangle ___?___.

    b. Prove Theorem 8-4.

    Given: $\triangle RST$; $\overline{RT}$ is the longest side; $l^2 < j^2 + k^2$
    Prove: $\triangle RST$ is an acute triangle.
    (*Hint*: Start by drawing right $\triangle UVW$ with legs j and k.
    Compare lengths l and n.)

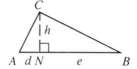

C 21. Given: $\overline{CN} \perp \overline{AB}$;
    h is the geometric mean between d and e.
    Prove: $\triangle ABC$ is a right triangle.

22. A frame in the shape of the simple *scissors truss* shown at the right
    below can be used to support a peaked roof. The weight of the roof
    compresses some parts of the frame (green), while other parts are in tension
    (blue). A frame made with s segments joined at j points is stable if
    $s \geq 2j - 3$. In the truss shown, 9 segments connect 6 points. Verify that
    the truss is stable. Then find the values of x and y.

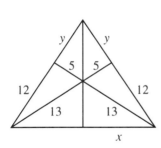

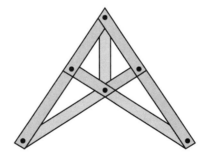

Explorations

**These exploratory exercises can be done using a computer with a program
that draws and measures geometric figures.**

The sides of a quadrilateral have lengths a, b, c, and d. The diagonals have
lengths e and f. For what kinds of quadrilaterals does

$$a^2 + b^2 + c^2 + d^2 = e^2 + f^2?$$

Draw various quadrilaterals including a parallelogram, rectangle, rhombus,
trapezoid, and a random quadrilateral.

♦ Computer Key-In

Suppose a, b, and c are positive integers such that $a^2 + b^2 = c^2$. Then the converse of the Pythagorean Theorem guarantees that a, b, and c are the lengths of the sides of a right triangle. Because of this, any such triple of integers is called a **Pythagorean triple.**

For example, 3, 4, 5 is a Pythagorean triple since $3^2 + 4^2 = 5^2$. Another triple is 6, 8, 10, since $6^2 + 8^2 = 10^2$. The triple 3, 4, 5 is called a *primitive* Pythagorean triple because no factor (other than 1) is common to all three integers. The triple 6, 8, 10 is *not* a primitive triple.

The following program in BASIC lists some Pythagorean triples.

```
10  FOR X = 2 TO 7
20    FOR Y = 1 TO X - 1
30      LET A = 2 * X * Y
40      LET B = X * X - Y * Y
50      LET C = X * X + Y * Y
60      PRINT A;",";B;",";C
70    NEXT Y
80  NEXT X
90  END
```

Exercises

1. Type and RUN the program. (If your computer uses a language other than BASIC, write and RUN a similar program.) What Pythagorean triples did it list? Which of these are primitive Pythagorean triples?

2. The program above uses a method for finding Pythagorean triples that was developed by Euclid around 320 B.C. His method can be stated as follows:

   If x and y are positive integers with $y < x$, then $a = 2xy$, $b = x^2 - y^2$, and $c = x^2 + y^2$ is a Pythagorean triple.

   To verify that Euclid's method is correct, show that the equation below is true.

$$(2xy)^2 + (x^2 - y^2)^2 = (x^2 + y^2)^2$$

3. Look at the primitive Pythagorean triples found in Exercise 1. List those triples that have an odd number as their lowest value. Do you notice a pattern in some of these triples?

   Another method for finding Pythagorean triples begins with an odd number. If n is any positive integer, $2n + 1$ is an odd number. A triple is given by: $a = 2n + 1$, $b = 2n^2 + 2n$, $c = (2n^2 + 2n) + 1$.
   For example, when $n = 3$, the triple is
   $$2(3) + 1 = 7, \quad 2(3^2) + 2(3) = 24, \quad 24 + 1 = 25.$$

   a. Use the formula to find another primitive triple with 33 as its lowest value. (*Hint:* $n = 16$)
   b. Use the Pythagorean Theorem to verify the method described.

8-4 *Special Right Triangles*

An isosceles right triangle is also called a 45°-45°-90° triangle, because the measures of the angles are 45, 45, and 90.

Theorem 8-6 *45°-45°-90° Theorem*
In a 45°-45°-90° triangle, the hypotenuse is $\sqrt{2}$ times as long as a leg.

Given: A 45°-45°-90° triangle

Prove: hypotenuse $= \sqrt{2} \cdot$ leg

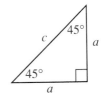

Plan for Proof: Let the sides of the given triangle be a, a, and c. Apply the Pythagorean Theorem and solve for c in terms of a.

Example 1 Find the value of x.

a.

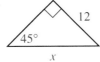

b.

Solution **a.** hyp $= \sqrt{2} \cdot$ leg

$x = \sqrt{2} \cdot 12$

$x = 12\sqrt{2}$

b. hyp $= \sqrt{2} \cdot$ leg

$8 = \sqrt{2} \cdot x$

$x = \dfrac{8}{\sqrt{2}} = \dfrac{8}{\sqrt{2}} \cdot \dfrac{\sqrt{2}}{\sqrt{2}} = \dfrac{8\sqrt{2}}{2}$

$x = 4\sqrt{2}$

Another special right triangle has acute angles measuring 30 and 60.

Theorem 8-7 *30°-60°-90° Theorem*
In a 30°-60°-90° triangle, the hypotenuse is twice as long as the shorter leg, and the longer leg is $\sqrt{3}$ times as long as the shorter leg.

Given: $\triangle ABC$, a 30°-60°-90° triangle

Prove: hypotenuse $= 2 \cdot$ shorter leg
longer leg $= \sqrt{3} \cdot$ shorter leg

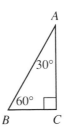

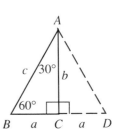

Plan for Proof: Build onto $\triangle ABC$ as shown. $\triangle ADC \cong \triangle ABC$, so $\triangle ABD$ is equiangular and equilateral with $c = 2a$. Since $\triangle ABC$ is a right triangle, $a^2 + b^2 = c^2$. By substitution, $a^2 + b^2 = 4a^2$, so $b^2 = 3a^2$ and $b = a\sqrt{3}$.

Example 2 Find the values of *x* and *y*.

a.

b.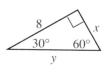

Solution **a.** hyp. = 2 · shorter leg

$$x = 2 \cdot 6$$

$$x = 12$$

longer leg = $\sqrt{3}$ · shorter leg

$$y = 6\sqrt{3}$$

b. longer leg = $\sqrt{3}$ · shorter leg

$$8 = \sqrt{3} \cdot x$$

$$x = \frac{8}{\sqrt{3}} = \frac{8\sqrt{3}}{3}$$

hyp. = 2 · shorter leg

$$y = 2 \cdot \frac{8\sqrt{3}}{3} = \frac{16\sqrt{3}}{3}$$

Classroom Exercises

Find the value of *x*.

1.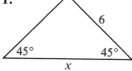

2.

3.

4.

5.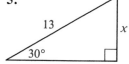

6.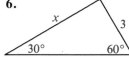

7.

8.

9.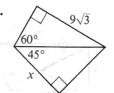

10. In regular hexagon *ABCDEF*, *AB* = 8. Find *AD* and *AC*.

11. Express *PQ*, *PS*, and *QR* in terms of *a*.

12. If the measures of the angles of a triangle are in the ratio 1:2:3, are the lengths of the sides in the same ratio? Explain.

Written Exercises

Copy and complete the tables.

A

	1.	2.	3.	4.	5.	6.	7.	8.
a	4	?	$\sqrt{5}$	?	?	?	?	?
b	?	$\frac{2}{3}$	?	?	?	?	$4\sqrt{2}$	?
c	?	?	?	$3\sqrt{2}$	6	$\sqrt{14}$	?	5

	9.	10.	11.	12.	13.	14.	15.	16.
d	7	$\frac{1}{4}$	?	?	?	?	?	?
e	?	?	$5\sqrt{3}$	6	?	?	3	?
f	?	?	?	?	10	13	?	$6\sqrt{3}$

17. Find the length of a diagonal of a square whose perimeter is 48.

18. A diagonal of a square has length 8. What is the perimeter of the square?

19. An altitude of an equilateral triangle has length $6\sqrt{3}$. What is the perimeter of the triangle?

20. Find the altitude of an equilateral triangle if each side is 10 units long.

Find the values of x and y in each diagram.

B 21.

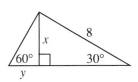

22.

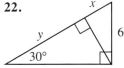

23.

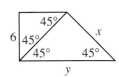

24.

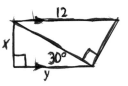

25.

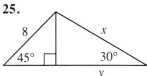

26.

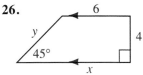

27. The diagram shows four 45°-45°-90° triangles. If $OA = 1$, find OB, OC, OD, and OE.

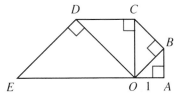

28. The diagonals of a rectangle are 8 units long and intersect at a 60° angle. Find the dimensions of the rectangle.

29. The perimeter of a rhombus is 64 and one of its angles has measure 120. Find the lengths of the diagonals.

30. Prove Theorem 8-6.

31. Explain why any triangle having sides in the ratio $1:\sqrt{3}:2$ must be a 30°-60°-90° triangle.

Find the lengths of as many segments as possible.

32.

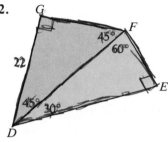

33.

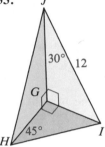

34.

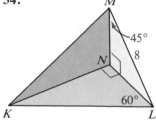

C 35. In quadrilateral $QRST$, $m \angle R = 60$, $m \angle T = 90$, $QR = RS$, $ST = 8$, and $TQ = 8$.
  a. How long is the longer diagonal of the quadrilateral?
  b. Find the ratio of RT to QS.

36. Find the perimeter of the triangle.

37. Find the length of the median of the trapezoid in terms of j.

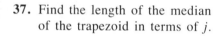
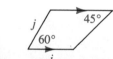

38. If the wrench just fits the hexagonal nut, what is the value of x?

★ 39. The six edges of the solid shown are 8 units long. A and B are midpoints of two edges as shown. Find AB.

Self-Test 1

1. Find the geometric mean between 3 and 15.

2. The diagram shows the altitude drawn to the hypotenuse of a right triangle.
   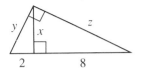
   a. $x = $ _____
   b. $y = $ _____
   c. $z = $ _____

3. The sides of a triangle are given. Is the triangle acute, right, or obtuse?
   a. 11, 60, 61   b. 7, 9, 11   c. 0.2, 0.3, 0.4

4. A rectangle has length 8 and width 4. Find the lengths of the diagonals.

5. Find the perimeter of a square that has diagonals 10 cm long.

6. The sides of an equilateral triangle are 12 cm long. Find the length of an altitude of the triangle.

7. How long is the altitude to the base of an isosceles triangle if the sides of the triangle are 13, 13, and 10?

Biographical Note	*Nikolai Lobachevsky*

Lobachevsky (1793–1856) was a Russian mathematician who brought a new insight to the study of geometry. He realized that Euclidean geometry is only one geometry, and that other geometric systems are possible.

A modern restatement of Euclid's fifth postulate, often called the Parallel Postulate, is "Through a point outside a line, there is exactly one line parallel to the given line." It is this postulate that defines *Euclidean* geometry. For 2000 years, mathematicians tried to prove this fifth postulate from the other four.

Lobachevsky tried a different approach. He created a geometric system where Euclid's first four postulates were the same but the fifth was changed to allow *more than one* parallel through a given point. The antique model at the left shows such a system. Other geometric systems based on a different fifth postulate followed. (See Extra: Non-Euclidean Geometries, page 233.)

Although Lobachevsky thought our universe was Euclidean, some physicists have decided the universe may be better described by Lobachevsky's system. Even so, over small regions Euclidean geometry is accurate. Similarly, although the surface of the Earth is a sphere, we treat small areas of it as flat.

Trigonometry

Objectives

1. Define the tangent, sine, and cosine ratios for an acute angle.
2. Solve right triangle problems by correct selection and use of the tangent, sine, and cosine ratios.

8-5 *The Tangent Ratio*

The word trigonometry comes from Greek words that mean "triangle measurement." In this book our study will be limited to the trigonometry of right triangles. In the right triangle shown, one acute angle is marked. The leg opposite this angle and the leg adjacent to this angle are labeled.

The following ratio of the lengths of the legs is called the *tangent ratio*.

tangent of $\angle A = \dfrac{\text{leg opposite } \angle A}{\text{leg adjacent to } \angle A}$

In abbreviated form: $\tan A = \dfrac{\text{opposite}}{\text{adjacent}}$

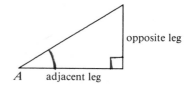

Example 1 Find $\tan X$ and $\tan Y$.

Solution $\tan X = \dfrac{\text{leg opposite } \angle X}{\text{leg adjacent to } \angle X} = \dfrac{12}{5}$

$\tan Y = \dfrac{\text{leg opposite } \angle Y}{\text{leg adjacent to } \angle Y} = \dfrac{5}{12}$

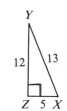

In the right triangles shown below, $m \angle A = m \angle R$. Then by the AA Similarity Postulate, the triangles are similar. We can write these proportions:

$\dfrac{a}{r} = \dfrac{b}{s}$ (Why?)

$\dfrac{a}{b} = \dfrac{r}{s}$ (A property of proportions)

$\tan A = \tan R$ (Def. of tangent ratio)

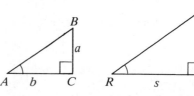

We have shown that if $m \angle A = m \angle R$, then $\tan A = \tan R$. Thus, we have shown that the value of the tangent of an angle depends only on the size of the angle, not on the size of the right triangle. It is also true that if $\tan A = \tan R$ for acute angles A and R, then $m \angle A = m \angle R$.

Since the tangent of an angle depends only on the measure of the angle, we can write tan 10°, for example, to stand for the tangent of any angle with a degree measure of 10. The table on page 311 lists the values of the tangents of some angles with measures between 0 and 90. Most of the values are approximations, rounded to four decimal places. Suppose you want the approximate value of tan 33°. Locate 33° in the angle column. Go across to the tangent column. Read .6494. You write tan 33° ≈ 0.6494, where the symbol ≈ means "is approximately equal to." You can also use a scientific calculator to find tan 33° ≈ 0.649407593. Your calculator may give more or fewer decimal places than the nine that are shown.

Example 2 Find the value of y to the nearest tenth.

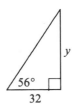

Solution $\tan 56° = \dfrac{y}{32}$

$y = 32(\tan 56°)$

$y \approx 32(1.4826)$

$y \approx 47.4432$, or 47.4

You can find the approximate degree measure of an angle with a given tangent by reading the table from the tangent column across to the angle column, or by using the inverse tangent key(s) of a calculator.

Example 3 The grade of a road is the ratio of its rise to its run and is usually given as a decimal or percent. Find the angle that the road makes with the horizontal if its grade is 4% ($\frac{4}{100}$ or 0.04).

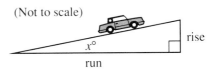

(Not to scale)

Solution $\tan x° = 0.0400$

$x° \approx 2°$

$$\text{grade} = \dfrac{\text{rise}}{\text{run}}$$

If you use the table on page 311, notice that 0.0400 falls between two values in the tangent column: tan 2° ≈ 0.0349 and tan 3° ≈ 0.0524. Since 0.0349 is closer to 0.0400, we use 2° as an approximate value for $x°$.

Classroom Exercises

In Exercises 1–3 express tan R as a ratio.

1.
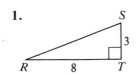

2.
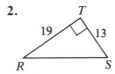

3.

4–6. Express tan S as a ratio for each triangle above.

7. Use the table on page 311 to complete the statements.
   a. tan 24° ≈ __?__
   b. tan 41° ≈ __?__
   c. tan 88° ≈ __?__
   d. tan __?__ ≈ 2.4751
   e. tan __?__ ≈ 0.3057
   f. tan __?__ ≈ 0.8098

8. Three 45°-45°-90° triangles are shown below.

   a. In each triangle, express tan 45° in simplified form.

   b. See the entry for tan 45° on page 311. Is the entry exact?

9. Three 30°-60°-90° triangles are shown below.

   a. In each triangle, express tan 60° in simplified radical form.

   b. Use $\sqrt{3} \approx 1.732051$ to find an approximate value for tan 60°.

   c. Is the entry for tan 60° on page 311 exact? Is it correct to four decimal places?

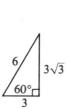   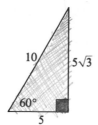   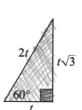

10. Notice that the tangent values increase rapidly toward the end of the table on page 311. Explain how you know that there is some angle with a tangent value equal to 1,000,000. Is there any upper limit to tangent values?

11. Two ways to find the value of x are started below.

Using tan 40°:

$$\tan 40° = \frac{27}{x}$$

$$0.8391 \approx \frac{27}{x}$$

Using tan 50°:

$$\tan 50° = \frac{x}{27}$$

$$1.1918 \approx \frac{x}{27}$$

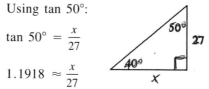

Which of the following statements are correct?

a. $x \approx 27 \cdot 0.8391$ **b.** $x \approx 27 \cdot 1.1918$

c. $x \approx \dfrac{27}{0.8391}$ **d.** $x \approx \dfrac{27}{1.1918}$

Which correct statement is easier to use for computing if you are *not* using a calculator for the arithmetic?

Written Exercises

Find the value of x to the nearest tenth. Use a calculator or the table on page 311.

A **1.**

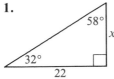

2.

3.

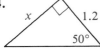

4.

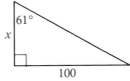

5.

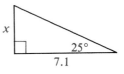

6.

Find $y°$ correct to the nearest degree.

7.

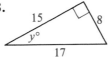

8.

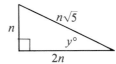

9.

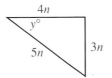

10.

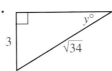

11.

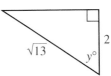

12.

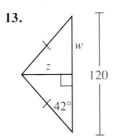

Find w, then z, correct to the nearest integer.

B **13.**

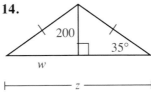

14.

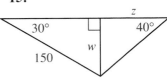

15.

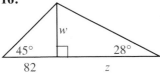

16.

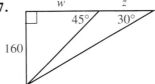

17.

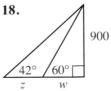

18.

19. The grade of a road is 7%. What angle does the road make with the horizontal?

20. A road climbs at an 8° angle with the horizontal. What is the grade of the road?

21. The base of an isosceles triangle is 70 cm long. The altitude to the base is 75 cm long. Find, to the nearest degree, the base angles of the triangle.

22. A rhombus has diagonals of length 4 and 10. Find the angles of the rhombus to the nearest degree.

23. The shorter diagonal of a rhombus with a 70° angle is 122 cm long. How long, to the nearest centimeter, is the longer diagonal?

24. A rectangle is 80 cm long and 20 cm wide. Find, to the nearest degree, the acute angle formed at the intersection of the diagonals.

25. A natural question to consider is the following:

$$\text{Does } \tan A + \tan B = \tan (A + B)?$$

Try substituting 35° for *A* and 25° for *B*.

   a. $\tan 35° + \tan 25° \approx \underline{\ ?\ } + \underline{\ ?\ } = \underline{\ ?\ }$
   b. $\tan (35° + 25°) = \tan \underline{\ ?\ }° \approx \underline{\ ?\ }$
   c. What is your answer to the general question raised in this exercise, *yes* or *no*?
   d. Do you think $\tan A - \tan B = \tan (A - B)$? Explain.

26. **a.** Given: $\triangle PQR$; $\angle R$ is a right angle.
       Prove: $\tan P \cdot \tan Q = 1$

   b. If $\tan 32° \approx \dfrac{5}{8}$, find $\tan 58°$
       without using a table or a calculator.

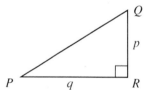

27. A rectangular box has length 4, width 3, and height 2.
   a. Find *BD*.
   b. Find $\angle GBD$ to the nearest degree.

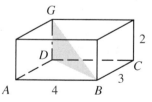

C 28. If the figure is a cube, find $\angle TQS$ to the nearest degree.

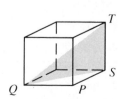

29. A person at window W, 40 ft above street level, sights points on a building directly across the street. H is chosen so that $\overline{WH}$ is horizontal. T is directly above H, and B is directly below. By measurement, $m \angle TWH = 61$ and $m \angle BWH = 37$. How far above street level is T?

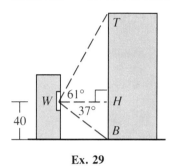

Ex. 29

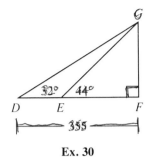

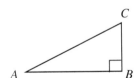

Ex. 30

30. Use the figure to find EF to the nearest integer.

Explorations

These exploratory exercises can be done using a computer with a program that draws and measures geometric figures.

As you will learn in the next section, two other trigonometric ratios are the *sine* and *cosine*. If $\triangle ABC$ has a right angle at B, then:

$$\sin A = \frac{\text{leg opposite } \angle A}{\text{hypotenuse}} = \frac{BC}{AC}$$

$$\cos A = \frac{\text{leg adjacent to } \angle A}{\text{hypotenuse}} = \frac{AB}{AC}$$

Using ASA, draw nine right triangles using nine values for $m \angle A$: 10, 20, 30, 40, 45, 50, 60, 70, and 80. Keep $m \angle B = 90$.

Compute and record $\sin A$, $\cos A$, and $\tan A$ for each measure of $\angle A$. What do you notice?

If you change the length of $\overline{AB}$ but keep the measures of $\angle A$ and $\angle B$ the same, do the sine, cosine, and tangent of $\angle A$ change?

Complete.

1. $\cos x° = \sin x°$ when $x =$ __?__

2. $\cos (90 - x)° = \sin$ __?__

3. $\sin (90 - x)° = \cos$ __?__

4. $\tan x° \cdot \tan (90 - x)° =$ __?__

5. For acute angles, what trigonometric ratios have values between 0 and 1?

6. What trigonometric ratio can have values greater than 1?

Table of Trigonometric Ratios

Angle	Sine	Cosine	Tangent	Angle	Sine	Cosine	Tangent
1°	.0175	.9998	.0175	46°	.7193	.6947	1.0355
2°	.0349	.9994	.0349	47°	.7314	.6820	1.0724
3°	.0523	.9986	.0524	48°	.7431	.6691	1.1106
4°	.0698	.9976	.0699	49°	.7547	.6561	1.1504
5°	.0872	.9962	.0875	50°	.7660	.6428	1.1918
6°	.1045	.9945	.1051	51°	.7771	.6293	1.2349
7°	.1219	.9925	.1228	52°	.7880	.6157	1.2799
8°	.1392	.9903	.1405	53°	.7986	.6018	1.3270
9°	.1564	.9877	.1584	54°	.8090	.5878	1.3764
10°	.1736	.9848	.1763	55°	.8192	.5736	1.4281
11°	.1908	.9816	.1944	56°	.8290	.5592	1.4826
12°	.2079	.9781	.2126	57°	.8387	.5446	1.5399
13°	.2250	.9744	.2309	58°	.8480	.5299	1.6003
14°	.2419	.9703	.2493	59°	.8572	.5150	1.6643
15°	.2588	.9659	.2679	60°	.8660	.5000	1.7321
16°	.2756	.9613	.2867	61°	.8746	.4848	1.8040
17°	.2924	.9563	.3057	62°	.8829	.4695	1.8807
18°	.3090	.9511	.3249	63°	.8910	.4540	1.9626
19°	.3256	.9455	.3443	64°	.8988	.4384	2.0503
20°	.3420	.9397	.3640	65°	.9063	.4226	2.1445
21°	.3584	.9336	.3839	66°	.9135	.4067	2.2460
22°	.3746	.9272	.4040	67°	.9205	.3907	2.3559
23°	.3907	.9205	.4245	68°	.9272	.3746	2.4751
24°	.4067	.9135	.4452	69°	.9336	.3584	2.6051
25°	.4226	.9063	.4663	70°	.9397	.3420	2.7475
26°	.4384	.8988	.4877	71°	.9455	.3256	2.9042
27°	.4540	.8910	.5095	72°	.9511	.3090	3.0777
28°	.4695	.8829	.5317	73°	.9563	.2924	3.2709
29°	.4848	.8746	.5543	74°	.9613	.2756	3.4874
30°	.5000	.8660	.5774	75°	.9659	.2588	3.7321
31°	.5150	.8572	.6009	76°	.9703	.2419	4.0108
32°	.5299	.8480	.6249	77°	.9744	.2250	4.3315
33°	.5446	.8387	.6494	78°	.9781	.2079	4.7046
34°	.5592	.8290	.6745	79°	.9816	.1908	5.1446
35°	.5736	.8192	.7002	80°	.9848	.1736	5.6713
36°	.5878	.8090	.7265	81°	.9877	.1564	6.3138
37°	.6018	.7986	.7536	82°	.9903	.1392	7.1154
38°	.6157	.7880	.7813	83°	.9925	.1219	8.1443
39°	.6293	.7771	.8098	84°	.9945	.1045	9.5144
40°	.6428	.7660	.8391	85°	.9962	.0872	11.4301
41°	.6561	.7547	.8693	86°	.9976	.0698	14.3007
42°	.6691	.7431	.9004	87°	.9986	.0523	19.0811
43°	.6820	.7314	.9325	88°	.9994	.0349	28.6363
44°	.6947	.7193	.9657	89°	.9998	.0175	57.2900
45°	.7071	.7071	1.0000				

8-6 *The Sine and Cosine Ratios*

Suppose you want to find the legs, x and y, in the triangle at the right. You can't easily find these values using the tangent ratio because the only side you know is the hypotenuse. The ratios that relate the legs to the hypotenuse are the *sine* and *cosine*.

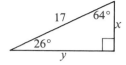

sine of $\angle A = \dfrac{\text{leg opposite } \angle A}{\text{hypotenuse}}$

cosine of $\angle A = \dfrac{\text{leg adjacent to } \angle A}{\text{hypotenuse}}$

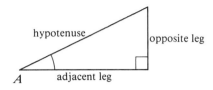

We now have three useful trigonometric ratios, given below in abbreviated form:

$$\tan A = \frac{\text{opposite}}{\text{adjacent}}$$

$$\sin A = \frac{\text{opposite}}{\text{hypotenuse}}$$

$$\cos A = \frac{\text{adjacent}}{\text{hypotenuse}}$$

Example 1 Find the values of x and y to the nearest integer.

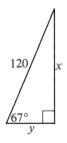

Solution
$\sin 67° = \dfrac{x}{120}$
$x = 120 \cdot \sin 67°$
$x \approx 120(0.9205)$
$x \approx 110.46$, or 110

$\cos 67° = \dfrac{y}{120}$
$y = 120 \cdot \cos 67°$
$y \approx 120(0.3907)$
$y \approx 46.884$, or 47

Example 2 Find the value of n to the nearest integer.

Solution
$\sin n° = \dfrac{22}{40}$
$\sin n° = 0.5500$
$n \approx 33$

Example 3 An isosceles triangle has sides 8, 8, and 6. Find the lengths of its three altitudes.

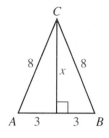

Solution The altitude to the base can be found using the Pythagorean Theorem.
$$x^2 = 8^2 - 3^2 = 55$$
$$x = \sqrt{55} \approx 7.4$$

Notice that $\cos B = \dfrac{3}{8}$ (so $m \angle B \approx 68$), and that the altitudes from A and B are congruent. (Why?) To find the length of the altitudes from A and B, use

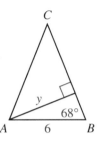

$$\sin B \approx \sin 68° \approx \frac{y}{6}.$$
$$y \approx 6 \cdot \sin 68°$$
$$y \approx 5.6$$

Classroom Exercises

In Exercises 1–3 express sin A, cos A, and tan A as fractions.

1.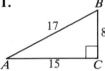

2.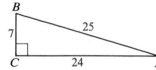

3.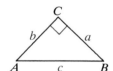

4–6. Using the triangles in Exercises 1–3, express sin B, cos B, and tan B as fractions.

7. Use the table on page 311 or a scientific calculator to complete the statements.
 a. sin 24° ≈ __?__
 b. cos 57° ≈ __?__
 c. sin 87° ≈ __?__
 d. cos __?__ ≈ 0.9659
 e. sin __?__ ≈ 0.1045
 f. cos __?__ ≈ 0.1500

State two different equations you could use to find the value of x.

8.

9.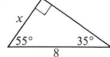

10.

11. The word *cosine* is related to the phrase "complement's sine." Explain the relationship by using the diagram to express the cosine of $\angle A$ and the sine of its complement, $\angle B$.

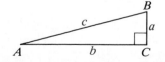

12. The table on page 311 lists 0.5000 as the value of sin 30°. This value is exact. Explain why.

13. Suppose sin $n° = \dfrac{5}{13}$. Find cos $n°$ and tan $n°$ without using a table or calculator.

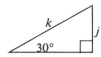

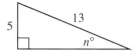

14. According to the table on page 311, sin 1° and tan 1° are both approximately 0.0175. Which is actually larger? How do you know?

15. a. Using the definition of sine, explain why the sine of an acute angle is always less than one.
    b. Is the cosine of an acute angle always less than one?

Written Exercises

In these exercises, use a scientific calculator or the table on page 311. Find lengths correct to the nearest integer and angles to the nearest degree.

In Exercises 1–12 find the values of the variables.

A **1.**

2.

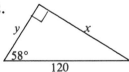

3.

4.

5.

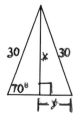

6.

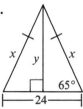

7.

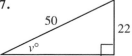

8.

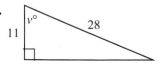

9.

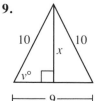

10.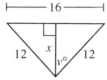

11.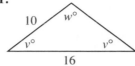

12.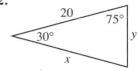

13. a. Use the Pythagorean Theorem to find the value of x in radical form.
   b. Use trigonometry to find the values of y, then x.
   c. Are the values of x from parts (a) and (b) in reasonable agreement?

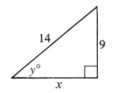

B **14.** A guy wire is attached to the top of a 75 m tower and meets the ground at a 65° angle. How long is the wire?

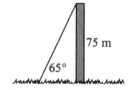

15. To find the distance from point A on the shore of a lake to point B on an island in the lake, surveyors locate point P with $m \angle PAB = 65$ and $m \angle APB = 25$. By measurement, $PA = 352$ m. Find AB.

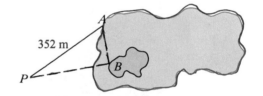

16. A certain jet is capable of a steady 20° climb. How much altitude does the jet gain when it moves 1 km through the air? Answer to the nearest 50 m.

17. A 6 m ladder reaches higher up a wall when placed at a 70° angle than when placed at a 60° angle. How much higher, to the nearest tenth of a meter?

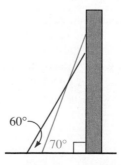

18. In $\triangle ABC$, $AB = AC = 13$ and $BC = 10$.
   a. Find the length of the altitude from A.
   b. Find the measures of the three angles of $\triangle ABC$.
   c. Find the length of the altitude from C.

19. In $\triangle ABC$, $m \angle B = m \angle C = 72$ and $BC = 10$.
   a. Find AB and AC.
   b. Find the length of the bisector of $\angle A$ to $\overline{BC}$.

20. In $\triangle PAL$, $m \angle A = 90$, $m \angle L = 24$ and median $\overline{AM}$ is 6 cm long. Find PA.

21. The diagonals of rectangle $ABCD$ are 18 cm long and intersect in a 34° angle. Find the length and width of the rectangle.

22. Points A, B, and C are three consecutive vertices of a regular decagon whose sides are 16 cm long. How long is diagonal $\overline{AC}$?

23. Points A, B, C, and D are consecutive vertices of a regular decagon with sides 20 cm long. $\overrightarrow{AB}$ and $\overrightarrow{DC}$ are drawn and intersect at X. Find BX.

For Exercises 24–26 write proofs in paragraph form.

C **24.** Prove that in any triangle with acute angles A and B, $\dfrac{a}{\sin A} = \dfrac{b}{\sin B}$. (*Hint*: Draw a perpendicular from the third vertex to $\overline{AB}$. Label it p.)

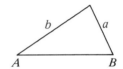

25. Prove: If R is any acute angle, $(\sin R)^2 + (\cos R)^2 = 1$. (*Hint*: From any point on one side of $\angle R$, draw a perpendicular to the other side.)

26. A rectangular card is 10 cm wide. The card is folded so that the vertex D falls at point D' on $\overline{AB}$ as shown. Crease $\overline{CE}$ with length k makes an $n°$ angle with $\overline{CD}$. Prove: $k = \dfrac{10}{\sin (2n)° \cos n°}$

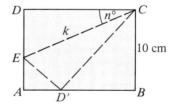

Challenge

The two blocks of wood have the same size and shape. It is possible to cut a hole in one block in such a way that you can pass the other block completely through the hole. How?

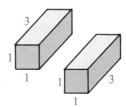

8-7 *Applications of Right Triangle Trigonometry*

Suppose an operator at the top of a lighthouse sights a sailboat on a line that makes a 2° angle with a horizontal line. The angle between the horizontal and the line of sight is called an **angle of depression.** At the same time, a person in the boat must look 2° above the horizontal to see the tip of the lighthouse. This is an **angle of elevation.**

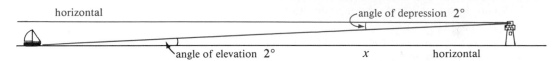

horizontal — angle of depression 2°

angle of elevation 2° — x — horizontal

If the top of the lighthouse is 25 m above sea level, the distance *x* between the boat and the base of the lighthouse can be found in these two ways:

Method 1	*Method 2*
$\tan 2° = \dfrac{25}{x}$	$\tan 88° = \dfrac{x}{25}$
$x = \dfrac{25}{\tan 2°}$	$x = 25(\tan 88°)$
$x \approx \dfrac{25}{0.0349}$	$x \approx 25(28.6363)$
$x \approx 716.3$	$x \approx 715.9$

Because the tangent values in the table are approximations, the two methods give slightly different answers. In practice, the angle measurement will not be exact, and the boat may be moving. In a case like this we cannot claim high accuracy for our answer. A good answer would be: The boat is roughly 700 m from the lighthouse.

Classroom Exercises

1. Two people at points *X* and *Y* sight an airplane at *A*.
   a. What is the angle of elevation from *X* to *A*?
   b. What is the angle of depression from *A* to *X*?
   c. What is the angle of depression from *A* to *Y*?
   d. What is the angle of elevation from *Y* to *A*?
   e. Is the measure of the angle of elevation from *Z* to *A* greater or less than 35?

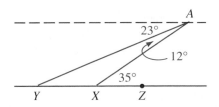

The lines shown are horizontal and vertical lines except for $\overleftrightarrow{HT}$ and $\overleftrightarrow{HG}$. Give the number of the angle and its special name when:

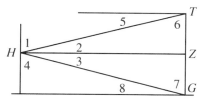

2. A person at *H* sights *T*.

3. A person at *H* sights *G*.

4. A person at *T* sights *H*.

5. A person at *G* sights *H*.

6. A driveway has a 15% grade.
  a. What is the angle of elevation of the driveway?
  b. If the driveway is 12 m long, about how much does it rise?

7. A toboggan travels from point *A* at the top of the hill to point *B* at the bottom. Because the steepness of the hill varies, the angle of depression from *A* to *B* is only an approximate measure of the hill's steepness. We can, however, think of this angle of depression as representing the average steepness.

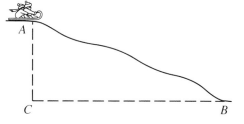

  a. If the toboggan travels 130 m from *A* to *B* and the vertical descent *AC* is 50 m, what is the approximate angle of depression?
  b. Why is your answer approximate?

Written Exercises

Express lengths correct to the nearest integer and angles correct to the nearest degree. Use a calculator or the table on page 311.

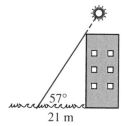

A 1. When the sun's angle of elevation is 57°, a building casts a shadow 21 m long. How high is the building?

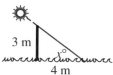

2. At a certain time, a vertical pole 3 m tall casts a 4 m shadow. What is the angle of elevation of the sun?

In Exercises 3–8 first draw a diagram.

3. A kite is flying at an angle of elevation of about 40°. All 80 m of string have been let out. Ignoring the sag in the string, find the height of the kite to the nearest 10 m.

4. An advertising blimp hovers over a stadium at an altitude of 125 m. The pilot sights a tennis court at an 8° angle of depression. Find the ground distance in a straight line between the stadium and the tennis court. (*Note:* In an exercise like this one, an answer saying *about . . . hundred meters* is sensible.)

5. An observer located 3 km from a rocket launch site sees a rocket at an angle of elevation of 38°. How high is the rocket at that moment?

6. To land, an airplane will approach an airport at a 3° angle of depression. If the plane is flying at 30,000 ft, find the ground distance from the airport to the point directly below the plane when the pilot begins descending. Give your answer to the nearest 10,000 feet.

B 7. Martha is 180 cm tall and her daughter Heidi is just 90 cm tall. Who casts the longer shadow, Martha when the sun is 70° above the horizon, or Heidi when the sun is 35° above the horizon? How much longer?

8. Two buildings on opposite sides of a street are 40 m apart. From the top of the taller building, which is 185 m high, the angle of depression to the top of the shorter building is 13°. Find the height of the shorter building.

9. Scientists can estimate the depth of craters on the moon by studying the lengths of their shadows in the craters. Shadows' lengths can be estimated by measuring them on photographs. Find the depth of a crater if the shadow is estimated to be 400 m long and the angle of elevation of the sun is 48°.

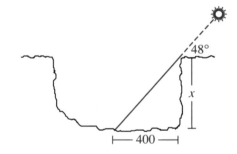

10. A road has a 10% grade.
    a. What is the angle of elevation of the road?
    b. If the road is 2 km long, how much does it rise?

11. A road 1.6 km long rises 400 m. What is the angle of elevation of the road?

12. The force of gravity pulling an object down a hill is its weight multiplied by the sine of the angle of elevation of the hill.
    a. With how many pounds of force is gravity pulling on a 3000 lb car on a hill with a 3° angle of elevation?
    b. Could you push against the car and keep it from rolling down the hill?

13. A soccer goal is 24 ft wide. Point *A* is 40 ft in front of the center of the goal. Point *B* is 40 ft in front of the right goal post.
  a. Which angle is larger, $\angle A$ or $\angle B$?
  b. From which point would you have a better chance of kicking the ball into the goal? Why?

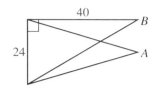

C 14. From the stage of a theater, the angle of elevation of the first balcony is 19°. The angle of elevation of the second balcony, 6.3 m directly above the first, is 29°. How high above stage level is the first balcony? (*Hint:* Use tan 19° and tan 29° to write two equations involving *x* and *d*. Solve for *d*, then find *x*.)

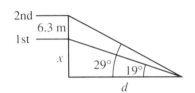

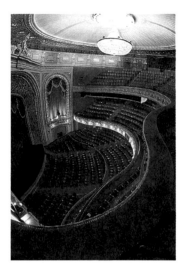

Self-Test 2

Exercises 1–5 refer to the diagram at the right.

1. $\tan E = \dfrac{?}{?}$ **2.** $\cos E = \dfrac{?}{?}$

3. $\sin E = \dfrac{?}{?}$ **4.** $\tan D = \dfrac{?}{?}$

5. To the nearest integer, $m \angle D = \underline{\ \ ?\ \ }$.

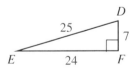

Find the value of *x* to the nearest integer.

6.   **7.** 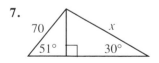 **8.**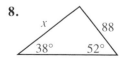

9. From a point on the ground 100 m from the foot of a cliff, the angle of elevation of the top of the cliff is 24°. How high is the cliff?

Application | *Passive Solar Design*

Passive solar homes are designed to let the sun heat the house during the winter but to prevent the sun from heating the house during the summer. Because the Earth's axis is not perpendicular to the *ecliptic* (the plane of the Earth's orbit around the sun), the sun is lower in the sky in the winter than it is in the summer.

From the latitude of the homesite the architect can determine the elevation angle of the sun (the angle at which a person has to look up from the horizontal to see the sun at noon) during the winter and during the summer. The architect can then design an overhang for windows that will let sunlight in the windows during the winter, but will shade the windows during the summer.

The Earth's axis makes an angle of $23\frac{1}{2}°$ with a perpendicular to the ecliptic plane. So for places in the northern hemisphere between the Tropic of Cancer and the Arctic Circle, the angle of elevation of the sun at noon on the longest day of the year, at the summer solstice, is $90°$ − the latitude + $23\frac{1}{2}°$. Its angle of elevation at noon on the shortest day, at the winter solstice, is $90°$ − the latitude − $23\frac{1}{2}°$. For example, in Terre Haute, Indiana, at latitude $39\frac{1}{2}°$ north, the angle of elevation of the sun at noon on the longest day is $74°$ $(90 − 39\frac{1}{2} + 23\frac{1}{2} = 74)$, and at noon on the shortest day it is $27°$ $(90 − 39\frac{1}{2} − 23\frac{1}{2} = 27)$.

Exercises

Find the angle of elevation of the sun at noon on the longest day and at noon on the shortest day in the following cities. The approximate north latitudes are in parentheses.

1. Seattle, Washington $(47\frac{1}{2}°)$

2. Chicago, Illinois $(42°)$

3. Houston, Texas $(30°)$

4. Los Angeles, California $(34°)$

5. Nome, Alaska $(64\frac{1}{2}°)$

6. Miami, Florida $(26°)$

7. For a city south of the Tropic of Cancer, such as San Juan, Puerto Rico $(18°N)$, the formula gives a summer solstice angle greater than $90°$. What does this mean?

8. For a place north of the Arctic Circle, such as Prudhoe Bay, Alaska $(70°N)$, the formula gives a negative value for the angle of elevation of the sun at noon at the winter solstice. What does this mean?

9. An architect is designing a passive solar house to be located in Terre Haute, Indiana. The diagram shows a cross-section of a wall that will face south. How long must the overhang x be to shade the entire window at noon at the summer solstice?

10. If the overhang has the length found in Exercise 9, how much of the window will be in the sun at noon at the winter solstice?

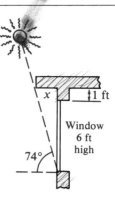

Chapter Summary

1. When $\dfrac{a}{x} = \dfrac{x}{b}$, x is the geometric mean between a and b.

2. A right triangle is shown with the altitude drawn to the hypotenuse.

   a. The two triangles formed are similar to the original triangle and to each other.

   $$\dfrac{x}{h} = \dfrac{h}{y} \qquad \dfrac{c}{b} = \dfrac{b}{x} \qquad \dfrac{c}{a} = \dfrac{a}{y}$$

   b. Pythagorean Theorem: $c^2 = a^2 + b^2$

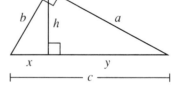

3. The longest side of the triangle shown is c.

   If $c^2 = a^2 + b^2$, then the triangle is a right triangle.
   If $c^2 > a^2 + b^2$, then the triangle is obtuse.
   If $c^2 < a^2 + b^2$, then the triangle is acute.

4. The sides of a 45°-45°-90° triangle and the sides of a 30°-60°-90° triangle are related as shown.

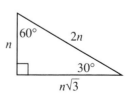

5. In the right triangle shown:

   $$\tan A = \dfrac{a}{b} \qquad \sin A = \dfrac{a}{c} \qquad \cos A = \dfrac{b}{c}$$

   The tangent, sine, and cosine ratios are useful in solving problems involving right triangles.

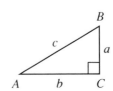

Chapter Review

1. Find the geometric mean between 12 and 3. **8–1**

2. $x = $ ___?___

3. $y = $ ___?___

4. $z = $ ___?___

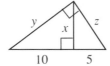

5. The legs of a right triangle are 3 and 6. Find the length of the hypotenuse. **8–2**

6. A rectangle has sides 10 and 8. Find the length of a diagonal.

7. The diagonal of a square has length 14. Find the length of a side.

8. The legs of an isosceles triangle are 10 units long and the altitude to the base is 8 units long. Find the length of the base.

Tell whether a triangle formed with sides having the lengths named is acute, right, or obtuse. If a triangle can't be formed, write *not possible*.

9. 4, 5, 6

10. 8, 8, 17 **8–3**

11. 11, 60, 61

12. $2\sqrt{3}$, $3\sqrt{2}$, 6

Find the value of x.

13.

14.

15. **8–4**

16. The legs of an isosceles right triangle have length 12. Find the lengths of the hypotenuse and the altitude to the hypotenuse.

Complete. Find angle measures and lengths correct to the nearest integer. Use a calculator or the table on page 311 if needed.

17.

18. 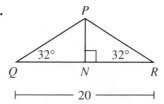 **8–5**

a. $\tan A = $ ___?___

b. $\tan B = $ ___?___

a. $QN = $ ___?___

c. $m \angle B \approx $ ___?___

b. $PN \approx $ ___?___

Complete. Find angle measures and lengths correct to the nearest integer.
Use a calculator or the table on page 311 if needed.

19.

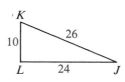

a. $\cos J = $ __?__

b. $\sin K = $ __?__

c. $m \angle K \approx $ __?__

20.

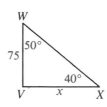

a. $WX \approx$ __?__

b. $VX \approx$ __?__

8–6

Find the values of x and y correct to the nearest integer.

21.

22.

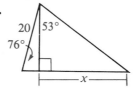

23.

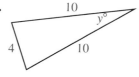

24. Lee, on the ground, looks up at Chong Ye in a hot air balloon at a 35°
angle of elevation. If Lee and Chong Ye are 500 ft apart, about how far
off the ground is Chong Ye?

8–7

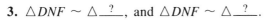

Chapter Test

Find the geometric mean between the numbers.

1. 5 and 20

2. 6 and 8

In the diagram, $\angle DNF$ is a right angle and $\overline{NE} \perp \overline{DF}$.

3. $\triangle DNF \sim \triangle$ __?__ , and $\triangle DNF \sim \triangle$ __?__ .

4. NE is the geometric mean between __?__ and __?__ .

5. NF is the geometric mean between __?__ and __?__ .

6. If $DE = 10$ and $EF = 15$, then $ND = $ __?__ .

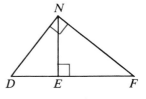

Find the values of x and y.

7.

8.

Tell whether a triangle formed with sides having the lengths named is acute, right, or obtuse. If a triangle can't be formed, write *not possible*.

9. 3, 4, 8

10. 11, 12, 13

11. 7, 7, 10

12. $\frac{3}{5}$, $\frac{4}{5}$, 1

Find the value of x.

13.

14.

15.

16.

17.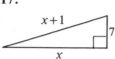

18.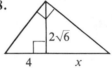

Find lengths correct to the nearest integer and angles correct to the nearest degree.

19.

20.

21.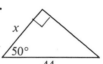

22.

23.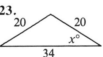

24.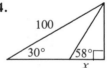

25. The sides of a rhombus are 4 units long and one diagonal has length 4. How long is the other diagonal?

26. In the diagram, $\angle RTS$ is a right angle; $\overline{RT}$, $\overline{RS}$, $\overline{VT}$ and $\overline{VS}$ have the lengths shown.
What is the measure of $\angle V$? Explain.

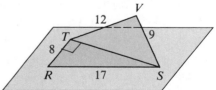

27. From the top of a lighthouse 18 m high, the angle of depression to sight a boat is 4°. What is the distance between the boat and the base of the lighthouse?

Preparing for College Entrance Exams

Strategy for Success

Problems in college entrance exams often involve right triangles. One thing you can do in preparing for the exams is to learn the common right-triangle lengths listed on page 295. These Pythagorean triples are often used on tests where calculators are not allowed. Also, keep in mind that if a, b, and c are the lengths of the sides of a right triangle, then for any $k > 0$, ak, bk, and ck are also lengths of sides of a right triangle.

Indicate the best answer by writing the appropriate letter.

1. In $\triangle ABC$, $m\angle A : m\angle B : m\angle C = 2:5:5$. $m\angle B =$
(A) 75   (B) 60   (C) 30   (D) 40   (E) 100

2. The proportion $\dfrac{t}{z} = \dfrac{m}{k}$ is *not* equivalent to:

(A) $\dfrac{t - z}{z} = \dfrac{m - k}{k}$ (B) $\dfrac{k}{z} = \dfrac{m}{t}$ (C) $\dfrac{t}{m} = \dfrac{k}{z}$ (D) $tk = mz$ (E) $\dfrac{z}{t} = \dfrac{k}{m}$

3. If $\triangle ABC \sim \triangle DEF$, which statement is not necessarily true?

(A) $\angle C \cong \angle F$ (B) $\overline{BC} \cong \overline{EF}$ (C) $\dfrac{AB}{BC} = \dfrac{DE}{EF}$

(D) $m\angle A + m\angle E = m\angle B + m\angle D$ (E) $AC \cdot DE = DF \cdot AB$

4. If $ZY = 2x + 9$, $ZM = 10$, $ZN = x + 3$, and $MW = x$, then $x =$
(A) $2 + \sqrt{34}$ (B) -12 (C) 12   (D) 5   (E) -5

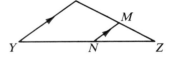

5. $\overrightarrow{BD}$ bisects $\angle ABC$ and D lies on $\overline{AC}$. If $AB = 6$, $BC = 14$, and $AC = 14$, find AD.
(A) 6   (B) 8.4   (C) 9.8   (D) 7   (E) 4.2

6. Find the geometric mean of $2x$ and $2y$.
(A) $2\sqrt{xy}$ (B) $\sqrt{2xy}$ (C) $2\sqrt{x + y}$ (D) $\sqrt{2(x + y)}$ (E) $4xy$

7. If $XY = 8$, $YZ = 40$, and $XZ = 41$, then:
(A) $\triangle XYZ$ is acute   (B) $\triangle XYZ$ is right   (C) $\triangle XYZ$ is obtuse
(D) $m\angle Y < m\angle Z$ (E) no $\triangle XYZ$ is possible

8. A rhombus contains a 120° angle. Find the ratio of the length of the longer diagonal to the length of the shorter diagonal.
(A) $\sqrt{3}:1$ (B) $\sqrt{3}:3$ (C) $\sqrt{2}:1$ (D) $\sqrt{2}:2$ (E) cannot be determined

9. $k =$

(A) $j \sin A$ (B) $j \tan A$ (C) $\dfrac{l}{\sin A}$

(D) $l \cos A$ (E) $l \tan A$

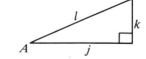

10. The legs of an isosceles triangle have length 4 and the base angles have measure 65. If $\sin 65° \approx 0.91$, $\cos 65° \approx 0.42$, and $\tan 65° \approx 2.14$, then the approximate length of the base of the triangle is:
(A) 1.7   (B) 1.9   (C) 3.4   (D) 3.6   (E) 4.4

Cumulative Review: Chapters 1–8

In Exercises 1–8, complete each statement.

1. If S is between R and T, then $RS + ST = RT$ by the __?__.

2. A statement that is accepted without proof is called a __?__.

3. A statement that can be proved easily by using a theorem is called a __?__.

4. To write an indirect proof, you assume temporarily that the __?__ is not true.

5. A conditional and its __?__ are always logically equivalent.

6. The sides of an obtuse triangle have lengths x, $2x + 2$, and $2x + 3$. __?__ $< x <$ __?__.

7. In an isosceles right triangle, the ratio of the length of a leg to the length of the hypotenuse is __?__.

8. If $\sin B = \dfrac{8}{17}$, then $\cos B =$ __?__.

9. Given: A triangle is equiangular only if it is isosceles.
   a. Write an if-then statement that is logically equivalent to the given conditional.
   b. State the converse. Sketch a diagram to disprove the converse.

10. Use inductive reasoning to guess the next two numbers in the sequence:
$$1, 2, 6, 15, 31, 56, \ldots$$

11. When two parallel lines are cut by a transversal, two corresponding angles have measures x^2 and $6x$. Find the measure of each angle.

12. In $\triangle XYZ$, $m\angle X : m\angle Y : m\angle Z = 3:3:4$.
    a. Is $\triangle XYZ$ scalene, isosceles, or equilateral?
    b. Is $\triangle XYZ$ acute, right, or obtuse?
    c. Name the longest side of $\triangle XYZ$.

13. If $AB = x - 5$, $BC = x - 2$, $CD = x + 4$, and $DA = x$, find the value of x.

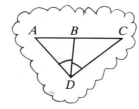

14. The diagonals of a rhombus have lengths 18 and 24. Find the length of one side.

15. Write a paragraph proof: If $\overline{AX}$ is a median and an altitude of $\triangle ABC$, then $\triangle ABC$ is isosceles.

16. Given: $NPQRST$ is a regular hexagon.
    Prove: $NPRS$ is a rectangle.
    (Begin by drawing a diagram.)

17. Given: $\angle WXY \cong \angle XZY$
    Prove: $(XY)^2 = WY \cdot ZY$

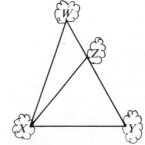

9 CIRCLES

One of the most useful applications of the circle is the wheel. From ancient times until today, the wheel has made an enormous contribution to progress in travel, transportation, industry, and other elements of civilization.

Tangents, Arcs, and Chords

Objectives

1. Define a circle, a sphere, and terms related to them.
2. Recognize circumscribed and inscribed polygons and circles.
3. Apply theorems that relate tangents and radii.
4. Define and apply properties of arcs and central angles.
5. Apply theorems about the chords of a circle.

9-1 Basic Terms

A **circle** is the set of points in a plane at a given distance from a given point in that plane. The given point is the **center** of the circle and the given *distance* is *the* **radius.** Any *segment* that joins the center to a point of the circle is called *a* radius. All radii of a circle are congruent. The rim of the Ferris wheel shown is a circle with center O ($\odot O$) and radius 10.

A **chord** is a segment whose endpoints lie on a circle. A **secant** is a line that contains a chord. A **diameter** is a chord that contains the center of a circle. (Like the word *radius*, the word *diameter* can refer to *the* length of a segment or to *a* segment.)

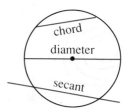

A **tangent** is a line in the plane of a circle that intersects the circle in exactly one point, called the **point of tangency.** The *tangent ray* $\overrightarrow{PA}$ and *tangent segment* $\overline{PA}$ are often called tangents.

$\overleftrightarrow{AP}$ is tangent to $\odot O$.

$\odot O$ is tangent to $\overleftrightarrow{AP}$.

A is the point of tangency.

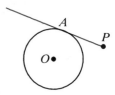

A **sphere** with center O and radius r is the set of all points in space at a distance r from point O. Many of the terms used with spheres are the same as those used with circles.

$\overline{OA}$, $\overline{OB}$, and $\overline{OD}$ are radii.

$\overline{BD}$ is a diameter.

$\overline{BC}$ is a chord.

$\overleftrightarrow{BC}$ is a secant.

$\overleftrightarrow{AT}$ is a tangent.

$\overline{AT}$ is a tangent segment.

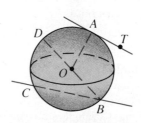

Congruent circles (or **spheres**) are circles (or spheres) that have congruent radii.

Concentric circles are circles that lie in the same plane and have the same center. The rings of the target illustrate concentric circles. **Concentric spheres** are spheres that have the same center.

A polygon is **inscribed in a circle** and the circle is **circumscribed about the polygon** when each vertex of the polygon lies on the circle.

   Inscribed polygons

Circumscribed circles

Classroom Exercises

1. Name three radii of ⊙*O*.
2. Name a diameter.
3. Consider $\overline{RS}$ and $\overleftrightarrow{RS}$. Which is a chord and which is a secant?
4. Why is $\overline{TK}$ not a chord?
5. Name a tangent to ⊙*O*.
6. What name is given to point *L*?
7. Name a line tangent to sphere *Q*.
8. Name a secant of the sphere and a chord of the sphere.
9. Name 4 radii. (None are drawn in the diagram.)
10. What is the diameter of a circle with radius 8? 5.2? $4\sqrt{3}$? *j*?
11. What is the radius of a sphere with diameter 14? 13? 5.6? 6*n*?

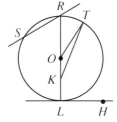

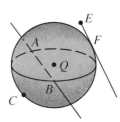

Written Exercises

A 1. Draw a circle and several parallel chords. What do you think is true of the midpoints of all such chords?
2. Draw a circle with center *O* and a line $\overleftrightarrow{TS}$ tangent to ⊙*O* at *T*. Draw $\overline{OT}$, and use a protractor to find *m*∠*OTS*.
3. a. Draw a right triangle inscribed in a circle.
   b. What do you know about the midpoint of the hypotenuse?
   c. Where is the center of the circle?
   d. If the legs of the right triangle are 6 and 8, find the radius of the circle.

4. Plane *Z* passes through the center of sphere *Q*.
   a. Explain why *QR* = *QS* = *QT*.
   b. Explain why the intersection of the plane and the sphere is a circle. (The intersection of a sphere with any plane passing through the center of the sphere is called a **great circle** of the sphere.)

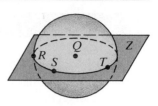

5. The radii of two concentric circles are 15 cm and 7 cm. A diameter $\overline{AB}$ of the larger circle intersects the smaller circle at *C* and *D*. Find two possible values for *AC*.

For each exercise draw a circle and inscribe the polygon in the circle.

6. A rectangle

7. A trapezoid

8. An obtuse triangle

9. A parallelogram

10. An acute isosceles triangle

11. A quadrilateral *PQRS*, with $\overline{PR}$ a diameter

For each exercise draw $\odot O$ with radius 12. Then draw radii $\overline{OA}$ and $\overline{OB}$ to form an angle with the measure named. Find the length of $\overline{AB}$.

B **12.** $m\angle AOB = 90$

13. $m\angle AOB = 180$

14. $m\angle AOB = 60$

15. $m\angle AOB = 120$

16. Draw two points *A* and *B* and several circles that pass through *A* and *B*. Locate the centers of these circles. On the basis of your experiment, complete the following statement:
The centers of all circles passing through *A* and *B* lie on __?__.
Write an argument to support your statement.

17. $\odot Q$ and $\odot R$ are congruent circles that intersect at *C* and *D*. $\overline{CD}$ is called the *common chord* of the circles.
   a. What kind of quadrilateral is *QDRC*? Why?
   b. $\overline{CD}$ must be the perpendicular bisector of $\overline{QR}$. Why?
   c. If *QC* = 17 and *QR* = 30, find *CD*.

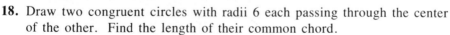

18. Draw two congruent circles with radii 6 each passing through the center of the other. Find the length of their common chord.

C **19.** $\odot P$ and $\odot Q$ have radii 5 and 7 and *PQ* = 6. Find the length of the common chord $\overline{AB}$. (*Hint: APBQ* is a kite and $\overline{PQ}$ is the perpendicular bisector of $\overline{AB}$. See Exercise 28, page 193. Let *N* be the intersection of $\overline{PQ}$ and $\overline{AB}$, and let *PN* = *x* and *AN* = *y*. Write two equations in terms of *x* and *y*.)

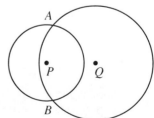

20. Draw a diagram similar to the one shown, but much larger. Carefully draw the perpendicular bisectors of $\overline{AB}$ and $\overline{BC}$.
   a. The perpendicular bisectors intersect in a point. Where does that point appear to be?
   b. Write an argument that justifies your answer to part (a).

Extra   *Networks*

The Pregel River flows through the old city of Koenigsberg, now Kaliningrad. Once, seven bridges joined the shores and the two islands in the river as shown in the diagram at the left below. A popular problem of that time was to try to walk across all seven bridges without crossing any bridge more than once. Can you find a way to do it?

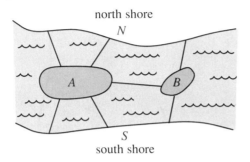

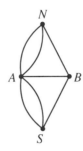

Mathematician Leonard Euler analyzed this problem using a diagram called a *network*, shown at the right above. He represented each land mass by a point (called a vertex) and each bridge by an arc. He then classified each vertex with an odd number of arcs coming from it as *odd* and each vertex with an even number of arcs as *even*. From here Euler discovered which networks can be traced without backtracking, that is, without drawing over an arc twice.

Exercises

Find the number of odd and even vertices in each network. Imagine traveling each network to see if it can be traced without backtracking.

1.   **2.** 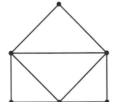   **3.**

The number of odd vertices will tell you whether or not a network can be traced without backtracking. Do you see how? If not, read on.

4. Suppose that a given network can be traced without backtracking.
   a. Consider a vertex that is neither the start nor end of a journey through this network. Is such a vertex odd or even?
   b. Now consider the two vertices at the start and finish of a journey through this network. Can both of these vertices be odd? even?
   c. Can just one of the start and finish vertices be odd?
5. Tell why it is impossible to walk across the seven bridges of Koenigsberg without crossing any bridge more than once.

9-2 *Tangents*

In Written Exercise 2 on page 330 you had the chance to preview the next theorem about tangents and radii.

Theorem 9-1

If a line is tangent to a circle, then the line is perpendicular to the radius drawn to the point of tangency.

Given: *m* is tangent to ⊙*O* at *T*.

Prove: $\overline{OT} \perp m$

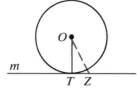

Proof:

Assume temporarily that $\overline{OT}$ is not perpendicular to *m*. Then the perpendicular segment from *O* to *m* intersects *m* in some other point *Z*. Draw $\overline{OZ}$. By Corollary 1, page 220, the perpendicular segment from *O* to *m* is the shortest segment from *O* to *m*, so $OZ < OT$. Because tangent *m* intersects ⊙*O* only in point *T*, *Z* lies outside ⊙*O*, and $OZ > OT$. The statements $OZ < OT$ and $OZ > OT$ are contradictory. Thus the temporary assumption must be false. It follows that $\overline{OT} \perp m$.

Corollary

Tangents to a circle from a point are congruent.

In the figure, $\overline{PA}$ and $\overline{PB}$ are tangent to the circle at *A* and *B*. By the corollary, $\overline{PA} \cong \overline{PB}$. For a proof, see Classroom Exercise 4.

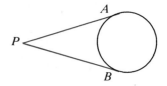

Theorem 9-2 is the converse of Theorem 9-1. Its proof is left as Exercise 22.

Theorem 9-2

If a line in the plane of a circle is perpendicular to a radius at its outer endpoint, then the line is tangent to the circle.

Given: Line *l* in the plane of ⊙*Q*;

  $l \perp$ radius $\overline{QR}$ at *R*

Prove: *l* is tangent to ⊙*Q*.

When each side of a polygon is tangent to a circle, the polygon is said to be **circumscribed about the circle** and the circle is **inscribed in the polygon.**

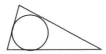

Circumscribed polygons

Inscribed circles

A line that is tangent to each of two coplanar circles is called a **common tangent.**

A common *internal* tangent intersects the segment joining the centers.

A common *external* tangent does *not* intersect the segment joining the centers.

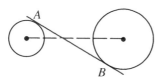

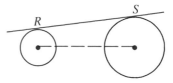

$\overleftrightarrow{AB}$ is a common internal tangent. Can you find another one that has not been drawn?

$\overleftrightarrow{RS}$ is a common external tangent. Can you find another one that has not been drawn?

A circle can be tangent to a line, but it can also be tangent to another circle. **Tangent circles** are coplanar circles that are tangent to the same line at the same point.

$\odot A$ and $\odot B$ are *externally* tangent.

$\odot C$ and $\odot D$ are *internally* tangent.

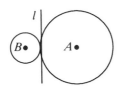

The ends of the plastic industrial pipes shown in the photograph illustrate externally tangent circles. Notice that when a circle is surrounded by tangent circles of the same radius, six of these circles fit exactly around the inner circle.

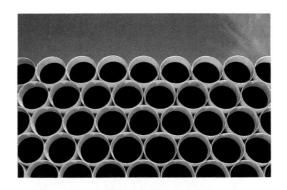

Classroom Exercises

1. How many common external tangents can be drawn to the two circles?

a.

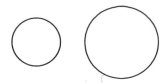

b.

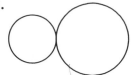

c.

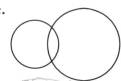

d.

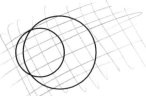

e.

f.

2. How many common internal tangents can be drawn to each pair of circles in Exercise 1 above?

3. a. Which pair of circles shown above are externally tangent?
   b. Which pair are internally tangent?

4. Given: $\overline{PA}$ and $\overline{PB}$ are tangents to $\odot O$.
   Use the diagram at the right to explain how the corollary on page 333 follows from Theorem 9-1.

5. In the diagram, which pairs of angles are congruent? Which pairs of angles are complementary? Which pairs of angles are supplementary?

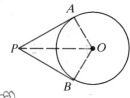

Written Exercises

$\overline{JT}$ **is tangent to** $\odot O$ **at** T. **Complete.**

A **1.** If $OT = 6$ and $JO = 10$, then $JT = \underline{\ ?\ }$.
   2. If $OT = 6$ and $JT = 10$, then $JO = \underline{\ ?\ }$.
   3. If $m \angle TOJ = 60$ and $OT = 6$, then $JO = \underline{\ ?\ }$.
   4. If $JK = 9$ and $KO = 8$, then $JT = \underline{\ ?\ }$.

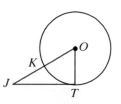

5. The diagram below shows tangent lines and circles. Find PD.

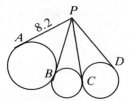

6. $\overline{RS}$ and $\overline{TU}$ are common internal tangents to the circles. If $RZ = 4.7$ and $ZU = 7.3$, find RS and TU.

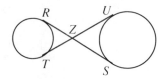

7. a. What do you think is true of common external
tangents $\overline{AB}$ and $\overline{CD}$? Prove it.
  b. Will your results in part (a) be true if the circles
  are congruent?

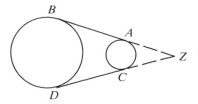

8. Given: $\overline{TR}$ and $\overline{TS}$ are tangents to $\odot O$ from T;
$m \angle RTS = 36$
  a. Copy the diagram. Draw $\overline{RS}$ and find $m \angle TSR$ and $m \angle TRS$.
  b. Draw radii $\overline{OS}$ and $\overline{OR}$ and find $m \angle ORS$ and $m \angle OSR$.
  c. Find $m \angle ROS$.
  d. Does your result in part (c) support one of your conclusions
  about angles in Classroom Exercise 5? Explain.

9. Draw $\odot O$ with perpendicular radii $\overline{OX}$ and $\overline{OY}$. Draw tangents to the circle
at X and Y.
  a. If the tangents meet at Z, what kind of figure is $OXZY$? Explain.
  b. If $OX = 5$, find OZ.

10. Given: $\overline{PT}$ is tangent to $\odot O$ at T; $\overline{TS} \perp \overline{PO}$
Complete the following statements.
  a. TS is the geometric mean between ___?___ and ___?___.
  b. TO is the geometric mean between ___?___ and ___?___.
  c. If $OS = 6$ and $SP = 24$, $TS =$ ___?___ and $TP =$ ___?___.

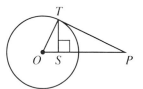

11. Given: $\overline{RS}$ is a common internal
tangent to $\odot A$ and $\odot B$.
Explain why $\dfrac{AC}{BC} = \dfrac{RC}{SC}$.

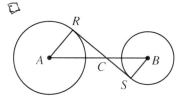

B **12.** Discover and prove a theorem about two lines tangent to a circle at the
endpoints of a diameter.

13. Is there a theorem about spheres related to the theorem in Exercise 12?
If so, state the theorem.

14. Quad. *ABCD* is circumscribed about
a circle. Discover and prove a relationship between $AB + DC$ and
$AD + BC$.

15. $\overline{PA}$, $\overline{PB}$, and $\overline{RS}$ are tangents.
Explain why $PR + RS + SP = PA + PB$.

16. $\overline{SR}$ is tangent to $\odot P$ and $\odot Q$.
$QT = 6$; $TR = 8$; $PR = 30$.
$PQ = \underline{\;?\;}$; $PS = \underline{\;?\;}$; $ST = \underline{\;?\;}$.

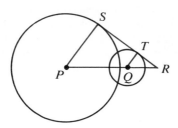

17. $\overline{JK}$ is tangent to $\odot P$ and $\odot Q$.
$JK = \underline{\;?\;}$ (*Hint*: What kind of quadrilateral is *JPQK*?)

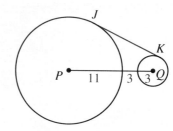

18. Circles P and Q have radii 6 and 2 and are tangent to each other. Find the length of their common external tangent $\overline{AB}$. (*Hint*: Draw $\overline{PQ}$, $\overline{PA}$, and $\overline{QB}$.)

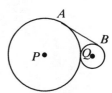

19. Given: Two tangent circles; $\overline{EF}$ is a common external tangent; $\overline{GH}$ is the common internal tangent.
   a. Discover and prove something interesting about point G.
   b. Discover and prove something interesting about $\angle EHF$.

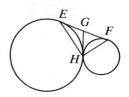

20. Three circles are shown. How many circles tangent to all three of the given circles can be drawn?

C 21. Suppose the three circles represent three spheres.
   a. How many planes tangent to each of the spheres can be drawn?
   b. How many spheres tangent to all three spheres can be drawn?

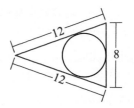

22. Prove Theorem 9-2. (*Hint*: Write an indirect proof.)

23. Find the radius of the circle inscribed in the triangle.

Mixed Review Exercises

Find *AB*. In Exercise 3, $\overline{CB}$ is tangent to $\odot A$.

1.

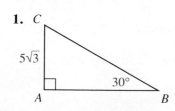

2.

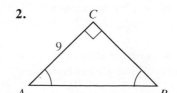

3.

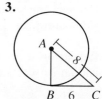

| **Biographical Note** | *Maria Gaetana Agnesi* |

Maria Gaetana Agnesi (1718–1799) was born in Milan, Italy. A child prodigy, she had mastered seven languages by the age of thirteen. Between the ages of twenty and thirty she compiled the works of the mathematicians of her time into two volumes on calculus, called *Analytical Institutions*. This was an enormous task, since the mathematicians had originally published their results in different languages and had used a variety of methods of approach.

Agnesi's volumes were praised as clear, methodical, and comprehensive. They were translated into English and French and were widely used as textbooks. One of the most famous aspects of Agnesi's volumes was an exercise in analytic geometry and the discussion of a curve called a *versirea*, shown at the left below. The name, derived from the Latin *vertere*, ''to turn,'' was apparently mistranslated into English texts as ''witch.'' Thus the curve is commonly known as the ''witch of Agnesi.''

Due to Agnesi's scholarship, she was elected to the Bologna Academy of Sciences and in 1750 she was appointed honorary professor in mathematics at the University of Bologna, shown at the left.

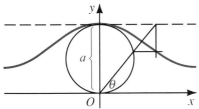

Explorations

These exploratory exercises can be done using a computer with a program that draws and measures geometric figures.

Draw parallelogram *ABCD*. Draw four circles as follows.

(1) Use *A*, *B*, and *D* to draw circle *E*.
(2) Use *A*, *D*, and *C* to draw circle *F*.
(3) Use *B*, *C*, and *D* to draw circle *G*.
(4) Use *A*, *B*, and *C* to draw circle *H*.

Connect the centers of the circles to get quad. *EFGH*.
Compare quad. *ABCD* with quad. *EFGH*. What do you notice?

Repeat on other types of quadrilaterals: a rhombus, a trapezoid, a rectangle, and an isosceles trapezoid. What do you notice?

9-3 *Arcs and Central Angles*

A **central angle** of a circle is an angle with its vertex at the center of the circle. In the diagrams below, $\angle YOZ$ is a central angle. An *arc* is an unbroken part of a circle. Two points Y and Z on a circle O are always the endpoints of two arcs. Y and Z and the points of $\odot O$ in the interior of $\angle YOZ$ form a **minor arc.** Y and Z and the remaining points of $\odot O$ form a **major arc.** If Y and Z are the endpoints of a diameter, then the two arcs are called **semicircles.** A minor arc is named by its endpoints: $\overset{\frown}{YZ}$ is read "arc YZ." You use three letters to name a semicircle or a major arc: $\overset{\frown}{YWZ}$ is read "arc YWZ."

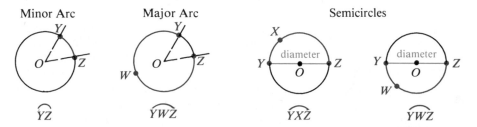

Minor Arc	Major Arc	Semicircles	
$\overset{\frown}{YZ}$	$\overset{\frown}{YWZ}$	$\overset{\frown}{YXZ}$	$\overset{\frown}{YWZ}$

The **measure of a minor arc** is defined to be the measure of its central angle. In the diagram at the left below, $m\overset{\frown}{YZ}$ represents the measure of minor arc YZ. In the middle diagram, can you see why the **measure of a major arc is 360 minus the measure of its minor arc?** The third diagram shows that the **measure of a semicircle** is 180.

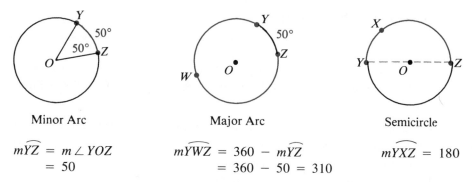

Minor Arc	Major Arc	Semicircle
$m\overset{\frown}{YZ} = m\angle YOZ$	$m\overset{\frown}{YWZ} = 360 - m\overset{\frown}{YZ}$	$m\overset{\frown}{YXZ} = 180$
$= 50$	$= 360 - 50 = 310$	

Adjacent arcs of a circle are arcs that have exactly one point in common. The following postulate can be used to find the measure of an arc formed by two adjacent arcs.

Postulate 16 *Arc Addition Postulate*

The measure of the arc formed by two adjacent arcs is the sum of the measures of these two arcs.

Applying the Arc Addition Postulate to the circle shown at the right, we have

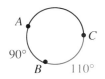

$$m\overarc{AB} + m\overarc{BC} = m\overarc{ABC}$$
$$90 + 110 = 200$$

Congruent arcs are arcs, in the same circle or in congruent circles, that have equal measures. In the diagram below, $\odot P$ and $\odot Q$ are congruent circles and $\overarc{AB} \cong \overarc{CD} \cong \overarc{EF}$. However, $\overarc{EF}$ is not congruent to $\overarc{RS}$ even though both arcs have the same degree measure, because $\odot Q$ is not congruent to $\odot O$.

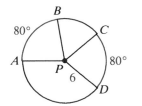 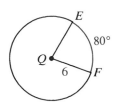

Notice that each of the congruent arcs above has an 80° central angle, so these congruent arcs have congruent central angles. The relationship between congruence of minor arcs and congruence of their central angles is stated in Theorem 9-3 below. This theorem follows immediately from the definition of congruent arcs.

Theorem 9-3

In the same circle or in congruent circles, two minor arcs are congruent if and only if their central angles are congruent.

Example The radius of the Earth is about 6400 km. The latitude of the Arctic Circle is 66.6° North. (That is, in the figure, $m\overarc{BE} = 66.6$.) Find the radius of the Arctic Circle.

Solution Let N be the North Pole and let $\overline{ON}$ intersect $\overline{AB}$ in M. Since $m\overarc{NE} = 90$, $m\overarc{NB} = 90 - 66.6 = 23.4$ and $m\angle NOB = 23.4$. Similarly, $m\angle NOA = 23.4$. Since $\triangle AOB$ is isosceles and $\overline{OM}$ bisects the vertex $\angle AOB$, (1) M is the midpoint of $\overline{AB}$ (and thus the center of the Arctic Circle) and (2) $\overline{OM} \perp \overline{AB}$. Using trigonometry in right $\triangle MOB$:

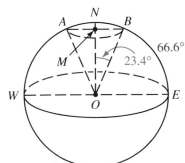

$$\sin 23.4° = \frac{MB}{OB}$$
$$MB = OB \cdot \sin 23.4°$$
$$MB \approx 6400(0.3971)$$
$$MB \approx 2500 \text{ km}$$

Classroom Exercises

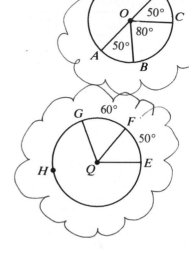

1. Using the letters shown in the diagram, name:
   a. two central angles   **b.** a semicircle
   c. two minor arcs   **d.** two major arcs

In Exercises 2–7 find the measure of the arc.

2. $\overset{\frown}{AB}$ 3. $\overset{\frown}{AC}$ 4. $\overset{\frown}{ABD}$
5. $\overset{\frown}{BAD}$ 6. $\overset{\frown}{CDA}$ 7. $\overset{\frown}{CDB}$

In Exercises 8–13 find the measure of the angle or the arc named.

8. $\angle GQF$ 9. $\angle EQF$ 10. $\angle GQE$
11. $\overset{\frown}{GE}$ 12. $\overset{\frown}{GHE}$ 13. $\overset{\frown}{EHF}$

Written Exercises

Find the measure of central $\angle 1$.

A

 1. 85°

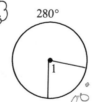 **2.** 280°

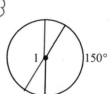 **3.** 150°

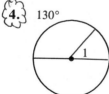 **4.** 130°

 5. 240° 68°

 6. 35°

7. At 11 o'clock the hands of a clock form an angle of __?__°.

8. The hands of a clock form a 120° angle at __?__ o'clock and at __?__ o'clock.

9. **a.** Draw a circle. Place points A, B, and C on it in such positions that $m\overset{\frown}{AB} + m\overset{\frown}{BC}$ does not equal $m\overset{\frown}{AC}$.
   b. Does your example in part (a) contradict Postulate 16?

Complete the tables in Exercises 10 and 11.

10.

$m\widehat{CB}$	60	70	?	?	?
$m\angle 1$	?	?	56	?	?
$m\angle 2$	?	?	?	25	x

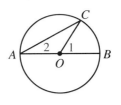

11.

$m\widehat{CB}$	70	60	66	60	p
$m\widehat{BD}$	30	28	?	?	q
$m\angle COD$	?	?	100	?	?
$m\angle CAD$	?	?	?	52	?

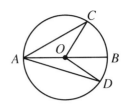

12. Use a compass to draw a large $\odot O$. Draw a central $\angle AOB$.
   a. Label three other points P, Q, and R that are on $\odot O$ but not on $\widehat{AB}$. Then draw $\angle APB$, $\angle AQB$, and $\angle ARB$.
   b. Use a protractor to find $m\angle AOB$, $m\angle APB$, $m\angle AQB$, and $m\angle ARB$.
   c. What is the relationship between $m\angle APB$, $m\angle AQB$, and $m\angle ARB$? What is the relationship between $m\angle AOB$ and $m\angle APB$?

13. a. Draw three large circles and inscribe a different-shaped quadrilateral $ABCD$ in each.
   b. Use a protractor to measure all the angles.
   c. Compute $m\angle A + m\angle C$ and $m\angle B + m\angle D$.
   d. What is the relationship between opposite angles of an inscribed quadrilateral?

B 14. Given: $\overline{WZ}$ is a diameter of $\odot O$; $\overline{OX} \parallel \overline{ZY}$
   Prove: $\widehat{WX} \cong \widehat{XY}$
   (*Hint*: Draw $\overline{OY}$.)

15. Given: $\overline{WZ}$ is a diameter of $\odot O$;
   $m\widehat{WX} = m\widehat{XY} = n$
   Prove: $m\angle Z = n$

16. $\overline{AC}$ is a diameter of $\odot O$.
   a. If $m\angle A = 35$, then $m\angle B = \underline{\ ?\ }$, $m\angle BOC = \underline{\ ?\ }$, and $m\widehat{BC} = \underline{\ ?\ }$.
   b. If $m\angle A = n$, then $m\widehat{BC} = \underline{\ ?\ }$.
   c. If $m\widehat{BC} = 6k$, then $m\angle A = \underline{\ ?\ }$.

In Exercises 17–20, the latitude of a city is given. Sketch the Earth and a circle of latitude through the city. Find the radius of this circle.

17. Milwaukee, Wisconsin; 43°N
18. Columbus, Ohio; 40°N
19. Sydney, Australia; 34°S
20. Rio de Janeiro; 23°S

C **21.** Given: ⊙O and ⊙Q intersect at R and S;
$m\widehat{RVS} = 60$; $m\widehat{RUS} = 120$
Prove: $\overline{OR}$ is tangent to ⊙Q;
$\overline{QR}$ is tangent to ⊙O.

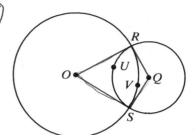

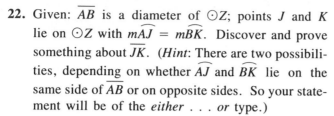

22. Given: $\overline{AB}$ is a diameter of ⊙Z; points J and K
lie on ⊙Z with $m\widehat{AJ} = m\widehat{BK}$. Discover and prove
something about $\overline{JK}$. (*Hint:* There are two possibilities, depending on whether $\widehat{AJ}$ and $\widehat{BK}$ lie on the
same side of $\overline{AB}$ or on opposite sides. So your statement will be of the *either . . . or* type.)

The diagram, not drawn to scale, shows satellite S above the Earth, represented
as sphere E. All lines tangent to the Earth from S touch the Earth at points
on a circle with center C. Any two points on the Earth's surface on or above
that circle can communicate with each other via S. X and Y are as far apart
as two communication points can be. The Earth distance between X and Y
equals the length of $\widehat{XTY}$, which equals $\dfrac{n}{360}$ · circumference of the Earth. That
circumference is approximately 40,200 km and the radius of the Earth is approximately 6400 km.

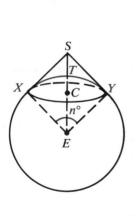

23. The photograph above shows the view from Gemini V looking north over
the Gulf of California toward Los Angeles. The orbit of Gemini V ranged
from 160 km to 300 km above the Earth. Take S to be 300 km above the
Earth. That is, $ST = 300$ km. Find the Earth distance, rounded to the
nearest 100 km, between X and Y. (*Hint:* Since you can find a value for
$\cos\dfrac{n°}{2}$ you can determine $n°$.)

24. Repeat Exercise 23, but with S twice as far from the Earth. Note that the
distance between X and Y is not twice as great as before.

9-4 *Arcs and Chords*

In $\odot O$ shown at the right, $\overline{RS}$ cuts off two arcs, $\overparen{RS}$ and $\overparen{RTS}$. We speak of $\overparen{RS}$, the minor arc, as being *the arc of chord* $\overline{RS}$.

Theorem 9-4

In the same circle or in congruent circles:

(1) Congruent arcs have congruent chords.

(2) Congruent chords have congruent arcs.

Here is a paragraph proof of part (1) for one circle. You will be asked to write a paragraph proof of part (2) in Written Exercise 16.

Given: $\odot O$; $\overparen{RS} \cong \overparen{TU}$
Prove: $\overline{RS} \cong \overline{TU}$

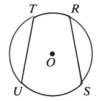 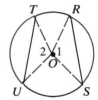

Proof:

Draw radii $\overline{OR}$, $\overline{OS}$, $\overline{OT}$, and $\overline{OU}$. $\overline{OR} \cong \overline{OT}$ and $\overline{OS} \cong \overline{OU}$ because they are all radii of the same circle. Since $\overparen{RS} \cong \overparen{TU}$, central angles 1 and 2 are congruent. Then $\triangle ROS \cong \triangle TOU$ by SAS and corresponding parts $\overline{RS}$ and $\overline{TU}$ are congruent.

A point Y is called the *midpoint* of $\overparen{XYZ}$ if $\overparen{XY} \cong \overparen{YZ}$. Any line, segment, or ray that contains Y bisects $\overparen{XYZ}$.

Theorem 9-5

A diameter that is perpendicular to a chord bisects the chord and its arc.

Given: $\odot O$; $\overline{CD} \perp \overline{AB}$
Prove: $\overline{AZ} \cong \overline{BZ}$; $\overparen{AD} \cong \overparen{BD}$

Plan for Proof: Draw $\overline{OA}$ and $\overline{OB}$. Then use the HL Theorem to prove that $\triangle OZA \cong \triangle OZB$. Then use corresponding parts of congruent triangles to show that $\overline{AZ} \cong \overline{BZ}$ and $\angle 1 \cong \angle 2$. Finally, apply the theorem that congruent central angles have congruent arcs.

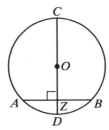 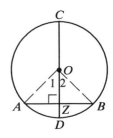

Example 1   Find the values of x and y.

Solution   Diameter $\overline{CD}$ bisects chord $\overline{AB}$, so $x = 5$.
(Theorem 9-5)

$\overline{AB} \cong \overline{EF}$, so $m\overset{\frown}{AB} = 86$.   (Theorem 9-4)

Diameter $\overline{CD}$ bisects $\overset{\frown}{AB}$, so $y = 43$.
(Theorem 9-5)

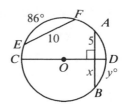

Recall (page 154) that the distance from a point to a line is the length of the perpendicular segment from the point to the line. This definition is used in the following example.

Example 2   Find the length of a chord that is a distance 5 from the center of a circle with radius 8.

Solution   Draw the perpendicular segment, $\overline{OP}$, from O to $\overline{AB}$.

$$x^2 + 5^2 = 8^2$$
$$x^2 + 25 = 64$$
$$x^2 = 39$$
$$x = \sqrt{39}$$

By Theorem 9-5, $\overline{OP}$ bisects $\overline{AB}$ so
$AB = 2 \cdot AP = 2x = 2\sqrt{39}$.

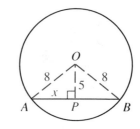

It should be clear that *all* chords in $\odot O$ above that are a distance 5 from center O will have length $2\sqrt{39}$. Thus, all such chords are congruent, as stated in part (1) of the next theorem. You will prove part (2) of the theorem as Classroom Exercise 6.

Theorem 9-6

In the same circle or in congruent circles:

(1) **Chords equally distant from the center (or centers) are congruent.**

(2) **Congruent chords are equally distant from the center (or centers).**

Example 3   Find the value of x.

Solution S is the midpoint of $\overline{RT}$, so $RT = 6$.
(Theorem 9-5)

$\overline{RT} \cong \overline{UV}$, so $x = 4$.   (Theorem 9-6)

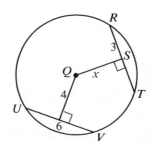

Classroom Exercises

1. If $\overline{PQ} \cong \overline{XY}$, can you conclude that $\overarc{PQ} \cong \overarc{XY}$? Why or why not?

2. If $\overline{PQ} \cong \overline{RS}$, can you conclude that $\overarc{PQ} \cong \overarc{RS}$? Why or why not?

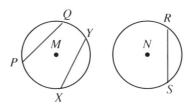

3. Study the diagram at the right and tell what theorem justifies each statement.
   a. $LK = 8$
   b. $OE = 3$
   c. $\overarc{LK} \cong \overarc{GH}$

4. $AB = 16$
   $OM = 6$
   radius = ___?___

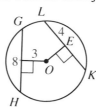

5. $PQ = 10$
   radius = 13
   $OM = $ ___?___

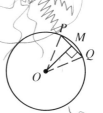

6. Supply reasons to complete a proof of Theorem 9-6, part (2), for one circle.
   Given: $\odot O$; $\overline{AB} \cong \overline{CD}$;
   $\overline{OY} \perp \overline{AB}$; $\overline{OZ} \perp \overline{CD}$
   Prove: $OY = OZ$

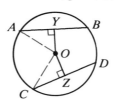

Proof:

Statements	Reasons
1. Draw radii $\overline{OA}$ and $\overline{OC}$.	1. ___?___
2. $\overline{OY} \perp \overline{AB}$; $\overline{OZ} \perp \overline{CD}$	2. ___?___
3. $\overline{AB} \cong \overline{CD}$, or $AB = CD$	3. ___?___
4. $\frac{1}{2}AB = \frac{1}{2}CD$	4. ___?___
5. $AY = \frac{1}{2}AB$; $CZ = \frac{1}{2}CD$	5. ___?___
6. $AY = CZ$, or $\overline{AY} \cong \overline{CZ}$	6. ___?___
7. $\overline{OA} \cong \overline{OC}$	7. ___?___
8. rt. $\triangle OYA \cong$ rt. $\triangle OZC$	8. ___?___
9. $\overline{OY} \cong \overline{OZ}$, or $OY = OZ$	9. ___?___

7. Suppose that in Theorem 9-6, the words "circle" and "circles" are replaced by "sphere" and "spheres." Is the resulting statement true?

Written Exercises

In the diagrams that follow, O is the center of the circle.

A **1.**

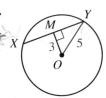

$XY = $ ___?___

2.

$PQ = 24; OM = $ ___?___

3.

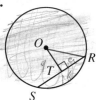

$OT = 9; RS = 18$
$OR = $ ___?___

4.

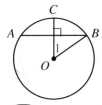

$m\widehat{ACB} = 110;$
$m\angle 1 = $ ___?___

5.

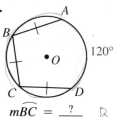

$m\widehat{BC} = $ ___?___

6.

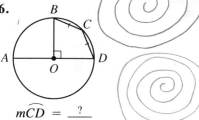

$m\widehat{CD} = $ ___?___

7.

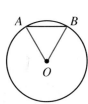

$m\angle AOB = 60;$
$AB = 24; OA = $ ___?___

8.

$OM = ON = 7;$
$CM = 6; EF = $ ___?___

9.

$AB = 18; OM = 12;$
$ON = 10; CD = $ ___?___

10. Sketch a circle with two noncongruent chords. Is the longer chord farther from the center or closer to the center than the shorter chord?

11. Sketch a circle O with radius 10 and chord $\overline{XY}$ 8 cm long. How far is the chord from O?

12. Sketch a circle Q with a chord $\overline{RS}$ that is 16 cm long and 2 cm from Q. What is the radius of $\odot Q$?

13. Sketch a circle P with radius 5 cm and chord $\overline{AB}$ that is 2 cm from P. Find the length of $\overline{AB}$.

14. Given: $\widehat{JZ} \cong \widehat{KZ}$
Prove: $\angle J \cong \angle K$

15. Prove the converse of Exercise 14.

B **16.** Write a paragraph proof of part (2) of Theorem 9-4. First list what is given and what is to be proved.

17.

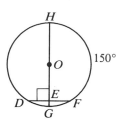

If $OJ = 10$, $JK = $ __?__ .

18.

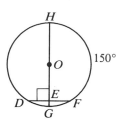

If $OE = 8\sqrt{3}$, $HG = $ __?__ .

19. A plane 5 cm from the center of a sphere intersects the sphere in a circle with diameter 24 cm. Find the diameter of the sphere.

20. A plane P cuts sphere O in a circle that has diameter 20. If the diameter of the sphere is 30, how far is the plane from O?

21. Use trigonometry to find the measure of the arc cut off by a chord 12 cm long in a circle of radius 10 cm.

22. In $\odot O$, $m\overset{\frown}{RS} = 70$ and $RS = 20$. Use trigonometry to find the radius of $\odot O$.

State and prove a theorem suggested by the figure.

C **23.**

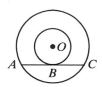

24.

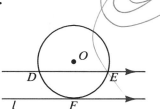

25. A, B, C are points on $\odot O$ such that $\triangle ABC$ is equilateral. If the radius of the circle is 6, what is the perimeter of $\triangle ABC$?

26. Investigate the possibility, given a circle, of drawing two chords whose lengths are in the ratio $1:2$ and whose distances from the center are in the ratio $2:1$. If the chords can be drawn, find the length of each in terms of the radius. If not, prove that the figure is impossible.

27. Three parallel chords of $\odot O$ are drawn as shown. Their lengths are 20, 16, and 12 cm. Find, to the nearest tenth of a centimeter, the length of chord $\overline{XY}$ (not shown).

Self-Test 1

1. Points *A*, *B*, and *C* lie on ⊙*Q*.
   a. Name two radii of ⊙*Q*.
   b. Name a diameter of ⊙*Q*.
   c. Name a chord and a secant of ⊙*Q*.

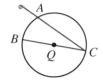

2. Sketch each of the following.
   a. △*ABC* inscribed in ⊙*O*   **b.** Quad. *LUMX* circumscribed about ⊙*Q*
3. $\overline{NP}$ is tangent to ⊙*O* at *P*. If *NO* = 25 and *NP* = 20, find *OP*.
4. A plane passes through the common center of two concentric spheres. Describe the intersection of the plane and the two spheres.
5. Find the length of a chord that is 3 cm from the center of a circle with radius 7 cm.
6. Points *E*, *F*, *G*, *H*, and *J* lie on ⊙*O*.
   a. $m\widehat{EF}$ = ___?___ and $m\widehat{EHF}$ = ___?___.
   b. Suppose $\overline{JH} \cong \overline{HG}$.   State the theorem that supports the conclusion that $\widehat{JH} \cong \widehat{HG}$.

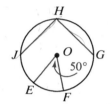

Angles and Segments

Objectives

1. Solve problems and prove statements involving inscribed angles.
2. Solve problems and prove statements involving angles formed by chords, secants, and tangents.
3. Solve problems involving lengths of chords, secant segments, and tangent segments.

9-5 *Inscribed Angles*

Angles 1 and 2 shown at the right are called *inscribed angles*. An **inscribed angle** is an angle whose vertex is on a circle and whose sides contain chords of the circle. We say that the angles at the right *intercept* the arcs shown in color. ∠ 1 intercepts a minor arc. ∠ 2 intercepts a major arc.

The next theorem compares the measure of an inscribed angle with the measure of its intercepted arc. Its proof requires us to consider three possible cases.

Theorem 9-7

The measure of an inscribed angle is equal to half the measure of its intercepted arc.

Given: $\angle ABC$ inscribed in $\odot O$

Prove: $m\angle ABC = \frac{1}{2}m\widehat{AC}$

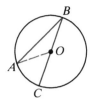

Case I:
Point O lies on $\angle ABC$.

Case II:
Point O lies inside $\angle ABC$.

Case III:
Point O lies outside $\angle ABC$.

Key steps of proof of Case I:

1. Draw radius $\overline{OA}$ and let $m\angle ABC = x$.
2. $m\angle A = x$ (Why?)
3. $m\angle AOC = 2x$ (Why?)
4. $m\widehat{AC} = 2x$ (Why?)
5. $m\angle ABC = \frac{1}{2}m\widehat{AC}$ (Substitution Prop.)

Now that Case I has been proved, it can be used to prove Case II and Case III. An auxiliary line will be used in those proofs, which are left as Classroom Exercises 12 and 13.

Example 1   Find the values of x and y in $\odot O$.

Solution $m\angle PTQ = \frac{1}{2}m\widehat{PQ}$, so
$$x = \frac{1}{2} \cdot 40 = 20.$$
$m\angle PSR = \frac{1}{2}m\widehat{PR}$, so
$$50 = \frac{1}{2}(40 + y) \text{ and } y = 60.$$

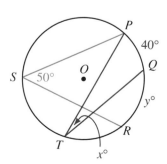

Proofs of the following three corollaries of Theorem 9-7 will be considered in Classroom Exercises 1–3.

Corollary 1

If two inscribed angles intercept the same arc, then the angles are congruent.

$\angle 1 \cong \angle 2$

Corollary 2

An angle inscribed in a semicircle is a right angle.

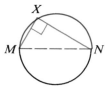

If $\overarc{MXN}$ is a semicircle,
then $\angle X$ is a right angle.

Corollary 3

If a quadrilateral is inscribed in a circle, then its opposite angles are supplementary.

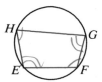

$\angle E$ is supp. to $\angle G$.
$\angle F$ is supp. to $\angle H$.

Example 2 Find the values of x, y, and z.

Solution $\angle ADB$ and $\angle ACB$ intercept the same arc, so
$x = 40$. (Corollary 1)

$\angle ABC$ is inscribed in a semicircle, so $\angle ABC$
is a right angle and $y = 90$. (Corollary 2)

$ABCD$ is an inscribed quadrilateral, so $\angle BAD$
and $\angle BCD$ are supplementary. (Corollary 3)

Therefore, $z = 180 - (x + 30)$
$z = 180 - (40 + 30) = 110.$

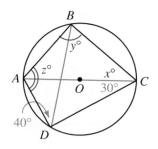

Study the diagrams below from left to right. Point *B* moves along the circle closer and closer to point *T*. Finally, in diagram (4), point *B* has merged with *T*, and one side of ∠*T* has become a tangent.

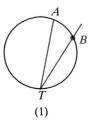
(1)

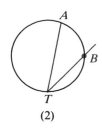
(2)

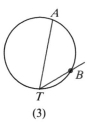
(3)

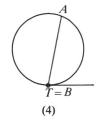
(4)

Apply Theorem 9-7 to diagrams (1), (2), and (3) and you have $m\angle T = \frac{1}{2}m\widehat{AB}$. As you might expect, this equation applies to diagram (4), too, since we say that ∠*T* intercepts $\widehat{AB}$ in this case as well. Diagram (4) suggests Theorem 9-8. In Exercises 13–15 you will prove the three cases of the theorem.

Theorem 9-8

The measure of an angle formed by a chord and a tangent is equal to half the measure of the intercepted arc.

For example, if $\overline{PT}$ is tangent to the circle and $\overline{AT}$ is a chord, then ∠*ATP* intercepts $\widehat{AT}$ and $m\angle ATP = \frac{1}{2}m\widehat{AT} = \frac{1}{2} \cdot 140 = 70$.

Classroom Exercises

1. Explain why Corollary 1 of Theorem 9-7 is true. That is, explain why ∠1 ≅ ∠2.

2. Explain why Corollary 2 is true. That is, explain how the fact that $\widehat{MXN}$ is a semicircle leads to a conclusion that ∠*X* is a right angle.

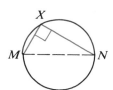

3. **a.** What is the sum of the measures of the red and blue arcs?
   b. Explain how part (a) allows you to deduce that
      $x + y = 180$.
   c. State the corollary of Theorem 9-7 that you have just
      proved.

Tangents and chords are shown. Find the values of x and y. In Exercise 5, O is the center of the circle.

4.

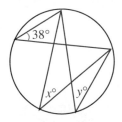

5.

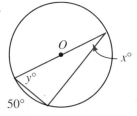

6.

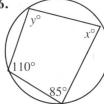

7.

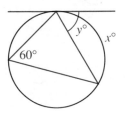

8.

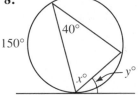

9.

10. **a.** State the contrapositive of Corollary 3.
    b. In quadrilateral $PQRS$, $m \angle P = 100$ and $m \angle R = 90$. Is it
       possible to circumscribe a circle about $PQRS$? Why or
       why not?

11. In the diagram, $m \angle AKB = m \angle CKD = n$.
    $m\widehat{AB} = \underline{\ ?\ }$ and $m\widehat{CD} = \underline{\ ?\ }$. State a
    theorem suggested by this exercise.

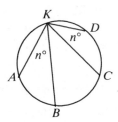

12. Outline a proof of Case II of Theorem 9-7. Use the diagram on page 350.
    (*Hint*: Draw the diameter from B and apply Case I.)

13. Repeat Exercise 12 for Case III.

14. Equilateral $\triangle ABC$ is inscribed in $\odot O$. Tangents to the circle at A and C
    meet at D. What kind of figure is $ABCD$?

Written Exercises

In the diagrams that follow, *O* is the center of the circle. In Exercises 1–9 find the values of *x*, *y*, and *z*.

A **1.**
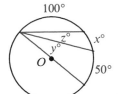

2.
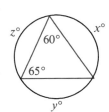

3.
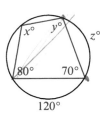

4.
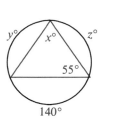

5.
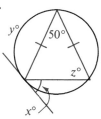

6.
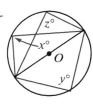

7.
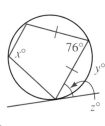

8.
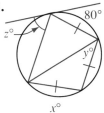

9.
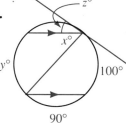

10. Prove: If two chords of a circle are parallel, the two arcs between the chords are congruent.

Given: $\overline{AB} \parallel \overline{CD}$
Prove: $\overparen{AC} \cong \overparen{BD}$

(*Hint*: Draw an auxiliary line.)

11. a. State the converse of the statement in Exercise 10.
 b. Is this converse true or false? If it is true, write a proof. If not, explain why it is false.

12. Prove: $\triangle UXZ \sim \triangle YVZ$

Exercises 13–15 prove the three possible cases of Theorem 9-8. In each case you are given chord $\overline{TA}$ and tangent $\overline{TP}$ of $\odot O$.

13. Supply reasons for the key steps of the proof that $m \angle ATP = \frac{1}{2}m\widehat{ANT}$ in Case I.

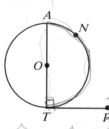

Case I: O lies on $\angle ATP$.
1. $\overline{TP} \perp \overline{TA}$ and $m \angle ATP = 90$.
2. $\widehat{ANT}$ is a semicircle and $m\widehat{ANT} = 180$.
3. $m \angle ATP = \frac{1}{2}m\widehat{ANT}$

In Case II and Case III, $\overline{AT}$ is not a diameter. You can draw diameter $\overline{TZ}$ and then use Case I, Theorem 9-7, and the Angle Addition and Arc Addition Postulates.

B **14.** Case II. O lies inside $\angle ATP$.
Prove $m \angle ATP = \frac{1}{2}m\widehat{ANT}$

15. Case III. O lies outside $\angle ATP$.
Prove $m \angle ATP = \frac{1}{2}m\widehat{ANT}$

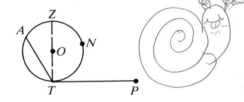
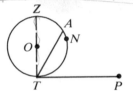

16. Prove that if one pair of opposite sides of an inscribed quadrilateral are congruent, then the other sides are parallel.

17. Draw an inscribed quadrilateral $ABCD$ and its diagonals intersecting at E. Name two pairs of similar triangles.

18. Draw an inscribed quadrilateral $PQRS$ with shortest side $\overline{PS}$. Draw its diagonals intersecting at T. Extend $\overrightarrow{QP}$ and $\overrightarrow{RS}$ to meet at V. Name two pairs of similar triangles such that each triangle has a vertex at V.

Exercises 19–21 refer to a quadrilateral $ABCD$ inscribed in a circle.

19. $m \angle A = x$, $m \angle B = 2x$, and $m \angle C = x + 20$. Find x and $m \angle D$.

20. $m \angle A = x^2$, $m \angle B = 9x - 2$, and $m \angle C = 11x$. Find x and $m \angle D$.

21. $m \angle D = 75$, $m\widehat{AB} = x^2$, $m\widehat{BC} = 5x$, and $m\widehat{CD} = 6x$. Find x and $m \angle A$.

22. Parallelogram $ABCD$ is inscribed in $\odot O$. Find $m \angle A$.

23. Equilateral $\triangle ABC$ is inscribed in a circle. P and Q are midpoints of $\widehat{BC}$ and $\widehat{CA}$, respectively. What kind of figure is quadrilateral $AQPB$? Justify your answer.

24. The diagram at the right shows a regular polygon with 7 sides.
    a. Explain why the numbered angles are all congruent. (*Hint:* You may assume that a circle can be circumscribed about any regular polygon.)
    b. Will your reasoning apply to a regular polygon with any number of sides?

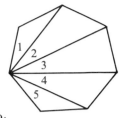

C 25. Given: Vertices *A*, *B*, and *C* of quadrilateral *ABCD* lie on $\odot O$;
    $m \angle A + m \angle C = 180$; $m \angle B + m \angle D = 180$.
    Prove: *D* lies on $\odot O$.

    (*Hint:* Use an indirect proof. Assume temporarily that *D* is not on $\odot O$. You must then treat two cases: (1) *D* is inside $\odot O$, and (2) *D* is outside $\odot O$. In each case let *X* be the point where $\overrightarrow{AD}$ intersects $\odot O$ and draw $\overline{CX}$. Show that what you can conclude about $\angle AXC$ contradicts the given information.)

26. Given: $\overline{PQ} \parallel \overline{SR}$
    Prove: $\overline{PS} \parallel \overline{QR}$

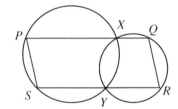

27. *Ptolemy's Theorem* states that in an inscribed quadrilateral, the sum of the products of its opposite sides is equal to the product of its diagonals. This means that for *ABCD* shown,

    $$AB \cdot CD + BC \cdot AD = AC \cdot BD.$$

    Prove the theorem by choosing point *Q* on $\overline{AC}$ so that $\angle ADQ \cong \angle BDC$. Then show $\triangle ADQ \sim \triangle BDC$ and $\triangle ADB \sim \triangle QDC$. Use these similar triangles to show that

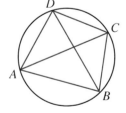

    $$AQ = \frac{BC \cdot AD}{BD} \quad \text{and} \quad QC = \frac{AB \cdot CD}{BD}.$$

    Add these two equations and complete the proof.

28. Equilateral $\triangle ABC$ is inscribed in a circle. *P* is any point on $\overarc{BC}$. Prove $PA = PB + PC$. (*Hint:* Use Ptolemy's Theorem.)

★ 29. Angle *C* of $\triangle ABC$ is a right angle. The sides of the triangle have the lengths shown. The smallest circle (not shown) through *C* that is tangent to $\overline{AB}$ intersects $\overline{AC}$ at *J* and $\overline{BC}$ at *K*. Express the distance *JK* in terms of *a*, *b*, and *c*.

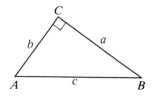

Mixed Review Exercises

1. Name a diameter of $\odot O$.
2. Name a secant of $\odot O$.
3. Name a tangent segment.
4. If $OQ = 7$, then $LM = \underline{\ ?\ }$.
5. If $m\widehat{MQ} = x$, express $m\widehat{QLM}$ in terms of x.
6. Find the geometric mean between 4 and 9.

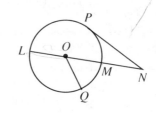

9-6 *Other Angles*

The preceding section dealt with angles that have their vertices on a circle. Theorem 9-9 deals with the angle formed by two chords that intersect inside a circle. Such an angle and its vertical angle intercept two arcs.

Theorem 9-9

The measure of an angle formed by two chords that intersect inside a circle is equal to half the sum of the measures of the intercepted arcs.

Given: Chords $\overline{AB}$ and $\overline{CD}$ intersect inside a circle.
Prove: $m\angle 1 = \frac{1}{2}(m\widehat{AC} + m\widehat{BD})$

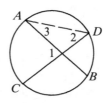

Proof:

Statements	Reasons
1. Draw chord $\overline{AD}$.	1. Through any two points there is exactly one line.
2. $m\angle 1 = m\angle 2 + m\angle 3$	2. The measure of an exterior $\angle$ of a $\triangle$ = the sum of the measures of the two remote interior $\angle$s.
3. $m\angle 2 = \frac{1}{2}m\widehat{AC}$; $m\angle 3 = \frac{1}{2}m\widehat{BD}$	3. The measure of an inscribed angle is equal to half the measure of its intercepted arc.
4. $m\angle 1 = \frac{1}{2}m\widehat{AC} + \frac{1}{2}m\widehat{BD}$ or $m\angle 1 = \frac{1}{2}(m\widehat{AC} + m\widehat{BD})$	4. Substitution (Step 3 in Step 2)

One case of the next theorem will be proved in Classroom Exercise 10, the other two cases in Exercises 25 and 26. Notice that angles formed by two secants, two tangents, or a secant and a tangent intercept two arcs.

Theorem 9-10

The measure of an angle formed by two secants, two tangents, or a secant and a tangent drawn from a point outside a circle is equal to half the difference of the measures of the intercepted arcs.

Case I: Two secants   Case II: Two tangents   Case III: A secant and a tangent

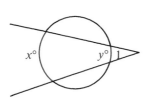   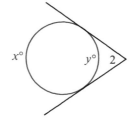   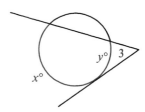

$$m \angle 1 = \tfrac{1}{2}(x - y) \qquad m \angle 2 = \tfrac{1}{2}(x - y) \qquad m \angle 3 = \tfrac{1}{2}(x - y)$$

Example 1 Find the measures of $\angle 1$ and $\angle 2$.

Solution $m \angle 1 = \tfrac{1}{2}(m\widehat{CE} + m\widehat{FD})$ (Theorem 9-9)
$m \angle 1 = \tfrac{1}{2}(70 + 50) = 60$

$m \angle 2 = \tfrac{1}{2}(m\widehat{CE} - m\widehat{BD})$ (Theorem 9-10)
$m \angle 2 = \tfrac{1}{2}(70 - 26) = 22$

Example 2 $\overline{BA}$ is a tangent. Find $m\widehat{BD}$ and $m\widehat{BC}$.

Solution $100 = \tfrac{1}{2}(m\widehat{BD} + 66)$ (Theorem 9-9)
$200 = m\widehat{BD} + 66$, so $m\widehat{BD} = 134$

$40 = \tfrac{1}{2}(m\widehat{BD} - m\widehat{BC})$ (Theorem 9-10)
$80 = 134 - m\widehat{BC}$, so $m\widehat{BC} = 54$

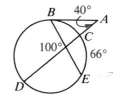

Classroom Exercises

Find the measure of each numbered angle.

1.

2.

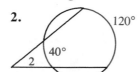

3.

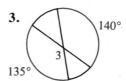

4.

260°

5.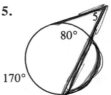

80°

170°

120
130
110
360

6.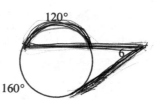

120°

160°

State an equation you could use to find the value of *x*. Then solve for *x*.

7.

x°

75°

100°

8.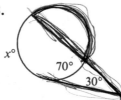

x°

70°

30°

9.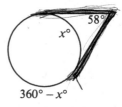

58°

x°

360° − *x*°

10. Supply reasons to complete a proof of Case I of Theorem 9-10.

Given: Secants $\overline{PA}$ and $\overline{PC}$

Prove: $m\angle 1 = \frac{1}{2}(m\widehat{AC} - m\widehat{BD})$

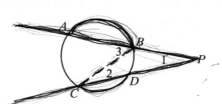

Proof:

1. Draw chord $\overline{BC}$.
2. $m\angle 1 + m\angle 2 = m\angle 3$
3. $m\angle 1 = m\angle 3 - m\angle 2$
4. $m\angle 3 = \frac{1}{2}m\widehat{AC}$; $m\angle 2 = \frac{1}{2}m\widehat{BD}$
5. $m\angle 1 = \frac{1}{2}m\widehat{AC} - \frac{1}{2}m\widehat{BD}$, or $m\angle 1 = \frac{1}{2}(m\widehat{AC} - m\widehat{BD})$

Written Exercises

A **1–10.** $\overleftrightarrow{BZ}$ is tangent to $\odot O$; $\overline{AC}$ is a diameter;
$m\widehat{BC} = 90$; $m\widehat{CD} = 30$; $m\widehat{DE} = 20$.
Draw your own large diagram so that you
can write arc measures alongside the arcs.
Find the measure of each numbered angle.

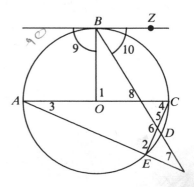

Complete.

11. If $m\widehat{RT} = 80$ and $m\widehat{US} = 40$, then $m\angle 1 = \underline{\quad?\quad}$.

12. If $m\widehat{RU} = 130$ and $m\widehat{TS} = 100$, then $m\angle 1 = \underline{\quad?\quad}$.

13. If $m\angle 1 = 50$ and $m\widehat{RT} = 70$, then $m\widehat{US} = \underline{\quad?\quad}$.

14. If $m\angle 1 = 52$ and $m\widehat{US} = 36$, then $m\widehat{RT} = \underline{\quad?\quad}$.

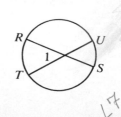

In Exercises 15–17 $\overline{AT}$ is a tangent.

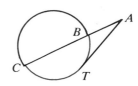

15. If $m\overset{\frown}{CT} = 110$ and $m\overset{\frown}{BT} = 50$, then $m\angle A = $ __?__.

16. If $m\angle A = 28$ and $m\overset{\frown}{BT} = 46$, then $m\overset{\frown}{CT} = $ __?__.

17. If $m\angle A = 35$ and $m\overset{\frown}{CT} = 110$, then $m\overset{\frown}{BT} = $ __?__.

In Exercises 18–21 $\overline{PX}$ and $\overline{PY}$ are tangents.

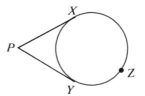

18. If $m\overset{\frown}{XZY} = 250$, then $m\angle P = $ __?__.

19. If $m\overset{\frown}{XY} = 90$, then $m\angle P = $ __?__.

20. If $m\overset{\frown}{XY} = t$, then $m\overset{\frown}{XZY} = $ __?__ and $m\angle P = $ __?__ in terms of t.

21. If $m\angle P = 65$, then $m\overset{\frown}{XY} = $ __?__.

B 22. A secant and a tangent to a circle intersect in a 42° angle. The two arcs of the circle intercepted by the secant and tangent have measures in a 7:3 ratio. Find the measure of the third arc.

23. A quadrilateral circumscribed about a circle has angles of 80°, 90°, 94°, and 96°. Find the measures of the four nonoverlapping arcs determined by the points of tangency.

24. In the inscribed quadrilateral $ABCD$, the sides $\overline{AB}$, $\overline{BC}$, and $\overline{CD}$ are congruent. $\overrightarrow{AB}$ and $\overrightarrow{DC}$ meet at a 32° angle. Find the measures of the angles of $ABCD$.

25. Prove Case II of Theorem 9-10. (*Hint:* See Classroom Exercise 10. Draw a figure like the second one shown below the theorem on page 358. Label your figure, and draw the chord joining the points of tangency.)

26. Prove Case III of Theorem 9-10.

27. Write an equation involving a, b, and c.   **28.** Find the ratio $x:y$.

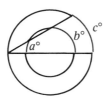   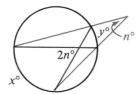

29. Isosceles $\triangle ABC$ with base $\overline{BC}$ is inscribed in a circle. P is a point on $\overset{\frown}{AC}$ and $\overrightarrow{AP}$ and $\overrightarrow{BC}$ meet at Q. Prove that $\angle ABP \cong \angle Q$.

C 30. $\overline{PT}$ is a tangent. It is known that $80 < m\overset{\frown}{RS} < m\overset{\frown}{ST} < 90$. State as much as you can about the measure of $\angle P$.

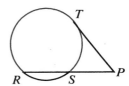

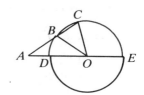

31. $\overline{AC}$ and $\overline{AE}$ are secants of $\odot O$. It is given that $\overline{AB} \cong \overline{OB}$. Discover and prove a relation between the measures of $\overset{\frown}{CE}$ and $\overset{\frown}{BD}$.

32. Take any point P outside a circle. Draw a tangent segment $\overline{PT}$ and a secant $\overline{PBA}$ with A and B points on the circle. Take K on $\overrightarrow{PA}$ so that $PK = PT$. Draw $\overrightarrow{TK}$. Let the intersection of $\overrightarrow{TK}$ with the circle be point X. Discover and prove a relationship between $\overset{\frown}{AX}$ and $\overset{\frown}{XB}$.

Explorations

These exploratory exercises can be done using a computer with a program that draws and measures geometric figures.

1. Draw a circle. Choose four points on the circle. Draw two intersecting chords using the points as endpoints.

Measure the lengths of the pieces of the chords and compute the products $w \cdot x$ and $y \cdot z$. What do you notice?

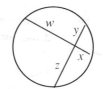

2. Draw any circle A. Choose two points B and C on the circle and a point D outside the circle. Draw secants $\overline{BD}$ and $\overline{CD}$. Label their intersections with the circle as E and F.

Measure and compute $DE \cdot DB$ and $DF \cdot DC$. What do you notice?

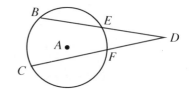

3. Draw any circle A. Choose three points B, C, and D on the circle. Draw a tangent to the circle through point B that intersects $\overleftrightarrow{CD}$ at a point E.

Measure and compute $(BE)^2$ and $ED \cdot CE$. What do you notice?

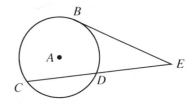

9-7 *Circles and Lengths of Segments*

You can use similar triangles to prove that lengths of chords, secants, and tangents are related in interesting ways.

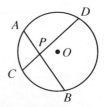

In the figure at the right, chords $\overline{AB}$ and $\overline{CD}$ intersect inside $\odot O$. $\overline{AP}$ and $\overline{PB}$ are called *the segments of chord $\overline{AB}$*. As we did with the terms "radius" and "diameter" we will use the phrase "segment of a chord" to refer to the length of a segment as well as to the segment itself.

Theorem 9-11

When two chords intersect inside a circle, the product of the segments of one chord equals the product of the segments of the other chord.

Given: $\overline{AB}$ and $\overline{CD}$ intersect at P.

Prove: $r \cdot s = t \cdot u$

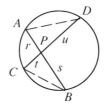

Proof:

Statements	Reasons
1. Draw chords $\overline{AD}$ and $\overline{CB}$.	1. Through any two points there is exactly one line.
2. $\angle A \cong \angle C$; $\angle D \cong \angle B$	2. If two inscribed angles intercept __?__.
3. $\triangle APD \sim \triangle CPB$	3. __?__
4. $\dfrac{r}{t} = \dfrac{u}{s}$	4. __?__
5. $r \cdot s = t \cdot u$	5. A property of proportions

For a proof of the following theorem, see Classroom Exercise 7. In the diagram for the theorem, $\overline{AP}$ and $\overline{CP}$ are *secant segments*. $\overline{BP}$ and $\overline{DP}$ are exterior to the circle and are referred to as *external segments*. The terms "secant segment" and "external segment" can refer to the length of a segment as well as to the segment itself.

Theorem 9-12

When two secant segments are drawn to a circle from an external point, the product of one secant segment and its external segment equals the product of the other secant segment and its external segment.

Given: $\overline{PA}$ and $\overline{PC}$ drawn to the circle from point P

Prove: $r \cdot s = t \cdot u$

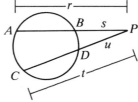

Study the diagrams at the top of the next page from left to right. As $\overline{PC}$ approaches a position of tangency, C and D move closer together until they merge. Then $\overline{PC}$ becomes a tangent, and $t = u$.

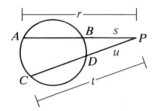 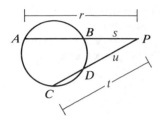 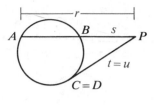

In the first two diagrams we know that $r \cdot s = t \cdot u$. In the third diagram, u and t both become equal to the length of the tangent segment, and the equation becomes $r \cdot s = t^2$. This result, stated below, will be proved more formally in Exercise 10. As with earlier terms, the term "tangent segment" can refer to the length of a segment as well as to the segment itself.

Theorem 9-13

When a secant segment and a tangent segment are drawn to a circle from an external point, the product of the secant segment and its external segment is equal to the square of the tangent segment.

Example 1 Find the value of x.

Solution $3x \cdot x = 6 \cdot 8$ (Theorem 9-11)
$3x^2 = 48,\ x^2 = 16,\ \text{and}\ x = 4$

Example 2 Find the values of x and y.

Solution $4(4 + 5) = 3(3 + x)$ (Theorem 9-12)
$36 = 3(3 + x),\ 12 = 3 + x,\ \text{and}\ x = 9$

$4(4 + 5) = y^2$ (Theorem 9-13)
$36 = y^2,\ \text{so}\ y = 6$

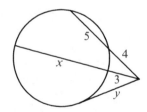

Classroom Exercises

Chords, secants, and tangents are shown. State the equation you would use to find the value of x. Then solve for x.

1.

2.

3.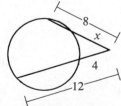

Chords, secants, and tangents are shown. **State the equation you would use to find the value of *x*. Then solve for *x*.**

4.

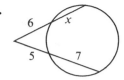

5.

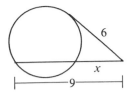

6.

7. Supply reasons to complete the proof of Theorem 9-12.

Given: $\overline{PA}$ and $\overline{PC}$ drawn to the circle from point P

Prove: $r \cdot s = t \cdot u$

Proof:

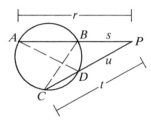

1. Draw chords $\overline{AD}$ and $\overline{BC}$.
2. $\angle A \cong \angle C$
3. $\angle P \cong \angle P$
4. $\triangle APD \sim \triangle CPB$
5. $\dfrac{r}{t} = \dfrac{u}{s}$
6. $r \cdot s = t \cdot u$

Written Exercises

Chords, secants, and tangents are shown. Find the value of *x*.

A **1.**

2.

3.

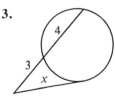

4.

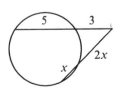

5.

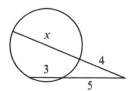

6.

7.

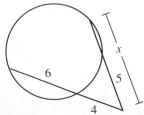

8.

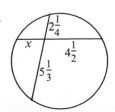

9.

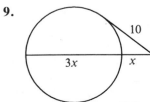

10. Write a proof of Theorem 9-13.

    Given: Secant segment $\overline{PA}$ and tangent segment $\overline{PC}$ drawn to the circle from P.

    Prove: $r \cdot s = t^2$

Plan for Proof: Draw chords $\overline{AC}$ and $\overline{BC}$. Show that $\angle A$ and $\angle PCB$ are congruent because they intercept the same arc. Then show that $\triangle PAC$ and $\triangle PCB$ are similar triangles and use the properties of proportions to complete the proof.

B 11. Given: $\odot O$ and $\odot P$ are tangent to $\overline{UT}$ at T.

    Prove: $UV \cdot UW = UX \cdot UY$

12. Given: $\overline{AB}$ is tangent to $\odot Q$; $\overline{AC}$ is tangent to $\odot S$.

    Prove: $\overline{AB} \cong \overline{AC}$

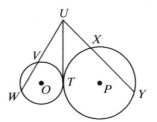
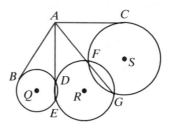

Chords $\overline{AB}$ and $\overline{CD}$ intersect at P. Find the lengths indicated.

Example $AP = 5$; $BP = 4$; $CD = 12$; $CP = \underline{\ ?\ }$

Solution   Let $CP = x$. Then $DP = 12 - x$.

$$x(12 - x) = 5 \cdot 4$$
$$12x - x^2 = 20$$
$$x^2 - 12x + 20 = 0$$
$$(x - 2)(x - 10) = 0$$
$$x = 2 \text{ or } x = 10$$
$$CP = 2 \text{ or } 10$$

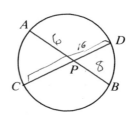

13. $AP = 6$; $BP = 8$; $CD = 16$; $DP = \underline{\ ?\ }$

14. $CD = 10$; $CP = 6$; $AB = 11$; $AP = \underline{\ ?\ }$

15. $AB = 12$; $CP = 9$; $DP = 4$; $BP = \underline{\ ?\ }$

16. $AP = 6$; $BP = 5$; $CP = 3 \cdot DP$; $DP = \underline{\ ?\ }$

$\overline{PT}$ **is tangent to the circle. Find the lengths indicated.**

17. $PT = 6$; $PB = 3$; $AB = \underline{\ ?\ }$

18. $PT = 12$; $CD = 18$; $PC = \underline{\ ?\ }$

19. $PD = 5$; $CD = 7$; $AB = 11$; $PB = \underline{\ ?\ }$

20. $PB = AB = 5$; $PD = 4$; $PT = \underline{\ ?\ }$ and $PC = \underline{\ ?\ }$

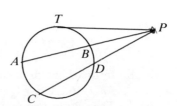

21. A secant, a radius, and a tangent of $\odot O$ are shown.
    a. Explain why $(r + h)^2 = r^2 + d^2$.
    b. Simplify the equation in part (a) to show that $d^2 = h(2r + h)$.
    c. You have proved a special case of a theorem. What theorem is this?

22. $\overleftrightarrow{PT}$ is tangent to $\odot O$. Secant $\overrightarrow{BA}$ is perpendicular to $\overrightarrow{PT}$ at P. If $TA = 6$ and $PA = 3$, find (a) AB, (b) the distance from O to $\overline{AB}$, and (c) the radius of $\odot O$.

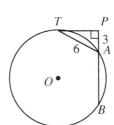

23. A bridge over a river has the shape of a circular arc. The span of the bridge is 24 meters. (The span is the length of the chord of the arc.) The midpoint of the arc is 4 meters higher than the endpoints. What is the radius of the circle that contains this arc?

24. A circle can be drawn through points X, Y, and Z.
    a. What is the radius of the circle?
    b. How far is the center of the circle from point W?

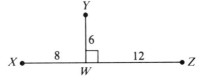

25. Draw two intersecting circles with common chord $\overline{PQ}$ and let X be any point on $\overline{PQ}$. Through X draw any chord $\overline{AB}$ of one circle. Also draw through X any chord $\overline{CD}$ of the other circle. Prove that $AX \cdot XB = CX \cdot XD$.

26. A line is tangent to two intersecting circles at P and Q. The common chord is extended to meet $\overline{PQ}$ at T. Prove that T is the midpoint of $\overline{PQ}$.

C 27. In the diagram at the left below, $\overrightarrow{PT}$ is tangent to $\odot O$ and $\overrightarrow{PN}$ intersects $\odot O$ at J. Find the radius of the circle.

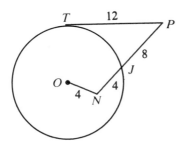

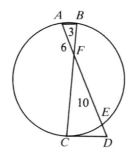

Ex. 27   Ex. 28

★ 28. In the diagram at the right above, $\overline{CD}$ is a tangent, $\overarc{AC} \cong \overarc{BC}$, $AB = 3$, $AF = 6$, and $FE = 10$. Find ED. (*Hint*: Let $ED = x$ and $CD = y$. Then write two equations in x and y.)

Self-Test 2

$\overset{\longleftrightarrow}{MD}$ **is tangent to the circle.**

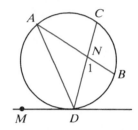

1. If $m\widehat{BD} = 80$, then $m\angle A = \underline{\quad?\quad}$.

2. If $m\angle ADM = 75$, then $m\widehat{AD} = \underline{\quad?\quad}$.

3. If $m\widehat{BD} = 80$ and $m\angle 1 = 81$, then $m\widehat{AC} = \underline{\quad?\quad}$.

4. If $AN = 12$, $BN = 6$, and $CN = 8$, then $DN = \underline{\quad?\quad}$.

5. $\overline{AB}$, $\overline{AC}$, and $\overline{DE}$ are tangents.
Find the values of x and y.

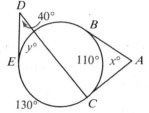

$\overline{PE}$ and $\overline{PF}$ **are secants and** $\overline{PJ}$ **is a tangent.**

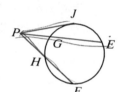

6. If $m\widehat{EF} = 100$ and $m\widehat{GH} = 30$, then $m\angle FPE = \underline{\quad?\quad}$.

7. If $PG = 4$, $PE = 15$, and $PH = 6$, then $PF = \underline{\quad?\quad}$.

8. If $PH = 8$ and $HF = 10$, then $PJ = \underline{\quad?\quad}$.

Application	*Distance to the Horizon*

If you look out over the surface of the Earth from a position at P, directly above point B on the surface, you see the horizon wherever your line of sight is tangent to the surface of the Earth. If the surface around B is smooth (say you are on the ocean on a calm day), the horizon will be a circle, and the higher your lookout is, the farther away this horizon circle will be.

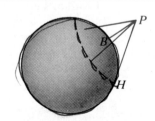

You can use Theorem 9-13 to derive a formula that tells how far you can see from any given height. The diagram at the right shows a section through the Earth containing P, H, and O, the center of the Earth. $\overline{PH}$ is tangent to circle O at H. $\overline{PA}$ is a secant passing through the center O. Theorem 9-13 says that:

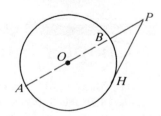

$$(PH)^2 = PA \cdot PB$$

In the formula $(PH)^2 = AP \cdot BP$, PH is the distance from the observer to the horizon, and BP is the observer's height above the surface of the Earth. If the height is small compared to the diameter, AB, of the Earth, then $AP \approx AB$ in the formula. Using 12,800,000 m for AB, you can rewrite the formula as:

$$(\text{distance})^2 \approx (12{,}800{,}000)(\text{height})$$

Taking square roots, you get:

$$\text{distance} \approx \sqrt{12{,}800{,}000} \cdot \sqrt{\text{height}} \approx 3600\sqrt{\text{height}}$$

So the approximate distance (in meters) to the horizon is 3600 times the square root of your height (in meters) above the surface of the Earth. If your height is less than 400 km, the error in this approximation will be less than one percent.

Exercises

In Exercises 1 and 2 give your answer to the nearest kilometer, in Exercises 3 and 5 to the nearest 10 km, and in Exercise 4 to the nearest meter.

1. If you stand on a dune with your eyes about 16 m above sea level, how far out to sea can you look?

2. A lookout climbs high in the rigging of a sailing ship to a point 36 m above the water line. About how far away is the horizon?

3. From a balloon floating 10 km above the ocean, how far away is the farthest point you can see on the Earth's surface?

4. How high must a lookout be to see an object on the horizon 8 km away?

5. You are approaching the coast of Japan in a small sailboat. The highest point on the central island of Honshu is the cone of Mount Fuji, 3776 m above sea level. Roughly how far away from the mountain will you be when you can first see the top? (Assume that the sky is clear!)

Chapter Summary

1. Many of the terms used with circles and spheres are discussed on pages 329 and 330.

2. If a line is tangent to a circle, then the line is perpendicular to the radius drawn to the point of tangency. The converse is also true.

3. Tangents to a circle from a point are congruent.

4. In the same circle or in congruent circles:
   a. Congruent minor arcs have congruent central angles.
   Congruent central angles have congruent arcs.
   b. Congruent arcs have congruent chords.
   Congruent chords have congruent arcs.
   c. Chords equally distant from the center are congruent.
   Congruent chords are equally distant from the center.

5. A diameter that is perpendicular to a chord bisects the chord and its arc.

6. If two inscribed angles intercept the same arc, then the angles are congruent.

7. An angle inscribed in a semicircle is a right angle.

8. If a quadrilateral is inscribed in a circle, then its opposite angles are supplementary.

9. Relationships expressed by formulas:

$m \angle 1 = k$

$m \angle 1 = \frac{1}{2}k$

$m \angle 1 = \frac{1}{2}k$

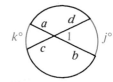

$m \angle 1 = \frac{1}{2}(k + j)$
$a \cdot b = c \cdot d$

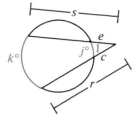

$m \angle 1 = \frac{1}{2}(k - j)$
$s \cdot e = r \cdot c$

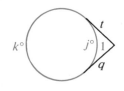

$m \angle 1 = \frac{1}{2}(k - j)$
$t = q$

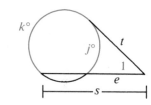

$m \angle 1 = \frac{1}{2}(k - j)$
$s \cdot e = t^2$

Chapter Review

Points A, B, and C lie on $\odot O$.

1. $\overline{AC}$ is called a __?__, while $\overleftrightarrow{AC}$ is called a __?__.

2. $\overline{OB}$ is called a __?__.

3. The best name for $\overline{AB}$ is __?__.

4. $\triangle ABC$ is _____?_____ $\odot O$.
   (inscribed in/circumscribed about)

5. $\overleftrightarrow{CD}$ intersects $\odot O$ in one point. $\overleftrightarrow{CD}$ is called a __?__.

9–1

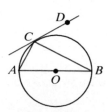

Lines $\overleftrightarrow{ZX}$ and $\overleftrightarrow{ZY}$ are tangent to $\odot P$.

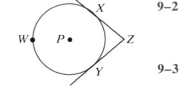

9–2

6. $\overline{PX}$, if drawn, would be __?__ to $\overleftrightarrow{XZ}$.

7. If the radius of $\odot P$ is 6 and $XZ = 8$, then $PZ = $ __?__.

8. If $m \angle Z = 90$ and $XZ = 13$, then $XY = $ __?__.

9. If $m \angle XPY = 100$, then $m\widehat{XY} = $ __?__.

9–3

10. If $m\widehat{XW} = 135$ and $m\widehat{WY} = 125$, then $m\widehat{XWY} = $ __?__.

11. If $\widehat{XW} \cong \widehat{WY}$, then $\angle XPW \cong $ __?__.

In $\odot X$, $m\widehat{AC} = 120$.

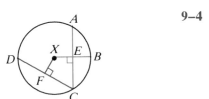

9–4

12. $m\widehat{AB} = $ __?__

13. If $\overline{AC} \cong \overline{CD}$, then $m\widehat{CD} = $ __?__.

14. If $XE = 5$ and $AC = 24$, then the radius = __?__.

15. If $\overline{AC} \cong \overline{DC}$, state the theorem that allows you to deduce that $XE = XF$.

$\overleftrightarrow{RS}$ is tangent to the circle at N.

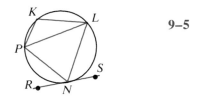

9–5

16. If $m \angle K = 105$, then $m \angle PNL = $ __?__.

17. If $m\widehat{PN} = 100$, then $m \angle PLN = $ __?__ and $m \angle PNR = $ __?__.

18. If $m \angle K = 110$, then $m\widehat{PNL} = $ __?__ and $m\widehat{PL} = $ __?__.

19. If $m\widehat{EF} = 120$ and $m\widehat{GH} = 90$, then $m \angle 1 = $ __?__.

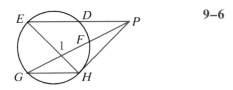

9–6

20. If $m\widehat{EG} = 100$ and $m\widehat{DF} = 40$, then $m \angle EPG = $ __?__.

21. If $\overline{PH}$ is a tangent, $m\widehat{GH} = 90$ and $m \angle GPH = 25$, then $m\widehat{FH} = $ __?__.

Chords, secants, and a tangent are shown. Find the value of x.

22.

23.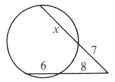

24.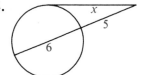

9–7

Chapter Test

Classify each statement as true or false.

1. Opposite angles of an inscribed quadrilateral must be congruent.
2. If a chord in one circle is congruent to a chord in another circle, the arcs of these chords must have congruent central angles.
3. A diameter that is perpendicular to a chord must bisect the chord.
4. If a line bisects a chord, that line must pass through the center of the circle.
5. If $\overrightarrow{GM}$ intersects a circle in just one point, $\overline{GM}$ must be tangent to the circle.
6. It is possible to draw two circles so that no common tangents can be drawn.
7. An angle inscribed in a semicircle must be a right angle.
8. When one chord is farther from the center of a circle than another chord, the chord farther from the center is the longer of the two chords.

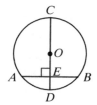

9. In $\odot O$, if $m\widehat{AB} = 100$, then $m\widehat{AC} = \underline{\quad?\quad}$.
10. If the radius of $\odot O$ is 17 and $AB = 30$, then $OE = \underline{\quad?\quad}$.

$\overline{DA}$ and $\overline{DB}$ are tangent to the circle.

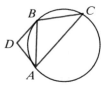

11. If $\overline{AB} \cong \overline{BC}$ and $m\widehat{BC} = 80$, then $m\angle ABC = \underline{\quad?\quad}$.
12. If $m\angle D = 110$, then $m\angle BCA = \underline{\quad?\quad}$.
13. Given: $m\widehat{BC} = m\widehat{AB}$
    Prove: $\overline{AC} \parallel \overline{DB}$

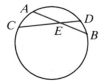

14. If $m\widehat{AC} = 40$ and $m\widehat{BD} = 28$, then $m\angle AEC = \underline{\quad?\quad}$.
15. If $AE = 10$, $EB = 9$, and $CE = 15$, then $ED = \underline{\quad?\quad}$.

$\overline{PT}$ is tangent to the circle.

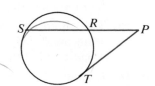

16. If $m\widehat{RS} = 120$ and $m\widehat{ST} = 160$, then $m\angle P = \underline{\quad?\quad}$.
17. If $PT = 12$ and $PS = 18$, then $PR = \underline{\quad?\quad}$.

18. Given: $\square ABCD$ is inscribed in a circle.
    Prove: $ABCD$ is a rectangle.

Cumulative Review: Chapters 1–9

A
1. If x, $x + 3$, and y are the lengths of the sides of a triangle, then
$$\underline{\quad ? \quad} < y < \underline{\quad ? \quad}.$$

2. Find the measure of an angle if the measures of a supplement and a complement of the angle have the ratio 5:2.

3. Given: $\overline{MN}$ is the median of a trapezoid $WXYZ$.
   Prove: $\overline{MN}$ bisects $\overline{WY}$.

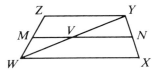

4. Prove: The diagonals of a rhombus divide the rhombus into four congruent triangles.

5. A 30°-60°-90° triangle is inscribed in a circle of radius 7. Find the length of each leg of the triangle.

6. Must three parallel lines be coplanar? Draw a diagram to illustrate your answer.

7. The measures of the angles of a triangle are in the ratio 1:9:10. Find the measure of each angle.

8. If a regular polygon has 18 sides, find the measure of each interior angle and the measure of each exterior angle.

9. If $ABCE$ is a square and $AC = 4$, find AB.

10. If the lengths of two sides of a right triangle are 6 and 10, find two possible lengths for the third side.

11. Given: $\angle 1 \cong \angle 2$; $\angle 2 \cong \angle 3$
    Prove: $\overline{AB} \cong \overline{DC}$

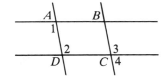

12. When the altitude to the hypotenuse of a certain right triangle is drawn, the altitude divides the hypotenuse into segments of lengths 8 and 10. Find the length of the shorter leg.

13. Write **(a)** the contrapositive and **(b)** the inverse of the following statement: If quad. $ABCD$ is a parallelogram, then $\angle A \cong \angle C$.

14. If $\overrightarrow{OB}$ bisects $\angle AOC$, $m\angle AOB = 5t - 7$, and $m\angle AOC = 8t + 10$, find the numerical measure of $\angle BOC$.

15. Two chords of a circle intersect inside a circle, dividing one chord into segments of length 15 and 12 and the other chord into segments of length 9 and t. Find the value of t.

16. If points R and S on a number line have coordinates -11 and 3, and $\overline{RS}$ has midpoint T, find RS and ST.

17. Complete with *outside*, *inside*, or *on*: In a right triangle, **(a)** the medians intersect __?__ the triangle, **(b)** the altitudes intersect __?__ the triangle, and **(c)** the perpendicular bisectors of the sides intersect __?__ the triangle.

18. In $\triangle RST$, the bisector of $\angle T$ meets $\overline{RS}$ at X. $RS = 15$, $ST = 27$, $TR = 18$. Find RX.

19. Given: All of Bill's sisters like to dance.
What can you conclude from each additional statement? If no conclusion is possible, write *no conclusion*.
  a. Janice is Bill's sister.   **b.** Holly loves to dance.
  c. Maureen is not Bill's sister.   **d.** Kim does not like to dance.

20. Suppose someone plans to write an indirect proof of the statement "In $\square ABCD$ if $\overline{AB} \perp \overline{BC}$, then $ABCD$ is a rectangle." Write a correct first sentence of the indirect proof.

Complete each statement with the words *always*, *sometimes*, or *never*.

21. A contrapositive of a true conditional statement is __?__ true.

22. The sides of a triangle are __?__ 14 cm, 17 cm, and 31 cm long.

23. In $\square ABCD$, if $m \angle A > m \angle B$, then $\angle D$ is __?__ an acute angle.

24. Two obtuse triangles are __?__ similar.

25. Two lines perpendicular to a third line are __?__ perpendicular to each other.

Complete.

26. If $\dfrac{7}{x} = \dfrac{9}{10}$, then __?__ = __?__, and $x =$ __?__.

B **27.** The sine of any acute angle must be greater than __?__ and less than __?__.

28. a. $\triangle RWZ \sim$ __?__
    b. $\dfrac{RW}{?} = \dfrac{ZR}{?} = \dfrac{WZ}{?}$
    c. $RW = 15$, $ZR = 10$, and $SZ = 8$.
      $WZ =$ __?__ and $RS =$ __?__

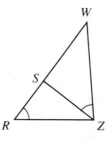

29. Given: $AB > AC$; $\overline{BD} \cong \overline{EC}$
    Prove: $BE > CD$

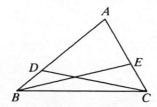

30. Given: $\dfrac{PR}{TR} = \dfrac{SR}{QR}$
    Prove: $\angle S \cong \angle Q$

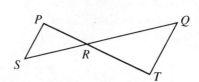

10 CONSTRUCTIONS AND LOCI

Engineers can test the efficiency of a car's design by observing the air currents in a wind tunnel. Strips of cloth attached to the body, as well as smoke patterns, indicate the flow of air around the car.

Basic Constructions

Objectives

1. Perform seven basic constructions.
2. Use these basic constructions in original construction exercises.
3. State and apply theorems involving concurrent lines.

10-1 *What Construction Means*

In Chapters 1–9 we have used rulers and protractors to draw segments with certain lengths and angles with certain measures. In this chapter we will *construct* geometric figures using only two instruments, a *straightedge* and a *compass*. (You may use a ruler as a straightedge as long as you do not use the marks on the ruler.)

Using a Straightedge in Constructions

Given two points A and B, we know from Postulate 6 that there is exactly one line through A and B. We agree that we can use a straightedge to draw $\overleftrightarrow{AB}$ or parts of the line, such as $\overline{AB}$ and $\overrightarrow{AB}$.

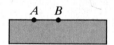

Using a Compass in Constructions

Given a point O and a length r, we know from the definition of a circle that there is exactly one circle with center O and radius r. We agree that we can use a compass to draw this circle or arcs of the circle.

Construction 1

Given a segment, construct a segment congruent to the given segment.

Given: $\overline{AB}$

Construct: A segment congruent to $\overline{AB}$

Procedure:

1. Use a straightedge to draw a line. Call it l.
2. Choose any point on l and label it X.
3. Set your compass for radius AB. Using X as center, draw an arc intersecting line l. Label the point of intersection Y.

$\overline{XY}$ is congruent to $\overline{AB}$.

Justification: Since you used AB for the radius of $\odot X$, $\overline{XY} \cong \overline{AB}$.

Construction 2

Given an angle, construct an angle congruent to the given angle.

Given: ∠ABC

Construct: An angle congruent to ∠ABC

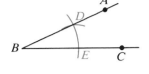

Procedure:

1. Draw a ray. Label it $\overrightarrow{RY}$.
2. Using *B* as center and any radius, draw an arc intersecting $\overrightarrow{BA}$ and $\overrightarrow{BC}$. Label the points of intersection *D* and *E*, respectively.
3. Using *R* as center and the same radius as in Step 2, draw an arc intersecting $\overrightarrow{RY}$. Label the arc $\overset{\frown}{XS}$, with *S* the point where the arc intersects $\overrightarrow{RY}$.

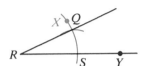

4. Using *S* as center and a radius equal to *DE*, draw an arc that intersects $\overset{\frown}{XS}$ at a point *Q*.
5. Draw $\overrightarrow{RQ}$.

∠QRS is congruent to ∠ABC.

Justification: If you draw $\overline{DE}$ and $\overline{QS}$, △DBE ≅ △QRS (SSS Postulate).
       Then ∠QRS ≅ ∠ABC.

Construction 3

Given an angle, construct the bisector of the angle.

Given: ∠ABC

Construct: The bisector of ∠ABC

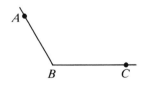

Procedure: *J*

1. Using *B* as center and any radius, draw an arc that intersects $\overrightarrow{BA}$ at *X* and $\overrightarrow{BC}$ at *Y*.
2. Using *X* as center and a suitable radius, draw an arc. Using *Y* as center and the same radius, draw an arc that intersects the arc with center *X* at a point *Z*.

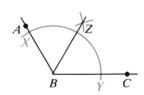

3. Draw $\overrightarrow{BZ}$.

$\overrightarrow{BZ}$ bisects ∠ABC.

Justification: If you draw $\overline{XZ}$ and $\overline{YZ}$, △XBZ ≅ △YBZ (SSS Postulate).
       Then ∠XBZ ≅ ∠YBZ and $\overrightarrow{BZ}$ bisects ∠ABC.

Example   Given ∠1 and ∠2, construct an angle whose
measure is equal to *m*∠1 + *m*∠2.

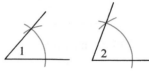

Solution   First use Construction 2 to construct ∠*LON*
congruent to ∠1. Then use the same method
to construct ∠*MOL* congruent to ∠2 (as
shown) so that *m*∠*MON* = *m*∠1 + *m*∠2.

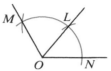

In construction exercises, you won't ordinarily have to write out the procedure
and the justification. However, you should be able to supply them when asked
to do so.

Classroom Exercises

1. Given: △*JKM*
   Explain how to construct a triangle that is
   congruent to △*JKM*.

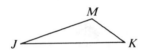

2. Draw any $\overline{AB}$.
   a. Construct $\overline{XY}$ so that *XY* = *AB*.
   b. Using *X* and *Y* as centers, and a radius equal to *AB*, draw arcs that
   intersect. Label the point of intersection *Z*.
   c. Draw $\overline{XZ}$ and $\overline{YZ}$.
   d. What kind of triangle is △*XYZ*?

3. Explain how you could construct a 30° angle.

4. Exercise 3 suggests that you could construct other angles with certain
   measures. Name some.

5. Suppose you are given the three lengths shown and are
   asked to construct a triangle whose sides have lengths
   r, *s*, and *t*. Can you do so? State the theorem from
   Chapter 6 that applies.

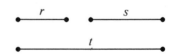

6. ∠1 and ∠2 are given. You see two attempts at constructing an angle
   whose measure is equal to *m*∠1 + *m*∠2. Are both constructions
   satisfactory?

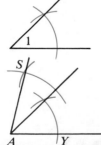

m∠*SAY* = *m*∠1 + *m*∠2

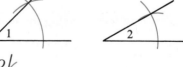

m∠*OUI* = *m*∠1 + *m*∠2

Written Exercises

On your paper, draw two segments roughly like those shown. Use these segments in Exercises 1–4 to construct a segment having the indicated length.

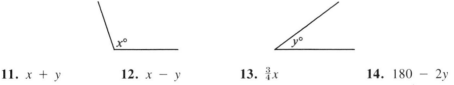

A   **1.** $a + b$ **2.** $b - a$ **3.** $3a - b$ **4.** $a + 2b$

   5. Using any convenient length for a side, construct an equilateral triangle.

   6. a. Construct a 30° angle.   **b.** Construct a 15° angle.

   7. Draw any acute $\triangle ACU$. Use a method based on the SSS Postulate to construct a triangle congruent to $\triangle ACU$.

   8. Draw any obtuse $\triangle OBT$. Use the SSS method to construct a triangle congruent to $\triangle OBT$.

   9. Repeat Exercise 7, but use the SAS method.

   10. Repeat Exercise 8, but use the ASA method.

On your paper, draw two angles roughly like those shown. Then for Exercises 11–14 construct an angle having the indicated measure.

11. $x + y$ **12.** $x - y$ **13.** $\frac{3}{4}x$ **14.** $180 - 2y$

B   **15. a.** Draw any acute triangle. Bisect each of the three angles.
   b. Draw any obtuse triangle. Bisect each of the three angles.
   c. What do you notice about the points of intersection of the bisectors in parts (a) and (b)?

   16. Construct a six-pointed star using the following procedure.
   1. Draw a ray, $\overrightarrow{AB}$. On $\overrightarrow{AB}$ mark off, in order, points C and D such that $AB = BC = CD$.
   2. Construct equilateral $\triangle ADG$.
   3. On $\overline{AG}$ mark off points E and F so that both AE and EF equal AB.
   4. On $\overline{GD}$ mark off points H and I so that both GH and HI equal AB.
   5. To complete the star, draw the three lines $\overleftrightarrow{FH}$, $\overleftrightarrow{EB}$, and $\overleftrightarrow{CI}$.

Construct an angle having the indicated measure.

17. 120   **18.** 150   **19.** 165   **20.** 45

21. Draw any $\triangle ABC$. Construct $\triangle DEF$ so that $\triangle DEF \sim \triangle ABC$ and $DE = 2AB$.

22. Construct a $\triangle RST$ such that $RS:ST:TR = 4:6:7$.

On your paper draw figures roughly like those shown. Use them in constructing the figures described in Exercises 23–25.

23. An isosceles triangle with a vertex angle of $n°$ and legs of length d

24. An isosceles triangle with a vertex angle of $n°$ and base of length s

C **25.** A parallelogram with an $n°$ angle, longer side of length s, and longer diagonal of length d

★ **26.** On your paper draw figures roughly like the ones shown. Then construct a triangle whose three angles are congruent to $\angle 1$, $\angle 2$, and $\angle 3$, and whose circumscribed circle has radius r.

Biographical Note *Grace Hopper*

In 1944 the Mark I, the first working computing machine, started operations at Harvard. It could do three additions per second; calculations that took six months by hand could now be done in a day. Today, computers are one *billion* times as fast, partly because software (programming) has become more efficient, but mostly because of advances in hardware (electronics) such as the development of integrated circuits and silicon chips.

Rear Adm. Grace Hopper, U.S. Navy (Ret.) worked on that first computing machine and many others since. After getting her Ph.D. in mathematics in 1934 from Yale and teaching for several years, Hopper joined the Navy in 1943 and was assigned to Harvard as a programmer of the Mark I. In 1957, her work on making programming faster and easier resulted in her language called Flowmatic, based on the novel idea of using English words in a computer language. The first machine-independent language, COBOL, was announced in 1960 and was based on her language. She continues today to promote computers and learning, saying computers are the "first tool to assist man's brain instead of his arm."

Mixed Review Exercises

Complete.

1. A median of a triangle is a segment from a vertex to the __?__ of the opposite side.

2. A quadrilateral with both pairs of opposite angles congruent is a __?__.

3. A parallelogram with congruent diagonals is a __?__.

4. A parallelogram with perpendicular diagonals is a __?__.

5. If a side of a square has length 5 cm, then a diagonal of the square has length __?__ cm.

6. The measure of each interior angle of a regular pentagon is __?__.

10-2 *Perpendiculars and Parallels*

The next three constructions are based on a theorem and postulate from earlier chapters. The theorem and postulate are repeated here for your use.

(1) If a point is equidistant from the endpoints of a segment, then the point lies on the perpendicular bisector of the segment.

(2) Through any two points there is exactly one line.

Construction 4

Given a segment, construct the perpendicular bisector of the segment.

Given: $\overline{AB}$

Construct: The perpendicular bisector of $\overline{AB}$

Procedure:

1. Using any radius greater than $\frac{1}{2}AB$, draw four arcs of equal radii, two with center A and two with center B. Label the points of intersections of these arcs X and Y.

2. Draw $\overleftrightarrow{XY}$.

$\overleftrightarrow{XY}$ is the perpendicular bisector of $\overline{AB}$.

Justification: Points X and Y are equidistant from A and B. Thus $\overleftrightarrow{XY}$ is the perpendicular bisector of $\overline{AB}$.

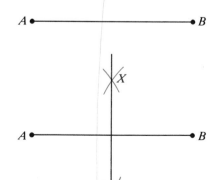

Note that you can use Construction 4 to find the midpoint of a segment.

Construction 5

Given a point on a line, construct the perpendicular to the line at the given point.

Given: Point C on line k

Construct: The perpendicular to k at C

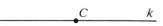

Procedure:

1. Using C as center and any radius, draw arcs intersecting k at X and Y.

2. Using X as center and a radius greater than CX, draw an arc. Using Y as center and the same radius, draw an arc intersecting the arc with center X at a point Z.

3. Draw $\overleftrightarrow{CZ}$.

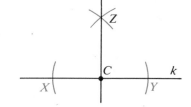

$\overleftrightarrow{CZ}$ is perpendicular to k at C.

Justification: You constructed points X and Y so that C is equidistant from X and Y. Then you constructed point Z so that Z is equidistant from X and Y. Thus $\overleftrightarrow{CZ}$ is the perpendicular bisector of $\overline{XY}$, and $\overleftrightarrow{CZ} \perp k$ at C.

Construction 6

Given a point outside a line, construct the perpendicular to the line from the given point.

Given: Point P outside line k

Construct: The perpendicular to k from P

Procedure:

1. Using P as center, draw two arcs of equal radii that intersect k at points X and Y.

2. Using X and Y as centers and a suitable radius, draw arcs that intersect at a point Z.

3. Draw $\overleftrightarrow{PZ}$.

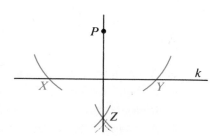

$\overleftrightarrow{PZ}$ is perpendicular to k.

Justification: Both P and Z are equidistant from X and Y. Thus $\overleftrightarrow{PZ}$ is the perpendicular bisector of $\overline{XY}$, and $\overleftrightarrow{PZ} \perp k$.

Construction 7

Given a point outside a line, construct the parallel to the given line through the given point.

Given: Point *P* outside line *k*

Construct: The line through *P* parallel to *k*

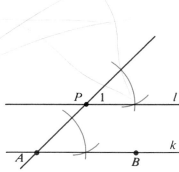

Procedure:

1. Let *A* and *B* be two points on line *k*. Draw $\overleftrightarrow{PA}$.

2. At *P*, construct ∠ 1 so that ∠ 1 and ∠ *PAB* are congruent corresponding angles. Let *l* be the line containing the ray you just constructed.

l is the line through *P* parallel to *k*.

Justification: If two lines are cut by a transversal and corresponding angles are congruent, then the lines are parallel. (Postulate 11)

Classroom Exercises

1. Suggest an alternative procedure for Construction 7 that uses Constructions 5 and 6.

Describe how you would construct each of the following.

2. The midpoint of $\overline{BC}$

3. The median of △*ABC* that contains vertex *B*

4. The altitude of △*ABC* that contains vertex *B*

5. The altitude of △*ABC* that contains vertex *A*

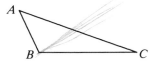

6. The perpendicular to $\overline{BC}$ at *C*

7. A square whose sides each have length *AC*

8. A square whose perimeter equals *AC*

9. A right triangle with hypotenuse and one leg equal to *AC* and *BC*, respectively

10. A triangle whose sides are in the ratio $1:2:\sqrt{5}$

Exercises 11–13 will analyze the following problem.

Given: Line *l*; points *X* and *Y*

Construct: A circle through *Y* and tangent to *l* at *X*

If the problem had been solved, we would have a diagram something like the one shown.

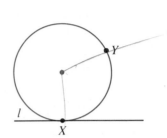

11. Where does the center of the circle lie with respect to line *l* and point *X*?

12. Where does the center of the circle lie with respect to $\overline{XY}$?

13. Explain how to carry out the construction of the circle.

Written Exercises

Draw a figure roughly like the one shown, but larger. Do the indicated construction clearly enough so that your method can be understood easily.

A   **1.** The perpendicular to *l* at *P*

2. The perpendicular to *l* from *S*

3. The perpendicular bisector of $\overline{JK}$

4. The parallel to *l* through *T*

5. The parallel to $\overleftrightarrow{ED}$ through *F*

6. The perpendicular to $\overleftrightarrow{BA}$ at *A*

7. The perpendicular to $\overleftrightarrow{HJ}$ from *G*

8. A complement of $\angle KMN$

Construct an angle with the indicated measure.

9. 45   **10.** 135

11. $22\frac{1}{2}$ **12.** 105

13. Draw a segment $\overline{AB}$. Construct a segment $\overline{XY}$ whose length equals $\frac{3}{4}AB$.

B **14. a.** Draw an acute triangle. Construct the perpendicular bisector of each side.
    b. Do the perpendicular bisectors intersect in one point?
    c. Repeat parts (a) and (b) using an obtuse triangle.

15. a. Draw an acute triangle. Construct the three altitudes.
    b. Do the lines that contain the altitudes intersect in one point?
    c. Repeat parts (a) and (b) using an obtuse triangle.

16. a. Draw a very large acute triangle. Construct the three medians.
    b. Do the lines that contain the medians intersect in one point?
    c. Repeat parts (a) and (b) using an obtuse triangle.

On your paper draw figures roughly like those shown. Use them in constructing the figures described in Exercises 17–24.

17. A parallelogram with an $n°$ angle and sides of lengths a and b

18. A rectangle with sides of lengths a and b

19. A square with perimeter $2a$

20. A rhombus with diagonals of lengths a and b

21. A square with diagonals of length b

22. A segment of length $\sqrt{a^2 + b^2}$

23. A square with diagonals of length $b\sqrt{2}$

24. A right triangle with hypotenuse of length a and one leg of length b

C **25.** Draw a segment and let its length be s. Construct a segment whose length is $s\sqrt{3}$.

26. Draw a diagram roughly like the one shown. Without laying your straightedge across any part of the lake, construct more of $\overrightarrow{RS}$.

27. Draw three noncollinear points R, S, and T. Construct a triangle whose sides have R, S, and T as midpoints. (*Hint*: How is $\overline{RT}$ related to the side of the triangle that has S as its midpoint?)

28. Draw a segment and let its length be 1.
    a. Construct a segment of length $\sqrt{5}$.
    b. Construct a segment of length $\dfrac{1}{2} + \dfrac{\sqrt{5}}{2}$, or $\dfrac{1 + \sqrt{5}}{2}$.
    c. Construct a *golden rectangle* (as discussed on page 253) whose sides are in the ratio $1 : \dfrac{1 + \sqrt{5}}{2}$.

Challenge

Given $\overline{AB}$, its midpoint M, and a point Z outside $\overline{AB}$, use only a straightedge (and *no* compass) to construct a line through Z parallel to $\overleftrightarrow{AB}$. (*Hint*: Use Ceva's Theorem, Exercise 33, page 273.)

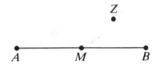

Explorations

These exploratory exercises can be done using a computer with a program that draws and measures geometric figures.

1. Draw any $\triangle ABC$. Draw the bisectors of the angles of the triangle. They should intersect in one point. Draw a perpendicular segment from this point to each of the sides. Measure the length of each perpendicular segment. What do you notice?

2. **a.** Draw any acute $\triangle ABC$. Draw the perpendicular bisector of each side of the triangle. They should intersect in one point. Measure the distance from this point of intersection to each of the vertices of the triangle. What do you notice?
    b. Repeat using an obtuse triangle and a right triangle. Is the same result true for these triangles as well?
    c. In a right triangle, the perpendicular bisectors of the sides intersect in what point?

3. Draw any $\triangle ABC$. Draw the three medians. They should intersect in one point, as shown in the diagram at the right. Find the ratios $\dfrac{AG}{AD}$, $\dfrac{BG}{BE}$, and $\dfrac{CG}{CF}$. What do you notice?

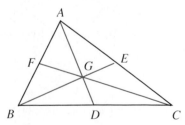

10-3 *Concurrent Lines*

When two or more lines intersect in one point, the lines are said to be **concurrent.** For example, as you saw in Exercise 15, page 378, the bisectors of the angles of a triangle are concurrent.

Theorem 10-1

The bisectors of the angles of a triangle intersect in a point that is equidistant from the three sides of the triangle.

Given: $\triangle ABC$; the bisectors of $\angle A$, $\angle B$, and $\angle C$

Prove: The angle bisectors intersect in a point; that point is equidistant from $\overline{AB}$, $\overline{BC}$, and $\overline{AC}$.

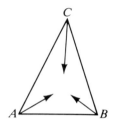

Proof:

The bisectors of $\angle A$ and $\angle B$ intersect at some point I. We will show that point I also lies on the bisector of $\angle C$ and that I is equidistant from $\overline{AB}$, $\overline{BC}$, and $\overline{AC}$.

Draw perpendiculars from I intersecting $\overline{AB}$, $\overline{BC}$, and $\overline{AC}$ at R, S, and T, respectively. Since any point on the bisector of an angle is equidistant from the sides of the angle (Theorem 4-7, page 154), $IT = IR$ and $IR = IS$. Thus $IT = IS$. Since any point equidistant from the sides of an angle is on the bisector of the angle (Theorem 4-8, page 154), I is on the bisector of $\angle C$. Since $IR = IS = IT$, point I is equidistant from $\overline{AB}$, $\overline{BC}$, and $\overline{AC}$.

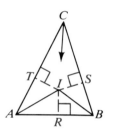

In Exercises 14–16, page 384, you discovered three other sets of concurrent lines related to triangles: the perpendicular bisectors of the sides, the lines containing the altitudes, and the medians. As you can see in the diagrams below, concurrent lines may intersect in a point outside the triangle. The intersection point may also lie on the triangle (see Classroom Exercise 4, page 388).

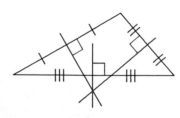

Perpendicular bisectors

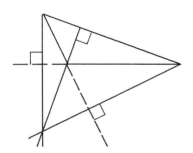

Lines containing altitudes

Theorem 10-2

The perpendicular bisectors of the sides of a triangle intersect in a point that is equidistant from the three vertices of the triangle.

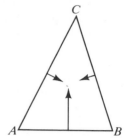

Given: △*ABC*; the ⊥ bisectors of $\overline{AB}$, $\overline{BC}$, and $\overline{AC}$

Prove: The ⊥ bisectors intersect in a point; that point is equidistant from *A*, *B*, and *C*.

Proof:

The perpendicular bisectors of $\overline{AC}$ and $\overline{BC}$ intersect at some point *O*. We will show that point *O* lies on the perpendicular bisector of $\overline{AB}$ and is equidistant from *A*, *B*, and *C*.

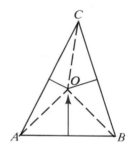

Draw $\overline{OA}$, $\overline{OB}$, and $\overline{OC}$. Since any point on the perpendicular bisector of a segment is equidistant from the endpoints of the segment (Theorem 4-5, page 153), *OA* = *OC* and *OC* = *OB*. Thus *OA* = *OB*. Since any point equidistant from the endpoints of a segment lies on the perpendicular bisector of the segment (Theorem 4-6, page 153), *O* is on the perpendicular bisector of $\overline{AB}$. Since *OA* = *OB* = *OC*, point *O* is equidistant from *A*, *B*, and *C*.

The following theorems will be proved in Chapter 13.

Theorem 10-3

The lines that contain the altitudes of a triangle intersect in a point.

Theorem 10-4

The medians of a triangle intersect in a point that is two thirds of the distance from each vertex to the midpoint of the opposite side.

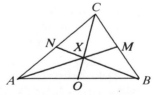

According to Theorem 10-4, if $\overline{AM}$, $\overline{BN}$, and $\overline{CO}$ are medians of △*ABC*, then:

$$AX = \tfrac{2}{3}AM$$
$$XN = \tfrac{1}{3}BN$$
$$CX:XO:CO = 2:1:3$$

The points of intersection described in the theorems in this section are sometimes called the *incenter* (point where the angle bisectors meet), *circumcenter* (point where the perpendicular bisectors meet), *orthocenter* (point where the altitudes meet), and *centroid* (point where the medians meet).

Classroom Exercises

1. Draw, if possible, a triangle in which the perpendicular bisectors of the sides intersect in a point with the location described.
   a. A point inside the triangle   **b.** A point outside the triangle
   c. A point on the triangle

2. Repeat Exercise 1, but work with angle bisectors.

3. Is there some kind of triangle such that the perpendicular bisector of each side is also an angle bisector, a median, and an altitude?

4. △*JAM* is a right triangle.
   a. Is $\overline{JM}$ an altitude of △*JAM*?
   b. Name another altitude shown.
   c. In what point do the three altitudes of △*JAM* meet?
   d. Where do the perpendicular bisectors of the sides of △*JAM* meet?
   e. Does your answer to (d) agree with Theorem 10-2?

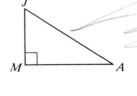

5. The medians of △*DEF* are shown. Find the lengths indicated.
   a. $EP = \underline{\ ?\ }$ **b.** $PR = \underline{\ ?\ }$
   c. If $FT = 9$, then $PT = \underline{\ ?\ }$ and $FP = \underline{\ ?\ }$.

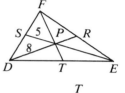

6. Given: $\overline{RJ}$ and $\overline{SK}$ are medians of △*RST*;
   X and Y are the midpoints of $\overline{RG}$ and $\overline{SG}$.
   a. How are $\overline{XY}$ and $\overline{RS}$ related? Why?
   b. How are $\overline{KJ}$ and $\overline{RS}$ related? Why?
   c. How are $\overline{KJ}$ and $\overline{XY}$ related? Why?
   d. What special kind of quadrilateral is *XYJK*? Why?
   e. Why does $XG = GJ$?
   f. Explain why $RG = \frac{2}{3}RJ$.

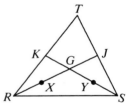

Written Exercises

A
1. Draw a triangle such that the lines containing the three altitudes intersect in a point with the location described.
   a. A point inside the triangle   **b.** A point outside the triangle
   c. A point on the triangle

Exercises 2–5 refer to the diagram in which the medians of a triangle are shown.

2. Find the values of *x* and *y*.

3. If $AB = 6$, then $BP = \underline{\ ?\ }$ and $AP = \underline{\ ?\ }$.

4. If $AB = 7$, then $BP = \underline{\ ?\ }$ and $AP = \underline{\ ?\ }$.

5. If $PB = 1.9$, then $AP = \underline{\ ?\ }$ and $AB = \underline{\ ?\ }$.

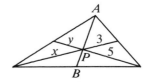

6. Use a ruler and a protractor to draw a regular pentagon. Then construct the perpendicular bisectors of the five sides.

7. Draw a regular pentagon as in Exercise 6. Construct the angle bisectors.

8. Draw any large $\triangle ABC$ and construct equilateral triangles on each of the sides as shown.

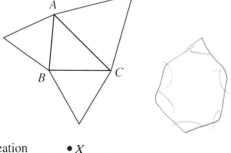

  a. In each of the three equilateral triangles, construct any two medians and find their point of intersection.
  b. Draw the three segments connecting these three points of intersection.
  c. What appears to be true about the triangle you drew in part (b)?

B 9. Three towns, located as shown, plan to build one recreation center to serve all three towns. They decide that the fair thing to do is to build the hall equidistant from the three towns. Comment about the wisdom of the plan.

•*X*

•*Y*   •*Z*

10. See Exercise 9. Locate three towns so that it isn't possible to find a spot equidistant from the three towns.

11. In the figure, $\overline{AD}$ and $\overline{BE}$ are congruent medians of $\triangle ABC$.

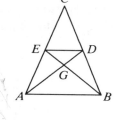

  a. Explain why $GD = GE$.
  b. $GA = \underline{\quad?\quad}$
  c. Name three angles congruent to $\angle GAB$.

$\overline{AU}$, $\overline{BV}$, and $\overline{CW}$ **are the medians of** $\triangle ABC$.

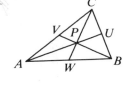

12. If $AP = x^2$ and $PU = 2x$, then $x = \underline{\quad?\quad}$.

13. If $BP = y^2 + 1$ and $PV = y + 2$, then $y = \underline{\quad?\quad}$ or $y = \underline{\quad?\quad}$.

14. If $CW = 2z^2 - 5z - 12$ and $CP = z^2 - 15$, then $z = \underline{\quad?\quad}$ and $PW = \underline{\quad?\quad}$.

15. $ABCD$ is a parallelogram with M the midpoint of $\overline{CD}$. If $\overline{BM}$ intersects $\overline{AC}$ at X, prove that $CX = \frac{1}{3}AC$. (*Hint*: Draw $\overline{BD}$.)

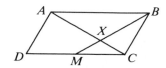

16. Prove that if two of the medians of a triangle are congruent, then the triangle is isosceles.

C 17. In the plane figure, point P is equidistant from R, S, and T. Describe the location of the following points in the plane.

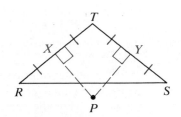

  a. Points farther from both R and S than from T.
  b. Points closer to both R and S than to T.

Self-Test 1

1. Draw any $\overline{CD}$. Construct the perpendicular bisector of $\overline{CD}$.
2. Construct a 60° angle, $\angle RST$, and its bisector, $\overrightarrow{SQ}$.
3. Draw a large acute $\triangle ABC$. Then construct altitude $\overline{AD}$ from vertex A.
4. Draw line t and choose any point P that is not on line t. Construct $\overleftrightarrow{PQ} \parallel t$.
5. Draw any $\overline{AB}$. Construct rectangle $JKLM$ so that $JK = 2AB$ and $KL = AB$.
6. Name four types of concurrent lines, rays, or segments that are associated with triangles.
7. The perpendicular bisectors of the sides of a right triangle intersect in a point located at __?__.
8. The medians of equilateral $\triangle ABC$ intersect at point X. If $\overline{AD}$ is a median and $AB = 12$, then $AX = $ __?__ and $XD = $ __?__.

Application   *Center of Gravity*

The *center of gravity* of an object is the point where the weight of the object is focused. If you lift or support an object, you can do this most easily under its center of gravity.

    A mobile is either hung or supported at its center of gravity. In planning a mobile, a sculptor must take into account the centers of gravity of the component parts.

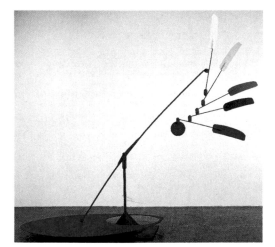

    If an object is not supported under its center of gravity, it becomes unstable. Suppose you hold a heavy bar in one hand. If you support it near the center of gravity, it will be easy to hold (Figure 1). To support it at one end requires more effort (Figure 2), since the pole tends to turn until the center of gravity is directly below the point of support (Figure 3).

    The center of gravity may be inside an object or outside of it. The center of gravity of an ice cube is in the middle of the ice, but the center of gravity of an automobile tire is not in a part of the tire itself.

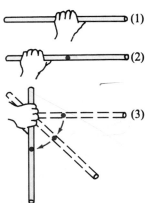

Exercises

1. For this experiment, cut out a large, irregularly shaped piece of cardboard.

   a. Near the edge, poke a hole just large enough to allow the cardboard to rotate freely when pinned through the hole.

   b. Pin the cardboard through the hole to a suitable wall surface. The piece of cardboard will position itself so that its center of gravity is as low as possible. This means that it will lie on a vertical line through the point of suspension. To find this line, tie a weighted string to the pin. Then draw on the cardboard the line determined by the string.

   c. Repeat parts (a) and (b) but use a different hole. The center of gravity of the cardboard ought to lie on both of the lines you have drawn and should therefore be their point of intersection. The cardboard should balance if supported at this point.

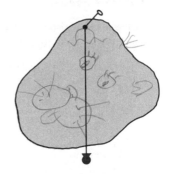

2. Cut out a piece of cardboard in the shape of a large scalene triangle.

   a. Follow the steps of Exercise 1 using three holes, one near each of the three vertices.

   b. If you worked carefully, all three lines drawn intersect in one point, the center of gravity of the cardboard. This point is also referred to as the *center of mass* or the *centroid* of the cardboard. Study the lines you have drawn and explain why in geometry the point of intersection of the medians of a triangle is called the *centroid of the triangle.*

3. Do you think that the center of gravity of a parallelogram is the point where the diagonals intersect? Use the technique of Exercise 1 to test this idea.

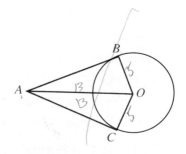

Mixed Review Exercises

$\overline{AB}$ **is tangent to** $\odot O$ **at** B. **Complete.**

1. If the radius of $\odot O$ is 5 and $AO = 13$, then $AB = \underline{\ ?\ }$.

2. If $m \angle ACO = 90$ and $AB = 10$, then $\overline{AC}$ is $\underline{\ ?\ }$ to $\odot O$ at C and $AC = \underline{\ ?\ }$.

3. A triangle circumscribed about a circle intersects the circle in how many points?

4. Quad. *QRST* is inscribed in a circle. If $m \angle Q = 39$, find $m \angle S$.

Explorations

These exploratory exercises can be done using a computer with a program that draws and measures geometric figures.

1. Inscribe a circle D inside a $\triangle ABC$. Draw $\overline{DA}$, $\overline{DB}$, and $\overline{DC}$. Compare the measures of $\angle ABD$ and $\angle ABC$, $\angle ACD$ and $\angle ACB$, $\angle BAD$ and $\angle BAC$. What do you notice? What type of lines intersect at the center of a circle inscribed in a triangle?

2. Circumscribe a circle D about a $\triangle ABC$. Draw perpendicular segments from D to $\overline{AB}$, $\overline{BC}$, and $\overline{CA}$, intersecting the sides at E, F, and G, respectively. Compare the lengths of $\overline{AE}$ and $\overline{AB}$, $\overline{BF}$ and $\overline{BC}$, and $\overline{CG}$ and $\overline{CA}$. What do you notice? What type of lines intersect at the center of a circle circumscribed about a triangle?

More Constructions

Objectives

1. Perform seven additional basic constructions.
2. Use the basic constructions in original construction exercises.

10-4 *Circles*

Construction 8

Given a point on a circle, construct the tangent to the circle at the given point.

Given: Point A on $\odot O$

Construct: The tangent to $\odot O$ at A

Procedure:

1. Draw $\overrightarrow{OA}$.

2. Construct the line perpendicular to $\overrightarrow{OA}$ at A. Call it t.

Line t is tangent to $\odot O$ at A.

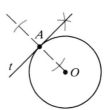

Justification: Because t is perpendicular to radius $\overline{OA}$ at A, t is tangent to $\odot O$.

Construction 9

Given a point outside a circle, construct a tangent to the circle from the given point.

Given: Point *P* outside $\odot O$

Construct: A tangent to $\odot O$ from *P*

Procedure:

1. Draw $\overline{OP}$.
2. Find the midpoint *M* of $\overline{OP}$ by constructing the perpendicular bisector of $\overline{OP}$.
3. Using *M* as center and *MP* as radius, draw a circle that intersects $\odot O$ in a point *X*.
4. Draw $\overrightarrow{PX}$.

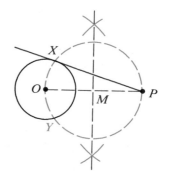

$\overrightarrow{PX}$ is tangent to $\odot O$ from *P*. $\overrightarrow{PY}$, not drawn, is the other tangent from *P*.

Justification: If you draw $\overline{OX}$, $\angle OXP$ is inscribed in a semicircle. Then $\angle OXP$ is a right angle and $\overrightarrow{PX} \perp \overline{OX}$. Because $\overrightarrow{PX}$ is perpendicular to radius $\overline{OX}$ at its outer endpoint, $\overrightarrow{PX}$ is tangent to $\odot O$.

Construction 10

Given a triangle, circumscribe a circle about the triangle.

Given: $\triangle ABC$

Construct: A circle passing through *A*, *B*, and *C*

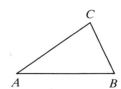

Procedure:

1. Construct the perpendicular bisectors of any two sides of $\triangle ABC$. Label the point of intersection *O*.
2. Using *O* as center and *OA* as radius, draw a circle.

Circle *O* passes through *A*, *B*, and *C*.

Justification: See Theorem 10-2 on page 387.

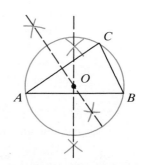

Construction 11

Given a triangle, inscribe a circle in the triangle.

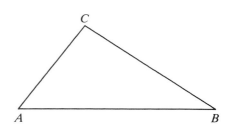

Given: △*ABC*

Construct: A circle tangent to $\overline{AB}$, $\overline{BC}$, and $\overline{AC}$

Procedure:

1. Construct the bisectors of ∠*A* and ∠*B*. Label the point of intersection *I*.

2. Construct a perpendicular from *I* to $\overline{AB}$, intersecting $\overline{AB}$ at a point *R*.

3. Using *I* as center and *IR* as radius, draw a circle.

Circle *I* is tangent to $\overline{AB}$, $\overline{BC}$, and $\overline{AC}$.

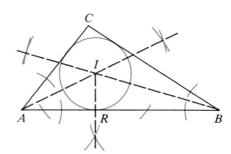

Justification: See Theorem 10-1 on page 386.

Classroom Exercises

1. Explain how to find the midpoint of $\overset{\frown}{AB}$.

2. Explain how to construct the center of the circle containing points *A*, *B*, and *C*.

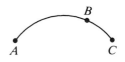

3. Explain how to find the line described.
   a. Parallel to $\overleftrightarrow{RS}$ and passing through *P*
   b. Parallel to $\overleftrightarrow{RS}$ and tangent to ⊙*P*

4. Here you see a common method for using just one compass setting for drawing a circle and dividing the circle into six congruent arcs. Explain how the method works.

5. Suppose a circle is given. Explain how you can use the method of Exercise 4 to inscribe an equilateral triangle in the circle.

6. Suppose the construction of Exercise 4 has been carried out. Explain how you can then inscribe a regular twelve-sided polygon in the circle.

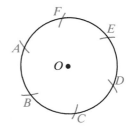

7. A student intends to inscribe a circle in $\triangle RST$. The center I has been found as shown. How should the student find the radius needed?

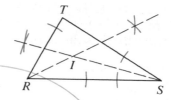

Written Exercises

In Exercises 1 and 2 draw a diagram similar to the one shown, but larger.

A **1.** Construct a tangent at A.

2. Construct two tangents from P.

3. Draw a large acute triangle. Construct the circumscribed circle.

4. Construct a large right triangle. Construct the circumscribed circle.

5. Draw a large obtuse triangle. Construct the circumscribed circle.

6. Draw a large acute triangle. Construct the inscribed circle.

7. Construct a large right triangle. Construct the inscribed circle.

8. Draw a large obtuse triangle. Construct the inscribed circle.

B **9.** Draw a circle. Inscribe an equilateral triangle in the circle.

10. Draw a circle. Inscribe a square in the circle.

11. a. Draw a circle. Inscribe a regular octagon in the circle.
  b. How would you use your construction in part (a) to create an eight-pointed star as shown at the right?

12. Draw a circle. Circumscribe a square about the circle.

13. Construct a square. Circumscribe a circle about the square.

14. Construct a square. Inscribe a circle in the square.

15. Draw a circle. Circumscribe an equilateral triangle about the circle.

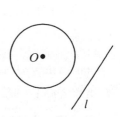

Ex. 11(b)

In each of Exercises 16 and 17 begin with a diagram roughly like the one shown, but larger.

16. Construct a line that is parallel to line l and tangent to $\odot O$.

17. Construct a line that is perpendicular to line l and tangent to $\odot O$.

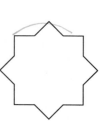

C **18.** Construct three congruent circles, each tangent to the other two circles. Then construct an equilateral triangle, each side of which is tangent to two of the circles.

In Exercises 19–21 begin with two circles P and Q such that $\odot P$ and $\odot Q$ do not intersect and Q is not inside $\odot P$. Let the radii of $\odot P$ and $\odot Q$ be p and q respectively, with $p > q$.

19. Construct a circle, with radius equal to PQ, that is tangent to $\odot P$ and $\odot Q$.

20. Construct a common external tangent to $\odot P$ and $\odot Q$. One method is suggested below.

    1. Draw a circle with center P and radius $p - q$.
    2. Construct a tangent to this circle from Q, and call the point of tangency Z.
    3. Draw $\overrightarrow{PZ}$. $\overrightarrow{PZ}$ intersects $\odot P$ in a point X.
    4. With center X and radius ZQ, draw an arc that intersects $\odot Q$ in a point Y.
    5. Draw $\overleftrightarrow{XY}$.

As a justification for this construction, you could begin by drawing $\overline{QY}$. Then show that $XZQY$ is a rectangle. The rest of the justification is easy.

21. Construct a common internal tangent to $\odot P$ and $\odot Q$. (*Hint:* Draw a circle with center P and radius $p + q$.)

10-5 *Special Segments*

Construction 12

Given a segment, divide the segment into a given number of congruent parts. (3 shown)

Given: $\overline{AB}$

A •————————————————• B

Construct: Points X and Y on $\overline{AB}$ so that
         $AX = XY = YB$

Procedure:

1. Choose any point Z not on $\overleftrightarrow{AB}$. Draw $\overrightarrow{AZ}$.

2. Using any radius, start with A as center and mark off R, S, and T so that $AR = RS = ST$.

3. Draw $\overline{TB}$.

4. At R and S construct lines parallel to $\overline{TB}$, intersecting $\overline{AB}$ in X and Y.

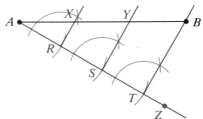

$\overline{AX}$, $\overline{XY}$, and $\overline{YB}$ are congruent parts of $\overline{AB}$.

Justification: Since the parallel lines you constructed cut off congruent segments on transversal $\overleftrightarrow{AZ}$, they cut off congruent segments on transversal $\overleftrightarrow{AB}$. (It may help you to think of the parallel to $\overline{TB}$ through A.)

Construction 13

Given three segments, construct a fourth segment so that the four segments are in proportion.

Given: Segments with lengths a, b, and c

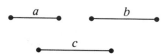

Construct: A segment of length x such that $\dfrac{a}{b} = \dfrac{c}{x}$

Procedure:

1. Draw an $\angle HIJ$.
2. On $\overrightarrow{IJ}$, mark off $IR = a$ and $RS = b$.
3. On $\overrightarrow{IH}$, mark off $IT = c$.
4. Draw $\overline{RT}$.
5. At S, construct a parallel to $\overrightarrow{RT}$, intersecting $\overrightarrow{IH}$ in a point U.

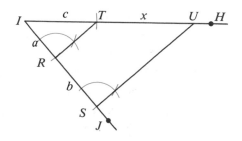

$\overline{TU}$ has length x such that $\dfrac{a}{b} = \dfrac{c}{x}$.

Justification: Because $\overline{RT}$ is parallel to side $\overline{SU}$ of $\triangle SIU$, $\overline{RT}$ divides the other

two sides of the triangle proportionally. Therefore, $\dfrac{a}{b} = \dfrac{c}{x}$.

Construction 14

Given two segments, construct their geometric mean.

Given: Segments with lengths a and b

Construct: A segment of length x such that $\dfrac{a}{x} = \dfrac{x}{b}$

           (or $x = \sqrt{ab}$)

Procedure:

1. Draw a line and mark off $RS = a$ and $ST = b$.
2. Locate the midpoint O of $\overline{RT}$ by constructing the perpendicular bisector of $\overline{RT}$.
3. Using O as center draw a semicircle with a radius equal to OR.
4. At S, construct a perpendicular to $\overline{RT}$. The perpendicular intersects the semicircle at a point Z.

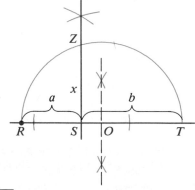

ZS, or x, is the geometric mean between a and b.

Justification: $\overparen{RZT}$ is a semicircle. If you draw $\overline{RZ}$ and $\overline{ZT}$, then $\triangle RZT$ is a right triangle. Since $\overline{ZS}$ is the altitude to the hypotenuse of rt. $\triangle RZT$, $\dfrac{a}{x} = \dfrac{x}{b}$.

Classroom Exercises

1. Given a segment, tell how to construct an equilateral triangle whose perimeter equals the length of the given segment.

Draw three segments and label their lengths a, b, and c.

2. Construct a segment of length x such that $\dfrac{c}{a} = \dfrac{b}{x}$.

3. Describe how to construct a segment of length x such that $x = \sqrt{2ab}$.

4. Describe how to construct a segment of length x such that $x = \sqrt{5ab}$.

5. Describe how to construct a segment of length x such that $x = \sqrt{4ab}$.

Exercises 6–11 will analyze the following problem.

Given: Line t; points A and B

Construct: A circle through A and B and tangent to t

If the problem had been solved, we would have a diagram something like the one shown.

6. Where does the center of the circle lie with respect to $\overline{AB}$?

7. Where does the center of the circle lie with respect to line t and K, the point of tangency?

Note that we don't have point K located in the given diagram. Hunting for ideas, we draw $\overleftrightarrow{AB}$. We now have a point J, which we can locate in the given diagram.

8. State an equation that relates JK to JA and JB.

9. Rewrite your equation in the form $\dfrac{?}{JK} = \dfrac{JK}{?}$.

10. What construction can we use to get the length JK?

In a *separate* diagram we can mark off the lengths JA and JB on some line l and then use Construction 14 to find x such that $\dfrac{JA}{x} = \dfrac{x}{JB}$. Once we have x, which equals JK, we return to the given diagram and draw an arc to locate K.

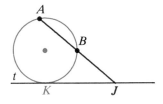

11. Explain how to complete the construction of the circle.

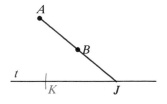

Written Exercises

In each of Exercises 1–4 begin by drawing $\overline{AB}$ roughly 15 cm long.

A **1.** Divide $\overline{AB}$ into three congruent segments.

  2. a. Use Construction 12 to divide $\overline{AB}$ into four congruent segments.

  b. Use Construction 4 to divide $\overline{AB}$ into four congruent segments.

  3. a. Use Construction 12 to divide $\overline{AB}$ into five congruent segments.

  b. Can Construction 4 be used to divide $\overline{AB}$ into five congruent segments?

  c. Divide $\overline{AB}$ into two segments that have the ratio $2:3$.

  4. Divide $\overline{AB}$ into two segments that have the ratio $3:4$.

On your paper draw four segments roughly as long as those shown below. Use your segments in Exercises 5–14. In each exercise construct a segment that has length x satisfying the given condition.

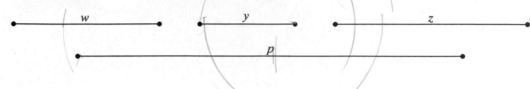

5. $\dfrac{y}{w} = \dfrac{z}{x}$ **6.** $\dfrac{w}{x} = \dfrac{x}{y}$ **7.** $x = \sqrt{yp}$ **8.** $3x = w + 2y$

B **9.** $zx = wy$ (*Hint:* First write a proportion that is equivalent to the given equation and has x as the last term.)

10. $x = \dfrac{yp}{z}$ **11.** $x = \frac{1}{3}\sqrt{yp}$ **12.** $x = \sqrt{3wz}$ **13.** $x = \sqrt{6yz}$

14. Construct $\overline{AB}$, with $AB = p$. Divide $\overline{AB}$ into two parts that have the ratio $w:y$.

15. Draw a segment like the one shown and let its length be 1. Use the segment to construct a segment of length $\sqrt{15}$.

16. a. If $x = a\sqrt{n}$, then x is the geometric mean between a and __?__.

  b. Draw a segment about 3 cm long. Call its length a. Use your results from part (a) to construct a segment of length $a\sqrt{n}$ for $n = 2, 3$, and 4.

C **17.** Draw $\overline{CD}$ about 20 cm long. Construct a triangle whose perimeter is equal to CD and whose sides are in the ratio $2:2:3$.

★ **18.** To trisect a general angle G, a student tried this procedure:

  1. Mark off $\overline{GA}$ congruent to $\overline{GB}$.

  2. Draw $\overline{AB}$.

  3. Divide $\overline{AB}$ into three congruent parts using Construction 12.

  4. Draw $\overline{GX}$ and $\overline{GY}$.

  Show that the student did not trisect $\angle G$. (*Hint:* Show that $GA > GY$. Then use an indirect proof to show that $m\angle 2 \neq m\angle 1$.)

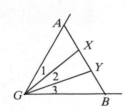

Career

Accountant

An accountant is a financial expert. Accountants study and analyze a company's or an organization's overall financial dealings. They prepare many different kinds of financial reports. These include profit-and-loss statements, which summarize the company's earnings for a given period of time, and balance sheets, which state the current net worth of the company. Another important accounting function is the preparation of tax reports and statements. A company's owners or managers rely on the accountant's reports

to determine whether the company is operating efficiently and profitably, and where improvements can be made.

Accountants are usually college graduates with a major in accounting. They often begin work as junior accountants. After sufficient work experience they take an examination in order to become certified. Upon passing this examination, an accountant becomes a certified public accountant, or CPA. Further

career advancement may lead to promotion to senior accountant, specializing in areas such as cost accounting or auditing. Advancement in a different direction may lead to a financial policy position such as that of controller or to starting an independent accounting company.

Self-Test 2

1. Draw a large $\odot O$. Choose a point A that is outside $\odot O$. Construct the two tangents to $\odot O$ from point A.

2. Draw a very large obtuse triangle. Construct the inscribed circle.

3. Draw a segment about half as long as the width of your paper. Then divide the segment by construction into two segments whose lengths have the ratio $2:1$.

4. Draw a large $\triangle ABC$. Then construct $\overline{DE}$ such that $\dfrac{AB}{BC} = \dfrac{AC}{DE}$.

5. Use $\triangle ABC$ drawn in Exercise 4 to construct a segment, $\overline{PQ}$, whose length is the geometric mean of AB and AC.

6. You are given $\odot S$ and diameter $\overline{PG}$. To construct parallel tangents to $\odot S$, you could construct a line that is ___?___ to $\overline{PG}$ at ___?___ and a line that is ___?___ to $\overline{PG}$ at ___?___.

7. You are given $\triangle TRI$. Describe the steps you would use to circumscribe a circle about $\triangle TRI$.

free thinker
…loli…

Locus

Objectives

1. Describe the locus that satisfies a given condition.
2. Describe the locus that satisfies more than one given condition.
3. Apply the concept of locus in the solution of construction exercises.

10-6 *The Meaning of Locus*

A radar system is used to determine the position, or *locus*, of airplanes relative to an airport. In geometry **locus** means a figure that is the set of all points, and only those points, that satisfy one or more conditions.

Suppose we have a line *k* in a plane and wish to picture the locus of points in the plane that are 1 cm from *k*. Several points are shown in the first diagram below.

All the points satisfying the given conditions are indicated in the next diagram. You see that the required locus is a pair of lines parallel to, and 1 cm from, *k*.

Suppose we wish to picture the locus of points 1 cm from *k* without requiring the points to be *in a plane*. The problem changes. Now you need to consider all the points in space that are 1 cm from line *k*. The required locus is a cylindrical surface with axis *k* and a 1 cm radius, as shown below. Of course, the surface will extend in both directions without end, just as line *k* does.

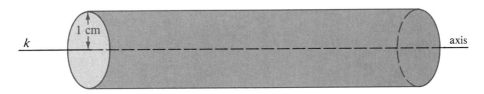

When you are solving a locus problem, always think in terms of three dimensions unless the statement of the problem restricts the locus to a plane.

Classroom Exercises

1. Draw a point *A* on the chalkboard.
   a. Draw several points on the chalkboard that are 20 cm from *A*.
   b. Draw all the points on the chalkboard that are 20 cm from *A*.
   c. Complete: The locus of all points on the chalkboard that are 20 cm from point *A* is __?__.
   d. Remove the restriction that the points must lie in the plane of the chalkboard. Now describe the locus.

2. Draw two parallel lines *k* and *l*.
   a. Draw several points that are in the plane containing *k* and *l* and are equidistant from *k* and *l*.
   b. Draw all the points that are in the plane containing *k* and *l* and are equidistant from *k* and *l*.
   c. Describe the locus of points that are in the plane of two parallel lines and equidistant from them.
   d. Remove the restriction that the points must lie in the plane of the two lines. Now describe the locus.

3. Draw an angle.
   a. Draw several points in the plane of the angle that are equidistant from the sides of the angle.
   b. Draw all the points in the plane of the angle that are equidistant from the sides of the angle.
   c. Describe the locus of points in the plane of a given angle that are equidistant from the sides of the angle.

4. What is the locus of points in your classroom that are equidistant from the ceiling and floor?

5. What is the locus of points in your classroom that are 1 m from the floor?

6. Choose a point *P* on the floor of the classroom.
   a. What is the locus of points, on the floor, that are 1 m from *P*?
   b. What is the locus of points, in the room, that are 1 m from *P*?

7. What is the locus of points in your classroom that are equidistant from the ceiling and floor and are also equidistant from two opposite side walls?

8. What is the locus of points in your classroom that are equidistant from the front and back walls and are also equidistant from the two side walls?

9. Describe the locus of points on a football field that are equidistant from the two goal lines.

10. Draw a circle with radius 6 cm. Use the following definition of *distance from a circle*: A point *P* is *x* cm from a circle if there is a point of the circle that is *x* cm from *P* but there is no point of the circle that is less than *x* cm from *P*.
    a. Draw all the points in the plane of the circle that are 2 cm from the circle.
    b. Complete: Given a circle with a 6 cm radius, the locus of all points in the plane of the circle and 2 cm from the circle is __?__.
    c. Remove the restriction that the points must lie in the plane of the circle. Now describe the locus.

11. Make up a locus problem for which the locus contains exactly one point.

12. Make up a locus problem for which the locus doesn't contain any points.

Written Exercises

Exercises 1–4 deal with figures in a plane. Draw a diagram showing the locus. Then write a description of the locus.

A

1. Given two points A and B, what is the locus of points equidistant from A and B?

2. Given a point P, what is the locus of points 2 cm from P?

3. Given a line h, what is the locus of points 2 cm from h?

4. Given $\odot O$, what is the locus of the midpoints of all radii of $\odot O$?

In Exercises 5–8 begin each exercise with a square $ABCD$ that has sides 4 cm long. Draw a diagram showing the locus of points on or inside the square that satisfy the given conditions. Then write a description of the locus.

5. Equidistant from $\overline{AB}$ and $\overline{CD}$

6. Equidistant from points B and D

7. Equidistant from $\overline{AB}$ and $\overline{BC}$

8. Equidistant from all four sides

Exercises 9–12 deal with figures in space.

9. Given two parallel planes, what is the locus of points equidistant from the two planes?

10. Given a plane, what is the locus of points 5 cm from the plane?

11. Given point E, what is the locus of points 3 cm from E?

12. Given points C and D, what is the locus of points equidistant from C and D?

Exercises 13–17 deal with figures in a plane. (*Note*: If a point in a segment or in an arc is not included in the locus, indicate the point by an open dot.)

B

13. **a.** Draw an angle HEX. Construct the locus of points equidistant from the sides of $\angle HEX$.

   b. Draw two intersecting lines j and k. Construct the locus of points equidistant from j and k.

14. Draw a segment $\overline{DE}$ and a line n. Construct the locus of points whose distance from n is DE.

15. Draw a segment $\overline{AB}$. Construct the locus of points P such that $\angle APB$ is a right angle.

16. Draw a segment $\overline{CD}$. Construct the locus of points Q such that $\triangle CQD$ is isosceles with base $\overline{CD}$.

17. Draw a segment $\overline{EF}$. Construct the locus of points G such that $\triangle EFG$ is isosceles with leg $\overline{EF}$.

Exercises 18–20 deal with figures in space.

18. Given a sphere, what is the locus of the midpoints of the radii of the sphere?

19. Given a square, what is the locus of points equidistant from the sides?

20. Given a scalene triangle, what is the locus of points equidistant from the vertices?

C **21.** A ladder leans against a house. As *A* moves up or down on the wall, *B* moves along the ground. What path is followed by midpoint *M*? (*Hint:* Experiment with a meter stick, a wall, and the floor.)

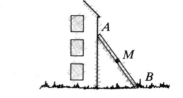

22. Given a segment $\overline{CD}$, what is the locus in space of points *P* such that $m\angle CPD = 90$?

23. A goat is tied to a square shed as shown. Using the scale 1:100, carefully draw a diagram that shows the region over which the goat can graze.

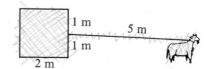

24. A tight wire $\overline{AC}$ is stretched between the tops of two vertical posts $\overline{AB}$ and $\overline{CD}$ that are 5 m apart and 2 m high. A ring, at one end of a 6 m leash, can slide along $\overline{AC}$. A dog is tied to the other end of the leash. Draw a diagram that shows the region over which the leashed dog can roam. Use the scale 1:100.

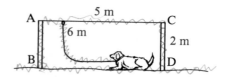

10-7 *Locus Problems*

The plural of *locus* is *loci*. The following problem involves intersections of loci.

Suppose you are given three noncollinear points, *A*, *B*, and *C*. In the plane of *A*, *B*, and *C*, what is the locus of points that are 1 cm from *A* and are, at the same time, equidistant from *B* and *C*?

You can analyze one part of the problem at a time.

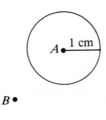

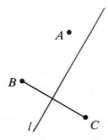

The locus of points 1 cm from *A* is $\odot A$ with radius 1 cm.

The locus of points equidistant from *B* and *C* is *l*, the perpendicular bisector of $\overline{BC}$.

The locus of points satisfying *both* conditions given on the previous page must lie on both circle *A* and line *l*. There are three possibilities, depending on the positions of *A*, *B*, and *C*, as shown below.

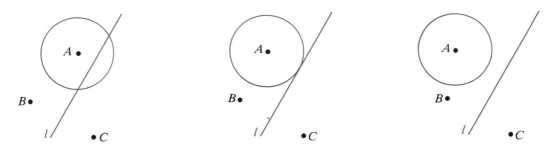

All three can be described in one sentence:

The locus is two points, one point, or no points, depending on the intersection of the circle with center *A* and radius 1 cm and the line that is the perpendicular bisector of $\overline{BC}$.

The example that follows deals with the corresponding problem in three dimensions.

Example   Given three noncollinear points *A*, *B*, and *C*, what is the locus of points 1 cm from *A* and equidistant from *B* and *C*?

Solution

The first locus is sphere *A* with radius 1 cm.

The second locus is plane *P*, the perpendicular bisector of $\overline{BC}$.

Possibilities:

The plane might cut the sphere in a circle.
The plane might be tangent to the sphere.
The plane might not have any points in common with the sphere.

Thus, the locus is a circle, one point, or no points, depending on the intersection of the sphere with center *A* and radius 1 cm and the plane which is the perpendicular bisector of $\overline{BC}$.

Classroom Exercises

Exercises 1–4 refer to coplanar figures. Describe the possible intersections of the figures named.

1. A line and a circle

2. Two circles

3. Two parallel lines and a circle

4. Two perpendicular lines and a circle

5. Consider the following problem: In a plane, what is the locus of points that are equidistant from the sides of $\angle A$ and are equidistant from two points B and C?
   a. The locus of points equidistant from the sides of $\angle A$ is __?__.
   b. The locus of points equidistant from B and C is __?__.
   c. Draw diagrams to show three possibilities with regard to points that satisfy both conditions (a) and (b).
   d. Describe the locus.

Exercises 6–9 refer to figures in space. Describe the possible intersections of the figures named.

6. A line and a plane

7. A line and a sphere

8. Two spheres

9. A plane and a sphere

10. Let C be the point in the center of your classroom (*not* the center of the floor). Describe the locus of points in the room that satisfy the given conditions.
    a. 3 m from C
    b. 3 m from C and equidistant from the ceiling and the floor
    c. 3 m from C and 1 m from either the ceiling or the floor

Written Exercises

Exercises 1 and 2 refer to plane figures.

A 1. Draw a new $\odot O$ for each part. Then place two points A and B outside $\odot O$ so that the locus of points on $\odot O$ and equidistant from A and B is:
   a. 2 points
   b. 0 points
   c. 1 point

2. Draw two parallel lines m and n. Then place two points R and S so that the locus of points equidistant from m and n and also equidistant from R and S is:
   a. 1 point
   b. 1 line
   c. 0 points

Exercises 3 and 4 refer to plane figures.

3. Consider the following problem: Given two points *D* and *E*, what is the locus of points 1 cm from *D* and 2 cm from *E*?
   a. The locus of points 1 cm from *D* is __?__.
   b. The locus of points 2 cm from *E* is __?__.
   c. Draw diagrams to show three possibilities with regard to points that satisfy both conditions (a) and (b).
   d. Give a one-sentence solution to the problem.

4. Consider the following problem: Given a point *A* and a line *k*, what is the locus of points 3 cm from *A* and 1 cm from *k*?
   a. The locus of points 3 cm from *A* is __?__.
   b. The locus of points 1 cm from *k* is __?__.
   c. Draw diagrams to show five possibilities with regard to points that satisfy both conditions (a) and (b).
   d. Give a one-sentence solution to the problem.

Exercises 5–10 refer to plane figures. Draw a diagram of the locus. Then write a description of the locus.

5. Point *P* lies on line *l*. What is the locus of points on *l* and 3 cm from *P*?

6. Point *Q* lies on line *l*. What is the locus of points 5 cm from *Q* and 3 cm from *l*?

7. Points *A* and *B* are 3 cm apart. What is the locus of points 2 cm from both *A* and *B*?

8. Lines *j* and *k* intersect in point *P*. What is the locus of points equidistant from *j* and *k*, and 2 cm from *P*?

9. Given ∠*A*, what is the locus of points equidistant from the sides of ∠*A* and 2 cm from vertex *A*?

10. Given △*RST*, what is the locus of points equidistant from $\overline{RS}$ and $\overline{RT}$ and also equidistant from *R* and *S*?

In Exercises 11–14 draw diagrams to show the possibilities with regard to points in a plane.

B 11. Given points *C* and *D*, what is the locus of points 2 cm from *C* and 3 cm from *D*?

12. Given point *E* and line *k*, what is the locus of points 3 cm from *E* and 2 cm from *k*?

13. Given a point *A* and two parallel lines *j* and *k*, what is the locus of points 30 cm from *A* and equidistant from *j* and *k*?

14. Given four points *P*, *Q*, *R*, and *S*, what is the locus of points that are equidistant from *P* and *Q* and equidistant from *R* and *S*?

Exercises 15–19 refer to figures in space. In each exercise tell what the locus is. You need not draw the locus or describe it precisely.

Example Given two parallel planes and a point *A*, what is the locus of points equidistant from the planes and 3 cm from *A*?

Solution The locus is a circle, a point, or no points.

15. Given plane *Z* and point *B* outside *Z*, what is the locus of points in *Z* that are 3 cm from *B*?

16. Given plane *Y* and point *P* outside *Y*, what is the locus of points 2 cm from *P* and 2 cm from *Y*?

17. Given $\overleftrightarrow{AB} \perp$ plane *Q*, what is the locus of points 2 cm from $\overleftrightarrow{AB}$ and 2 cm from *Q*?

18. Given square *ABCD*, what is the locus of points equidistant from the vertices of the square?

19. Given point *A* in plane *Z*, what is the locus of points 5 cm from *A* and *d* cm from *Z*? (More than 1 possibility)

20. Given three points, each 2 cm from the other two, draw a diagram to show the locus of points that are in the plane of the given points and are not more than 2 cm away from any of them.

21. Points *R*, *S*, *T*, and *W* are not coplanar and no three of them are collinear.
    a. The locus of points equidistant from *R* and *S* is __?__.
    b. The locus of points equidistant from *R* and *T* is __?__.
    c. The loci found in parts (a) and (b) intersect in a __?__, and all points in this __?__ are equidistant from points *R*, *S*, and *T*.
    d. The locus of points equidistant from *R* and *W* is __?__.
    e. The intersection of the figures found in (c) and (d) is a __?__. This __?__ is equidistant from the four given points.

C 22. Can you locate four points *J*, *K*, *L*, and *M* so that the locus of points equidistant from *J*, *K*, *L*, and *M* is named below? If the answer is *yes*, describe the location of the points *J*, *K*, *L*, and *M*.
    a. a point
    b. a line
    c. a plane
    d. no points

23. Assume that the Earth is a sphere. How many points are there on the Earth's surface that are equidistant from
    a. Houston and Toronto?
    b. Houston, Toronto, and Los Angeles?
    c. Houston, Toronto, Los Angeles, and Mexico City?

24. A mini-radio transmitter has been secured to a bear. Rangers D, E, and F are studying the bear's movements. Rangers D and E can receive the bear's beep at distances up to 10 km, ranger F at distances up to 15 km.

Draw a diagram showing where the bear might be at these times:

a. When all three rangers can receive the signal

b. When ranger F suddenly detects the signal after a period of time during which only rangers D and E could receive the signal

c. When ranger D is off duty, and ranger F begins to detect the signal just as ranger E loses it

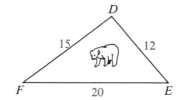

Challenge

Given $\overline{AB}$, it is possible to construct the midpoint M of $\overline{AB}$ using only a compass (and *no* straightedge). Study the diagram until you understand the procedure. Then draw $\overline{AB}$, about 10 cm long, construct its midpoint M as shown, and prove that M is the midpoint.

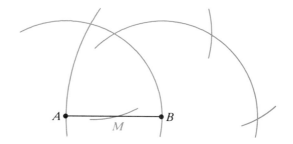

10-8 *Locus and Construction*

Sometimes the solution to a construction problem depends on finding a point that satisfies more than one condition. To locate the point, you may have to begin by constructing a locus of points satisfying *one* of the conditions.

Example   Given the angle and the segments shown, construct $\triangle ABC$ with $m \angle A = n$, $AB = r$, and the altitude to $\overleftrightarrow{AB}$ having length s.

Solution   It is easy to construct $\angle A$ and side $\overline{AB}$. Point C must satisfy two conditions: C must lie on $\overrightarrow{AZ}$, and C must be s units from $\overleftrightarrow{AB}$. The locus of points s units from $\overleftrightarrow{AB}$ is a pair of parallel lines. Only the upper parallel will intersect $\overrightarrow{AZ}$. We construct that parallel to $\overleftrightarrow{AB}$ as follows:

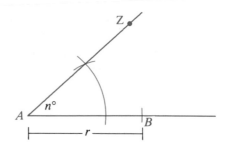

1. Construct the perpendicular to $\overleftrightarrow{AB}$ at any convenient point X.

2. Mark off s units on the perpendicular to locate point Y.

3. Construct the perpendicular to $\overrightarrow{XY}$ at Y. Call it $\overleftrightarrow{YW}$.

Note that all points on $\overleftrightarrow{YW}$ are s units from $\overleftrightarrow{AB}$. Thus the intersection of $\overleftrightarrow{YW}$ and $\overrightarrow{AZ}$ is the desired point C. To complete the solution, we simply draw $\overline{CB}$.

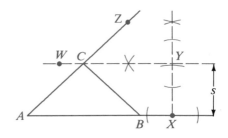

Classroom Exercises

1. The purpose of this exercise is to analyze the following construction problem:

   Given a circle and a segment with length k, inscribe in the circle an isosceles triangle RST with base $\overline{RS}$ k units long.

   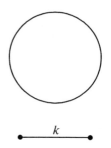

   a. Suppose R has been chosen. Where must S lie so that RS equals k? (In other words, what is the locus of points k units from R?)

   b. Now suppose $\overline{RS}$ has been drawn. Where must T lie so that $RT = ST$? (In other words, what is the locus of points equidistant from R and S?)

   c. Explain the steps of the construction shown.

   (1) 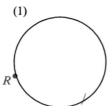   (2) 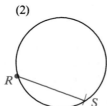   (3)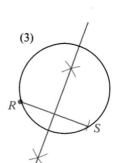

   d. Explain two different ways to finish the construction.

2. Two different solutions, both correct, are shown for the following construction problem. Analyze the diagrams and explain the solutions.

Given segments with lengths *r* and *s*, construct △*ABC* with *m* ∠ *C* = 90, *AC* = *r*, and *AB* = *s*.

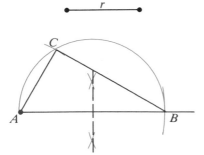

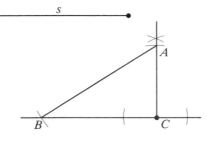

First solution   Second solution

Written Exercises

Exercises 1–4 refer to plane figures.

A   1. Draw any $\overline{AB}$ and a segment with length *h*. Use the following steps to construct the locus of points *P* such that for every △*APB* the altitude from *P* to $\overleftrightarrow{AB}$ would equal *h*.
   a. Construct a perpendicular to $\overline{AB}$.
   b. Construct two lines parallel to $\overline{AB}$, *h* units from $\overline{AB}$.

2. Begin each part of this exercise by drawing any $\overline{CD}$. Then construct the locus of points *P* that meet the given condition.
   a. ∠ *CDP* is a right angle.
   b. ∠ *CPD* is a right angle. (*Hint*: See Classroom Exercise 2.)

On your paper draw a segment roughly as long as the one shown. Use it in Exercises 3 and 4.

3. Draw an angle *XYZ*. Construct a circle, with radius *a*, that is tangent to the sides of ∠ *XYZ*. (*Hint*: The center of the circle will be *a* units from the sides of ∠ *XYZ*.)

4. Draw a figure roughly like the one shown. Then construct a circle, with radius *a*, that passes through *N* and is tangent to line *k*. (*Hint*: Construct the locus of points that would, as centers, be the correct distance from *k*. Also construct the locus of points that would, as centers, be the correct distance from *N*.)

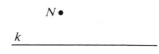

On your paper draw an angle and three segments roughly like those shown. Use them in Exercises 5–19. You may find it helpful to begin with a sketch.

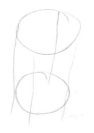

5. Construct $\overline{AB}$ so that $AB = t$. Then construct the locus of all points C so that in $\triangle ABC$ the altitude from C has length r.

6. Construct $\overline{AB}$ so that $AB = t$. Then construct the locus of all points C so that in $\triangle ABC$ the median from C has length s.

B 7. Construct isosceles $\triangle ABC$ so that $AB = AC = t$ and so that the altitude from A has length s.

8. Construct an isosceles trapezoid $ABCD$ with $\overline{AB}$ the shorter base, with $AB = AD = BC = t$, and with an altitude of length r.

9. Construct $\triangle ABC$ so that $AB = t$, $AC = s$, and the median to $\overline{AB}$ has length r.

10. Construct $\triangle ABC$ so that $m\angle A = m\angle B = n$, and the altitude to $\overline{AB}$ has length s.

11. Construct $\triangle ABC$ so that $m\angle C = 90$, $m\angle A = n$, and the altitude to $\overline{AB}$ has length s.

12. Construct $\triangle ABC$ so that $AB = s$, $AC = t$, and the altitude to $\overline{AB}$ has length r.

13. Construct $\triangle ABC$ so that $AB = t$, and the median to $\overline{AB}$ and the altitude to $\overline{AB}$ have lengths s and r, respectively.

14. Construct a right triangle such that the altitude to the hypotenuse and the median to the hypotenuse have lengths r and s, respectively.

15. Construct both an acute isosceles triangle and an obtuse isosceles triangle such that each leg has length s and each altitude to a leg has length r.

C 16. Construct a square whose sides each have length $4s$. A segment of length $3s$ moves so that its endpoints are always on the sides of the square. Construct the locus of the midpoint of the moving segment.

17. Construct a right triangle such that the bisector of the right angle divides the hypotenuse into segments whose lengths are r and s.

18. Construct an isosceles right triangle such that the radius of the inscribed circle is r.

19. Construct $\overline{AB}$ so that $AB = t$. Then construct the locus of points P such that $m\angle APB = n$.

Self-Test 3

Describe briefly the locus of points that satisfy the conditions.

1. In the plane of two intersecting lines *j* and *k*, and equidistant from the lines

2. In space and *t* units from point *P*

3. In space and equidistant from points *W* and *X* that are 10 cm apart

4. In the plane of ∠*DEF*, equidistant from the sides of the angle, and 4 cm from $\overrightarrow{EF}$

5. In the plane of two parallel lines *s* and *t*, equidistant from *s* and *t*, and 4 cm from a particular point *A* in the plane (three possibilities)

6. Construct a large isosceles △*RST*. Then construct the locus of points that are equidistant from the vertices of △*RST*.

7. Draw a long segment, $\overline{BC}$, and an acute angle, ∠1. Construct a right triangle with an acute angle congruent to ∠1 and hypotenuse congruent to $\overline{BC}$.

Extra   *The Nine-Point Circle*

Given any △*ABC*, let *H* be the intersection of the three altitudes. There is a circle that passes through these nine special points:

> midpoints *L*, *M*, *N* of the three sides
>
> points *R*, *S*, *T*, where the three altitudes of the triangle meet the sides
>
> midpoints *X*, *Y*, *Z* of $\overline{HA}$, $\overline{HB}$, $\overline{HC}$

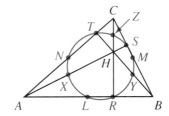

Key steps of proof:

1. *XYMN* is a rectangle.

2. The circle circumscribed about *XYMN* has diameters $\overline{MX}$ and $\overline{NY}$.

3. Because ∠*XSM* and ∠*YTN* are right angles, the circle contains points *S* and *T* as well as *X*, *Y*, *M*, and *N*.

4. *XLMZ* is a rectangle.

5. The circle circumscribed about *XLMZ* has diameters $\overline{MX}$ and $\overline{LZ}$.

6. Because ∠*XSM* and ∠*ZRL* are right angles, the circle contains points *S* and *R* as well as *X*, *L*, *M*, and *Z*.

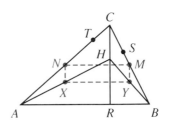

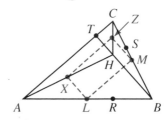

7. The circle of Steps 1–3 and the circle of Steps 4–6 must be the same circle, because $\overline{MX}$ is a diameter of both circles.

8. There is a circle that passes through the nine points, L, M, N, R, S, T, X, Y, and Z. (See Steps 3 and 6.)

One way to locate the center of the circle is to locate points X and M, then the midpoint of $\overline{XM}$.

Exercises

1. Test your mechanical skill by constructing the nine-point circle for an acute triangle. (The larger the figure, the better.)

2. Repeat Exercise 1, but use an obtuse triangle.

3. Repeat Exercise 1, but use an equilateral triangle. What happens to some of the nine points?

4. Repeat Exercise 1, but use a right triangle. How many of the nine points are at the vertex of the right angle?

5. Prove that $XYMN$ is a rectangle. Use the diagram shown for Steps 1–3 of the key steps of proof. (*Hint*: Compare $\overline{NM}$ with $\overline{AB}$ and $\overline{NX}$ with $\overline{CR}$.)

6. What is the ratio of the radius of the nine-point circle to the radius of the circumscribed circle?

Chapter Summary

1. Geometric constructions are diagrams that are drawn using only a straightedge and a compass.

2. Basic constructions:
    (1) A segment congruent to a given segment, page 375
    (2) An angle congruent to a given angle, page 376
    (3) The bisector of a given angle, page 376
    (4) The perpendicular bisector of a given segment, page 380
    (5) A line perpendicular to a given line at a given point on the line, page 381
    (6) A line perpendicular to a given line from a given point outside the line, page 381
    (7) A line parallel to a given line through a given point outside the line, page 382
    (8) A tangent to a given circle at a given point on the circle, page 392
    (9) A tangent to a given circle from a given point outside the circle, page 393
    (10) A circle circumscribed about a given triangle, page 393
    (11) A circle inscribed in a given triangle, page 394

(12) Division of a given segment into any number of congruent parts, page 396

(13) A segment of length x such that $\dfrac{a}{b} = \dfrac{c}{x}$ when segments of lengths a, b, and c are given, page 397

(14) A segment whose length is the geometric mean between the lengths of two given segments, page 397

3. Every triangle has these concurrency properties:
    (1) The bisectors of the angles intersect in a point that is equidistant from the three sides of the triangle.
    (2) The perpendicular bisectors of the sides intersect in a point that is equidistant from the three vertices of the triangle.
    (3) The lines that contain the altitudes intersect in a point.
    (4) The medians intersect in a point that is two thirds of the distance from each vertex to the midpoint of the opposite side.

4. A locus is the set of all points, and only those points, that satisfy one or more conditions.

5. A locus that satisfies more than one condition is found by considering all possible intersections of the loci for the separate conditions.

Chapter Review

In Exercises 1–3 draw a diagram that is similar to, but larger than, the one shown. Then do the constructions.

1. Draw any line m. On m construct $\overline{ST}$ such that $ST = 3XY$.
2. Construct an angle with measure equal to $m \angle X + m \angle Z$.
3. Bisect $\angle Y$.

10–1

Use a diagram like the one below for Exercises 4–7.

4. Construct the perpendicular bisector of $\overline{AB}$.
5. Construct the perpendicular to $\overleftrightarrow{AC}$ at C.
6. Construct the perpendicular to $\overleftrightarrow{AC}$ from D.
7. Construct the parallel to $\overleftrightarrow{AC}$ through E.

10–2

8. The __?__ of a triangle intersect in a point that is equidistant from the vertices of the triangle.

9. The __?__ of a triangle intersect in a point that is equidistant from the sides of the triangle.

10. If $MR = 12$, then $MP = $ __?__.

11. $QR:RO = $ __?__ (numerical answer)

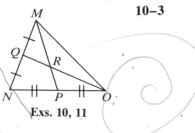

Exs. 10, 11

10–3

Draw a large ⊙O. Label a point F on ⊙O and a point G outside ⊙O.

12. Construct the tangent to ⊙O at F.   10–4

13. Construct a tangent to ⊙O from G.

14. Draw a large acute triangle. Find, by construction, the center of the circle that could be inscribed in the triangle.

15. Draw a large obtuse triangle. Construct a circle that circumscribes the triangle.

Draw segments about as long as those shown below. In each exercise, construct a segment with the required length t.

16. $t^2 = bc$ **17.** $at = bc$ **18.** $t = \frac{1}{3}(a + b)$ 10–5

19. Given two parallel lines l and m, what is the locus of points in their plane and equidistant from them?   10–6

20. Given two points A and B, what is the locus of points, in space, equidistant from A and B?

21. What is the locus of points in space equidistant from two parallel planes?

22. What is the locus of points in space that are equidistant from the vertices of equilateral $\triangle HJK$?

23. Points P and Q are 6 cm apart. What is the locus of points in a plane that are equidistant from P and Q and are 8 cm from P? Sketch the locus.   10–7

24. Point R is on line l. What is the locus in space of points that are 8 cm from l and 8 cm from R?

25. What is the locus of points in space that are 1 m from plane Q and 2 m from point Z not in Q? (There is more than one possibility.)

Use the segments with lengths a, b, and c that you drew for Exercises 16–18.

26. Construct an isosceles right triangle with hypotenuse of length a.   10–8

27. Construct a $\triangle RST$ with $RS = a$, $RT = c$, and the median to $\overline{RS}$ of length b.

Chapter Test

Begin by drawing segments and an angle roughly like those shown.

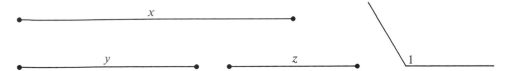

1. Construct an isosceles triangle with vertex angle congruent to ∠ 1 and legs of length z.
2. Construct a 30°-60°-90° triangle with shorter leg of length y.
3. Construct a segment of length $\sqrt{xy}$.
4. Construct a segment of length $\frac{2}{3}(y + 2z)$.
5. Construct a segment of length n such that $\dfrac{x}{z} = \dfrac{y}{n}$.
6. Draw a large circle and a point K not on the circle. Using K as one vertex, construct any triangle that is circumscribed about the circle.
7. Draw a large triangle and construct the circle inscribed in the triangle.
8. In a right triangle **(a)** the __?__ of the triangle intersect at a point on the hypotenuse, **(b)** the __?__ intersect at a point inside the triangle, and **(c)** the altitudes of the triangle intersect at a __?__ of the triangle.
9. An isosceles triangle has sides of length 5, 5, and 8.
   a. What is the length of the median to the base?
   b. When the three medians are drawn, the median to the base is divided into segments with lengths __?__ and __?__.
10. Given points R and S in plane Z, what is the locus of points **(a)** in Z and equidistant from R and S and **(b)** in space and equidistant from R and S?
11. Given points T and U 8 units apart, what is the locus of points, in space, that are 6 units from T and 4 units from U?
12. Draw a line l and a point A on it. Using y and z from Exercises 1–5, construct the locus of points z units from l and y units from A.

Algebra Review: *Evaluating Formulas*

Evaluate each expression for the given values of the variables.

Example $\frac{1}{2}bh$ when $b = 12$ and $h = 6\sqrt{3}$ **Solution** $\frac{1}{2}(12)(6\sqrt{3}) = 36\sqrt{3}$

1. Area of a square: s^2 when $s = 1.3$
2. Length of hypotenuse of a right triangle: $\sqrt{a^2 + b^2}$ when $a = 15$ and $b = 20$
3. Perimeter of parallelogram: $2x + 2y$ when $x = \frac{5}{3}$ and $y = \frac{3}{2}$
4. Perimeter of triangle: $a + b + c$ when $a = 11.5$, $b = 7.2$, and $c = 9.9$
5. Area of a rectangle: lw when $l = 2\sqrt{6}$ and $w = 3\sqrt{3}$
6. Perimeter of isosceles trapezoid: $2r + s + t$ when $r = \frac{4}{7}$, $s = 1$, and $t = \frac{13}{7}$

7. πr^2 when $r = 30$ (Use 3.14 for π.)   8. lwh when $l = 8$, $w = 6\frac{1}{4}$, and $h = 3\frac{1}{2}$

9. $2(lw + wh + lh)$ when $l = 4.5, w = 3$, and $h = 1$ 10. $\dfrac{x - 3}{y + 2}$ when $x = 3$ and $y = -4$

11. $\dfrac{x + 5}{y - 2}$ when $x = -2$ and $y = -4$ 12. $mx + b$ when $x = -6$, $m = \frac{5}{2}$, and $b = -2$

13. $6t^2$ when $t = 3$ 14. $(6t)^2$ when $t = 3$
15. $\frac{1}{2}h(a + b)$ when $h = 3$, $a = 3\sqrt{2}$, and $b = 7\sqrt{2}$ 16. $\sqrt{(x - 5)^2 + (y - 3)^2}$ when $x = 1$ and $y = 0$
17. $\frac{1}{3}x^2h$ when $x = 4\sqrt{3}$ and $h = 6$ 18. $2s^2 + 4sh$ when $s = \sqrt{6}$ and $h = \frac{5}{2}\sqrt{6}$

Use the given information to rewrite each expression.

Example Bh when $B = \frac{1}{2}rs$ **Solution** $Bh = (\frac{1}{2}rs)h = \frac{1}{2}rsh$

19. $c(x + y)$ when $x + y = d$ 20. $\frac{1}{3}Bh$ when $B = \pi r^2$ 21. $\frac{1}{2}pl$ when $p = 2\pi r$
22. $2(l + w)$ when $l = s$ and $w = s$ 23. $4\pi r^2$ when $r = \frac{1}{2}d$ 24. $n(\frac{1}{2}sa)$ when $ns = p$

Solve each formula for the variable shown in color.

Example $y = mx + b$ **Solution** $y - b = mx$;
$$x = \frac{y - b}{m}, \ m \neq 0$$

25. $ax + by = c$ 26. $C = \pi d$ 27. $S = (n - 2)180$ 28. $x^2 + y^2 = r^2$
29. $\dfrac{x}{h} = \dfrac{h}{y}$ 30. $a^2 + b^2 = (a\sqrt{2})^2$ 31. $A = \frac{1}{2}bh$ 32. $m = \dfrac{y + 4}{x - 2}$

Preparing for College Entrance Exams

Strategy for Success
Often the answer to a question can be found by writing an equation or inequality and solving it. When a complete solution is time-consuming, you may find that the fastest way to answer the question is to test the suggested answers in your equation or inequality.

Indicate the best answer by writing the appropriate letter.

1. $\overline{AB}$ and $\overline{AC}$ are tangent to $\odot O$ at B and C. If $m\overset{\frown}{BC} = x$, then $m\angle BAC =$
   (A) x **(B)** $180 - x$ **(C)** $360 - x$ **(D)** $180 + x$ **(E)** $\frac{1}{2}x$

2. If quadrilateral $JKLM$ is inscribed in a circle and $\angle J$ and $\angle K$ are supplementary angles, then $\angle J$:
   (A) must be congruent to $\angle L$ **(B)** must be a right angle
   (C) must be congruent to $\angle M$ **(D)** must be an acute angle
   (E) must be supplementary to $\angle M$

3. In $\odot M$, chords $\overline{RS}$ and $\overline{TU}$ intersect at X. If $RX = 15$, $XS = 18$, and $TX:XU = 3:10$, then $XU =$
   (A) 3   **(B)** 9   **(C)** $20\frac{10}{13}$ **(D)** $25\frac{5}{13}$ **(E)** 30

4. If $m\overset{\frown}{XW} = 60$, $m\overset{\frown}{WZ} = 70$, and $m\overset{\frown}{ZY} = 70$, then $m\angle 1 =$
   (A) 45   **(B)** 50   **(C)** 60   **(D)** 65   **(E)** 70

5. If $VW = 10$, $WX = 6$, and $VZ = 8$, then $ZY =$
   (A) 4.8   **(B)** 12   **(C)** 7.5   **(D)** 20   **(E)** 16

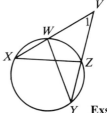

Exs. 4, 5

6. Given $\triangle ABC$, you can find the locus of points in the plane of $\triangle ABC$ and equidistant from $\overline{AB}$, $\overline{BC}$, and $\overline{AC}$ by constructing:
   (A) two medians   **(B)** two altitudes   **(C)** two angle bisectors
   (D) the perpendicular bisectors of two sides   **(E)** the circumscribed circle

7. To construct a tangent to $\odot R$ from a point S outside $\odot R$, you need to construct:
   (A) the perpendicular bisector of $\overline{RS}$
   (B) a perpendicular to $\overline{RS}$ at the point where $\overline{RS}$ intersects $\odot R$
   (C) a diameter that is perpendicular to $\overline{RS}$
   (D) a perpendicular to $\overline{RS}$ through point S
   (E) a 30°-60°-90° triangle with vertex S

8. The locus of points 6 cm from plane P and 10 cm from a given point J *cannot* be:
   (A) no points   **(B)** one point   **(C)** a line   **(D)** a circle   **(E)** two circles

9. The locus of the midpoints of all 8 cm chords in a circle of radius 5 cm is:
   (A) a point   **(B)** a segment   **(C)** a line   **(D)** a ray   **(E)** a circle

Cumulative Review: Chapters 1–10

Write *always*, *sometimes*, or *never* to complete each statement.

A **1.** A quadrilateral __?__ has four obtuse angles.

2. Two isosceles right triangles with congruent hypotenuses are __?__ congruent.

3. If $\overarc{AC}$ on $\odot O$ and $\overarc{BD}$ on $\odot P$ have the same measure, then $\overarc{AC}$ is __?__ congruent to $\overarc{BD}$.

4. If two consecutive sides of a parallelogram are perpendicular, then the diagonals are __?__ perpendicular.

5. If the lengths of the sides of two triangles are in proportion, then the corresponding angles are __?__ congruent.

6. The tangent of an angle is __?__ greater than 1.

7. A triangle with sides of length $2x$, $3x$, and $4x$, with $x > 0$, is __?__ acute.

8. Given a plane containing points A and B, the locus of points in the plane that are equidistant from A and B and are 10 cm from A is __?__ one point.

Complete each statement in Exercises 9–12.

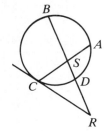

9. If $m\overarc{AB} = 80$, $m\overarc{CD} = 66$, and $m\overarc{DA} = 70$, then $m\angle ASD = $ __?__.

10. If $BS = 12$, $SD = 6$, and $AS = 8$, then $SC = $ __?__.

11. If $RD = 9$ and $DB = 16$, then $RC = $ __?__.

12. If $m\overarc{AB} = 80$, $m\overarc{CD} = 66$, and $m\overarc{DA} = 70$, then $m\angle R = $ __?__.

13. Draw a large $\triangle MNP$. Construct a $\triangle XYZ$ congruent to $\triangle MNP$.

B **14.** Describe the locus of points in space that are 4 cm from plane X and 8 cm from point J.

15. $\triangle DEF$ is a right triangle with hypotenuse $\overline{DF}$. $DE = 6$ and $EF = 8$.
  a. If $\overline{EX} \perp \overline{DF}$ at X, find DX.
  b. If Y lies on $\overline{DF}$ and $\overrightarrow{EY}$ bisects $\angle DEF$, find DY.

16. If each interior angle of a regular polygon has measure 160, how many sides does the polygon have?

17. Given: $\odot O$; $m\angle 1 = 45$
  Prove: $\triangle OPQ$ is a 45°-45°-90° $\triangle$.

18. Use the given diagram to prove that $WX \cdot YV = XV \cdot ZY$.

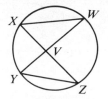

19. Draw $\overline{AB}$. Construct any rectangle with a diagonal congruent to $\overline{AB}$.

11 AREAS OF PLANE FIGURES

Notice how planes are featured in the design of this modern chair. The designer used geometric principles to create an attractive piece of furniture that can be mass-produced from machine-made materials.

Milwaukee Art Museum, Purchase, Layton Art Collection

Areas of Polygons

Objectives

1. Understand what is meant by the area of a polygon.
2. Understand the area postulates.
3. Know and use the formulas for the areas of rectangles, parallelograms, triangles, rhombuses, trapezoids, and regular polygons.

11-1 *Areas of Rectangles*

In everyday conversation people often refer to the *area* of a rectangle when what they really mean is the area of a rectangular region.

Rectangle   Rectangular region

We will continue this common practice to simplify our discussion. Thus, when we speak of the area of a triangle, we will mean the area of the triangular region that includes the triangle *and* its interior.

In Chapter 1 we accepted postulates that enable us to express the lengths of segments and the measures of angles as positive numbers. Similarly, the areas of figures are positive numbers with properties given by area postulates.

Postulate 17

The area of a square is the square of the length of a side. ($A = s^2$)

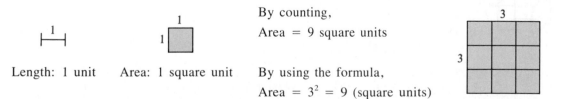

Length: 1 unit   Area: 1 square unit

By counting,
Area = 9 square units

By using the formula,
Area = 3^2 = 9 (square units)

Postulate 18   *Area Congruence Postulate*
If two figures are congruent, then they have the same area.

Postulate 19 *Area Addition Postulate*

The area of a region is the sum of the areas of its non-overlapping parts.

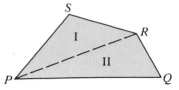

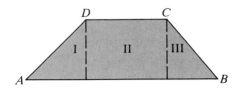

Area of *PQRS* = Area I + Area II   Area of *ABCD* = Area I + Area II + Area III

Any side of a rectangle or other parallelogram can be considered to be a **base.** The length of a base will be denoted by *b*. In this text the term *base* will be used to refer either to the line segment or to its length. An **altitude** to a base is any segment perpendicular to the line containing the base from any point on the opposite side. The length of an altitude is called the **height** (*h*). All the altitudes to a particular base have the same length.

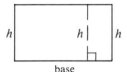 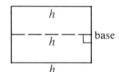 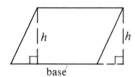 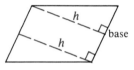

Theorem 11-1

The area of a rectangle equals the product of its base and height. ($A = bh$)

Given: A rectangle with base *b* and height *h*

Prove: $A = bh$

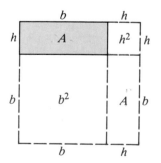

Proof:

Building onto the given shaded rectangle, we can draw a large square consisting of these non-overlapping parts:

> the given rectangle with area A
> a congruent rectangle with area A
> a square with area b^2
> a square with area h^2

Area of big square = $2A + b^2 + h^2$ (Area Addition Postulate)
Area of big square = $(b + h)^2 = b^2 + 2bh + h^2$ ($A = s^2$)
$\quad 2A + b^2 + h^2 = b^2 + 2bh + h^2$ (Substitution Prop.)
$\quad\quad\quad\quad 2A = 2bh$ (Subtraction Prop. of =)
$\quad\quad\quad\quad\ A = bh$ (Division Prop. of =)

Areas are always measured in square units. Some common units of area are the square centimeter (cm^2) and the square meter (m^2). In part (b) of the example below, notice that the unit of length and the unit of area are understood to be "units" and "square units," respectively. It is important to remember that the implied units for length and area are different.

Example   Find the area of each figure.

a. A rectangle with base 3.5 cm and height 2 cm

b.

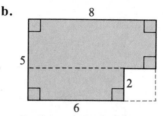

Solution   **a.** $A = 3.5(2) = 7$ (cm^2)

b. *Method 1* (see blue lines) $A = (8 \cdot 5) - (2 \cdot 2) = 40 - 4 = 36$
Method 2 (see red line) $A = (8 \cdot 3) + (6 \cdot 2) = 24 + 12 = 36$

Classroom Exercises

1. Tell what each letter represents in the formula $A = s^2$.

2. Tell what each letter represents in the formula $A = bh$.

3. Find the area and perimeter of a square with sides 5 cm long.

4. The perimeter of a square is 28 cm. What is the area?

5. The area of a square is 64 cm^2. What is the perimeter?

Exercises 6–13 refer to rectangles. Complete the table.

	6.	**7.**	**8.**	**9.**	**10.**	**11.**	**12.**	**13.**
b	8 cm	4 cm	12 m	?	$3\sqrt{2}$	$4\sqrt{2}$	$5\sqrt{3}$	$x + 3$
h	3 cm	1.2 cm	?	5 cm	2	$\sqrt{2}$	$2\sqrt{3}$	x
A	?	?	36 m^2	55 cm^2	?	?	?	?

14. a. What is the converse of the Area Congruence Postulate?
b. Is this converse true or false? Explain.

15. a. Draw three noncongruent rectangles, each with perimeter 20 cm. Find the area of each rectangle.
b. Of all rectangles having perimeter 20 cm, which one do you think has the greatest area? (Give its length and width.)

Written Exercises

Exercises 1–16 refer to rectangles. Complete the tables. p is the perimeter.

A

	1.	2.	3.	4.	5.	6.	7.	8.
b	12 cm	8.2 cm	16 cm	?	$3\sqrt{2}$	$\sqrt{6}$	$2x$	$4k-1$
h	5 cm	4 cm	?	8 m	$4\sqrt{2}$	$\sqrt{2}$	$x-3$	$k+2$
A	?	?	80 cm²	120 m²	?	?	?	?

	9.	10.	11.	12.	13.	14.	15.	16.
b	9 cm	40 cm	16 cm	$x+5$	$a+3$	$k+7$	x	?
h	4 cm	?	?	x	$a-3$	?	?	y
A	?	?	?	?	?	?	x^2-3x	y^2+7y
p	?	100 cm	42 cm	?	?	$4k+20$	?	?

Consecutive sides of the figures are perpendicular. Find the area of each figure.

B 17.

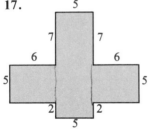

18.

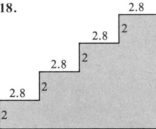

19.

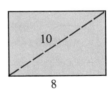

20.

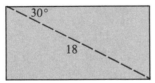

21.

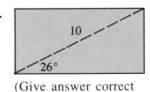

(Give answer correct
to the nearest tenth.)

22.

In Exercises 23 and 24 find each area in terms of the variables.

23.

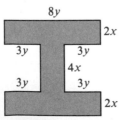

24.

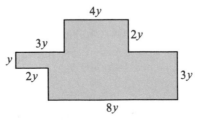

25. Find the area of a square with diagonals of length d.

26. The length of a rectangle is 12 cm more than its width. Find the area of the rectangle if its perimeter is 100 cm.

27. A path 2 m wide surrounds a rectangular garden 20 m long and 12 m wide. Find the area of the path.

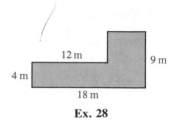

28. How much will it cost to blacktop the driveway shown if blacktopping costs $11.00 per square meter?

Ex. 28

29. A room 28 ft long and 20 ft wide has walls 8 ft high.
  a. What is the total wall area?
  b. How many gallon cans of paint should be bought to paint the walls if 1 gal of paint covers 300 ft^2?

30. A wooden fence 6 ft high and 220 ft long is to be painted on both sides.
  a. What is the total area to be painted?
  b. A gallon of a certain type of paint will cover only 200 ft^2 of area for the first coat, but on the second coat a gallon of the same paint will cover 300 ft^2. If the fence is to be given two coats of paint, how many gallons of paint should be bought?

31. A rectangle having area 392 m^2 is twice as long as it is wide. Find its dimensions.

32. The lengths of the sides of three squares are s, $s + 1$, and $s + 2$. If their total area is 365 cm^2, find their total perimeter.

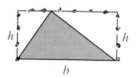

33. Derive a formula for the area of the triangle shown by using the formula for the area of a rectangle.

Ex. 33

34. The diagonals of a rectangle are 18 cm long and intersect at a 60° angle. Find the area of the rectangle.

35. a. Suppose you have 40 m of fencing with which to make a rectangular pen for a dog. If one side of the rectangle is x m long, explain why the other side is $(20 - x)$ m long.

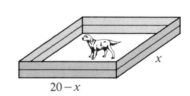

  b. Express the area of the pen in terms of x.
  c. Find the area of the pen for each value of x: 0, 2, 4, 6, 8, 10, 12, 14, 16, 18, 20. Record your answers on a set of axes like the one shown.
  d. Give the dimensions of the pen with the greatest area.

C **36.** A farmer has 100 m of fencing with which to make a rectangular corral. A side of a barn will be used as one side of the corral, as shown in the overhead view.

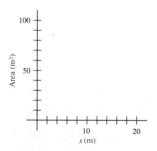

  a. If the width of the corral is x, express the length and the area in terms of x.
  b. Make a graph showing values of x on the horizontal axis and the corresponding areas on the vertical axis.
  c. What dimensions give the corral the greatest possible area?

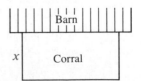

37. Draw a rectangle. Then construct a square with equal area.

◆ Computer Key-In

The shaded region shown is bounded by the graph of $y = x^2$, by the x-axis, and by the vertical line through the points $(1, 0)$ and $(1, 1)$. You can approximate the area of the shaded region by drawing ten rectangles having base vertices at $x = 0, 0.1, 0.2, 0.3, \ldots, 1.0$, as shown, and computing the sum of the areas of the ten rectangles. The base of each rectangle is 0.1, and the height of each rectangle is given by $y = x^2$.

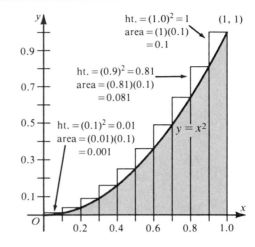

The following computer program will compute and add the areas of the ten rectangles shown in the diagram. In line 30, Y is the height of each rectangle. In line 40, A gives the current total of all the areas.

```
10  LET X = 0.1
20  FOR N = 1 TO 10
30  LET Y = X ↑ 2
40  LET A = A + Y * 0.1
50  LET X = X + 0.1
60  NEXT N
70  PRINT "AREA IS APPROXIMATELY ";A
80  END
```

If the program is run, the computer will print

```
AREA IS APPROXIMATELY 0.385
```

Exercises

1. A better approximation can be found by using 100 smaller rectangles with base vertices at $0, 0.01, 0.02, 0.03, \ldots, 1.00$. Change lines 10, 20, 40, and 50 as follows:

```
10  LET X = 0.01
20  FOR N = 1 TO 100
40  LET A = A + Y * 0.01
50  LET X = X + 0.01
```

   RUN the program to approximate the area of the shaded region.

2. Modify the given program so that it will use 1000 rectangles with base vertices at $0, 0.001, 0.002, 0.003, \ldots, 1.000$ to approximate the area of the shaded region. RUN the program.

3. Is the actual area of the shaded region more or less than the value given by the computer program? Explain.

11-2 *Areas of Parallelograms, Triangles, and Rhombuses*

Although proofs of most area formulas are easy to understand, detailed formal proofs are long and time consuming. Therefore, we will show the key steps of each proof.

Theorem 11-2

The area of a parallelogram equals the product of a base and the height to that base. $(A = bh)$

Given: $\square PQRS$

Prove: $A = bh$

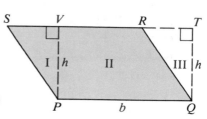

Key steps of proof:

1. Draw altitudes $\overline{PV}$ and $\overline{QT}$, forming two rt. $\triangle$.
2. Area I = Area III ($\triangle PSV \cong \triangle QRT$ by HL or AAS)
3. Area of $\square PQRS$ = Area II + Area I
   $\qquad\qquad\quad$ = Area II + Area III
   $\qquad\qquad\quad$ = Area of rect. $PQTV$
   $\qquad\qquad\quad$ = bh

Example 1 Find the area of the parallelogram shown.

Solution Notice the 45°-45°-90° triangle.

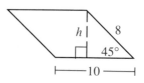

$$h = \frac{8}{\sqrt{2}} = \frac{8}{\sqrt{2}} \cdot \frac{\sqrt{2}}{\sqrt{2}} = \frac{8\sqrt{2}}{2} = 4\sqrt{2}$$

$$A = bh = 10 \cdot 4\sqrt{2} = 40\sqrt{2}$$

Theorem 11-3

The area of a triangle equals half the product of a base and the height to that base. $(A = \frac{1}{2}bh)$

Given: $\triangle XYZ$

Prove: $A = \frac{1}{2}bh$

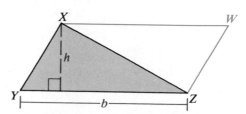

Key steps of proof:

1. Draw $\overline{XW} \parallel \overline{YZ}$ and $\overline{ZW} \parallel \overline{YX}$, forming $\square WXYZ$.
2. $\triangle XYZ \cong \triangle ZWX$ (SAS or SSS)
3. Area of $\triangle XYZ = \frac{1}{2} \cdot$ Area of $\square WXYZ$
   $\qquad\qquad\qquad = \frac{1}{2}bh$

Example 2 Find the area of a triangle with sides 8, 8, and 6.

Solution Draw the altitude to the base shown. Since the triangle is isosceles, this altitude bisects the base.

$$h^2 + 3^2 = 8^2 \quad \text{(Pythagorean Theorem)}$$
$$h^2 = 64 - 9 = 55$$
$$h = \sqrt{55}$$
$$A = \tfrac{1}{2}bh = \tfrac{1}{2} \cdot 6 \cdot \sqrt{55} = 3\sqrt{55}$$

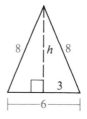

Example 3 Find the area of an equilateral triangle with side 6.

Solution Draw an altitude. Two 30°-60°-90° triangles are formed.

$$h = 3\sqrt{3}$$
$$A = \tfrac{1}{2}bh = \tfrac{1}{2} \cdot 6 \cdot 3\sqrt{3} = 9\sqrt{3}$$

Theorem 11-4

The area of a rhombus equals half the product of its diagonals.
$(A = \tfrac{1}{2}d_1d_2)$

Given: Rhombus $ABCD$ with diagonals d_1 and d_2

Prove: $A = \tfrac{1}{2}d_1d_2$

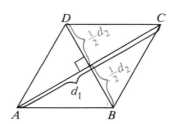

Key steps of proof:

1. $\triangle ADC \cong \triangle ABC$ (SSS)
2. Since $\overline{DB} \perp \overline{AC}$, the area of $\triangle ADC = \tfrac{1}{2}bh = \tfrac{1}{2} \cdot d_1 \cdot \tfrac{1}{2}d_2 = \tfrac{1}{4}d_1d_2$.
3. Area of rhombus $ABCD = 2 \cdot \tfrac{1}{4}d_1d_2 = \tfrac{1}{2}d_1d_2$

Classroom Exercises

1. The area of the parallelogram can be found in two ways:
   a. $A = 8 \cdot \underline{\ ?\ } = \underline{\ ?\ }$
   b. $A = 4 \cdot \underline{\ ?\ } = \underline{\ ?\ }$

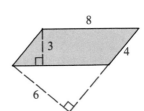

2. Find the areas of $\triangle ABC$, $\triangle DBC$, and $\triangle EBC$.

3. Give two formulas that can be used to find the area of a rhombus. (*Hint:* Every rhombus is also a $\underline{\ ?\ }$.)

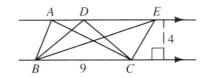

Find the area of each figure.

4.

5.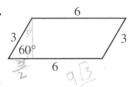

6.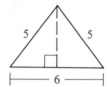

7.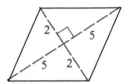

8.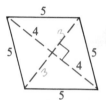

9.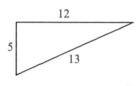

Written Exercises

In Exercises 1–20 find the area of each figure.

A   **1.** A triangle with base 5.2 m and corresponding height 11.5 m

     2. A triangle with sides 3, 4, and 5

     3. A parallelogram with base $3\sqrt{2}$ and corresponding height $2\sqrt{2}$

     4. A rhombus with diagonals 4 and 6

     5. An equilateral triangle with sides 8 ft

     6. An isosceles triangle with sides 10, 10, and 16

7.

8.

9.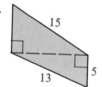

     10. An isosceles triangle with base 10 and perimeter 36

     11. An isosceles right triangle with hypotenuse 8

     12. An equilateral triangle with perimeter 18

B   **13.** A parallelogram with a 45° angle and sides 6 and 10

     14. A rhombus with a 120° angle and sides 6 cm

     15. A 30°-60°-90° triangle with hypotenuse 10

     16. An equilateral triangle with height 9

     17. A rhombus with perimeter 68 and one diagonal 30

     18. A regular hexagon with perimeter 60

     19. A square inscribed in a circle with radius *r*

     20. A rectangle with length 16 inscribed in a circle with radius 10

In Exercises 21–24 use a calculator or the trigonometry table on page 311 to find the area of each figure to the nearest tenth.

21.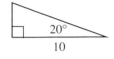

22.

23.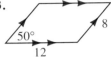

24. An isosceles triangle with a 32° vertex angle and a base of 8 cm

25. $\overline{FG}$ is the altitude to the hypotenuse of $\triangle DEF$. Name three similar triangles and find their areas. (*Hint*: See Theorem 8-1 and Corollary 1 on pages 285–286.)

26. If the area of $\square PQRS$ is 36 and T is a point on $\overline{PQ}$, find the area of $\triangle RST$. (*Hint*: Draw a diagram.)

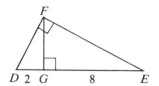

In Exercises 27 and 28, $\overline{AM}$ is a median of $\triangle ABC$.

27. If $BC = 16$ and $h = 5$, find the areas of $\triangle ABC$ and $\triangle AMB$.

28. Prove: Area of $\triangle AMB = \frac{1}{2} \cdot$ Area of $\triangle ABC$

29. a. Find the ratio of the areas of $\triangle QRT$ and $\triangle QTS$.
  b. If the area of $\triangle QRS$ is 240, find the length of the altitude from S to $\overleftrightarrow{QR}$.

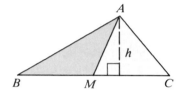

30. An isosceles triangle has sides that are 5 cm, 5 cm, and 8 cm long. Find its area and the lengths of the three altitudes.

31. a. Find the area of the right triangle in terms of a and b.
  b. Find the area of the right triangle in terms of c and h.
  c. Solve for h in terms of the other variables.
  d. A right triangle has legs 6 and 8. Find the lengths of the altitude and the median to the hypotenuse.

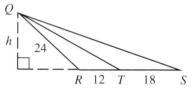

32. Use the diagram at the right.
  a. Find the area of $\square PQRS$.
  b. Find the area of $\triangle PSR$.
  c. Find the area of $\triangle OSR$. (*Hint*: Refer to $\triangle PSR$ and use Exercise 28.)
  d. What is the area of $\triangle PSO$?
  e. What must the area of $\triangle POQ$ be? Why? What must the area of $\triangle OQR$ be?
  f. State what you have shown in parts (a)–(e) about how the diagonals of a parallelogram divide the parallelogram.

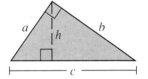

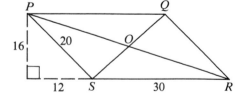

33. a. An equilateral triangle has sides of length s. Show that its area is $\frac{s^2}{4}\sqrt{3}$.

  b. Find the area of an equilateral triangle with side 7.

34. Think of a parallelogram made with cardboard strips and hinged at each vertex so that the measure of ∠C will vary. Find the area of the parallelogram for each measure of ∠C given in parts (a)–(e).

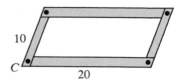

    a. 30   **b.** 45   **c.** 60   **d.** 90   **e.** 120

    f. Approximate your answers to parts (b), (c), and (e) by using $\sqrt{2} \approx 1.4$ and $\sqrt{3} \approx 1.7$. Then record your answers to parts (a)–(e) on a set of axes like the one below.

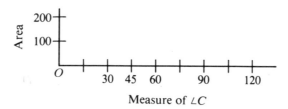

35. The area of a rhombus is 100. Find the length of the two diagonals if one is twice as long as the other.

36. The base of a triangle is 1 cm longer than its altitude. If the area of the triangle is 210 cm², how long is the altitude?

C **37.** Find the area of quadrilateral *ABCD* given *A*(2, −2), *B*(6, 4), *C*(−1, 5), and *D*(−5, 2).

38. Two squares each with sides 12 cm are placed so that a vertex of one lies at the center of the other. Find the area of the shaded region.

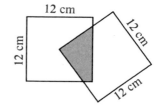

39. The diagonals of a parallelogram are 82 cm and 30 cm. One altitude is 18 cm long. Find the two possible values for the area.

For Exercises 40–42, draw a scalene triangle *ABC*.

40. Construct an isosceles triangle whose area is equal to the area of △*ABC*.

41. Construct an isosceles right triangle whose area is equal to the area of △*ABC*.

42. Construct an equilateral triangle whose area is equal to the area of △*ABC*.

Explorations

These exploratory exercises can be done using a computer with a program that draws and measures geometric figures.

Draw any quadrilateral and connect the midpoints of its sides. You should get a parallelogram (see Exercise 11, page 186). Compare the area of the original quadrilateral and the area of this parallelogram. What do you notice? Can you explain why this is true?

◆ Calculator Key-In

More than 2000 years ago, Heron, a mathematician from Alexandria, Egypt, derived a formula for finding the area of a triangle when the lengths of its sides are known. This formula is known as **Heron's Formula.** To find the area of $\triangle ABC$ using this formula:

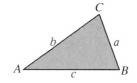

Step 1   Find the *semiperimeter* $s = \frac{1}{2}(a + b + c)$.
Step 2   Area $= A = \sqrt{s(s - a)(s - b)(s - c)}$

Example   If $a = 5$, $b = 6$, and $c = 7$, find the area of $\triangle ABC$.

Solution
Step 1 $s = \frac{1}{2}(5 + 6 + 7) = 9$
Step 2 $A = \sqrt{s(s - a)(s - b)(s - c)}$
$= \sqrt{9(9 - 5)(9 - 6)(9 - 7)}$
$= \sqrt{9 \cdot 4 \cdot 3 \cdot 2}$
$= 6\sqrt{6}$

It is convenient to use a calculator when evaluating areas by using Heron's Formula. A calculator gives 14.7 as the approximate area of the triangle in the example above.

Exercises

The lengths of the sides of a triangle are given. Use a calculator to find the area and the three heights of the triangle, each correct to three significant digits. (*Hint:* $h = \dfrac{2A}{b}$.)

1. 9, 10, 11   **2.** 5, 7, 8   **3.** 6, 11, 13   **4.** 15, 16, 17

5. 6.3, 7.2, 10.1   **6.** 68, 77, 105   **7.** 5.5, 6.5, 10   **8.** 12, 18, 27

Use two different methods to find the exact area of each triangle whose sides are given.

9. 3, 4, 5   **10.** 6, 6, 6   **11.** 13, 13, 10   **12.** 29, 29, 42

13. Something strange happens when Heron's Formula is used with $a = 47$, $b = 38$, and $c = 85$. Why does this occur?

14. Heron also derived the following formula for the area of an inscribed quadrilateral with sides a, b, c, and d:

$A = \sqrt{(s - a)(s - b)(s - c)(s - d)}$,
where the semiperimeter $s = \frac{1}{2}(a + b + c + d)$

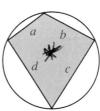

Use this formula to find the area of an isosceles trapezoid with sides 10, 10, 10, and 20 that is inscribed in a circle.

11-3 *Areas of Trapezoids*

An **altitude** of a trapezoid is any segment perpendicular to a line containing a base from a point on the opposite base. Since the bases are parallel, all altitudes have the same length, called the *height* (*h*) of the trapezoid.

Theorem 11-5

The area of a trapezoid equals half the product of the height and the sum of the bases. ($A = \frac{1}{2}h(b_1 + b_2)$)

Key steps of proof:

1. Draw diagonal $\overline{BD}$ of trap. *ABCD*, forming two triangular regions, I and II, each with height *h*.
2. Area of trapezoid = Area I + Area II

$$= \tfrac{1}{2}b_1h + \tfrac{1}{2}b_2h$$
$$= \tfrac{1}{2}h(b_1 + b_2)$$

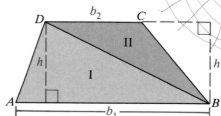

Example 1 Find the area of a trapezoid with height 7 and bases 12 and 8.

Solution $A = \frac{1}{2}h(b_1 + b_2) = \frac{1}{2} \cdot 7 \cdot (12 + 8) = 70$

Example 2 Find the area of an isosceles trapezoid with legs 5 and bases 6 and 10.

Solution When you draw the two altitudes shown, you get a rectangle and two congruent right triangles. The segments of the lower base must have lengths 2, 6, and 2. First find *h*:

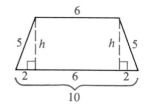

$$h^2 + 2^2 = 5^2$$
$$h^2 = 21$$
$$h = \sqrt{21}$$

Then find the area: $A = \frac{1}{2}h(b_1 + b_2) = \frac{1}{2}\sqrt{21}(10 + 6) = 8\sqrt{21}$

Classroom Exercises

Find the area of each trapezoid and the length of the median.

1.

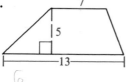

2.

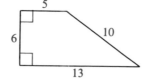

3.

4. Use your answers from Exercises 1–3 to explain why the area of a trapezoid can be given by the formula

$$Area = height \times median.$$

5. Does the median of a trapezoid divide it into two regions of equal area?

6. Does the segment joining the midpoints of the parallel sides of a trapezoid divide it into two regions of equal area?

7. a. If the congruent trapezoids shown are slid together, what special quadrilateral is formed?
  b. Use your answer in part (a) to derive the formula $A = \frac{1}{2}h(b_1 + b_2)$.

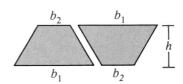

Find the area of each trapezoid.

8. 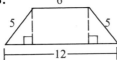 **9.** 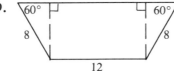 **10.**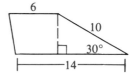

Written Exercises

Exercises 1–8 refer to trapezoids and m is the length of the median. Complete the table.

A

	1.	2.	3.	4.	5.	6.	7.	8.
b_1	12	6.8	$3\frac{1}{6}$	45	27	3	7	?
b_2	8	3.2	$4\frac{1}{3}$	15	9	?	?	$3k$
h	7	6.1	$1\frac{3}{5}$	?	?	3	$9\sqrt{2}$	$5k$
A	?	?	?	300	90	12	$36\sqrt{2}$	$45k^2$
m	?	?	?	?	?	?	?	?

9. A trapezoid has area 54 and height 6. How long is its median?

In Exercises 10–18, find the area of each trapezoid.

10.   **11.** 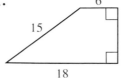 **12.**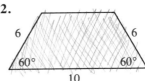

13.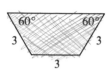

14.

15.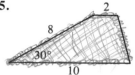

16. An isosceles trapezoid with legs 13 and bases 10 and 20

17. An isosceles trapezoid with legs 10 and bases 10 and 22

18. A trapezoid with bases 8 and 18 and 45° base angles

Use a calculator or the trigonometry table on page 311 to find the area of each trapezoid to the nearest tenth.

B **19.**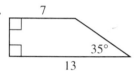

20.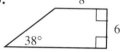

21.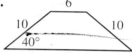

22. The legs of an isosceles trapezoid are 10 cm. The bases are 9 cm and 21 cm. Find the area of the trapezoid and the lengths of the diagonals.

23. An isosceles trapezoid has bases 12 and 28. The area is 300. Find the height and the perimeter.

24. *ABCD* is a trapezoid with bases 4 cm and 12 cm, as shown. Find the ratio of the areas of:
   a. △*ABD* and △*ABC*
   b. △*AOD* and △*BOC*
   c. △*ABD* and △*ADC*

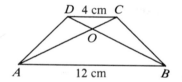

25. *ABCDEF* is a regular hexagon with side 12. Find the areas of the three regions formed when diagonals $\overline{AC}$ and $\overline{AD}$ are drawn.

26. An isosceles trapezoid with bases 12 and 16 is inscribed in a circle of radius 10. The center of the circle lies in the interior of the trapezoid. Find the area of the trapezoid.

27. A trapezoid of area 100 cm^2 has bases of 5 cm and 15 cm. Find the areas of the two triangles formed by extending the legs until they intersect.

C **28.** Draw a non-isosceles trapezoid. Then construct an isosceles trapezoid with equal area.

Find the exact area of each trapezoid. In Exercise 31, $\odot O$ is inscribed in quadrilateral *ABCD*.

29.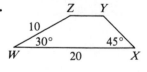

30.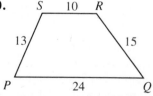

31.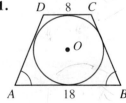

32. President James Garfield discovered a proof of the Pythagorean Theorem in 1876 that used a diagram like the one at the right. Refer to the diagram and write your own proof of the Pythagorean Theorem. (*Hint:* Express the area of quad. *MNOP* in two ways.)

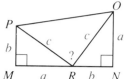

33. Show that the area of square *ABCD* equals the area of rectangle *EFGD*.

★ **34.** If *NS* = 16, find the area of ▱*MNOP*.

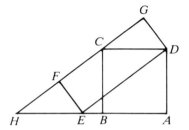

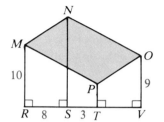

◆ Computer Key-In

The shaded region shown below is bounded by the graph of $y = x^2$, the *x*-axis, and the vertical lines $x = 1$ and $x = 2$. The area of this region can be approximated by drawing rectangles. (See the Computer Key-In, page 428.) This area can also be approximated by drawing trapezoids. The curve $y = x^2$ has been exaggerated slightly to better show the trapezoids. The diagrams below suggest that you can obtain a closer approximation for the area by using trapezoids than by using rectangles.

Let us approximate the area using five rectangles and five trapezoids. The base of each rectangle is 0.2 and the height is given by $y = x^2$.

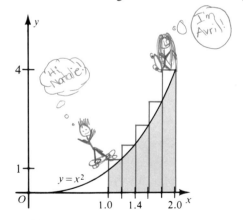

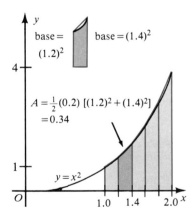

For each trapezoid in the diagram at the right above, the parallel bases are vertical segments from the *x*-axis to the curve $y = x^2$. The altitude is a horizontal segment with length 0.2. For example, in the second trapezoid, the bases are $(1.2)^2$ and $(1.4)^2$, respectively, and the height is 0.2.

The area of the shaded region is first approximated by the sum of the areas of the five rectangles and then by the sum of the areas of the five trapezoids. Compare these approximations. (Calculus can be used to prove that the exact area is $\frac{7}{3}$. Note that $\frac{7}{3} \approx 2.33$.)

Area approximated by five rectangles:
$$A \approx (1.2)^2(0.2) + (1.4)^2(0.2) + (1.6)^2(0.2) + (1.8)^2(0.2) + (2.0)^2(0.2)$$
$$A \approx 2.64$$

Area approximated by five trapezoids:
$$A \approx \tfrac{1}{2}(0.2)[(1.0)^2 + (1.2)^2] + \tfrac{1}{2}(0.2)[(1.2)^2 + (1.4)^2] + \tfrac{1}{2}(0.2)[(1.4)^2 + (1.6)^2]$$
$$+ \tfrac{1}{2}(0.2)[(1.6)^2 + (1.8)^2] + \tfrac{1}{2}(0.2)[(1.8)^2 + (2.0)^2]$$
$$A \approx 2.34$$

The following computer program will compute and add the areas of the five trapezoids shown in the diagram on the preceding page.

```
10  LET X = 1
20  FOR N = 1 TO 5
30  LET B1 = X ↑ 2
40  LET B2 = (X + 0.2) ↑ 2
50  LET A = A + 0.5 * 0.2 * (B1 + B2)
60  LET X = X + 0.2
70  NEXT N
80  PRINT "AREA IS APPROXIMATELY ";A
90  END
```

Exercises

1. A better approximation can be found by using 100 smaller trapezoids with base vertices at 1.00, 1.01, 1.02, . . . , 1.99, 2.00. Change lines 20, 40, 50, and 60 as follows:

   ```
   20  FOR N = 1 TO 100
   40  LET B2 = (X + 0.01) ↑ 2
   50  LET A = A + 0.5 * 0.01 * (B1 + B2)
   60  LET X = X + 0.01
   ```

   RUN the program to approximate the area of the shaded region.

2. Modify the given computer program so that it will use 1000 trapezoids with base vertices at 1.000, 1.001, 1.002, . . . , 2.000 to approximate the area of the shaded region. RUN the program.

3. Modify the given computer program so that it will use ten trapezoids to approximate the area of the region that is bounded by the graph of $y = x^2$, the x-axis, and *the vertical lines $x = 0$ and $x = 1$*. RUN the program. Compare your answer with that obtained on page 428, where ten rectangles were used. (*Note*: Calculus can be used to prove that the exact area is $\frac{1}{3}$.)

Mixed Review Exercises

Complete.

1. In $\odot O$, if the measure of central angle AOB is 52, then the measure of arc AB is ___?___.

2. In $\odot P$, if the measure of inscribed angle RST is 73, then the measure of arc RT is ___?___.

3. The measure of each interior angle of a regular octagon is ___?___.

4. If the measure of each exterior angle of a regular polygon is 20, then the polygon has ___?___ sides.

5. In a 45°-45°-90° triangle with legs 20 cm long, the length of the altitude to the hypotenuse is ___?___.

6. In a 30°-60°-90° triangle with hypotenuse 30 cm long, the lengths of the legs are ___?___ and ___?___.

7. In an isosceles triangle with vertex angle of 60° and legs 10 m long, the length of the base is ___?___.

8. In $\triangle ABC$ if $m \angle C = 90$, $AC = 8$, and $AB = 17$, then $\cos B = $ ___?___.

11-4 *Areas of Regular Polygons*

The beautifully symmetrical designs of kaleidoscopes are produced by mirrors that reflect light through loose particles of colored glass. Since the body of a kaleidoscope is a tube, the designs always appear to be inscribed in a circle. The photograph of a kaleidoscope pattern at the right suggests a regular hexagon.

Given any circle, you can inscribe in it a regular polygon of any number of sides. The diagrams below show how this can be done.

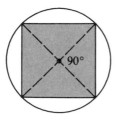

Square in circle: draw four 90° central angles.

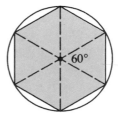

Regular hexagon in circle: draw six 60° central angles.

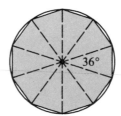

Regular decagon in circle: draw ten 36° central angles.

It is also true that if you are given any regular polygon, you can circumscribe a circle about it. This relationship between circles and regular polygons leads to the following definitions:

The **center of a regular polygon** is the center of the circumscribed circle.

The **radius of a regular polygon** is the distance from the center to a vertex.

A **central angle of a regular polygon** is an angle formed by two radii drawn to consecutive vertices.

The **apothem of a regular polygon** is the (perpendicular) distance from the center of the polygon to a side.

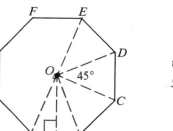

Center of regular octagon: O

Radius: OA, OB, OC, and so on

Central angle: $\angle AOB$, $\angle BOC$, and so on

Measure of central angle: $\dfrac{360}{8} = 45$

Apothem: OX

If you know the apothem and the perimeter of a regular polygon, you can use the next theorem to find the area of the polygon.

Theorem 11-6

The area of a regular polygon is equal to half the product of the apothem and the perimeter. $(A = \frac{1}{2}ap)$

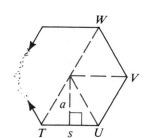

Given: Regular n-gon $TUVW \ldots$; apothem a; side s;
    perimeter p; area A

Prove: $A = \frac{1}{2}ap$

Key steps of proof:

1. If all radii are drawn, n congruent triangles are formed.
2. Area of each $\triangle = \frac{1}{2}sa$
3. $A = n(\frac{1}{2}sa) = \frac{1}{2}a(ns)$
4. Since $ns = p$, $A = \frac{1}{2}ap$.

Example 1   Find the area of a regular hexagon with apothem 9.

Solution   Use 30°-60°-90° $\triangle$ relationships.

$$\tfrac{1}{2}s = \frac{9}{\sqrt{3}} = 3\sqrt{3}$$
$$s = 6\sqrt{3}; \quad p = 36\sqrt{3}$$
$$A = \tfrac{1}{2}ap = \tfrac{1}{2} \cdot 9 \cdot 36\sqrt{3}$$
$$= 162\sqrt{3}$$

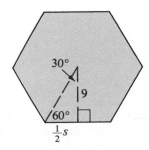

Example 2 Find the area of a regular polygon with 9 sides inscribed in a circle with radius 10.

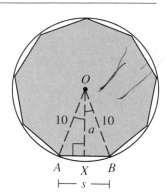

Solution $m \angle AOB = \dfrac{360}{9} = 40; \ m \angle AOX = 20$

Use trigonometry to find a and s:

$$\cos 20° = \frac{a}{10} \qquad\qquad \sin 20° = \frac{\frac{1}{2}s}{10}$$

$$a = 10 \cdot \cos 20° \qquad \tfrac{1}{2}s = 10 \sin 20°$$

$$a \approx 10(0.9397) \qquad s \approx 20(0.3420)$$

$$a \approx 9.397 \qquad\qquad s \approx 6.840$$

To find the area of the polygon, use either of two methods:

Method 1 Area of polygon $= 9 \cdot$ area of $\triangle AOB$

$$= 9 \cdot \tfrac{1}{2}sa$$

$$\approx \tfrac{9}{2}(6.840)(9.397)$$

$$\approx 289$$

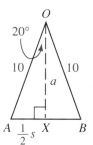

Method 2 Area of polygon $= \tfrac{1}{2}ap$

$$\approx \tfrac{1}{2}(9.397)(9 \cdot 6.840)$$

$$\approx 289$$

Classroom Exercises

1. Find the measure of a central angle of a regular polygon with **(a)** 10 sides, **(b)** 15 sides, **(c)** 360 sides, and **(d)** n sides.

Find the perimeter and the area of each regular polygon described.

2. A regular octagon with side 4 and apothem a

3. A regular pentagon with side s and apothem 3

4. A regular decagon with side s and apothem a

5. Explain why the apothem of a regular polygon must be less than the radius.

For each regular polygon shown, find (a) the perimeter, (b) the measure of a central angle, (c) the apothem a, (d) the radius r, and (e) the area A.

6.

7.

8.

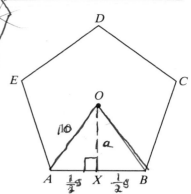

9. *ABCDE* is a regular pentagon with radius 10.
   a. Find the measure of $\angle AOB$.
   b. Explain why $m \angle AOX = 36$.

   Note: For parts (c)–(e), use a calculator or the table on page 311.

   c. $\cos 36° = \dfrac{a}{?}$. To the nearest tenth, $a \approx$ __?__.

   d. $\sin 36° = \dfrac{\frac{1}{2}s}{?}$. To the nearest tenth, $s \approx$ __?__.

   e. Find the perimeter and area of the pentagon.

Written Exercises

Copy and complete the tables for the regular polygons shown. In these tables, *p* represents the perimeter and *A* represents the area.

	r	*a*	*A*
1.	$8\sqrt{2}$	?	?
2.	?	5	?
3.	?	?	49
4.	?	$\sqrt{6}$	?

	r	*a*	*p*	*A*
5.	6	?	?	?
6.	?	4	?	?
7.	?	?	12	?
8.	?	?	$9\sqrt{3}$	?

	r	*a*	*p*	*A*
9.	4	?	?	?
10.	?	$5\sqrt{3}$	?	?
11.	?	6	?	?
12.	?	?	$12\sqrt{3}$	?

Find the area of each polygon.

13. An equilateral triangle with radius $4\sqrt{3}$

14. A square with radius $8k$

15. A regular hexagon with perimeter 72

16. A regular hexagon with apothem 4

17. A regular decagon is shown inscribed in a circle with radius 1.
   a. Explain why $m \angle AOX = 18$.
   b. Use a calculator or the table on page 311 to evaluate *OX* and *AX* below.

   $$\sin 18° = \frac{AX}{1}, \text{ so } AX \approx \underline{\quad?\quad}.$$

   $$\cos 18° = \frac{?}{?}, \text{ so } OX \approx \underline{\quad?\quad}.$$

   c. Perimeter of decagon $\approx$ __?__
   d. Area of $\triangle AOB \approx$ __?__
   e. Area of decagon $\approx$ __?__

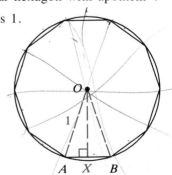

Three regular polygons are inscribed in circles with radii 1. Find the apothem, the perimeter, and the area of each polygon. Use $\sqrt{3} \approx 1.732$ and $\sqrt{2} \approx 1.414$.

18. 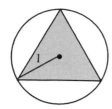 **19.** 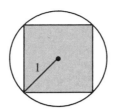 **20.**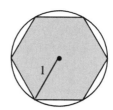

21. Find the perimeter and area of a regular dodecagon (12 sides) inscribed in a circle with radius 1. Use the procedure suggested by Exercise 17.

C 22. A regular polygon with n sides is inscribed in a circle with radius 1.

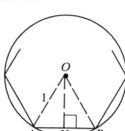

    a. Explain why $m \angle AOX = \dfrac{180}{n}$.

    b. Show that $AX = \sin\left(\dfrac{180}{n}\right)^{\circ}$.

    c. Show that $OX = \cos\left(\dfrac{180}{n}\right)^{\circ}$.

    d. Show that the perimeter of the polygon is $p = 2n \cdot \sin\left(\dfrac{180}{n}\right)^{\circ}$.

    e. Show that the area of the polygon is $A = n \cdot \sin\left(\dfrac{180}{n}\right)^{\circ} \cdot \cos\left(\dfrac{180}{n}\right)^{\circ}$.

Self-Test 1

Find the area of each polygon.

1. A square with diagonal $9\sqrt{2}$
2. A rectangle with base 12 and diagonal 13
3. A parallelogram with sides 8 and 10 and an angle of measure 60
4. An equilateral triangle with perimeter 12 cm
5. An isosceles triangle with sides 7 cm, 7 cm, and 12 cm
6. A rhombus with diagonals 8 and 10
7. An isosceles trapezoid with legs 5 and bases 9 and 17
8. A regular hexagon with sides 10
9. A regular decagon with sides x and apothem y
10. The quadrilateral shown at the right

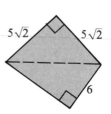

Ex. 10

♦ Calculator Key-In

If a regular polygon with n sides is inscribed in a circle with radius 1, then its perimeter and area are given by the formulas derived in Exercise 22 on the preceding page.

$$\text{Perimeter} = 2n \cdot \sin\left(\frac{180}{n}\right)^{\circ} \qquad \text{Area} = n \cdot \sin\left(\frac{180}{n}\right)^{\circ} \cdot \cos\left(\frac{180}{n}\right)^{\circ}$$

Exercises

1. Use the formulas and a calculator to complete the table at the right.

2. Use your answers in Exercise 1 to suggest approximations to the perimeter and the area of a *circle* with radius 1.

Number of sides	Perimeter	Area
18	?	?
180	?	?
1800	?	?
18000	?	?

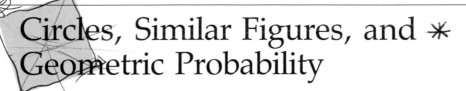

Circles, Similar Figures, and ✳ Geometric Probability

Objectives

1. Know and use the formulas for the circumferences and areas of circles.
2. Know and use the formulas for arc lengths and the areas of sectors of a circle.
3. Find the ratio of the areas of two triangles.
4. Understand and apply the relationships between scale factors, perimeters, and areas of similar figures.
5. Use lengths and areas to solve problems involving geometric probability.

11-5 *Circumferences and Areas of Circles*

When you think of the perimeter of a figure, you probably think of the distance around the figure. Since the word "around" is not mathematically precise, perimeter is usually defined in other ways. For example, the perimeter of a polygon is defined as the sum of the lengths of its sides. Since a circle is not a polygon, the perimeter of a circle must be defined differently.

First consider a sequence of regular polygons inscribed in a circle with radius *r*. Four such polygons are shown below. Imagine that the number of sides of the regular polygons continues to increase. As you can see in the diagrams, the more sides a regular polygon has, the closer it approximates (or "fits") the curve of the circle.

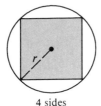
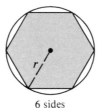
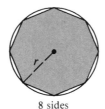
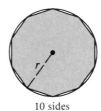

| 4 sides | 6 sides | 8 sides | 10 sides |

Now consider the perimeters and the areas of this sequence of regular polygons. The table below contains values that are approximations (using trigonometry) of the perimeters and the areas of regular polygons in terms of the radius, *r*.

As the table suggests, these perimeters give us a sequence of numbers that get closer and closer to a limiting number. This limiting number is defined to be the perimeter, or **circumference,** of the circle.

The area of a circle is defined in a similar way. The areas of the inscribed regular polygons get closer and closer to a limiting number, defined to be the **area** of the circle.

The results in the table suggest that the circumference and the area of a circle with radius *r* are *approximately* $6.28r$ and $3.14r^2$.

Number of Sides of Polygon	Perimeter	Area
4	$5.66r$	$2.00r^2$
6	$6.00r$	$2.60r^2$
8	$6.12r$	$2.83r^2$
10	$6.18r$	$2.93r^2$
20	$6.26r$	$3.09r^2$
30	$6.27r$	$3.12r^2$
100	$6.28r$	$3.14r^2$

The exact values are given by the formulas below. (Proofs are suggested in Classroom Exercises 13 and 14 and Written Exercise 33.)

Circumference, C, of circle with radius r:	$C = 2\pi r$
Circumference, C, of circle with diameter d:	$C = \pi d$
Area, A, of circle with radius r:	$A = \pi r^2$

These formulas involve a famous number denoted by the Greek letter π (*pi*), which is the first letter in a Greek word that means "measure around." The number π is the ratio of the circumference of a circle to the diameter. This ratio is a constant for *all* circles. Because π is an irrational number, there isn't any decimal or fraction that expresses the constant number π exactly. Here are some common approximations for π:

$$3.14 \qquad \frac{22}{7} \qquad 3.1416 \qquad 3.14159$$

When you calculate the circumference and area of a circle, leave your answers in terms of π unless you are told to replace π by an approximation.

Example 1 Find the circumference and area of a circle with radius 6 cm.

Solution $C = 2\pi r = 2\pi \cdot 6 = 12\pi$ (cm)
$A = \pi r^2 = \pi \cdot 6^2 = 36\pi$ (cm^2)

Example 2 The photograph shows land that is supplied with water by an irrigation system. This system consists of a moving arm that sprinkles water over a circular region. If the arm is 430 m long, what is the area, correct to the nearest thousand square meters, of the irrigated region? (Use $\pi \approx 3.14$.)

Solution $A = \pi r^2 = \pi \cdot 430^2$
$A \approx 3.14 \cdot 184,900 = 580,586$
$A \approx 581,000$ m^2 (to the nearest 1000 m^2)

Example 3 Find the circumference of a circle if the area is 25π.

Solution Since $\pi r^2 = 25\pi$, $r^2 = 25$ and $r = 5$.
Then $C = 2\pi r = 2\pi \cdot 5 = 10\pi$.

Classroom Exercises

Complete the table. Leave answers in terms of π.

	1.	**2.**	**3.**	**4.**	**5.**	**6.**	**7.**	**8.**
Radius	3	4	0.8	?	?	?	?	?
Circumference	?	?	?	10π	18π	?	?	?
Area	?	?	?	?	?	36π	49π	144π

Find the circumference and area to the nearest tenth. Use $\pi \approx 3.14$.

9. $r = 2$ **10.** $r = 6$ **11.** $r = \dfrac{3}{2}$ **12.** $r = 1.2$

13. The number π is defined to be the ratio of the circumference of a circle to the diameter. This ratio is the same for all circles. Supply the missing reasons for the key steps of proof below.

Given: $\odot O$ and $\odot O'$ with circumferences
     C and C' and diameters d and d'

Prove: $\dfrac{C}{d} = \dfrac{C'}{d'}$

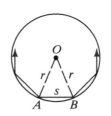

Key steps of proof:

1. Inscribe in each circle a regular polygon of n sides. Let p and p' be the perimeters.

2. $p = ns$ and $p' = ns'$ (Why?)

3. $\dfrac{p}{p'} = \dfrac{ns}{ns'} = \dfrac{s}{s'}$ (Why?)

4. $\triangle AOB \sim \triangle A'O'B'$ (Why?)

5. $\dfrac{s}{s'} = \dfrac{r}{r'} = \dfrac{d}{d'}$ (Why?)

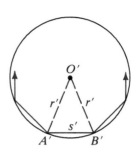

6. Thus $\dfrac{p}{p'} = \dfrac{d}{d'}$.   (Steps 3 and 5)

7. Steps 2–5 hold for any number of sides n. We can let n be so large that p is practically the same as C, and p' is practically the same as C'. In advanced courses, you learn that C and C' can be substituted for p and p' in Step 6. This gives $\dfrac{C}{C'} = \dfrac{d}{d'}$, or $\dfrac{C}{d} = \dfrac{C'}{d'}$.

   (This constant ratio is the number π. Then, since $\dfrac{C}{d} = \pi$, $C = \pi d$.)

14. Use the formula $C = \pi d$ to derive the formula $C = 2\pi r$.

Written Exercises

Complete the table. Leave answers in terms of π.

A

	1.	2.	3.	4.	5.	6.	7.	8.
Radius	7	120	$\frac{5}{2}$	$6\sqrt{2}$	?	?	?	?
Circumference	?	?	?	?	20π	12π	?	?
Area	?	?	?	?	?	?	25π	50π

9. Use $\pi \approx \frac{22}{7}$ to find the circumference and area of a circle when the diameter is **(a)** 42 and **(b)** $14k$.

10. Use $\pi \approx 3.14$ to find the circumference and area of a circle when the diameter is **(a)** 8 (Answer to the nearest tenth.) and **(b)** $4t$.

11. A basketball rim has diameter 18 in. Find the circumference of the rim and the area it encloses. Use $\pi \approx 3.14$.

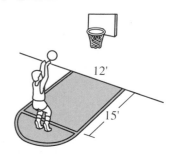

12. When a basketball player is shooting a free throw, the other players must stay out of the shaded region shown. This region consists of a semicircle together with a rectangle. Find the area of this region to the nearest square foot (ft^2). Use $\pi \approx 3.14$.

13. If 6 oz of dough are needed to make an 8-in. pizza, how much dough will be needed to make a 16-in. pizza of the same thickness? (*Hint*: Compare the areas of the pizza tops.)

14. One can of pumpkin pie mix will make a pie of diameter 8 in. If two cans of pie mix are used to make a larger pie of the same thickness, find the diameter of that pie. Use $\sqrt{2} \approx 1.414$.

15. A school's wrestling mat is a square with 40 ft sides. A circle 28 ft in diameter is painted on the mat. No wrestling is allowed outside the circle. Find the area of the part of the mat that is *not* used for wrestling. Use $\pi \approx \frac{22}{7}$.

16. An advertisement states that a Roto-Sprinkler can water a circular region with area 1000 ft^2. Find the diameter of this region to the nearest foot. Use $\pi \approx 3.14$.

17. Which is the better buy, a 10-in. pizza costing \$5 or a 15-in. pizza costing \$9? Use $\pi \approx 3.14$.

18. Semicircles are constructed on the sides of the right triangle shown at the right. If $a = 6$ and $b = 8$, show that
$$\text{Area I} + \text{Area II} = \text{Area III.}$$

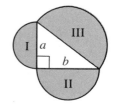

B 19. Repeat Exercise 18 if the right triangle has legs a and b and hypotenuse c.

Exs. 18, 19

20. A Ferris wheel has diameter 42 ft. How far will a rider travel during a 4-min ride if the wheel rotates once every 20 seconds? Use $\pi \approx \frac{22}{7}$.

21. The tires of a racing bike are approximately 70 cm in diameter.
  a. How far does a bike racer travel in 5 min if the wheels are turning at a speed of 3 revolutions per second? Use $\pi \approx \frac{22}{7}$.
  b. How many revolutions does a wheel make in a 22 km race? Use $\pi \approx \frac{22}{7}$.

22. A slide projector casts a circle of light with radius 2 ft on a screen that is 10 ft from the projector. If the screen is removed, the projector shines an even larger circle of light on the wall that was 10 ft behind the screen. Find the circumferences and areas of the circles of light on the screen and on the wall. Leave answers in terms of π.

23. A target consists of four concentric circles with radii 1, 2, 3, and 4.
    a. Find the area of the bull's eye and of each ring of the target.
    b. Find the area of the *n*th ring if the target contains *n* rings and a bull's eye.

Ex. 24

24. The shaded region in the diagram at the right above is formed by drawing four quarter-circles within a square of side 8. Find the area of the shaded region. (*Hint*: It is possible to give the answer without using pencil and paper or a calculator.)

25. The figure at the right consists of semicircles within a circle. Find the area of each shaded region.

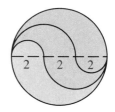

Ex. 25

Find the area of each shaded region. In Exercise 28, leave your answer in terms of *r*.

26.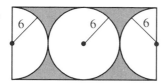

27.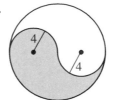

28.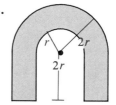

29. Draw a square and its inscribed and circumscribed circles. Find the ratio of the areas of these two circles.

30. Draw an equilateral triangle and its inscribed and circumscribed circles. Find the ratio of the areas of these two circles.

C 31. The diagram shows part of a regular polygon of 12 sides inscribed in a circle with radius *r*. Find the area enclosed between the circle and the polygon in terms of *r*. Use $\pi \approx 3.14$.

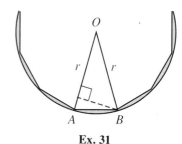

Ex. 31

32. A regular octagon is inscribed in a circle with radius *r*. Find the area enclosed between the circle and the octagon in terms of *r*. Use $\pi \approx 3.14$ and $\sqrt{2} \approx 1.414$.

33. A regular polygon with apothem *a* is inscribed in a circle with radius *r*.
    a. Complete: As the number of sides increases, the value of *a* gets nearer to __?__ and the perimeter of the polygon gets nearer to $2\pi r$.
    b. In the formula $A = \frac{1}{2}ap$, replace *a* by *r*, and *p* by $2\pi r$. What formula do you get?

34. Find the circumference of a circle inscribed in a rhombus with diagonals 12 cm and 16 cm.

35. Draw any circle *O* and any circle *P*. Construct a circle whose area equals the sum of the areas of circle *O* and circle *P*.

◆ Calculator Key-In

The number π is an irrational number. It cannot be expressed exactly as the ratio of two integers. Decimal *approximations* of π have been computed to thousands of decimal places. We can easily look up values in reference books, but such was not always the case. In the past, mathematicians had to rely on their cleverness to compute an approximate value of π. One of the earliest approximations was that of Archimedes, who found that $3\frac{1}{7} > \pi > 3\frac{10}{71}$.

Exercises

1. Find decimal approximations of $3\frac{1}{7}$ and $3\frac{10}{71}$. Did Archimedes approximate π correct to hundredths?

In Exercises 2–4, find approximations for π. The more terms or factors you use, the better your approximations will be.

2. $\pi \approx 2\sqrt{3}\left(1 - \dfrac{1}{3 \cdot 3} + \dfrac{1}{3^2 \cdot 5} - \dfrac{1}{3^3 \cdot 7} + \dfrac{1}{3^4 \cdot 9} - \dfrac{1}{3^5 \cdot 11} + \cdots\right)$

(Sharpe, 18th century)

3. $\pi \approx 2 \cdot \dfrac{2}{1} \cdot \dfrac{2}{3} \cdot \dfrac{4}{3} \cdot \dfrac{4}{5} \cdot \dfrac{6}{5} \cdot \dfrac{6}{7} \cdot \dfrac{8}{7} \cdot \dfrac{8}{9} \cdots$

(Wallis, 17th century)

4. This exercise is for calculators that have a square root function and a memory.

$$\pi \approx 2 \div \left(\sqrt{0.5} \cdot \sqrt{0.5 + 0.5\sqrt{0.5}} \cdot \sqrt{0.5 + 0.5\sqrt{0.5 + 0.5\sqrt{0.5}}} \cdots\right)$$

(Vieta, 16th century)

Algebra Review: *Evaluating Expressions*

Find the value of each expression using the given values of the variables.

Example $2\pi r$ when $r = \dfrac{5}{4}$

Solution $2 \cdot \pi \cdot \dfrac{5}{4} = \left(2 \cdot \dfrac{5}{4}\right)\pi = \dfrac{5}{2}\pi$

1. πr^2 when $r = \dfrac{2}{3}\sqrt{3}$

2. $\pi r l$ when $r = 4\dfrac{1}{5}$ and $l = 15$

3. $\dfrac{1}{3}\pi r^2 h$ when $r = 2\sqrt{6}$ and $h = 4$

4. $\dfrac{4}{3}\pi r^3$ when $r = 6$

5. $\pi r\sqrt{r^2 + h^2}$ when $r = h = \sqrt{5}$

6. $2\pi r^2 + 2\pi r h$ when $r = 10$ and $h = 6$

7. $\pi r^2 + \pi r\sqrt{r^2 + h^2}$ when $r = 2$ and $h = 2\sqrt{3}$

8. $\pi(r_1^2 - r_2^2)$ when $r_1 = 6$ and $r_2 = 3\sqrt{2}$

11-6 *Arc Lengths and Areas of Sectors*

A *pie chart* is often used to analyze data or to help plan business strategy. The radii of a pie chart divide the interior of the circle into regions called sectors, whose areas represent the relative sizes of particular items. A **sector of a circle** is a region bounded by two radii and an arc of the circle. The shaded region of the diagram at the right below is called sector *AOB*. The unshaded region is also a sector.

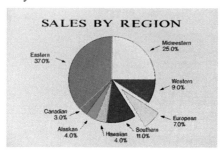

SALES BY REGION

The length of $\overparen{AB}$ in circle O is part of the circumference of the circle. Since $m\overparen{AB} = 60$ and $\dfrac{60}{360} = \dfrac{1}{6}$, the length of $\overparen{AB}$ is $\dfrac{1}{6}$ of the circumference. Thus,

$$\text{Length of } \overparen{AB} = \frac{1}{6}(2\pi \cdot 5) = \frac{5}{3}\pi.$$

Similarly, the area of sector AOB is $\dfrac{1}{6}$ of the area of the circle. Thus,

$$\text{Area of sector } AOB = \frac{1}{6}(\pi \cdot 5^2) = \frac{25}{6}\pi.$$

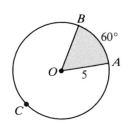

In general, if $m\overparen{AB} = x$:

$$\text{Length of } \overparen{AB} = \frac{x}{360} \cdot 2\pi r$$

$$\text{Area of sector } AOB = \frac{x}{360} \cdot \pi r^2$$

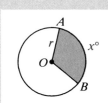

Example 1 In $\odot O$ with radius 9, $m\angle AOB = 120$. Find the lengths of the arcs $\overparen{AB}$ and $\overparen{ACB}$ and the areas of the two sectors shown.

Solution $m\overparen{AB} = 120$, and $m\overparen{ACB} = 240$.

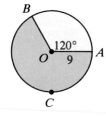

Minor arc $\overparen{AB}$:

$$\text{Length of } \overparen{AB} = \frac{120}{360} \cdot (2\pi \cdot 9) = \frac{1}{3}(18\pi) = 6\pi$$

$$\text{Area of sector } AOB = \frac{120}{360} \cdot (\pi \cdot 9^2) = \frac{1}{3}(81\pi) = 27\pi$$

Major arc $\overparen{ACB}$:

$$\text{Length of } \overparen{ACB} = \frac{240}{360} \cdot (2\pi \cdot 9) = \frac{2}{3}(18\pi) = 12\pi$$

$$\text{Area of sector} = \frac{240}{360} \cdot (\pi \cdot 9^2) = \frac{2}{3}(81\pi) = 54\pi$$

Example 2 Find the area of the shaded region bounded by $\overline{XY}$ and $\overset{\frown}{XY}$.

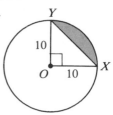

Solution Area of sector $XOY = \dfrac{90}{360} \cdot \pi \cdot 10^2 = 25\pi$

Area of $\triangle XOY = \dfrac{1}{2} \cdot 10 \cdot 10 = 50$

Area of shaded region $= 25\pi - 50$

Classroom Exercises

Find the arc length and area of each shaded sector.

1.   **2.** 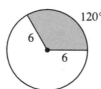   **3.**   **4.**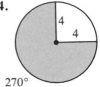

5. In a circle with radius 6, $m\overset{\frown}{AB} = 60$. Make a sketch and find the area of the region bounded by $\overline{AB}$ and $\overset{\frown}{AB}$.

6. A circle has area 160π cm^2. If a sector of the circle has area 40π cm^2, find the measure of the arc of the sector.

7. Compare the areas of two sectors if
  a. they have the same central angle, but the radius of one is twice as long as the radius of the other.
  b. they have the same radius, but the central angle of one is twice as large as the central angle of the other.

Written Exercises

Sector AOB is described by giving $m\angle AOB$ and the radius of circle O. Make a sketch and find the length of $\overset{\frown}{AB}$ and the area of sector AOB.

A

	1.	**2.**	**3.**	**4.**	**5.**	**6.**	**7.**	**8.**	**9.**	**10.**
$m\angle AOB$	30	45	120	240	180	270	40	320	108	192
radius	12	4	3	3	1.5	0.8	$\dfrac{9}{2}$	$1\dfrac{1}{5}$	$5\sqrt{2}$	$3\sqrt{3}$

11. The area of sector AOB is 10π and $m\angle AOB = 100$. Find the radius of circle O.

12. The area of sector AOB is $\dfrac{7\pi}{2}$ and $m\angle AOB = 315$. Find the radius of circle O.

Find the area of each shaded region. Point *O* marks the center of a circle.

B **13.**

14.

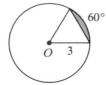

15.

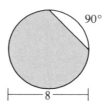

16.

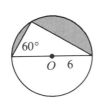

17.

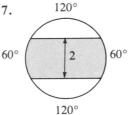

18.

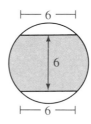

19. A rectangle with length 16 cm and width 12 cm is inscribed in a circle. Find the area of the region inside the circle but outside the rectangle.

20. From point *P*, $\overline{PA}$ and $\overline{PB}$ are drawn tangent to circle *O* at points *A* and *B*. If the radius of the circle is 6 and $m \angle APB = 60$, find the area of the region outside the circle but inside quadrilateral *AOBP*.

You may wish to use a calculator for Exercises 21–23. Use $\pi \approx 3.14$.

21. Chord *AB* is 18 cm long and the radius of the circle is 12 cm.
   a. Use trigonometry to find the measures of $\angle AOX$ and $\angle AOB$, correct to the nearest integer.
   b. Find the area of the shaded region to the nearest square centimeter. Use $\sqrt{7} \approx 2.646$.

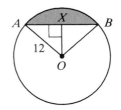

22. The diagram shows some dimensions in a baseball stadium. *H* represents home plate. Approximate the ratio of the areas of fair territory (shaded region) and foul territory (nonshaded region).

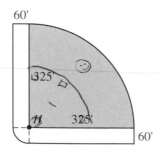

23. A cow is tied by a 25 m rope to the corner of a barn as shown. A fence keeps the cow out of the garden. Find, to the nearest square meter, the grazing area. Use $\sqrt{2} \approx 1.414$.

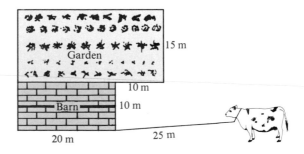

24. *ABCD* is a square with sides 8 cm long. Two circles each with radius 8 cm are drawn, one with center *A* and the other with center *C*. Find the area of the region inside both circles.

25. Two circles have radii 6 cm and their centers are 6 cm apart. Find the area of the region common to both circles.

26. **a.** Draw a square. Then construct the figure shown at the right.
b. If the radius of the square is 2, find the area of the shaded region.

C **27.** **a.** Using only a compass, construct the six-pointed figure shown at the right.
b. If the radius of the circle is 6, find the area of the shaded region.

28. Three circles with radii 6 are tangent to each other. Find the area of the region enclosed between them.

★ **29.** Circles *X* and *Y*, with radii 6 and 2, are tangent to each other. $\overline{AB}$ is a common external tangent. Find the area of the shaded region. (*Hint*: What kind of figure is *AXYB*? What is the measure of ∠*AXY*?)

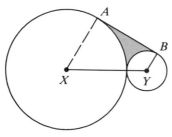

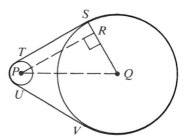

Ex. 29   **Ex. 30**

★ **30.** The diagram at the right above shows a belt tightly stretched over two wheels with radii 5 cm and 25 cm. The distance between the centers of the wheels is 40 cm. Find the length of the belt.

Challenge

Here $\overline{XY}$ has been divided into five congruent segments and semicircles have been drawn. But suppose $\overline{XY}$ were divided into millions of congruent segments and semicircles were drawn. What would the sum of the lengths of the arcs be?

Sarah says, "*XY*, because all the points would be so close to $\overline{XY}$." Mike says, "A really large number, because there would be so many arc lengths to add up." What do you say?

11-7 *Ratios of Areas*

In this section you will learn to compare the areas of figures by finding ratios.

Example 1 Find the ratios of the areas of two triangles:
  a. with equal heights
  b. with equal bases
  c. that are similar

Solution **a.** $\dfrac{\text{area of } \triangle ABD}{\text{area of } \triangle DBC} = \dfrac{\frac{1}{2}ah}{\frac{1}{2}bh} = \dfrac{a}{b}$ **b.** $\dfrac{\text{area of } \triangle ABC}{\text{area of } \triangle ADC} = \dfrac{\frac{1}{2}bh}{\frac{1}{2}bk} = \dfrac{h}{k}$

  ratio of areas = ratio of bases   ratio of areas = ratio of heights

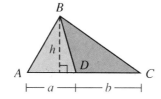

c. $\dfrac{\text{area of } \triangle ABC}{\text{area of } \triangle DEF} = \dfrac{\frac{1}{2}bh}{\frac{1}{2}ek} = \dfrac{bh}{ek} = \dfrac{b}{e} \cdot \dfrac{h}{k}$

It follows from Exercise 25 on
page 259 that if

$\triangle ABC \sim \triangle DEF$, then $\dfrac{h}{k} = \dfrac{b}{e}$.

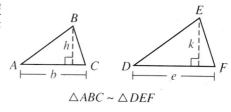

$\triangle ABC \sim \triangle DEF$

Thus, $\dfrac{\text{area of } \triangle ABC}{\text{area of } \triangle DEF} = \dfrac{b}{e} \cdot \dfrac{h}{k} = \dfrac{b}{e} \cdot \dfrac{b}{e} = \left(\dfrac{b}{e}\right)^2 = (\text{scale factor})^2.$

ratio of areas = square of scale factor

Example 1 justifies the following properties.

Comparing Areas of Triangles

1. If two triangles have equal heights, then the ratio of their areas equals the ratio of their bases.
2. If two triangles have equal bases, then the ratio of their areas equals the ratio of their heights.
3. If two triangles are similar, then the ratio of their areas equals the square of their scale factor.

Example 2   *ABCD* is a trapezoid. Find the ratio of the areas of:

   a. △*COD* and △*AOB*
   b. △*COD* and △*AOD*
   c. △*OAB* and △*DAB*

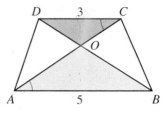

Solution   △*COD* ~ △*AOB* by the AA Similarity Postulate, with a scale factor of 3:5. Thus each of the corresponding sides and heights of these triangles has a 3:5 ratio.

   a. Since △*COD* ~ △*AOB*,

$$\frac{\text{area of } \triangle COD}{\text{area of } \triangle AOB} = \left(\frac{3}{5}\right)^2 = \frac{9}{25}.$$

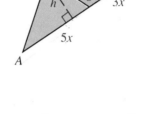

   b. Since △*COD* and △*AOD* have the same height, *h*, their area ratio equals their base ratio.

$$\frac{\text{area of } \triangle COD}{\text{area of } \triangle AOD} = \frac{CO}{AO} = \frac{3x}{5x} = \frac{3}{5}$$

   c. Since △*OAB* and △*DAB* have the same base, $\overline{AB}$, their area ratio equals their height ratio. Notice that the height of △*DAB* is 3*y* + 5*y*, or 8*y*.

$$\frac{\text{area of } \triangle OAB}{\text{area of } \triangle DAB} = \frac{5y}{8y} = \frac{5}{8}$$

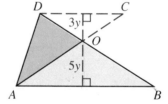

   You know that the ratios of the perimeters and areas of two similar triangles are related to their scale factor. These relationships can be generalized to any two similar figures.

Theorem 11-7

If the scale factor of two similar figures is *a*:*b*, then

(1) the ratio of the perimeters is *a*:*b*.

(2) the ratio of the areas is $a^2:b^2$.

Example 3   Find the ratio of the perimeters and the ratio of the areas of the two similar figures.

Solution   The scale factor is 8:12, or 2:3. Therefore, the ratio of the perimeters is 2:3. The ratio of the areas is $2^2:3^2$, or 4:9.

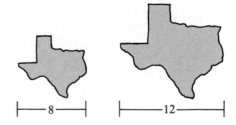

Classroom Exercises

Find the ratio of the areas of △ABC and △ADC.

1.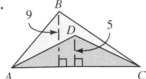

2.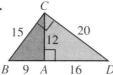

3.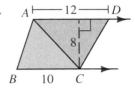

The table refers to similar figures. Complete the table.

	4.	**5.**	**6.**	**7.**	**8.**	**9.**	**10.**	**11.**
Scale factor	1:3	1:5	3:4	2:3	?	?	?	?
Ratio of perimeters	?	?	?	?	4:5	3:5	?	?
Ratio of areas	?	?	?	?	?	?	16:49	36:25

12. a. Are all circles similar?
  b. If two circles have radii 9 and 12, what is the ratio of the circumferences? of the areas?

13. a. Are regions I and II similar?
  b. Name two similar triangles.
  c. What is the ratio of their areas?
  d. What is the ratio of the areas of regions I and II?

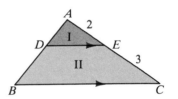

Find the ratio of the areas of triangles (a) I and II and (b) I and III.

14.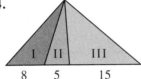

15.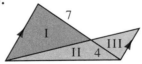

Written Exercises

The table refers to similar figures. Copy and complete the table.

A

	1.	**2.**	**3.**	**4.**	**5.**	**6.**	**7.**	**8.**
Scale factor	1:4	3:2	$r:2s$	?	?	?	?	?
Ratio of perimeters	?	?	?	9:5	3:13	?	?	?
Ratio of areas	?	?	?	?	?	25:1	9:64	2:1

9. On a map of California, 1 cm corresponds to 50 km. Find the ratio of the map's area to the actual area of California.

10. The areas of two circles are 36π and 64π. What is the ratio of the diameters? of the circumferences?

11. *L*, *M*, and *N* are the midpoints of the sides of $\triangle ABC$. Find the ratio of the perimeters and the ratio of the areas of $\triangle LMN$ and $\triangle ABC$.

12. The lengths of two similar rectangles are x^2 and xy, respectively. What is the ratio of the areas?

Name two similar triangles and find the ratio of their areas. Then find *DE*.

13.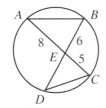

14.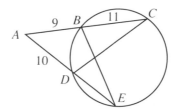

15. A quadrilateral with sides 8 cm, 9 cm, 6 cm, and 5 cm has area 45 cm². Find the area of a similar quadrilateral whose longest side is 15 cm.

16. A pentagon with sides 3 m, 4 m, 5 m, 6 m, and 7 m has area 48 m². Find the perimeter of a similar pentagon whose area is 27 m².

Find the ratio of the areas of triangles (a) I and II and (b) I and III. In Exercise 19(b), use the fact that Area I + Area II = Area III.

B 17.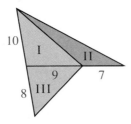

18.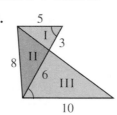

19.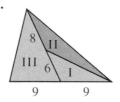

20. In the diagram below, *PQRS* is a parallelogram. Find the ratio of the areas for each pair of triangles.
    a. $\triangle TOS$ and $\triangle QOP$ **b.** $\triangle TOS$ and $\triangle TQR$

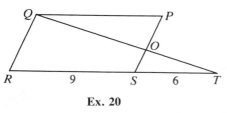

Ex. 20

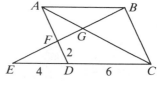

Ex. 21

21. In the diagram above, *ABCD* is a parallelogram. Name *four pairs* of similar triangles and give the ratio of the areas for each pair.

22. The area of parallelogram $ABCD$ is 48 cm^2 and $DE = 2 \cdot EC$. Find the area of:

    a. $\triangle ABE$ b. $\triangle BEC$ c. $\triangle ADE$

    d. $\triangle CEF$ e. $\triangle DEF$ f. $\triangle BEF$

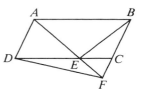

23. $ABCD$ is a parallelogram. Find each ratio.

    a. $\dfrac{\text{Area of } \triangle DEF}{\text{Area of } \triangle ABF}$

    b. $\dfrac{\text{Area of } \triangle DEF}{\text{Area of } \triangle CEB}$

    c. $\dfrac{\text{Area of } \triangle DEF}{\text{Area of trap. } DEBA}$

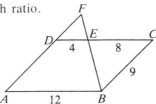

The figures in Exercises 24 and 25 are trapezoids. Find the ratio of the areas of (a) $\triangle$ I and $\triangle$ III, (b) $\triangle$ I and $\triangle$ II, (c) $\triangle$ I and $\triangle$ IV, (d) $\triangle$ II and $\triangle$ IV, and (e) $\triangle$ I and the trapezoid.

24.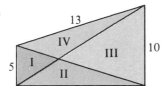

25.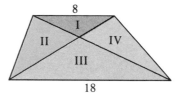

Find the ratio of the areas of regions I and II.

26.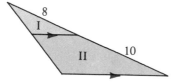

27.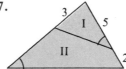

28.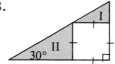

C 29. G is the intersection point of the medians of $\triangle ABC$. A line through G parallel to $\overline{BC}$ divides the triangle into two regions. What is the ratio of their areas? (*Hint:* See Theorem 10-4, page 387.)

30. In $\triangle LMN$, altitude $\overline{LK}$ is 12 cm long. Through point J of $\overline{LK}$ a line is drawn parallel to $\overline{MN}$, dividing the triangle into two regions with equal areas. Find LJ.

31. If you draw the three medians of a triangle, six small triangles are formed. Prove whatever you can about the areas of these six triangles.

★ 32. $\triangle ABC$ is equilateral; $\dfrac{AP}{PB} = \dfrac{BQ}{QC} = \dfrac{CR}{RA} = \dfrac{2}{1}$.

    Prove: Area of $\triangle XYZ = \dfrac{1}{7}(\text{area of } \triangle ABC)$

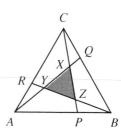

Ex. 32

11-8 *Geometric Probability*

The geometric probability problems in this section can be solved by using one of the following two principles.

1. Suppose a point P of $\overline{AB}$ is picked at random. Then:

$$\text{probability that } P \text{ is on } \overline{AC} = \frac{\text{length of } \overline{AC}}{\text{length of } \overline{AB}}$$

2. Suppose a point P of region S is picked at random. Then:

$$\text{probability that } P \text{ is in region } R = \frac{\text{area of } R}{\text{area of } S}$$

Example 1 Every ten minutes a bus pulls up to a hotel and waits for two minutes while passengers get on and off. Then the bus leaves. If a person walks out of the hotel front door at a random time, what is the probability that a bus is there?

Solution Think of a time line in which the colored segments represent times when the bus is at the hotel. For *any* ten-minute period, a two-minute subinterval is colored. Thus:

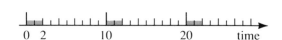

$$\text{probability that a bus will be there} = \frac{\text{length of colored segment}}{\text{length of whole segment}}$$

$$= \frac{2}{10} = \frac{1}{5}$$

Example 2 A person who is just beginning archery lessons misses the target frequently. And when a beginner hits the target, each spot is as likely to be hit as another. If a beginner shoots an arrow and it hits the target, what is the probability that the arrow hits the red bull's eye?

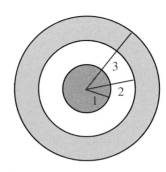

Solution probability arrow hits bull's eye if it hits target =

$$\frac{\text{area of bull's eye}}{\text{area of target}} = \frac{\pi \cdot 1^2}{\pi \cdot 3^2} = \frac{1}{9}$$

Example 3 At a carnival game, you can toss a coin on a large table that has been divided into squares 30 mm on a side. If the coin comes to rest without touching any line, you win. Otherwise you lose your coin. What are your chances of winning on one toss of a dime? (A dime has a radius of 9 mm.)

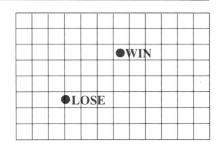

Solution Although there are many squares on the board, it is only necessary to consider the square in which the center of the dime lands. In order for the dime to avoid touching a line, the dime's center must be more than 9 mm from each side of the square. Its center must land in the shaded square shown. Thus the probability that the dime does not touch a line is equal to the probability that the center of the dime lies in the shaded region.

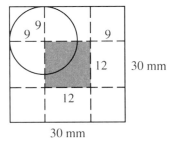

$$\text{probability of winning} = \frac{\text{area of shaded square}}{\text{area of larger square}} = \frac{12^2}{30^2} = 0.16$$

Note: In practice, your probability of winning would be less than 16% for several reasons. For example, a carnival might require you to toss a quarter instead of the smaller dime, as discussed in Written Exercise 9.

Classroom Exercises

1. A point P is picked at random on $\overline{RW}$. What is the probability that P is on:
   a. $\overline{RS}$? **b.** $\overline{SV}$? **c.** $\overline{SW}$?
   d. $\overline{RW}$? **e.** $\overline{XY}$? **f.** $\overline{RY}$?

2. A friend promises to call you sometime between 4:00 and 4:30 P.M. If you are not home to receive the call until 4:10, what is the probability that you miss the first call that your friend makes to you?

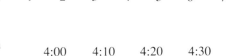

3. A dart lands at a random point on the square dartboard shown. What is the probability that the dart is within the outer circle? within the bull's eye? Use $\pi \approx 3.14$.

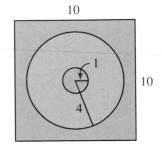

4. A ship is known to have sunk in the ocean in a square region 100 mi on a side. A salvage vessel anchors at a random spot in this square. Divers search 1 mi in all directions from the point on the ocean floor directly below the vessel. What is the approximate probability that they locate the sunken ship on the first try? Use $\pi \approx 3.14$.

Written Exercises

If a value of π is required in the following exercises, use $\pi \approx 3.14$.

A **1.** M is the midpoint of $\overline{AB}$ and Q is the midpoint of $\overline{MB}$. If a point of $\overline{AB}$ is picked at random, what is the probability that the point is on $\overline{MQ}$? (*Hint:* Make a sketch.)

2. In the diagram, $AC = CB$, $CD = DB$, and $DE = EB$. If a point X is selected at random from $\overline{AB}$, what is the probability that:
   a. X is between A and C?
   b. X is between D and B?
   c. X is between C and E?

3. A friend promises to call you at home sometime between 3 P.M. and 4 P.M. At 2:45 P.M. you must leave your house unexpectedly for half an hour. What is the probability you miss the first call?

4. **a.** At a subway stop, a train arrives every six minutes, waits one minute, and then leaves. If you arrive at a random time, what is the probability there will be a train waiting?
   b. If you arrive and there is no train waiting, what is the probability that you will wait no more than two minutes before one arrives?

5. A circular dartboard has diameter 40 cm. Its bull's eye has diameter 8 cm.
   a. If an amateur throws a dart and it hits the board, what is the probability that the dart hits the bull's eye?
   b. After many throws, 75 darts have hit the target. Estimate the number hitting the bull's eye.

6. Several hundred darts are thrown at the square dartboard shown. About what percentage of those hitting the board will land in the location described?
   a. Inside the inner square
   b. Outside the inner square but inside the circle

7. A dart is thrown at a board 12 m long and 5 m wide. Attached to the board are 30 balloons, each with radius 10 cm. Assuming each balloon lies entirely on the board, find the probability that a dart that hits the board will also hit a balloon.

8. Parachutists jump from an airplane and land in the rectangular field shown. What is the probability that a parachutist avoids the two trees represented by circles in the diagram? (Assume that the person is unable to control the landing point.)

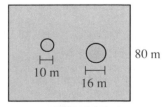

9. Refer to Example 3. Suppose that a quarter, instead of a dime, is tossed and lands on the table shown on the preceding page. What is the probability of winning on one toss? (The radius of a quarter is 12 mm.)

10. Repeat Exercise 9, using a nickel instead of a quarter. The radius of a nickel is 11 mm.

B **11.** A piece of wire 6 in. long is cut into two pieces at a random point. What is the probability that both pieces of wire will be at least 1 in. long?

12. A piece of string 8 cm long is cut at a random point. What is the probability that:
a. each piece is at least 2 cm long?
b. the lengths of the two pieces differ by no more than 2 cm?
c. the lengths of the two pieces total 8 cm?

13. Darts are thrown at a 1-meter square which contains an irregular red region. Of 100 darts thrown, 80 hit the square. Of these, 10 hit the red region. Estimate the area of this region.

1 m

1 m

14. A carnival game has a white dartboard 5 m long and 2 m wide on which 100 red stars are painted. Each player tries to hit a star with a dart. Before trying it, you notice that only 3 shots out of the previous 50 hit a star. Estimate the area of one star.

15. a. Suppose that a coin with radius R is tossed and lands on the table shown on page 462. Show that the probability the coin does not touch a line is $\left(\dfrac{30 - 2R}{30}\right)^2$.
b. Find the value of R for which the probability is 0.25.

16. A and B are the endpoints of a diameter. If C is a point chosen at random from the points on the circle (excluding A and B), what is the probability that:
a. $\triangle ABC$ is a right triangle?   **b.** $m \angle CAB \leq 30$?

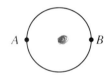

C **17.** A researcher was tape recording birdcalls. The tape recorder had a 1-hour tape in it. Eight minutes after the recorder was turned on, a 5-minute birdcall began. Unfortunately, the researcher accidentally erased 10 min of the tape. What is the probability that:
a. part of the birdcall was erased?   **b.** all of the birdcall was erased?
(*Hint:* Draw a time line from 0 to 60 min and locate on this line when the birdcall took place. Also consider the possible starting times when the erasure could have occurred.)

Challenge

Three segments through point P and parallel to the sides of $\triangle XYZ$ divide the whole region into six subregions. The three triangular subregions have the areas shown. Find the area of $\triangle XYZ$.

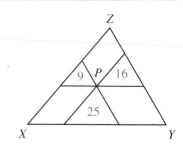

Self-Test 2

Leave your answers in terms of π unless you are told to use an approximation.

1. Find the circumference and area of a circle with radius 14. Use $\pi \approx \frac{22}{7}$.
2. The circumference of a circle is 18π. What is its area?
3. In $\odot O$ with radius 12, $m\overset{\frown}{AB} = 90$.
   a. Find the length of $\overset{\frown}{AB}$.
   b. Find the area of sector *AOB*.
   c. Find the area of the region bounded by $\overline{AB}$ and $\overset{\frown}{AB}$.
4. Find the ratio of the areas of two circles with radii 4 and 7.
5. The areas of two similar triangles are 36 and 81. Find the ratio of their perimeters.
6. *PQRS* is a parallelogram. Find the ratio of the areas of:
   a. $\triangle PTO$ and $\triangle RSO$ b. $\triangle RPS$ and $\triangle TPS$

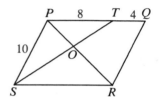

Each polygon is a regular polygon. Find the area of the shaded region.

7.

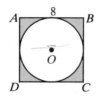

$\odot O$ is inscribed in square *ABCD*.

8.

$\triangle ABC$ is inscribed in $\odot P$.

9. Refer to $\square PQRS$ in Exercise 6. A point is randomly chosen on $\overline{PR}$. Find the probability that the point is on $\overline{OR}$.
10. Suppose that the figure in Exercise 7 is a dartboard. Imagine that someone with poor aim throws a dart and you *hear* it hit the dartboard. What is the probability that the dart landed inside the circle?

Extra   *Congruence and Area*

The SAS Postulate tells us that a triangle is *determined*, or fixed in size and shape, when two sides and the included angle are fixed. This means that the other parts of the triangle and its area can be determined from the given SAS information. Similarly, the area of a triangle can be determined when given ASA, SSS, AAS, or HL information. Computing the area of a triangle can often be simplified by using a calculator.

Example 1   Given the SAS information shown for $\triangle ABC$, find its area.

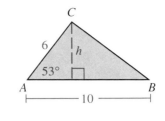

Solution   Draw the altitude from C.

Then $\dfrac{h}{6} = \sin 53° \approx 0.7986;\ h \approx 4.79.$

Area $= \frac{1}{2}bh \approx \frac{1}{2}(10)(4.79) \approx 24.0$

Example 2   Given the ASA information shown for $\triangle ABC$, find its area.

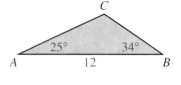

Solution

Step 1   Draw the altitude from C.

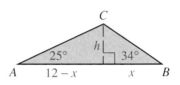

Then $\tan 25° = \dfrac{h}{12 - x}$ and $\tan 34° = \dfrac{h}{x}.$

$(12 - x)\tan 25° = h$ and $x \tan 34° = h$

$(12 - x)\tan 25° = x \tan 34°$

$(12 - x)(0.4663) \approx x(0.6745)$

$5.5956 - 0.4663x \approx 0.6745x$

$5.5956 \approx 1.1408x$

$4.905 \approx x$

Step 2   Knowing x, we can find h:

$h = x \tan 34° \approx (4.905)(0.6745) \approx 3.308$

Step 3   Area $= \frac{1}{2}bh \approx \frac{1}{2}(12)(3.308) \approx 19.8$

Exercises

Use the given information to find the approximate area of $\triangle ABC$. In Exercises 6 and 7 the altitude from C lies outside the triangle.

1. (SAS) $AB = 8,\ m \angle B = 67,\ BC = 15$

2. (HL) $m \angle C = 90,\ AB = 30,\ BC = 20$ (Use $\sqrt{5} \approx 2.236.$)

3. (SSS) $AB = 10,\ BC = 12,\ CA = 8$ (*Hint*: Use Heron's Formula.)

4. (ASA) $m \angle A = 28,\ AB = 10,\ m \angle B = 42$

5. (AAS) $m \angle A = 36,\ m \angle B = 80,\ BC = 10$ (*Hint*: Find the measure of $\angle C$. Then proceed as in Example 2.)

6. (SAS) $AB = 12,\ m \angle A = 118,\ AC = 20$

7. (ASA) $m \angle A = 107,\ AB = 20,\ m \angle B = 35$

★ **8.** The two triangles shown have two pairs of congruent corresponding sides and one pair of congruent corresponding non-included angles (SSA). Of course, they are *not* congruent. Find the area of each triangle.

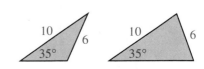

Application   *Space Shuttle Landings*

The space shuttle is launched vertically as a rocket but lands horizontally as a glider with no power and no second chance at the runway. NASA studied many different guidance systems for the final portion of entry and landing. The system that was selected for the first shuttle flights used a cylinder called the *Heading Alignment Cylinder* (HAC), shown in the diagram below. Notice that the projection of the flight path onto the Earth's surface is called the *ground track*.

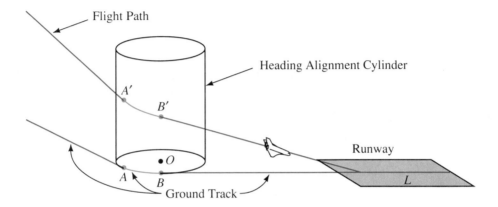

The shuttle followed a straight-line flight path to *A'* and then it followed a curved path along the Heading Alignment Cylinder to point *B'*. The shuttle continued to lose altitude so that it was closer to the Earth's surface at *B'* than at *A'*. From *B'* to landing at *L* the shuttle followed a straight path aligned with the center of the runway.

The points A' and B' where the shuttle's turn begins and ends can be determined by looking at the ground track. The figure at the right shows what you would see if you were high above the ground looking straight down on the ground track and runway. A' and B' are directly above A and B, which are located as follows: Extend the HAC acquisition line and the center line of the runway to meet at M. Bisect the angle at M and choose point O on the bisector so that a circle with center O and radius

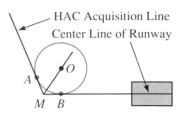

20,000 ft will be tangent to the sides of the angle. Call the points of tangency A and B. $\odot O$ is the base of the Heading Alignment Cylinder and A' and B' are on the cylinder directly above A and B.

Normally the shuttle approached the cylinder at 800 ft/s and turned along its surface by lowering one wing tip so that the wings formed an angle of about 45° with the horizontal (called the *bank angle*). Under high-speed conditions the shuttle would have approached the cylinder at 1000 ft/s. This would have required a bank angle of about 57° to follow the surface of the cylinder. Unfortunately, at that bank angle, the shuttle would have lost lift, and the astronauts would have lost some of their control capability.

NASA has refined the guidance system so that the shuttle can be safely landed even under these adverse circumstances. Now, instead of following the surface of a cylinder, it spirals along the surface of a cone called the *Heading Alignment Cone*. Once every second during this part of the landing the shuttle's computers recompute the radius of turn necessary to keep the shuttle on the surface of the cone. Now even under most high-speed conditions the bank angle will not exceed approximately 42°. At point Q the shuttle is heading directly toward the runway; it leaves the cone and continues along a straight course to touchdown.

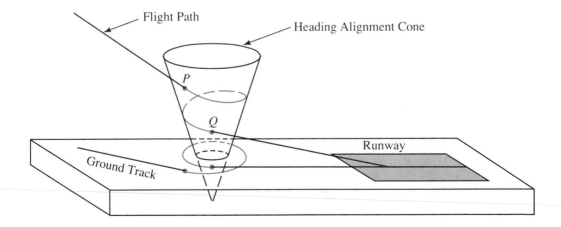

Exercises

1. Let T be any point on the bisector of $\angle AMB$. Show that if $\odot T$ is drawn tangent to $\overleftrightarrow{MA}$ it will also be tangent to $\overleftrightarrow{MB}$.

2. The radius of the Heading Alignment Cylinder was 20,000 ft, and a typical measure for $\angle AMB$ was 120. How long was the curved portion of the ground track, $\overset{\frown}{AB}$? (Use 3.1416 for π.)

3. The shuttle's turning radius changes as it moves along the surface of the Heading Alignment Cone. Is the radius larger near P or near Q?

4. A good approximation of the detailed landing procedure uses a Heading Alignment Cone with vertex below the surface of the Earth. A typical radius of the cone at a height of 30,000 ft above the Earth's surface is 20,000 ft. At a height of 12,000 ft, which is a typical height for Q, the radius of the cone is 14,000 ft.
   a. How far below the surface of the Earth is the vertex of the cone?
   b. What is the radius of the cone at a height of 15,000 ft?
   c. At what height is the radius of the cone equal to 12,000 ft?

Chapter Summary

1. If two figures are congruent, then they have the same area.

2. The area of a region is the sum of the areas of its non-overlapping parts.

3. The list below gives the formulas for areas of polygons.

Square:	$A = s^2$	Rectangle:	$A = bh$
Parallelogram:	$A = bh$	Triangle:	$A = \frac{1}{2}bh$
Rhombus:	$A = \frac{1}{2}d_1d_2$	Trapezoid:	$A = \frac{1}{2}h(b_1 + b_2)$
Regular polygon:	$A = \frac{1}{2}ap$, where a is the apothem and p is the perimeter		

4. The list below gives the formulas related to circles.

   $$C = 2\pi r \qquad \text{Length of arc} = \frac{x}{360} \cdot 2\pi r$$
   $$C = \pi d$$
   $$A = \pi r^2 \qquad \text{Area of sector} = \frac{x}{360} \cdot \pi r^2$$

   where x is the measure of the arc

5. If two triangles have equal heights, then the ratio of their areas equals the ratio of their bases. If two triangles have equal bases, then the ratio of their areas equals the ratio of their heights.

6. If the scale factor of two similar figures is $a:b$, then
   (1) the ratio of the perimeters is $a:b$.
   (2) the ratio of the areas is $a^2:b^2$.

7. Two principles used in geometric probability problems are stated and illustrated on page 461.

Chapter Review

1. Find the area of a square with perimeter 32.   11–1

2. Find the area of a rectangle with length 4 and diagonal 6.

3. Find the area of a square with side $3\sqrt{2}$ cm.

4. Find the area of a rhombus with side 17 and longer diagonal 30.   11–2

5. A parallelogram has sides 8 and 12. The shorter altitude is 6. Find the length of the other altitude.

6. Find the perimeter and the area of the triangle shown.

7. Find the height of a trapezoid with median 12 and area 84.   11–3

8. Find the area of an isosceles trapezoid with legs 5 and bases 4 and 12.

9. Find the perimeter and the area of the figure shown.

10. Find the area of a square with apothem 3 m.   11–4

11. Find the area of an equilateral triangle with radius $2\sqrt{3}$.

12. Find the area of a regular hexagon with perimeter 12 cm.

13. Find the circumference and area of a circle with radius 30. Use $\pi \approx 3.14$.   11–5

14. The area of a circle is 121π cm². Find the diameter.

15. A square with side 8 is inscribed in a circle. Find the circumference and the area of the circle.

16. Find the length of a 135° arc in a circle with radius 24.   11–6

Find the area of each shaded region.

17. 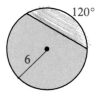   18.   19.

20. If $AB = 9$ and $CD = 12$, find the ratio of the areas of: 11–7
    a. $\triangle AEB$ and $\triangle DEC$ **b.** $\triangle AED$ and $\triangle DEC$

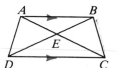

21. Two regular octagons have perimeters 16 cm and 32 cm, respectively. What is the ratio of their areas?

22. Two similar polygons have the scale factor 7:5. The area of the large polygon is 147. Find the area of the smaller polygon.

23. A point is randomly chosen inside the larger circle of Exercise 19. What is the probability that the point is inside the smaller circle?   11–8

Chapter Test

Find the area of each figure described.

1. A circle with diameter 10

2. A square with diagonal 4 cm

3. An isosceles right triangle with hypotenuse $6\sqrt{2}$

4. A circle with circumference 30π m

5. A rhombus with diagonals 5 and 4

6. An isosceles trapezoid with legs 10 and bases 6 and 22

7. A parallelogram with sides 6 and 10 that form a 30° angle

8. A regular hexagon with apothem $2\sqrt{3}$ cm

9. Sector AOB of $\odot O$ with radius 4 and $m\widehat{AB} = 45$

10. A rectangle with length 12 inscribed in a circle with radius 7.5

11. A sector of a circle with radius 12 and arc length 10π

12. A square with radius 9

$24\pi + 9\sqrt{3}$

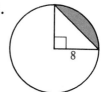

Find the area of each shaded region.

13.
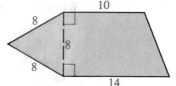

14.
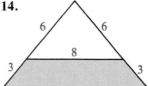

15.

16. The areas of two circles are 100π and 36π. Find the ratio of their radii and the ratio of their circumferences.

17. Two regular pentagons have sides of 14 m and 3.5 m, respectively. Find their scale factor and the ratio of their areas.

18. In the diagram of $\odot Q$, $m\widehat{ABC} = 288$ and $QA = 10$.
    a. Find the circumference of $\odot Q$.
    b. Find the length of $\widehat{AC}$.
    c. Find the area of sector AQC.

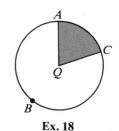

Ex. 18

19. A point is randomly chosen on $\overline{AD}$. Find the probability that the point is on $\overline{AE}$.

20. A point is randomly chosen inside $\triangle ABC$. What is the probability that the point is inside:
    a. $\triangle ABD$?   **b.** $\triangle BDE$?

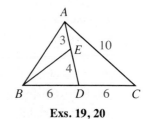

Exs. 19, 20

Cumulative Review: Chapters 1–11

In Exercises 1–12 classify each statement as true or false.

A 1. If point A lies on $\overrightarrow{BC}$, but not on $\overline{BC}$, then B is between A and C.

2. A true conditional always has a true converse.

3. The statement "If $ac = bc$, then $a = b$" is true for all real numbers a, b, and c.

4. If two parallel lines are cut by transversal t and t is perpendicular to one of the lines, then t must also be perpendicular to the other line.

5. If $\triangle ABC \cong \triangle DEF$ and $\angle A \cong \angle B$, then $\overline{DE} \cong \overline{EF}$.

6. If the opposite sides of a quadrilateral are congruent and the diagonals are perpendicular, then the quadrilateral must be a square.

7. If $\triangle GBS \sim \triangle JFK$, then $\dfrac{JF}{JK} = \dfrac{GB}{GS}$.

8. The length of the altitude to the hypotenuse of a right triangle is always the geometric mean between the lengths of the legs.

9. In any right triangle, the sine of one acute angle is equal to the cosine of the other acute angle.

10. If an angle inscribed in a circle intercepts a major arc, then the measure of the angle must be between 180 and 360.

11. The angle bisectors of an obtuse triangle intersect at a point that is equidistant from the three vertices.

12. If $JK = 10$, then the locus of points in space that are 4 units from J and 5 units from K is a circle.

13. Two lines that do not intersect are either __?__ or __?__ .

14. In $\triangle RST$, $m \angle R = 2x + 10$, $m \angle S = 3x - 10$, and $m \angle T = 4x$.
   a. Find the numerical measure of each angle.
   b. Is $\triangle RST$ scalene, isosceles, or right? Why?

15. Use inductive thinking to guess the next number: 10, 9, 5, -4, -20, __?__ .

16. If a diagonal of an equilateral quadrilateral is drawn, what method could be used to show that the two triangles formed are congruent?

17. A trapezoid has bases with lengths $x + 3$ and $3x - 1$ and a median of length 11. Find the value of x.

18. If 4, 7, and x are the lengths of the sides of a triangle and x is an integer, list the possible values for x.

19. Describe the locus of points in space that are 4 cm or less from a given point P.

20. Two similar rectangles have diagonals of $6\sqrt{3}$ and 9. Find the ratio of their perimeters and the ratio of their areas.

B **21.** Given: $\overline{AB} \perp \overline{BC}$; $\overline{DC} \perp \overline{BC}$; $\overline{AC} \cong \overline{BD}$
Prove: $\triangle BCE$ is isosceles.

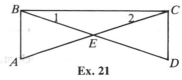

Ex. 21

22. Given: Quad. *EFGH*; $\overline{EF} \cong \overline{HG}$; $\overline{EF} \parallel \overline{HG}$
Prove: $\angle EHF \cong \angle GFH$

23. Use an indirect proof to show that no triangle has sides of length x, y, and $x + y$.

24. The legs of a right triangle are 4 cm and 8 cm long. What is the length of the median to the hypotenuse?

25. If a 45°-45°-90° triangle has legs of length $5\sqrt{2}$, find the length of the altitude to the hypotenuse.

26. The altitude to the hypotenuse of a 30°-60°-90° triangle divides the hypotenuse into segments with lengths in the ratio ___?___ : ___?___.

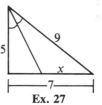

Ex. 27

27. Find the value of x in the diagram.

28. In $\triangle DEF$, $m\angle F = 42$, $m\angle E = 90$, and $DE = 12$. Find *EF* to the nearest integer. (Use the table on page 311.)

29. In right $\triangle XYZ$ with hypotenuse $\overline{XZ}$ if $\cos X = \dfrac{7}{10}$ and $XZ = 24$, then to the nearest integer $XY = $ ___?___.

30. If a tree is 20 m high and the distance from point *P* on the ground to the base of the tree is also 20 m, then the angle of elevation of the top of the tree from point *P* is ___?___.

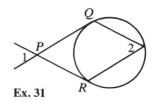

Ex. 31

31. If $\overline{PQ}$ and $\overline{PR}$ are tangents to the circle and $m\angle 1 = 58$, find $m\angle 2$.

32. $\triangle ABC$ is an isosceles right triangle with hypotenuse $\overline{AC}$ of length $2\sqrt{2}$. If medians $\overline{AD}$ and $\overline{BE}$ intersect at *M*, find *AD* and *AM*.

33. Draw two segments and let their lengths be x and y. Construct a segment of length t such that $t = \dfrac{2x^2}{y}$.

34. An equilateral triangle has perimeter 12 cm. Find its area.

35. Find the area of an isosceles trapezoid with legs 7 and bases 11 and 21.

36. **a.** Find the length of a 200° arc in a circle with diameter 24.
b. Find the area of the sector determined by this arc.

37. *B* and *E* are the respective midpoints of $\overline{AC}$ and $\overline{AD}$. Given that $AB = 9$, $BE = 6$, and $AE = 8$, find:
a. the perimeter of $\triangle ACD$
b. the ratio of the areas of $\triangle ABE$ and $\triangle ACD$

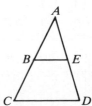

12 AREAS AND VOLUMES OF SOLIDS

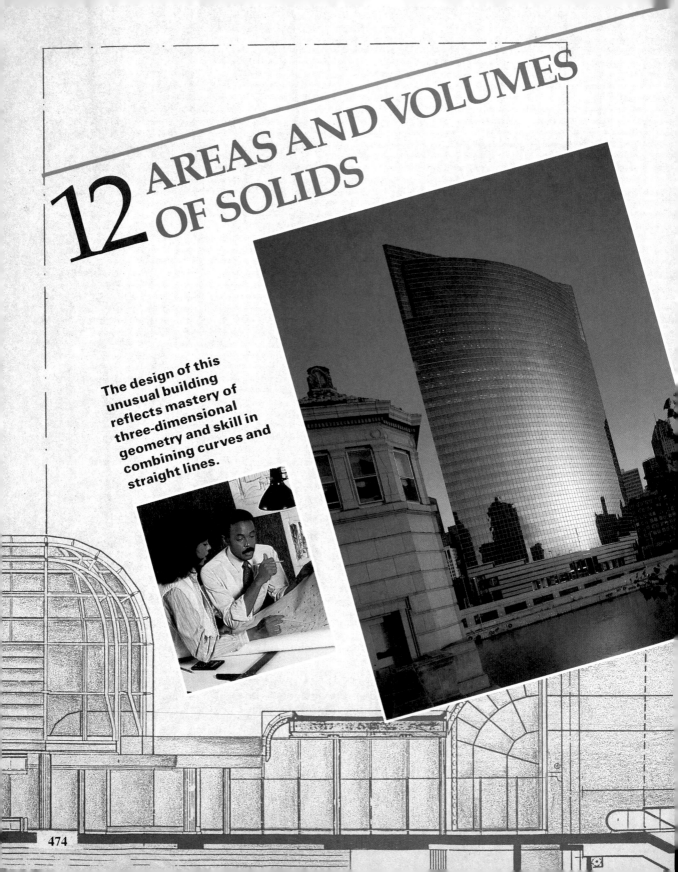

The design of this unusual building reflects mastery of three-dimensional geometry and skill in combining curves and straight lines.

Important Solids

Objectives

1. Identify the parts of prisms, pyramids, cylinders, and cones.
2. Find the lateral areas, total areas, and volumes of right prisms and regular pyramids.
3. Find the lateral areas, total areas, and volumes of right cylinders and right cones.

12-1 *Prisms*

In this chapter you will be calculating surface areas and volumes of special solids. It is possible to begin with some postulates and then prove as theorems the formulas for areas and volumes of solids, as we did for plane figures. Instead, the formulas for solids will be stated as theorems, and informal arguments will be given to show you that the formulas are reasonable.

The first solid we will study is the **prism.** The two shaded faces of the prism shown are its **bases.** Notice that the bases are congruent polygons lying in parallel planes. An **altitude** of a prism is a segment joining the two base planes and perpendicular to both. The length of an altitude is the *height, h,* of the prism.

The faces of a prism that are not its bases are called **lateral faces.** Adjacent lateral faces intersect in parallel segments called **lateral edges.**

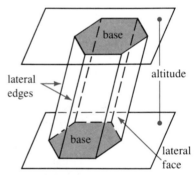

The lateral faces of a prism are parallelograms. If they are rectangles, the prism is a **right prism.** Otherwise the prism is an **oblique prism.** The diagrams below show that a prism is also classified by the shape of its bases. Note that in a right prism, the lateral edges are also altitudes.

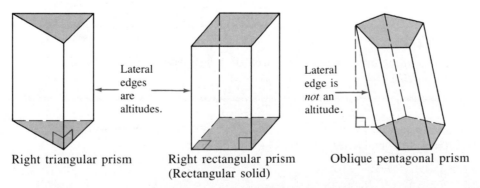

Right triangular prism

Right rectangular prism
(Rectangular solid)

Oblique pentagonal prism

The surface area of a solid is measured in square units. The **lateral area** (L.A.) of a prism is the sum of the areas of its lateral faces. The **total area** (T.A.) is the sum of the areas of all its faces. Using B to denote the area of a base, we have the following formula.

$$\text{T.A.} = \text{L.A.} + 2B$$

If a prism is a right prism, the next theorem gives us an easy way to find the lateral area.

Theorem 12-1

The lateral area of a right prism equals the perimeter of a base times the height of the prism. (L.A. = ph)

The formula for lateral area applies to any right prism. The right pentagonal prism can be used to illustrate the development of the formula:

$$
\begin{aligned}
\text{L.A.} &= ah + bh + ch + dh + eh \\
&= (a + b + c + d + e)h \\
&= \text{perimeter} \cdot h \\
&= ph
\end{aligned}
$$

Prisms have *volume* as well as area. A rectangular solid with square faces is a **cube.** Since each edge of the blue cube shown is 1 unit long, the cube is said to have a volume of 1 cubic unit. The larger rectangular solid has 3 layers of cubes, each layer containing $(4 \cdot 2)$ cubes. Hence its volume is $(4 \cdot 2) \cdot 3$, or 24 cubic units.

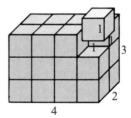

$$
\begin{aligned}
\text{Volume} &= \text{Base area} \times \text{height} \\
&= (4 \cdot 2) \cdot 3 \\
&= 24 \text{ cubic units}
\end{aligned}
$$

The same sort of reasoning is used to find the volume of any right prism.

Theorem 12-2

The volume of a right prism equals the area of a base times the height of the prism. ($V = Bh$)

Volume is measured in cubic units. Some common units for measuring volume are the cubic centimeter (cm^3) and the cubic meter (m^3).

Example 1   A right trapezoidal prism is shown. Find the **(a)** lateral area,
(b) total area, and **(c)** volume.

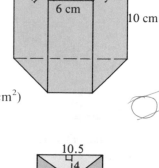

12 cm

4 cm

5 cm

5 cm

6 cm

10 cm

Solution   **a.** First find the perimeter of a base.
$p = 5 + 6 + 5 + 12 = 28$ (cm)

Now use the formula for lateral area.
L.A. $= ph = 28 \cdot 10 = 280$ (cm^2)

b. First find the area of a base.
$B = \frac{1}{2} \cdot 4 \cdot (12 + 6) = 36$ (cm^2)

Now use the formula for total area.
T.A. $=$ L.A. $+ 2B = 280 + 2 \cdot 36 = 352$ (cm^2)

c. $V = Bh = 36 \cdot 10 = 360$ (cm^3)

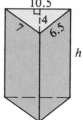

Example 2   A right triangular prism is shown. The volume
is 315. Find the total area.

10.5

7

4

6.5

h

Solution   First find the height of the prism.
$$V = Bh$$
$$315 = (\tfrac{1}{2} \cdot 10.5 \cdot 4)h$$
$$315 = 21h$$
$$15 = h$$

Next find the lateral area.
L.A. $= ph = (10.5 + 6.5 + 7) \cdot 15 = 24 \cdot 15 = 360$

Now use the formula for total area.
T.A. $=$ L.A. $+ 2B = 360 + 2 \cdot 21 = 402$

Classroom Exercises

Exercises 1–6 refer to the right prism shown.

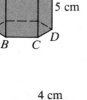

U   T

P   S

Q   R

5 cm

A   D

B   C

1. The prism is called a right ___?___ prism.

2. How many lateral faces are there?

3. What kind of figure is each lateral face?

4. Name two lateral edges and an altitude.

5. The length of an altitude is called the ___?___ of the prism.

6. Suppose the bases are regular hexagons with 4 cm edges.
   a. Find the lateral area.   **b.** Find the base area.
   c. Find the total area.   **d.** Find the volume.

4 cm

7. Can a prism have lateral faces that are triangles?

8. What is the minimum number of faces a prism can have?

9. If two prisms have equal volumes, must they also have equal total areas?

10. **a.** Since 1 yd $= 3$ ft, 1 yd$^2 =$ ___?___ ft^2 and 1 yd$^3 =$ ___?___ ft^3.
    b. Since 1 ft $=$ ___?___ in., 1 ft$^2 =$ ___?___ in.^{2} and 1 ft$^3 =$ ___?___ in.^{3}
    c. Since 1 m $=$ ___?___ cm, 1 m$^2 =$ ___?___ cm^2 and 1 m$^3 =$ ___?___ cm^3.

Written Exercises

Exercises 1–6 refer to rectangular solids with dimensions *l*, *w*, and *h*. Complete the table.

A

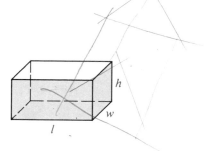

	1.	2.	3.	4.	5.	6.
l	6	50	6	?	9	5x
w	4	30	3	8	?	4x
h	2	15	?	5	2	3x
L.A.	?	?	?	?	60	?
T.A.	?	?	?	?	?	?
V	?	?	54	360	?	?

Exercises 7–12 refer to cubes with edges of length *e*. Complete the table.

	7.	8.	9.	10.	11.	12.
e	3	e	?	?	?	2x
T.A.	?	?	?	?	150	?
V	?	?	1000	64	?	?

13. Find the lateral area of a right pentagonal prism with height 13 and base edges 3.2, 5.8, 6.9, 4.7, and 9.4.

14. A right triangular prism has lateral area 120 cm². If the base edges are 4 cm, 5 cm, and 6 cm long, find the height of the prism.

15. If the edge of a cube is doubled, the total area is multiplied by __?__ and the volume is multiplied by __?__ .

16. If the length, width, and height of a rectangular solid are all tripled, the lateral area is multiplied by __?__ , the total area is multiplied by __?__ , and the volume is multiplied by __?__ .

Facts about the base of a right prism and the height of the prism are given. Sketch each prism and find its lateral area, total area, and volume.

17. Equilateral triangle with side 8; *h* = 10

18. Triangle with sides 9, 12, 15; *h* = 10

B **19.** Isosceles triangle with sides 13, 13, 10; *h* = 7

20. Isosceles trapezoid with bases 10 and 4 and legs 5; *h* = 20

21. Rhombus with diagonals 6 and 8; *h* = 9

22. Regular hexagon with side 8; *h* = 12

23. The container shown has the shape of a rectangular solid. When a rock is submerged, the water level rises 0.5 cm. Find the volume of the rock.

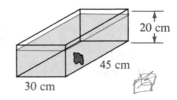

24. A driveway 30 m long and 5 m wide is to be paved with blacktop 3 cm thick. How much will the blacktop cost if it is sold at the price of $175 per cubic meter?

25. A brick with dimensions 20 cm, 10 cm, and 5 cm weighs 1.2 kg. A second brick of the same material has dimensions 25 cm, 15 cm, and 4 cm. What is its weight?

26. A drinking trough for horses is a right trapezoidal prism with dimensions shown below. If it is filled with water, about how much will the water weigh? (*Hint*: 1 m³ of water weighs 1 metric ton.)

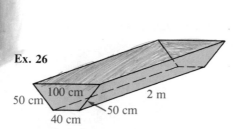

Ex. 26

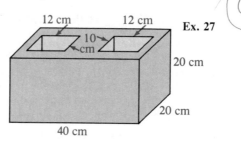

Ex. 27

27. Find the weight, to the nearest kilogram, of the cement block shown. Cement weighs 1700 kg/m³.

28. Find the weight, to the nearest 10 kg, of the steel **I**-beam shown below. Steel weighs 7860 kg/m³.

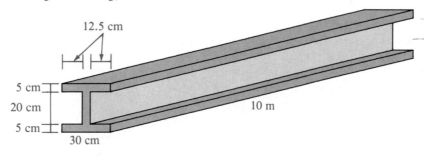

Find the volume and the total surface area of each solid in terms of the given variables.

29.

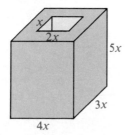

30.

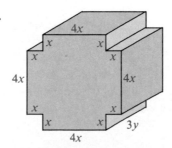

31. The length of a rectangular solid is twice the width, and the height is three times the width. If the volume is 162 cm³, find the total area of the solid.

32. A right prism has square bases with edges that are three times as long as the lateral edges. The prism's total area is 750 m². Find the volume.

33. A diagonal of a box forms a 35° angle with a diagonal of the base, as shown. Use trigonometry to approximate the volume of the box.

34. Refer to Exercise 33. Suppose another box has a base with dimensions 8 by 6 and a diagonal that forms a 70° angle with a diagonal of a base. Show that the ratio of the volumes of the two boxes is $\frac{\tan 35°}{\tan 70°}$.

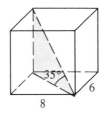

C 35. A right prism has height x and bases that are equilateral triangles with sides x. Show that the volume is $\frac{1}{4}x^3\sqrt{3}$.

36. A right prism has height h and bases that are regular hexagons with sides s. Show that the volume is $\frac{3}{2}s^2h\sqrt{3}$.

37. A rectangular beam of wood 3 m long is cut into six pieces, as shown. Find the volume of each piece in cubic centimeters.

38. A diagonal of a cube joins two vertices not in the same face. If the diagonals are $4\sqrt{3}$ cm long, what is the volume?

39. All nine edges of a right triangular prism are congruent. Find the length of these edges if the volume is $54\sqrt{3}$ cm³.

40. If the length and width of a rectangular solid are each decreased by 20%, by what percent must the height be increased for the volume to remain unchanged? Give your answer to the nearest whole percent.

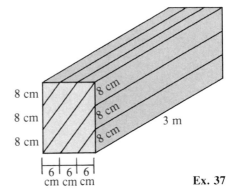

Ex. 37

Challenge

1. Given two rectangles, find one line that divides each rectangle into two parts of equal area.

2. Given three rectangular solids, tell how to find one plane that divides each of these solids into two parts of equal volume.

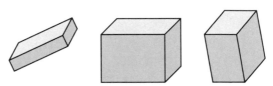

♦ Computer Key-In

A manufacturing company produces metal boxes of different sizes by cutting out square corners from rectangular pieces of metal that measure 9 in. by 12 in. The metal is then folded along the dashed lines to form a box without a top. If a customer requests the box with the greatest possible volume, what dimensions should be used?

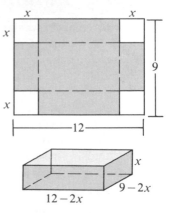

The volume, V, of the box can be expressed in terms of x.

$$V = \text{length} \cdot \text{width} \cdot \text{height}$$
$$= (12 - 2x) \cdot (9 - 2x) \cdot x$$

To form a box, the possible values for x are $0 < x < \frac{9}{2}$.

The following computer program finds the volumes of the boxes produced for ten values of x from 0 to 4.5.

```
10  PRINT "X", "VOLUME"
15  PRINT
20  FOR X = 0 TO 4.5 STEP 0.5
30  LET V = (12 - 2 * X) * (9 - 2 * X) * X
40  PRINT X, V
50  NEXT X
60  END
```

X	VOLUME
0	0
.5	44
1	70
1.5	81
2	80
2.5	70
3	54
3.5	35
4	16
4.5	0

The print-out at the right shows that the maximum volume of the box probably occurs when the value of x is between 1 and 2. Also, the print-out shows that the maximum volume is about 81 in.[3]

Exercises

1. To find a more accurate value for x, change line 20 to:

     FOR X = 1 TO 2 STEP 0.1

Between what values of x does the maximum volume occur?

2. Modify line 20 to find the maximum volume, correct to the nearest tenth of a cubic inch. What are the length, width, and height, correct to the nearest tenth of an inch, of the box with maximum volume?

3. Suppose the manufacturing company cuts square corners out of pieces of metal that measure 8 in. by 15 in.
  a. Express the volume in terms of x.
  b. Find the maximum volume, correct to the nearest tenth of a cubic inch.
  c. What are the length, width, and height of the box with maximum volume? Give each correct to the nearest tenth of an inch.

12-2 *Pyramids*

The diagram shows the pentagonal **pyramid** *V-ABCDE*. Point *V* is the **vertex** of the pyramid and pentagon *ABCDE* is the **base**. The segment from the vertex perpendicular to the base is the **altitude** and its length is the *height*, *h*, of the pyramid.

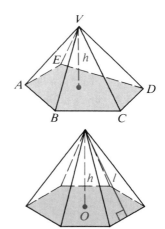

The five triangular faces with *V* in common, such as △*VAB*, are **lateral faces**. These faces intersect in segments called **lateral edges**.

Most of the pyramids you'll study will be **regular pyramids**. These are pyramids with the following properties:

(1) The base is a regular polygon.
(2) All lateral edges are congruent.
(3) All lateral faces are congruent isosceles triangles. The height of a lateral face is called the **slant height** of the pyramid. It is denoted by *l*.
(4) The altitude meets the base at its center, *O*.

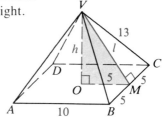

Regular hexagonal pyramid

Example 1   A regular square pyramid has base edges 10 and lateral edges 13. Find **(a)** its slant height and **(b)** its height.

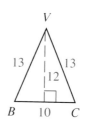

Solution   Use the Pythagorean Theorem.

a. In rt. △*VMC*,
$$l = \sqrt{13^2 - 5^2} = 12.$$

b. In rt. △*VOM*,
$$h = \sqrt{12^2 - 5^2} = \sqrt{119}.$$

Example 2   Find the lateral area of the pyramid in Example 1.

Solution   The four lateral faces are congruent.

area of △*VBC* $= \frac{1}{2} \cdot 10 \cdot 12 = 60$

lateral area = area of 4 lateral faces
= 4 · area of △*VBC*
= 4 · 60 = 240

Example 2 illustrates a simple method for finding the lateral area of a regular pyramid. It is Method 1, summarized below.

To find the lateral area of a **regular** pyramid with *n* lateral faces:

Method 1   Find the area of one lateral face and multiply by *n*.

Method 2   Use the formula L.A. $= \frac{1}{2}pl$, stated as the next theorem.

Theorem 12-3

The lateral area of a regular pyramid equals half the perimeter of the base times the slant height. (L.A. = $\frac{1}{2}pl$)

This formula is developed using Method 1 on the previous page. The area of one lateral face is $\frac{1}{2}bl$. Then:

$$\begin{aligned} \text{L.A.} &= (\tfrac{1}{2}bl)n \\ &= \tfrac{1}{2}(nb)l \end{aligned}$$

Since $nb = p$,   L.A. $= \frac{1}{2}pl$

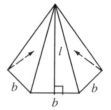

The prism and pyramid below have congruent bases and equal heights. Since the volume of the prism is Bh, the volume of the pyramid must be less than Bh. In fact, it is exactly $\frac{1}{3}Bh$. This result is stated as Theorem 12-4. Although no proof will be given, Classroom Exercise 1 and the Computer Key-In on pages 488–489 help justify the formula.

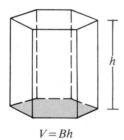

$V = Bh$

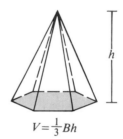

$V = \frac{1}{3}Bh$

Theorem 12-4

The volume of a pyramid equals one third the area of the base times the height of the pyramid. ($V = \frac{1}{3}Bh$)

Example 3   Suppose the regular hexagonal pyramid shown at the right above Theorem 12-4 has base edges 6 and height 12. Find its volume.

Solution   Find the area of the hexagonal base.

Divide the base into six equilateral triangles. Find the area of one triangle and multiply by 6.

Base area $= B = 6(\frac{1}{2} \cdot 6 \cdot 3\sqrt{3}) = 54\sqrt{3}$

Then $V = \frac{1}{3}Bh = \frac{1}{3} \cdot 54\sqrt{3} \cdot 12 = 216\sqrt{3}$

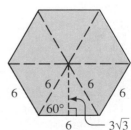

Example 4   A regular triangular pyramid has lateral edge 10 and
height 6. Find the **(a)** lateral area and **(b)** volume.

Solution   **a.** In rt. $\triangle VOA$, $AO = \sqrt{10^2 - 6^2} = \sqrt{64} = 8$.

Since $AO = \frac{2}{3}AM$ (why?), $\frac{2}{3}AM = 8$,
$AM = 12$, and $OM = 4$.
$$l = \sqrt{6^2 + 4^2} = \sqrt{52} = 2\sqrt{13}$$
In $30°\text{-}60°\text{-}90°$ $\triangle AMC$, $CM = \dfrac{12}{\sqrt{3}} = \dfrac{12\sqrt{3}}{3} = 4\sqrt{3}$.

Base edge $= BC = 2 \cdot 4\sqrt{3} = 8\sqrt{3}$
L.A. $= \frac{1}{2}pl = \frac{1}{2}(3 \cdot 8\sqrt{3}) \cdot 2\sqrt{13} = 24\sqrt{39}$

b. Area of base $= B = \frac{1}{2} \cdot BC \cdot AM = \frac{1}{2} \cdot 8\sqrt{3} \cdot 12 = 48\sqrt{3}$
$V = \frac{1}{3}Bh = \frac{1}{3} \cdot 48\sqrt{3} \cdot 6 = 96\sqrt{3}$

Classroom Exercises

1. The diagonals of a cube intersect to divide the cube into six
   congruent pyramids as shown. The base of each pyramid is a
   face of the cube, and the height of each pyramid is $\frac{1}{2}e$.
   a. Use the formula for the volume of a cube to explain why the
   volume of each pyramid is $V = \frac{1}{6}e^3$.
   b. Use the formula in part (a) to show that $V = \frac{1}{3}Bh$. (*Note:*
   This exercise shows that $V = \frac{1}{3}Bh$ gives the correct result
   for *these* pyramids.)

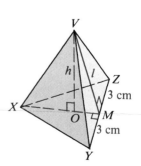

V-ABCD **is a regular square pyramid. Find numerical answers.**

2. $OM = \underline{\quad?\quad}$ 3. $l = \underline{\quad?\quad}$

4. Area of $\triangle VBC = \underline{\quad?\quad}$ 5. L.A. $= \underline{\quad?\quad}$

6. Volume $= \underline{\quad?\quad}$ 7. $VC = \underline{\quad?\quad}$

Each edge of pyramid *V-XYZ* is 6 cm. Find numerical answers.

8. $XM = \underline{\quad?\quad}$ 9. $XO = \underline{\quad?\quad}$

10. $h = \underline{\quad?\quad}$ 11. Base area $= \underline{\quad?\quad}$

12. Volume $= \underline{\quad?\quad}$ 13. Slant height $= \underline{\quad?\quad}$

14. L.A. $= \underline{\quad?\quad}$ 15. T.A. $= \underline{\quad?\quad}$

16. Can the height of a regular pyramid be greater than the slant
    height? Explain.

17. Can the slant height of a regular pyramid be greater than the
    length of a lateral edge? Explain.

18. Can the area of the base of a regular pyramid be greater than
    the lateral area? Explain.

Written Exercises

Copy and complete the table below for the regular square pyramid shown.

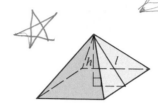

A

	1.	2.	3.	4.	5.	6.
height, h	4	12	24	?	?	6
slant height, l	5	13	?	12	5	?
base edge	?	?	14	?	8	?
lateral edge	?	?	?	15	?	10

You can use the following three steps to sketch a square pyramid.

   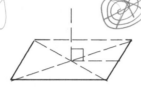   

(1) Draw a parallelogram for the base and sketch the diagonals.

(2) Draw a vertical segment at the point where the diagonals intersect.

(3) Join the vertex to the base vertices.

Sketch each pyramid, as shown above. Then find its lateral area.

7. A regular triangular pyramid with base edge 4 and slant height 6

8. A regular pentagonal pyramid with base edge 1.5 and slant height 9

9. A regular square pyramid with base edge 12 and lateral edge 10

10. A regular hexagonal pyramid with base edge 10 and lateral edge 13

For Exercises 11–14 sketch each square pyramid described. Then find its lateral area, total area, and volume.

11. base edge = 6, height = 4

12. base edge = 16, slant height = 10

13. height = 12, slant height = 13

14. base edge = 16, lateral edge = 17

15. A pyramid has a base area of 16 cm² and a volume of 32 cm³. Find its height.

16. A regular octagonal pyramid has base edge 3 m and lateral area 60 m². Find its slant height.

B **17.** *V-ABCD* is a pyramid with a rectangular base 18 cm long and 10 cm wide. *O* is the center of the rectangle. The height, *VO*, of the pyramid is 12 cm.
   a. Find *VX* and *VY*.
   b. Find the lateral area of the pyramid. (Why can't you use the formula L.A. = $\frac{1}{2}pl$?)

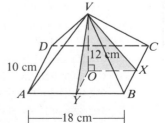

18. Find the height and the volume of a regular hexagonal pyramid with lateral edges 10 ft and base edges 6 ft.

19. The shaded pyramid in the diagram is cut from a rectangular solid. How does the volume of the pyramid compare with the volume of the rectangular solid?

20. A pyramid and a prism both have height 8.2 cm and congruent hexagonal bases with area 22.3 cm². Give the ratio of their volumes. (*Hint*: You do *not* need to calculate their volumes.)

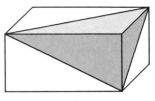

Ex. 19

Exercises 21–25 refer to the regular triangular pyramid shown below.

21. If $AM = 9$ and $VA = 10$, find h and l.

22. **a.** If $BC = 6$, find AM and AO.
    b. If $BC = 6$ and $VA = 4$, find h and l.

23. **a.** If $h = 4$ and $l = 5$, find OM, OA, and BC.
    b. Find the lateral area and the volume.

24. If $VA = 5$ and $h = 3$, find the slant height, the lateral area, and the volume.

25. If $AB = 12$ and $VA = 10$, find the lateral area and the volume.

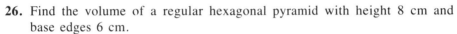

26. Find the volume of a regular hexagonal pyramid with height 8 cm and base edges 6 cm.

27. Use trigonometry to find the volume of the regular pyramid below to the nearest cubic unit.

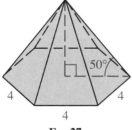

Ex. 27

Ex. 28

28. Show that the ratio of the volumes of the two regular square pyramids shown above is $\dfrac{\tan 40°}{\tan 80°}$.

C 29. All the edges of a regular triangular pyramid are x units long. Find the volume of the pyramid in terms of x.

30. The base of a pyramid is a regular hexagon with sides y cm long. The lateral edges are $2y$ cm long. Find the volume of the pyramid in terms of y.

31. Use a calculator and trigonometry to find the volume of the regular square pyramid shown to the nearest cubic unit.

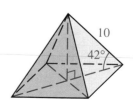

Ex. 31

★ **32.** Different pyramids are inscribed in two identical cubes, as shown below.
    a. Which pyramid has the greater volume?
    b. Which pyramid has the greater total area?

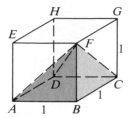

Pyramid *F-ABCD*

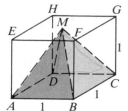

Pyramid *M-ABCD* has vertex *M* at the center of square *EFGH*.

Challenge

1. Accurately draw or construct a large equilateral triangle. Choose any point inside the triangle and carefully measure the distances x, y, z, and h. Then find $x + y + z$.

2. Now choose another point on or inside the triangle and find $x + y + z$. What do you notice? Why does this happen?

3. Use your answers in Exercises 1 and 2 to complete the following statement: From any point inside an equilateral triangle, the sum of the __?__ equals the __?__.

4. Generalize the statement in Exercise 3 from two dimensions to three dimensions.

Mixed Review Exercises

Copy and complete the table for circles.

	1.	2.	3.	4.	5.	6.	7.	8.
Radius	6	11	$\frac{1}{2}$	$3\sqrt{3}$	?	?	?	?
Circumference	?	?	?	?	10π	18π	?	?
Area	?	?	?	?	?	?	49π	15π

Draw a diagram for each exercise.

9. A circle is inscribed in a square with sides 24 mm. Find **(a)** the area of the circle and **(b)** the area of the square.

10. A square is inscribed in a circle with diameter $8\sqrt{2}$. Find **(a)** the perimeter of the square and **(b)** the circumference of the circle.

◆ Calculator Key-In

The Great Pyramid of King Cheops has a
square base with sides 755 feet long. The
original height was 481 feet, but the top
part of the pyramid, which was 31 feet in
height, has been destroyed. Approximately
what percent of the original volume remains?
Answer to the nearest hundredth of a
percent.

◆ Computer Key-In

The earliest pyramids, which were built
about 2750 B.C., are called *step pyramids*
because the lateral faces are not triangles
but a series of great stone steps. To find
the volume of such a pyramid it is necessary
to find the sum of the volumes of the steps,
or layers. Each layer is a rectangular solid
with a square base.

Let us consider a pyramid with base
edges 10 and height 10. Suppose that this
pyramid is made up of 10 steps with equal
heights. The top layer is a cube (base edges
equal the height), and the base edge for each succeeding layer increases by an
amount equal to the height of a layer. As the left side of the diagram at the
bottom of this page shows, the height of each step is $\frac{10}{10} = 1$, and the volume
of the top layer is $V_1 = Bh = (1^2) \cdot 1 = 1$. The volumes of the second and
third layers are $V_2 = (2^2) \cdot 1 = 4$ and $V_3 = (3^2) \cdot 1 = 9$, respectively. Con-
tinuing in this way, the total volume of the pyramid is:

$$V = 1^2 \cdot 1 + 2^2 \cdot 1 + 3^2 \cdot 1 + 4^2 \cdot 1 + \cdots + 9^2 \cdot 1 + 10^2 \cdot 1 = 385$$

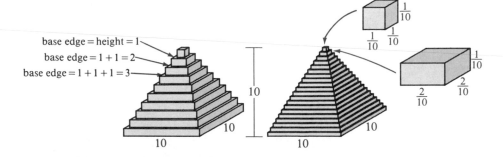

Now consider another pyramid with the same base and height but having 100 steps instead of 10 steps. The height of each layer is $\dfrac{10}{100} = \dfrac{1}{10}$, and the volume of each layer is computed using the formula $V = Bh$:

$$\text{Volume of top layer} \quad = \left(\frac{1}{10}\right)^2 \cdot \frac{1}{10}$$

$$\text{Volume of second layer} \quad = \left(2 \cdot \frac{1}{10}\right)^2 \cdot \frac{1}{10} = \left(\frac{2}{10}\right)^2 \cdot \frac{1}{10}$$

$$\text{Volume of third layer} \quad = \left(3 \cdot \frac{1}{10}\right)^2 \cdot \frac{1}{10} = \left(\frac{3}{10}\right)^2 \cdot \frac{1}{10}$$

$$\vdots \qquad\qquad \vdots$$

$$\text{Volume of 99th layer} \quad = \left(99 \cdot \frac{1}{10}\right)^2 \cdot \frac{1}{10} = \left(\frac{99}{10}\right)^2 \cdot \frac{1}{10}$$

$$\text{Volume of 100th layer} \quad = \left(100 \cdot \frac{1}{10}\right)^2 \cdot \frac{1}{10} = \left(\frac{100}{10}\right)^2 \cdot \frac{1}{10}$$

Thus, the volume of the pyramid is:

$$V = \left(\frac{1}{10}\right)^2 \cdot \frac{1}{10} + \left(\frac{2}{10}\right)^2 \cdot \frac{1}{10} + \left(\frac{3}{10}\right)^2 \cdot \frac{1}{10} + \cdots + \left(\frac{99}{10}\right)^2 \cdot \frac{1}{10} + \left(\frac{100}{10}\right)^2 \cdot \frac{1}{10}$$

The following computer program finds the total volume for the given pyramid with N steps.

```
10  LET V = 0
20  PRINT "HOW MANY STEPS ARE THERE";
30  INPUT N
40  LET H = 10/N
50  FOR X = 1 TO N
60  LET V = V + (X * H) ↑2 * H
70  NEXT X
80  PRINT "VOLUME OF PYRAMID WITH ";N;" STEPS IS ";V
90  END
```

Exercises

1. RUN the given program to verify the volume of the 10-step pyramid and to find the volume of the 100-step pyramid.

2. **a.** Suppose that another pyramid with the same base and height has 1000 steps. RUN the program to find the volume.
   b. Make a chart that shows the volume for the given number of steps: 10, 100, 500, 750, 900, 1000.
   c. As the number of steps increases, what value do the volumes seem to be approaching?
   d. What is the volume of a regular square pyramid with base edge of 10 and height 10?
   e. What can you conclude from comparing the answers to parts (b)–(d)?

12-3 *Cylinders and Cones*

A **cylinder** is like a prism except that its bases are circles instead of polygons. In a **right cylinder,** the segment joining the centers of the circular bases is an **altitude.** The length of an altitude is called the *height*, h, of the cylinder. A radius of a base is also called a **radius,** r, of the cylinder.

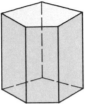
Right prism

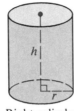
Right cylinder

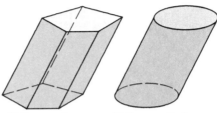
Oblique prism   Oblique cylinder

The diagrams above show the relationship between prisms and cylinders. In the discussion and exercises that follow, the word "cylinder" will always refer to a right cylinder.

It is not surprising that the formulas for cylinders are related to those for prisms: L.A. $= ph$ and $V = Bh$. Since the base of a cylinder is a circle, we substitute $2\pi r$ for p and πr^2 for B and get the following formulas.

Theorem 12-5

The lateral area of a cylinder equals the circumference of a base times the height of the cylinder. (L.A. $= 2\pi rh$)

Theorem 12-6

The volume of a cylinder equals the area of a base times the height of the cylinder. ($V = \pi r^2 h$)

A **cone** is like a pyramid except that its base is a circle instead of a polygon. The relationship between pyramids and cones is shown in the diagrams below.

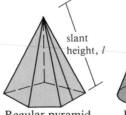
Regular pyramid

Right cone

Oblique pyramid

Oblique cone

Note that "slant height" applies only to a regular pyramid and a right cone. We will use the word "cone" to refer to a right cone.

The formulas for cones are related to those for pyramids: L.A. $= \frac{1}{2}pl$ and $V = \frac{1}{3}Bh$. Since the base of a cone is a circle, we again substitute $2\pi r$ for p and πr^2 for B and get the following formulas.

Theorem 12-7

The lateral area of a cone equals half the circumference of the base times the slant height. (L.A. $= \frac{1}{2} \cdot 2\pi r \cdot l$, or L.A. $= \pi rl$)

Theorem 12-8

The volume of a cone equals one third the area of the base times the height of the cone. ($V = \frac{1}{3}\pi r^2 h$)

So far our study of solids has not included formulas for oblique solids. The volume formulas for cylinders and cones, but *not* the area formulas, can be used for the corresponding oblique solids. (See the Extra on pages 516–517.)

Example 1   A cylinder has radius 5 cm and height 4 cm. Find the **(a)** lateral area, **(b)** total area, and **(c)** volume of the cylinder.

Solution   **a.** L.A. $= 2\pi rh = 2\pi \cdot 5 \cdot 4 = 40\pi$ (cm²)

b. T.A. $=$ L.A. $+ 2B$
$= 40\pi + 2(\pi \cdot 5^2) = 90\pi$ (cm²)

c. $V = \pi r^2 h = \pi \cdot 5^2 \cdot 4 = 100\pi$ (cm³)

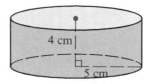

Example 2   Find the **(a)** lateral area, **(b)** total area, and **(c)** volume of the cone shown.

Solution   **a.** First use the Pythagorean Theorem to find l.
$l = \sqrt{6^2 + 3^2} = \sqrt{45} = 3\sqrt{5}$
L.A. $= \pi rl = \pi \cdot 3 \cdot 3\sqrt{5} = 9\pi\sqrt{5}$

b. T.A. $=$ L.A. $+ B = 9\pi\sqrt{5} + \pi \cdot 3^2 = 9\pi\sqrt{5} + 9\pi$

c. $V = \frac{1}{3}\pi r^2 h = \frac{1}{3}\pi \cdot 3^2 \cdot 6 = 18\pi$

Classroom Exercises

1. a. When the label of a soup can is cut off and laid flat, it is a rectangular piece of paper. (See the diagram below.) How are the length and width of this rectangle related to r and h?

   b. What is the area of this rectangle?

2. **a.** Find the lateral areas of cylinders I, II, and III.
   b. Notice that the height of II is twice the height of I.
   Is the lateral area of II twice the lateral area of I?
   c. Notice that the radius of III is twice the radius of I.
   Is the lateral area of III twice the lateral area of I?

3. **a.** Find the total areas of cylinders I, II, and III.
   b. Are the ratios of the total areas the same as those of
   the lateral areas in Exercise 2?

4. **a.** Find the volumes of cylinders I, II, and III.
   b. Notice that the height of II is twice the height of I.
   Is the volume of II twice the volume of I?
   c. Notice that the radius of III is twice the radius of I.
   Is the volume of III twice the volume of I?

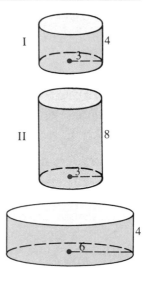

Complete the table for the cone shown.

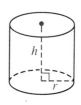

	r	h	l	**L.A.**	**T.A.**	V
5.	3	4	?	?	?	?
6.	?	12	13	?	?	?
7.	6 cm	?	10 cm	?	?	?

8. Describe the intersection of a plane and a cone if the plane is
   the perpendicular bisector of the altitude of the cone.

Written Exercises

You can use the following three steps to sketch a cylinder.

(1) Draw two congruent ovals, one above the other.

(2) Join the ovals with two vertical segments.

(3) Draw in the altitude and a radius.

Sketch each cylinder. Then find its lateral area, total area, and volume.

A **1.** $r = 4$; $h = 5$ **2.** $r = 8$; $h = 10$ **3.** $r = 4$; $h = 3$ **4.** $r = 8$; $h = 15$

5. The volume of a cylinder is 64π. If $r = h$, find r.
6. The lateral area of a cylinder is 18π. If $h = 6$, find r.
7. The volume of a cylinder is 72π. If $h = 8$, find the lateral area.
8. The total area of a cylinder is 100π. If $r = h$, find r.

You can use the following three steps to sketch a cone.

(1) Draw an oval and a vertex point over the center of the oval.

(2) Join the vertex to the oval, as shown.

(3) Draw in the altitude and a radius.

Sketch each cone. Copy and complete the table.

	r	h	l	L.A.	T.A.	V
9.	4	3	?	?	?	?
10.	8	6	?	?	?	?
11.	12	?	13	?	?	?
12.	?	2	6	?	?	?
13.	?	?	15	180π	?	?
14.	21	?	?	609π	?	?
15.	15	?	?	?	?	600π
16.	9	?	?	?	?	324π

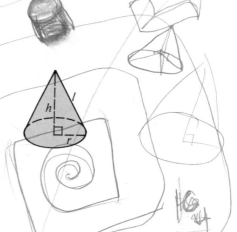

17. In Exercises 9 and 10, the ratio of the radii is $\frac{4}{8}$, or $\frac{1}{2}$, and the ratio of the heights is $\frac{3}{6}$, or $\frac{1}{2}$. Use the answers you found for these two exercises to determine the ratios of the following:
   a. lateral areas   **b.** total areas   **c.** volumes

18. A manufacturer needs to decide which container to use for packaging a product. One container is twice as wide as another but only half as tall. Which container holds more, or do they hold the same amount? Guess first and then calculate the ratio of their volumes.

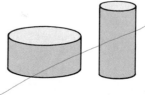

19. A cone and a cylinder both have height 48 and radius 15. Give the ratio of their volumes without calculating the two volumes.

B **20. a.** Guess which contains more, the can or the bottle. (Assume that the top part of the bottle is a complete cone.)
   b. See if your guess is right by finding the volumes of both.

21. A solid metal cylinder with radius 6 cm and height 18 cm is melted down and recast as a solid cone with radius 9 cm. Find the height of the cone.

Ex. 20

22. A pipe is 2 m long and has inside radius 5 cm and outside radius 6 cm. Find the volume of metal contained in the pipe to the nearest cubic centimeter. Use $\pi \approx 3.14$.

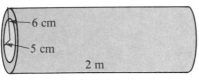

Ex. 22

Ex. 23

23. Water is pouring into a conical (cone-shaped) reservoir at the rate of 1.8 m³ per minute. Find, to the nearest minute, the number of minutes it will take to fill the reservoir. Use $\pi \approx 3.14$.

24. Two water pipes of the same length have inside diameters of 6 cm and 8 cm. These two pipes are replaced by a single pipe of the same length, which has the same capacity as the smaller pipes combined. What is the inside diameter of the new pipe?

25. The total area of a cylinder is 40π. If $h = 8$, find r.

26. The total area of a cylinder is 90π. If $h = 12$, find r.

27. In rectangle $ABCD$, $AB = 10$ and $AD = 6$.
   a. The rectangle is rotated in space about $\overline{AB}$. Describe the solid that is formed and find its volume.
   b. Answer part (a) if the rectangle is rotated about $\overline{AD}$.

28. a. The segment joining (0, 0) and (4, 3) is rotated about the x-axis, forming the lateral surface of a cone. Find the lateral area and the volume of this cone.
   b. Sketch the cone that would be formed if the segment had been rotated about the y-axis. Find the lateral area and the volume of this cone.
   c. Are your answers to parts (a) and (b) the same?

29. Each prism shown below is inscribed in a cylinder with height 10 and radius 6. Find the volume and lateral area of each prism.

a.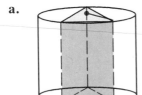

Base is an equilateral triangle.

b.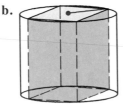

Base is a square.

c.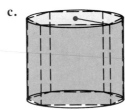

Base is a regular hexagon.

30. An equilateral triangle with 6 cm sides is rotated about an altitude. Draw a diagram and find the volume of the solid formed.

31. A square is rotated in space about a side s. Describe the solid formed and find its volume in terms of s.

32. A cylinder with height 10 and radius 6 is inscribed in a square prism. Make a sketch. Then find the volume of the prism.

33. A regular square pyramid with base edge 4 cm is inscribed in a cone with height 6 cm. What is the volume of the cone?

34. A regular square pyramid is inscribed in a cone with radius 4 cm and height 4 cm.
   a. What is the volume of the pyramid?
   b. Find the slant heights of the cone and the pyramid.

Exs. 33, 34

35. A cone is inscribed in a regular square pyramid with slant height 9 cm and base edge 6 cm. Make a sketch. Then find the volume of the cone.

36. The lateral area of a cone is three-fifths the total area. Find the ratio of the radius and the slant height.

C 37. A regular hexagonal pyramid with base edge 6 and height 8 is inscribed in a cone. Find the lateral areas of the cone and the pyramid.

38. A 120° sector is cut out of a circular piece of tin with radius 6 in. and bent to form the lateral surface of a cone. What is the volume of the cone?

39. In $\triangle ABC$, $AB = 15$, $AC = 20$, and $BC = 25$. The triangle is rotated in space about $\overline{BC}$. Find the volume of the solid formed.

40. An equilateral triangle with sides of length s is rotated in space about one side. Show that the volume of the solid formed is $\frac{1}{4}\pi s^3$.

Challenge

A piece of wood contains a square hole, a circular hole, and a triangular hole, as shown. Explain how one block of wood in the shape of a cube with 2 cm edges can be cut down so that it can pass through, and exactly fit, all three holes.

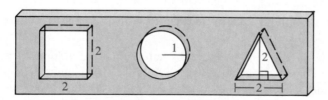

Self-Test 1

For Exercises 1–5 find the lateral area, total area, and volume of each solid.

1. A rectangular solid with length 10, width 8, and height 4.5

2. A regular square pyramid with base edge 24 and slant height 13

3. A cylinder with radius 10 in. and height 7 in.

4. A right hexagonal prism with height 5 cm and base edge 6 cm

5. A cone with height 12 and radius 9

6. The total area of a cube is 2400 m^2. Find the volume.

7. A solid metal cylinder with radius 2 and height 2 is recast as a solid cone with radius 2. Find the height of the cone.

8. A prism with height 2 m and a pyramid with height 5 m have congruent triangular bases. Find the ratio of their volumes.

♦ Calculator Key-In

1. A cylinder has radius 10 and height 12. Suppose that the lateral surface of the cylinder is covered with a thin coat of paint having thickness 0.1. The volume of the paint can be calculated approximately or exactly.
   a. Use the diagrams below to explain the following formula.

   Approximate volume = (lateral area of cylinder) · (thickness of paint)
   $$V \approx (2\pi rh) \cdot (t)$$

   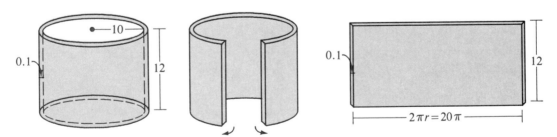

   b. Why is this formula only an approximation of the volume?

2. Use a calculator and the formula to find the approximate volume of paint for each thickness: 0.1, 0.01, 0.001.

3. The exact volume of paint can be found by subtracting the volume of the inner cylinder (the given cylinder) from the volume of the outer cylinder (the given cylinder plus paint). Use a calculator to evaluate the exact volume of paint for each thickness: 0.1, 0.01, 0.001.

4. Compare the values for the approximate volume and exact volume for each thickness in Exercises 2 and 3. What can you conclude?

Similar Solids

Objectives

1. Find the area and the volume of a sphere.
2. State and apply the properties of similar solids.

12-4 *Spheres*

Recall (page 329) that a sphere is the set of all points that are a given distance from a given point. The sphere has many useful applications. One recent application is the development of a spherical blimp. An experimental model of the blimp is shown in the photograph. A spherical shape was selected for this blimp because a sphere gives excellent mobility, stability, hovering capabilities, and lift. The rotation of the top of a sphere away from the direction in which the sphere is traveling provides lifting power.

The surface area and the volume of a sphere are given by the formulas below. After some examples showing how these formulas are used, justifications of the formulas will be presented.

Theorem 12-9

The area of a sphere equals 4π times the square of the radius. $(A = 4\pi r^2)$

Theorem 12-10

The volume of a sphere equals $\frac{4}{3}\pi$ times the cube of the radius. $(V = \frac{4}{3}\pi r^3)$

Example 1 Find the area and the volume of a sphere with radius 2 cm.

Solution
$$A = 4\pi r^2 = 4\pi \cdot 2^2 = 16\pi \ (\text{cm}^2)$$
$$V = \frac{4}{3}\pi r^3 = \frac{4}{3}\pi \cdot 2^3 = \frac{32\pi}{3} \ (\text{cm}^3)$$

Example 2   The area of a sphere is 256π. Find the volume.

Solution   To find the volume, first find the radius.

$$(1)\ A = 256\pi = 4\pi r^2 \qquad (2)\ V = \frac{4}{3}\pi r^3 = \frac{4}{3}\pi \cdot 8^3$$
$$64 = r^2$$
$$8 = r \qquad\qquad\qquad\qquad\qquad = \frac{2048\pi}{3}$$

Example 3   A plane passes 4 cm from the center of a sphere with radius 7 cm. Find the area of the circle of intersection.

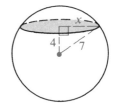

Solution   Let x = radius of the circle.
$$x = \sqrt{7^2 - 4^2} = \sqrt{33}$$
$$\text{Area} = \pi x^2 = \pi(\sqrt{33})^2 = 33\pi \ (\text{cm}^2)$$

Justification of the Volume Formula (Optional)

Any solid can be approximated by a stack of thin circular discs of equal thickness, as shown by the sphere drawn at the right. Each disc is actually a cylinder with height h.

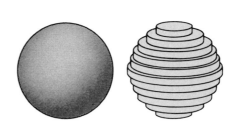

   The sphere, the cylinder, and the double cone below all have radius r and height $2r$. Look at the disc that is x units above the center of each solid.

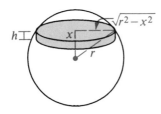
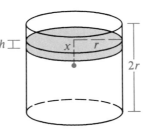
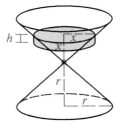

Disc volume:
$$\pi(\sqrt{r^2 - x^2})^2 h = \pi(r^2 - x^2)h$$
$$= \pi r^2 h - \pi x^2 h$$

Disc volume: $\pi r^2 h$

Disc volume: $\pi x^2 h$

   Note from the calculations above that no matter what x is, the volume of the first disc equals the difference between the volumes of the other two discs.

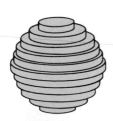

Total volume of
discs in sphere

=

Total volume of
discs in cylinder

−

Total volume of
discs in double cone

The relationship on the bottom of the previous page holds if there are just a few discs approximating each solid or very many discs. If there are very many discs, their total volume will be practically the same as the volume of the solid. Thus:

$$\begin{aligned}\text{Volume of sphere} &= \text{volume of cylinder} - \text{volume of double cone} \\ &= \pi r^2 \cdot 2r - 2(\tfrac{1}{3}\pi r^2 \cdot r) \\ &= 2\pi r^3 - \tfrac{2}{3}\pi r^3 \\ &= \tfrac{4}{3}\pi r^3\end{aligned}$$

Justification of the Area Formula (Optional)

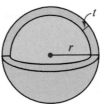

Imagine a rubber ball with inner radius r and rubber thickness t. To find the volume of the rubber, we can use the formula for the volume of a sphere. We just subtract the volume of the inner sphere from the volume of the outer sphere.

$$\begin{aligned}\text{Exact volume of rubber} &= \tfrac{4}{3}\pi(r + t)^3 - \tfrac{4}{3}\pi r^3 \\ &= \tfrac{4}{3}\pi[(r + t)^3 - r^3] \\ &= \tfrac{4}{3}\pi[r^3 + 3r^2t + 3rt^2 + t^3 - r^3] \\ &= 4\pi r^2 t + 4\pi r t^2 + \tfrac{4}{3}\pi t^3\end{aligned}$$

The volume of the rubber can be found in another way as well. If we think of a small piece of the rubber ball, its approximate volume would be its outer area A times its thickness t. The same thing is true for the whole ball.

$$\text{Volume of rubber} \approx \text{Surface area} \cdot \text{thickness}$$
$$V \approx A \cdot t$$

Now we can equate the two formulas for the volume of the rubber:

$$A \cdot t \approx 4\pi r^2 t + 4\pi r t^2 + \tfrac{4}{3}\pi t^3$$

If we divide both sides of the equation by t, we get the following result:

$$A \approx 4\pi r^2 + 4\pi r t + \tfrac{4}{3}\pi t^2$$

This approximation for A gets better and better as the layer of rubber gets thinner and thinner. As t approaches zero, the last two terms in the formula for A also approach zero. As a result, the surface area gets closer and closer to $4\pi r^2$. Thus:

$$A = 4\pi r^2$$

This is exactly what we would expect, since the surface area of a ball clearly depends on the size of the radius, not on the thickness of the rubber.

Classroom Exercises

Copy and complete the table for spheres.

	1.	2.	3.	4.	5.	6.
Radius	1	8	$3t$	?	?	?
Area	?	?	?	36π	100π	?
Volume	?	?	?	?	?	$\dfrac{4000\pi}{3}$

A plane passes h cm from the center of a sphere with radius r cm. Find the area of the circle of intersection, shaded in the diagram, for the given values.

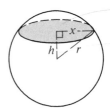

7. $r = 5$
$\quad h = 3$

8. $r = 17$
$\quad h = 8$

9. $r = 7$
$\quad h = 6$

Written Exercises

Copy and complete the table for spheres.

A

	1.	2.	3.	4.	5.	6.	7.	8.
Radius	7	5	$\frac{1}{2}$	$\frac{3}{4}k$	?	?	$\sqrt{2}$	?
Area	?	?	?	?	64π	324π	?	?
Volume	?	?	?	?	?	?	?	288π

9. If the radius of a sphere is doubled, the area of the sphere is multiplied by __?__ and the volume is multiplied by __?__.

10. Repeat Exercise 9 if the radius of the sphere is tripled.

11. The area of a sphere is π cm². Find the diameter of the sphere.

12. The volume of a sphere is 36π m³. Find its area.

13. Find the area of the circle formed when a plane passes 2 cm from the center of a sphere with radius 5 cm.

14. Find the area of the circle formed when a plane passes 7 cm from the center of a sphere with radius 8 cm.

15. A sphere has radius 2 and a hemisphere (''half'' a sphere) has radius 4. Compare their volumes.

16. A scoop of ice cream with diameter 6 cm is placed in an ice-cream cone with diameter 5 cm and height 12 cm. Is the cone big enough to hold all the ice cream if it melts?

17. Approximately 70% of the Earth's surface is covered by water. Use a calculator to find the area covered by water to the nearest million square kilometers. (The radius of the Earth is approximately 6380 km.)

Ex. 16

18. **a.** Find the volume, correct to the nearest cubic centimeter, of a sphere inscribed in a cube with edges 6 cm long. Use $\pi \approx 3.14$.
    b. Find the volume of the region inside the cube but outside the sphere.

B 19. A silo of a barn consists of a cylinder capped by a hemisphere, as shown. Find the volume of the silo.

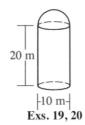

20. About two cans of paint are needed to cover the hemispherical dome of the silo shown. Approximately how many cans are needed to paint the rest of the silo's exterior?

Exs. 19, 20

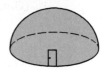

21. An experimental one-room house is a hemisphere with a floor. If three cans of paint are needed to cover the floor, how many cans will be needed to paint the ceiling? (Ignore door and windows.)

22. A hemispheric bowl with radius 25 contains water whose depth is 10. What is the area of the water's surface?

Ex. 21

23. A solid metal ball with radius 8 cm is melted down and recast as a solid cone with the same radius.
    a. What is the height of the cone?
    b. Use a calculator to show that the lateral area of the cone is about 3% more than the area of the sphere.

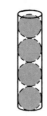

24. Four solid metal balls fit snugly inside a cylindrical can. A geometry student claims that two extra balls of the same size can be put into the can, provided all six balls can be melted and the molten liquid poured into the can. Is the student correct? (*Hint*: Let the radius of each ball be r.)

Ex. 24

25. A sphere with radius r is inscribed in a cylinder. Find the volume of the cylinder in terms of r.

26. A sphere is inscribed in a cylinder. Show that the area of the sphere equals the lateral area of the cylinder.

Exs. 25, 26

27. A double cone is inscribed in the cylinder shown. Find the volume of the space inside the cylinder but outside the double cone.

28. A hollow rubber ball has outer radius 11 cm and inner radius 10 cm.
    a. Find the exact volume of the rubber. Then evaluate the volume to the nearest cubic centimeter.
    b. The volume of the rubber can be approximated by the formula:
    $$V \approx \text{inner surface area} \cdot \text{thickness of rubber}$$
    Use this formula to approximate V. Compare your answer with the answer in part (a).
    c. Is the approximation method used in part (b) better for a ball with a thick layer of rubber or a ball with a thin layer?

Ex. 27

Ex. 28

29. A circle with diameter 9 in. is rotated about a diameter. Find the area and volume of the solid formed.

30. A cylinder with height 12 is inscribed in a sphere with radius 10. Find the volume of the cylinder.

C **31.** A cylinder with height $2x$ is inscribed in a sphere with radius 10.
  a. Show that the volume of the cylinder, V, is
  $$2\pi x(100 - x^2).$$
  b. By using calculus, one can show that V is maximum when $x = \dfrac{10\sqrt{3}}{3}$. Substitute this value for x to find the maximum volume V.
  c. (Optional) Use a calculator or a computer to evaluate $V = 2\pi x(100 - x^2)$ for various values of x between 0 and 10. Show that the maximum volume V occurs when $x \approx 5.77$.

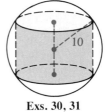

Exs. 30, 31

32. A cone is inscribed in a sphere with radius 10, as shown.
  a. Show that the volume of the cone, V, is
  $$\tfrac{1}{3}\pi(100 - x^2)(10 + x).$$
  b. By using calculus, one can show that V is maximum when $x = \frac{10}{3}$. Substitute this value for x to find the maximum volume V.
  c. (Optional) Use a calculator or a computer to evaluate the volume for various values of x between 0 and 10. Show that the maximum volume V occurs when $x = \frac{10}{3}$.

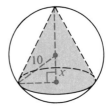

33. Sketch two intersecting spheres with radii 15 cm and 20 cm, respectively. The centers of the spheres are 25 cm apart. Find the area of the circle that is formed by the intersection. (*Hint*: Use Exercise 42 on page 289.)

34. A sphere is inscribed in a cone with radius 6 cm and height 8 cm. Find the volume of the sphere.

Challenge

In a training exercise, two research submarines are assigned to take any two positions in the ocean at a depth between 100 m and 500 m and within a square area with 6000 m sides. If one submarine takes its position in the center of the square area at the 100-meter depth, what is the probability that the other submarine is within 400 m? Use $\pi \approx 3.14$. (*Hint*: Consider each submarine to be a point in a rectangular solid. Use the ideas of geometric probability.)

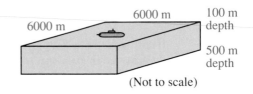

(Not to scale)

♦ Calculator Key-In

1. The purpose of this exercise is to suggest the following statement: Of all figures in a plane with a fixed perimeter, the circle has the greatest possible area. If each regular polygon below has perimeter 60 mm and the circle has circumference 60 mm, find the area of each to the nearest square millimeter.

a.

$s = \underline{\ \ ?\ \ }$
$A \approx \underline{\ \ ?\ \ }$

b.

$x = \underline{\ \ ?\ \ }$
$A = \underline{\ \ ?\ \ }$

c.

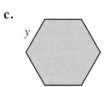

$y = \underline{\ \ ?\ \ }$
$A \approx \underline{\ \ ?\ \ }$

d.

$r \approx \underline{\ \ ?\ \ }$
$A \approx \underline{\ \ ?\ \ }$

2. The regular pyramid, the cube, and the sphere below all have total surface area 600 mm³. Find the volume of each to the nearest cubic millimeter.

a.

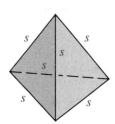

$$\text{T.A.} = 4\left(\frac{s^2\sqrt{3}}{4}\right) = 600 \text{ mm}^2$$

$s \approx \underline{\ \ ?\ \ }$

$$V = \frac{s^3\sqrt{2}}{12} \approx \underline{\ \ ?\ \ }$$

(See Ex. 29, page 486.)

b.

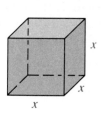

$$\text{T.A.} = 6x^2 = 600 \text{ mm}^2$$

$x = \underline{\ \ ?\ \ }$

$$V = x^3 = \underline{\ \ ?\ \ }$$

c.

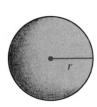

$$\text{T.A.} = 4\pi r^2 = 600 \text{ mm}^2$$

$r \approx \underline{\ \ ?\ \ }$

$$V = \frac{4}{3}\pi r^3 \approx \underline{\ \ ?\ \ }$$

   d. Use the results of parts (a)–(c) to complete the following statement, which is similar to the one in Exercise 1: Of all solid figures with a fixed __?__, the __?__ has the __?__.

★ 3. Suppose the plane figures in Exercise 1 all have area 900 cm². Find the perimeter of each polygon and the circumference of the circle to the nearest centimeter. What do your answers suggest about all plane figures with a fixed area?

★ 4. Suppose the solid figures in Exercise 2 all have volume 1000 cm³. Find the total surface area of each to the nearest square centimeter. What do your answers suggest about all solid figures with a fixed volume?

♦ Computer Key-In

The volume of a sphere with radius 10 can be approximated by cylindrical discs with equal heights, as discussed on page 498. It is convenient to work with the upper half of the sphere, then double the result.

Suppose you use ten discs to approximate the upper hemisphere, as shown at the left below.

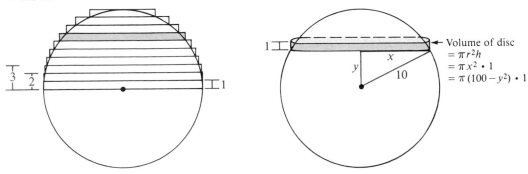

The diagram at the right above shows that the volume of a disc y units from the center of the sphere is $V = \pi(100 - y^2)$. You can substitute $y = 0, 1, 2, \ldots, 9$ to compute the volumes of the ten discs.

Now suppose you use n discs to approximate the upper hemisphere. Then the height of each disc equals $\dfrac{10}{n}$, and the volume of a disc y units from the center of the sphere is $V = \pi(100 - y^2) \cdot \dfrac{10}{n}$. The following computer program adds the volumes of the n discs, then doubles the result. Note that line 80 calculates the volume of the sphere using the formula $V = \frac{4}{3}\pi r^3$.

```
10  LET Y = 0
15  LET V = 0
20  PRINT "HOW MANY DISCS";
25  INPUT N
30  FOR I = 1 TO N
40  LET Y = (I - 1) * 10/N
50  LET V = V + 3.14159 * (100 - Y↑2) * (10/N)
60  NEXT I
70  PRINT "VOLUME OF DISCS IS ";2 * V
80  PRINT "VOLUME OF SPHERE IS ";4/3 * 3.14159 * 10↑3
90  END
```

Exercises

1. Use 10 for N and RUN the program. By about what percent does the disc method overapproximate the volume of the sphere?

2. RUN the program to find the total volume of n discs for each value of n.
   a. 20   **b.** 50   **c.** 100   **d.** 1000

   As n increases, does the approximate volume approach the actual volume?

3. The program uses discs that extend outside the sphere, so it yields approximations greater than the actual volume. To use discs that are inside the sphere, replace line 40 with: LET Y = I * (10/N)
   a. RUN the new program for N = 100.
   b. Find the average of the result in part (a) and the result in Exercise 2, part (c). Is the average close to the actual volume of the sphere?

Application *Geodesic Domes*

A spherical dome is an efficient way of enclosing space, since a sphere holds a greater volume than any other container with the same surface area. (See Calculator Exercise 2 on page 503.) In 1947, R. Buckminster Fuller patented the *geodesic dome*, a framework made by joining straight pieces of steel or aluminum tubing in a network of triangles. A thin cover of aluminum or plastic is then attached to the tubing.

The segments forming the network are of various lengths, but the vertices are all equidistant from the center of the dome, so that they lie on a sphere. When we follow a chain of segments around the dome, we find that they approximate a circle on this sphere, often a great circle. It is this property that gives the dome design its name: A *geodesic* on any surface is a path of minimum length between two points on the surface, and on a sphere these shortest paths are arcs of great circles.

Though the geodesic dome is very light and has no internal supports, it is very strong, and standardized parts make construction of the dome relatively easy. Domes have been used with success for theaters, exhibition halls, sports arenas, and greenhouses.

The United States Pavilion that Fuller designed for Expo '67 in Montreal uses two domes linked together. The design of this structure is illustrated at the right. The red triangular network is the outer dome, the black hexagons form the inner dome, and the blue segments represent the trusses that tie the two domes together. The arrows mark one of the many chains of segments that form arcs of circles on the dome. You can see all of these features of the structure in the photograph at the right, which shows a view from inside the dome.

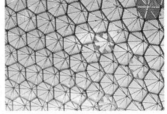

Although a grid of hexagons will interlock nicely to cover the plane, they cannot interlock to cover a sphere unless twelve of the hexagons are changed to pentagons. (The reason for this is given in the exercises on the next page.)

Exercises

For any solid figure with polygons for faces, Euler's formula, $F + V - E = 2$, must hold. In this formula, F, V, and E stand for the number of faces, vertices, and edges, respectively, that the figure has.

1. Verify Euler's formula for each figure below.

    a. Cube   **b.** Octahedron   **c.** Icosahedron (20 faces)

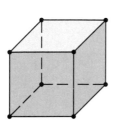 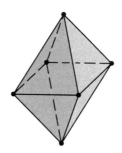 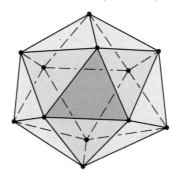

2. If each edge of the icosahedron above is trisected and the trisection points are "popped out" to the surface of the circumscribing sphere, one of the many possible geodesic domes is formed. By subdividing the edges of the icosahedron into more than three parts, the resulting geodesic dome is even more spherelike, as shown. In the diagram, find a group of equilateral triangles that cluster to form **(a)** a hexagon, and **(b)** a pentagon.

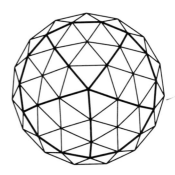

3. In this exercise, Euler's formula will be used to show that **(a)** the dome's framework *cannot* consist of hexagons only, and **(b)** the framework *can* consist of hexagons plus exactly 12 pentagons.

    a. Use an indirect proof and assume that the framework has n faces, all hexagons. Thus $F = n$. To find V, the number of vertices on the framework, notice that each hexagon contributes 6 vertices, but each vertex is shared by 3 hexagons. Thus $V = \dfrac{6n}{3}$. To find E, the number of edges of the framework, notice that each hexagon contributes 6 edges, but each edge is shared by 2 hexagons. Thus $E = \dfrac{6n}{2}$. According to Euler's Formula: $F + V - E$ must equal 2. Does it? What does this contradiction tell you?

    b. Suppose that 12 of the n faces of the framework are pentagons. Show that $V = \dfrac{6n - 12}{3}$ and that $E = \dfrac{6n - 12}{2}$. Then use algebra to show that $F + V - E = 2$. Since Euler's formula is satisfied, a dome framework can be constructed when n faces consist of 12 pentagons and $n - 12$ hexagons.

Biographical Note   *R. Buckminster Fuller*

The early curiosity shown by R. Buckminster Fuller (1895–1983) about the
world around him led to a life of in-
vention and philosophy. As a mathematician he made many contributions
to the fields of engineering, architecture,
and cartography. His ultimate goal was
always "to do more with less." Thus his discoveries
often had economic and ecological implications.

Fuller's inventions include the geodesic dome (see
pages 505 and 506), the 3-wheeled Dymaxion car, and
the Dymaxion Air-ocean World Map on which he was
able to project the spherical earth as a flat surface without
any visible distortions. He also designed other structures
that were based upon triangles and circles instead of the
usual rectangular surfaces.

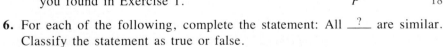

Mixed Review Exercises

Trapezoid *ABCD* is similar to trapezoid *PQRS*.

1. Find the scale factor of the trapezoids.

2. Draw an altitude from point *B* and use the Pythagorean
Theorem to find the value of *w*.

3. Find the values of *x*, *y*, and *z*.

4. a. Find the perimeter of each trapezoid.
   b. Find the ratio of the perimeters.
   c. Compare the ratio of the perimeters to the scale
factor you found in Exercise 1.

5. a. Find the area of each trapezoid.
   b. Find the ratio of the areas.
   c. Compare the ratio of the areas to the scale factor
you found in Exercise 1.

6. For each of the following, complete the statement: All __?__ are similar.
Classify the statement as true or false.

   a. squares
   c. circles
   e. right triangles
   g. regular pentagons

   b. rectangles
   d. rhombuses
   f. equilateral triangles
   h. isosceles trapezoids

12-5 *Areas and Volumes of Similar Solids*

One of the best-known attractions in The Hague, the Netherlands, is a unique miniature city, Madurodam, consisting of five acres of carefully crafted reproductions done on a scale of 1:25. Everything in this model city works, including the two-mile railway network, the canal locks, the harbor fireboats, and the nearly 50,000 tiny street lights. In this section you will study the relationship between scale factors of *similar solids* and their areas and volumes.

Similar solids are solids that have the same shape but not necessarily the same size. It's easy to see that all spheres are similar. To decide whether two other solids are similar, determine whether bases are similar and corresponding lengths are proportional.

Right cylinders

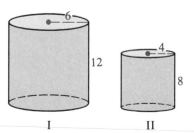

The bases are similar because all circles are similar. The lengths are proportional because $\dfrac{6}{4} = \dfrac{12}{8}$.

So the solids are similar with scale factor $\dfrac{3}{2}$.

Regular square pyramids

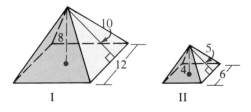

The bases are similar because all squares are similar. The lengths are proportional because $\dfrac{12}{6} = \dfrac{8}{4} = \dfrac{10}{5}$.

So the solids are similar with scale factor $\dfrac{2}{1}$.

The table below shows the ratios of the perimeters, areas, and volumes for both pairs of similar solids shown on page 508. Notice the relationship between the scale factor and the ratios in each column.

	Cylinders I and II	**Pyramids I and II**
Scale factor	$\dfrac{3}{2}$	$\dfrac{2}{1}$
$\dfrac{\text{Base perimeter (I)}}{\text{Base perimeter (II)}}$	$\dfrac{2\pi \cdot 6}{2\pi \cdot 4} = \dfrac{6}{4}$, or $\dfrac{3}{2}$	$\dfrac{4 \cdot 12}{4 \cdot 6} = \dfrac{12}{6}$, or $\dfrac{2}{1}$
$\dfrac{\text{L.A. (I)}}{\text{L.A. (II)}}$	$\dfrac{2\pi \cdot 6 \cdot 12}{2\pi \cdot 4 \cdot 8} = \dfrac{9}{4}$, or $\dfrac{3^2}{2^2}$	$\dfrac{\frac{1}{2} \cdot 48 \cdot 10}{\frac{1}{2} \cdot 24 \cdot 5} = \dfrac{4}{1}$, or $\dfrac{2^2}{1^2}$
$\dfrac{\text{Volume (I)}}{\text{Volume (II)}}$	$\dfrac{\pi \cdot 6^2 \cdot 12}{\pi \cdot 4^2 \cdot 8} = \dfrac{27}{8}$, or $\dfrac{3^3}{2^3}$	$\dfrac{\frac{1}{3} \cdot 12^2 \cdot 8}{\frac{1}{3} \cdot 6^2 \cdot 4} = \dfrac{8}{1}$, or $\dfrac{2^3}{1^3}$

The results shown in the table above are generalized in the following theorem. (See Exercises 22–27 for proofs.)

Theorem 12-11

If the scale factor of two similar solids is $a:b$, then

(1) the ratio of corresponding perimeters is $a:b$.

(2) the ratio of the base areas, of the lateral areas, and of the total areas is $a^2:b^2$.

(3) the ratio of the volumes is $a^3:b^3$.

Example For the similar solids shown, find the ratios of the **(a)** base perimeters, **(b)** lateral areas, and **(c)** volumes.

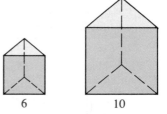

Solution The scale factor is $6:10$, or $3:5$.
    a. Ratio of base perimeters $= 3:5$
    b. Ratio of lateral areas $= 3^2:5^2 = 9:25$
    c. Ratio of volumes $= 3^3:5^3 = 27:125$

Theorem 12-11 is the three-dimensional counterpart of Theorem 11-7 on page 457. (Take a minute to compare these theorems.) There is a similar relationship between the two cases shown below.

In two dimensions:
If $\overline{XY} \parallel \overline{AB}$, then
$\triangle VXY \sim \triangle VAB$.

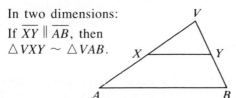

In three dimensions:
If plane $XYZ \parallel$ plane ABC,
then $V\text{-}XYZ \sim V\text{-}ABC$.

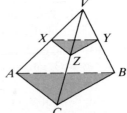

Classroom Exercises

Tell whether the solids in each pair are similar. Explain your answer.

1.

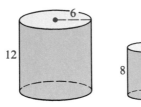

Right cylinders

2.

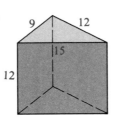

Right prisms

3. For the prisms in Exercise 2, find the ratios of:
   a. the lateral areas   **b.** the total areas   **c.** the volumes

4. Two spheres have diameters 24 cm and 36 cm.
   a. What is the ratio of the areas?   **b.** What is the ratio of the volumes?

5. Two spheres have volumes 2π m^3 and 16π m^3. Find the ratios of:
   a. the volumes   **b.** the diameters   **c.** the areas

Complete the table below, which refers to two similar cones.

	6.	7.	8.	9.	10.	11.
Scale factor	3:4	5:7	?	?	?	?
Ratio of base circumferences	?	?	2:1	?	?	?
Ratio of slant heights	?	?	?	1:6	?	?
Ratio of lateral areas	?	?	?	?	4:9	?
Ratio of total areas	?	?	?	?	?	?
Ratio of volumes	?	?	?	?	?	8:125

12. Plane *PQR* is parallel to the base of the pyramid and bisects the altitude. Find the following ratios.
   a. The perimeter of $\triangle PQR$ to the perimeter of $\triangle ABC$
   b. The lateral area of the top part of the pyramid to the lateral area of the whole pyramid
   c. The lateral area of the top part of the pyramid to the lateral area of the bottom part
   d. The volume of the top part of the pyramid to the volume of the bottom part

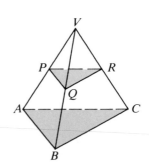

13. Find each ratio in Exercise 12 if the height of the top pyramid is 3 cm and the height of the whole pyramid is 5 cm.

Written Exercises

A 1. Two cones have radii 6 cm and 9 cm. The heights are 10 cm and 15 cm, respectively. Are the cones similar?

2. The heights of two right prisms are 18 ft and 30 ft. The bases are squares with sides 8 ft and 15 ft, respectively. Are the prisms similar?

3. Two similar cylinders have radii 3 and 4. Find the ratios of the following:
   a. heights   **b.** base circumferences   **c.** lateral areas   **d.** volumes

4. Two similar pyramids have heights 12 and 18. Find the ratios of the following:
   a. base areas   **b.** lateral areas   **c.** total areas   **d.** volumes

5. Assume that the Earth and the moon are smooth spheres with diameters 12,800 km and 3,200 km, respectively. Find the ratios of the following:
   a. lengths of their equators   **b.** areas   **c.** volumes

6. Two similar cylinders have lateral areas 81π and 144π. Find the ratios of:
   a. the heights   **b.** the total areas   **c.** the volumes

7. Two similar cones have volumes 8π and 27π. Find the ratios of:
   a. the radii   **b.** the slant heights   **c.** the lateral areas

8. Two similar pyramids have volumes 3 and 375. Find the ratios of:
   a. the heights   **b.** the base areas   **c.** the total areas

9. The package of a model airplane kit states that the scale is $1:200$. Compare the amounts of paint required to cover the model and the actual airplane. (Assume the paint on the model is as thick as that on the actual airplane.)

10. The scale for a certain model freight train is $1:48$. If the model hopper car (usually used for carrying coal) will hold 90 in.^{3} of coal, what is the capacity in cubic feet of the actual hopper car? (*Hint*: See Exercise 10, page 477.)

11. Two similar cones have radii of 4 cm and 6 cm. The total area of the smaller cone is 36π cm^2. Find the total area of the larger cone.

B 12. A diagonal of one cube is 2 cm. A diagonal of another cube is $4\sqrt{3}$ cm. The larger cube has volume 64 cm^3. Find the volume of the smaller cube.

13. Two balls made of the same metal have radii 6 cm and 10 cm. If the smaller ball weighs 4 kg, find the weight of the larger ball to the nearest 0.1 kg.

14. A snow man is made using three balls of snow with diameters 30 cm, 40 cm, and 50 cm. If the head weighs about 6 kg, find the total weight of the snow man. (Ignore the arms, eyes, nose and mouth.)

15. A certain kind of string is sold in a ball 6 cm in diameter and in a ball 12 cm in diameter. The smaller ball costs $1.00 and the larger one costs $6.50. Which is the better buy?

16. Construction engineers know that the strength of a column is proportional to the area of its cross section. Suppose that the larger of two similar columns is three times as high as the smaller column.
    a. The larger column is __?__ times as strong as the smaller column.
    b. The larger column is __?__ times as heavy as the smaller column.
    c. Which can support more, *per pound of column material*, the larger or the smaller column?

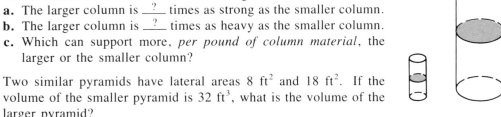

17. Two similar pyramids have lateral areas 8 ft^2 and 18 ft^2. If the volume of the smaller pyramid is 32 ft^3, what is the volume of the larger pyramid?

18. Two similar cones have volumes 12π and 96π. If the lateral area of the smaller cone is 15π, what is the lateral area of the larger cone?

19. A plane parallel to the base of a cone divides the cone into two pieces. Find the ratios of the following:
    a. The areas of the shaded circles
    b. The lateral area of the top part of the cone to the lateral area of the whole cone
    c. The lateral area of the top part of the cone to the lateral area of the bottom part
    d. The volume of the top part of the cone to the volume of the whole cone
    e. The volume of the top part of the cone to the volume of the bottom part

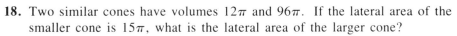

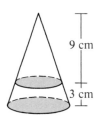

9 cm

3 cm

20. Redraw the figure for Exercise 19, changing the 9 cm and 3 cm dimensions to 10 cm and 4 cm, respectively. Then find the five ratios described in Exercise 19.

21. A pyramid with height 15 cm is separated into two pieces by a plane parallel to the base and 6 cm above it. What are the volumes of these two pieces if the volume of the original pyramid is 250 cm^3?

The purpose of Exercises 22–27 is to prove Theorem 12-11 for some similar solids.

22. Two spheres have radii a and b. Prove that the ratio of the areas is $a^2 : b^2$.

23. Two spheres have radii a and b. Prove that the ratio of the volumes is $a^3 : b^3$.

24. Two similar cones have radii r_1 and r_2 and heights h_1 and h_2. Prove that the ratio of the volumes is $h_1^3 : h_2^3$.

25. Two similar cones have radii r_1 and r_2 and slant heights l_1 and l_2. Prove that the ratio of the lateral areas is $r_1^2 : r_2^2$.

26. The bases of two similar right prisms are regular pentagons with base edges e_1 and e_2 and base areas B_1 and B_2. The heights are h_1 and h_2. Prove that the ratio of the lateral areas is $e_1^2 : e_2^2$.

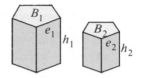

27. Refer to Exercise 26. Prove that the ratio of the volumes of the prisms is $e_1^3 : e_2^3$.

C 28. The purpose of this exercise is to prove that if plane *XYZ* is parallel to plane *ABC*, then *V-XYZ* ~ *V-ABC*. To do this, suppose that $VA = k \cdot VX$ and show that every edge of *V-ABC* is *k* times as long as the corresponding edge of *V-XYZ*. (*Hint*: Use Theorem 3-1.)

29. A plane parallel to the base of a pyramid separates the pyramid into two pieces with equal volumes. If the height of the pyramid is 12, find the height of the top piece.

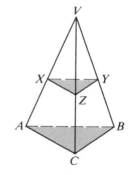

Self-Test 2

1. Find the area and volume of a sphere with diameter 6 cm.

2. The volume of a sphere is $\frac{32}{3}\pi$ m³. Find the area of the sphere.

3. The students of a school decide to bury a time capsule consisting of a cylinder capped by two hemispheres. Find the volume of the time capsule shown.

4. Find the area of the circle formed when a plane passes 12 cm from the center of a sphere with radius 13 cm.

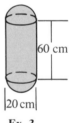

60 cm

20 cm

Ex. 3

5. One regular triangular pyramid has base edge 8 and height 6. A similar pyramid has height 4.
   a. Find the base edge of the smaller pyramid.
   b. Find the ratio of the total areas of the pyramids.

6. The base areas of two similar prisms are 32 and 200, respectively.
   a. Find the ratio of their heights. **b.** Find the ratio of their volumes.

Challenge

A pattern for a model is shown. Can you tell what it is? To build it, make a large copy of the pattern on stiff paper. Cut along the solid lines, fold along the dashed lines, and tape the edges together.

   If you want to make a pattern for a figure, think about the number of faces, their shapes, and how the edges are related. Try to create and build models for a triangular prism, a triangular pyramid, and a cone.

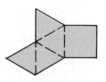

♦ Calculator Key-In

Each diagram shows a rectangle inscribed in an isosceles triangle with legs 5 and base 6. There are many more such rectangles. Which one has the greatest area?

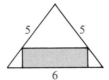 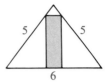

To solve the problem, let *CDEF* represent any rectangle inscribed in isosceles △*ABV* with legs 5 and base 6. If we let *OD* = *x* and *ED* = *y*, then the area of the rectangle is 2*xy*. Our goal is to express this area in terms of *x* alone. Then we can find out how the area changes as *x* changes.

1. In right △*VOB*, *OB* = 3 and *VB* = 5. Thus *VO* = 4 by the Pythagorean Theorem.

2. △*EDB* ~ △*VOB*   (Why?)

3. $\dfrac{ED}{VO} = \dfrac{DB}{OB}$ (Why?)

4. $\dfrac{y}{4} = \dfrac{3 - x}{3}$ (By substitution in Step 3)

5. $y = \dfrac{4}{3}(3 - x)$ (Multiplication Property of =)

6. Area of rectangle: $A = 2xy = 2x \cdot \dfrac{4}{3}(3 - x) = \dfrac{8x(3 - x)}{3}$

Use the formula in Step 6 and a calculator to find the area for many values of *x*. Calculate 3 − *x* first, then multiply by *x*, then multiply by 8, and divide by 3.

x	Area
0	0
0.25	1.83333
0.5	3.33333
0.75	4.5
1	5.33333
1.25	5.83333
1.5	6
1.75	5.83333
2	5.33333
2.25	4.5
2.50	3.33333
2.75	1.83333
3	0

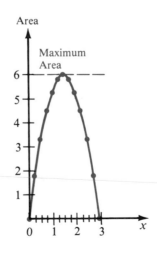

The table was used to make a graph showing how the area varies with *x*. Both the table and the graph suggest that the greatest area, 6 square units, occurs when *x* = 1.5.

Exercises

Suppose the original triangle had sides 5, 5, and 8 instead of 5, 5, and 6.

1. Draw a diagram. Then show that $A = \dfrac{3x(4 - x)}{2}$.

2. Find the value of x for which the greatest area occurs.

◆ Computer Key-In

A rectangle is inscribed in an isosceles triangle with legs 5 and base 6 and the triangle is rotated in space about the altitude to the base. The resulting figure is a cylinder inscribed in a cone with diameter 6 and slant height 5, as shown below. Which of the cylinders such as these has the greatest volume?

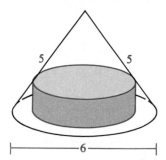 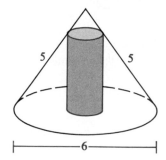 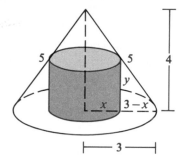

The diagram at the right above shows a typical inscribed cylinder. Using similar triangles, we have the proportion $\dfrac{y}{4} = \dfrac{3 - x}{3}$. Thus $y = \dfrac{4}{3}(3 - x)$.

The volume of the cylinder is found as follows:

$$V = \pi x^2 y = \pi x^2 \cdot \tfrac{4}{3}(3 - x) \approx \tfrac{4}{3}(3.14159)x^2(3 - x)$$

The following program in BASIC evaluates V for various values of x.

```
10  PRINT "X", "VOLUME"
20  FOR X = 0 TO 3 STEP 0.25
30  LET V = 4/3 * 3.14159 * X↑2 * (3 - X)
40  PRINT X, V
50  NEXT X
60  END
```

Exercises

1. RUN the program. Make a graph that shows how the volume varies with x. For what value of x did you find the greatest volume?

2. Suppose the original triangle has sides 5, 5, and 8 instead of 5, 5, and 6. Rotate the triangle in space about the altitude to the base.
   a. Draw a diagram. Show that $V = \tfrac{3}{4}\pi x^2(4 - x)$.
   b. Change lines 20 and 30 of the program and RUN the revised program to find the value of x for which the greatest volume occurs.

Extra	*Cavalieri's Principle*

Suppose you have a right rectangular prism and divide it horizontally into thin rectangular slices. The base of each rectangular slice, or *cross section*, has the same area as the base of the prism. If you rearrange the slices, the total volume of the slices does not change.

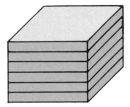 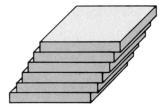 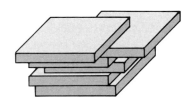

Bonaventura Cavalieri (1598–1647), an Italian mathematician, used this idea to compare the volumes of solids. His conclusion is known as *Cavalieri's Principle*.

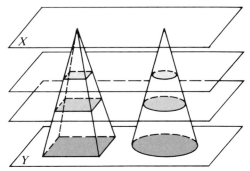

Cavalieri's Principle

If two solids lying between parallel planes have equal heights and all cross sections at equal distances from their bases have equal areas, then the solids have equal volumes.

Using Cavalieri's Principle you can find the volume of an oblique prism. Consider a right triangular prism and an oblique prism that have the same base and height.

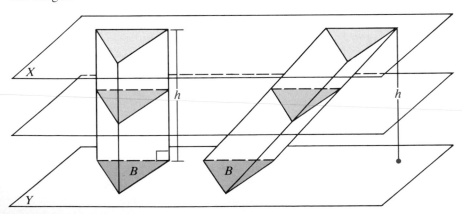

By Theorem 12-2, the volume of the right prism is $V = Bh$. Every cross section of each prism has the same area as that prism's base. Since the base areas are equal, the corresponding cross sections of the two prisms have equal areas. Therefore by Cavalieri's Principle, the volume of the oblique prism also is $V = Bh$.

You can use similar reasoning to show that the volume formulas given for a regular pyramid, right cylinder, and right cone hold true for the corresponding oblique solids.

$$V = Bh \text{ for } any \text{ prism or cylinder}$$
$$V = \tfrac{1}{3}Bh \text{ for } any \text{ pyramid or cone}$$

Exercises

Find the volume of the solid shown with the given altitude.

1.

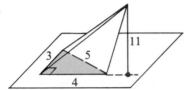

2.

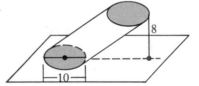

3. Find the volume of an oblique cone with radius 4 and height 3.5.

4. The oblique square prism shown below has base edge 3. A lateral edge that is 15 makes a 60° angle with the plane containing the base. Find the exact volume.

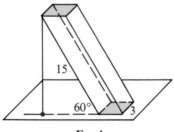

Ex. 4

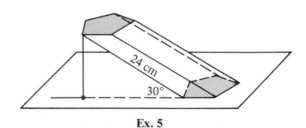

Ex. 5

5. The volume of the oblique pentagonal prism shown above is 96 cm³. A lateral edge that is 24 cm makes a 30° angle with the plane containing the base. Find the area of the base.

6. Refer to the justification of the formula for the volume of a sphere given on pages 498–499. How does Cavalieri's Principle justify the statement that the volume of the sphere is equal to the difference between the volumes of the cylinder and the double cone?

Chapter Summary

1. The list below summarizes area and volume formulas for solids. The cylinder formulas are special cases of the prism formulas with $p = 2\pi r$ and $B = \pi r^2$. Also the cone formulas are special cases of the pyramid formulas with the same substitutions for p and B. To find the total area of each of the four solids, add lateral area to the area of the base(s).

Right prism	L.A. $= ph$	$V = Bh$
Right cylinder	L.A. $= 2\pi rh$	$V = \pi r^2 h$
Regular pyramid	L.A. $= \frac{1}{2}pl$	$V = \frac{1}{3}Bh$
Right cone	L.A. $= \pi rl$	$V = \frac{1}{3}\pi r^2 h$
Sphere	$A = 4\pi r^2$	$V = \frac{4}{3}\pi r^3$

2. If the scale factor of two similar solids is $a:b$, then
   a. the ratio of corresponding perimeters is $a:b$.
   b. the ratio of corresponding areas is $a^2:b^2$.
   c. the ratio of the volumes is $a^3:b^3$.

Chapter Review

1. In a right prism, each __?__ is also an altitude. 12–1

2. Find the lateral area of a right octagonal prism with height 12 and base edge 7.

3. Find the total area and volume of a rectangular solid with dimensions 8, 6, and 5.

4. A right square prism has base edge 9 and volume 891. Find the total area.

5. Find the volume of a regular triangular pyramid with base edge 8 and height 10. 12–2

6. A regular pentagonal pyramid has base edge 6 and lateral edge 5. Find the slant height and the lateral area.

A regular square pyramid has base edge 30 and total area 1920.

7. Find the area of the base, the lateral area, and the slant height.

8. Find the height and the volume of the pyramid.

9. Find the lateral area and the total area of a cylinder with radius 4 and height 3. 12–3

10. Find the lateral area, total area, and volume of a cone with radius 6 cm and slant height 10 cm.

11. A cone has volume 8π cm^3 and height 6 cm. Find its slant height.

12. The radius of a cylinder is doubled and its height is halved. How does the volume change?

13. A sphere has radius 7 m. Use $\pi \approx \frac{22}{7}$ to find the approximate area of the sphere. **12–4**

14. Find, in terms of π, the volume of a sphere with diameter 12 ft.

15. Find the volume of a sphere with area 484π cm^2.

Plane *RST* ∥ **plane** *XYZ* **and** *VS* : *VY* = 1 : 3.

16. $\dfrac{\text{perimeter of } \triangle RST}{\text{perimeter of } \triangle XYZ} = \underline{\ \ ?\ \ }$ **12–5**

17. $\dfrac{\text{total area of small pyramid}}{\text{total area of large pyramid}} = \underline{\ \ ?\ \ }$

18. $\dfrac{\text{volume of small pyramid}}{\text{volume of bottom part}} = \underline{\ \ ?\ \ }$

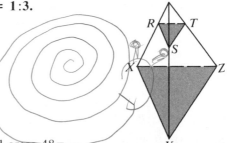

19. Two similar cylinders have lateral areas 48π and 27π. Find the ratio of their volumes.

Chapter Test

1. Find the volume and the total area of a cube with edge $2k$.

2. A regular square pyramid has base edge 3 cm and volume 135 cm^2. Find the height.

3. A cone has radius 8 and height 6. Find the volume.

4. Find the lateral area and the total area of the cone in Exercise 3.

5. A right triangular prism has height 20 and base edges 5, 12, and 13. Find the total area.

6. Find the volume of the prism in Exercise 5.

7. A cylinder has radius 6 cm and height 4 cm. Find the lateral area.

8. Find the volume of the cylinder in Exercise 7.

9. A regular square pyramid has lateral area 60 m^2 and base edge 6 m. Find the volume.

10. A sphere has radius 6 cm. Find the area and the volume.

11. Two cones have radii 12 cm and 18 cm, and have slant heights 18 cm and 24 cm. Are the cones similar? Explain.

12. A regular pyramid has height 18 and total area 648. A similar pyramid has height 6. Find the total area of the smaller pyramid.

13. The volumes of two similar rectangular solids are 1000 cm^3 and 64 cm^3. What is the ratio of their lateral areas?

14. A cone and a cylinder each have radius 3 and height 4. Find the ratio of their volumes and of their lateral areas.

15. Find the volume of a sphere with area 9π.

16. A cylinder with radius 7 has total area 168π cm^2. Find its height.

Preparing for College Entrance Exams

Strategy for Success

Questions on college entrance exams often require knowledge of areas and volumes. Be sure that you know all the important formulas developed in Chapters 11 and 12. To avoid doing unnecessary calculations, be sure to read the directions to find out whether answers may be expressed in terms of π.

Indicate the best answer by writing the appropriate letter.

1. A cone has volume 320π and height 15. Find the total area.
   (A) 200π (B) 368π (C) 264π (D) 136π (E) 320π

2. Two equilateral triangles have perimeters 6 and $9\sqrt{3}$. The ratio of their areas is:
   (A) $2:3\sqrt{3}$ (B) $2\sqrt{3}:9$ (C) $4:27$ (D) $4:9$ (E) $8:81\sqrt{3}$

3. A sphere has volume 288π. Its diameter is:
   (A) $12\sqrt{6}$ (B) $6\sqrt{2}$ (C) 6   (D) $12\sqrt{2}$ (E) 12

4. *RSTW* is a rhombus with $m \angle R = 60$ and $RS = 4$. If *X* is the midpoint of $\overline{RS}$, find the area of trapezoid *SXWT*.
   (A) 12   (B) 16   (C) $8\sqrt{3}$ (D) $6\sqrt{3}$ (E) $16 - 2\sqrt{2}$

5. If *ABCD* is a square and $AE = y$, the area of *ABCDE* is
   (A) $\frac{5}{4}y^2$ (B) $\frac{5}{2}y^2$ (C) $3y^2$
   (D) $(4 + \frac{1}{2}\sqrt{3})y^2$ (E) $(\frac{1}{2} + \sqrt{2})y^2$

Compare the quantity in Column A with that in Column B. Select:
(A) if the quantity in Column A is greater;
(B) if the quantity in Column B is greater;
(C) if the two quantities are equal;
(D) if the relationship cannot be determined from the information given.

Column A	Column B

6. volume of square pyramid   volume of square prism

7.   area of triangle   area of sector

Cumulative Review: Chapters 1–12

For Exercises 1–9 classify each statement as true or false.

A 1. No more than one plane contains two given intersecting lines.

2. The conditional "*p* only if *q*" is equivalent to "if *p*, then *q*."

3. If the vertex angle of an isosceles triangle has measure *j*, then the measure of a base angle is $180 - 2j$.

4. In $\triangle RST$, if $m \angle R = 48$ and $m \angle S = 68$, then $RT > RS$.

5. If right $\triangle JEH$ has hypotenuse $\overline{JE}$, then $\tan J = \dfrac{JH}{EH}$.

6. It is possible to construct an angle of measure 105.

7. The area of a triangle with sides 3, 3, and 2 is $4\sqrt{2}$.

8. When a square is circumscribed about a circle, the ratio of the areas is $4 : \pi$.

9. A triangle with sides of length $\sqrt{3}$, 2, and $\sqrt{7}$ is a right triangle.

B 10. In $\square JKLM$, $m \angle J = \frac{3}{2}x$ and $m \angle L = x + 17$. Find the numerical measure of $\angle K$.

11. Given: $\overline{WZ} \perp \overline{ZY}$; $\overline{WX} \perp \overline{XY}$; $\overline{WX} \cong \overline{YZ}$
Prove: $\overline{WZ} \parallel \overline{XY}$

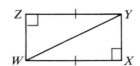

12. Prove: If the diagonals of a parallelogram are perpendicular, then the parallelogram must be a rhombus.

13. For $\triangle JKL$ and $\triangle XYZ$ use the following statement:
"If $\angle J \cong \angle X$ and $\angle K \cong \angle Y$, then $\triangle JKL \sim \triangle XYZ$."
a. Name the postulate or theorem that justifies the statement.
b. Write the converse of the statement. Is the converse true or false?

14. Find the value of *x* in the diagram at the right.

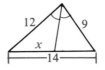

15. $\overline{AB}$ and $\overline{CD}$ are chords of $\odot P$ intersecting at *X*. If $AX = 7.5$, $BX = 3.2$, $CD = 11$, and $CX > DX$, find CX.

16. Describe each possibility for the locus of points in space that are equidistant from the sides of a $\triangle ABC$ and 4 cm from *A*.

17. $\overparen{AB}$ lies on $\odot O$ with $m\overparen{AB} = 60$. $\odot O$ has radius 8. Find AB.

18. A regular square pyramid has base edge 10 and height 12. Find its total area and volume.

19. A cylinder has a radius equal to its height. The total area of the cylinder is 100π cm^2. Find its volume.

20. A sphere has a diameter of 1.8 cm. Find its surface area to the nearest square centimeter. (Use $\pi \approx 3.14$.)

13 COORDINATE GEOMETRY

Every eighteen seconds, a plane lands at or departs O'Hare International Airport in Chicago. Modern instruments and a knowledge of coordinate geometry enable air traffic controllers to track and direct these planes safely and efficiently.

Geometry and Algebra

Objectives

1. State and apply the distance formula.
2. State and apply the general equation of a circle.
3. State and apply the slope formula.
4. Determine whether two lines are parallel, perpendicular, or neither.
5. Understand the basic properties of vectors.
6. State and apply the midpoint formula.

13-1 *The Distance Formula*

Some of the terms you have used in your study of graphs are reviewed below.

Origin: Point O

Axes: x-axis and y-axis

Quadrants: Regions I, II, III, and IV

Coordinate plane: The plane of the x-axis and the y-axis

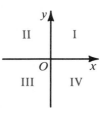

The arrowhead on each axis shows the positive direction.

You can easily find the distance between two points that lie on a horizontal line or on a vertical line.

The distance between A and B is 4.
Using the x-coordinates of A and B:

$|3 - (-1)| = 4$, or $|(-1) - 3| = 4$

The distance between C and D is 3.
Using the y-coordinates of C and D:

$|1 - (-2)| = 3$, or $|(-2) - 1| = 3$

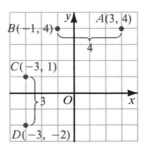

When two points do not lie on a horizontal or vertical line, you can find the distance between the points by using the Pythagorean Theorem.

Example 1 Find the distance between points $A(4, -2)$ and $B(1, 2)$.

Solution Draw the horizontal and vertical segments shown. The coordinates of T are $(1, -2)$. Then $AT = 3$, $BT = 4$, $(AB)^2 = 3^2 + 4^2 = 25$, and $AB = 5$.

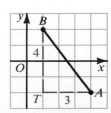

Using a method suggested by Example 1, you can find a formula for the distance between points $P(x_1, y_1)$ and $Q(x_2, y_2)$. First draw a right triangle as shown. The coordinates of R are (x_2, y_1).

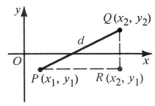

$$PR = |x_2 - x_1|; \quad QR = |y_2 - y_1|$$
$$d^2 = (PR)^2 + (QR)^2$$
$$= |x_2 - x_1|^2 + |y_2 - y_1|^2$$
$$= (x_2 - x_1)^2 + (y_2 - y_1)^2$$
$$d = \sqrt{(x_2 - x_1)^2 + (y_2 - y_1)^2}$$

Since d represents distance, d cannot be negative.

Theorem 13-1   *The Distance Formula*

The distance d between points (x_1, y_1) and (x_2, y_2) is given by:

$$d = \sqrt{(x_2 - x_1)^2 + (y_2 - y_1)^2}$$

Example 2   Find the distance between points $(-4, 2)$ and $(2, -1)$.

Solution 1   Draw a right triangle. The legs have lengths 6 and 3.

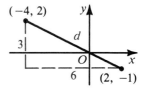

$$d^2 = 6^2 + 3^2 = 36 + 9 = 45$$
$$d = \sqrt{45} = \sqrt{9} \cdot \sqrt{5} = 3\sqrt{5}$$

Solution 2   Let (x_1, y_1) be $(-4, 2)$ and (x_2, y_2) be $(2, -1)$.
Then $d = \sqrt{(x_2 - x_1)^2 + (y_2 - y_1)^2}$
$$= \sqrt{(2 - (-4))^2 + ((-1) - 2)^2}$$
$$= \sqrt{6^2 + (-3)^2} = \sqrt{36 + 9} = \sqrt{45} = 3\sqrt{5}$$

You can use the distance formula to find an equation of a circle. Example 3 shows how to do this.

Example 3   Find an equation of a circle with center $(5, 6)$ and radius 4.

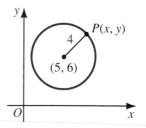

Solution   Let $P(x, y)$ represent any point on the circle. Since the distance from P to the center is 4,
$$\sqrt{(x - 5)^2 + (y - 6)^2} = 4,$$
$$\text{or } (x - 5)^2 + (y - 6)^2 = 16.$$

Either of these two equations is an equation of the circle, but the second equation is the one usually used.

Example 3 can be generalized to give the theorem at the top of the next page.

Theorem 13-2

An equation of the circle with center (a, b) and radius r is

$$(x - a)^2 + (y - b)^2 = r^2.$$

Example 4 Find the center and the radius of the circle with equation $(x - 1)^2 + (y + 2)^2 = 9$. Sketch the graph.

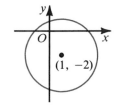

Solution $(x - 1)^2 + (y - (-2))^2 = 3^2$

The center is point $(1, -2)$ and the radius is 3.
The graph is shown at the right.

Classroom Exercises

1. What is the x-coordinate of every point that lies on a vertical line through C?

2. Which of the following points lie on a horizontal line through C?

   (2, 4)   (2, −4)   (0, 4)
   (4, 3)   (15, 4)   (−4, 3)

3. Find OD and BF.

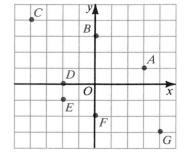

In Exercises 4–9 state: a. the coordinates of T
                              b. the lengths of the legs of the right triangle
                              c. the length of the segment shown

4.
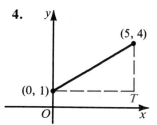

5.
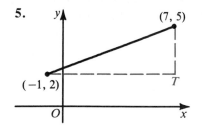

6.
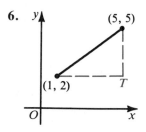

7.
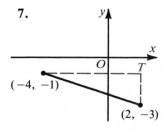

8.
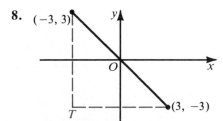

9.
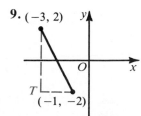

10. Find the distance between the points named. Give all answers in simplest form.
    a. $(0, 0)$ and $(5, -3)$
    b. $(3, -2)$ and $(-5, -2)$
    c. $(4, 4)$ and $(-3, -3)$

11. Find the center and the radius of each circle.
    a. $(x - 2)^2 + y^2 = 1$
    b. $(x + 2)^2 + (y - 8)^2 = 16$
    c. $x^2 + (y + 5)^2 = 112$
    d. $(x + 3)^2 + (y + 7)^2 = 14$

12. Find an equation of the circle that has the given center and radius.
    a. Center $(2, 5)$; radius 3
    b. Center $(-2, 0)$; radius 5
    c. Center $(-2, 3)$; radius 10
    d. Center (j, k); radius n

Written Exercises

Find the distance between the two points. If necessary, you may draw graphs but you shouldn't need to use the distance formula.

A   **1.** $(-2, -3)$ and $(-2, 4)$ **2.** $(3, 3)$ and $(-2, 3)$
      3. $(3, -4)$ and $(-1, -4)$ **4.** $(0, 0)$ and $(3, 4)$

Use the distance formula to find the distance between the two points.

      5. $(-6, -2)$ and $(-7, -5)$ **6.** $(3, 2)$ and $(5, -2)$
      7. $(-8, 6)$ and $(0, 0)$ **8.** $(12, -1)$ and $(0, -6)$

Find the distance between the points named. Use any method you choose.

      9. $(5, 4)$ and $(1, -2)$ **10.** $(-2, -2)$ and $(5, 7)$
   11. $(-2, 3)$ and $(3, -2)$ **12.** $(-4, -1)$ and $(-4, 3)$

Given points A, B, and C. Find AB, BC, and AC. Are A, B, and C collinear? If so, which point lies between the other two?

   13. $A(0, 3)$, $B(-2, 1)$, $C(3, 6)$ **14.** $A(5, -5)$, $B(0, 5)$, $C(2, 1)$
   15. $A(-5, 6)$, $B(0, 2)$, $C(3, 0)$ **16.** $A(3, 4)$, $B(-3, 0)$, $C(-1, 1)$

Find the center and the radius of each circle.

   17. $(x + 3)^2 + y^2 = 49$ **18.** $(x + 7)^2 + (y - 8)^2 = \dfrac{36}{25}$
   19. $(x - j)^2 + (y + 14)^2 = 17$ **20.** $(x + a)^2 + (y - b)^2 = c^2$

Write an equation of the circle that has the given center and radius.

   21. $C(3, 0)$; $r = 8$ **22.** $C(0, 0)$; $r = 6$
   23. $C(-4, -7)$; $r = 5$ **24.** $C(-2, 5)$; $r = \dfrac{1}{3}$

   25. Sketch the graph of $(x - 3)^2 + (y + 4)^2 = 36$.
   26. Sketch the graph of $(x - 2)^2 + (y - 5)^2 \le 9$.

In Exercises 27–32 find and then compare lengths of segments.

B **27.** Show that the triangle with vertices $A(-3, 4)$, $M(3, 1)$, and $Y(0, -2)$ is isosceles.

28. Quadrilateral *TAUL* has vertices $T(4, 6)$, $A(6, -4)$, $U(-4, -2)$, and $L(-2, 4)$. Show that the diagonals are congruent.

29. Triangles *JAN* and *RFK* have vertices $J(-2, -2)$, $A(4, -2)$, $N(2, 2)$, $R(8, 1)$, $F(8, 4)$, and $K(6, 3)$. Show that $\triangle JAN$ is similar to $\triangle RFK$.

30. The vertices of $\triangle KAT$ and $\triangle IES$ are $K(3, -1)$, $A(2, 6)$, $T(5, 1)$, $I(-4, 1)$, $E(-3, -6)$, and $S(-6, -1)$. What word best describes the relationship between $\triangle KAT$ and $\triangle IES$?

31. Find the area of the rectangle with vertices $B(8, 0)$, $T(2, -9)$, $R(-1, -7)$, and $C(5, 2)$.

32. Show that the triangle with vertices $D(0, 0)$, $E(3, 1)$, and $F(-2, 6)$ is a right triangle, then find the area of the triangle.

33. There are twelve points, each with integer coordinates, that are 10 units from the origin. List the points. (*Hint:* Recall the 6, 8, 10 right triangle.)

34. a. List twelve points, each with integer coordinates, that are 5 units from $(-8, 1)$.
 b. Find an equation of the circle containing these points.

In Exercises 35–38 find an equation of the circle described and sketch the graph.

35. The circle has center $(0, 6)$ and passes through point $(6, 14)$.

36. The circle has center $(-2, -4)$ and passes through point $(3, 8)$.

37. The circle has diameter $\overline{RS}$ where *R* is $(-3, 2)$ and *S* is $(3, 2)$.

38. The circle has center (p, q) and is tangent to the *x*-axis.

39. a. Find the radii of the circles
$$x^2 + y^2 = 25 \text{ and } (x - 9)^2 + (y - 12)^2 = 100.$$
 b. Find the distance between the centers of the circles.
 c. Explain why the circles must be externally tangent.
 d. Sketch the graphs of the circles.

40. a. Find the radii of the circles
$$x^2 + y^2 = 2 \text{ and } (x - 3)^2 + (y - 3)^2 = 32.$$
 b. Find the distance between the centers of the circles.
 c. Explain why the circles must be internally tangent.
 d. Sketch the graphs of the circles.

41. Discover and prove something about the quadrilateral with vertices $R(-1, -6)$, $A(1, -3)$, $Y(11, 1)$, and $J(9, -2)$.

42. Discover and prove two things about the triangle with vertices $K(-3, 4)$, $M(3, 1)$, and $J(-6, -2)$.

C **43.** It is known that $\triangle GHM$ is isosceles. *G* is point $(-2, -3)$, *H* is point $(-2, 7)$, and the *x*-coordinate of *M* is 4. Find all five possible values for the *y*-coordinate of *M*.

44. Find the coordinates of the point that is equidistant from $(-2, 5)$, $(8, 5)$, and $(6, 7)$.

45. Find the center and the radius of the circle $x^2 + 4x + y^2 - 8y = 16$.
    (*Hint*: Express the given equation in the form
    $$(x - a)^2 + (y - b)^2 = r^2.)$$

♦ Computer Key-In

The graph shows a quarter-circle inscribed in a square with area 1. If points are picked at random inside the square, some of them will also be inside the quarter-circle. Let n be the number of points picked inside the square and let q be the number of these points that fall inside the quarter-circle. If many, many points are picked at random inside the square, the following ratios are approximately equal:

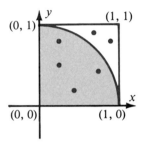

$$\frac{\text{Area of quarter-circle}}{\text{Area of square}} \approx \frac{q}{n}$$

$$\frac{\text{Area of quarter-circle}}{1} \approx \frac{q}{n}$$

$$\text{Area of whole circle} \approx 4 \times \frac{q}{n}$$

Any point (x, y) in the square region has coordinates such that $0 < x < 1$ and $0 < y < 1$. (Note that this restriction excludes points on the boundaries of the square.) A computer can pick a random point inside the unit square by choosing two random numbers x and y between 0 and 1. We let d be the distance from O to the point (x, y). By the Pythagorean Theorem, $d = \sqrt{x^2 + y^2}$. Do you see that if $d < 1$, the point lies inside the quarter-circle?

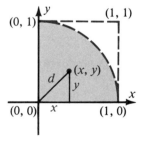

Exercises

1. Write a computer program to do all of the following:
    a. Choose n random points (x, y) inside the unit square.
    b. Using the distance formula test each point chosen to see whether it lies inside the quarter-circle.
    c. Count the number of points (q) which *do* lie inside the quarter-circle.
    d. Print out the value of $4 \times \frac{q}{n}$.

2. RUN your program for $n = 100$, $n = 500$, and $n = 1000$.

3. Calculate the area of the circle, using the formula given on page 446. Compare this result with your computer approximations.

13-2 *Slope of a Line*

The effect of steepness, or *slope*, must be considered in a variety of everyday situations. Some examples are the grade of a road, the pitch of a roof, the incline of a wheelchair ramp, and the tilt of an unloading platform, such as the one at a paper mill in Maine shown in the photograph at the right. In this section, the informal idea of steepness is generalized and made precise by the mathematical concept of *slope of a line through two points*.

Informally, slope is the ratio of the *change in y* (vertical change) to the *change in x* (horizontal change). The **slope,** denoted by m, of the nonvertical line through the points (x_1, y_1) and (x_2, y_2) is defined as follows:

$$\text{slope } m = \frac{y_2 - y_1}{x_2 - x_1}$$

$$= \frac{\text{change in } y}{\text{change in } x}$$

When you are given several points on a line you can use any two of them to compute the slope. Furthermore, the slope of a line does not depend on the order in which the points are chosen because $\dfrac{y_2 - y_1}{x_2 - x_1} = \dfrac{y_1 - y_2}{x_1 - x_2}$.

Example 1 Find the slope of each segment.

a.

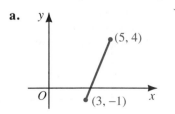

b.

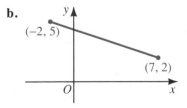

Solution

a. $\dfrac{y_2 - y_1}{x_2 - x_1} = \dfrac{4 - (-1)}{5 - 3} = \dfrac{5}{2}$

b. $\dfrac{y_2 - y_1}{x_2 - x_1} = \dfrac{2 - 5}{7 - (-2)} = \dfrac{-3}{9} = -\dfrac{1}{3}$

Example 2 Sketch each line described, showing several points on the line.

a. The line passes through (1, 2) and has slope $\frac{3}{5}$.

b. The line passes through (0, 5) and has slope $-\frac{2}{3}$.

Solution **a.** Since $\dfrac{\text{change in } y}{\text{change in } x} = \dfrac{3}{5}$, every horizontal change of 5 units is matched by a vertical change of 3 units. Start at (1, 2), move 5 units to the right and 3 units up.

b. Since $\dfrac{\text{change in } y}{\text{change in } x} = -\dfrac{2}{3} = \dfrac{-2}{3}$, every horizontal change of 3 units is matched by a vertical change of 2 units. Start at (0, 5), move 3 units to the right and 2 units down.

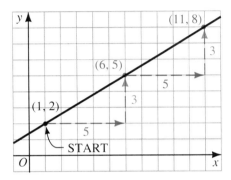

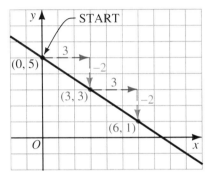

The examples above and the diagrams below illustrate the following facts.

> Lines with positive slope rise to the right.
>
> Lines with negative slope fall to the right.
>
> The greater the absolute value of a line's slope, the steeper the line.

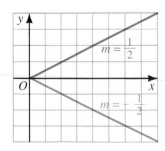

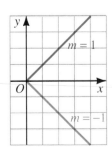

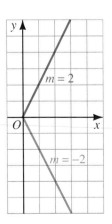

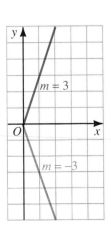

The diagrams below explain why the following are true.

The slope of a horizontal line is zero.

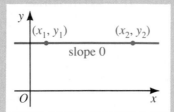

Since $y_1 = y_2$,

$$\frac{y_2 - y_1}{x_2 - x_1} = \frac{0}{x_2 - x_1} = 0.$$

The slope of a vertical line is not defined.

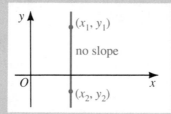

Since $x_1 = x_2$,

$$\frac{y_2 - y_1}{x_2 - x_1} = \frac{y_2 - y_1}{0}, \text{ which is}$$

not defined.

Classroom Exercises

Find the slope of the line.

1.

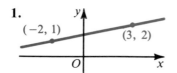

2.

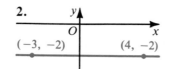

3.

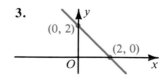

Tell whether each expression is positive or negative for the line shown:

a. $y_2 - y_1$ b. $x_2 - x_1$ c. $\dfrac{y_2 - y_1}{x_2 - x_1}$

4.

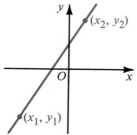

5.

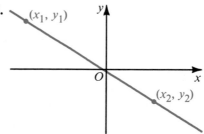

6. Does the slope of the line appear to be positive, negative, zero, or not defined?

a.

b.

c.

d. _____

7. a. Find the slope of $\overleftrightarrow{AB}$.
   b. Find tan $n°$.
   c. Consider the statement: If a line with positive slope makes an acute angle of $n°$ with the x-axis, then the slope of the line is tan $n°$. Do you think this statement is true or false? Explain.

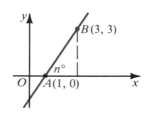

8. This exercise provides a geometric method of justifying the fact that you can use any two points on a line to determine the slope of the line. Horizontal and vertical segments have been drawn as shown. Supply the reason for each step.

Key steps of proof:

1. $\angle B \cong \angle A$
2. $\angle 1 \cong \angle 2$
3. $\triangle LBN \sim \triangle DAJ$
4. $\dfrac{BN}{AJ} = \dfrac{LB}{DA}$, or $\dfrac{BN}{LB} = \dfrac{AJ}{DA}$

5. The slope of $\overline{LN}$ equals $\dfrac{BN}{LB}$, and

   the slope of $\overline{DJ}$ equals $\dfrac{AJ}{DA}$.
6. Slope of $\overline{LN}$ = slope of $\overline{DJ}$

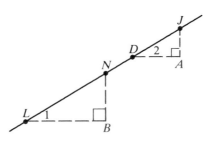

Written Exercises

A **1.** Name each line in the figure whose slope is:
   a. positive
   b. negative
   c. zero
   d. not defined

2. What can you say about the slope of **(a)** the x-axis? and **(b)** the y-axis?

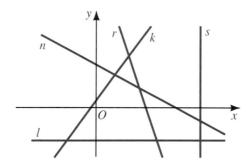

Find the slope of the line through the points named. If the slope is not defined, write *not defined*.

3. $(1, 2)$; $(3, 4)$ **4.** $(1, 2)$; $(-2, -5)$ **5.** $(1, 2)$; $(-2, 5)$

6. $(0, 0)$; $(5, 1)$ **7.** $(7, 2)$; $(2, 7)$ **8.** $(3, 3)$; $(3, 7)$

9. $(6, -6)$; $(-6, -6)$ **10.** $(6, -6)$; $(4, 3)$ **11.** $(-4, -3)$; $(-6, -6)$

Find the slope and length of $\overline{AB}$.

12. $A(3, -1)$, $B(5, -7)$ **13.** $A(-3, -2)$, $B(7, -6)$

14. $A(8, -7)$, $B(-3, -5)$ **15.** $A(0, -9)$, $B(8, -3)$

In Exercises 16–19 a point *P* on a line and the slope of the line are given. Sketch the line and find the coordinates of two other points on the line.

16. $P(-2, 1)$; slope $= \dfrac{1}{3}$ **17.** $P(-3, 0)$; slope $= \dfrac{2}{5}$

18. $P(2, 4)$; slope $= -\dfrac{3}{2}$ **19.** $P(0, -5)$; slope $= -\dfrac{1}{4}$

In Exercises 20 and 21 show that points *P*, *Q*, and *R* are collinear by showing that $\overline{PQ}$ and $\overline{QR}$ have the same slope.

20. $P(-1, 3)$ $Q(2, 7)$ $R(8, 15)$ **21.** $P(-8, 6)$ $Q(-5, 5)$ $R(4, 2)$

B   **22.** A wheelchair ramp is to be built at the town library. If the entrance to the library is 18 in. above ground, and the slope of the ramp is $\dfrac{1}{15}$, how far out from the building will the ramp start?

Complete.

23. A line with slope $\dfrac{3}{4}$ passes through points $(2, 3)$ and $(10, \underline{})$.

24. A line with slope $-\dfrac{5}{2}$ passes through points $(7, -4)$ and $(\underline{}, 6)$.

25. A line with slope *m* passes through points (p, q) and $(r, \underline{})$.

26. a. Find the slopes of $\overline{OD}$ and $\overline{NF}$.
    b. Why is $\triangle OCD \cong \triangle NEF$?
    c. Why is $\angle DOC \cong \angle FNE$?
    d. Why is $\overline{OD} \parallel \overline{NF}$?
    e. What do you think is true about the slopes of parallel lines?

27. a. Show that $\triangle OAB \cong \triangle ORS$.
    b. Why is $\overline{OB} \perp \overline{OS}$?
    c. Find the product of the slopes of $\overline{OB}$ and $\overline{OS}$.

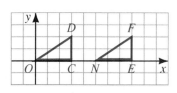

Ex. 26

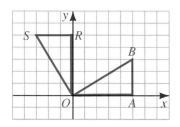

Ex. 27

In Exercises 28 and 29, (a) find the lengths of the sides of $\triangle RST$, (b) use the converse of the Pythagorean Theorem to show that $\triangle RST$ is a right triangle, and (c) find the product of the slopes of $\overline{RT}$ and $\overline{ST}$.

28. $R(4, 2)$, $S(-1, 7)$, $T(1, 1)$ **29.** $R(4, 3)$, $S(-3, 6)$, $T(2, 1)$

30. a. Show that $\tan \angle A$ = slope of $\overline{AC}$.
    b. Use trigonometry to find $m \angle A$.

31. A line intersects the x-axis at a 45° angle. What is its slope?

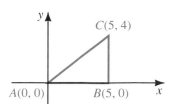

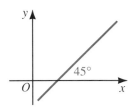

C **32.** A line passes through points $(-2, -1)$ and $(4, 3)$. Where does the line intersect the x-axis? the y-axis?

33. A line through $H(3, 1)$ and $J(5, a)$ has positive slope and makes a 60° angle measured counterclockwise with the positive x-axis. Find the value of a.

34. Find two values of k such that the points $(-3, 4)$, $(0, k)$, and $(k, 10)$ are collinear.

Algebra Review: *Exponents*

Rules of Exponents

When a and b are nonzero real numbers and m and n are integers:

(1) $a^0 = 1$ **Examples** $5^0 = 1$

(2) $a^m \cdot a^n = a^{m+n}$ $x^2 \cdot x^4 = x^{2+4} = x^6$

(3) $\dfrac{a^m}{a^n} = a^{m-n}$ $\dfrac{b^7}{b^3} = b^{7-3} = b^4$

(4) $(a^m)^n = a^{mn}$ $(y^3)^4 = y^{3 \cdot 4} = y^{12}$

(5) $a^{-m} = \dfrac{1}{a^m}$ $6^{-2} = \dfrac{1}{6^2} = \dfrac{1}{36}$

Simplify.

1. $(-6)^3$ **2.** $(-5)^4$ **3.** 3^{-2} **4.** 2^{-3}

5. $(-4)^{-3}$ **6.** $\left(\dfrac{2}{3}\right)^{-2}$ **7.** $\left(\dfrac{5}{3}\right)^{-3}$ **8.** 15^0

9. $(-1)^{20}$ **10.** $(-1)^{99}$ **11.** $2^3 \cdot 2^2 \cdot 2^{-4}$ **12.** $4^2 \cdot 3^3 \cdot 2^{-3}$

Simplify. Use only positive exponents in your answers.

13. $r^5 \cdot r^8$ **14.** $x^{-1} \cdot x^{-2}$ **15.** $\dfrac{r^9}{r^4}$ **16.** $\dfrac{t^3}{t^5}$

17. $a \cdot a^{-1}$ **18.** $(x^2)^{-2}$ **19.** $(b^4)^2$ **20.** $(s^5)^3$

21. $(3y^2)(2y^4)$ **22.** $(4x^3y^2)(-3xy)$ **23.** $(5a^2b^3)(a^{-2}b)$ **24.** $(-2st^5)(-4st^{-3})$

13-3 *Parallel and Perpendicular Lines*

When you look at two parallel lines, you probably believe that the lines have equal slopes. This idea is illustrated by the photograph below. The parallel beams shown are needed to support a roof with a fixed pitch.

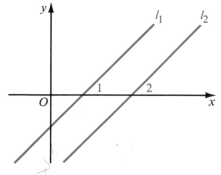

You can use trigonometry and properties of parallel lines to show the following for two nonvertical lines l_1 and l_2 (see the diagram at the right above):

1. $l_1 \parallel l_2$ if and only if $\angle 1 \cong \angle 2$
2. $\angle 1 \cong \angle 2$ if and only if $\tan \angle 1 = \tan \angle 2$
3. $\tan \angle 1 = \tan \angle 2$ if and only if slope of l_1 = slope of l_2

Therefore $l_1 \parallel l_2$ if and only if slope of l_1 = slope of l_2.

Although the diagram shows two lines with positive slope, this result can also be proved for two lines with negative slope. When the lines are parallel to the *x*-axis, both have slope zero.

Theorem 13-3

Two nonvertical lines are parallel if and only if their slopes are equal.

In Exercises 27–29 of the preceding section, you may have noticed that perpendicular lines, too, have slopes that are related in a special way. See Classroom Exercise 11 and Written Exercise 23 for proofs of the following theorem.

Theorem 13-4

Two nonvertical lines are perpendicular if and only if the product of their slopes is −1.

$$m_1 \cdot m_2 = -1, \text{ or } m_1 = -\frac{1}{m_2}$$

Example Given points $S(5, -1)$ and $T(-3, 3)$, find the slope of every line
(a) parallel to $\overleftrightarrow{ST}$ and (b) perpendicular to $\overleftrightarrow{ST}$.

Solution Slope of $\overleftrightarrow{ST} = \dfrac{3 - (-1)}{-3 - 5} = \dfrac{4}{-8} = -\dfrac{1}{2}$

a. Any line parallel to $\overleftrightarrow{ST}$ has slope $-\dfrac{1}{2}$. (Theorem 13-3)

b. Any line perpendicular to $\overleftrightarrow{ST}$ has slope

$$-\dfrac{1}{-\frac{1}{2}} = -1 \cdot (-2) = 2. \quad \text{(Theorem 13-4)}$$

Classroom Exercises

1. Given: $l \perp n$. Find the slope of line n if the slope of line l is:

   a. 2   **b.** $\dfrac{4}{5}$ **c.** -4 **d.** not defined   **e.** 0

The slopes of two lines are given. Are the lines parallel, perpendicular, or neither?

2. $\dfrac{3}{4}; \dfrac{12}{16}$ 3. $1; -1$ 4. $3; -3$ 5. $-\dfrac{3}{4}; \dfrac{4}{3}$

6. $3; \dfrac{-1}{3}$ 7. $\dfrac{-2}{3}; \dfrac{2}{-3}$ 8. $0; -1$ 9. $\dfrac{5}{6}; \dfrac{6}{5}$

10. State two conditionals that are combined in the biconditional of Theorem 13-3.

11. The purpose of this exercise is to prove the statement: If two nonvertical lines are perpendicular, then the product of their slopes is -1. Supply the reason for each step.

   Given: l_1 has slope m_1;
   $\quad\quad\quad l_2$ has slope m_2;
   $\quad\quad\quad l_1 \perp l_2$
   Prove: $m_1 \cdot m_2 = -1$

   Key steps of proof:

   1. Draw the vertical segment shown.

   2. $\dfrac{u}{v} = \dfrac{v}{w}$

   3. $m_1 = \dfrac{v}{u}$

   4. $m_2 = -\dfrac{v}{w}$

   5. $m_1 \cdot m_2 = \left(\dfrac{v}{u}\right) \cdot \left(-\dfrac{v}{w}\right)$

   6. $m_1 \cdot m_2 = \left(\dfrac{v}{u}\right) \cdot \left(-\dfrac{u}{v}\right) = -1$

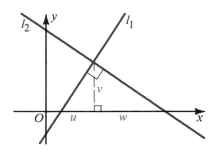

Written Exercises

Find the slope of (a) $\overleftrightarrow{AB}$, (b) any line parallel to $\overleftrightarrow{AB}$, and (c) any line perpendicular to $\overleftrightarrow{AB}$.

A **1.** $A(-2, 0)$ and $B(4, 4)$ **2.** $A(-3, 1)$ and $B(2, -1)$

   3. In the diagram at the left below, *OEFG* is a parallelogram. What is the slope of $\overline{OE}$? of $\overline{GF}$? of $\overline{OG}$? of $\overline{EF}$?

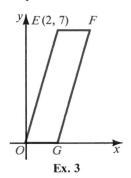

Ex. 3

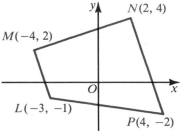

Ex. 4

   4. In the diagram at the right above, *HIJK* is a rectangle. What is the slope of $\overline{HI}$? of $\overline{JK}$? of $\overline{IJ}$? of $\overline{KH}$?

   5. a. What is the slope of $\overline{LM}$? of $\overline{PN}$?
     b. Why is $\overline{LM} \parallel \overline{PN}$?
     c. What is the slope of $\overline{MN}$? of $\overline{LP}$?
     d. Why is $\overline{MN}$ not parallel to $\overline{LP}$?
     e. What special kind of quadrilateral is *LMNP*?

   6. Quadrilateral *RSTV* is known to be a parallelogram.
     a. What is the slope of $\overline{RV}$? of $\overline{TV}$?
     b. Why is $\overline{RV} \perp \overline{TV}$?
     c. Why is $\square RSTV$ a rectangle?
     d. Find the coordinates of *S*.

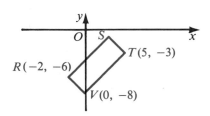

Find the slope of each side and each altitude of $\triangle ABC$.

   7. $A(0, 0)$ $B(7, 3)$ $C(2, -5)$ **8.** $A(1, 4)$ $B(-1, -3)$ $C(4, -5)$

Use slopes to show that $\triangle RST$ is a right triangle.

   9. $R(-3, -4)$ $S(2, 2)$ $T(14, -8)$ **10.** $R(-1, 1)$ $S(2, 4)$ $T(5, 1)$

B **11.** Given the points $A(-6, -4)$, $B(4, 2)$, $C(6, 8)$, and $D(-4, 2)$ show that *ABCD* is a parallelogram using two different methods.
     a. Show that opposite sides are parallel.   **b.** Show that opposite sides are congruent.

12. Given: Points $E(-4, 1)$, $F(2, 3)$, $G(4, 9)$, and $H(-2, 7)$
    a. Show that *EFGH* is a rhombus.
    b. Use slopes to verify that the diagonals are perpendicular.

13. Given: Points $R(-4, 5)$, $S(-1, 9)$, $T(7, 3)$ and $U(4, -1)$
    a. Show that *RSTU* is a rectangle.
    b. Use the distance formula to verify that the diagonals are congruent.

14. Given: Points $N(-1, -5)$, $O(0, 0)$, $P(3, 2)$, and $Q(8, 1)$
    a. Show that *NOPQ* is an isosceles trapezoid.
    b. Show that the diagonals are congruent.

Decide what special type of quadrilateral *HIJK* is. Then prove that your answer is correct.

15. $H(0, 0)$ $I(5, 0)$ $J(7, 9)$ $K(1, 9)$

16. $H(0, 1)$ $I(2, -3)$ $J(-2, -1)$ $K(-4, 3)$

17. $H(7, 5)$ $I(8, 3)$ $J(0, -1)$ $K(-1, 1)$

18. $H(-3, -3)$ $I(-5, -6)$ $J(4, -5)$ $K(6, -2)$

19. Point $N(3, -4)$ lies on the circle $x^2 + y^2 = 25$. What is the slope of the line that is tangent to the circle at N? (*Hint:* Recall Theorem 9-1.)

20. Point $P(6, 7)$ lies on the circle $(x + 2)^2 + (y - 1)^2 = 100$. What is the slope of the line that is tangent to the circle at P?

In Chapter 3 parallel lines are defined as coplanar lines that do not intersect. It is also possible to define parallel lines *algebraically* as follows:

Lines *a* and *b* are *parallel* if and only if slope of *a* = slope of *b* (or both *a* and *b* are vertical).

21. Use the algebraic definition to classify each statement as true or false.
    a. For any line *l* in a plane, $l \parallel l$.
    b. For any lines *l* and *n* in a plane, if $l \parallel n$, then $n \parallel l$.
    c. For any lines *k*, *l*, and *n* in a plane, if $k \parallel l$ and $l \parallel n$, then $k \parallel n$.

22. Refer to Exercise 21. Is parallelism of lines an equivalence relation? (See Exercise 15, page 43.) Explain.

C   23. This exercise shows another way to prove Theorem 13-4.
    a. Use the Pythagorean Theorem to prove:

       If $\overleftrightarrow{TU} \perp \overleftrightarrow{US}$, then the product of the slopes of $\overleftrightarrow{TU}$ and $\overleftrightarrow{US}$ equals -1. That is, prove

$$\left(-\frac{c}{a}\right) \cdot \left(-\frac{c}{b}\right) = -1.$$

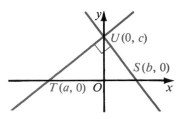

    b. Use the converse of the Pythagorean Theorem to prove:

$$\text{If } \left(-\frac{c}{a}\right) \cdot \left(-\frac{c}{b}\right) = -1, \text{ then } \overleftrightarrow{TU} \perp \overleftrightarrow{US}.$$

13-4 *Vectors*

The journey of a boat or airplane can be described by giving its speed and direction, such as 50 km/h northeast. Any quantity such as force, velocity, or acceleration, that has both *magnitude* (size) and *direction*, is a **vector.**

When a boat moves from point A to point B, its journey can be represented by drawing an arrow from A to B, $\overrightarrow{AB}$ (read "*vector AB*"). If $\overrightarrow{AB}$ is drawn in the coordinate plane, then the journey can also be represented as an ordered pair.

$$\overrightarrow{AB} = (\text{change in } x, \text{change in } y)$$

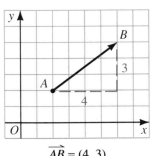

$$\overrightarrow{AB} = (4, 3)$$

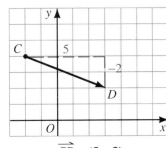

$$\overrightarrow{CD} = (5, -2)$$

The **magnitude** of a vector $\overrightarrow{AB}$ is the length of the arrow from point A to point B and is denoted by the symbol $|\overrightarrow{AB}|$. You can use the Pythagorean Theorem or the Distance Formula to find the magnitude of a vector. In the diagrams above,

$$|\overrightarrow{AB}| = \sqrt{4^2 + 3^2} = 5$$
$$\text{and} \quad |\overrightarrow{CD}| = \sqrt{5^2 + 2^2} = \sqrt{29}.$$

Example 1 Given: Points $P(-5, 4)$ and $Q(1, 2)$
  a. Sketch $\overrightarrow{PQ}$.
  b. Find $\overrightarrow{PQ}$.
  c. Find $|\overrightarrow{PQ}|$.

Solution **a.**

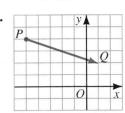

b. $\overrightarrow{PQ} = (1 - (-5), 2 - 4) = (6, -2)$
c. $|\overrightarrow{PQ}| = \sqrt{6^2 + (-2)^2} = \sqrt{40} = 2\sqrt{10}$

The symbol $2\overrightarrow{PQ}$ represents a vector that has twice the magnitude of $\overrightarrow{PQ}$ and has the same direction. If $\overrightarrow{PQ} = (3, 2)$, it should not surprise you that $2\overrightarrow{PQ} = (2 \cdot 3, 2 \cdot 2) = (6, 4)$. In general, if the vector $\overrightarrow{PQ} = (a, b)$, then $k\overrightarrow{PQ} = (ka, kb)$; $k\overrightarrow{PQ}$ is called a **scalar multiple** of $\overrightarrow{PQ}$. Multiplying a vector by a real number k multiplies the length of the vector by $|k|$. If $k < 0$, the direction of the vector reverses as well. This is illustrated in the diagrams below. What ordered pair represents $-3\overrightarrow{PQ}$?

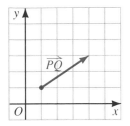 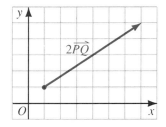 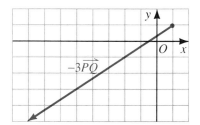

Two vectors are *perpendicular* if the arrows representing them have perpendicular directions. Two vectors are *parallel* if the arrows representing them have the same direction or opposite directions. All the vectors shown at the right are parallel. Notice that $\overrightarrow{OA}$ and $\overrightarrow{BC}$ are parallel even though the points O, A, B, and C are collinear.

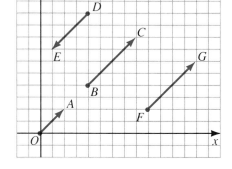

Two vectors are **equal** if they have the same magnitude and the same direction. In the diagram, $\overrightarrow{BC} = \overrightarrow{FG}$.

You can tell by using slopes whether nonvertical vectors are parallel or perpendicular. Example 2 shows how.

Example 2   **a.** Show that $(9, -6)$ and $(-6, 4)$ are parallel.
  b. Show that $(9, -6)$ and $(2, 3)$ are perpendicular.

Solution   **a.** Slope of $(9, -6)$ is $\dfrac{-6}{9} = -\dfrac{2}{3}$.

Slope of $(-6, 4) = \dfrac{4}{-6} = -\dfrac{2}{3}$.

Since the slopes are equal, the vectors are parallel.

b. Slope of $(9, -6)$ is $\dfrac{-6}{9} = -\dfrac{2}{3}$.

Slope of $(2, 3)$ is $\dfrac{3}{2}$.

Since $\dfrac{-2}{3} \cdot \dfrac{3}{2} = -1$, the vectors are perpendicular.

Vectors can be added by the following simple rule:

$$(a, b) + (c, d) = (a + c, b + d)$$

To see an application of adding vectors, suppose that a jet travels from P to Q and then from Q to R. The jet could have made the same journey by flying directly from P to R. $\overrightarrow{PR}$ is the **sum** of $\overrightarrow{PQ}$ and $\overrightarrow{QR}$. We abbreviate this fact by writing

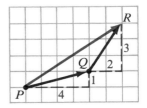

$$\overrightarrow{PQ} + \overrightarrow{QR} = \overrightarrow{PR}$$
$$(4, 1) + (2, 3) = (6, 4)$$

Classroom Exercises

Exercises 1–4 refer to the figure at the right.

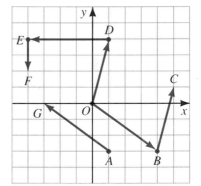

1. Name each vector as an ordered pair.
   a. $\overrightarrow{OB}$ b. $\overrightarrow{OD}$ c. $\overrightarrow{DE}$
   d. $\overrightarrow{EF}$ e. $\overrightarrow{BC}$ f. $\overrightarrow{AG}$

2. Find the magnitude of each vector in Exercise 1.

3. a. Is $\overrightarrow{BC}$ parallel to $\overrightarrow{OD}$? Explain.
   b. Is $\overrightarrow{BC} = \overrightarrow{OD}$? Explain.
   c. What kind of figure is $OBCD$? Explain.

4. a. Is $\overrightarrow{AG}$ parallel to $\overrightarrow{OB}$? Explain.
   b. Is $\overrightarrow{AG} = \overrightarrow{OB}$? Explain.

5. Refer to the diagram. Find $|\overrightarrow{ST}|$ and $\tan \angle S$.

6. Find each sum.
   a. $(3, 1) + (5, 6)$
   b. $(0, -6) + (7, 4)$
   c. $(-3, 10) + (-5, -12)$

7. Find each scalar multiple.
   a. $2(3, 1)$ b. $3(-5, 1)$ c. $-\frac{1}{2}(-6, 0)$

8. If $\overrightarrow{PQ}$ represents a wind blowing 45 km/h from the north, state two ways you could name the vector representing a wind blowing 45 km/h from the south.

Written Exercises

In Exercises 1–9 points A and B are given. Make a sketch. Then find $\overrightarrow{AB}$ and $|\overrightarrow{AB}|$.

A 1. $A(1, 1)$, $B(5, 4)$ 2. $A(2, 0)$, $B(8, 8)$ 3. $A(6, 1)$, $B(4, 3)$
   4. $A(0, 5)$, $B(-3, 2)$ 5. $A(3, 5)$, $B(-1, 7)$ 6. $A(4, -2)$, $B(0, 0)$
   7. $A(0, 0)$, $B(5, -9)$ 8. $A(-3, 5)$, $B(3, 0)$ 9. $A(-1, -1)$, $B(-4, -7)$

Use a grid and draw arrows to represent the following vectors. You can choose any starting point you like for each vector.

10. (3, 5) and 2(3, 5)

11. (4, −1) and 3(4, −1)

12. (−8, 4) and $\frac{1}{2}$(−8, 4)

13. (−6, −9) and $\frac{1}{3}$(−6, −9)

14. (4, 1) and −3(4, 1)

15. (6, −4) and −$\frac{1}{2}$(6, −4)

16. Name two vectors parallel to (3, −8).

17. The vectors (8, 6) and (12, k) are parallel. Find the value of k.

18. Show that (4, −5) and (15, 12) are perpendicular.

19. The vectors (8, k) and (9, 6) are perpendicular. Find the value of k.

Find each vector sum. Then illustrate each sum with a diagram like that on page 541.

20. (2, 1) + (3, 6)

21. (3, −5) + (4, 5)

22. (−8, 2) + (4, 6)

23. (−3, −3) + (7, 7)

24. (1, 4) + 2(3, 1)

25. (7, 2) + 3(−1, 0)

B 26. Two forces $\overrightarrow{AB}$ and $\overrightarrow{AC}$ are pulling an object at point A. The single force $\overrightarrow{AD}$ that has the same effect as these two forces is their sum $\overrightarrow{AB} + \overrightarrow{AC}$. This sum can be found by completing parallelogram $ABDC$ as shown. Explain why the diagonal $\overrightarrow{AD}$ is the sum of $\overrightarrow{AB}$ and $\overrightarrow{AC}$.

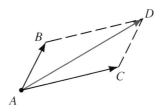

27. Make a drawing showing an object being pulled by the two forces $\overrightarrow{KX} = (−1, 5)$ and $\overrightarrow{KY} = (7, 3)$. What single force has the same effect as the two forces acting together? What is the magnitude of this force?

28. Repeat Exercise 27 for the forces $\overrightarrow{KX} = (2, −3)$ and $\overrightarrow{KY} = (−2, 3)$.

29. In the diagram, M is the midpoint of $\overline{AB}$ and T is a trisector point of $\overline{AB}$.
   a. Complete: $\overrightarrow{AB} = (\underline{\ ?\ }, \underline{\ ?\ })$, $\overrightarrow{AM} = (\underline{\ ?\ }, \underline{\ ?\ })$ and $\overrightarrow{AT} = (\underline{\ ?\ }, \underline{\ ?\ })$.
   b. Find the coordinates of M and T.

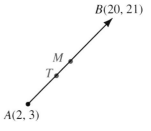

30. Repeat Exercise 29 given the points $A(−10, 9)$ and $B(20, −15)$.

31. Use algebra to prove $|(ka, kb)| = |k| \cdot |(a, b)|$.

C 32. a. Use definitions I and II below to prove that
$$k[(a, b) + (c, d)] = k(a, b) + k(c, d).$$
   I. Definition of scalar multiple $k(a, b) = (ka, kb)$
   II. Definition of vector addition $(a, b) + (c, d) = (a + c, b + d)$
   b. Make a diagram illustrating what you proved in part (a).

33. a. Given: $\overrightarrow{AB} = 2\overrightarrow{DB}$ and $\overrightarrow{BC} = 2\overrightarrow{BE}$

Supply the reasons for each step.

1. $\overrightarrow{AC} = \overrightarrow{AB} + \overrightarrow{BC}$
2. $\quad = 2\overrightarrow{DB} + 2\overrightarrow{BE}$
3. $\quad = 2(\overrightarrow{DB} + \overrightarrow{BE})$ (*Hint*: See Exercise 32.)
4. $\quad = 2\overrightarrow{DE}$

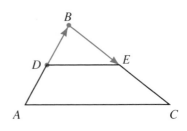

b. What theorem about midpoints does part (a) prove?

34. Suppose two nonvertical vectors (a, b) and (c, d) are perpendicular.

a. Use slopes to show that $\dfrac{bd}{ac} = -1$.

b. Show that $ac + bd = 0$.

c. The number $ac + bd$ is called the **dot product** of vectors (a, b) and (c, d). Complete: If (a, b) and (c, d) are perpendicular vectors, then their dot product ___?___ .

d. Verify the statement in part (c) for the vectors in Example 2(b) on page 540 and in Exercise 18 on page 542.

Mixed Review Exercises

1. On a number line, point A has coordinate -11 and B has coordinate 7. Find the coordinate of the midpoint of $\overline{AB}$.

2. If M is the midpoint of the hypotenuse $\overline{AB}$ of right triangle ABC, and $AM = 6$, find MB and MC.

3. The lengths of the bases of a trapezoid are 12 and 20. Find the length of the median.

4. If the length of one side of an equilateral triangle is $2a$, find the length of an altitude.

5. Find the measure of each interior angle of a regular hexagon.

6. Each side of a regular hexagon $ABCDEF$ has length x. Find AD and AC.

7. Find the measure of each exterior angle of a regular octagon.

8. Find the coordinates of the fourth vertex of a rectangle that has three vertices at $(-3, -2)$, $(2, -2)$, and $(2, 5)$.

9. The vertices of quad. $ABCD$ are $A(2, 0)$, $B(7, 0)$, $C(7, 5)$, and $D(2, 5)$. Find the area of quad. $ABCD$.

10. The vertices of $\triangle PQR$ are $P(0, 0)$, $Q(-6, 0)$, and $R(-6, 6)$. Find the area of $\triangle PQR$.

11. $\triangle DEF$ has vertices $D(-5, 1)$, $E(-2, -3)$, and $F(6, 3)$.

a. Use the distance formula to show that $\triangle DEF$ is a right triangle.

b. Use slopes to show that $\triangle DEF$ is a right triangle.

12. $\triangle ABC$ has vertices $A(6, 0)$, $B(4, 8)$, and $C(2, 6)$.

a. Find the slope of the altitude from B to $\overline{AC}$.

b. Find the slope of the perpendicular bisector of $\overline{AB}$.

13-5 *The Midpoint Formula*

On a number line, if points A and B have coordinates x_1 and x_2, then the midpoint of $\overline{AB}$ has coordinate $\dfrac{x_1 + x_2}{2}$, the average of x_1 and x_2. (See Exercise 19 on page 47.)

This idea can be used to find the midpoint of any horizontal or vertical segment.

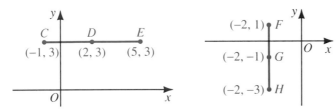

If a segment $\overline{PQ}$ is neither horizontal nor vertical, then the coordinates of its midpoint M can be found by drawing horizontal and vertical auxiliary lines as shown.

Since M is the midpoint of $\overline{PQ}$ and $\overline{MS} \parallel \overline{QR}$, S is the midpoint of $\overline{PR}$ (Theorem 5-10). Thus both S and M have x-coordinate $\dfrac{x_1 + x_2}{2}$.

Similarly, $\overline{MT} \parallel \overline{PR}$, so T is the midpoint of $\overline{QR}$.

Thus both T and M have y-coordinate $\dfrac{y_1 + y_2}{2}$.

This discussion leads to the following theorem.

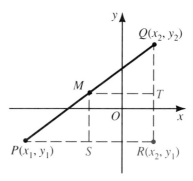

Theorem 13-5 *The Midpoint Formula*

The midpoint of the segment that joins points (x_1, y_1) and (x_2, y_2) is the point

$$\left(\frac{x_1 + x_2}{2}, \frac{y_1 + y_2}{2} \right).$$

Example 1 Find the midpoint of the segment that joins $(-11, 3)$ and $(8, -7)$.

Solution The x-coordinate of the midpoint is

$$\frac{x_1 + x_2}{2} = \frac{-11 + 8}{2} = \frac{-3}{2}, \text{ or } -\frac{3}{2}.$$

The y-coordinate of the midpoint is

$$\frac{y_1 + y_2}{2} = \frac{3 - 7}{2} = \frac{-4}{2} = -2.$$

The midpoint is $\left(-\dfrac{3}{2}, -2 \right)$.

Example 2 Given points $A(2, 1)$ and $B(8, 5)$, show that $P(3, 6)$ is on the perpendicular bisector of $\overline{AB}$.

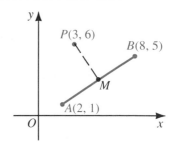

Solution 1 Join P to M, the midpoint of $\overline{AB}$ and show that $\overline{PM} \perp \overline{AB}$.

Step 1 $M = \left(\dfrac{2 + 8}{2}, \dfrac{1 + 5}{2}\right) = (5, 3)$

Step 2 Slope of $\overline{AB} = \dfrac{5 - 1}{8 - 2} = \dfrac{4}{6} = \dfrac{2}{3}$

Slope of $\overline{PM} = \dfrac{3 - 6}{5 - 3} = \dfrac{-3}{2}$

Step 3 Since the product of the slopes of $\overline{AB}$ and $\overline{PM}$ is -1, $\overline{PM} \perp \overline{AB}$.

Solution 2 Show that P is equidistant from A and B and apply Theorem 4-6, page 153.

Step 1 $PA = \sqrt{(3 - 2)^2 + (6 - 1)^2} = \sqrt{26}$
$PB = \sqrt{(3 - 8)^2 + (6 - 5)^2} = \sqrt{26}$

Step 2 Since $PA = PB$, P is on the perpendicular bisector of $\overline{AB}$.

Classroom Exercises

Find the coordinates of the midpoint of the segment that joins the given points.

1. $(3, 5)$ and $(7, 5)$
2. $(0, 4)$ and $(4, 3)$
3. $(-2, 2)$ and $(6, 4)$
4. $(-3, 7)$ and $(-7, -5)$
5. $(-1, -3)$ and $(-3, 6)$
6. $(2b, 3)$ and $(4, -5)$
7. $(t, 2)$ and $(t + 4, -4)$
8. (a, n) and (d, p)

9. $M(3, 5)$ is the midpoint of $\overline{P_1P_2}$, where P_1 has coordinates $(0, 1)$. Find the coordinates of P_2.

10. Point $(1, -1)$ is the midpoint of $\overline{AB}$, where A has coordinates $(-1, 3)$. Find the coordinates of B.

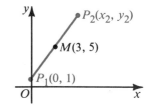

Written Exercises

Find the coordinates of the midpoint of the segment that joins the given points.

A **1.** $(0, 2)$ and $(6, 4)$
2. $(-2, 6)$ and $(4, 3)$
3. $(6, -7)$ and $(-6, 3)$
4. $(a, 4)$ and $(a + 2, 0)$
5. $(2.3, 3.7)$ and $(1.5, -2.9)$
6. (a, b) and (c, d)

Find the length, slope, and midpoint of $\overline{PQ}$.

7. $P(3, -8)$, $Q(-5, 2)$ **8.** $P(-3, 4)$, $Q(7, 8)$ **9.** $P(-7, 11)$, $Q(1, -4)$

In Exercises 10–12, M is the midpoint of $\overline{AB}$, where the coordinates of A are given. Find the coordinates of B.

10. $A(4, -2)$; $M(4, 4)$ **11.** $A(1, -3)$; $M(5, 1)$ **12.** $A(r, s)$; $M(0, 2)$

B   **13.** Given points $A(0, 0)$ and $B(8, 4)$, show that $P(2, 6)$ is on the perpendicular bisector of $\overline{AB}$ by using both of the methods in Example 2.

14. a. Given points $R(1, 0)$, $S(7, 4)$, and $T(11, -2)$, show that $\triangle RST$ is isosceles.
     b. The altitude from the vertex meets the base at K. Find the coordinates of K.

15. Find the midpoints of the legs, then the length of the median of the trapezoid with vertices $C(-4, -3)$, $D(-1, 4)$, $E(4, 4)$, and $F(7, -3)$.

16. Find the length of the longest median of the triangle with vertices $X(-2, 3)$, $Y(6, -3)$, and $Z(4, 7)$.

17. a. Verify that $\overline{OQ}$ and $\overline{PR}$ have the same midpoint.
     b. Part (a) shows that the diagonals of $OPQR$ bisect each other. Therefore $OPQR$ is a __?__.
     c. Use slopes to verify that the opposite sides of $OPQR$ are parallel.
     d. Use the distance formula to verify that the opposite sides are congruent.

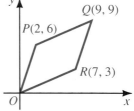

18. Graph the points $A(-5, 0)$, $B(3, 2)$, $C(5, 6)$, and $D(-3, 4)$. Then show that $ABCD$ is a parallelogram by two different methods.
     a. Show that one pair of opposite sides are both congruent and parallel.
     b. Show that the diagonals bisect each other (have the same midpoint).

19. In right $\triangle OAT$, M is the midpoint of $\overline{AT}$.
     a. M has coordinates (__?__, __?__).
     b. Find, and compare, the lengths MA, MT, and MO.
     c. State a theorem from Chapter 5 suggested by this exercise.
     d. Find an equation of the circle that circumscribes $\triangle OAT$.

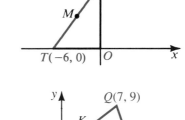

20. Given points $A(1, 1)$, $B(13, 9)$, and $C(3, 7)$. D is the midpoint of $\overline{AB}$, and E is the midpoint of $\overline{AC}$.
     a. Find the coordinates of D and E.
     b. Use slopes to show that $\overline{DE} \parallel \overline{BC}$.
     c. Use the distance formula to show that $DE = \frac{1}{2}BC$.

21. a. Find the coordinates of the midpoints J, K, L, and M.
     b. What kind of figure is $JKLM$? Prove it.

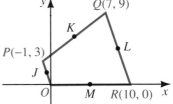

Ex. 21

22. Suppose E is on $\overline{PQ}$ and $PE = \frac{1}{4}PQ$. If $P = (x_1, y_1)$ and $Q = (x_2, y_2)$, where $x_1 < x_2$, show that $E = \left(\frac{3}{4}x_1 + \frac{1}{4}x_2, \frac{3}{4}y_1 + \frac{1}{4}y_2\right)$.

C 23. Suppose F is on $\overline{PQ}$ and $PF = \frac{3}{8}PQ$. If $P = (x_1, y_1)$ and $Q = (x_2, y_2)$, where $x_1 < x_2$, find the coordinates of F. (*Hint*: See Exercise 22.)

24. Given points $P(2, 1)$ and $D(7, 11)$, find the coordinates of a point T on $\overline{PD}$ such that $\dfrac{PT}{TD} = \dfrac{2}{3}$.

Self-Test 1

For each pair of points find (a) the distance between the two points and (b) the midpoint of the segment that joins the two points.

1. $(5, 1)$ and $(3, 1)$ **2.** $(8, -6)$ and $(0, 0)$

3. $(-2, 7)$ and $(8, -3)$ **4.** $(-3, 2)$ and $(-5, 7)$

Write an equation of the circle described.

5. Center at the origin; radius 9

6. Center $(-1, 2)$; radius 5

7. Find the center and the radius of the circle $(x + 2)^2 + (y - 3)^2 = 36$.

Find the slope of the line through the points named.

8. $(0, 0)$ and $(7, 4)$ **9.** $(-4, 2)$ and $(1, -1)$

10. For which is slope *not* defined, a horizontal line or a vertical line?

11. Given $P(3, -2)$ and $Q(5, 2)$, find:

    a. the slope of any line parallel to $\overleftrightarrow{PQ}$

    b. the slope of any line perpendicular to $\overleftrightarrow{PQ}$

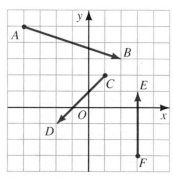

12. Name each vector as an ordered pair.

    a. $\overrightarrow{AB}$ **b.** $\overrightarrow{CD}$ **c.** $\overrightarrow{FE}$

13. Find the magnitude of each vector in Exercise 12.

14. Complete.

    a. $(-3, 2) + (7, -11) = \underline{\ ?\ }$

    b. $3(4, -1) + (-2)(-5, 3) = \underline{\ ?\ }$

15. If $M(-3, 7)$ is the midpoint of $\overline{PQ}$, where P has coordinates $(9, -4)$, find the coordinates of Q.

Exs. 12, 13

Lines and Coordinate Geometry Proofs

Objectives

1. Identify the slope and y-intercept of the line specified by a given equation.
2. Draw the graph of the line specified by a given equation.
3. Write an equation of a line when given either one point and the slope of the line, or two points on the line.
4. Determine the intersection of two lines.
5. Given a polygon, choose a convenient placement of coordinate axes and assign appropriate coordinates.
6. Prove statements by using coordinate geometry methods.

13-6 *Graphing Linear Equations*

A **linear equation** is an equation whose graph is a line. As you will learn in this section and the next, linear equations can be written in different forms: *standard form*, *slope-intercept form*, and *point-slope form*. We state a theorem for the standard form, but omit the proof.

Theorem 13-6 *Standard Form*

The graph of any equation that can be written in the form

$$Ax + By = C$$

where A and B are not both zero, is a line.

The advantage of the standard form is that it is easy to determine the points where the line crosses the x-axis and the y-axis. If a line intersects the x-axis at the point $(a, 0)$, then its *x-intercept* is a; if it intersects the y-axis at the point $(0, b)$, then its *y-intercept* is b.

Example 1   Graph the line $2x - 3y = 12$.

Solution   Since two points determine a line, begin by plotting two convenient points, such as the points where the line crosses the axes. Then draw the line through the points.

To find the *x*-intercept, let $y = 0$.

$$2x - 3(0) = 12$$
$$x = 6$$

Thus $(6, 0)$ is a point on the line.

To find the *y*-intercept, let $x = 0$.

$$2(0) - 3y = 12$$
$$y = -4$$

Thus $(0, -4)$ is a point on the line.

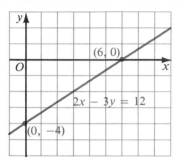

Example 2 Use algebra to find the intersection of the lines $2x - 3y = 9$ and $4x + y = 4$. Illustrate by drawing the graphs of the two lines.

Solution

$$\begin{array}{ll} 2x - 3y = 9 & \text{(First equation)} \\ \underline{12x + 3y = 12} & \text{(Second equation} \times 3) \\ 14x = 21 & \text{(Add to eliminate } y.) \\ x = 1.5 & \\ 4(1.5) + y = 4 & \text{(Substitution)} \\ y = -2 & \end{array}$$

The point of intersection is $(1.5, -2)$.

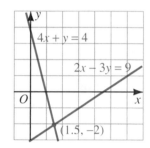

The equations in Examples 1 and 2 are all written in standard form. These equations can also be written in the *slope-intercept form* $y = mx + b$. This form tells you at a glance what the line's slope and *y*-intercept are.

standard form	*slope-intercept form*	*slope*	*y-intercept*
$2x - 3y = 12$	$y = \dfrac{2}{3}x - 4$	$\dfrac{2}{3}$	-4
$2x - 3y = 9$	$y = \dfrac{2}{3}x - 3$	$\dfrac{2}{3}$	-3
$4x + y = 4$	$y = -4x + 4$	-4	4

Theorem 13-7 *Slope-Intercept Form*

A line with the equation $y = mx + b$ has slope m and y-intercept b.

Proof:

When $x = 0$, $y = b$. So b is the y-intercept.

When $x = 1$, $y = m + b$.

Let $(x_1, y_1) = (0, b)$ and $(x_2, y_2) = (1, m + b)$.

Then the slope is $\dfrac{y_2 - y_1}{x_2 - x_1} = \dfrac{(m + b) - b}{1 - 0} = m$.

Example 3   Graph the line $y = -\dfrac{3}{4}x + 6$.

Solution   The slope is $\dfrac{-3}{4}$ and the y-intercept is 6.

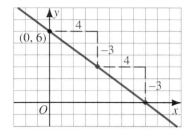

Step 1   Start at the point (0, 6).

Step 2   Use $\dfrac{\text{change in } y}{\text{change in } x} = \dfrac{-3}{4}$ to find other points of the line. (See Example 2, page 530.)

Classroom Exercises

1. Which points lie on the line $3x - 2y = 12$?

    a. (0, 4)   **b.** (2, −3)   **c.** $\left(3, \dfrac{3}{2}\right)$ **d.** (0, −6)

2. Which point is the intersection of $x + 2y = 8$ and $2x + 3y = 10$?

    a. (−2, 5)   **b.** (−4, 6)   **c.** (2, 3)   **d.** (−1, 4)

Find the x- and y-intercepts of each line.

3. $2x + 3y = 6$ **4.** $3x - 5y = 15$ **5.** $-4x + 3y = 24$

6. $x + 3y = 9$ **7.** $y = 5x - 10$ **8.** $y = 2x + 5$

Find the slope and y-intercept of each line.

9. $y = \dfrac{2}{5}x - 9$ **10.** $2x + y = 8$ **11.** $3x - 4y = 6$

12. What is the slope of the line $y = 4$? Name three points that lie on the line.

13. The graph of $x = 5$ is a vertical line through (5, 0). Name three other points on the line and check to see if their coordinates satisfy the equation.

Written Exercises

A   **1.** On the same axes, graph $y = mx$ for $m = 2, -2, \dfrac{1}{2},$ and $-\dfrac{1}{2}$.

    2. On the same axes, graph $y = mx + 2$ for $m = 3, -3, \dfrac{1}{3},$ and $-\dfrac{1}{3}$.

    3. On the same axes, graph $y = \dfrac{1}{2}x + b$ for $b = 0, 2, 4, -2,$ and -4.

    4. On the same axes, graph $y = -\dfrac{2}{3}x + b$ for $b = 0, 3, 6, -3,$ and -6.

    5. On the same axes, graph the lines $y = 0$, $y = 3$, and $y = -3$.

    6. On the same axes, graph the lines $x = 0$, $x = 2$, and $x = -2$.

Find the *x*-intercept and *y*-intercept of each line. Then graph the equation.

7. $3x + y = -21$ **8.** $4x - 5y = 20$ **9.** $3x + 2y = 12$

10. $3x - 2y = 12$ **11.** $5x + 8y = 20$ **12.** $3x + 4y = -18$

Find the slope and *y*-intercept of each line. Plot the *y*-intercept. Then, using the slope, plot one more point. Finally, graph the line.

13. $y = 2x - 3$ **14.** $y = 2x + 3$ **15.** $y = -4x$

16. $y = \dfrac{3}{4}x + 1$ **17.** $y = -\dfrac{2}{3}x - 4$ **18.** $y = \dfrac{5}{3}x - 2$

Find the slope and *y*-intercept of each line.

Example $x + 3y = -6$

Solution   Write the equation in slope-intercept form.
$$3y = -x - 6$$
$$y = -\frac{1}{3}x - 2$$
The slope is $-\dfrac{1}{3}$. The *y*-intercept is -2.

19. $4x + y = 10$ **20.** $2x - y = 5$ **21.** $5x - 2y = 10$

22. $3x + 4y = 12$ **23.** $x - 4y = 6$ **24.** $4x + 3y = 8$

Solve each pair of equations algebraically. Then draw the graphs of the equations and label their intersection point.

25. $x + y = 3$ **26.** $2x + y = 7$ **27.** $x + 2y = 10$
      $x - y = -1$ $3x + y = 9$ $3x - 2y = 6$

28. $3x + 2y = -30$ **29.** $4x + 5y = -7$ **30.** $3x + 2y = 8$
      $y = x$ $2x - 3y = 13$ $-x + 3y = 12$

B

31. a. Find the slopes of the lines $6x + 3y = 10$ and $y = -2x + 5$.
   b. Do the lines intersect?
   c. What happens when you solve these equations algebraically?

32. Give a geometric reason and an algebraic reason why the lines $y = 3x - 5$ and $y = 3x + 5$ do not intersect.

33. a. Find the slopes of the lines $2x - y = 7$ and $x + 2y = 4$.
   b. What can you conclude about the lines? State the theorem that supports your answer.

34. a. On the same axes, graph
$$y = -2, x = -3, \text{ and } 2x + 3y = 6.$$
   b. Find the coordinates of the three points where the lines intersect.
   c. Find the area of the triangle determined by the three lines.

35. **a.** On the same axes, graph
$$y = \frac{1}{2}x - 2, \quad y = -2x + 3, \quad \text{and} \quad y = 3x + 8.$$
   b. Find the coordinates of the three points where the lines intersect.
   c. Find the area of the triangle determined by the three lines.

36. Find the area of the region inside the circle $x^2 + y^2 = 2$ *and* above the line $y = 1$.

C 37. Use algebra to find each point at which the line $x - 2y = -5$ intersects the circle $x^2 + y^2 = 25$. Graph both equations to verify your answer.

38. **a.** Verify that the point $P(4, -2)$ is on the line $2x - y = 10$ and on the circle $x^2 + y^2 = 20$.
   b. Show that the segment joining the center of the circle to P is perpendicular to the line.
   c. What do parts (a) and (b) tell you about the line and the circle?

39. Graph each equation.
   a. $|x| = |y|$ **b.** $|x| + |y| = 6$ **c.** $|x| + 2|y| = 4$

Explorations

These exploratory exercises can be done using a graphing calculator.

Graph the lines $y = 2x$, $y = 2x + 1$, and $y = 2x + 3$ on the same screen. What do you notice about these lines?

What theorem does this illustrate?

Use what you have observed to write an equation of the line whose y-intercept is 7 and that is parallel to $y = 2x$.

Graph the lines $y = 2x$ and $y = -\frac{1}{2}x$ on the same screen.

Graph the lines $y = \frac{2}{3}x$ and $y = -\frac{3}{2}x$ on the same screen.

What do you notice about both pairs of lines?

What theorem does this illustrate?

Use what you have observed to write an equation of the line through the origin that is perpendicular to $y = \frac{4}{5}x$.

Challenge

Draw segments that divide an obtuse triangle into acute triangles.

13-7 *Writing Linear Equations*

In the previous section, you were given a linear equation and asked to draw its graph. In this section you will be given information about a graph and asked to find an equation of the line described.

Example 1   Find an equation of each line described.

      a. Slope $= -\dfrac{5}{3}$, y-intercept $= 4$

      b. x-intercept $= -6$, y-intercept $= 3$

Solution   **a.** $y = mx + b$

$$y = -\frac{5}{3}x + 4$$

      b. Because the y-intercept is 3, you have $b = 3$.
      Because the points $(-6, 0)$ and $(0, 3)$ lie on the line,

$$\text{slope} = \frac{3 - 0}{0 - (-6)} = \frac{3}{6} = \frac{1}{2}.$$

      Since the slope is $\dfrac{1}{2}$, you have $m = \dfrac{1}{2}$.
      Now substitute into the equation $y = mx + b$ to get

$$y = \frac{1}{2}x + 3.$$

    Both linear equations in Example 1 were written in slope-intercept form. This form is very easy to use if the y-intercept is given. If the y-intercept is not given, the *point-slope form* can be used.

Theorem 13-8   *Point-Slope Form*

An equation of the line that passes through the point (x_1, y_1) and has slope m is

$$y - y_1 = m(x - x_1).$$

Proof:

Let (x, y) be any point on the line. Since the line also contains the point (x_1, y_1) the slope must, by definition, equal $\dfrac{y - y_1}{x - x_1}.$

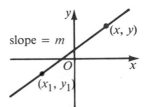

From $m = \dfrac{y - y_1}{x - x_1}$,

we get $y - y_1 = m(x - x_1).$

Example 2 Find an equation of each line described.
   a. The line through $(1, 2)$ and parallel to the line $y = 3x - 7$
   b. The line through $(1, 2)$ and perpendicular to the line $y = 3x - 7$
   c. The line through the points $(-3, 0)$ and $(1, 8)$

Solution **a.** If the line is parallel to the line $y = 3x - 7$, its slope must be 3.
      Substituting in $y - y_1 = m(x - x_1)$ gives $y - 2 = 3(x - 1)$,
      or $y = 3x - 1$.
   b. The required line has slope $-\dfrac{1}{3}$. (Why?) Thus an equation in point-slope

   form is $y - 2 = -\dfrac{1}{3}(x - 1)$, or $y = -\dfrac{1}{3}x + \dfrac{7}{3}$, or $x + 3y = 7$.

   c. First find the slope: $m = \dfrac{8 - 0}{1 - (-3)} = 2$

   Then use the point-slope form with *either* given point.
   Using $(-3, 0)$, the equation is $y - 0 = 2[x - (-3)]$, or $y = 2x + 6$.
   Using $(1, 8)$, the equation is $y - 8 = 2(x - 1)$, or $y = 2x + 6$.

Classroom Exercises

Give an equation of each line described.

1. Slope $= -\dfrac{1}{2}$; y-intercept $= 5$

2. Slope $= \dfrac{3}{7}$; y-intercept $= 8$

3. x-intercept $= 2$; y-intercept $= 4$

4. x-intercept $= 2$; y-intercept $= -6$

5. The x-axis

6. The y-axis

7. y-intercept $= -3$; parallel to $y = -\dfrac{4}{5}x + 2$

8. y-intercept $= 0$; perpendicular to $y = -\dfrac{7}{4}x + 9$

9. Slope $= \dfrac{5}{8}$; passes through $(3, 4)$

10. Slope $= -2$; passes through $(8, 6)$

State the slope of the line and name two points on the line.

11. $y = -(x + 7)$

12. $y + 2 = \dfrac{1}{2}(x - 5)$

13. $y - c = \dfrac{a}{b}(x - d)$

14. Line l is tangent to $\odot O$ at point $P(3, 4)$.
   a. Find the radius of the circle.
   b. Give an equation of the circle.
   c. Find the slope of line l.
   d. Give an equation of line l.

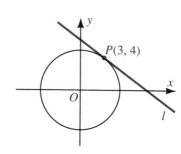

Written Exercises

Give an equation of each line described. Use the form specified by your teacher.

A

	1.	2.	3.	4.	5.	6.
slope	2	-3	$\dfrac{1}{2}$	$\dfrac{3}{4}$	$-\dfrac{7}{5}$	$-\dfrac{3}{2}$
y-intercept	5	6	-8	-9	8	-7

	7.	8.	9.	10.
x-intercept	8	9	-8	-5
y-intercept	2	-3	4	-2

	11.	12.	13.	14.	15.	16.
point	$(1, 2)$	$(3, 8)$	$(-3, 5)$	$(6, -6)$	$(-4, 0)$	$(-10, 3)$
slope	5	4	$\dfrac{1}{3}$	$-\dfrac{2}{3}$	$-\dfrac{1}{2}$	$-\dfrac{2}{5}$

17. line through $(1, 1)$ and $(4, 7)$

18. line through $(-1, -3)$ and $(2, 1)$

19. line through $(-3, 1)$ and $(3, 3)$

20. line through $(-2, -1)$ and $(-6, -5)$

21. vertical line through $(2, -5)$

22. horizontal line through $(3, 1)$

23. line through $(5, -3)$ and parallel to the line $x = 4$

24. line through $(-8, -2)$ and parallel to the line $x = 5$

B **25.** line through $(5, 7)$ and parallel to the line $y = 3x - 4$

26. line through $(-1, 3)$ and parallel to the line $3x + 5y = 15$

27. line through $(-3, -2)$ and perpendicular to the line $8x - 5y = 0$

28. line through $(8, 0)$ and perpendicular to the line $3x + 4y = 12$

29. perpendicular bisector of the segment joining $(0, 0)$ and $(10, 6)$

30. perpendicular bisector of the segment joining $(-3, 7)$ and $(5, 1)$

31. the line through $(5, 5)$ that makes a 45° angle measured counterclockwise from the positive x-axis

32. the line through the origin that makes a 135° angle measured counterclockwise from the positive x-axis

33. Find each value of k for which the lines $y = 9kx - 1$ and $kx + 4y = 12$ are perpendicular.

34. Quad. *BECK* is known to be a rhombus. Two of the vertices are $B(3, 5)$ and $C(7, -3)$.

a. Find the slope of diagonal $\overline{EK}$.

b. Find an equation of $\overleftrightarrow{EK}$.

35. Find the center of the circle that passes through (2, 10), (10, 6), and (−6, −6).

Exercises 36–39 refer to △QRS with vertices Q(−6, 0), R(12, 0), and S(0, 12).

C **36. a.** Find the equations of the three lines that contain the medians.
  b. Show that the three medians meet in a point G (called the *centroid*). (*Hint:* Solve two equations simultaneously and show that their solution satisfies the third equation.)
  c. Show that the length QG is $\frac{2}{3}$ of the length of the median from Q.

37. a. Find the equations of the three perpendicular bisectors of the sides of △QRS.
  b. Show that the three perpendicular bisectors meet in a point C (called the *circumcenter*). (See the hint from Exercise 36(b).)
  c. Show that C is equidistant from Q, R, and S by using the distance formula.
  d. Find the equation of the circle that can be circumscribed about △QRS.

38. a. Find the equations of the three lines that contain the altitudes of △QRS.
  b. Show that the three altitudes meet in a point H (called the *orthocenter*).

39. a. Refer to Exercises 36, 37, and 38. Use slopes to show that the points C, G, and H are collinear. (The line through these points is called *Euler's Line.*)
  b. Show that $GH = 2GC$.

13-8 *Organizing Coordinate Proofs*

We will illustrate coordinate geometry methods by proving Theorem 5-15:

The midpoint of the hypotenuse of a right triangle is equidistant from the three vertices.

Proof:

Let $\overleftrightarrow{OP}$ and $\overleftrightarrow{OR}$ be the *x*-axis and *y*-axis.
Let P and R have the coordinates shown.

Then the coordinates of M are (a, b).

$MO = \sqrt{(a - 0)^2 + (b - 0)^2} = \sqrt{a^2 + b^2}$
$MP = \sqrt{(a - 2a)^2 + (b - 0)^2} = \sqrt{a^2 + b^2}$
Thus $MO = MP$.

By the definition of midpoint, $MP = MR$.

Hence $MO = MP = MR$.

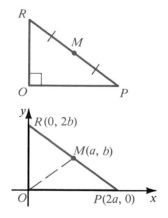

Notice that $2a$ and $2b$ are convenient choices for coordinates since they lead to expressions that do not contain fractions for the coordinates of M.

If you have a right triangle, such as $\triangle POR$ on page 556, the most convenient place to put the x-axis and y-axis is usually along the legs of the triangle. If a triangle is not a right triangle, the two most convenient ways to place your axes are shown below. Notice that these locations for the axes maximize the number of times zero is a coordinate of a vertex.

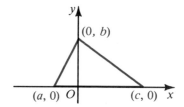
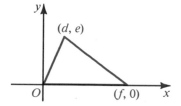

Some common ways of placing coordinate axes on other special figures are shown below.

$\triangle COD$ is isosceles; $CO = CD$. Then C can be labeled (a, b).

$\triangle EFG$ is isosceles; $EF = EG$. Then F can be labeled $(-a, 0)$.

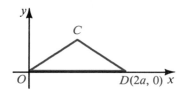
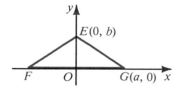

$HOJK$ is a rectangle. Then K can be labeled (a, b).

$MONP$ is a parallelogram. Then P can be labeled $(a + b, c)$.

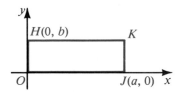
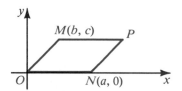

$ROST$ is a trapezoid. Then T can be labeled (d, c).

$UOVW$ is an isosceles trapezoid. Then W can be labeled $(a - b, c)$.

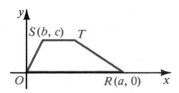
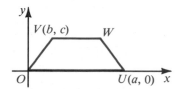

Classroom Exercises

Supply the missing coordinates without introducing any new letters.

1. *POST* is a square.

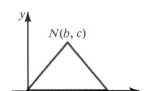

2. △*MON* is isosceles.

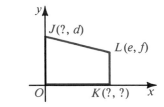

3. *JOKL* is a trapezoid.

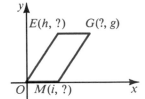

4. *GEOM* is a parallelogram.

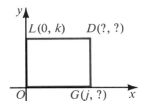

5. *GOLD* is a rectangle.

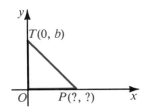

6. Rt. △*TOP* is isosceles.

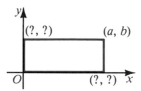

Written Exercises

Copy the figure. Supply the missing coordinates without introducing any new letters.

A   **1.** Rectangle

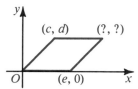

2. Parallelogram

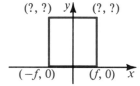

3. Square

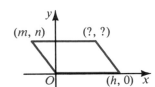

4. Isosceles triangle

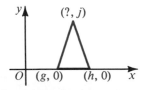

5. Parallelogram

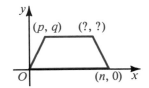

6. Isosceles trapezoid

B **7.** An equilateral triangle is shown below. Express the missing coordinates in terms of *s*.

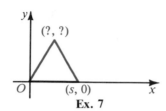

Ex. 7

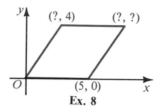

Ex. 8

8. A rhombus is shown above. Find the missing coordinates.

9. Rhombus *OABC* is shown at the right. Express the missing coordinates in terms of *a* and *b*. (*Hint*: See Exercise 8.)

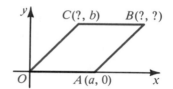

10. Supply the missing coordinates to prove: The segments that join the midpoints of opposite sides of any quadrilateral bisect each other. Let *H*, *E*, *A*, and *R* be the midpoints of the sides of quadrilateral *SOMK*. Choose axes and coordinates as shown.

a. *R* has coordinates (__?__, __?__).
b. *E* has coordinates (__?__, __?__).
c. The midpoint of $\overline{RE}$ has coordinates (__?__, __?__).
d. *A* has coordinates (__?__, __?__).
e. *H* has coordinates (__?__, __?__).
f. The midpoint of $\overline{AH}$ has coordinates (__?__, __?__).
g. Because (__?__, __?__) is the midpoint of both $\overline{RE}$ and $\overline{AH}$, $\overline{RE}$ and $\overline{AH}$ bisect each other.

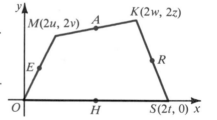

Draw the figure named. Select axes and label the coordinates of the vertices in terms of a single letter.

C **11.** a regular hexagon   **12.** a regular octagon

13. Given isosceles trapezoid *HOJK* and the axes and coordinates shown, use the definition of an isosceles trapezoid to prove that $e = c$ and $d = a - b$.

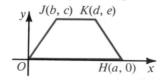

13-9 *Coordinate Geometry Proofs*

It is easy to verify that $\triangle OPQ$ is an isosceles triangle. Knowing this, we can deduce that medians $\overline{OR}$ and $\overline{QS}$ are congruent by using the midpoint and distance formulas.

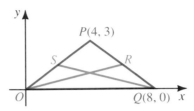

In order to give a coordinate proof that the medians to the legs are congruent for *any* isosceles triangle, and not just for the specific isosceles triangle above, you could use the figure below. Compare the general coordinates given in the figure below with the specific coordinates given for the triangle above. A coordinate proof follows.

Example 1

Prove that the medians to the legs of an isosceles triangle are congruent.

Proof:

Let OPQ be any isosceles triangle with $PO = PQ$. Choose convenient axes and coordinates as shown.

By the midpoint formula,

S has coordinates $\left(\dfrac{a}{2}, \dfrac{b}{2}\right)$ and R has coordinates $\left(\dfrac{3a}{2}, \dfrac{b}{2}\right)$.

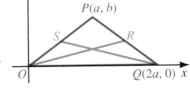

By the distance formula,

$$OR = \sqrt{\left(\frac{3a}{2} - 0\right)^2 + \left(\frac{b}{2} - 0\right)^2}$$

$$= \sqrt{\frac{9a^2}{4} + \frac{b^2}{4}}$$

and $\quad QS = \sqrt{\left(\frac{a}{2} - 2a\right)^2 + \left(\frac{b}{2} - 0\right)^2}$

$$= \sqrt{\frac{9a^2}{4} + \frac{b^2}{4}} \qquad \text{Therefore, } \overline{OR} \cong \overline{QS}.$$

It is possible to prove many theorems of geometry by using coordinate methods rather than the noncoordinate methods involving congruent triangles and angles formed by parallel lines. Coordinate proofs are sometimes, but not always, much easier than noncoordinate proofs. For example, compare the proof in Example 2 with the proof of Theorem 5-11 on page 178.

Example 2

Prove that the segment joining the midpoints of two sides of a triangle is parallel to the third side and is half as long as the third side.

Proof:

Let *OPQ* be any triangle. Choose convenient axes and coordinates as shown. By the midpoint formula, *M* has coordinates $\left(\dfrac{a}{2}, \dfrac{b}{2}\right)$ and *N* has coordinates $\left(\dfrac{a+c}{2}, \dfrac{b}{2}\right)$.

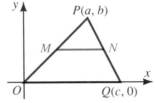

Slope of $\overline{MN} = 0$ and slope of $\overline{OQ} = 0$. (Why?)
Since $\overline{MN}$ and $\overline{OQ}$ have equal slopes, $\overline{MN} \parallel \overline{OQ}$.

Since $MN = \dfrac{a+c}{2} - \dfrac{a}{2} = \dfrac{c}{2}$ and $OQ = c - 0 = c$, $MN = \frac{1}{2}OQ$.

Classroom Exercises

In Exercises 1–4 use the diagram at the right.

1. What kind of figure is quad. *OQRS*? Why?

2. Show that $\overline{OR} \cong \overline{QS}$.

3. Show that $\overline{OR} \perp \overline{QS}$.

4. Show that $\overline{OR}$ bisects $\overline{QS}$.

5. The purpose of this exercise is to prove that the lines that contain the altitudes of a triangle intersect in a point (called the *orthocenter*).

Given $\triangle ROM$, with lines *j*, *k*, and *l* containing the altitudes, we choose axes and coordinates as shown.

a. The equation of line *k* is ___?___.

b. Since the slope of $\overline{MR}$ is $\dfrac{c}{b-a}$, the slope of line *l* is ___?___.

c. Show that an equation of line *l* is $y = \left(\dfrac{a-b}{c}\right)x$.

d. Show that lines *k* and *l* intersect where $x = b$ and $y = \dfrac{ab - b^2}{c}$.

e. Since the slope of $\overline{OM} = $ ___?___, the slope of line *j* is ___?___.

f. Show that an equation of line *j* is $y = -\dfrac{b}{c}(x - a)$.

g. Show that lines *k* and *j* intersect where $x = b$ and $y = \dfrac{ab - b^2}{c}$.

h. From parts (d) and (g) we see that the three altitude lines intersect in a point. Name the coordinates of that point.

Written Exercises

Use coordinate geometry to prove each statement. First draw a figure and choose convenient axes and coordinates.

A 1. The diagonals of a rectangle are congruent. (Theorem 5-12)

2. The diagonals of a parallelogram bisect each other. (Theorem 5-3)

3. The diagonals of a rhombus are perpendicular. (Theorem 5-13)
   (*Hint*: Let the vertices be $(0, 0)$, $(a, 0)$, $(a + b, c)$, and (b, c). Show that $c^2 = a^2 - b^2$.)

Exercises 4–6 refer to trapezoid *MNOP* at the right.

4. Prove that the median of a trapezoid:
   a. is parallel to the bases.
   b. has a length equal to the average of the base lengths.
      (Theorem 5-19)

5. Prove that the segment joining the midpoints of the diagonals of a trapezoid is parallel to the bases and has a length equal to half the difference of the lengths of the bases.

6. Assume that $a = b + d$.
   a. Show that the trapezoid is isosceles.
   b. Prove that its diagonals are congruent.

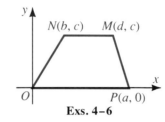

Exs. 4–6

B 7. Prove that the figure formed by joining, in order, the midpoints of the sides of quadrilateral *ROST* is a parallelogram.

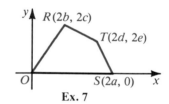

Ex. 7

8. Prove that the quadrilateral formed by joining, in order, the midpoints of the sides of an isosceles trapezoid is a rhombus.

9. Prove that an angle inscribed in a semicircle is a right angle. (*Hint*: The coordinates of *C* must satisfy the equation of the circle.)

10. Prove that the sum of the squares of the lengths of the sides of a parallelogram is equal to the sum of the squares of the lengths of the diagonals.

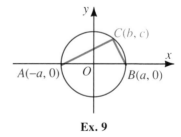

Ex. 9

C 11. Use axes and coordinates as shown to prove: The medians of a triangle intersect in a point (called the *centroid*) that is two thirds of the distance from each vertex to the midpoint of the opposite side. (*Hint*: Find the coordinates of the midpoints, then the slopes of the medians, then the equations of the lines containing the medians.)

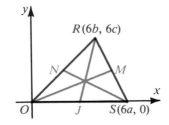

Exercises 12, 13, and 14 refer to the diagram at the right.

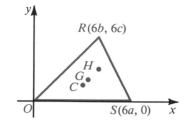

12. Prove that the perpendicular bisectors of the three sides of $\triangle ROS$ meet in a point C (called the *circumcenter*) whose coordinates are $\left(3a, \dfrac{3b^2 + 3c^2 - 3ab}{c}\right)$.

13. Prove that the lines containing the altitudes of $\triangle ROS$ intersect in a point $H\left(6b, \dfrac{6ab - 6b^2}{c}\right)$. (*Hint*: Use the procedure of Classroom Exercise 5.)

14. G, the intersection point of the medians of $\triangle ROS$, has coordinates $(2a + 2b, 2c)$. (See Exercise 11.)
Prove each statement.

   a. Points C, G, and H are collinear. The line containing these points is called *Euler's Line*. (*Hint*: One way to prove this is to show that slope of $\overline{CG}$ = slope of $\overline{GH}$.)

   b. $CG = \frac{1}{3}CH$

Self-Test 2

1. Find the slope and y-intercept of the line $2x - 5y = 20$.

2. Graph the line $2x + 3y = 6$.

3. Write an equation of the line through $(1, 2)$ and $(5, 0)$.

4. Write an equation of the horizontal line through $(-2, 5)$.

5. Find the intersection point of the lines $y = 3x - 4$ and $5x - 2y = 7$.

State the coordinates of point J without introducing any new letters.

6. Isosceles triangle

7. Parallelogram

8. Isosceles trapezoid

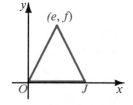

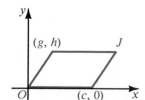

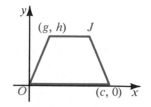

9. The vertices of a quadrilateral are $G(4, -1)$, $O(0, 0)$, $L(2, 6)$, and $D(6, 5)$. Show that quadrilateral *GOLD* is a parallelogram.

| **Application** | *Steiner's Problem* |

Four villages plan to build a system of roads of minimum length that will connect them all. Shown below are some plans for how to build the roads. Which plan shows the shortest road? Is there another way to connect the villages by an even shorter system of roads?

This problem was first investigated by the German mathematician Jacob Steiner (1796–1863), and carried his name, *Steiner's problem.*

Because a soap film automatically minimizes its surface area you can build a model that will help you solve Steiner's problem. You will need:

    a sheet of clear plastic
    8 split-pin paper fasteners
    a drinking straw cut into four 3-cm long pieces

Bend the sheet of plastic without creasing it. Cut four small slits (to represent the location of the four villages) through both layers of the sheet. Insert the paper fasteners through all eight slits. Slip two fasteners through each of four straws, so that your model looks like the figure at the right. The halves of the plastic sheet should be parallel, and the straws perpendicular to them.

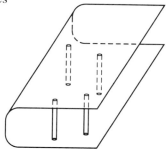

Dip the model in a soap solution and carefully lift it out. You should see a system of vertical soap films between the two sheets of plastic and joining the straws, revealing the solution to the problem. (Should any soap film adhere to the curved part of the plastic sheet, wet a drinking straw with the soap solution and push the straw through the soap films. You can suck air out through the straw to allow the films to form the minimum connection.)

Exercises

1. Gently place a protractor on top of the model and measure the angles where the soap films meet. What are the measures of these angles?

Make other models to find the shortest connection between the vertices of the following polygons. In each model, find the measures of the angles where the soap films meet.

2. Triangle   3. Square   4. Pentagon

Extra   *Points in Space*

To locate points in three-dimensional space, three coordinate axes are needed. Think of the *y-axis* and *z-axis* as lying in the plane of the paper with the *x-axis* perpendicular to the plane of the paper. The axes intersect at the *origin*, or zero point, of each axis. The arrowhead on each axis indicates the positive direction.

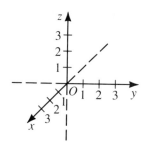
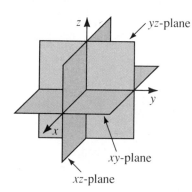
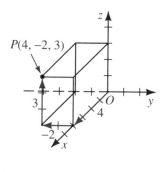

The coordinate axes determine three *coordinate planes*, as shown in the middle diagram above. Each point in space has three coordinates: the *x-coordinate*, *y-coordinate*, and *z-coordinate*. For example, point *P* in the diagram at the right above, has coordinates (**4, −2, 3**). The red arrows in the figure show that to *graph P* you start at *O*, move **4** units in the positive direction on the *x*-axis, **−2** units parallel to the *y*-axis (that is 2 units in the negative direction parallel to the *y*-axis), and **3** units in the positive direction parallel to the *z*-axis.

Exercises

On which axis or axes does each point lie?

1. (0, 7, 0)   **2.** (0, 0, −9)   **3.** (5, 0, 0)   **4.** (0, 0, 0)

On which coordinate plane or planes does each point lie?

5. (1, −3, 0)   **6.** (−7, 0, −1)   **7.** (0, 8, 5)   **8.** (0, 0, 0)

Graph each point on a coordinate system in space.

9. (−1, 4, 0)   **10.** (2, 3, 1)   **11.** (−2, −3, 4)   **12.** (0, 1, −5)

Sketch the triangle in space whose vertices have the given coordinates.

13. (4, 0, 0), (0, 8, 0), (0, 0, 2)   **14.** (1, 0, 0), (0, −5, 0), (0, 0, −5)

15. (−3, 0, 0), (0, −4, 0), (0, 0, 6)   **16.** (0, 0, 0), (3, 0, 3), (0, −4, 5)

Chapter Summary

1. The distance between points (x_1, y_1) and (x_2, y_2) is
$$\sqrt{(x_2 - x_1)^2 + (y_2 - y_1)^2}.$$
The midpoint of the segment joining these points is the point
$$\left(\frac{x_1 + x_2}{2}, \frac{y_1 + y_2}{2} \right).$$

2. The circle with center (a, b) and radius r has the equation
$$(x - a)^2 + (y - b)^2 = r^2.$$

3. The slope m of a line through two points (x_1, y_1) and (x_2, y_2), $x_1 \neq x_2$, is defined as follows: $m = \dfrac{y_2 - y_1}{x_2 - x_1}$. The slope of a horizontal line is zero. Slope is not defined for vertical lines.

4. Two nonvertical lines with slopes m_1 and m_2 are:
   a. parallel if and only if $m_1 = m_2$.
   b. perpendicular if and only if $m_1 \cdot m_2 = -1$.

5. Any quantity that has both magnitude and direction is called a vector. A vector can be represented by an arrow or by an ordered pair. The magnitude of $\overrightarrow{AB}$ equals the length of $\overline{AB}$. Two vectors are perpendicular if the arrows representing them are perpendicular. Two vectors are parallel if the arrows representing them have the same or opposite directions. Two vectors are equal if they have the same magnitude and direction.

6. Two operations with vectors were discussed: multiplication of a vector by a real number, and addition of vectors.

7. The graph of any equation that can be written in the form $Ax + By = C$, with A and B not both zero, is a line. An equation of the line through point (x_1, y_1) with slope m is $y - y_1 = m(x - x_1)$. An equation of the line with slope m and y-intercept b is $y = mx + b$. The coordinates of the point of intersection of two lines can be found by solving their equations simultaneously.

8. To prove theorems using coordinate geometry, proceed as follows:
   a. Place x- and y-axes in a convenient position with respect to a figure.
   b. Use known properties to assign coordinates to points of the figure.
   c. Use the distance formula, the midpoint formula, and the slope properties of parallel and perpendicular lines to prove theorems.

Chapter Review

Exercises 1 and 2 refer to points $X(-2, -4)$, $Y(2, 4)$, **and** $Z(2, -6)$.

1. Graph X, Y, and Z on one set of axes, then find XY, YZ, and XZ. 13–1

2. Use the distance formula to show that $\triangle XYZ$ is a right triangle.

Find the center and radius of each circle.

3. $(x + 3)^2 + y^2 = 100$ 4. $(x - 5)^2 + (y + 1)^2 = 49$

5. Write an equation of the circle that has center $(-6, -1)$ and radius 3.

6. Find the slope of the line through $(-5, -1)$ and $(15, -6)$. 13–2

7. A line with slope $\frac{2}{3}$ passes through $(9, -13)$ and $(0, \underline{\ ?\ })$.

8. A line through $(0, -2)$ has slope 5. Find three other points on the line.

9. What is the slope of a line that is parallel to the x-axis?

10. Show that $QRST$ is a trapezoid. 13–3

11. Since the slope of $\overline{QT}$ is $\underline{\ ?\ }$, the slope of an altitude to $\overline{QT}$ is $\underline{\ ?\ }$.

12. If U is a point on $\overline{QT}$ such that $\overline{UR} \parallel \overline{ST}$, then U has coordinates $(\underline{\ ?\ }, \underline{\ ?\ })$.

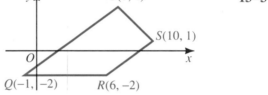

13. Given points $P(3, -2)$ and $Q(7, 1)$, find **(a)** $\overrightarrow{PQ}$, **(b)** $|\overrightarrow{PQ}|$, and **(c)** $-2\overrightarrow{PQ}$. 13–4

14. Find the vector sum $(2, 6) + 3(1, -2)$ and illustrate with a diagram.

Find the coordinates of the midpoint of the segment that joins the given points.

15. $(7, -2)$ and $(1, -1)$ 16. $(-4, 5)$ and $(2, -5)$ 17. (a, b) and $(-a, b)$ 13–5

18. $M(0, 5)$ is the midpoint of $\overline{RS}$. If S has coordinates $(11, -1)$, then R is point $(\underline{\ ?\ }, \underline{\ ?\ })$.

19. Graph the line $y = 2x - 3$. 20. Graph the line $x + 2y = 4$. 13–6

21. Find the point of intersection of the two lines in Exercises 19 and 20.

22. Find an equation of the line with slope 4 and y-intercept 7. 13–7

23. Find an equation of the line through $(-1, 2)$ and $(3, 10)$.

24. If $OPQR$ is a parallelogram, what are the coordinates of Q? 13–8

25. Let M be the midpoint of $\overline{RQ}$ and N be the midpoint of $\overline{OP}$. Use coordinate geometry to prove that $ONQM$ is a parallelogram. 13–9

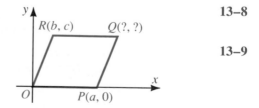

Chapter Test

Given: Points $M(-2, 1)$ and $N(2, 4)$

1. Find **(a)** MN, **(b)** the slope of $\overline{MN}$, and **(c)** the midpoint of $\overline{MN}$.
2. Write an equation of $\overleftrightarrow{MN}$.
3. Write an equation of a circle with center M and radius MN.
4. If M is the midpoint of $\overline{NZ}$, what are the coordinates of Z?

In Exercises 5–8 write an equation of each line described.

5. The line with slope $-\dfrac{3}{2}$ and y-intercept 4
6. The line with y-intercept 5 and x-intercept 3
7. The line through $(-2, 5)$ and parallel to $3x + y = 6$
8. The line with y-intercept 7 and perpendicular to $y = -2x + 3$

9. Given points $P(-2, 5)$ and $Q(4, 1)$, find **(a)** $\overrightarrow{PQ}$ and **(b)** $|\overrightarrow{PQ}|$.
10. The vectors $(3, 6)$ and $(-2, k)$ are parallel. Find the value of k.
11. The vectors $(3, -5)$ and $(c, 6)$ are perpendicular. Find the value of c.
12. Evaluate the vector sum $(5, -3) + 4(-2, 1)$.
13. Find the point of intersection of the lines $x + 2y = 8$ and $3x - y = 3$.

Draw the graph of each equation.

14. $2x - 3y = 6$ 15. $y = 5$

16. Name 3 points on the line through $(2, 2)$ with slope $\dfrac{4}{3}$.

17. An isosceles trapezoid is shown. Give the missing coordinates without introducing any new letters.

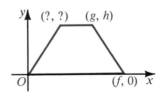

Use points $J(-12, 0)$, $K(0, 6)$, and $L(-3, -3)$.

18. Show that $\triangle JKL$ is isosceles.
19. Use slopes to show that $\triangle JKL$ is a right triangle.

Use coordinate geometry to prove each statement.

20. The diagonals of a rectangle bisect each other.
21. The segments joining the midpoints of consecutive sides of a rectangle form a rhombus.

Cumulative Review: Chapters 1–13

A 1. $\overrightarrow{BD}$ bisects $\angle ABC$, $m\angle ABC = 5x - 4$, and $m\angle CBD = \frac{3}{2}x + 21$.
Is $\angle ABC$ acute, obtuse, or right?

2. Name five ways to prove that two lines are parallel.

3. If the diagonals of a quadrilateral are congruent and perpendicular, must the quadrilateral be a square? a rhombus? Draw a diagram to illustrate your answer.

4. Write "$x = 1$ only if $x \neq 0$" in if-then form. Then write the contrapositive and classify the contrapositive as true or false.

5. Refer to the diagram.
   a. Show that $\angle B \cong \angle D$.
   b. Find the value of x.
   c. Find the ratio of the areas of the triangles.

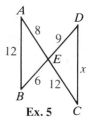

Ex. 5

6. Is a triangle with sides of lengths 12, 35, and 37 acute, right, or obtuse?

7. In $\triangle ABC$, $\overline{AB} \perp \overline{BC}$, $AB = 1$, and $AC = 3$. Find:
   a. $\cos A$ **b.** $\sin C$ **c.** $\tan A$ **d.** $\cos C$

8. Find the perimeter and area of a regular hexagon with apothem $\sqrt{3}$ cm.

9. Find the total area and volume of a cylinder with radius 10 and height 8.2.

10. Describe the locus of the centers of all circles tangent to each of two given parallel lines.

Find the value of x.

11.

12.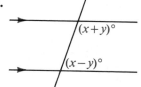

13.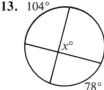

B 14. If x is the length of a tangent segment in the diagram, find the values of x and y.

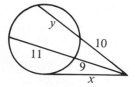

15. Prove: If the ray that bisects an angle of a triangle is perpendicular to the side that it intersects, then the triangle is an isosceles triangle.

16. Draw an obtuse triangle. Construct a circumscribed circle about the triangle.

17. Use coordinate geometry to prove that the median of a trapezoid is parallel to each base.

14 TRANSFORMATIONS

This striking color photograph shows a repeated design, identical balconies on one face of a building. A mathematical operation that changes the position of a figure without changing its shape is called a *transformation*.

Some Basic Mappings

Objectives

1. Recognize and use the terms *image*, *preimage*, *mapping*, *one-to-one mapping*, *transformation*, *isometry*, and *congruence mapping*.
2. Locate images of figures by reflection, translation, glide reflection, rotation, and dilation.
3. Recognize the properties of the basic mappings.

14-1 *Mappings and Functions*

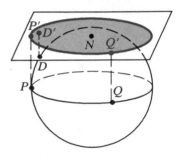

Have you ever wondered how maps of the round Earth can be made on flat paper? The diagram illustrates the idea behind a *polar map* of the northern hemisphere. A plane is placed tangent to a globe of the Earth at its North Pole N. Every point P of the globe is projected straight upward to exactly one point, called P', in the plane. P' is called the **image** of P, and P is called the **preimage** of P'. The diagram shows the images of two points P and Q on the globe's equator. It also shows D', the image of a point D not on the equator.

This correspondence between points of the globe's northern hemisphere and points in the plane is an example of a *mapping*. If we call this mapping M, then we could indicate that M maps P to P' by writing $M:P \rightarrow P'$. Notice that since the North Pole N is mapped to itself, we can write $M:N \rightarrow N$.

The word *mapping* is used in geometry as the word *function* is used in algebra. While a **mapping** is a correspondence between sets of points, a **function** is a correspondence between sets of numbers. Each number in the first set corresponds to exactly one number in the second set. For example, the squaring function f maps each real number x to its square x^2. We can write $f:x \rightarrow x^2$. Another way to indicate that the value of the function at x is x^2 is to write $f(x) = x^2$ (read "f of x equals x^2"). Similarly, for the mapping M, above, we can write $M(P) = P'$ to indicate that the image of P is P'. With all of these similarities, it should not surprise you that mathematicians often use the words *function* and *mapping* interchangeably.

A mapping (or a function) from set A to set B is called a **one-to-one mapping** (or a one-to-one function) if every member of B has exactly one preimage in A. The polar projection illustrated at the top of the page is a one-to-one mapping of the northern hemisphere of the globe onto a circular region in the tangent plane (the shaded area in the diagram). However, the squaring function $f:x \rightarrow x^2$ is *not* one-to-one because, for example, 9 has two preimages, 3 and -3.

Example 1   Function *g* maps every number to a number that is six more than its double.

    a. Express this fact using function notation.

    b. Find the image of 7.

    c. Find the preimage of 8.

Solution   **a.** $g : x \rightarrow 2x + 6$, or $g(x) = 2x + 6$

    b. $g : 7 \rightarrow 2 \cdot 7 + 6 = 20$. Thus the image of 7 is 20.

    c. $g : x \rightarrow 2x + 6 = 8$. Therefore $x = 1$, so 1 is the preimage of 8.

Example 2   Mapping *G* maps each point (x, y) to the point $(2x, y - 1)$.

    a. Express this fact using mapping notation.

    b. Find P' and Q', the images of $P(3, 0)$ and $Q(1, 4)$.

    c. Decide whether *G* maps *M*, the midpoint of $\overline{PQ}$, to M', the midpoint of $\overline{P'Q'}$.

    d. Decide whether $PQ = P'Q'$.

Solution   **a.** $G : (x, y) \rightarrow (2x, y - 1)$

    b. $G : (3, 0) \rightarrow (2 \cdot 3, 0 - 1) = (6, -1) = P'$

       $G : (1, 4) \rightarrow (2 \cdot 1, 4 - 1) = (2, 3) = Q'$

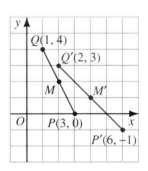

    c. $M = \left(\dfrac{3 + 1}{2}, \dfrac{0 + 4}{2} \right) = (2, 2)$

       $M' = \left(\dfrac{6 + 2}{2}, \dfrac{-1 + 3}{2} \right) = (4, 1)$

       $G : (2, 2) \rightarrow (2 \cdot 2, 2 - 1) = (4, 1)$

       Thus *G* does map midpoint *M* to midpoint M'.

    d. Use the distance formula to show that

$$PQ = \sqrt{(1 - 3)^2 + (4 - 0)^2}$$
$$= \sqrt{(-2)^2 + 4^2} = \sqrt{20} = 2\sqrt{5}$$
$$P'Q' = \sqrt{(2 - 6)^2 + (3 - (-1))^2}$$
$$= \sqrt{(-4)^2 + 4^2} = \sqrt{32} = 4\sqrt{2}$$

       Thus $PQ \neq P'Q'$.

    Although the diagram for Example 2 shows only points of $\overline{PQ}$ and their image points, you should understand that mapping *G* maps *every* point of the plane to an image point. Also, every point of the plane has a preimage point. A one-to-one mapping from the whole plane to the whole plane is called a **transformation.** Moreover, if a transformation maps every segment to a congruent segment, it is called an **isometry.** The transformation in Example 2 is *not* an isometry because $PQ \neq P'Q'$.

    By definition, an isometry maps any segment to a congruent segment, so we can say that an isometry *preserves* distance. The next theorem states that an isometry also maps any triangle to a congruent triangle. For this reason, an isometry is sometimes called a **congruence mapping.**

Theorem 14-1

An isometry maps a triangle to a congruent triangle.

Given: Isometry $T: \triangle ABC \rightarrow \triangle A'B'C'$

Prove: $\triangle ABC \cong \triangle A'B'C'$

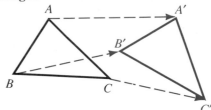

Proof:

Statements	Reasons
1. $\overline{AB} \cong \overline{A'B'}$, $\overline{BC} \cong \overline{B'C'}$, $\overline{AC} \cong \overline{A'C'}$	1. Definition of isometry
2. $\triangle ABC \cong \triangle A'B'C'$	2. SSS Postulate

Corollary 1

An isometry maps an angle to a congruent angle.

Corollary 2

An isometry maps a polygon to a polygon with the same area.

Example 3 Mapping R maps each point (x, y) to an image point $(-x, y)$.

a. Decide if $BA = B'A'$, $CB = C'B'$, and $CA = C'A'$.

b. Does R *appear* to be an isometry? Does part (a) *prove* that R is an isometry? Explain.

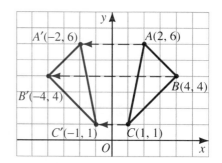

Solution **a.** Use the distance formula to show that

$$BA = \sqrt{(2-4)^2 + (6-4)^2} = \sqrt{(-2)^2 + 2^2} = \sqrt{8} = 2\sqrt{2}$$
$$B'A' = \sqrt{(-2-(-4))^2 + (6-4)^2} = \sqrt{2^2 + 2^2} = \sqrt{8} = 2\sqrt{2}$$
$$CB = \sqrt{(4-1)^2 + (4-1)^2} = \sqrt{3^2 + 3^2} = \sqrt{18} = 3\sqrt{2}$$
$$C'B' = \sqrt{(-4-(-1))^2 + (4-1)^2} = \sqrt{(-3)^2 + 3^2} = \sqrt{18} = 3\sqrt{2}$$
$$CA = \sqrt{(2-1)^2 + (6-1)^2} = \sqrt{1^2 + 5^2} = \sqrt{26}$$
$$C'A' = \sqrt{(-2-(-1))^2 + (6-1)^2} = \sqrt{(-1)^2 + 5^2} = \sqrt{26}$$

We have $BA = B'A'$, $CB = C'B'$, and $CA = C'A'$.

b. R *appears* to be an isometry because part (a) shows that three segments are mapped to congruent segments. Part (a) *does not prove* that R is an isometry because a proof must show that the image of *every* segment is a congruent segment.

Classroom Exercises

1. Explain why each of the correspondences pictured below is not a one-to-one mapping from set *A* to set *B*.

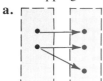

a.

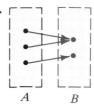

b.

c.

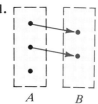

d.

2. **a.** If $f : x \rightarrow |x|$, find the images of -3, 6, and -6.
   b. Is f a one-to-one function? Explain.

3. **a.** If mapping $M : (x, y) \rightarrow (2x, 2y)$, find the images of *P* and *Q* in the diagram.
   b. Is *M* a transformation?
   c. Does *M* appear to be an isometry?
   d. Decide whether *M* maps the midpoint of $\overline{PQ}$ to the midpoint of $\overline{P'Q'}$.

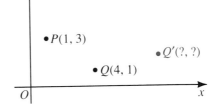

4. **a.** If $g(x) = 2x - 1$, find $g(8)$ and $g(-8)$.
   b. Find the image of 5.
   c. Find the preimage of 7.

5. Use the transformation $T : (x, y) \rightarrow (x + 1, y + 2)$ in this exercise.
   a. Plot the following points and their images on the chalkboard: $A(0, 0)$, $B(3, 4)$, $C(5, 1)$, and $D(-1, -3)$.
   b. Find AB, $A'B'$, CD, and $C'D'$.
   c. Does this transformation appear to be an isometry?
   d. What is the preimage of $(0, 0)$? of $(4, 5)$?

Exercises 6–8 refer to the globe shown on page 571.

6. What is the image of point *N*?

7. Is the distance between *N* and *P* on the globe the same as the corresponding distance on the polar map?

8. Does the polar map preserve or distort distances?

9. Explain how Corollary 1 follows from Theorem 14-1.

10. Explain how Corollary 2 follows from Theorem 14-1.

Written Exercises

A
1. If function $f : x \rightarrow 5x - 7$, find the image of 8 and the preimage of 13.
2. If function $g : x \rightarrow 8 - 3x$, find the image of 5 and the preimage of 0.
3. If $f(x) = x^2 + 1$, find $f(3)$ and $f(-3)$. Is f a one-to-one function?
4. If $h(x) = 6x + 1$, find $h(\frac{1}{2})$. Is h a one-to-one function?

For each transformation given in Exercises 5–10:
a. **Plot the three points $A(0, 4)$, $B(4, 6)$, and $C(2, 0)$ and their images A', B', and C' under the transformation.**
b. **State whether the transformation appears to be an isometry.**
c. **Find the preimage of $(12, 6)$.**

5. $T:(x, y) \to (x + 4, y - 2)$ **6.** $S:(x, y) \to (2x + 4, 2y - 2)$

7. $D:(x, y) \to (3x, 3y)$ **8.** $H:(x, y) \to (-x, -y)$

9. $M:(x, y) \to (12 - x, y)$ **10.** $G:(x, y) \to (-\tfrac{1}{2}x, -\tfrac{1}{2}y)$

11. O is a point equidistant from parallel lines l_1 and l_2. A mapping M maps each point P of l_1 to the point P' where $\overrightarrow{PO}$ intersects l_2.

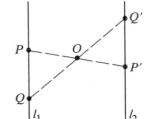

  a. Is the mapping a one-to-one mapping of l_1 onto l_2?
  b. Does this mapping preserve or distort distance?
  c. If l_1 and l_2 were not parallel, would the mapping preserve distance? Illustrate your answer with a sketch.

12. $\triangle XYZ$ is isosceles with $\overline{XY} \cong \overline{XZ}$. Describe a way of mapping each point of $\overline{XY}$ to a point of $\overline{XZ}$ so that the mapping is an isometry.

B   **13.** $ABCD$ is a trapezoid. Describe a way of mapping each point of $\overline{DC}$ to a point of $\overline{AB}$ so that the mapping is one-to-one. Is your mapping an isometry?

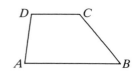

14. The red and blue squares are congruent and have the same center O. A mapping maps each point P of the red square to the point P' where $\overrightarrow{OP}$ intersects the blue square.

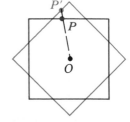

  a. Is this mapping one-to-one?
  b. Copy the diagram and locate a point X that is its own image.
  c. Locate two points R and S on the red square and their images R' and S' on the blue square that have the property that $RS \neq R'S'$.
  d. Does this mapping preserve distance?
  e. Describe a mapping from the red square onto the blue square that *does* preserve distance.

15. The transformation $T:(x, y) \to (x + y, y)$ preserves areas of figures even though it does not preserve distances. Illustrate this by drawing a square with vertices $A(2, 3)$, $B(4, 3)$, $C(4, 5)$, and $D(2, 5)$ and its image $A'B'C'D'$. Find the area and perimeter of each figure.

A piece of paper is wrapped around a globe of the Earth to form a cylinder as shown. *O* is the center of the Earth and a point *P* of the globe is projected along $\overrightarrow{OP}$ to a point *P'* of the cylinder.

16. Describe the image of the globe's equator.

17. Is the image of the Arctic Circle congruent to the image of the equator?

18. Are distances near the equator distorted more than or less than distances near the Arctic Circle?

19. Does the North Pole (point *N*) have an image?

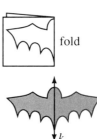

20. Consider the mapping *S*: $(x, y) \to (x, 0)$.
    a. Plot the points $P(4, 5)$, $Q(-3, 2)$, and $R(-3, -1)$ and their images.
    b. Does *S* appear to be an isometry? Explain.
    c. Is *S* a transformation? Explain.

21. Mapping *M* maps points *A* and *B* to the same image point. Explain why the mapping *M* does not preserve distance.

22. Fold a piece of paper. Cut a design connecting the top and bottom point of the fold, as shown. Unfold the shape. Consider a mapping *M* of the points in the gray region to the corresponding points in the red region.
    a. Does *M* appear to be an isometry?
    b. If a point *P* is on line *k*, what is the image of *P*?
    c. If a point *Q* is not on line *k*, and $M(Q) = Q'$, what is the relationship between line *k* and $\overleftrightarrow{QQ'}$?

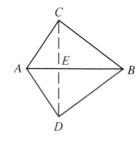

C 23. a. Plot the points $A(6, 1)$, $B(3, 4)$, and $C(1, -3)$ and their images A', B', and C' under the transformation $R:(x, y) \to (-x, y)$.
    b. Prove that *R* is an isometry. (*Hint*: Let $P(x_1, y_1)$ and $Q(x_2, y_2)$ be any two points. Find P' and Q', and use the distance formula to show that $PQ = P'Q'$.)

Explorations

These exploratory exercises can be done using a computer with a program that draws and measures geometric figures.

As you will learn in the next lesson, a *reflection* is a mapping in the plane across a mirror line, just as your reflection in a mirror is a mapping in space across a mirror plane.

Draw any $\triangle ABC$. Reflect *C* in $\overleftrightarrow{AB}$ to locate point *D*. Draw $\overline{AD}$ and $\overline{BD}$. What do you notice about $\triangle ABC$ and $\triangle ABD$?

Draw $\overline{CD}$. Label the intersection of $\overleftrightarrow{AB}$ and $\overleftrightarrow{CD}$ as *E*. Compare *CE* and *DE*. What do you notice? Measure the angles with vertex *E*. What do you notice?

Repeat the construction with other types of triangles.

14-2 *Reflections*

When you stand before a mirror, your image appears to be as far behind the mirror as you are in front of it. The diagram shows a transformation in which a line acts like a mirror. Points P and Q are reflected in line m to their images P' and Q'. This transformation is called a *reflection*. Line m is called the *line of reflection* or the mirror line.

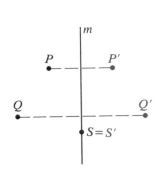

A **reflection** in line m maps every point P to a point P' such that:

(1) If P is not on the line m, then m is the perpendicular bisector of $\overline{PP'}$.
(2) If P is on line m, then $P' = P$.

To abbreviate *reflection in line m*, we write R_m. To abbreviate the statement R_m maps P to P', we write $R_m: P \rightarrow P'$ or $R_m(P) = P'$. This may also be read as *P is reflected in line m to P'*.

Theorem 14-2

A reflection in a line is an isometry.

To prove Theorem 14-2 by using coordinates, we assign coordinates in the plane so that the line of reflection becomes the y-axis. Then in coordinate terms the reflection is $R:(x_1, y_1) \rightarrow (-x_1, y_1)$. In Exercise 23 on page 576 the distance formula was used to prove that $PQ = P'Q'$. Although the diagram shows P and Q on the same side of the y-axis, you should realize that the coordinates x_1, y_1, x_2, and y_2 can be positive, negative, or zero, thereby covering all cases.

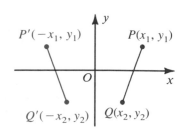

Theorem 14-2 can also be proved without the use of coordinates. If coordinates are not used, we must show that $PQ = P'Q'$ for all choices of P and Q. Four of the possible cases are shown below. In Written Exercises 18–20 you will prove Theorem 14-2 for Cases 2–4, using the fact that the line of reflection, m, is the perpendicular bisector of $\overline{PP'}$ and $\overline{QQ'}$.

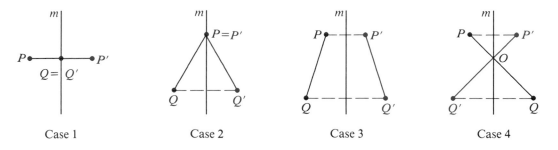

Case 1   Case 2   Case 3   Case 4

Since a reflection is an isometry, it preserves distance, angle measure, and the area of a polygon. Another way to say this is that distance, angle measure, and area are *invariant* under a reflection. On the other hand, the orientation of a figure is *not* invariant under a reflection because a reflection changes a clockwise orientation to a counterclockwise one, as shown at the right.

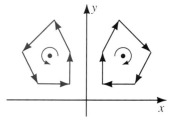

Example   Find the image of point $P(2, 4)$ and $\triangle ABC$ under each reflection.
  a. The line of reflection is the x-axis.
  b. The line of reflection is the line $y = x$.

Solution   The images are shown in red.
  a.

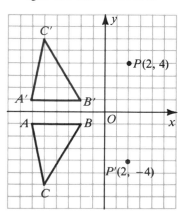

  b.

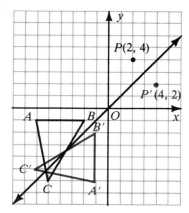

Notice that under reflection in the line $y = x$, the point (x, y) is mapped to the point (y, x).

Classroom Exercises

Complete the following. Assume points D, C, U, W, X, and Y are obtained by reflection in line k or j.

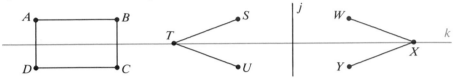

1. R_k stands for ___?___.

2. $R_k : A \rightarrow$ ___?___

3. $R_k(B) =$ ___?___

4. $R_k : \overline{AB} \rightarrow$ ___?___

5. $R_k(C) =$ ___?___

6. $R_k : T \rightarrow$ ___?___

7. $R_k : \overline{BC} \rightarrow$ ___?___

8. $R_k : \angle STU \rightarrow$ ___?___

9. $R_j(S) =$ ___?___

10. $R_j : \overline{ST} \rightarrow$ ___?___

11. $R_j($ ___?___ $) = \overline{XY}$

12. $R_j : \text{line } k \rightarrow$ ___?___

Points A–D are reflected in the x-axis. Points E–H are reflected in the y-axis. State the coordinates of the images.

13.

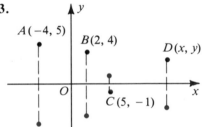

14.

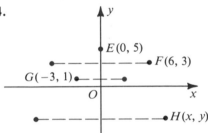

Sketch each figure on the chalkboard. With a different color, sketch its image, using the dashed line as the line of reflection.

15.

16.

17.

18.

19. Under a reflection, is an angle always mapped to a congruent angle? Is a polygon always mapped to a polygon with the same area? Explain.

20. Explain in your own words the meaning of each phrase.
  a. An isometry preserves distance.
  b. Area is invariant under a reflection.
  c. Orientation is not invariant under a reflection.

Written Exercises

Copy each figure on graph paper. Then draw the image by reflection in line *k*.

A

1.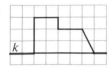

2.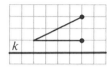

3.

4.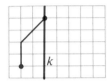

5.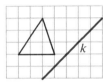

6.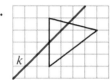

Write the coordinates of the image of each point by reflection in (a) the *x*-axis, (b) the *y*-axis, and (c) the line $y = x$. (*Hint*: Refer to the Example on page 578.)

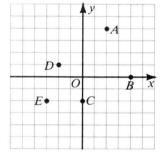

Exs. 7–12

7. *A*   **8.** *B*   **9.** *C*

10. *D*   **11.** *E*   **12.** *O*

13. When the word MOM is reflected in a vertical line, the image is still MOM. Can you think of other words that are unchanged when reflected in a vertical line?

14. When the word HIDE is reflected in a horizontal line, the image is still HIDE. Can you think of other words that are unchanged when reflected in a horizontal line?

B **15.** Draw a triangle and a line *m* such that R_m maps the triangle to itself. What kind of triangle did you use?

16. Draw a pentagon and a line *n* such that R_n maps the pentagon to itself.

17. The sketch illustrates a *reflection in plane X*. Write a definition of this reflection similar to the definition of a reflection in line *m* on page 577.

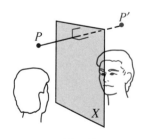

Ex. 17

In Exercises 18–20, refer to the diagrams on page 578. Given the reflection $R_m : \overline{PQ} \to \overline{P'Q'}$, write the key steps of a proof that $PQ = P'Q'$ for each case.

18. Case 2   **19.** Case 3   **20.** Case 4

21. Draw a line *t* and a point *A* not on *t*. Then use a straightedge and compass to construct the image of *A* under R_t.

22. Draw any two points *B* and *B'*. Then use a straightedge and compass to construct the line of reflection *j* so that $R_j(B) = B'$.

23. If a transformation maps two parallel lines to two image lines that are also parallel, we say that parallelism is invariant under the transformation. Is parallelism invariant under a reflection?

The photograph shows a reflected beam of laser light. Exercises 24–28 deal with the similar reflected path of a golf ball bouncing off the walls of a miniature golf layout. These exercises show how the geometry of reflections can be used to solve the problem of aiming a reflected path at a particular target.

24. A rolling ball that does not have much spin will bounce off a wall so that the two angles that the path forms with the wall are congruent. Thus, to roll the ball from *B* off the wall shown and into hole *H*, you need to aim the ball so that $\angle 1 \cong \angle 2$.

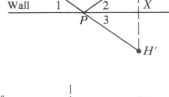

    a. Let *H'* be the image of *H* by reflection in the wall. $\overline{BH'}$ intersects the wall at *P*. Why is $\angle 1 \cong \angle 3$? Why is $\angle 3 \cong \angle 2$? Why is $\angle 1 \cong \angle 2$? You can conclude that if you aim for *H'*, the ball will roll to *H*.

    b. Show that the distance traveled by the ball equals the distance *BH'*.

25. In the two-wall shot illustrated at the right, a reflection in one wall maps *H* to *H'*, and a reflection in a second wall (extended) maps *H'* to *H"*. To roll the ball from *B* to *H*, you aim for *H"*. Show that the total distance traveled by the ball equals the distance *BH"*.

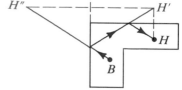

26. Show how to score a hole in one on the fifth hole of the golf course shown by rolling the ball off one wall.

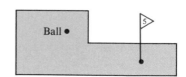

27. Repeat Exercise 26 but roll the ball off two walls.

28. Repeat Exercise 26 but roll the ball off three walls.

29. A ball rolls at a 45° angle away from one side of a billiard table that has a coordinate grid on it. If the ball starts at the point (0, 1) it will eventually return to its starting point. Would this happen if the ball started from other points on the *y*-axis between (0, 0) and (0, 4)?

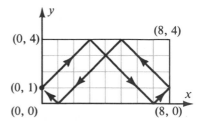

30. The line with equation $y = 2x + 3$ is reflected in the *y*-axis. Find an equation of the image line.

31. The line with equation $y = x + 5$ is reflected in the *x*-axis. Find an equation of the image line.

In each exercise $R_k : A \rightarrow A'$. Find an equation of line k.

	32.	33.	34.	35.	36.	37.
A	(5, 0)	(1, 4)	(4, 0)	(5, 1)	(0, 2)	(−1, 2)
A'	(9, 0)	(3, 4)	(4, 6)	(1, 5)	(4, 6)	(4, 5)

C **38.** Draw the x- and y-axes and the line l with equation $y = -x$. Plot several points and their images under R_l. What is the image of (a, b)?

39. Draw the x- and y-axes and the vertical line j with equation $x = 5$. Find the images under R_j of the following points.

    a. (4, 3)   **b.** (0, −2)   **c.** (−3, 1)   **d.** (x, y)

40. Repeat Exercise 39 letting j be the horizontal line with equation $y = 6$.

Application   *Mirrors*

If a ray of light strikes a mirror at an angle of 40°, it will be reflected off the mirror at an angle of 40° also. The angle between the mirror and the reflected ray is always congruent to the angle between the mirror and the initial light ray. In the diagram at the left below, $\angle 2 \cong \angle 1$.

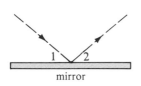

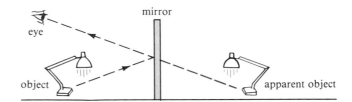

We see objects in a mirror when the reflected light ray reaches the eye. The object appears to lie behind the mirror as shown in the diagram at the right above.

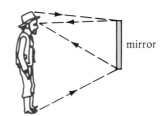

You don't need a full-length mirror to see all of yourself. A mirror that is only half as tall as you are will do if the mirror is in a position as shown. You see the top of your head at the top of the mirror and your feet at the bottom of the mirror. If the mirror is too high or too low, you will not see your entire body.

A periscope uses mirrors to enable a viewer to see above the line of sight. The diagram at the right is a simple illustration of the principle used in a periscope. It has two mirrors, parallel to each other, at the top and at the bottom. The mirrors are placed at an angle of 45° with the horizontal. Horizontal light rays from an object entering at the top are reflected down to the mirror at the bottom. They are then reflected to the eye of the viewer.

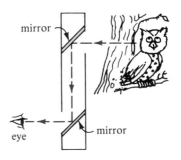

Exercises

1. What are the measures of the angles that the initial light ray and the reflected light rays make with the mirrors in the diagram of the periscope on the previous page?

2. If you can see the eyes of someone when you look into a mirror, can the other person see your eyes in that same mirror?

3. A person with eyes at A, 150 cm above the floor, faces a mirror 1 m away. The mirror extends 30 cm above eye level. How high can the person see on a wall 2 m behind point A?

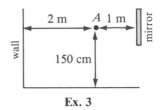

Ex. 3

4. Prove that you can see all of yourself in a mirror that is only half as tall as you are. (*Hint*: Study the diagram on page 582.)

5. Prove that the point D which is as far behind the mirror as the object A is in front of the mirror lies on $\overleftrightarrow{BC}$. (*Hint*: Show that $\angle CBE$ and $\angle EBD$ are supplementary.)

6. Show that the light ray follows the shortest possible path from A to C via the mirror by proving that for any point E on the mirror (other than B) $AE + EC > AB + BC$. (*Hint*: See the Application: Finding the Shortest Path, on page 224.)

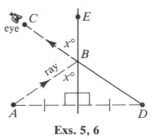

Exs. 5, 6

14-3 *Translations and Glide Reflections*

The photograph at the right suggests the transformation called a *translation*, or *glide*. The skate blades of the figure-skating pair move in identical ways when the pair is skating together. A transformation that glides all points of the plane the same distance in the same direction is called a **translation.**

If a transformation maps A to A', B to B', and C to C', points A, B, and C glide along parallel or collinear segments and $AA' = BB' = CC'$. Any of the vectors $\overrightarrow{AA'}$, $\overrightarrow{BB'}$, or $\overrightarrow{CC'}$ could describe this translation.

Each vector has the same magnitude of $\sqrt{1^2 + 3^2}$, or $\sqrt{10}$, and each vector has the same direction as indicated by its slope of $\frac{1}{3}$. Note that we don't need to know the coordinates of points A, B, or C to describe the translation. All that is important is the change in the x-coordinate and y-coordinate of each point.

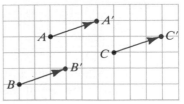

Consider a translation in which every point glides 8 units right and 2 units up. We could use the vector (8, 2) to indicate such a translation, or we could use the coordinate expression $T:(x, y) \rightarrow (x + 8, y + 2)$. The following diagram shows how $\triangle PQR$ is mapped by T to $\triangle P'Q'R'$. You can use the distance formula to check that each segment is mapped to a congruent segment so that T is an isometry.

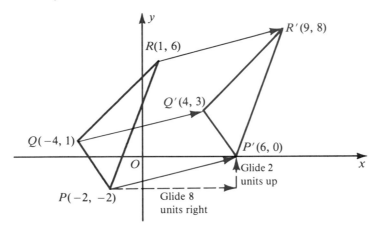

The illustration just presented should help you to understand why we use the following definition of a translation when working in the coordinate plane. A **translation,** or glide, in a plane is a transformation T which maps any point (x, y) to the point $(x + a, y + b)$ where a and b are constants. This definition makes it possible to give a simple proof of the following theorem.

Theorem 14-3

A translation is an isometry.

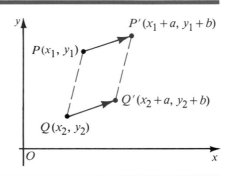

Plan for Proof: Label two points P and Q and their images P' and Q' as shown in the diagram. To show that T is isometry, we need to show that $PQ = P'Q'$. Use the distance formula to show that:

$$PQ = P'Q' = \sqrt{(x_2 - x_1)^2 + (y_2 - y_1)^2}$$

Of course, since a translation is an isometry, we know by the corollaries of Theorem 14-1, page 573, that a translation preserves angle measure and area.

A glide and a reflection can be carried out one after the other to produce a transformation known as a *glide reflection*. A **glide reflection** is a transformation in which every point P is mapped to a point P'' by these steps:

1. A glide maps P to P'.
2. A reflection in a line parallel to the glide line maps P' to P''.

A glide reflection combines two isometries to produce a new transformation, which is itself an isometry. The succession of footprints shown illustrates a glide reflection. Note that the reflection line is parallel to the direction of the glide.

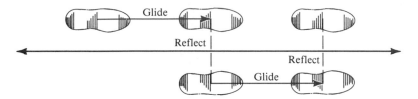

As long as the glide is parallel to the line of reflection, it doesn't matter whether you glide first and then reflect, or reflect first and then glide. For other combinations of mappings, the order in which you perform the mappings will affect the result. We will look further at such combinations of mappings in Section 14-6.

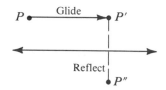

Classroom Exercises

1. Complete each statement for the translation $T:(x, y) \rightarrow (x + 3, y - 1)$.
 a. T glides points _?_ units right and 1 unit _?_.
 b. The image of (4, 6) is (_?_, _?_).
 c. The preimage of (2, 3) is (_?_, _?_).

Describe each translation in words, as in Exercise 1(a), and give the image of (4, 6) and the preimage of (2, 3).

2. $T:(x, y) \rightarrow (x - 5, y + 4)$ **3.** $T:(x, y) \rightarrow (x + 1, y)$

Each diagram shows a point P on the coordinate plane and its image P' under a translation T. Complete the statement $T:(x, y) \rightarrow$ (_?_, _?_).

4. 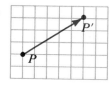 **5.** 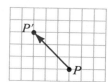 **6.**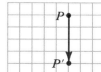

7. For a given translation, the image of the origin is (5, 7). What is the preimage of the origin?

8. A glide reflection has the glide translation $T:(x, y) \rightarrow (x + 2, y + 2)$. The line of reflection is line m with equation $y = x$.
 a. Find the image, point S', of $S(-1, 3)$ under T.
 b. Find the image, point S'', of S' under R_m. (*Hint*: Recall from the example on page 578 that $R_m:(x, y) \rightarrow (y, x)$.)
 c. Under the glide reflection, (x, y) is first mapped to (_?_, _?_) and then to (_?_, _?_).

Written Exercises

In Exercises 1 and 2 a translation *T* is described. For each:
a. Graph △*ABC* and its image △*A'B'C'*. Is △*ABC* ≅ △*A'B'C'*?
b. In color, draw arrows from *A* to *A'*, *B* to *B'*, and *C* to *C'*.
c. Are your arrows the same length? Are they parallel?

A

1. $T:(x, y) \rightarrow (x - 2, y + 6)$
   $A(-2, 0)$, $B(0, 4)$, $C(3, -1)$

2. $T:(x, y) \rightarrow (x - 3, y - 6)$
   $A(3, 6)$, $B(-3, 6)$, $C(-1, -2)$

3. If $T:(0, 0) \rightarrow (5, 1)$, then $T:(3, 3) \rightarrow (\underline{\ ?\ }, \underline{\ ?\ })$.

4. If $T:(1, 1) \rightarrow (3, 0)$, then $T:(0, 0) \rightarrow (\underline{\ ?\ }, \underline{\ ?\ })$.

5. If $T:(-2, 3) \rightarrow (2, 6)$, then $T:(\underline{\ ?\ }, \underline{\ ?\ }) \rightarrow (0, 0)$.

6. The image of $P(-1, 5)$ under a translation is $P'(5, 7)$. What is the pre-image of *P*?

In each exercise a glide reflection is described. Graph △*ABC* and its image under the glide, △*A'B'C'*. Also graph △*A"B"C"*, the image of △*A'B'C'* under the reflection.

7. Glide: All points move up 4 units.
   Reflection: All points are reflected in the *y*-axis.
   $A(1, 0)$, $B(4, 2)$, $C(5, 6)$

8. Glide: All points move left 7 units.
   Reflection: All points are reflected in the *x*-axis.
   $A(4, 2)$, $B(7, 0)$, $C(9, -3)$

B

9. Where does the glide reflection in Exercise 7 map (x, y)?

10. Where does the glide reflection in Exercise 8 map (x, y)?

11. Which of the following properties are invariant under a translation?
    a. distance   **b.** angle measure   **c.** area   **d.** orientation

12. Which of the properties listed in Exercise 11 are invariant under a glide reflection?

In Exercises 13 and 14 translations *R* and *S* are described. *R* maps point *P* to *P'*, and *S* maps *P'* to *P"*. Find *T*, the translation that maps *P* to *P"*.

13. $R:(x, y) \rightarrow (x + 1, y + 2)$
    $S:(x, y) \rightarrow (x - 5, y + 7)$
    $T:(x, y) \rightarrow (\underline{\ ?\ }, \underline{\ ?\ })$

14. $R:(x, y) \rightarrow (x - 5, y - 3)$
    $S:(x, y) \rightarrow (x + 4, y - 6)$
    $T:(x, y) \rightarrow (\underline{\ ?\ }, \underline{\ ?\ })$

15. If a translation *T* maps *P* to *P'*, then *T* can be described by the vector $\overrightarrow{PP'}$. Suppose a translation *T* is described by the vector $(3, -4)$ because it glides all points 3 units right and 4 units down.
    a. Graph points $A(-1, 2)$, $B(0, 6)$, *A'*, and *B'*, where $T(A) = A'$ and $T(B) = B'$.
    b. What kind of figure is *AA'B'B*? What is its perimeter?

16. a. Graph $\triangle POQ$ with vertices $P(0, 3)$, $O(0, 0)$, and $Q(6, 0)$.

    b. $T_1:(x, y) \rightarrow (x + 2, y - 4)$ and $T_2:(x, y) \rightarrow (x + 5, y + 6)$. If $T_1:\triangle POQ \rightarrow \triangle P'O'Q'$ and $T_2:\triangle P'O'Q' \rightarrow \triangle P''O''Q''$, graph $\triangle P'O'Q'$ and $\triangle P''O''Q''$.

    c. Find T_3, a translation that maps $\triangle POQ$ directly to $\triangle P''O''Q''$.

    d. Because T_1 glides all points 2 units right and 4 units down, the translation can be described by the vector $\overrightarrow{T_1} = (2, -4)$. Describe T_2 and T_3 by vectors. How are these three vectors related?

17. A glide reflection maps $\triangle ABC$ to $\triangle A'B'C'$. Copy the diagram and locate the midpoints of $\overline{AA'}$, $\overline{BB'}$, and $\overline{CC'}$. What seems to be true about these midpoints? Try to prove your conjecture.

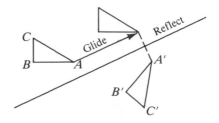

18. Copy the figure and use the result of Exercise 17 to construct the reflecting line of the glide reflection that maps $\triangle ABC$ to $\triangle A'B'C'$. Also construct the glide image of $\triangle ABC$.

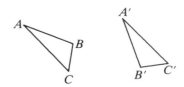

19. Explain why a glide reflection is an isometry.

20. Given $\odot A$ and $\odot B$ and $\overline{CD}$, construct a segment $\overline{XY}$ parallel to and congruent to $\overline{CD}$ and having X on $\odot A$ and Y on $\odot B$. (*Hint:* Translate $\odot A$ along a path parallel to and congruent to $\overline{CD}$.)

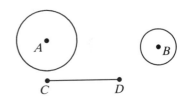

C 21. Describe how you would construct points X and Y, one on each of the lines shown, so that $\overline{XY}$ is parallel to and congruent to $\overline{EF}$.

22. Show by example that if a glide is not parallel to a line of reflection, then the image of a point when the glide is followed by the reflection will be different from the image of the same point when the reflection is followed by the glide.

23. Prove Theorem 14-3 (page 584).

14-4 *Rotations*

A *rotation* is a transformation suggested by rotating a paddle wheel. When the wheel moves, each paddle rotates to a new position. When the wheel stops, the new position of a paddle (P') can be referred to mathematically as the image of the initial position of the paddle (P).

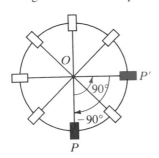

For the counterclockwise rotation shown about point O through $90°$, we write $\mathcal{R}_{O,\,90}$. A counterclockwise rotation is considered positive, and a clockwise rotation is considered negative. If the red paddle is rotated about O clockwise until it moves into the position of the black paddle, the rotation is denoted by $\mathcal{R}_{O,\,-90}$. (Note that to avoid confusion with the R used for reflections we use a script $\mathcal{R}$ for rotations.)

A full revolution, or $360°$ rotation about point O, rotates any point P around to itself so that $P' = P$. The diagram at the left below shows a rotation of $390°$ about O. Since $390°$ is $30°$ more than one full revolution, the image of any point P under a $390°$ rotation is the same as its image under a $30°$ rotation, and the two rotations are said to be equal. Similarly, the diagram at the right below shows that a $90°$ counterclockwise rotation is equal to a $270°$ clockwise rotation because both have the same effect on any point P.

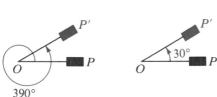

$$\mathcal{R}_{O,\,390} = \mathcal{R}_{O,\,30}$$
Notice: $390 - 360 = 30$

$$\mathcal{R}_{O,\,90} = \mathcal{R}_{O,\,-270}$$
Notice: $90 - 360 = -270$

In the following definition of a rotation, the angle measure x can be positive or negative and can be more than 180 in absolute value.

A **rotation** about point O through $x°$ is a transformation such that:

(1) If a point P is different from O, then $OP' = OP$ and $m\angle POP' = x$.
(2) If point P is the point O, then $P' = P$.

Theorem 14-4

A rotation is an isometry.

Given: $\mathscr{R}_{O,\,x}$ maps P to P' and Q to Q'.
Prove: $\overline{PQ} \cong \overline{P'Q'}$

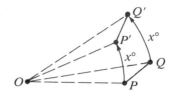

Key steps of proof:

1. $OP = OP'$, $OQ = OQ'$ (Definition of rotation)
2. $m \angle POP' = m \angle QOQ' = x$ (Definition of rotation)
3. $m \angle POQ = m \angle P'OQ'$ (Subtraction Property of = : subtract $m \angle QOP'$.)
4. $\triangle POQ \cong \triangle P'OQ'$ (SAS Postulate)
5. $\overline{PQ} \cong \overline{P'Q'}$ (Corr. parts of $\cong$ $\triangle$ are $\cong$.)

A rotation about point O through $180°$ is called a **half-turn** about O and is usually denoted by H_O. The diagram shows $\triangle PQR$ and its image $\triangle P'Q'R'$ by H_O. Notice that O is the midpoint of $\overline{PP'}$, $\overline{QQ'}$, and $\overline{RR'}$.

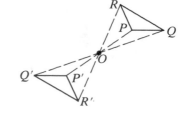

Using coordinates, a half-turn H_O about the origin can be written

$$H_O:(x,\ y) \rightarrow (-x,\ -y).$$

Classroom Exercises

State another name for each rotation.

1. $\mathscr{R}_{O,\,50}$ **2.** $\mathscr{R}_{O,\,-40}$ **3.** $\mathscr{R}_{O,\,-90}$ **4.** $\mathscr{R}_{O,\,400}$ **5.** $\mathscr{R}_{O,\,-180}$

In the diagram for Exercises 6–11, O is the center of equilateral $\triangle PST$. State the images of points P, S, and T for each rotation.

6. $\mathscr{R}_{O,\,120}$ **7.** $\mathscr{R}_{O,\,-120}$ **8.** $\mathscr{R}_{O,\,360}$

Name each image point.

9. $\mathscr{R}_{T,\,60}(S)$ **10.** $\mathscr{R}_{T,\,-60}(P)$ **11.** $\mathscr{R}_{O,\,240}(P)$

Exs. 6–11

12. Draw a coordinate grid on the chalkboard. Plot the origin and $A(4,\ 1)$. Give the coordinates of **(a)** $H_O\ (A)$, **(b)** $\mathscr{R}_{O,\,90}\ (A)$, and **(c)** $\mathscr{R}_{O,\,-90}\ (A)$.

13. Repeat Exercise 12 if A has coordinates $(-3,\ 5)$.

14. Is congruence invariant under a half-turn mapping? Explain.

15. Read each expression aloud.
    a. $R_k(A) = A'$ **b.** $H_O:(-2,\ 0) \rightarrow (2,\ 0)$
    c. $T:(x,\ y) \rightarrow (x - 1,\ y + 3)$ **d.** $\mathscr{R}_{P,\,10}$

Written Exercises

State another name for each rotation.

A **1.** $\mathcal{R}_{O,\ 80}$ **2.** $\mathcal{R}_{O,\ -15}$ **3.** $\mathcal{R}_{A,\ 450}$ **4.** $\mathcal{R}_{B,\ -720}$ **5.** H_O

The diagonals of regular hexagon $ABCDEF$ form six equilateral triangles as shown. Complete each statement below.

6. $\mathcal{R}_{O,\ 60}: E \to$ __?__   **7.** $\mathcal{R}_{O,\ -60}: D \to$ __?__

8. $\mathcal{R}_{O,\ 120}: F \to$ __?__   **9.** $\mathcal{R}_{D,\ 60}:$ __?__ $\to O$

10. $\mathcal{R}_{B,\ -60}(O) =$ __?__   **11.** $H_O(A) =$ __?__

12. A reflection in $\overleftrightarrow{FC}$ maps B to __?__ and D to __?__.

13. If k is the perpendicular bisector of $\overline{FE}$, then $R_k(A) =$ __?__.

14. If a translation maps A to B, then it also maps O to __?__ and E to __?__.

Exs. 6–14

State whether the specified triangle is mapped to the other triangle by a reflection, translation, rotation, or half-turn.

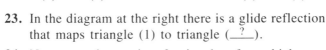

15. $\triangle(1)$ to $\triangle(2)$ **16.** $\triangle(1)$ to $\triangle(3)$

17. $\triangle(1)$ to $\triangle(4)$ **18.** $\triangle(1)$ to $\triangle(5)$

19. $\triangle(2)$ to $\triangle(4)$ **20.** $\triangle(2)$ to $\triangle(7)$

21. $\triangle(4)$ to $\triangle(6)$ **22.** $\triangle(4)$ to $\triangle(8)$

B **23.** In the diagram at the right there is a glide reflection that maps triangle (1) to triangle (__?__).

24. Name another pair of triangles for which one triangle is mapped to another by a glide reflection.

Exs. 15–24

25. Which of the following properties are invariant under a half-turn?
   a. distance   **b.** angle measure   **c.** area   **d.** orientation

26. Which of the properties listed in Exercise 25 are invariant under the rotation $\mathcal{R}_{O,\ 90}$?

Copy the figure on graph paper. Draw the image by the specified rotation.

27. $\mathcal{R}_{O,\ 90}$ **28.** $\mathcal{R}_{O,\ -90}$ **29.** H_O

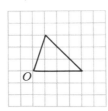 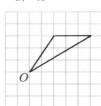 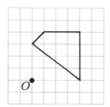

30. If $H_C:(1,\ 1) \to (7,\ 3)$, find the coordinates of C.

31. A rotation maps A to A' and B to B'. Construct the center of the rotation. (*Hint*: If the center is O, then $OA = OA'$ and $OB = OB'$.)

32. a. Draw a coordinate grid with origin O and plot the points $A(0, 3)$ and $B(4, 1)$.
   b. Plot A' and B', the images of A and B by $\mathscr{R}_{O, 90}$.
   c. Compare the slopes of $\overleftrightarrow{AB}$ and $\overleftrightarrow{A'B'}$. What does this tell you about these lines?
   d. Without using the distance formula, you know that $A'B' = AB$. State the theorem that tells you this.
   e. What reason supports the conclusion that $\triangle AOB$ and $\triangle A'OB'$ have the same area?
   f. Use your graph to find the image of (x, y) by $\mathscr{R}_{O, 90}$.

33. Repeat Exercise 32 using $\mathscr{R}_{O, 270}$.

34. A half-turn about $(3, 2)$ maps P to P'. Where does this half-turn map the following points?

   a. P' **b.** $(0, 0)$ **c.** $(3, 0)$
   d. $(1, 4)$ **e.** $(-2, 1)$ **f.** (x, y)

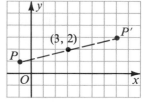

35. The rotation $\mathscr{R}_{O, x}$ maps line l to line l'. (You can think of rotating $\overline{OF}$, the perpendicular from O to l, through $x°$. Its image will be $\overline{OF'}$.) Show that one of the angles between l and l' has measure x.

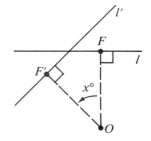

36. $\triangle ABC$ and $\triangle DCE$ are equilateral.
   a. What rotation maps A to B and D to E?
   b. Why does $AD = BE$?
   c. Find the measure of an acute angle between $\overleftrightarrow{AD}$ and $\overleftrightarrow{BE}$. (*Hint*: See Exercise 35.)

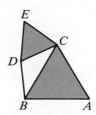

37. $\triangle ABC$ and $\triangle DEC$ are isosceles right triangles.
   a. What rotation maps B to A and E to D?
   b. Why does $AD = BE$?
   c. Explain why $\overline{AD} \perp \overline{BE}$. (*Hint*: See Exercise 35.)

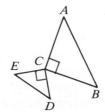

C 38. Given: Parallel lines *l* and *k* and point *A*.

    a. Construct an equilateral $\triangle ABC$ with *B* on *k* and *C* on *l* using the following method.

      Step 1. Rotate *l* through 60° about *A* and let *B* be the point on *k* where the image of *l* intersects *k*. (The diagram for Exercise 35 may be helpful in rotating *l*.)

      Step 2. Let point *C* on *l* be the preimage of *B*.

    b. Explain why $\triangle ABC$ is equilateral.

    c. Are there other equilateral triangles with vertices at *A* and on *l* and *k*?

  39. Given the figure for Exercise 38, construct a square *AXYZ* with *X* on *k* and *Z* on *l*.

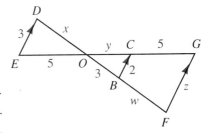

Exs. 38, 39

Mixed Review Exercises

Given $\overline{ED} \parallel \overline{BC} \parallel \overline{FG}$. Complete the statements.

1. $\triangle OBC$ is similar to $\triangle\underline{\ ?\ }$ and $\triangle\underline{\ ?\ }$.

2. The scale factor of $\triangle OBC$ to $\triangle ODE$ is $\underline{\ ?\ }$.

3. Find the values of *x*, *y*, *z*, and *w*.

4. The scale factor of $\triangle ODE$ to $\triangle OFG$ is $\underline{\ ?\ }$.

5. The ratio of the areas of $\triangle OBC$ and $\triangle ODE$ is $\underline{\ ?\ }$.

6. The ratio of the areas of $\triangle ODE$ and $\triangle OFG$ is $\underline{\ ?\ }$.

7. The ratio of the areas of $\triangle OBC$ and $\triangle OFG$ is $\underline{\ ?\ }$.

14-5 *Dilations*

Reflections, translations, glide reflections, and rotations are isometries, or *congruence* mappings. In this section we consider a transformation related to *similarity* rather than congruence. It is called a **dilation.** The dilation $D_{O,k}$ has *center O* and nonzero *scale factor k*. $D_{O,k}$ maps any point *P* to a point *P'* determined as follows:

(1) If $k > 0$, *P'* lies on $\overrightarrow{OP}$ and $OP' = k \cdot OP$.

(2) If $k < 0$, *P'* lies on the ray opposite $\overrightarrow{OP}$ and $OP' = |k| \cdot OP$.

(3) The center *O* is its own image.

If $|k| > 1$, the dilation is called an **expansion.**
If $|k| < 1$, the dilation is called a **contraction.**

   A developing leaf undergoes an expansion, keeping approximately the same shape as it grows in size.

Example 1   Find the image of $\triangle ABC$ under the expansion $D_{O, 2}$.

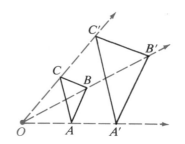

Solution $D_{O, 2}: \triangle ABC \rightarrow \triangle A'B'C'$

$OA' = 2 \cdot OA$
$OB' = 2 \cdot OB$
$OC' = 2 \cdot OC$

Example 2   Find the image of $\triangle RST$ under the contraction $D_{O, \frac{2}{3}}$.

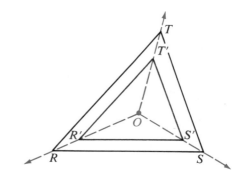

Solution $D_{O, \frac{2}{3}}: \triangle RST \rightarrow \triangle R'S'T'$

$OR' = \frac{2}{3} \cdot OR$
$OS' = \frac{2}{3} \cdot OS$
$OT' = \frac{2}{3} \cdot OT$

In the examples above, can you prove that the two triangles are similar? How are the areas of each pair of triangles related?

Example 3   Find the image of figure F under the contraction $D_{O, -\frac{1}{2}}$.

Solution $D_{O, -\frac{1}{2}}:$ figure $F \rightarrow$ figure F'
$\overrightarrow{OP}$ is opposite to $\overrightarrow{OP'}$.
$OP' = |-\frac{1}{2}| \cdot OP = \frac{1}{2} \cdot OP$

If the scale factor in Example 3 was -1 instead of $-\frac{1}{2}$, the figure F' would be congruent to the figure F, and the transformation would be an isometry, equivalent to a half-turn. In general, however, as these examples illustrate, dilations do not preserve distance. Therefore a dilation is not an isometry (unless $k = 1$ or $k = -1$).

But a dilation always maps any geometric figure to a similar figure. In the examples above, $\triangle ABC \sim \triangle A'B'C'$, $\triangle RST \sim \triangle R'S'T'$ and the figure F is similar to the figure F'. For this reason, a dilation is an example of a **similarity mapping**.

Theorem 14-5

A dilation maps any triangle to a similar triangle.

Given: $D_{O,k}: \triangle ABC \rightarrow \triangle A'B'C'$

Prove: $\triangle ABC \sim \triangle A'B'C'$

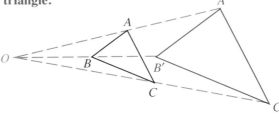

Key steps of proof:

1. $OA' = |k| \cdot OA$, $OB' = |k| \cdot OB$ (Definition of dilation)

2. $\triangle OAB \sim \triangle OA'B'$ (SAS Similarity Theorem)

3. $\dfrac{A'B'}{AB} = \dfrac{OA'}{OA} = |k|$ (Corr. sides of $\sim \triangle$ are in proportion.)

4. Similarly, $\dfrac{B'C'}{BC} = \dfrac{A'C'}{AC} = |k|$ (Repeat Steps 1–3 for $\triangle OBC$ and $\triangle OB'C'$ and for $\triangle OAC$ and $\triangle OA'C'$.)

5. $\triangle ABC \sim \triangle A'B'C'$ (SSS Similarity Theorem)

Corollary 1

A dilation maps an angle to a congruent angle.

Corollary 2

A dilation $D_{O,k}$ maps any segment to a parallel segment $|k|$ times as long.

Corollary 3

A dilation $D_{O,k}$ maps any polygon to a similar polygon whose area is k^2 times as large.

The diagram for the theorem above shows the case in which $k > 0$. You should draw the diagram for $k < 0$ and convince yourself that the proof is the same.

Theorem 14-5 can also be proved by using coordinates (see Exercise 28). To do this, you set the center of dilation at the origin, and describe $D_{O,k}$ in terms of coordinates by writing $D_{O,k}: (x, y) \rightarrow (kx, ky)$. You can see that this description satisfies the definition of a dilation because O, P, and P' are collinear (use slopes) and $OP' = |k| \cdot OP$ (use the distance formula).

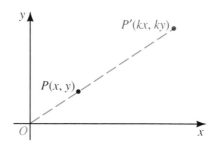

Classroom Exercises

Sketch each triangle on the chalkboard. Then sketch its image under the given dilation.

1. $D_{O,\,3}$

2. $D_{S,\,\frac{1}{2}}$

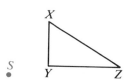

3. $D_{E,\,-2}$

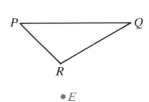

4. $D_{N,\,-\frac{1}{3}}$

5. Find the coordinates of the images of points A, B, and C under the dilation $D_{O,\,2}$.

6. Find the image of $(x,\,y)$ under $D_{O,\,2}$.

7. What dilation with center O maps A to B?

8. What dilation with center O maps C to the point $(-6,\,0)$?

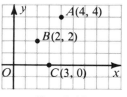

Exs. 5–10

9. Find the coordinates of the image of A under $D_{B,\,2}$.

10. Find the coordinates of the image of B under $D_{C,\,3}$.

11. Match each scale factor in the first column with the name of the corresponding dilation in the second column.

Scale factor	Transformation
$\frac{2}{5}$	Half-turn
-4	Contraction
-1	Expansion

12. Describe the dilation $D_{O,\,1}$.

13. If $\odot S$ has radius 4, describe the image of $\odot S$ under $D_{S,\,5}$ and under $D_{S,\,-1}$.

14. If point A is on line k, what is the image of line k under $D_{A,\,2}$?

15. The dilation $D_{O,\,3}$ maps P to P' and Q to Q'.
  a. If $OQ = 2$, find OQ'.
  b. If $PQ = 7$, find $P'Q'$.
  c. If $PP' = 10$, find OP.

16. Explain how Corollary 1 follows from Theorem 14-5.

17. Explain how Corollary 3 follows from Corollaries 1 and 2.

Written Exercises

Find the coordinates of the images of *A*, *B*, and *C* by the given dilation.

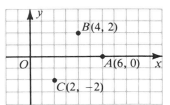

A
1. $D_{O, 2}$ 2. $D_{O, 3}$ 3. $D_{O, \frac{1}{2}}$ 4. $D_{O, -\frac{1}{2}}$
5. $D_{O, -2}$ 6. $D_{O, 1}$ 7. $D_{A, -\frac{1}{2}}$ 8. $D_{A, 2}$

A dilation with the origin, *O*, as center maps the given point to the image point named. Find the scale factor of the dilation. Is the dilation an expansion or a contraction?

9. $(2, 0) \rightarrow (8, 0)$ 10. $(2, 3) \rightarrow (4, 6)$ 11. $(3, 9) \rightarrow (1, 3)$
12. $(4, 10) \rightarrow (-2, -5)$ 13. $(0, \frac{1}{6}) \rightarrow (0, \frac{2}{3})$ 14. $(-6, 2) \rightarrow (18, -6)$

B 15. Which of the following properties are invariant under any dilation?
 a. distance   **b.** angle measure   **c.** area   **d.** orientation

16. Is parallelism invariant under a dilation? (*Hint:* See Exercise 23 on page 581.)

17. If A', B', C', and D' are the images of any four points *A*, *B*, *C*, and *D*, then we say the ratio of distances is invariant under the transformation if $\dfrac{AB}{CD} = \dfrac{A'B'}{C'D'}$. For which of the following transformations is the ratio of distances invariant?
 a. reflection   **b.** rotation   **c.** dilation

Graph quad. *PQRS* and its image by the dilation given. Find the ratio of the perimeters and the ratio of the areas of the two quadrilaterals.

18. $P(-1, 1)$ $Q(0, -1)$ $R(4, 0)$ $S(2, 2)$ $D_{O, 3}$
19. $P(12, 0)$ $Q(0, 15)$ $R(-9, 6)$ $S(3, -9)$ $D_{O, \frac{2}{3}}$
20. $P(3, 0)$ $Q(3, 4)$ $R(6, 6)$ $S(5, -1)$ $D_{O, -2}$
21. $P(-2, -2)$ $Q(0, 0)$ $R(4, 0)$ $S(6, -2)$ $D_{O, -\frac{1}{2}}$

22. The diagram illustrates a dilation of three-dimensional space. $D_{O, 2}$ maps the smaller cube to the larger cube.
 a. What is the ratio of the surface areas of these cubes?
 b. What is the ratio of the volumes of these cubes?

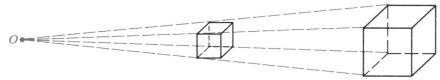

23. A dilation with scale factor $\frac{3}{4}$ maps a sphere with center *C* to a concentric sphere.
 a. What is the ratio of the surface areas of these spheres?
 b. What is the ratio of the volumes of these spheres?

24. *G* is the intersection of the medians of $\triangle XYZ$. Complete the following statements. (*Hint*: Use Theorem 10-4 on page 387.)

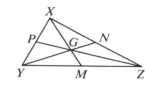

a. $\dfrac{XG}{XM} = \underline{\ ?\ }$ **b.** $\dfrac{GM}{GX} = \underline{\ ?\ }$

c. What dilation maps *X* to *M*?

d. What is the image under this dilation of *Y*? of *Z*?

25. $D_{O,\,k}$ maps $\overline{PQ}$ to $\overline{P'Q'}$.

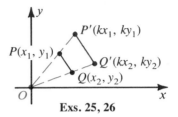

a. Show that the slopes of $\overline{PQ}$ and $\overline{P'Q'}$ are equal.

b. Part (a) proves that $\overline{PQ}$ and $\overline{P'Q'}$ are $\underline{\ ?\ }$.

C 26. Use the distance formula to show that
$$P'Q' = |k|\sqrt{(x_1 - x_2)^2 + (y_1 - y_2)^2} = |k| \cdot PQ.$$

Exs. 25, 26

27. A dilation with center (a, b) and scale factor k maps $A(3, 4)$ to $A'(1, 8)$, and $B(3, 2)$ to $B'(1, 2)$. Find the coordinates of the center (a, b) and the value of k.

28. Prove Theorem 14-5 using the coordinate definition of a dilation, $D_{O,\,k}:(x, y) \to (kx, ky)$. (*Hint*: Let *A*, *B*, and *C* have coordinates (p, q), (r, s), and (t, u) respectively.)

Self-Test 1

1. Define an isometry.

2. If $f(x) = 3x - 7$, find the image of 2 and the preimage of 2.

3. If $T:(x, y) \to (x + 1, y - 2)$, find the image and preimage of the origin.

4. Find the image of (3, 5) when reflected in each line.
 a. the *x*-axis   **b.** the *y*-axis   **c.** the line $y = x$.

5. A dilation with scale factor 3 maps $\triangle ABC$ to $\triangle A'B'C'$. Which of the following are true?

 a. $\overline{AB} \parallel \overline{A'B'}$ **b.** $\dfrac{A'B'}{AB} = 3$ **c.** $\dfrac{\text{area of } \triangle A'B'C'}{\text{area of } \triangle ABC} = 3$

6. Give two other names for the rotation $\mathcal{R}_{O,\,-30}$.

Complete. R_x and R_y denote reflections in the *x*- and *y*-axes, respectively.

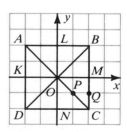

7. $R_y:A \to \underline{\ ?\ }$ **8.** $R_x:B \to \underline{\ ?\ }$

9. $R_x:\overline{DC} \to \underline{\ ?\ }$ **10.** $R_y:\underline{\ ?\ } \to \overline{OA}$

11. $H_O:K \to \underline{\ ?\ }$ **12.** $H_O:\underline{\ ?\ } \to \overline{CO}$

13. $\mathcal{R}_{O,\,90}$ maps *M* to $\underline{\ ?\ }$.   **14.** $\mathcal{R}_{O,\,-90}$ maps $\triangle MCO$ to $\triangle \underline{\ ?\ }$.

15. $D_{O,\,2}$ maps *P* to $\underline{\ ?\ }$.   **16.** $D_{M,\,-\frac{1}{2}}$ maps *B* to $\underline{\ ?\ }$.

17. A translation that maps *A* to *L* maps *N* to $\underline{\ ?\ }$.

18. The glide reflection in $\overleftrightarrow{BD}$ that maps *K* to *M* maps *N* to $\underline{\ ?\ }$.

Computer Animation Programmer

One problem computer animation programmers have encountered is how to produce natural-looking landscapes. The structures of trees, mountains, clouds, and coastlines are complex. To create them in a computer landscape can require storing a great deal of information. Also, since animations often include moving through space, data about the landscape features needs to be provided at many levels of detail. (If you specified the appearance of a mountain from only one viewpoint, say, then "zooming in" for a close-up would reveal that details are missing, a problem known as *loss of resolution*.)

One new approach involves using fractals. A *fractal* is a complex shape that looks more or less the same at all magnifications. Fractals are made by following simple rules, called *algorithms*. The snowflake shape in the diagram at right is an example. Its algorithm is: Start with an equilateral triangle. Divide each side of the polygon in thirds; add an equilateral triangle to each center third; repeat. No matter how much you magnify a piece of this polygon, it will retain the overall pattern and complexity of the original. When you

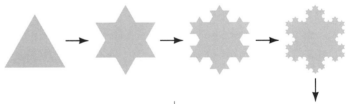

"zoom in" on a fractal shape, there is no loss of resolution.

Computer programmers are taking advantage of this property of fractals to approximate many items in nature, such as the mountains in the photograph above. Programming a computer to follow these algorithms uses less memory than specifying the exact shape of each element from many different viewpoints and distances.

Composition and Symmetry

Objectives

1. Locate the images of figures by composites of mappings.
2. Recognize and use the terms *identity* and *inverse* in relation to mappings.
3. Describe the symmetry of figures and solids.

14-6 *Composites of Mappings*

Suppose a transformation T maps point P to P' and then a transformation S maps P' to P''. Then T and S can be combined to produce a new transformation that maps P directly to P''. This new transformation is called the **composite** of S and T and is written $S \circ T$. Notice in the diagram that $P'' = S(P') = S(T(P)) = (S \circ T)(P)$.

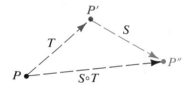

We reduce the number of parentheses needed to indicate that the composite of S and T maps P to P'' by writing $S \circ T : P \to P''$. Notice that T, the transformation that is applied first, is written closer to P, and S, the transformation that is applied second, is written farther from P. For this reason, the composite $S \circ T$ is often read "S after T," or "T followed by S."

The operation that combines two mappings (or functions) to produce the composite mapping (or composite function) is called *composition*. We shall see that composition has many characteristics similar to multiplication, but there is one important exception. For multiplication, it makes no difference in which order you multiply two numbers. For composition, however, the order of the mappings or functions usually *does* make a difference. Examples 1 and 2 illustrate this.

Example 1 If $f(x) = x^2$ and $g(x) = 2x$, find **(a)** $(g \circ f)(x)$ and **(b)** $(f \circ g)(x)$.

Solution **a.** $(g \circ f)(x) = g(f(x))$ **b.** $(f \circ g)(x) = f(g(x))$
$\qquad\qquad = g(x^2) \qquad\qquad\qquad\qquad = f(2x)$
$\qquad\qquad = 2x^2 \qquad\qquad\qquad\qquad\ = (2x)^2, \text{ or } 4x^2$

In mapping notation, we could write that $g \circ f : x \to 2x^2$ and $f \circ g : x \to 4x^2$. Note that since $2x^2 \neq 4x^2$, $g \circ f \neq f \circ g$.

Example 2   Show that $H_O \circ R_j \neq R_j \circ H_O$.

Solution   Study the two diagrams below.

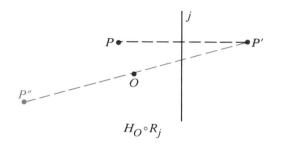
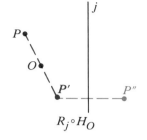

Here R_j, the reflection of P in line j, is carried out first, mapping P to P'. Then H_O maps P' to P''. Thus P'' is the image of P under the composite $H_O \circ R_j$.

With the order changed in the composite, the half-turn is carried out first, followed by the reflection in line j. The image point P'' is now in a different place.

Notice that the two composites map P to different image points, so the composites are not equal.

Example 2 shows that the order in a composite of transformations can be very important, but this is not always true. For example, if S and T are two translations, then order is not important, since $S \circ T = T \circ S$ (see Exercise 10).

Example 2 above shows the effect of a composite of mappings on a single point P. The diagram below shows a composite of reflections acting on a whole figure, F. F is reflected in line j to F', and F' is reflected in line k to F''. Thus $R_k \circ R_j$ maps F to F''. Again notice that the first reflection, R_j, is written on the right.

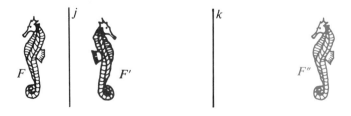

The final image F'' is the same size and shape as F. Also, F'' is the image of F under a translation. This illustrates our next two theorems. First, the composite of any two isometries is an isometry. Second, the composite of reflections in two parallel lines is a translation.

Theorem 14-6

The composite of two isometries is an isometry.

Theorem 14-7

A composite of reflections in two parallel lines is a translation. The translation glides all points through twice the distance from the first line of reflection to the second.

Although we will not present a formal proof of Theorem 14-7, the following argument should convince you that it is true. Assume that $j \parallel k$ and that R_j maps P to P' and Q to Q', and that R_k maps P' to P'' and Q' to Q''. To show that the composite $R_k \circ R_j$ is a translation we will demonstrate that $PP'' = QQ''$ and that $\overleftrightarrow{PP''}$ and $\overleftrightarrow{QQ''}$ are parallel.

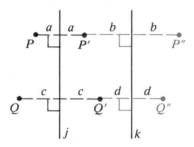

The letters a, b, c, and d in the diagram label pairs of distances that are equal according to the definition of a reflection. P, P', and P'' are collinear and

$$PP'' = 2a + 2b = 2(a + b)$$

Similarly,

$$QQ'' = 2c + 2d = 2(c + d)$$

But $(a + b) = (c + d)$, since by Theorem 5-8, the distance between the parallel lines j and k is constant. Therefore $PP'' = QQ'' =$ twice the distance from j to k.

That $\overleftrightarrow{PP''}$ and $\overleftrightarrow{QQ''}$ are parallel follows from the fact that both lines are perpendicular to j and k. Theorem 3-7 guarantees that if two lines in a plane are perpendicular to the same line, then the two lines are parallel.

You should make diagrams for the case when P is on j or k, when P is located between j and k, and when P is to the right of k. Convince yourself that $PP'' = 2(a + b)$ in these cases also. In every case, the glide is perpendicular to j and k and goes in the direction from j to k (that is, from the first line of reflection toward the second line of reflection).

Theorem 14-7 shows that when lines j and k are parallel, $R_k \circ R_j$ translates points through twice the distance between the lines. If lines j and k intersect, $R_k \circ R_j$ rotates points through twice the measure of the angle between the lines. This is our next theorem.

Theorem 14-8

A composite of reflections in two intersecting lines is a rotation about the point of intersection of the two lines. The measure of the angle of rotation is twice the measure of the angle from the first line of reflection to the second.

Given: j intersects k, forming an angle of measure y at O.

Prove: $R_k \circ R_j = \mathcal{R}_{O,\, 2y}$

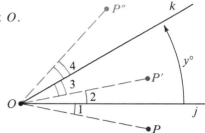

Proof:

The diagram shows an arbitrary point P and its image P' by reflection in j. The image of P' by reflection in k is P''. According to the definition of a rotation we must prove that $OP = OP''$ and $m \angle POP'' = 2y$.

R_j and R_k are isometries, so they preserve both distance and angle measure. Therefore $OP = OP'$, $OP' = OP''$, $m \angle 1 = m \angle 2$, and $m \angle 3 = m \angle 4$. Thus $OP = OP''$ and the measure of the angle of rotation equals

$$m \angle 1 + m \angle 2 + m \angle 3 + m \angle 4 = 2m \angle 2 + 2m \angle 3 = 2y.$$

Corollary

A composite of reflections in perpendicular lines is a half-turn about the point where the lines intersect.

Classroom Exercises

1. If $f(x) = x + 1$ and $g(x) = 3x$, find the following.
   - **a.** $f(4)$
   - **b.** $(g \circ f)(4)$
   - **c.** $(g \circ f)(x)$
   - **d.** $g(2)$
   - **e.** $(f \circ g)(2)$
   - **f.** $(f \circ g)(x)$
2. Repeat Exercise 1 if $f:x \rightarrow \sqrt{x}$ and $g:x \rightarrow x + 7$.

Complete the following. R_x and R_y are reflections in the x- and y-axes.

3. $R_x \circ R_y : A \rightarrow$ _?_

4. $R_x \circ R_y : D \rightarrow$ _?_

5. $H_O \circ R_y : B \rightarrow$ _?_

6. $R_y \circ H_O : B \rightarrow$ _?_

7. $H_O \circ H_O : A \rightarrow$ _?_

8. $R_y \circ R_y : C \rightarrow$ _?_

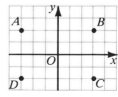

Copy the figure on the chalkboard and find its image by $R_k \circ R_j$. Then copy the figure again and find its image by $R_j \circ R_k$.

9.

10.

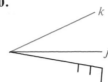

11. Prove Theorem 14-6. (*Hint*: Let S and T be isometries. Consider a $\overline{PQ}$ under $S \circ T$.)

12. Explain how the Corollary follows from Theorem 14-8.

Written Exercises

A **1.** If $f(x) = x^2$ and $g(x) = 2x - 7$, evaluate the following.
   a. $(g \circ f)(2)$ **b.** $(g \circ f)(x)$ **c.** $(f \circ g)(2)$ **d.** $(f \circ g)(x)$

2. Repeat Exercise 1 if $f(x) = 3x + 1$ and $g(x) = x - 9$.

3. If $h: x \to \dfrac{x + 1}{2}$ and $k: x \to x^3$, complete the following.
   a. $k \circ h: 3 \to \underline{\ ?\ }$ **b.** $k \circ h: 5 \to \underline{\ ?\ }$ **c.** $k \circ h: x \to \underline{\ ?\ }$
   d. $h \circ k: 3 \to \underline{\ ?\ }$ **e.** $h \circ k: 5 \to \underline{\ ?\ }$ **f.** $h \circ k: x \to \underline{\ ?\ }$

4. Repeat Exercise 3 if $h: x \to x^2 - 1$ and $k: x \to 2x + 7$.

Copy each figure and find its image under $R_k \circ R_j$. Then copy the figure again and find its image under $R_j \circ R_k$.

5.

6.

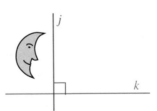

Copy each figure twice and show the image of the red flag under each of the composites given.

7. a. $H_B \circ H_A$ **8. a.** $R_j \circ H_C$ **9. a.** $H_E \circ D_{E, \frac{1}{3}}$
   b. $H_A \circ H_B$ **b.** $H_C \circ R_j$ **b.** $D_{E, \frac{1}{3}} \circ H_E$

10. Given $A(4, 1)$, $B(1, 5)$, and $C(0, 1)$. S and T are translations. $S:(x, y) \to (x + 1, y + 4)$ and $T:(x, y) \to (x + 3, y - 1)$. Draw $\triangle ABC$ and its images under $S \circ T$ and $T \circ S$.
   a. Does $S \circ T$ appear to be a translation?
   b. Is $S \circ T$ equal to $T \circ S$?
   c. $S \circ T:(x, y) \to (\underline{\quad?\quad}, \underline{\quad?\quad})$ and $T \circ S:(x, y) \to (\underline{\quad?\quad}, \underline{\quad?\quad})$

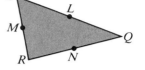

11. L, M, and N are midpoints of the sides of $\triangle QRS$.
   a. $H_N \circ H_M:S \to \underline{\quad?\quad}$ b. $H_M \circ H_N:Q \to \underline{\quad?\quad}$
   c. $D_{S, \frac{1}{2}} \circ H_N:Q \to \underline{\quad?\quad}$ d. $H_N \circ D_{S, 2}:M \to \underline{\quad?\quad}$
   e. $H_L \circ H_M \circ H_N:Q \to \underline{\quad?\quad}$

   Exs. 11, 12

B 12. If T is a translation that maps R to N, then:
   a. $T:M \to \underline{\quad?\quad}$ b. $T \circ D_{S, \frac{1}{2}}:R \to \underline{\quad?\quad}$ c. $T \circ T:R \to \underline{\quad?\quad}$

In Exercises 13–16 tell which of the following properties are invariant under the given transformation.
a. distance   b. angle measure   c. area   d. orientation

13. The composite of a reflection and a dilation   14. The composite of two reflections

15. The composite of a rotation and a translation   16. The composite of two dilations

For each exercise draw a grid and find the coordinates of the image point. O is the origin and A is the point (3, 1). R_x and R_y are reflections in the x- and y-axes.

17. $R_x \circ R_y:(3, 1) \to (\underline{\quad?\quad}, \underline{\quad?\quad})$ 18. $R_y \circ H_O:(1, -2) \to (\underline{\quad?\quad}, \underline{\quad?\quad})$

19. $H_A \circ H_O:(3, 0) \to (\underline{\quad?\quad}, \underline{\quad?\quad})$ 20. $H_O \circ H_A:(1, 1) \to (\underline{\quad?\quad}, \underline{\quad?\quad})$

21. $R_x \circ D_{O, 2}:(2, 4) \to (\underline{\quad?\quad}, \underline{\quad?\quad})$ 22. $\mathcal{R}_{O, 90} \circ R_y:(-2, 1) \to (\underline{\quad?\quad}, \underline{\quad?\quad})$

23. $\mathcal{R}_{A, 90} \circ \mathcal{R}_{O, -90}:(-1, -1) \to (\underline{\quad?\quad}, \underline{\quad?\quad})$ 24. $D_{O, -\frac{1}{3}} \circ D_{A, 4}:(3, 0) \to (\underline{\quad?\quad}, \underline{\quad?\quad})$

25. Let R_l be a reflection in the line $y = x$ and R_y be a reflection in the y-axis. Draw a grid and label the origin O.
   a. Plot the point $P(5, 2)$ and its image Q under the mapping $R_y \circ R_l$.
   b. According to Theorem 14-8, $m \angle POQ = \underline{\quad?\quad}$.
   c. Use the slopes of $\overline{OP}$ and $\overline{OQ}$ to verify that $\overline{OP} \perp \overline{OQ}$.
   d. Find the images of (x, y) under $R_y \circ R_l$ and $R_l \circ R_y$.

26. Let R_k be a reflection in the line $y = -x$ and R_x be a reflection in the x-axis.
   a. Plot $P(-6, -2)$ and its image Q under the mapping $R_k \circ R_x$.
   b. Use slopes to show that $m \angle POQ = 90$ where O is the origin. (Do you see that this result agrees with Theorem 14-8?)
   c. Find the images of (x, y) under $R_k \circ R_x$ and $R_x \circ R_k$.

C 27. Explain how you would construct line j so that $R_k \circ R_j:A \to B$.

28. The figure shows that $H_B \circ H_A : P \to P''$.

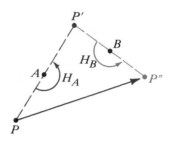

    a. Copy the figure and verify by measuring that $PP'' = 2 \cdot AB$. What theorem about the midpoints of the sides of a triangle does this suggest?

    b. Choose another point Q and carefully locate Q'', the image of Q under $H_B \circ H_A$. Does $QQ'' = 2 \cdot AB$?

    c. Measure PQ and $P''Q''$. Are they equal? What kind of transformation does $H_B \circ H_A$ appear to be?

29. $D_{A,\,2} : \overline{PQ} \to \overline{P'Q'}$ and $D_{B,\,\frac{1}{2}} : \overline{P'Q'} \to \overline{P''Q''}$. What kind of transformation is the composite $D_{B,\,\frac{1}{2}} \circ D_{A,\,2}$? Explain.

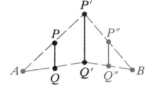

30. The point P is called a *fixed point* of the transformation T if $T : P \to P$.

    a. How many fixed points does each of the following have: $\mathcal{R}_{O,\,90}$? R_y? $D_{O,\,3}$? the translation $T : (x, y) \to (x - 3, y + 2)$?

    b. O is the origin and A is the point $(1, 0)$. Find the coordinates of a fixed point of the composite $D_{O,\,2} \circ D_{A,\,\frac{1}{4}}$.

14-7 *Inverses and the Identity*

Suppose that the pattern below continues indefinitely to both the left and the right. The translation T glides each runner one place to the right. The translation that glides each runner one place to the *left* is called the *inverse* of T, and is denoted T^{-1}. Notice that T followed by T^{-1} keeps *all* points fixed:

$$T^{-1} \circ T : P \to P$$

The composite $T \circ T$, also written $T \cdot T$, and usually denoted by T^2, glides each runner two places to the right.

    The mapping that maps every point to itself is called the **identity** transformation I. The words "identity" and "inverse" are used for mappings in much the same way that they are used for numbers. In fact, the composite of two mappings is very much like the product of two numbers. For this reason, the composite $S \circ T$ is often called the **product** of S and T.

Relating Algebra and Geometry

For products of numbers	*For composites of mappings*
1 is the identity.	I is the identity.
$a \cdot 1 = a$ and $1 \cdot a = a$	$S \circ I = S$ and $I \circ S = S$
The inverse of a is written a^{-1}, or $\dfrac{1}{a}$.	The inverse of S is written S^{-1}.
$a \cdot a^{-1} = 1$ and $a^{-1} \cdot a = 1$	$S \circ S^{-1} = I$ and $S^{-1} \circ S = I$

In general, the **inverse** of a transformation T is defined as the transformation S such that $S \circ T = I$. The inverses of some other transformations are illustrated below.

Example 1   Find the inverses of **(a)** translation $T:(x, y) \to (x + 5, y - 4)$,
(b) rotation $\mathcal{R}_{O, x}$, and **(c)** dilation $D_{O, 2}$.

Solution

a. $T^{-1}:(x, y) \to (x - 5, y + 4)$

$T:(0, 6) \to (5, 2)$
$T^{-1}:(5, 2) \to (0, 6)$

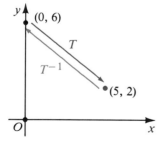

b. The inverse of $\mathcal{R}_{O, x}$ is $\mathcal{R}_{O, -x}$.

$\mathcal{R}_{O, x}:F \to G$
$\mathcal{R}_{O, -x}:G \to F$

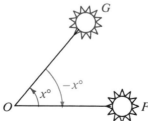

c. The inverse of $D_{O, 2}$ is $D_{O, \frac{1}{2}}$.

$D_{O, 2}:(3, 2) \to (6, 4)$
$D_{O, \frac{1}{2}}:(6, 4) \to (3, 2)$

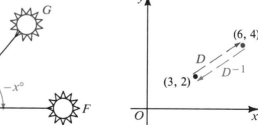

Example 2   What is the inverse of R_j?
(Refer to the diagram at right.)

Solution   Since $R_j \circ R_j = I$, the inverse of R_j is R_j itself. In symbols, $R_j^{-1} = R_j$. Do you see that the inverse of any reflection is that same reflection?

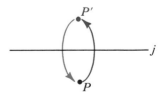

Classroom Exercises

The symbol 2^{-1} stands for the multiplicative inverse of 2, or $\frac{1}{2}$. Give the value of each of the following.

1. 3^{-1} **2.** 7^{-1} **3.** $\left(\frac{4}{5}\right)^{-1}$ **4.** $(2^{-1})^{-1}$

The translation T maps all points five units to the right. Describe each of the following transformations.

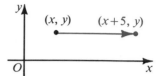

5. T^2 **6.** T^3 **7.** T^{-1}

8. T^{-2} **9.** $T \circ T^{-1}$ **10.** $(T^{-1})^{-1}$

The rotation $\mathcal{R}$ maps all points 120° about G, the center of equilateral $\triangle ABC$. Give the image of A under each rotation.

11. $\mathcal{R}$ **12.** $\mathcal{R}^2$ **13.** $\mathcal{R}^3$

14. $\mathcal{R}^6$ **15.** $\mathcal{R}^{-1}$ **16.** $\mathcal{R}^{-2}$

17. $\mathcal{R}^2 \circ \mathcal{R}^{-2}$ **18.** $\mathcal{R}^2 \circ \mathcal{R}^{-3}$ **19.** $\mathcal{R}^{100}$

20. What number is the identity for multiplication?

21. The product of any number t and the identity for multiplication is ___?___.

22. The product of any transformation T and the identity is ___?___.

23. State the inverse of each transformation.
     a. R_l **b.** $\mathcal{R}_{O,\,30}$ **c.** $T:(x, y) \rightarrow (x - 4, y + 1)$ **d.** $D_{O,\,-1}$

24. Name an important difference between products of numbers and products of transformations.

Written Exercises

Give the value of each of the following.

A   **1.** 4^{-1} **2.** 9^{-1} **3.** $\left(\frac{2}{3}\right)^{-1}$ **4.** $(5^{-1})^{-1}$

The rotation $\mathcal{R}$ maps all points 90° about O, the center of square $ABCD$. Give the image of A under each rotation.

5. $\mathcal{R}^2$ **6.** $\mathcal{R}^3$ **7.** $\mathcal{R}^4$

8. $\mathcal{R}^{-1}$ **9.** $\mathcal{R}^{-2}$ **10.** $\mathcal{R}^{-3}$

11. $\mathcal{R}^{-3} \circ \mathcal{R}^3$ **12.** $\mathcal{R}^5$ **13.** $\mathcal{R}^{50}$

Complete.

14. By definition, the identity mapping I maps every point P to ___?___.

15. $H_O{}^2$ is the same as the mapping ___?___.

16. The inverse of H_O is ___?___.

17. $H_O{}^3$ is the same as the mapping ___?___.

18. If $T:(x, y) \rightarrow (x + 2, y)$, then $T^2:(x, y) \rightarrow (\underline{\quad?\quad}, \underline{\quad?\quad})$.

19. If $T:(x, y) \rightarrow (x + 3, y - 4)$, then $T^2:(x, y) \rightarrow (\underline{\quad?\quad}, \underline{\quad?\quad})$.

20. If R_x is reflection in the x-axis, then $(R_x)^2:P \rightarrow \underline{\quad?\quad}$.

In each exercise, a rule is given for a mapping S. Write the rule for S^{-1}.

B **21.** $S:(x, y) \rightarrow (x + 5, y + 2)$ **22.** $S:(x, y) \rightarrow (x - 3, y - 1)$

23. $S:(x, y) \rightarrow (3x, -\frac{1}{2}y)$ **24.** $S:(x, y) \rightarrow (\frac{1}{4}x, \frac{1}{4}y)$

25. $S:(x, y) \rightarrow (x - 4, 4y)$ **26.** $S:(x, y) \rightarrow (y, x)$

27. If $S:(x, y) \rightarrow (x + 12, y - 3)$, find a translation T such that $T^6 = S$.

28. Find a transformation S (other than the identity) for which $S^5 = I$.

C **29. a.** j and k are vertical lines 1 unit apart. According to Theorem 14-7, $R_k \circ R_j$ and $R_j \circ R_k$ are both translations. Describe in words the distance and direction of each translation.

     b. Show that $R_k \circ R_j$ and $R_j \circ R_k$ are inverses by showing that their composite is I. Note: Forming composites of transformations is an associative operation, so $(R_k \circ R_j) \circ (R_j \circ R_k) = R_k \circ (R_j \circ R_j) \circ R_k$.

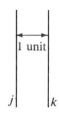

30. The blue lines in the diagram illustrate the statement $H_B \circ H_A =$ translation T. The red lines show that $H_A \circ H_B =$ translation S.

     a. How is translation S related to translation T?

     b. Prove your answer correct by showing that $(H_A \circ H_B) \circ (H_B \circ H_A) = I$. (*Hint:* See Exercise 29, part (b).)

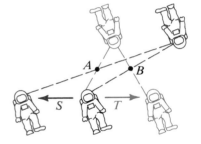

31. Complete the proof by giving a reason for each step.

Given: $l_1 \perp l_2$; $l_3 \perp l_2$; R_1, R_2, and R_3 denote reflections in l_1, l_2, and l_3.

Prove: $H_B \circ H_A$ is a translation.

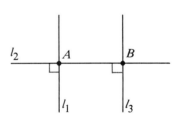

Proof:

Statements	Reasons
1. $H_A = R_2 \circ R_1$	1. $\underline{\quad?\quad}$
2. $H_B = R_3 \circ R_2$	2. $\underline{\quad?\quad}$
3. $H_B \circ H_A = (R_3 \circ R_2) \circ (R_2 \circ R_1)$	3. $\underline{\quad?\quad}$
4. $H_B \circ H_A = (R_3 \circ (R_2 \circ R_2)) \circ R_1$	4. Composition is associative.
5. $H_B \circ H_A = (R_3 \circ I) \circ R_1$	5. $\underline{\quad?\quad}$
6. $H_B \circ H_A = R_3 \circ R_1$	6. $\underline{\quad?\quad}$
7. $H_B \circ H_A$ is a translation.	7. $\underline{\quad?\quad}$

14-8 *Symmetry in the Plane and in Space*

A figure in the plane has **symmetry** if there is an isometry, other than the identity, that maps the figure onto itself. We call such an isometry a *symmetry* of the figure.

Both of the figures below have **line symmetry.** This means that for each figure there is a symmetry line k such that the reflection R_k maps the figure onto itself. The pentagon at the left has one symmetry line. The regular pentagon at the right has five symmetry lines.

 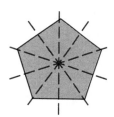

Each figure below has **point symmetry.** This means that for each figure there is a symmetry point O such that the half-turn H_O maps the figure onto itself.

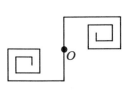 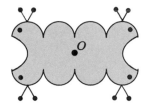 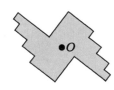

Besides having a symmetry point, the middle figure above has a vertical symmetry line and a horizontal symmetry line.

A third kind of symmetry is **rotational symmetry.** The figure below has the four rotational symmetries listed. Each symmetry has center O and rotates the figure onto itself. Note that 180° rotational symmetry is another name for point symmetry.

(1) 90° rotational symmetry: $\mathcal{R}_{O,\,90}$
(2) 180° rotational symmetry: $\mathcal{R}_{O,\,180}$ (or H_O)
(3) 270° rotational symmetry: $\mathcal{R}_{O,\,270}$
(4) 360° rotational symmetry: the identity I

The identity mapping always maps a figure onto itself, and we usually include the identity when listing the symmetries of a figure. However, we do not call a figure *symmetric* if the identity is its only symmetry.

A figure can also have **translational symmetry** if there is a translation that maps the figure onto itself. For example, imagine that the design at the right extends in all directions to fill the plane. If you consider the distance between the eyes of adjacent blue fish as a unit, then a translation through one or more units right, left, up, or down maps the whole pattern onto itself. Do you see that you can also translate the pattern along diagonal lines?

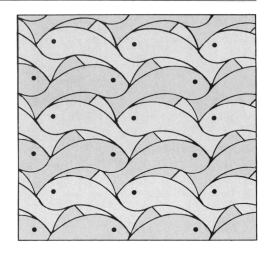

It is also possible to map the blue fish, which all face to the left, onto the right-facing green fish by translating the whole pattern a half unit up and then reflecting it in a vertical line. Thus, if we ignore color differences, the pattern has **glide reflection symmetry.**

A design like this pattern of fish, in which congruent copies of a figure completely fill the plane without overlapping, is called a *tessellation*. Tessellations can have any of the kinds of symmetry we have discussed. Here are two more examples.

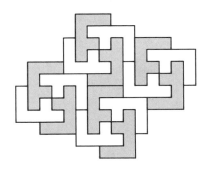

A tessellation of the letter *F*. This pattern has point symmetry and translational symmetry.

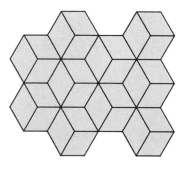

This tessellation has line, point, rotational, translational, and glide reflection symmetry.

Coloring a tessellation often changes its symmetries. For example, if the green were removed from the tessellation of the letter *F*, the pattern would also have 90° and 270° rotational symmetry.

A figure in space has **plane symmetry** if there is a symmetry plane *X* such that reflection in the plane maps the figure onto itself. (See Exercise 17, page 580.) Most living creatures have a single plane of symmetry. Such symmetry is called *bilateral symmetry*. The photographs on the next page illustrate bilateral symmetry.

Copy the figure shown. Then complete the figure so that it has the specified symmetries.

12.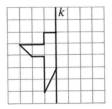

symmetry in line *k*

13.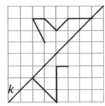

symmetry in line *k*

14.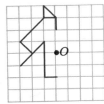

symmetry in point *O*

Copy the figure shown. Then complete the figure so that it has the specified symmetries.

B **15.**

60°, 120°, and 180° rotational symmetry

16.

90°, 180°, and 270° rotational symmetry

17.

2 symmetry lines and 1 symmetry point

18. a. An octopus has one symmetry. Describe it.
   b. If you disregard the eyes and mouth of an octopus, it has many symmetries. Describe them.

19. a. Describe the symmetries of the ellipse shown.
   b. If the ellipse is rotated in space about one of its symmetry lines, an ellipsoid (an egg-like figure) is formed. Its volume is $V = \frac{4}{3}\pi a^2 b$. Interpret this formula when $a = b$.
   c. Describe the symmetries of an ellipsoid.

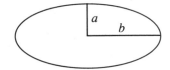

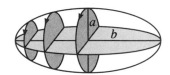

20. Tell whether a tessellation can be made with the given figure.
   a. A regular hexagon
   b. A scalene triangle
   c. A regular pentagon
   d. A nonisosceles trapezoid

In Exercises 21–23 draw the figure if there is one that meets the conditions. Otherwise write *not possible*.

21. A trapezoid with **(a)** no symmetry, **(b)** one symmetry line, **(c)** a symmetry point.

22. A parallelogram with **(a)** four symmetry lines, **(b)** just two symmetry lines, **(c)** just one symmetry line.

23. An octagon with **(a)** eight rotational symmetries, **(b)** just four rotational symmetries, **(c)** only point symmetry.

24. If you use tape to hinge together two pocket mirrors as shown and place the mirrors at a 120° angle, then a coin placed between the mirrors will be reflected, giving a pattern with 120° and 240° rotational symmetry.

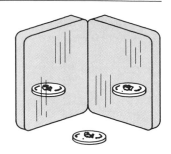

    a. What kinds of symmetries occur when the mirrors are at a right angle?

    b. Experiment by forming various angles with two mirrors. Be sure to try 60°, 45°, and 30° angles. Record the number of coins you see, including the actual coin.

25. You can make a tessellation by tracing around *any* quadrilateral, placing copies of the quadrilateral systematically as shown.

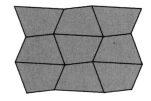

    a. The tessellation shown has many symmetry points but none of these are at vertices of the quadrilateral. Where are they?

    b. What other kind of symmetry does this mosaic have?

26. A figure has 60° rotational symmetry. What other rotational symmetries *must* it have? Explain your answer.

C 27. Show that if a hexagon has point symmetry, then its opposite sides must be parallel.

28. A figure has 50° rotational symmetry. What other rotational symmetries *must* it have? Explain your answer.

★ 29. Tell how many planes of symmetry and axes of rotation each solid has.

    a. a right circular cone
    b. a cube

    c. a regular tetrahedron (a pyramid formed by four equilateral triangles)

Challenges

1. A mouse moves along $\overline{AJ}$. For any position M of the mouse, X and Y are such that $\overline{AX} \perp \overline{AJ}$ with $AX = AM$, and $\overline{JY} \perp \overline{AJ}$ with $JY = JM$. The cat is at C, the midpoint of $\overline{XY}$. Describe the locus of the cat as the mouse moves from A to J.

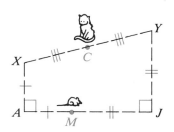

2. Points O, A, B, and C lie on a number line with coordinates 0, 8, 12, and 26. Take any point P not on the line. Draw $\overline{PA}$ and label its midpoint Q. Draw $\overline{QB}$ and label its midpoint R. Draw $\overline{PC}$ and label its midpoint S. Draw $\overrightarrow{SR}$. What is the coordinate of the point where $\overrightarrow{SR}$ intersects the number line?

Self-Test 2

For Exercises 1–6, refer to the figure.

1. $R_x \circ \mathcal{R}_{O,\,90}:B \to$ ____?____
2. $R_x \circ H_O:A \to$ ____?____
3. $\mathcal{R}_{O,\,110} \circ \mathcal{R}_{O,\,70}:C \to$ ____?____
4. $D_{O,\,\frac{1}{2}} \circ D_{R,\,\frac{1}{2}}:P \to$ ____?____
5. What is the symmetry line of $\triangle ABC$?
6. Does $\triangle ABC$ have point symmetry?

7. For any transformation T, $T^{-1} \circ T:P \to$ __?__ .
8. The composite of any transformation T and the identity is __?__ .
9. If line a is parallel to line b, then the composite $R_a \circ R_b$ is a __?__ .
10. Give the inverse of each transformation.
    a. $D_{O,\,5}$ **b.** $\mathcal{R}_{O,\,-70}$ **c.** R_y **d.** $S:(x,\ y) \to (x + 2,\ y - 3)$
11. How many lines of symmetry does a regular hexagon have?

Symmetry Groups

Cut out a cardboard or paper rectangle and color each corner with a color of its own on both front and back. Also on the front and back draw symmetry lines j and k and label symmetry point O. The rectangle has four symmetries: I, R_j, R_k, and H_O. The effect of each of these on the original rectangle is shown below.

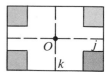

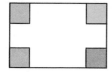

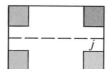

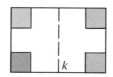

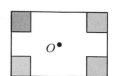

| Effect of I: Rectangle unchanged | Effect of R_j | Effect of R_k | Effect of H_O |

Our goal is to see how the four symmetries of the rectangle combine with each other. For example, if the original rectangle is mapped first by R_j and then by H_O, the images look like this:

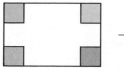 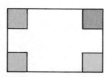 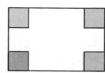

Mapping the rectangle by R_j and then by H_O has the same effect as the single symmetry R_k, so $H_O \circ R_j = R_k$. We can record this fact in a table resembling a multiplication table. Follow the row for R_j to where it meets the column for H_O, and enter the product $H_O \circ R_j$, which is R_k.

$\circ$	I	R_j	R_k	H_O
I				
R_j				R_k
R_k				
H_O				

We can determine other products of symmetries in the same way, but sometimes short cuts can be used. For example, we know that

(1) $R_j \circ R_j = I$ and $R_k \circ R_k = I$ (Why?)
(2) $H_O \circ H_O = I$ (Why?)
(3) $R_j \circ R_k = H_O$ and $R_k \circ R_j = H_O$
   (Corollary to Theorem 14-8)

$\circ$	I	R_j	R_k	H_O
I	I	R_j	R_k	H_O
R_j	R_j	I	H_O	R_k
R_k	R_k	H_O	I	R_j
H_O	H_O	R_k	R_j	I

Also we know that the product of any symmetry and the identity is that same symmetry. The completed table is shown at the right.

By studying the table you can see that the symmetries of the rectangle have these four properties, similar to the properties of nonzero real numbers under multiplication:

(1) The product of two symmetries is another symmetry.
(2) The set of symmetries contains the identity.
(3) Each symmetry has an inverse that is also a symmetry. (In this example each symmetry is its own inverse.)
(4) Forming products of transformations is an associative operation:
   $A \circ (B \circ C) = (A \circ B) \circ C$ for any three symmetries A, B, and C.

A set of symmetries with these four properties is called a symmetry *group*. Symmetry groups are used in crystallography, and more general groups are important in physics and advanced mathematics. The exercises that follow illustrate the fact that the symmetries of any figure form a group.

Exercises

1. An isosceles triangle has just two symmetries, including the identity. Make a 2 by 2 group table showing how these symmetries combine.

2. **a.** List the four symmetries of the rhombus shown. (Include the identity.)
   b. Make a group table showing all products of two symmetries.
   c. Is your table in part (b) identical to the table of symmetries for the rectangle?

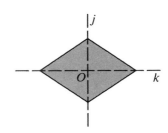

3. Make a group table for the three symmetries of this figure.

4. Make a group table for the four symmetries of this figure.

5. A transformation that is its own inverse is called a *self-inverse*.
  a. How many of the four symmetries of the figure in Exercise 4 are self-inverses?
  b. How many of the four symmetries of the rectangle are self-inverses?

6. A symmetry group is called *commutative* if $A \circ B = B \circ A$ for every pair of symmetries A and B in the group. The symmetry group of the rectangle is commutative, as you can see from the completed table. (For example, $H_O \circ R_j$ and $R_j \circ H_O$ are both equal to R_k.) Tell whether the groups in Exercises 3 and 4 are commutative or not.

7. An equilateral triangle has three rotational symmetries (I, $\mathcal{R}_{O, 120}$, and $\mathcal{R}_{O, 240}$) and three line symmetries (R_j, R_k, and R_l).
  a. Make a group table for these six symmetries.
  b. Give an example which shows that this group is *not* commutative.

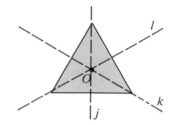

8. A square has four rotational symmetries (including the identity) and four line symmetries. Make a group table for these eight symmetries. Is this a commutative group?

9. The four rotational symmetries of the square satisfy the four requirements for a group, and so they are called a *subgroup* of the full symmetry group. (Notice that the identity is one of these rotational symmetries and that the product of two rotations is another rotation in the subgroup.)
  a. Do the four line symmetries of the square form a subgroup?
  b. Does the symmetry group of the equilateral triangle have a subgroup?
  c. Which two symmetries of the figure in Exercise 4 form a subgroup?

10. The tessellation with fish on page 610 has translational symmetry. Let S be the horizontal translation mapping each fish to the fish of the same color to its right, and let T be the vertical translation mapping each fish to the fish of the same color directly above.
  a. Describe the mapping S^3. Is it a symmetry of the pattern?
  b. Describe T^{-1}. Is it a symmetry?
  c. Describe $S \circ T$. Is it a symmetry?
  d. How many symmetries does the tessellation have?
  e. Does this set of symmetries satisfy the four requirements for a group?

Chapter Summary

1. A transformation is a one-to-one mapping from the whole plane to the whole plane. If the transformation S maps P to P', we write $S:P \rightarrow P'$ or $S(P) = P'$.

2. The word "mapping" is used in geometry as the word "function" is used in algebra. If the function f maps every number to its square we write $f:x \rightarrow x^2$ or $f(x) = x^2$.

3. An isometry is a transformation that preserves distance. An isometry maps any figure to a congruent figure.

4. Some basic isometries are:

   Reflection in a line. R_j is a reflection in line j.

   Translation or glide. $T:(x, y) \rightarrow (x + a, y + b)$ is a translation.

   Rotation about a point. $\mathcal{R}_{O, x}$ is a rotation counterclockwise about O through $x°$. H_O is a half-turn about O.

   Glide reflection. A glide followed by a reflection in a line parallel to the glide yields a glide reflection.

5. A dilation maps any figure to a similar figure. $D_{O, k}$ is a dilation with center O and nonzero scale factor k. A dilation is an isometry if $|k| = 1$.

6. Properties of figures that are preserved by a transformation are said to be invariant under that transformation. Invariant properties are checked in the table below.

	Distance	Angle Measure	Parallelism	Ratio of distances	Area
Isometry:	✓	✓	✓	✓	✓
Dilation:		✓	✓	✓	

7. The combination of one mapping followed by another is called a composite or product of mappings. The mapping A followed by B is written $B \circ A$.

8. A composite of isometries is an isometry.
   A composite of reflections in two parallel lines is a translation.
   A composite of reflections in two intersecting lines is a rotation.
   A composite of reflections in two perpendicular lines is a half-turn.

9. The identity transformation I keeps all points fixed. A transformation S followed by its inverse S^{-1} is equal to the identity.

10. A symmetry of a figure is an isometry that maps the figure onto itself. Figures can have line symmetry, point symmetry, and rotational symmetry. A tessellation, or covering of the plane with congruent figures, may also have translational and glide reflection symmetry. Solid figures in space can have planes of symmetry and rotational symmetry about an axis.

Chapter Review

1. If isometry S maps A to A' and B to B', then $\overline{AB}$ __?__ $\overline{A'B'}$. 14–1

2. If $f(x) = 3x$, find the image and preimage of 6.

3. **a.** If $S:(x, y) \rightarrow (2x, y - 2)$, find the image and preimage of $(3, 3)$.
   b. Is S an isometry?

4. Find the image of $(-7, 5)$ when reflected in **(a)** the x-axis, **(b)** the y-axis, and **(c)** the line $y = x$. 14–2

5. Draw the line $y = 2x + 1$ and its image under reflection in the y-axis.

6. **a.** If translation $T:(5, 5) \rightarrow (7, 1)$, then $T:(x, y) \rightarrow (\underline{\ ?\ }, \underline{\ ?\ })$ 14–3
   b. Is distance invariant under T?
   c. Is angle measure invariant under T?
   d. Is area invariant under T?

7. Find the image of $(7, -2)$ under the glide reflection that moves all points 5 units to the right and then reflects all points in the x-axis.

8. Plot the points $A(3, 2)$, $B(-1, 1)$, and $C(1, -3)$. Label the origin O. Draw $\triangle ABC$ and its images under **(a)** $\mathcal{R}_{O, 90}$ and **(b)** H_O. 14–4

9. Which of the given rotations are equal to $\mathcal{R}_{O, 140}$?
   a. $\mathcal{R}_{O, 500}$ **b.** $\mathcal{R}_{O, -140}$ **c.** $\mathcal{R}_{O, -220}$

10. If O is the origin then the dilation $D_{O, 2}:(3, -2) \rightarrow (\underline{\ ?\ }, \underline{\ ?\ })$. 14–5

11. Find the image of $(3, 1)$ under a dilation with center $(0, 4)$ and scale factor $\frac{1}{3}$.

12. Find the image of $(3, 1)$ under the following transformations: 14–6
    a. $R_x \circ R_y$ **b.** $R_y \circ H_O$ **c.** $R_x \circ \mathcal{R}_{O, -90}$

Complete.

13. If $T:(x, y) \rightarrow (x - 1, y + 6)$, then $T^{-1}:(x, y) \rightarrow (\underline{\ ?\ }, \underline{\ ?\ })$. 14–7

14. The inverse of $D_{O, 4}$ is $D_{?, ?}$.

15. $R_j \circ R_j = \underline{\ ?\ }$ 16. $\mathcal{R}_{O, 75} \circ \mathcal{R}_{O, ?} = I$

17. Does a scalene triangle have line symmetry? 14–8

18. Does a rectangle have point symmetry?

19. Does a regular octagon have 90° rotational symmetry?

20. Name a figure that has 72° rotational symmetry.

Chapter Test

State whether the transformation mapping the black triangle to the red triangle is a reflection, a translation, a glide reflection, or a rotation.

1.   **2.**   **3.**   **4.**

5. If $f(x) = \frac{1}{2}x + 3$, find the image and preimage of 4.

Give the coordinates of the image of point P under the transformation specified.

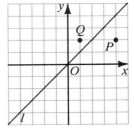

6. R_l

7. $\mathcal{R}_{O,\,-90}$

8. $D_{O,\,\frac{1}{2}}$

9. $H_O \circ R_x$

10. $\mathcal{R}_{O,\,90} \circ \mathcal{R}_{O,\,90}$

11. $D_{Q,\,\frac{1}{3}}$

12. $R_l \circ D_{Q,\,-2}$

13. $R_l \circ (R_y \circ R_x)$

Give the inverse of each transformation.

14. H_O **15.** R_x **16.** $D_{O,\,-2}$

T **is the translation mapping (4, 1) to (6, 2). Find the coordinates of the image of the origin under each mapping.**

17. T **18.** T^3 **19.** T^{-1}

Classify each statement as true or false.

20. All regular polygons have rotational symmetry.

21. 180° rotational symmetry is the same as point symmetry.

22. All regular n-gons have exactly n symmetry lines.

23. A figure that has two intersecting lines of symmetry must have rotational symmetry.

24. a. Is a half-turn a transformation? Is it an isometry?
  b. Name three properties that are invariant under a half-turn.

25. A line has slope 2. What is the slope of the image of the line under a:
  a. reflection in the x-axis?
  b. reflection in the line $y = x$?
  c. dilation $D_{O,\,3}$?

Preparing for College Entrance Exams

Strategy for Success

Try to work quickly and accurately on exam questions. Do not take time to double-check your answers unless you finish all the questions before the deadline. Skip questions that are too difficult for you, and spend no more than a few minutes on each question.

Indicate the best answer by writing the appropriate letter.

1. Find an equation of the perpendicular bisector of the segment joining $(3, -1)$ and $(-1, 7)$.
   (A) $x + 2y = 7$ **(B)** $x - 2y = -5$ **(C)** $2x + y = -5$
   (D) $2x + y = 5$ **(E)** $2x - y = -1$

2. A circle has a diameter with endpoints $(0, -8)$ and $(-6, -16)$. An equation of the circle is:
   (A) $(x + 3)^2 + (y + 12)^2 = 25$ **(B)** $(x + 3)^2 + (y + 12)^2 = 100$
   (C) $(x - 3)^2 + (y - 12)^2 = 25$ **(D)** $(x - 3)^2 + (y - 12)^2 = 100$
   (E) $(x + 6)^2 + (y + 24)^2 = 100$

3. The point $(\frac{1}{2}, -\frac{1}{2})$ lies on line t. Which of the following allow you to find an equation for t?
   I. slope of t is -3 II. x-intercept of t is 7   III. t is parallel to $4x - 5y = 7$
   (A) I only   **(B)** III only   **(C)** I and III only   **(D)** II only   **(E)** I, II, and III

4. Given $A(-3, 5)$, $B(0, -4)$, $C(2, 5)$, and $D(-6, -1)$, find the intersection point of $\overleftrightarrow{AB}$ and $\overleftrightarrow{CD}$.
   (A) $(6, 23)$ **(B)** $(2, -10)$ **(C)** $(-2, 2)$ **(D)** $(-18, 5)$ **(E)** cannot be determined

5. What is the best name for quadrilateral $WXYZ$ with vertices $W(-3, -2)$, $X(-5, 2)$, $Y(1, 5)$, and $Z(3, 1)$?
   (A) isosceles trapezoid   **(B)** parallelogram   **(C)** rectangle
   (D) rhombus   **(E)** square

6. Two vertices of an isosceles right triangle are $(0, 0)$ and $(j, 0)$. The third vertex cannot be:
   (A) $(0, j)$ **(B)** $(0, -j)$ **(C)** (j, j) **(D)** $\left(\frac{j}{2}, \frac{j}{2}\right)$ **(E)** $\left(\frac{j}{2}, j\right)$

7. What is the image of $(-2, 3)$ under reflection in the line $y = x$?
   (A) $(3, -2)$ **(B)** $(2, 3)$ **(C)** $(-2, -3)$ **(D)** $(2, -3)$ **(E)** $(-3, 2)$

8. Find the preimage of $(0, 0)$ under $D_{P, \frac{1}{4}}$, where P is the point $(-1, 1)$.
   (A) $(-4, 4)$ **(B)** $(-\frac{3}{4}, \frac{3}{4})$ **(C)** $(-\frac{1}{4}, \frac{1}{4})$ **(D)** $(4, -4)$ **(E)** $(3, -3)$

9. A regular pentagon does *not* have:
   (A) line symmetry   **(B)** point symmetry   **(C)** 360° rotational symmetry
   (D) 216° rotational symmetry   **(E)** 72° rotational symmetry

10. If $CDEF$ is a square with vertices labeled counterclockwise, then $\mathcal{R}_{C, -450} : \overline{CF} \rightarrow \underline{}$.
    (A) $\overline{FE}$ **(B)** $\overline{ED}$ **(C)** $\overline{CF}$ **(D)** $\overline{CD}$ **(E)** none of these

Cumulative Review: Chapters 1-14

True-False Exercises

Classify each statement as true or false.

A

1. Three given points are always coplanar.

2. Each interior angle of a regular n-gon has measure $\dfrac{(n-2)180}{n}$.

3. If $\triangle RST \cong \triangle RSV$, then $\angle SRT \cong \angle SRV$.

4. The contrapositive of a true conditional is sometimes false.

5. Corresponding parts of similar triangles must be congruent.

6. An acute angle inscribed in a circle must intercept a minor arc.

7. In a plane the locus of points equidistant from M and N is the midpoint of $\overline{MN}$.

8. If a cylinder and a right prism have equal base areas and equal heights, then they have equal volumes.

9. A triangle with vertices $(a, 0)$, $(-a, 0)$, and $(0, a)$ is equilateral.

10. If the slopes of two lines have opposite signs, the lines are perpendicular.

11. $R_k \circ R_k = I$

12. If a figure has 90° rotational symmetry, then it also has point symmetry.

B

13. A point lies on the bisector of $\angle ABC$ if and only if it is equidistant from A and C.

14. In $\triangle RST$, if $RS < ST$, then $\angle R$ must be the largest angle of the triangle.

15. A triangle with sides of length $2x$, $3x$, and $4x$ must be obtuse.

16. In a right triangle, the altitude to the hypotenuse is always the shortest of the three altitudes.

17. Given a segment of length t, it is possible to construct a segment of length $t\sqrt{3}$.

18. If an equilateral triangle and a regular hexagon are inscribed in a circle, then the ratio of their areas is $1:2$.

19. The lateral area of a cone can be equal to the area of the base of the cone.

20. The circle $(x + 3)^2 + (y - 2)^2 = 4$ is tangent to the line $x = -1$.

Multiple-Choice Exercises

Write the letter that indicates the best answer.

A

1. The measure of an interior angle of a regular decagon is:
   a. 36   **b.** 108   **c.** 72   **d.** 144

2. Which of the following is *not* a method for proving two triangles congruent?
   a. HL   **b.** AAS   **c.** SSA   **d.** SAS

3. The median to the hypotenuse of a right triangle divides the triangle into two triangles that are both:

    a. similar   **b.** isosceles   **c.** scalene   **d.** right

4. Which proportion is *not* equivalent to $\dfrac{a}{b} = \dfrac{c}{d}$?

    a. $\dfrac{a}{c} = \dfrac{b}{d}$ **b.** $\dfrac{b}{a+b} = \dfrac{d}{c+d}$ **c.** $\dfrac{b}{a} = \dfrac{d}{c}$ **d.** $\dfrac{a}{d} = \dfrac{c}{b}$

5. For every acute angle X:

    a. $\cos X < \sin X$ **b.** $\cos X > \tan X$ **c.** $\tan X > 1$ **d.** $\cos X < 1$

B   **6.** If A, B, and C are points on $\odot O$, $\overline{AC}$ is a diameter, and $m\angle AOB = 60$, then $m\angle ACB =$

    a. 30   **b.** 60   **c.** 90   **d.** 120

7. A rectangle with perimeter 30 and area 44 has length:

    a. $2\sqrt{11}$ **b.** 8   **c.** 11   **d.** 10

8. A regular hexagon with perimeter 24 has area:

    a. $24\sqrt{3}$ **b.** $16\sqrt{3}$ **c.** $48\sqrt{3}$ **d.** $32\sqrt{3}$

9. In $\odot O$, $m\overset{\frown}{AB} = 90$ and $OA = 6$. The region bounded by $\overline{AB}$ and $\overset{\frown}{AB}$ has area:

    a. $3\pi - 6$ **b.** $9\pi - 36$ **c.** $9\pi - 18$ **d.** $36\pi - 6\sqrt{2}$

10. Two regular octagons have sides of length $6\sqrt{3}$ and 9. The ratio of their areas is:

    a. $2\sqrt{3}{:}3$ **b.** $4{:}3$ **c.** $2{:}3$ **d.** $8\sqrt{3}{:}9$

11. If F is the point $(-3, 5)$ and G is the point $(0, -4)$, then an equation of $\overleftrightarrow{FG}$ is:

    a. $y = -\frac{1}{3}x + 4$ **b.** $y = -3x - 4$ **c.** $y = \frac{1}{3}x + 4$ **d.** $y = -3x + 4$

Completion Exercises

Write the correct word, number, phrase, or expression.

A   **1.** If $5x - 1 = 14$, then the statement $5x = 15$ is justified by the __?__.

2. If two parallel lines are cut by a transversal, then __?__ angles are congruent, __?__ angles are congruent, and __?__ angles are supplementary.

3. The measures of two angles of a triangle are 56 and 62. The measure of the largest exterior angle of the triangle is __?__.

4. In $\triangle BEV$ with $m\angle B = 53$ and $m\angle E = 63$, the longest side is __?__.

5. The area of a triangle with vertices $(-2, 0)$, $(9, 0)$, and $(3, 6)$ is __?__.

6. The distance between $(-5, -2)$ and $(1, -6)$ is __?__.

7. If $j \perp k$ and line j has slope $\frac{2}{3}$, then k has slope __?__.

8. If A is $(-8, 3)$ and B is $(-4, -1)$, then the midpoint of $\overline{AB}$ is $(\underline{\ ?\ }, \underline{\ ?\ })$.

B   **9.** If O is the origin, then $R_x \circ \mathcal{R}_{O,\,90}:(-2, 5) \to (\underline{\ ?\ }, \underline{\ ?\ })$.

10. If $RX = 18$, $XS = 10$, and $RT = 35$, then $YT = \underline{\ ?\ }$.

11. If $RX = 16$, $XS = 8$, and $XY = 15$, then $ST = \underline{\ ?\ }$.

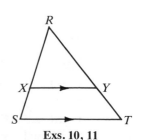

Exs. 10, 11

12. If a dart thrown at a 64-square checkerboard lands on the board, the probability that it lands on a black square is __?__. The probability that it lands on one of the four central squares is __?__.

13. In $\triangle ABC$, $\overline{AB} \perp \overline{BC}$, $AB = 15$, and $BC = 8$. Then the exact value of $\sin C$ is __?__.

14. A tree 5 m tall casts a shadow 8 m long. To the nearest degree, the angle of elevation of the sun is __?__. (Use the table on page 311.)

15. A trapezoid with sides 8, 8, 8, and 10 has area __?__.

16. A circle with area 100π has circumference __?__.

17. A cone with radius 9 and slant height 15 has volume __?__ and lateral area __?__.

18. A sphere with surface area 144π cm^2 has volume __?__.

19. If $B = (2, 0)$, then $D_{B, -2}:(1, 1) \rightarrow (\underline{}, \underline{})$.

C 20. If each edge of a regular triangular pyramid is 6 cm, then the pyramid has total area __?__ and volume __?__.

21. A plane parallel to the base of a cone and bisecting the altitude divides the cone into two parts whose volumes have the ratio __?__.

Always-Sometimes-Never Exercises

Write A, S, or N to indicate your answer.

A 1. Vertical angles are __?__ adjacent angles.

2. If J is a point outside $\odot P$ and $\overline{JA}$ and $\overline{JB}$ are tangents to $\odot P$ with A and B on $\odot P$, then $\triangle JAB$ is __?__ scalene.

3. A conclusion based on inductive reasoning is __?__ correct.

4. Two right triangles with congruent hypotenuses are __?__ congruent.

5. If the diagonals of a quadrilateral are perpendicular bisectors of each other, then the quadrilateral is __?__ a rhombus.

6. If $\triangle RST$ is a right triangle with hypotenuse $\overline{RS}$, then $\sin R$ and $\cos S$ are __?__ equal.

7. A circle __?__ contains three collinear points.

8. A lateral edge of a regular pyramid is __?__ longer than the slant height.

9. Transformations are __?__ isometries.

10. Under a half-turn about point O, point O is __?__ mapped onto itself.

B 11. If the measures of three consecutive angles of a quadrilateral are 58, 122, and 58, then the diagonals __?__ bisect each other.

12. A triangle with sides of length x, $x + 2$, and $x + 4$ is __?__ an acute triangle.

13. If $\overset{\frown}{RS}$ and $\overset{\frown}{XY}$ are arcs of $\odot O$ and $m\overset{\frown}{RS} < m\overset{\frown}{XY}$, then RS and XY are __?__ equal.

14. The center of the circle that can be circumscribed about a given triangle is ⟶?⟵ outside the triangle.

15. Given two segments with lengths r and s, it is ⟶?⟵ possible to construct a segment of length $\frac{3}{4}\sqrt{2rs}$.

16. A median of a triangle ⟶?⟵ separates the triangle into two triangles with equal areas.

17. A composite of reflections in two lines is ⟶?⟵ a translation.

Construction Exercises

A 1. Construct an angle of measure $22\frac{1}{2}$.

2. Draw a circle O and choose a point T on $\odot O$. Construct the tangent to $\odot O$ at T.

3. Draw a large triangle. Inscribe a circle in the triangle.

For Exercises 4–7, draw two long segments. Let their lengths be x and y, with $x > y$.

4. Construct a segment of length $\frac{1}{2}(x + y)$.

B 5. Construct a rectangle with width y and diagonal x.

6. Construct any triangle with area xy.

7. Construct a segment with length $\sqrt{3xy}$.

8. Draw a very long $\overline{AB}$. Construct a rectangle with perimeter AB and sides in the ratio $3:2$.

Proof Exercises

A 1. Given: $\overline{PQ} \parallel \overline{RS}$
Prove: $\dfrac{PO}{RO} = \dfrac{PQ}{RS}$

2. Given: $\overline{PR} \perp \overline{QS}$; $\overline{PS} \cong \overline{QR}$; $\overline{OS} \cong \overline{OR}$
Prove: $\angle PSO \cong \angle QRO$

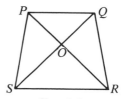
Exs. 1–3

B 3. Given: $\angle OSR \cong \angle ORS$; $\angle OPQ \cong \angle OQP$
Prove: $\triangle PSR \cong \triangle QRS$

4. Prove: The diagonals of a rectangle intersect to form four congruent segments.

5. Use coordinate geometry to prove that the triangle formed by joining the midpoints of the sides of an isosceles triangle is an isosceles triangle.

C 6. Use an indirect proof to show that a trapezoid cannot have two pairs of congruent sides.

7. Prove: If two coplanar circles intersect in two points, then the line joining those points bisects a common tangent segment.

Examinations

Chapter 1

Indicate the best answer by writing the appropriate letter.

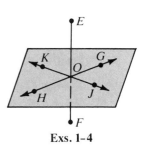

Exs. 1–4

1. Which of the following sets of points are *not* coplanar?
   a. E, H, O, G **b.** K, O, G, E
   c. E, O, F, J **d.** H, K, O, J

2. Which of the following sets of points are contained in *more* than one plane?
   a. G, O, J **b.** E, O, G
   c. H, E, G **d.** G, O, H

3. How many planes contain point E and $\overleftrightarrow{JK}$?
   a. 0 **b.** exactly 1
   c. unlimited **d.** unknown

4. If $\overleftrightarrow{GH}$ bisects $\overline{EF}$, which statement is *not necessarily* true?
   a. O is the midpoint of $\overline{GH}$. **b.** $\overline{EO} \cong \overline{OF}$
   c. E, F, G, H, and O are coplanar. **d.** $GO + OH = GH$

5. Points A, B, C are collinear, but they do not necessarily lie on a line in the order named. If $AB = 5$ and $BC = 3$, what is the length of $\overline{AC}$?
   a. either 2 or 8 **b.** either 2 or 4 **c.** 2 **d.** 8

6. On a number line, point R has coordinate -5 and point S has coordinate 3. Point X lies on $\overrightarrow{SR}$ and $SX = 5$. Find the coordinate of X.
   a. -10 **b.** -2 **c.** 8 **d.** 0

7. Which angle appears to be obtuse?
   a. $\angle AEB$ **b.** $\angle DEB$
   c. $\angle CEA$ **d.** $\angle AED$

8. If $\overrightarrow{EC}$ bisects $\angle DEB$, $\overrightarrow{EB}$ bisects $\angle DEA$, and $m\angle BEC = 28$, find the measure of $\angle CEA$.
   a. 28 **b.** 56
   c. 84 **d.** 112

9. Which two angles are adjacent angles?
   a. $\angle DEB$ and $\angle BEA$ **b.** $\angle DEB$ and $\angle CEA$
   c. $\angle DEC$ and $\angle BEA$ **d.** $\angle DEA$ and $\angle DEC$

Exs. 7–9

10. M is the midpoint of $\overline{YZ}$. If $YM = r + 3$ and $YZ = 3r - 1$, find MZ.
    a. 7 **b.** 10 **c.** 20 **d.** 4

11. Which of the following is *not always* true when lines j and k intersect?
    a. Exactly one plane contains line j.
    b. The lines intersect in exactly one point.
    c. All points on j and k are coplanar points.
    d. Given any point P on j and any point Q on k, P and Q are collinear points.

Chapter 2

Indicate the best answer by writing the appropriate letter.

1. If $m \angle 1 = 60$ and $m \angle 2 = 30$, then $\angle 1$ and $\angle 2$ *cannot* be which of the following?
   a. acute ⩘
   b. adjacent ⩘
   c. vertical ⩘
   d. complementary ⩘

2. Given: If q, then r. Which of the following is the converse of the given conditional?
   a. r implies q.
   b. r if q.
   c. q only if r.
   d. r if and only if q.

3. What are basic mathematical assumptions called?
   a. theorems   **b.** postulates   **c.** conditionals   **d.** conclusions

4. Which of the following *cannot* be used as a reason in a proof?
   a. a definition
   b. a postulate
   c. yesterday's theorem
   d. tomorrow's theorem

5. $\angle A$ and $\angle B$ are supplements, $m \angle A = 2x - 14$, and $m \angle B = x + 8$. Find the measure of $\angle B$.
   a. 62   **b.** 30   **c.** 40   **d.** 70

6. If $\angle 1$ and $\angle 2$ are complements, $\angle 2$ and $\angle 3$ are complements, and $\angle 3$ and $\angle 4$ are supplements, what are $\angle 1$ and $\angle 4$?
   a. supplements   **b.** complements   **c.** congruent angles   **d.** can't be determined

7. The statement "If $m \angle A = m \angle B$ and $m \angle D = m \angle A + m \angle C$, then $m \angle D = m \angle B + m \angle C$" is justified by what property?
   a. Transitive   **b.** Substitution   **c.** Symmetric   **d.** Reflexive

8. If $\overline{TQ} \perp \overline{QR}$, which angles *must* be complementary angles?
   a. $\angle 2$ and $\angle 3$
   b. $\angle 3$ and $\angle 4$
   c. $\angle 5$ and $\angle 8$
   d. $\angle 3$ and $\angle 7$

9. If $m \angle 8 = x + 80$, what is the measure of $\angle 9$?
   a. $100 - x$
   b. $100 + x$
   c. $x - 80$
   d. $x - 180$

10. If $\overline{QT} \perp \overline{PS}$, which statement is *not* always true?
    a. $\angle 8 \cong \angle 9$
    b. $\angle 2 \cong \angle 3$
    c. $\angle 8$ is a rt. $\angle$.
    d. $\angle 8$ and $\angle 9$ are supp. ⩘.

11. If $\overrightarrow{SQ}$ bisects $\angle RST$, which statement *must* be true?
    a. $2 \cdot m \angle 6 = m \angle RST$
    b. $\frac{1}{2} m \angle 7 = m \angle RST$
    c. $\angle 4 \cong \angle 6$
    d. $\angle RST \cong \angle RQT$

Exs. 8–11

Chapter 3

Indicate the best answer by writing the appropriate letter.

1. If $\overrightarrow{BE}$ bisects $\angle ABC$, what is the measure of $\angle AEB$?
   a. 30   **b.** 35   **c.** 40   **d.** 45

2. If $m\angle ABE = 40$, what is the measure of $\angle DEB$?
   a. 140   **b.** 40   **c.** 75   **d.** 135

3. If $\overline{AB} \parallel \overline{DC}$, what is the measure of $\angle D$?
   a. 70   **b.** 80   **c.** 90   **d.** 100

4. Which of the following would allow you to conclude that $\overline{AD} \parallel \overline{BC}$?
   a. $\angle DEC \cong \angle BCE$ **b.** $\angle ABE \cong \angle BEC$
   c. $\angle BEC \cong \angle BCE$ **d.** $m\angle A + m\angle AEC = 180$

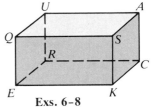

Exs. 1–4

5. What is the measure of each interior angle of a regular octagon?
   a. 150   **b.** 144   **c.** 140   **d.** 135

6. The plane containing Q, S, A, U appears to be parallel to the plane containing which points?
   a. Q, E, K, S **b.** E, K, C, R
   c. R, E, Q, U **d.** U, R, C, A

7. Which of the following appear to be skew lines?
   a. $\overleftrightarrow{QE}$ and $\overleftrightarrow{AC}$ **b.** $\overleftrightarrow{QU}$ and $\overleftrightarrow{KC}$
   c. $\overleftrightarrow{AC}$ and $\overleftrightarrow{UR}$ **d.** $\overleftrightarrow{QS}$ and $\overleftrightarrow{AC}$

Exs. 6–8

8. $\overleftrightarrow{EK}$ does *not* appear to be parallel to the plane containing which points?
   a. U, A, C **b.** Q, U, A **c.** Q, U, R **d.** Q, S, C

9. The sum of the measures of the interior angles of a certain polygon is the same as the sum of the measures of its exterior angles. How many sides does the polygon have?
   a. four   **b.** six   **c.** eight   **d.** ten

10. What is the next number in the sequence 1, 2, 4, 7, 11, . . . ?
    a. 17   **b.** 13   **c.** 16   **d.** 15

11. $\overline{AC}$ is a diagonal of regular pentagon $ABCDE$. What is the measure of $\angle ACD$?
    a. 36   **b.** 54   **c.** 72   **d.** 108

12. A, B, C, and D are coplanar points. $\overleftrightarrow{AB} \parallel \overleftrightarrow{CD}$, $\overleftrightarrow{AB} \perp \overleftrightarrow{AC}$, and $m\angle ACD = 2x + 8$. Find the value of x.
    a. 41   **b.** 49   **c.** 90   **d.** 180

13. What is the *principal* basis for inductive reasoning?
    a. definitions   **b.** previously proved theorems
    c. postulates   **d.** past observations

Chapter 4

In Exercises 1–8 write a method (SSS, SAS, ASA, AAS, or HL) that can be used to prove the two triangles congruent.

1.

2.

3.

4.

5.

6.

7. Given: $\overline{PO} \perp$ plane X; $OT = OS$

8. Given: $\overline{PO} \perp$ plane X; $PT = PS$

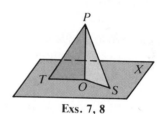

Exs. 7, 8

Indicate the best answer by writing the appropriate letter.

9. In $\triangle RXT$, $\angle R \cong \angle T$, $RT = 2x + 5$, $RX = 5x - 7$, and $TX = 2x + 8$. What is the perimeter of $\triangle RXT$?

    a. 5   **b.** 15   **c.** 18   **d.** 51

10. If $\triangle DEF \cong \triangle PRS$, which of these congruences *must* be true?

    a. $\overline{DF} \cong \overline{PS}$ **b.** $\overline{EF} \cong \overline{PR}$ **c.** $\angle E \cong \angle S$ **d.** $\angle F \cong \angle R$

11. In $\triangle ABC$, $AB = AC$, $m\angle A = 46$, and $\overline{BD}$ is an altitude. What is the measure of $\angle CBD$?

    a. 23   **b.** 44   **c.** 67   **d.** 134

12. An equiangular triangle *cannot* be which of the following?

    a. equilateral   **b.** isosceles   **c.** scalene   **d.** acute

13. Point X is equidistant from vertices T and N of scalene $\triangle TEN$. Point X *must* lie on which of the following?

    a. bisector of $\angle E$ **b.** perpendicular bisector of $\overline{TN}$

    c. median to $\overline{TN}$ **d.** the altitude to $\overline{TN}$

14. Given: $\overline{AB} \parallel \overline{DC}$; $\overline{AB} \cong \overline{CD}$; $\angle 1 \cong \angle 2$

To prove that $\overline{DE} \cong \overline{BF}$, what would you prove first?

    a. $\triangle ADE \cong \triangle CBF$ **b.** $\triangle ABF \cong \triangle CDE$

    c. $\triangle ABC \cong \triangle CDA$ **d.** cannot be proved

Chapter 5

Indicate the best answer by writing the appropriate letter.

1. Both pairs of opposite sides of a quadrilateral are parallel. Which special kind of quadrilateral *must* it be?

   a. parallelogram   **b.** rectangle   **c.** rhombus   **d.** trapezoid

2. The diagonals of a certain quadrilateral are congruent. Which term could *not* be used to describe the quadrilateral?

   a. isosceles trapezoid   **b.** rectangle

   c. rhombus   **d.** parallelogram with a 60° angle

3. M is the midpoint of hypotenuse $\overline{TK}$ of right $\triangle TAK$. $AM = 13$. What is the length of $\overline{TK}$?

   a. 26   **b.** $19\frac{1}{2}$ **c.** 13   **d.** none of these

4. In $\square$ $WXYZ$, $WX = 10$. What does ZW equal?

   a. 16   **b.** YZ **c.** WY **d.** none of these

5. A diagonal of a parallelogram bisects one of its angles. Which special kind of parallelogram *must* it be?

   a. rectangle   **b.** rhombus

   c. square   **d.** parallelogram with a 60° angle

6. The lengths of the bases of a trapezoid are 18 and 26. What is the length of the median?

   a. 8   **b.** 22   **c.** 44   **d.** 34

7. In quad. $PQRS$, $PQ = SR$, $QR = PS$, and $m \angle P = m \angle Q$. Which of the following is *not necessarily* true?

   a. $\overline{PR} \perp \overline{QS}$ **b.** $\overline{PR} \cong \overline{QS}$ **c.** $\angle P \cong \angle R$ **d.** $\angle R \cong \angle S$

8. In $\triangle ABC$, $AB = 8$, $BC = 10$, and $AC = 12$. M is the midpoint of $\overline{AB}$, and N is the midpoint of $\overline{BC}$. What is the length of $\overline{MN}$?

   a. 4   **b.** 5   **c.** 6   **d.** 9

9. If $EFGH$ is a parallelogram, which of the following *must* be true?

   a. $\angle E \cong \angle F$ **b.** $\angle F \cong \angle H$

   c. $\overline{FG} \parallel \overline{GH}$ **d.** $m \angle E + m \angle G = 180$

10. Which information does *not* prove that quad. $ABCD$ is a parallelogram?

    a. $\overline{AC}$ and $\overline{BD}$ bisect each other.   **b.** $\overline{AD} \parallel \overline{BC}$; $\overline{AD} \cong \overline{BC}$

    c. $\overline{AB} \parallel \overline{CD}$; $\overline{AD} \cong \overline{BC}$ **d.** $\angle A \cong \angle C$; $\angle B \cong \angle D$

11. In the figure, $\overline{RU} \cong \overline{US}$ and $\angle 1 \cong \angle 2$. Which of the following *cannot* be proved?

    a. $\angle 3 \cong \angle 4$ **b.** $\overline{RV} \cong \overline{VT}$

    c. $\overline{US} \cong \overline{VT}$ **d.** $ST = 2 \cdot UV$

12. Which of the following *must* be true for any trapezoid?

    a. Any two consecutive angles are supplementary.

    b. At least one angle is obtuse.

    c. The diagonals bisect each other.

    d. The median bisects each base.

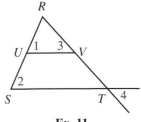

Ex. 11

Chapter 6

Indicate the best answer by writing the appropriate letter.

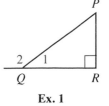

Ex. 1

1. Which of the following statements *must* be false?
   a. $QR + PR > PQ$
   b. $m \angle 2 > m \angle P + m \angle R$
   c. $\frac{1}{2} m \angle 2 > \frac{1}{2} m \angle 1$
   d. $PQ > PR$

2. You don't need a figure to do this exercise. Given that $m \angle A = m \angle B$, you want to prove that $m \angle 3 = m \angle 4$. To write an an indirect proof, you should begin by temporarily assuming which statement?
   a. $m \angle A \neq m \angle B$
   b. $m \angle A = m \angle B$
   c. $m \angle 3 = m \angle 4$
   d. $m \angle 3 \neq m \angle 4$

3. In quadrilateral $MNPQ$, $MN = 5$, $NP = 6$, $PQ = 7$, and $QM = 9$. Which of the following might possibly be the length of $\overline{NQ}$?
   a. 12.5
   b. 14
   c. 2
   d. all of these

4. Given: (1) If A is white, then B is red.
   (2) B is not red.
   Which of the following *must* be true?
   a. B is white.
   b. B is not white.
   c. A is not white.
   d. A is red.

5. If a conditional is known to be true, then which of the following *must* also be true?
   a. its converse
   b. its contrapositive
   c. its inverse
   d. none of these

6. In $\triangle DEF$, $m \angle D = 50$, and an exterior angle with vertex F has measure 120. What is the longest side of $\triangle DEF$?
   a. $\overline{DE}$
   b. $\overline{EF}$
   c. $\overline{DF}$
   d. unknown

7. In $\triangle MNP$, $MN = 8$ and $NP = 10$. Which of these *must* be true?
   a. $MP > 2$
   b. $MP < 2$
   c. $MP > 10$
   d. $MP < 10$

8. What is the inverse of "If $x = 3$, then $x > 0$"?
   a. If $x > 0$, then $x = 3$.
   b. If $x \neq 3$, then $x \leq 0$.
   c. If $x \leq 0$, then $x \neq 3$.
   d. If $x = 3$, then $x \leq 0$.

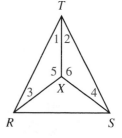

9. If $RT = ST$ and $RX > SX$, what can you conclude?
   a. $m \angle 1 > m \angle 2$
   b. $m \angle XRS > m \angle XSR$
   c. $m \angle 3 = m \angle 4$
   d. $m \angle 5 > m \angle 6$

10. If $RX = SX$ and $m \angle 5 > m \angle 6$, what can you conclude?
    a. $m \angle TRS < m \angle TSR$
    b. $RT < ST$
    c. $m \angle 1 > m \angle 2$
    d. $m \angle 3 > m \angle 4$

11. Which of the following is an important part of an indirect proof?
    a. Proving that the hypothesis *cannot* be deduced from the conclusion
    b. Proving that the temporary assumption must be true
    c. Assuming temporarily that the conclusion must be true
    d. Finding a contradiction of a known fact

Chapter 7

Indicate the best answer by writing the appropriate letter.

1. If the measures of the angles of a triangle are in the ratio $3:3:4$, what is the measure of the largest angle of the triangle?
   - **a.** 40
   - **b.** 54
   - **c.** 72
   - **d.** 90

2. If $\triangle ABC \sim \triangle JOT$, which of these is a correct proportion?
   - **a.** $\dfrac{BC}{AC} = \dfrac{JT}{OT}$
   - **b.** $\dfrac{AB}{JT} = \dfrac{AC}{JO}$
   - **c.** $\dfrac{AB}{BC} = \dfrac{OT}{JT}$
   - **d.** $\dfrac{AC}{JT} = \dfrac{BC}{OT}$

3. If $\dfrac{a}{b} = \dfrac{x}{y}$, what does $\dfrac{y}{b}$ equal?
   - **a.** $\dfrac{x}{a}$
   - **b.** $\dfrac{a}{x}$
   - **c.** $\dfrac{y}{x}$
   - **d.** $\dfrac{b}{y}$

4. $\triangle ABC \sim \triangle DEF$, $AB = 8$, $BC = 12$, $AC = 16$, and $DE = 12$. What is the perimeter of $\triangle DEF$?
   - **a.** 36
   - **b.** 40
   - **c.** 48
   - **d.** 54

5. Which of the following pairs of polygons *must* be similar?
   - **a.** two rectangles
   - **b.** two regular hexagons
   - **c.** two isosceles triangles
   - **d.** two parallelograms with a 60° angle

6. Quad. $GHJK \sim$ quad. $RSTU$, $GH = JK = 10$, $HJ = KG = 14$, and $RS = TU = 16$. What is the scale factor of quad. $GHJK$ to quad. $RSTU$?
   - **a.** $\dfrac{5}{7}$
   - **b.** $\dfrac{5}{8}$
   - **c.** $\dfrac{7}{8}$
   - **d.** $\dfrac{16}{10}$

7. Which of the following can you use to prove that the two triangles are similar?
   - **a.** SAS Similarity Theorem
   - **b.** AA Similarity Postulate
   - **c.** SSS Similarity Theorem
   - **d.** Def. of similar triangles

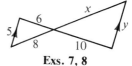

Exs. 7, 8

8. Which statement is correct?
   - **a.** $\dfrac{6}{10} = \dfrac{8}{x}$
   - **b.** $\dfrac{6}{8} = \dfrac{x}{10}$
   - **c.** $6 \cdot 10 = 8x$
   - **d.** $\dfrac{5}{y} = \dfrac{8}{10}$

9. What is the value of u?
   - **a.** 8
   - **b.** 10
   - **c.** 16
   - **d.** 25

10. What is the value of z?
   - **a.** 25
   - **b.** 28
   - **c.** $\dfrac{28}{3}$
   - **d.** $\dfrac{70}{3}$

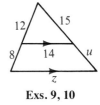

Exs. 9, 10

11. In $\triangle APC$, the bisector of $\angle P$ meets $\overline{AC}$ at B. $PA = 30$, $PC = 50$, and $AB = 12$. What is the length of $\overline{BC}$?
   - **a.** $\dfrac{36}{5}$
   - **b.** 12
   - **c.** 20
   - **d.** 32

12. If $\triangle RST \sim \triangle XYZ$, what is the ratio of $m \angle S$ to $m \angle Y$?
   - **a.** $m \angle R : m \angle Z$
   - **b.** $1:1$
   - **c.** $RS:XY$
   - **d.** not enough information

Chapter 8

Indicate the best answer by writing the appropriate letter.

1. The shorter leg of a 30°-60°-90° triangle has length 7. Find the length of the hypotenuse.
   - **a.** 14
   - **b.** $7\sqrt{2}$
   - **c.** $7\sqrt{3}$
   - **d.** $\sqrt{14}$

2. The altitude to the hypotenuse of a right triangle divides the hypotenuse into segments 25 cm and 30 cm long. How long is the altitude?
   - **a.** $15\sqrt{3}$ cm
   - **b.** $15\sqrt{5}$ cm
   - **c.** $5\sqrt{30}$ cm
   - **d.** $5\sqrt{55}$ cm

3. The hypotenuse and one leg of a right triangle have lengths 61 and 11. Find the length of the other leg.
   - **a.** 36
   - **b.** $5\sqrt{2}$
   - **c.** 60
   - **d.** $\sqrt{3842}$

4. Each side of an equilateral triangle has length 12. Find the length of an altitude.
   - **a.** 6
   - **b.** 12
   - **c.** $6\sqrt{2}$
   - **d.** $6\sqrt{3}$

5. One side of a square has length s. Find the length of a diagonal.
   - **a.** $2\sqrt{s}$
   - **b.** $s\sqrt{2}$
   - **c.** $\frac{s}{2}\sqrt{3}$
   - **d.** $s\sqrt{3}$

6. What kind of triangle has sides of lengths 12, 13, and 18?
   - **a.** an obtuse triangle
   - **b.** a right triangle
   - **c.** an acute triangle
   - **d.** an impossibility

7. In $\triangle RST$, $m \angle S = 90$. What is the value of $\sin T$?
   - **a.** $\dfrac{ST}{RT}$
   - **b.** $\dfrac{RS}{ST}$
   - **c.** $\dfrac{RS}{RT}$
   - **d.** $\dfrac{RT}{RS}$

8. What is the geometric mean between 2 and 24?
   - **a.** 48
   - **b.** $16\sqrt{3}$
   - **c.** $4\sqrt{6}$
   - **d.** $4\sqrt{3}$

9. One acute angle of a certain right triangle has measure n. If $\sin n° = \dfrac{3}{5}$, what is the value of $\tan n°$?
   - **a.** $\dfrac{4}{3}$
   - **b.** $\dfrac{4}{5}$
   - **c.** $\dfrac{3}{4}$
   - **d.** $\dfrac{5}{3}$

10. Which equation could be used to find the value of x?
    - **a.** $\cos 58° = \dfrac{x}{18.9}$
    - **b.** $\sin 32° = \dfrac{x}{16}$
    - **c.** $\cos 44° = \dfrac{x}{10.4}$
    - **d.** $\tan 46° = \dfrac{x}{10.4}$

    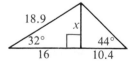

11. In rt. $\triangle ABC$, $\overline{AB} \perp \overline{BC}$, $\overline{BD} \perp \overline{AC}$ at point D, $BC = 9$, and $AC = 12$. Find the ratio of AD to DC.
    - **a.** $\dfrac{9}{16}$
    - **b.** $\dfrac{16}{9}$
    - **c.** $\dfrac{7}{9}$
    - **d.** $\dfrac{9}{7}$

12. For what value(s) of x is a triangle with sides of lengths x, $x + 7$, and $x + 8$ a right triangle?
    - **a.** $x = -7$
    - **b.** $x = 5$
    - **c.** $x = -7$ or $x = 5$
    - **d.** $-7 < x < 5$

Chapter 9

Indicate the best answer by writing the appropriate letter.

In Exercises 1–3, $\overline{PT}$ is tangent to $\odot M$ at T.

1. If $m \angle TMA = 80$, what is the measure of $\overset{\frown}{TBA}$?
   a. 100   b. 80   c. 280   d. 145

2. If $m \angle M = 80$, $m \angle P = 50$, what is the measure of $\angle MAP$?
   a. 140   b. 150   c. 160   d. 170

3. If $PA = 9$ and $AB = 16$, what does PT equal?
   a. 12   b. $\dfrac{25}{2}$ c. 15   d. 20

4. Suppose $\overline{PS}$ were drawn tangent to $\odot M$ at point S. If $m \angle SPT = 62$, find $m\overset{\frown}{ST}$.
   a. 62   b. 236   c. 118   d. 242

5. How many common tangents can be drawn to two circles that are externally tangent?
   a. one   b. two   c. three   d. four

6. Points A, B, and C lie on a circle in the order named. $m\overset{\frown}{AB} = 110$ and $m\overset{\frown}{BC} = 120$. What is the measure of $\angle BAC$?
   a. 130   b. 65   c. 60   d. 55

7. Refer to Exercise 6. If point D lies on $\overset{\frown}{AC}$, what is the sum of the measures of $\angle ABC$ and $\angle ADC$?
   a. 180   b. 170   c. 160   d. 130

8. R and S are points on a circle. $\overline{RS}$ could be which of these?
   a. radius   b. diameter   c. secant   d. tangent

9. If $m\overset{\frown}{BC} = 120$ and $m\overset{\frown}{AD} = 50$, what is the measure of $\angle X$?
   a. 25   b. 35   c. 60   d. 70

10. If $m\overset{\frown}{BC} = 120$ and $m\overset{\frown}{AD} = 50$, what is the measure of $\angle 1$?
    a. 60   b. 85   c. 90   d. 95

11. If $AY = j$, $YC = k$, and $YD = 7$, what does BY equal?
    a. $\dfrac{jk}{7}$ b. $\dfrac{7j}{k}$ c. $\dfrac{7k}{j}$ d. $\dfrac{k}{7j}$

In Exercises 12–14, $\overleftrightarrow{XA}$ is tangent to $\odot O$ at X.

12. Which of these equals $m \angle AXZ$?
    a. $m\overset{\frown}{XYZ}$ b. $m \angle OXM$ c. $\frac{1}{2}m\overset{\frown}{XY}$ d. $\frac{1}{2}m\overset{\frown}{XZ}$

13. If the radius of $\odot O$ is 13 and $XZ = 24$, what is the distance from O to chord $\overline{XZ}$?
    a. 5   b. 8   c. 11   d. $\sqrt{407}$

14. If $OM = 8$ and $MY = 9$, what does XZ equal?
    a. $6\sqrt{2}$ b. $2\sqrt{17}$ c. $\sqrt{145}$ d. 30

Exs. 1–4

Exs. 9–11

Exs. 12–14

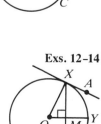

Chapter 10

Indicate the best answer by writing the appropriate letter.

1. In a plane, what is the locus of points equidistant from two given points?
   a. a point
   b. a circle
   c. a line
   d. a pair of lines

2. Point P lies on line l in a plane. What is the locus of points, in that plane, that lie 8 cm from P and 2 cm from l?
   a. no points
   b. two points
   c. three points
   d. four points

3. To inscribe a circle in a triangle, what should you construct first?
   a. two medians
   b. two angle bisectors
   c. two altitudes
   d. the perpendicular bisectors of two sides

4. The lengths of two segments are r and s, with $r > s$. It is *not* possible to construct a segment with which of these lengths?
   a. $\frac{1}{5}(r + s)$
   b. rs
   c. $\sqrt{3rs}$
   d. $\sqrt{r^2 + s^2}$

5. It *is* possible to construct an angle with which of these measures?
   a. 10
   b. 20
   c. 30
   d. 40

6. You are to construct a tangent to a given $\odot O$ from a point P outside the circle. In the process, it would be useless to construct which of these?
   a. $\overline{OP}$
   b. the perpendicular bisector of $\overline{OP}$
   c. a circle with O and P on it
   d. a line parallel to $\overline{OP}$

7. You are given points R and S. Which of the following could *not* be the locus of points in space that are equidistant from R and S and also 4 cm from point S?
   a. a pair of circles
   b. a circle
   c. a point
   d. the empty set

8. Where *must* the perpendicular bisectors of the sides of a triangle meet?
   a. inside the triangle
   b. on the triangle
   c. outside the triangle
   d. none of these

9. In space, what is the locus of points 3 cm from a given point A?
   a. a line
   b. a plane
   c. a circle
   d. a sphere

10. In a plane, what is the locus of points equidistant from the sides of a square?
    a. a square
    b. a line
    c. a circle
    d. a point

11. In $\triangle ABC$, $\overline{AD}$ and $\overline{BE}$ are medians. If $AG = 8$, find AD.
    a. 12
    b. 32
    c. 4
    d. 16

12. You are to construct a perpendicular to a line l at a given point X on l. In how many places on l will you need to position the point of your compass in order to do this construction?
    a. one
    b. two
    c. three
    d. four

Chapter 11

Indicate the best answer by writing the appropriate letter.

1. One side of a rectangle is 14 and the perimeter is 44. What is the area?
   a. 112   **b.** 210   **c.** 224   **d.** 420

2. What is the area of a square inscribed in a circle with radius 8?
   a. 32   **b.** 64   **c.** $64\sqrt{2}$ **d.** 128

3. The area of a circle is 25π. What is its circumference?
   a. 5π **b.** 10π **c.** 12.5π **d.** 50π

4. What is the area of a trapezoid with bases 7 and 8 and height 6?
   a. 90   **b.** 336   **c.** 45   **d.** 168

5. A parallelogram and a triangle have equal areas. The base and height of the parallelogram are 12 and 9. If the base of the triangle is 36, find its height.
   a. 3   **b.** 6   **c.** 9   **d.** 12

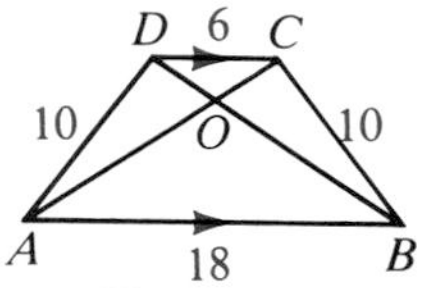

6. What is the area of trapezoid *ABCD*?
   a. 96   **b.** 120   **c.** 144   **d.** 192

7. What is the ratio of the areas of $\triangle AOB$ and $\triangle COD$?
   a. $\sqrt{3}:1$ **b.** $\sqrt{3}:3$ **c.** $3:1$ **d.** $9:1$

8. What is the ratio of the areas of $\triangle AOB$ and $\triangle AOD$?
   a. $\sqrt{3}:1$ **b.** $3:1$ **c.** $9:1$ **d.** cannot be determined

9. What is the area of a regular hexagon inscribed in a circle with radius 8?
   a. $16\sqrt{3}$ **b.** $96\sqrt{3}$ **c.** $128\sqrt{3}$ **d.** $192\sqrt{3}$

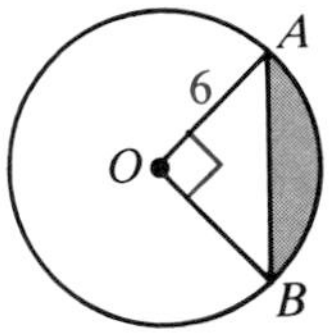

10. In the diagram, what is the length of $\overarc{AB}$?
    a. $6\sqrt{2}$ **b.** 6π **c.** 3π **d.** 36π

11. In the diagram, what is the area of the shaded region?
    a. $9\pi - 36$ **b.** $12\pi - 36$ **c.** $9\pi - 18$ **d.** $12\pi - 18$

12. If a point is chosen at random in the interior of $\odot O$, what is the probability that the point is inside $\triangle AOB$?
    a. $\dfrac{2}{\pi}$ **b.** $\dfrac{1}{4}$ **c.** $\dfrac{3}{2\pi}$ **d.** $\dfrac{1}{2\pi}$

13. A rhombus has diagonals 6 and 8. What is its area?
    a. 12   **b.** 24   **c.** 36   **d.** 48

14. What is the area of a circle with diameter 12?
    a. $24\pi^2$ **b.** 12π **c.** 144π **d.** 36π

15. What is the area of an equilateral triangle with perimeter 24?
    a. $64\sqrt{3}$ **b.** $32\sqrt{3}$ **c.** $\dfrac{32\sqrt{3}}{3}$ **d.** $16\sqrt{3}$

16. What is the area of a triangle with sides 15, 15, and 24?
    a. 54   **b.** 108   **c.** 180   **d.** 216

Chapter 12

Indicate the best answer by writing the appropriate letter.

1. What is the volume of a rectangular solid with dimensions 12, 9, and 6?
   a. 108   **b.** 216   **c.** 432   **d.** 648

2. What is the total surface area of the solid in Exercise 1?
   a. 234   **b.** 468   **c.** 252   **d.** 360

3. Two similar cones have heights 5 and 20. What is the ratio of their volumes?
   a. 1:64   **b.** 1:4   **c.** 1:16   **d.** 4:16

4. What is the volume of a regular square pyramid with base edge 16 and height 6?
   a. 128   **b.** 256   **c.** 512   **d.** 1536

5. What is the lateral area of the pyramid in Exercise 4?
   a. 256   **b.** 320   **c.** 576   **d.** 640

6. A sphere has area 16π. What is its volume?
   a. $\dfrac{8\pi}{3}$ **b.** $\dfrac{32\pi}{3}$ **c.** $\dfrac{64\pi}{3}$ **d.** $\dfrac{256\pi}{3}$

7. A cone has radius 5 and height 12. A cylinder with radius 10 has the same volume as the cone. What is the cylinder's height?
   a. 1   **b.** 2   **c.** 3   **d.** 4

8. A cube is inscribed in a cylinder with radius 5. What is the volume of the cube?
   a. $15\sqrt{2}$ **b.** $250\sqrt{2}$ **c.** 125   **d.** 100

9. A plane passes 2 cm from the center of a sphere with radius 4 cm. What is the area of the circle of intersection?
   a. 12π cm^2 **b.** 16π cm^2 **c.** 18π cm^2 **d.** 20π cm^2

10. Find the total surface area of a cylinder with radius 4 and height 6.
    a. 16π **b.** 32π **c.** 48π **d.** 80π

11. Two similar pyramids have volumes 27 and 125. If the smaller has lateral area 18, what is the lateral area of the larger?
    a. 30   **b.** $83\frac{1}{3}$ **c.** 50   **d.** 25

12. The base of a right prism is a regular hexagon with side 4. The height of the prism is 6. What is the volume of the prism?
    a. $144\sqrt{3}$ **b.** $72\sqrt{3}$ **c.** $48\sqrt{3}$ **d.** $36\sqrt{3}$

13. What is the lateral area of the prism in Exercise 12?
    a. 24   **b.** 36   **c.** 72   **d.** 144

14. Find the total surface area of a cone with radius 9 and siant height 12.
    a. 108π **b.** 189π **c.** $81\pi\sqrt{7}$ **d.** 216π

Chapter 13

Indicate the best answer by writing the appropriate letter.

1. Given $P(-2, 0)$ and $Q(2, 5)$, find $\overrightarrow{PQ}$.
   a. $(0, 2.5)$ **b.** $(0, 5)$ **c.** $(-4, -5)$ **d.** $(4, 5)$

2. Refer to Exercise 1. Find $|\overrightarrow{PQ}|$.
   a. 5 **b.** 3 **c.** $\sqrt{29}$ **d.** $\sqrt{41}$

3. A line with slope $\dfrac{2}{5}$ passes through point $(1, 4)$. What is an equation of the line?

   a. $y - 4 = \dfrac{2}{5}(x - 1)$ **b.** $y - 4 = \dfrac{5}{2}(x - 1)$

   c. $y + 4 = \dfrac{2}{5}(x + 1)$ **d.** $y - 1 = \dfrac{5}{2}(x - 4)$

4. The midpoint of $\overline{AB}$ is $(3, 4)$. If the coordinates of B are $(6, 6)$, what are the coordinates of A?
   a. $(9, 10)$ **b.** $(4.5, 5)$ **c.** $(0, 2)$ **d.** $(9, 10)$

5. What is an equation of the line through $(-4, 7)$ and perpendicular to $y = \dfrac{2}{3}x + 5$?

   a. $y = \dfrac{3}{4}x + 10$ **b.** $y = -\dfrac{3}{2}x - 5$ **c.** $y = -\dfrac{7}{4}x$ **d.** $y = -\dfrac{3}{2}x + 1$

6. What is an equation of the circle with center $(3, 0)$ and radius 8?
   a. $x^2 + y^2 = 64$ **b.** $(x - 3)^2 + y^2 = 64$
   c. $(x + 3)^2 + y^2 = 8$ **d.** $(x - 3)^2 + y^2 = 8$

7. Find an equation of the line through points $(-3, 5)$ and $(2, 8)$.
   a. $5x + 3y = 16$ **b.** $3x - 5y = -34$
   c. $5x - 3y = -30$ **d.** $5x + 3y = 0$

8. Three consecutive vertices of a parallelogram are $(j, 5)$, $(0, 0)$, and $(7, 0)$. Which is the fourth vertex?
   a. $(7, 5)$ **b.** $(5, 7)$ **c.** $(j + 7, 5)$ **d.** $(j + 5, 7)$

9. Points $(2, 2)$ and $(8, v)$ lie on a line with slope $\dfrac{1}{2}$. What is the value of v?
   a. -10 **b.** -1 **c.** 5 **d.** 14

10. What is the *best* term for a triangle with vertices $(1, -3)$, $(6, 2)$, and $(0, 4)$?
    a. isosceles triangle **b.** equilateral triangle
    c. right triangle **d.** none of these

11. Which point is the intersection of lines $3x + 2y = 17$ and $x - 4y = 1$?
    a. $(1, 5)$ **b.** $(5, 1)$ **c.** $(-1, 5)$ **d.** $\left(\dfrac{33}{5}, \dfrac{7}{5}\right)$

12. $\triangle ABC$ is equilateral with vertices $A(r, 0)$ and $B(-r, 0)$. Which of the following could be the coordinates of point C?
    a. $(r\sqrt{3}, 0)$ **b.** $(0, r\sqrt{3})$ **c.** $(0, r)$ **d.** $(0, 2r)$

Chapter 14

Indicate the best answer by writing the appropriate letter.

1. A regular hexagon does *not* have which symmetry?
   a. line   **b.** point   **c.** 30° rotational   **d.** 120° rotational

2. What is the image of the point (2, 3) by reflection in the x-axis?
   a. (3, 2)   **b.** $(-2, 3)$ **c.** $(2, -3)$ **d.** $(-2, -3)$

3. $T:(x, y) \rightarrow (x, y - 2)$. What is the preimage of (3, 5)?
   a. (5, 7)   **b.** (3, 7)   **c.** (3, 3)   **d.** (5, 3)

4. If O is the point (0, 0), what is the image of (3, 6) by $D_{O,\frac{1}{3}}$?
   a. (9, 18)   **b.** (2, 4)   **c.** (1, 2)   **d.** $(-1, -2)$

5. What is the image of $(-1, 3)$ by a half-turn about (1, 2)?
   a. (3, 1)   **b.** $(1, -3)$ **c.** $(-1, -2)$ **d.** $(3, -1)$

6. $T:(x, y) \rightarrow (x, y - 2)$. What is the image of (5, 3) by T^{-1}?
   a. (3, 3)   **b.** (5, 1)   **c.** (3, 5)   **d.** (5, 5)

7. Isometry $S: \square ABCD \rightarrow \square JKLM$. Which statement *must* be true?
   a. $\angle DAB \cong \angle JKL$ **b.** $AC = JL$
   c. $S:C \rightarrow M$ **d.** $CD = MJ$

8. What is the line of reflection for a transformation that maps $(-2, 1)$ to (2, 1)?
   a. the x-axis   **b.** the line $y = x$
   c. the y-axis   **d.** the origin

9. How many lines of symmetry does a rhombus with a 60° angle have?
   a. none   **b.** one   **c.** two   **d.** four

10. If k is the line $y = x$, find the image of J by $R_k \circ R_y$.
    a. J **b.** K **c.** L **d.** M

11. T is a translation that maps K to N. What is the image of J under T?
    a. K **b.** O **c.** N **d.** L

12. What is the image of J under $R_x \circ H_O$?
    a. J **b.** K **c.** L **d.** M

13. What is the image of $\triangle LMJ$ by $\mathcal{R}_{O,90}$?
    a. $\triangle JKL$ **b.** $\triangle KLM$ **c.** $\triangle LMJ$ **d.** $\triangle MJK$

Exs. 10–13

14. Which mapping is *not* an isometry?
    a. glide reflection   **b.** translation
    c. dilation   **d.** the identity transformation

15. Which of the following is *not* invariant under a glide reflection?
    a. angle measure   **b.** orientation of points
    c. parallelism of lines   **d.** areas of polygons

College Entrance Exams

If you are planning to attend college, you will probably be required to take college entrance exams. Some of these exams test your knowledge of specific subject areas; others are more general exams that attempt to measure the extent to which your verbal and mathematical reasoning abilities have been developed. These abilities are ones that can be improved through study and practice. Generally the best preparation for college entrance exams is to follow a strong academic program in high school and to read as extensively as possible.

At the end of every even-numbered chapter in this book are tests called ''Preparing for College Entrance Exams.'' These tests contain questions similar to the questions asked on college entrance exams.

The following test-taking strategies may be useful:

- Familiarize yourself with the test you will be taking well in advance of the test date. Sample tests, with accompanying explanatory material, are available for many standardized tests. By working through this sample material, you become comfortable with the types of questions and directions that will appear on the test and you develop a feeling for the pace at which you must work in order to complete the test.

- Find out how the test is scored so that you know whether it is advantageous to guess.

- Skim sections of the test before starting to answer the questions, to get an overview of the questions. You may wish to answer the easiest questions first. Correctly answering the easy questions earns you the same credit as correctly answering the difficult ones, so do not waste time on questions you do not understand; go on to those that you do.

- Mark your answer sheet carefully, checking the numbering on the answer sheet about every five questions to avoid errors caused by misplaced answer markings.

- Write in the test booklet if it is helpful; for example, cross out incorrect alternatives and do mathematical calculations. Do *not* make extra marks on the answer sheet.

- Work carefully, but do not take time to double-check your answers unless you finish before the deadline and have extra time.

- Arrive at the test center early and come well prepared with any necessary supplies such as sharpened pencils and a watch.

College entrance exams that test general reasoning abilities, such as the Scholastic Aptitude Test, frequently include questions relating to basic geometric concepts and skills. The following topics often appear on such exams. For each topic, page references have been given to the places in your textbook where the topic is discussed.

Properties of Parallel and Perpendicular Lines (pages 56, 73–74, 78–79, 535)

If two parallel lines are cut by a transversal, then alternate interior angles are congruent, corresponding angles are congruent, and same-side interior angles are supplementary.

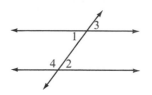

$$m \angle 1 = m \angle 2$$
$$m \angle 3 = m \angle 2$$
$$m \angle 1 + m \angle 4 = 180$$

If two lines are perpendicular, they form congruent adjacent angles.

Angle Measure Relationships (pages 51, 94–95, 102, 204)

Vertical angles are congruent.

$$m \angle 3 = m \angle 5$$

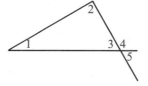

The sum of the measures of the angles of a triangle is 180.

$$m \angle 1 + m \angle 2 + m \angle 3 = 180$$

The measure of an exterior angle of a triangle equals the sum of the measures of the two remote interior angles.

$$m \angle 4 = m \angle 1 + m \angle 2$$

The measure of an exterior angle of a triangle is greater than the measure of either remote interior angle.

The sum of the measures of the angles of a convex polygon with n sides is $(n - 2)180$.

For example, the sum of the measures of the angles of the pentagon at the right is $3 \cdot 180 = 540$.

Triangle Side Relationships (pages 219–220, 228–229)

The sum of the lengths of any two sides of a triangle is greater than the length of the third side.

For example, $AB + BC > AC$.

If one side of a triangle is longer than a second side, then the angle opposite the first side is larger than the angle opposite the second side. The converse is also true.

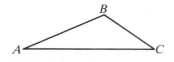

For example, if $AC > BC$, then $m \angle B > m \angle A$;
if $m \angle C < m \angle B$, then $AB < AC$.

Special Triangle Relationships (pages 93, 135–136, 290, 295, 300)

Isosceles Triangle

At least 2 sides are congruent.

Angles opposite congruent sides are congruent.

Equilateral Triangle

All sides are congruent.

All angles are congruent.

By the Pythagorean Theorem, in $\triangle ABC$
$$c^2 = a^2 + b^2.$$

Since $\angle C$ is a right angle,
$$m \angle A + m \angle B = 90.$$

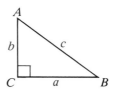

45°-45°-90° Triangle

$a = b$
$c = \sqrt{2}\, a$
$\quad = \sqrt{2}\, b$

Legs are congruent.

Hypotenuse $= \sqrt{2} \cdot$ leg

30°-60°-90° Triangle

$c = 2a$
$b = \sqrt{3}\, a$

Hypotenuse $= 2 \cdot$ shorter leg

Longer leg $= \sqrt{3} \cdot$ shorter leg

Perimeter, Area, and Volume Formulas (pages 424, 429, 447, 469, 476, 518)

Rectangle

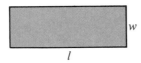

Perimeter $= 2l + 2w$

Area $= lw$

Triangle

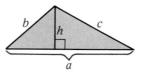

Perimeter $= a + b + c$

Area $= \frac{1}{2}(\text{base} \times \text{height})$
$\quad\quad = \frac{1}{2}ah$

Circle   **Rectangular Solid**

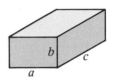

Circumference $= 2\pi r$ Total area $= 2ab + 2bc + 2ac$

Area $= \pi r^2$ Volume $= abc$

Locating Points on a Grid (pages 113, 523–525)

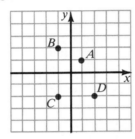

The points shown are $A(1, 1)$, $B(-1, 2)$, $C(-1, -2)$, and $D(2, -2)$.

The distance d between points (x_1, y_1) and (x_2, y_2) is given by
$$d = \sqrt{(x_2 - x_1)^2 + (y_2 - y_1)^2}.$$

An equation of the circle with center at the origin and radius r is $x^2 + y^2 = r^2$.

Logic

Statements and Truth Tables

In algebra, you have used letters to represent numbers. In logic, letters are used to represent statements that are either true or false. For example, p might represent the statement "Paris is the capital city of France," and q might represent the statement "The moon is made of green cheese." Deciding whether statements are true or false involves investigating the real world, not the "logic" of an argument.

Statements can be joined to form **compound statements.** Two important compound statements are defined below.

A **conjunction** is a compound statement composed of two statements joined by the word "and." The symbol $\wedge$ is used to represent the word "and."

A **disjunction** is a compound statement composed of two statements joined by the word "or." The symbol $\vee$ is used to represent the word "or."

Example 1 Statements: p Inez plays the flute.
                           q Sue Yin plays the cello.
         Conjunction: $p \wedge q$ Inez plays the flute and Sue Yin plays the cello.
         Disjunction: $p \vee q$ Inez plays the flute or Sue Yin plays the cello.

The table at the left below is called a **truth table.** It tells you the conditions under which a conjunction is a true statement. "T" stands for "true" and "F" for "false." The first row of the table shows that when statement p is true and statement q is true, the conjunction $p \wedge q$ is true. The other rows of the table show that $p \wedge q$ is false when either of its statements is false.

Truth table for conjunction

p	q	$p \wedge q$
T	T	T
T	F	F
F	T	F
F	F	F

Truth table for disjunction

p	q	$p \vee q$
T	T	T
T	F	T
F	T	T
F	F	F

A disjunction is true if either of its statements is true or both are true. This corresponds to what is called the *inclusive use* of "or." (The *exclusive use* of "or" would imply that one of the statements is true, but not both. We deal only with the inclusive use of "or" in this course.) The first row of the truth table for disjunction shows that when both p and q are true, $p \vee q$ is true. The next two rows show that the compound statement $p \vee q$ is true when either of its statements is true. The last row shows that a disjunction is false when both of its statements are false.

In addition to the words "and" and "or," the word "not" is an important word in logic. If p is a statement, then the statement "p is not true," usually shortened to "not p" and written $\sim p$, is called the **negation** of p.

Example 2 Statement: p Will is sleeping in class.
              Negation: $\sim p$ It is not true that Will is sleeping in class.
              or $\sim p$ Will is not sleeping in class.

Truth table for negation

p	$\sim p$
T	F
F	T

The truth table for negation shows that when p is true, $\sim p$ is false. When p is false, $\sim p$ is true. Note that it is impossible for a statement and its negation to be both true or both false at the same time. The conjunction $p \wedge \sim p$ would have Fs in both rows of its truth table. Such a statement is called a *contradiction*.

An example will show how to make truth tables for some other compound statements.

Example 3 Make a truth table for $\sim p \vee \sim q$.

p	q	$\sim p$	$\sim q$	$\sim p \vee \sim q$
T	T	F	F	F
T	F	F	T	T
F	T	T	F	T
F	F	T	T	T

Solution

1. Make a column for p and a column for q. Write all possible combinations of T and F in the standard pattern shown.

2. Since $\sim p$ is a part of the given statement, add a column for $\sim p$. To fill out this column, use the first column and refer to the truth table for negation above. Similarly, add a column for $\sim q$.

3. Using the columns for $\sim p$ and $\sim q$, refer to the truth table for disjunction on the preceding page in order to fill out the column for $\sim p \vee \sim q$. Remember that a disjunction is false only when both of its statements are false.

To make a truth table for a compound statement involving three simple statements p, q, and r, you would need an eight-row table to show all possible combinations of T and F. The standard pattern across the three columns headed p, q, and r is as follows: TTT, TTF, TFT, TFF, FTT, FTF, FFT, FFF.

Exercises

Suppose p stands for "I like the city," and q stands for "You like the country." Express in words each of the following statements.

1. $p \wedge q$ **2.** $\sim p$ **3.** $\sim q$ **4.** $p \vee q$ **5.** $p \vee \sim q$

6. $\sim(p \wedge q)$ **7.** $\sim p \vee \sim q$ **8.** $\sim p \wedge q$ **9.** $\sim(p \vee q)$ **10.** $\sim p \wedge \sim q$

Suppose p stands for "Hawks swoop," and q stands for "Gulls glide." Express in symbolic form each of the following statements.

11. Hawks swoop or gulls glide.   **12.** Gulls do not glide.

13. It is not true that "Hawks swoop or gulls glide."

14. Hawks do not swoop and gulls do not glide.

15. It is not true that "Hawks swoop and gulls glide."

16. Hawks do not swoop or gulls do not glide.

17. Do the statements in Exercises 13 and 14 mean the same thing?

18. Do the statements in Exercises 15 and 16 mean the same thing?

Make a truth table for each of the following statements.

19. $p \lor \sim q$ **20.** $\sim p \lor q$

21. $\sim(\sim p)$ **22.** $\sim(p \land q)$

23. $p \lor \sim p$ **24.** $p \land \sim p$

25. $p \land (q \lor r)$ **26.** $(p \land q) \lor (p \land r)$

Truth Tables for Conditionals

Truth table for conditionals

The conditional statement "If p, then q," which is discussed in Lesson 2–1, is symbolized as $p \rightarrow q$. This is also read as "p implies q" and as "q follows from p." The truth table for $p \rightarrow q$ is shown at the right. Notice that the only time a conditional is false is when the hypothesis p is true and the conclusion q is false. The example below will show why this is a reasonable way to make out the truth table.

p	q	$p \rightarrow q$
T	T	T
T	F	F
F	T	T
F	F	T

Example   Mom promises, "If I catch the early train home I'll take you swimming." Consider the four possibilities of the truth table.
1. Mom catches the early train home and takes you swimming. She kept her promise; her statement was *true*.
2. Mom catches the early train home but does not take you swimming. She broke her promise; her statement was *false*.
3. Mom does not catch the early train home but still takes you swimming. She has not broken her promise; her statement was *true*.
4. Mom does not catch the early train home and does not take you swimming. She has not broken her promise; her statement was *true*.

The tables on the next page show the converse and contrapositive of $p \rightarrow q$. Make sure that you understand how these tables were made. Notice that the last column of the table for the contrapositive $\sim q \rightarrow \sim p$ is identical with the last column of the table for the conditional on this page. In other words, the contrapositive of a statement is true (or false) if and only if the statement itself is true (or false). This is what we mean when we say that a statement and its contrapositive are logically equivalent (see Lesson 6–2). On the other hand, a statement and its converse are not logically equivalent. Can you see why?

Converse of $p \rightarrow q$		
p	q	$q \rightarrow p$
T	T	T
T	F	T
F	T	F
F	F	T

Contrapositive of $p \rightarrow q$				
p	q	$\sim q$	$\sim p$	$\sim q \rightarrow \sim p$
T	T	F	F	T
T	F	T	F	F
F	T	F	T	T
F	F	T	T	T

Exercises

Suppose p represents "You like to paint," q represents "You are an artist," and r represents "You draw landscapes." Express in words each of the following statements.

1. $p \rightarrow q$ **2.** $q \rightarrow r$ **3.** $\sim q \rightarrow \sim r$ **4.** $\sim(p \rightarrow q)$

5. $(p \wedge q) \rightarrow r$ **6.** $p \wedge (q \rightarrow r)$ **7.** $(r \vee q) \rightarrow p$ **8.** $r \vee (q \rightarrow p)$

Let b, s, and k represent the following statements.
b: Bonnie bellows. s: Sheila shouts. k: Keiko cackles.
Express in symbolic form each of the following statements.

9. If Bonnie bellows, then Keiko cackles.

10. If Keiko cackles, then Sheila does not shout.

11. If Bonnie does not bellow or Keiko does not cackle, then Sheila shouts.

12. Sheila shouts, and if Bonnie bellows, then Keiko cackles.

13. It is not true that Sheila shouts if Bonnie bellows.

14. If Bonnie does not bellow, then Keiko cackles and Sheila shouts.

15. a. Make a truth table for $\sim p \rightarrow \sim q$ (the inverse of $p \rightarrow q$). Your first two columns should be the same as the first two columns of the table for $p \rightarrow q$. The last columns of the two tables should be different. Are they? Is $\sim p \rightarrow \sim q$ logically equivalent to $p \rightarrow q$?

    b. Compare the truth table for $\sim p \rightarrow \sim q$ (the inverse of $p \rightarrow q$) with the truth table for $q \rightarrow p$ (the converse of $p \rightarrow q$). Are the last columns the same? Are the inverse and the converse logically equivalent?

Make truth tables for the following statements.

16. $p \rightarrow \sim q$ **17.** $\sim(p \rightarrow q)$ **18.** $p \wedge \sim q$

19. By comparing the truth tables in Exercises 16–18, you should find that two of the three statements are logically equivalent. Which two?

20. The biconditional statement "p if and only if q" is defined as $(p \rightarrow q) \wedge (q \rightarrow p)$. Make a truth table for this statement.

Some Rules of Inference

Four rules for making logical inferences are symbolized below. A horizontal line separates the given information, or premises, from the conclusion. If you accept the given statement or statements as true, then you must accept as true the conclusions shown.

1. Modus Ponens

$$p \rightarrow q$$
$$\underline{p \qquad\qquad}$$
Therefore, q

2. Modus Tollens

$$p \rightarrow q$$
$$\underline{\sim q \qquad\qquad}$$
Therefore, $\sim p$

3. Simplification

$$p \wedge q$$
$$\underline{\qquad\qquad\qquad}$$
Therefore, p

4. Disjunctive Syllogism

$$p \vee q$$
$$\underline{\sim p \qquad\qquad}$$
Therefore, q

You should convince yourself that these rules make good sense. For example, Rule 4 says that if you know that ''p or q'' is true and then you find out that p is not true, you must conclude that q is true.

Example 1 If today is Tuesday, then tomorrow is Wednesday.
Today is Tuesday.
Therefore, tomorrow is Wednesday. (Rule 1)

Example 2 If a figure is a triangle, then it is a polygon.
This figure is not a polygon.
Therefore, this figure is not a triangle. (Rule 2)

Example 3 It is Tuesday and it is April.
Therefore, it is Tuesday. (Rule 3)

Example 4 It is a square or it is a triangle.
It is not a square.
Therefore, it is a triangle. (Rule 4)

Example 5 Given: $p \rightarrow q$; $p \vee r$; $\sim q$
Prove: r

Proof:

Statements	Reasons
1. $p \rightarrow q$	1. Given
2. $\sim q$	2. Given
3. $\sim p$	3. Steps 1 and 2 and Modus Tollens
4. $p \vee r$	4. Given
5. r	5. Steps 3 and 4 and Disj. Syllogism

Exercises

Supply the reasons to complete each proof.

1. Given: $p \wedge q$; $p \to s$
Prove: s

Statements

1. $p \wedge q$
2. p
3. $p \to s$
4. s

2. Given: $r \to s$; r; $s \to t$
Prove: t

Statements

1. $r \to s$
2. r
3. s
4. $s \to t$
5. t

Write two-column proofs for the following.

3. Given: $p \vee q$; $\sim p$; $q \to s$
Prove: s

4. Given: $a \to b$; $a \vee c$; $\sim b$
Prove: c

5. Given: $a \wedge b$; $a \to \sim c$; $c \vee d$
Prove: d

6. Given: $p \wedge q$; $p \to \sim s$; $r \to s$
Prove: $\sim r$

Symbolize the statements using the letters indicated, accept the statements as true, and write two-column proofs.

7. If Jorge wins the marathon, then he will receive a gold medal. If Jorge receives a gold medal, then his country will be proud. Jorge wins the marathon and Yolanda wins the javelin contest. Prove that Jorge's country will be proud.

(Use the letter w for "Jorge wins the marathon," g for "Jorge receives a gold medal," p for "Jorge's country will be proud," and y for "Yolanda wins the javelin contest.")

8. The sides of *ABCD* are not all the same length, and *ABCD* is a plane figure. *ABCD* is a square or a rectangle. If the sides of *ABCD* are not all the same length, then it is not a square.

Prove that *ABCD* is a rectangle. (Use the letters l, p, s, and r.)

Valid Arguments and Mistaken Premises

A statement whose truth table contains only Ts in the last column is called a *tautology*. An example is the disjunction $p \vee \sim p$ ("p or not p"). This is always true, no matter whether p is true (and $\sim p$ is false) or p is false (in which case $\sim p$ is true).

Tautology

p	$\sim p$	$p \vee \sim p$
T	F	T
F	T	T

Valid argument

p	p	$p \to p$
T	T	T
F	F	T

A conditional whose truth table contains only Ts in the last column is a tautology that represents a **valid argument.** A valid argument is true no matter what the truth or falsity of the components is. Its validity is independent of the real world. For example, the conditional "If p, then p" is always true, even if p stands for "Unicorns exist."

The geometrical theorems in this text are all valid (if you accept the Postulates), and you can use them with confidence in any logical argument. Every theorem in this text can be written as a tautology in the form of a conditional whose hypothesis is a conjunction of givens, definitions, and theorems, and whose conclusion is the statement that is to be proved.

$$\text{Theorem:} \quad [a \wedge b \wedge (c \to d) \wedge (e \to f)] \to g$$
$$\text{givens} + \text{definition} + \text{theorem} = \text{conclusion}$$

In everyday situations, though, you must be careful to inspect the logic of arguments. Even though the reasoning is logically correct, the conclusion may be wrong. The problem is usually that some of the given premises or conditionals are wrong.

Example   The following is a logically valid argument. Is the conclusion true?

    1. The weather is sunny.
    2. If the weather is sunny, the plane will arrive on time.
    3. If the plane arrives on time, we will be able to ski today.
    4. Therefore, we will be able to ski today.

Solution   We cannot evaluate the truth of the conclusion unless we investigate all the premises. The first statement, a given, may not be accurate. Perhaps it is cloudy. Also, one or more of the remaining conditional statements might be wrong. Perhaps the plane will malfunction and be late even though the weather is sunny. Perhaps we won't be able to get to the ski area, even if the plane lands on time. Or maybe we don't even know how to ski! It is important to investigate the truth of every premise before you can draw meaningful conclusions.

Exercises

1. Make a truth table for each statement. Which is a tautology?

    a. $(p \vee q) \to p$ **b.** $(p \to q) \to p$
    c. $(p \wedge q) \to p$ **d.** $(p \to q) \to q$

2. Show that $(p \vee q \vee r) \vee \sim(p \wedge q \wedge r)$ is a tautology by making an eight-row truth table.

3. Show that the argument $(p \wedge q \wedge r) \to (p \vee q \vee r)$ is valid by making an eight-row truth table.

4. A chain of deductive reasoning is often used in geometric proofs. For example, if you are given three premises: p; $p \rightarrow q$; and $q \rightarrow r$, then you can conclude r. Prove the validity of this argument by filling in the truth table for $[p \wedge (p \rightarrow q) \wedge (q \rightarrow r)] \rightarrow r$.

5. The following argument is not logically valid, because it is missing a premise.

"I have \$5.00 to spend for lunch. If the sandwich I want to buy costs \$3.50, then I'll have enough money left over to buy a beverage. If milk costs less than \$1.50, I'll buy it. The price of milk is \$1.00. Therefore, I'll buy milk with my sandwich."

a. Add a premise that would make this argument complete and valid.

b. Can you think of any reasons why I still might not buy milk?

6. The following argument is logically valid. But the conclusion that two equals one is nonsensical. Can you find the mistake in one of the conditionals below?

1. Let $x = 1$.   (Given.)
2. If $x = 1$, then $x - 2 = -1$.   (Subtract 2.)
3. If $x - 2 = -1$, then $x^2 + x - 2 = x^2 - 1$.   (Add x^2.)
4. If $x^2 + x - 2 = x^2 - 1$, then $(x + 2)(x - 1) = (x + 1)(x - 1)$.   (Factor.)
5. If $(x + 2)(x - 1) = (x + 1)(x - 1)$, then $(x + 2) = (x + 1)$.   (Divide by $x - 1$.)
6. If $(x + 2) = (x + 1)$, then $2 = 1$.   (Subtract x.)
7. Therefore, if $x = 1$, then $2 = 1$.

Some Rules of Replacement

The symbol $\equiv$ means "is logically equivalent to." Thus Rule 5 below states that the conditional statement $p \rightarrow q$ is logically equivalent to its contrapositive, $\sim q \rightarrow \sim p$. Rules 6–10 give other logical equivalences. These can be verified by comparing the truth tables of the statements on both sides of the $\equiv$ sign.

5. Contrapositive Rule

$p \rightarrow q \equiv \sim q \rightarrow \sim p$

6. Double Negation

$\sim(\sim p) \equiv p$

7. Commutative Rules

$p \wedge q \equiv q \wedge p$
$p \vee q \equiv q \vee p$

8. Associative Rules

$(p \wedge q) \wedge r \equiv p \wedge (q \wedge r)$
$(p \vee q) \vee r \equiv p \vee (q \vee r)$

9. Distributive Rules

$p \wedge (q \vee r) \equiv (p \wedge q) \vee (p \wedge r)$
$p \vee (q \wedge r) \equiv (p \vee q) \wedge (p \vee r)$

10. DeMorgan's Rules

$\sim(p \wedge q) \equiv \sim p \vee \sim q$
$\sim(p \vee q) \equiv \sim p \wedge \sim q$

Any logically equivalent expressions can replace each other wherever they occur in a proof.

Example   Given: $p \wedge q$; $q \rightarrow \sim(r \vee s)$
                Prove: $\sim r \wedge \sim s$

Proof:

Statements	Reasons
1. $p \wedge q$	1. Given
2. $q \wedge p$	2. Step 1 and Commutative Rule
3. q	3. Step 2 and Simplification
4. $q \rightarrow \sim(r \vee s)$	4. Given
5. $\sim(r \vee s)$	5. Steps 3 and 4 and Modus Ponens
6. $\sim r \wedge \sim s$	6. Step 5 and DeMorgan's Rule

Exercises

Supply the reasons to complete each proof.

1. Given: $a \rightarrow \sim b$; b
Prove: $\sim a$

Statements
1. b
2. $\sim(\sim b)$
3. $a \rightarrow \sim b$
4. $\sim a$

2. Given: $a \vee (b \wedge c)$; $\sim b$
Prove: a

Statements
1. $a \vee (b \wedge c)$
2. $(a \vee b) \wedge (a \vee c)$
3. $a \vee b$
4. $b \vee a$
5. $\sim b$
6. a

Write two-column proofs for the following.

3. Given: $a \wedge (b \wedge c)$
Prove: c

4. Given: $(p \wedge q) \rightarrow s$; $\sim s$
Prove: $\sim p \vee \sim q$

5. Given: $p \vee \sim q$; q
Prove: p

6. Given: $\sim q \rightarrow \sim p$; $q \rightarrow r$; p
Prove: r

7. Given: $p \vee (q \wedge s)$
Prove: $p \vee s$

8. Given: $t \vee (r \vee s)$; $\sim r \wedge \sim s$
Prove: t

Assume the given statements are true, symbolize the statements, and write a two-column proof.

9. If solid X is a cube, then it has twelve edges. If solid X is not a cube, then it does not have all square faces. Solid X has all square faces. Prove that solid X has twelve edges. (Use the letters c, t, and s.)

10. Pat loves me or Jean loves me. Pat sent me a valentine or Pat sent Kevin a valentine. If Pat sent me a valentine, I would have received it by now. If Pat sent Kevin a valentine, then Pat doesn't love me. I have received no valentines.
Prove that Jean loves me. (Use the letters p, j, v, k, r.)

Application of Logic to Circuits

The diagram at the right represents part of an electrical circuit. When switch p is open, the electricity that is flowing from A will not reach B. When switch p is closed, as in the second diagram, the electricity flows through the switch to B.

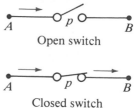

Open switch

Closed switch

The diagram at the left below represents two switches p and q that are *connected in series*. Notice that current will flow if and only if both switches are closed. The diagram at the right below represents the switches p and q *connected in parallel*. Notice that current will flow if either switch is closed or if both switches are closed. If switches p and q are both open, the current cannot flow.

Series circuit Parallel circuit

In order to understand how circuits are related to truth tables, let us do the following:

(1) If a switch is closed, label it T. If it is open, label it F.

(2) If current will flow in a circuit, label the circuit T. If the current will not flow, label the circuit F.

With these agreements, we can use truth tables to show what happens in the series circuit and the parallel circuit illustrated above.

Series circuit

p	q	Circuit
T	T	T
T	F	F
F	T	F
F	F	F

Parallel circuit

p	q	Circuit
T	T	T
T	F	T
F	T	T
F	F	F

The truth table for the series circuit is just like the truth table for $p \wedge q$. Also, the truth table for the parallel circuit is just like the table for $p \vee q$.

Now study the circuit shown at the right. Notice that one of the switches is labeled $\sim q$. This means that this switch is open if switch q is closed, and vice versa. The circuit shown is basically a parallel circuit, but in each branch of the circuit there are two switches connected in series. This explains why the circuit is labeled $(p \wedge q) \vee (p \wedge \sim q)$. A truth table for this circuit is given on the next page.

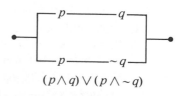

$(p \wedge q) \vee (p \wedge \sim q)$

p	q	$\sim q$	$p \wedge q$	$p \wedge \sim q$	$(p \wedge q) \vee (p \wedge \sim q)$
T	T	F	T	F	T
T	F	T	F	T	T
F	T	F	F	F	F
F	F	T	F	F	F

Notice that the first and last columns of the truth table are identical. This means that the complicated circuit shown can be replaced by a simpler circuit that contains just switch p! In other words, logic can be used to replace a complex electrical circuit by a simpler one.

Exercises

Symbolize each circuit using $\wedge$, $\vee$, $\sim$, and letters given for the switches in each diagram.

1.

2.

3.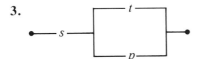

4.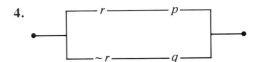

5.

6.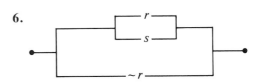

7. Draw a diagram for the circuit $p \wedge \sim p$; also for the circuit $p \vee \sim p$. Electricity can always pass through one of these circuits and can never pass through the other. Which is which?

8. According to the commutative rule, $p \wedge q \equiv q \wedge p$. This means that the circuit $p \wedge q$ does the same thing as the circuit $q \wedge p$. Make a diagram of each circuit.

9. According to the associative rule, $(p \vee q) \vee r \equiv p \vee (q \vee r)$. Draw diagrams for each circuit.

10. The distributive rule says that $p \wedge (q \vee r) \equiv (p \wedge q) \vee (p \wedge r)$. Draw diagrams for each circuit.

11. Make both a diagram and a truth table for the circuit $(p \vee q) \vee \sim q$. Notice that the last column of your table is always T so that current always flows. This means that all of the switches could be eliminated.

12. Make both a diagram and a truth table for the circuit $(p \vee q) \wedge (p \vee \sim q)$. Describe a simpler circuit equivalent to this circuit.

Flow Proofs

Proofs can be written in a variety of forms, including: (1) two-column form, (2) paragraph form, and (3) flow form. In a *flow proof*, a diagram with implication arrows (→) shows the logical flow of the statements of a proof. The statements in the diagram are numbered and the reasons for each are given below the flow diagram.

Example 1   Given: $\angle 1 \cong \angle 2$;
$\qquad\qquad\qquad \angle 3 \cong \angle 4$
$\qquad\qquad$ Prove: $\angle 5 \cong \angle 6$

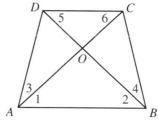

Flow Proof:

1. $\angle 1 \cong \angle 2 \rightarrow$ 2. $\overline{AO} \cong \overline{BO}$
$\qquad\qquad\qquad$ 3. $\angle 3 \cong \angle 4 \qquad\qquad\Big\} \rightarrow$ 5. $\triangle AOD \cong \triangle BOC$
$\qquad\qquad\qquad$ 4. $\angle AOD \cong \angle BOC$

6. $\overline{DO} \cong \overline{CO} \rightarrow$ 7. $\angle 5 \cong \angle 6$

Reasons

1. Given
2. If 2 $\angle$s of a $\triangle$ are $\cong$, the sides opp. them are $\cong$.
3. Given
4. Vertical $\angle$s are $\cong$.
5. ASA Postulate
6. Corr. parts of $\cong$ $\triangle$s are $\cong$.
7. Isosceles $\triangle$ Theorem

Because this flow proof is long, we have drawn an arrow connecting steps 5 and 6 to show that the proof continues below. You can do this or turn your paper sideways to accommodate a long proof.

One advantage of flow proof is that it shows clearly which steps depend on other steps. In the example above, for instance, we see that step 5 (whose justification is ASA) depends on steps 2, 3, and 4, each of which provides one of the three congruences needed for ASA. The next example shows how a complex proof can be understood more easily by organizing it into a flow proof.

Example 2   Given: $\overline{QS} \perp \overline{PR}$;

               $\overrightarrow{PR}$ bisects $\angle QPS$.

       Prove: $\overrightarrow{RP}$ bisects $\angle QRS$.

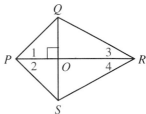

Flow Proof:

1. $\overline{QS} \perp \overline{PR}$ →   3. $\angle POQ \cong \angle POS$ ⎫

2. $\overrightarrow{PR}$ bis. $\angle QPS$ → 4. $\angle 1 \cong \angle 2$ ⎬ → 6. $\triangle POQ \cong \triangle POS$

                 5. $\overline{PO} \cong \overline{PO}$ ⎭

      7. $\overline{PQ} \cong \overline{PS}$ ⎫

      8. $\overline{PR} \cong \overline{PR}$ ⎬ → 10. $\triangle PQR \cong \triangle PSR$ → 11. $\angle 3 \cong \angle 4$ → 12. $\overrightarrow{RP}$ bis. $\angle QRS$.

      9. $\angle 1 \cong \angle 2$ ⎭

Reasons

1. Given
2. Given
3. If two lines are $\perp$, then they form $\cong$ adj. $\angle\!s$.
4. Def. of $\angle$ bisector
5. Reflexive Prop.
6. ASA Postulate
7. Corr. parts of $\cong$ $\triangle$ are $\cong$.
8. Reflexive Prop.
9. See Step 4. (Repeating this makes Step 10 easier to follow.)
10. SAS Postulate
11. Corr. parts of $\cong$ $\triangle$ are $\cong$.
12. Def. of $\angle$ bisector

     When you are working on a flow proof, you may find it helpful to wait until you have completed the structure of the proof before numbering the steps. You might work backwards from the statement you wish to prove, for example, filling in the intermediate steps until you arrive at given statements. Then you can number the steps and give the reasons underneath. Sometimes there will be more than one correct way to do this; just make sure that the number at the head of each arrow is always greater than the number(s) at the tail of the arrow.

Exercises

Write a flow proof for each exercise referred to below.

1. Exercise 3, page 130   **2.** Exercise 16, page 125   **3.** Exercise 15, page 145

4. Exercise 13, page 81   **5.** Exercise 17, page 126   **6.** Exercise 2, page 143

7. Exercise 7, page 149   **8.** Exercise 24, page 82   **9.** Exercise 20, page 145

Handbook for Integrating Coordinate and Transformational Geometry

The purpose of the following sections is to enable students to study coordinate geometry and transformational geometry throughout the year, interspersed with their study of traditional (synthetic) geometry. Teachers will need to present Lessons 13-1 through 13-7 after Chapter 4 and Lessons 14-1 through 14-4 after Chapter 5. (Complete instructions are provided in the Teacher's Edition.) The sections presented here are to follow each of Chapters 3 through 11, and a final section, called "Deciding Which Method to Use in a Problem," provides guidance in selecting the approach that best suits a problem. Each section is intended for use *after* the chapter listed with its title.

Translation and Rotation (Chapter 3)

Objective: Study the effect of the basic transformations of translation and rotation upon polygons.

You may have noticed from putting a jigsaw puzzle together that the size and shape of a piece do not change if you slide or spin it around on the table. The same is true of polygons in the plane. Consider the pattern of identical triangles shown below. A pattern of identical shapes that fills the plane in this way is called a *tiling*, or *tessellation*.

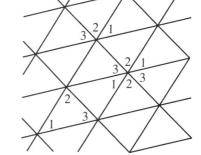

You could create this pattern by first sliding a triangle and leaving copies of it across the plane. If you then put your finger on a vertex of one triangle and spin the triangle 180 degrees, the result is an "upside-down" triangle that fills in the rest of the pattern. Notice that the angle measures of the triangles are not altered by these movements, so we can number the angles in the diagram.

The movements of sliding and spinning are known in geometry as *translations* and *rotations*. They are examples of *transformations* that preserve length and angle measure.

Example   a. In the diagram above, find three numbered angles whose measures add up to a straight angle.

b. What does this tell you about the sum of the measures of the angles of a triangle?

c. There are 3 sets of lines that *look* parallel. What theorem or postulate tells you that they *must be* parallel?

Solution   **a.** Every straight angle is made up of one ∠1, one ∠2, and one ∠3.

b. Since each triangle in the diagram is also made up of one ∠1, one ∠2, and one ∠3, the sum of the measures of a triangle must equal the measure of a straight angle, or 180.

c. Postulate 11

Exercises

1. The repetitive pattern at the right is a tiling of the plane with identical quadrilaterals (four-sided polygons). What does it tell you about the sum of the angle measures in a quadrilateral? Explain.

2. In the tessellation shown in Exercise 1, it is possible to slide, or glide, the lower left quadrilateral so it fits exactly on top of the lower right quadrilateral. Is it possible to glide the lower left quadrilateral so it fits exactly on any of the other quadrilaterals?

3. The figure shows a tiling of the plane by parallelograms, which are quadrilaterals having parallel opposite sides.
   a. Copy the diagram and mark all angles that must be congruent to ∠1 if you reason from Postulate 10.
   b. Mark any additional angles on the diagram that are also congruent to ∠1, and tell why they are congruent to ∠1.
   c. This exercise suggests a theorem about the opposite angles of a parallelogram. State this theorem.

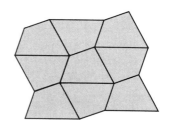

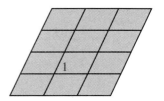

4. Draw a triangle *ABC* and place a pencil at *A* as shown in Figure 1. Glide the pencil along $\overline{AB}$ until the eraser is at *B* and then rotate it about *B* as shown in Figure 2. Now glide it to *C* and rotate it about *C* (Figure 3). Finally glide it back to *A* and rotate it about *A* (Figure 4).
   a. Through how many degrees has your pencil rotated?
   b. What theorem is suggested by this exercise?

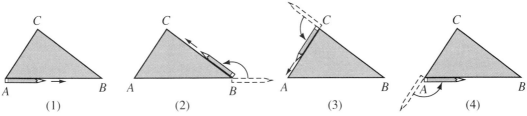

(1)   (2)   (3)   (4)

5. Suppose you follow these instructions: Walk 1 m north, then walk 1 m northeast. As the diagram shows, you turn through a 45° angle.
   a. Copy the diagram and continue according to these additional instructions: Walk 1 m east, 1 m southeast, 1 m south, 1 m southwest, 1 m west, and 1 m northwest.
   b. What is the total number of degrees through which you turned?

6. Create another closed path as in Exercise 5, but this time use a variety of different distances. What is the total number of degrees through which you turned?

7. Create a third closed path, but this time vary the angles at which you turn. What is the total number of degrees through which you turned? What theorem does this exercise suggest?

Reflection and Symmetry (Chapter 4)

Objective: Study the concepts of reflection and symmetry by using paper cutouts.

Flipping a jigsaw puzzle piece over exposes the cardboard back and makes it unusable in a puzzle. In geometry, however, the front and back of a polygon share many important features. We say that the original and the flipped versions are *congruent*. The two figures are called *mirror images* of each other because each looks like the other's reflection in a mirror. In fact, this type of transformation is called a *reflection*.

You can investigate reflections by folding graph paper along one of its lines and cutting along a path that begins on the fold line and ends somewhere else on the fold line. The resulting cutouts will have *mirror symmetry* across the fold line. The exercises will help you develop your skills in visualizing mirror symmetry.

Exercises

For each figure, (a) sketch on graph paper the figure you would obtain if you were to cut out the figure along the red lines shown and open it up, and (b) test your predictions by actually cutting out the shapes.

1.   **2.**   **3.**   **4.**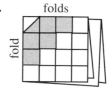

5. a. Draw a line segment $\overline{AB}$ on a sheet of paper and then fold the paper so that A lies on top of B.

    b. Cut the paper so that when it is opened an isosceles triangle is formed with $\overline{AB}$ as its base.

    c. What does your cut-out triangle tell you about the angles at A and B? Explain.

6. Lines l and m are perpendicular at O. Figure I is reflected across line l to figure II and figure II is reflected across line m to figure III.

    a. What single transformation moves figure I directly to figure III?

    b. Join the eye of figure I to the eye of figure III. How is O related to these two points?

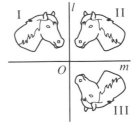

7. Draw two parallel lines l and m and figure I as shown. Then reflect figure I in line l to figure II and reflect figure II in line m to figure III.

    a. What single transformation moves figure I directly to figure III?

    b. How does the distance between corresponding points in figures I and III compare with the distance between l and m?

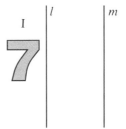

8. If you were to fan fold a piece of paper as shown and then cut out the figure shown, can you predict what you would get if you then opened the paper? Try it.

★ **9.** Can you modify the design cut out in Exercise 8 to create four figures all facing in the same direction?

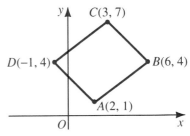

Quadrilaterals (Chapter 5)

Objective: Investigate the properties of various quadrilaterals and classify them using coordinate methods. (Requires understanding of Lessons 13-1 through 13-7.)

At the bottom of page 172, five ways are shown for proving that a quadrilateral is a parallelogram. At this stage of your learning, coordinate geometry methods can be used for all but the fourth of these methods.

Example   Prove that quad. $ABCD$ is a $\square$.

Solution

Method 1   *Show that both pairs of opposite sides are congruent.*

$$AB = \sqrt{(6-2)^2 + (4-1)^2} = \sqrt{4^2 + 3^2} = \sqrt{25} = 5$$
$$DC = \sqrt{(3-(-1))^2 + (7-4)^2} = \sqrt{4^2 + 3^2} = \sqrt{25} = 5$$
$$AD = \sqrt{(-1-2)^2 + (4-1)^2} = \sqrt{(-3)^2 + 3^2} = \sqrt{18} = 3\sqrt{2}$$
$$BC = \sqrt{(3-6)^2 + (7-4)^2} = \sqrt{(-3)^2 + 3^2} = \sqrt{18} = 3\sqrt{2}$$

Method 2   *Show that both pairs of opposite sides are parallel.*

Slope of $\overline{AB} = \dfrac{4-1}{6-2} = \dfrac{3}{4}$;

slope of $\overline{DC} = \dfrac{7-4}{3-(-1)} = \dfrac{3}{4}$.

Slope of $\overline{AD} = \dfrac{4-1}{-1-2} = \dfrac{3}{-3} = -1$;

slope of $\overline{BC} = \dfrac{7-4}{3-6} = \dfrac{3}{-3} = -1$.

Method 3   *Show that the diagonals bisect each other.*

Midpoint of $\overline{AC} = \left(\dfrac{3+2}{2}, \dfrac{1+7}{2}\right) = \left(\dfrac{5}{2}, 4\right)$;

midpoint of $\overline{BD} = \left(\dfrac{-1+6}{2}, \dfrac{4+4}{2}\right) = \left(\dfrac{5}{2}, 4\right)$.

Since the diagonals have the same midpoint, they bisect each other.

You could also use the work in Methods 1 and 2 to show that one pair of opposite sides is both congruent and parallel.

Exercises

The coordinates of the vertices of quadrilateral *ABCD* are given. Show that *ABCD* is a parallelogram by using each of the three methods shown in the example.

1. $A(5, 7)$, $B(0, 3)$, $C(1, -3)$, $D(6, 1)$ **2.** $A(-2, 6)$, $B(-3, 2)$, $C(2, -4)$, $D(3, 0)$

Decide whether quadrilateral *DEFG* is a parallelogram.

3. $D(3, 5)$, $E(5, 7)$, $F(3, 4)$, $G(0, 1)$ **4.** $D(3, -2)$, $E(-2, 5)$, $F(5, 6)$, $G(10, -1)$

The coordinates of three vertices of $\square PQRS$ are given. Find the coordinates of the missing vertex.

5. $P(0, 0)$, $Q(5, 2)$, $R(8, 4)$ **6.** $P(-2, 0)$, $Q(2, 1)$, $S(0, 5)$

7. a. Draw the triangle with vertices $O(0, 0)$, $I(4, 2)$, and $J(2, 6)$.
  b. Find the coordinates of M and N, the midpoints of $\overline{OJ}$ and $\overline{IJ}$, respectively.
  c. Find the slopes of $\overline{MN}$ and $\overline{OI}$. What do your results tell you about $\overline{MN}$ and $\overline{OI}$? What kind of quadrilateral is *OMNI*?

8. Repeat Exercise 7 for the general triangle with vertices $O(0, 0)$, $I(a, 0)$, and $J(b, c)$.

9. Given the quadrilateral *ABCD* with vertices $A(-4, 5)$, $B(4, -1)$, $C(7, 3)$, and $D(-1, 9)$.
  a. Use slopes to show that opposite sides are parallel and adjacent sides are perpendicular.
  b. What kind of quadrilateral is *ABCD*?

10. Given the quadrilateral *RSTU* with vertices $R(5, -3)$, $S(9, 0)$, $T(3, 8)$, and $U(-1, 5)$.

   a. Show that *RSTU* is a rectangle.
   b. Use the distance formula to verify that the diagonals are congruent.

11. Given the quad. *DEFG* with vertices $D(-4, 1)$, $E(2, 3)$, $F(4, 9)$, and $G(-2, 7)$.

   a. Use the distance formula to show that *DEFG* is a rhombus.
   b. Use slopes to verify that the diagonals are perpendicular.

12. Suppose two congruent trapezoids glide together as shown. Explain how you can deduce the length of the median, shown in red, of a trapezoid.

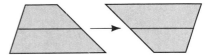

13–34. Work Exercises 28 and 41 on page 527; Exercises 3–6 and 11–18 on pages 537–538; Classroom Exercise 3 on page 541 and Exercises 26, 33 on pages 542–543; Exercises 15, 18, 19, and 21 on page 546; and Exercise 34 on page 555.

Minimal Paths (Chapter 6)

Objective: Solve "shortest distance" problems using translations and coordinate geometry. (Requires understanding of Lessons 13-1 through 13-7 and 14-1 through 14-4.)

The Application found on page 224 shows how to find the shortest path from point *A* to line *l* to point *B*. The solution is found by using a reflection of *B* in line *l*. Example 1 shows a different kind of shortest-path problem. It is solved by another kind of transformation: a *translation*.

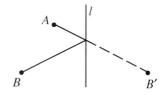

Example 1   Where should a bridge perpendicular to two parallel river banks be built if the total distance from *A* to *B*, including the distance across the bridge, is to be minimum?

Solution   Translate *B* toward the river a distance equal to the width of the river. Draw $\overline{AB'}$ and build the bridge at the point *X* where $\overline{AB'}$ intersects the river on *A*'s side.

Here is why this method works: We want to minimize $AX + XY + YB$, but since XY is fixed, we need to minimize $AX + YB$, which equals $AX + XB'$ because $XYBB'$ is a parallelogram. This sum is minimum when *X* is on $\overline{AB'}$. In effect, translating *B* to B' "sews up" the gap of the river.

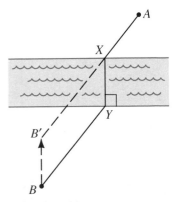

Example 2   Find the shortest distance from $P(0, 5)$ to the line l with equation $y = 2x$.

Solution   The shortest segment is the perpendicular $\overline{PQ}$.
Its length can be found in three steps.

Step 1   Find the equation of $\overleftrightarrow{PQ}$.

Slope of line $y = 2x$ is 2.

Then the slope of $\overleftrightarrow{PQ}$ is $-\dfrac{1}{2}$.

The equation of $\overleftrightarrow{PQ}$ is $y = -\dfrac{1}{2}x + 5$.

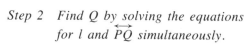

*Step 2   Find Q by solving the equations
for l and $\overleftrightarrow{PQ}$ simultaneously.*

$y = 2x$ and $y = -\dfrac{1}{2}x + 5$

$2x = -\dfrac{1}{2}x + 5$

$\dfrac{5}{2}x = 5$

$x = 2$

If $x = 2$, then $y = 2x = 4$. Thus Q is $(2, 4)$.

Step 3   Find PQ by using the distance formula.

$$PQ = \sqrt{(0 - 2)^2 + (5 - 4)^2} = \sqrt{5}$$

Exercises

**In Exercises 1–5 assume that each bridge must be perpendicular to the two
river banks it joins.**

1. Copy the figure shown and find the location of a bridge across the river that will allow the path from A to B to be minimum.

⋆ **2.** Two bridges are to be built over the parallel rivers shown. Find where they should be built if the total distance from C to D, including the distances across the bridges, is to be minimum.

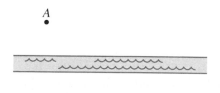

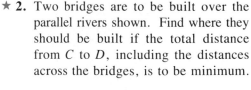

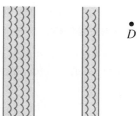

3. A river flows between the lines $x = 3$ and $x = 4$. Where should a bridge be constructed to minimize the path from $O(0, 0)$ to $P(5, 4)$?

4. A river flows between the lines $y = x$ and $y = x + 2$. Where should a bridge be constructed to minimize the path from $Q(0, 6)$ to $R(8, 5)$?

★ **5.** Two bridges are to be built over the river forks shown. Find where they should be built if the total distance from E to F via the island, including the distances across the bridges, is to be minimum. What do you notice about the three non-bridge portions of your path?

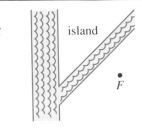

Find the shortest distance from the given point to the line whose equation is given.

6. $N(0, 10)$; $y = \frac{1}{3}x$ **7.** $O(0, 0)$; $y = 2x + 5$ ★ **8.** $P(2, -1)$; $y = \frac{2}{3}x + 2$

Dilations and Similarity (Chapter 7)

Objective: Understand the close relationship between dilation transformations and the similar figures they produce. (Requires understanding of Lessons 13-6, 13-7, and 14-1 through 14-5.)

If a transformation maps a figure to a similar figure, it is called a similarity mapping. Every similarity mapping can be broken into two components: (1) a dilation and (2) a congruence mapping.

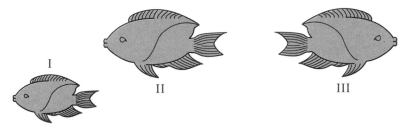

In the figure, a dilation maps fish I to the similar fish II, and a reflection maps fish II to a congruent fish III. If you perform these two mappings consecutively, the result is a similarity mapping that maps fish I to fish III.

Exercises

Give the scale factor for the dilation that maps figure I to figure II, and tell what kind of congruence mapping maps figure II to figure III.

1.   **2.**

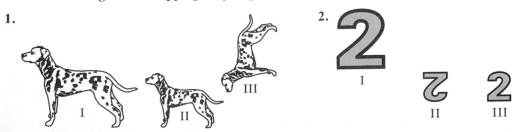

3. $D_{O, 2}$ maps $\triangle RST$ to $\triangle R'S'T'$. How are $\overrightarrow{RS}$ and $\overrightarrow{R'S'}$ related?

4 $D_{O, -3}$ maps $\triangle JKL$ to $\triangle J'K'L'$. How are $\overrightarrow{JK}$ and $\overrightarrow{J'K'}$ related?

5. A dilation maps square I to square II, and another dilation maps square II to square III. If the areas of these three squares are 25, 100, and 900, respectively, find the scale factor of each dilation.

6. The square $WXYZ$ has three vertices on the sides of $\triangle ABC$. How can you find a square with all four vertices on $\triangle ABC$? (*Hint:* Consider a dilation.)

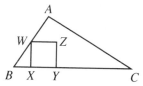

7. O is the origin and the dilation $D_{O, 3}$ maps line l to line l'. If the equation of l is $y = 2x + 1$, what is the equation of l'?

8. **a.** Draw a figure like the one shown, in which the dilation $D_{P, 2}$ maps $\triangle ABC$ to $\triangle A'B'C'$ and the dilation $D_{Q, 2}$ maps $\triangle A'B'C'$ to $\triangle A''B''C''$.

   b. There is a single dilation that will map $\triangle ABC$ directly to $\triangle A''B''C''$. What is its scale factor?

   c. Draw lines $\overleftrightarrow{AA''}$, $\overleftrightarrow{BB''}$, and $\overleftrightarrow{CC''}$ to locate the center of the single dilation. How is this point related to points P and Q?

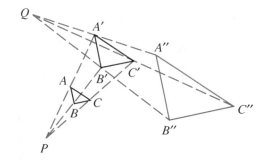

9–10. Work Exercise 29 on page 527 and Classroom Exercise 8 on page 532.

Right Triangles (Chapter 8)

Objective: Explore the properties of right triangles and other special triangles by using coordinate, vector, and transformational approaches. (Requires understanding of Lessons 13-1 through 13-7 and 14-1 through 14-5.)

The example below shows how to use coordinates, the distance formula, and Theorems 8-3, 8-4, and 8-5 to tell whether a triangle is acute, right, or obtuse.

Example   Is $\triangle OPQ$ acute, right, or obtuse?

Solution   First find the lengths of the sides of the triangle by using the distance formula.

$OP = \sqrt{6^2 + 3^2} = \sqrt{45}$

$OQ = \sqrt{4^2 + 5^2} = \sqrt{41}$

$PQ = \sqrt{(6 - 4)^2 + (3 - 5)^2} = \sqrt{8}$

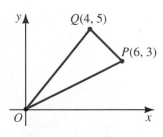

(*Solution continued on next page.*)

Since $\overline{OP}$ is the longest side of the triangle, we compare as follows:

$(OP)^2 \underline{\quad ? \quad} (OQ)^2 + (PQ)^2$

$(\sqrt{45})^2 \underline{\quad ? \quad} (\sqrt{41})^2 + (\sqrt{8})^2$

$\quad 45 \underline{\quad ? \quad} 41 + 8$

$\quad 45 \; < \; 49$

Thus the triangle is acute by Theorem 8-4.

Exercises

The vertices of a triangle are given. Decide whether the triangle is acute, right, or obtuse.

1. $O(0, 0)$, $P(8, 4)$, $Q(6, 8)$

2. $O(0, 0)$, $R(-1, 5)$, $S(-6, 0)$

3. $A(1, 1)$, $B(3, -2)$, $C(8, 1)$

4. $D(-5, 0)$, $E(0, 7)$, $F(3, -3)$

5. Given vertices $A(3, 0)$, $B(5, 3)$, $C(-1, 7)$. Show that $\triangle ABC$ is a right triangle
    a. by using slopes.   **b.** by using Theorem 8-3.

6. In Exercise 5, $\triangle ABC$ has a right angle at B.
    a. Find the coordinates of M, the midpoint of the hypotenuse.
    b. Use the distance formula to show that $MB = \frac{1}{2}AC$.
    c. What theorem does part (b) illustrate?

7. Describe how to map $\triangle PNQ$ to $\triangle QNR$ with a rotation followed by a dilation. Give the scale factor of the dilation.

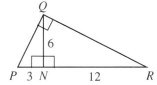

8. a. Draw $\triangle OPQ$ where O is the origin, $\overrightarrow{OP} = (4, 3)$ and $\overrightarrow{PQ} = (-2, 11)$.
    b. Find $\overrightarrow{OQ}$.
    c. Find $|\overrightarrow{OP}|$, $|\overrightarrow{PQ}|$, and $|\overrightarrow{OQ}|$.
    d. Is $\triangle OPQ$ a right triangle? Explain.

9. Four right triangles and a small square are arranged to form a larger square as shown at the left below. What is the area of the large square? If two of the triangles are rotated about P and Q as shown, we get the figure at the right below. Show that this figure can be considered as the sum of two squares. Find the dimensions of the two squares. What theorem does this exercise suggest?

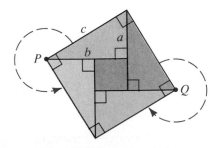

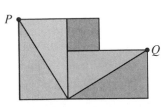

10. The inner square in the diagram is formed by connecting the midpoints of the outer square. It is possible to map the outer square to the inner square by performing a rotation followed by a dilation.

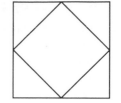

   a. Give the center of the rotation and the amount of the rotation.
   b. Give the center of the dilation and the scale factor.

11–21. Work Exercises 32, 42 on page 527; Classroom Exercise 7 on page 532; Exercises 28–31, 33 on pages 533–534; Classroom Exercise 11 on page 536; Exercise 23 on page 538; and Classroom Exercise 5 on page 541.

Circles (Chapter 9)

Objective: Write the equations of tangent lines, and make observations about circles and their symmetry. (Requires understanding of Lessons 13-1 through 13-7 and 14-1 through 14-5.)

Many relationships among circles and their chords and tangents can be investigated by using coordinates and transformations. Before studying the example below, you should understand how the equation of a circle is used in Examples 3 and 4 of Lesson 13-1.

Example
   a. Sketch the circle $x^2 + y^2 = 10$ and the line $y = 3x + 10$.
   b. Solve the two equations simultaneously and show that there is just one solution for x.
   c. Find the corresponding value for y. Label the solution point, T, on your sketch. What does this tell you about the line and the circle?
   d. Use the slopes to show that $\overline{OT}$ is perpendicular to the line $y = 3x + 10$.

Solution
   a. The circle has center O and radius $\sqrt{10}$.

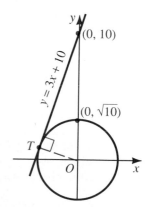

   b.
   $$x^2 + y^2 = 10$$
   $$x^2 + (3x + 10)^2 = 10$$
   $$x^2 + 9x^2 + 60x + 100 = 10$$
   $$10x^2 + 60x + 90 = 0$$
   $$x^2 + 6x + 9 = 0$$
   $$(x + 3)(x + 3) = 0$$
   $$x + 3 = 0$$
   $$x = -3$$

   c. $y = 3x + 10$; $y = 3(-3) + 10$; $y = 1$. Point T is $(-3, 1)$. The line is tangent to the circle at point T.

   d. The slope of $\overline{OT}$ is $\dfrac{1 - 0}{-3 - 0} = -\dfrac{1}{3}$. The slope of $y = 3x + 10$ is 3. Since the slopes are negative reciprocals, the lines are perpendicular.

Exercises

1. a. What is the equation of a circle with center $(6, 0)$ and radius 5?
  b. Is the point $Q(2, 3)$ on the circle?
  c. Plot $P(12, 8)$. Is $\overline{PQ}$ tangent to the circle?

In Exercises 2–4, (a) verify that points A and B lie on circle O, (b) make a sketch and find M, the midpoint of $\overline{AB}$, and (c) use slopes to verify that $\overline{OM} \perp \overline{AB}$.

2. Circle O has radius 5. The points are $A(0, 5)$ and $B(4, 3)$.

3. Circle O has radius 10. The points are $A(6, 8)$ and $B(-8, 6)$.

4. Circle O has radius $5\sqrt{2}$. The points are $A(5, 5)$ and $B(-7, 1)$.

5. Sketch the circles $x^2 + y^2 = 225$ and $(x - 6)^2 + (y - 8)^2 = 25$ and explain why the circles must be internally tangent. (*Hint:* Find the two radii and the distance between the centers of the circles.)

6. a. Sketch the circle $x^2 + y^2 = 25$ and the line $y = 2x - 5$.
  b. Solve the two equations simultaneously by substituting $2x - 5$ for y in the equation $x^2 + y^2 = 25$. Solve the resulting quadratic equation by factoring. For each value of x, find the corresponding value of y by substituting into the equation $y = 2x - 5$.
  c. Your two solutions in part (b) correspond to two points on the circle. Show them on your sketch.

7. $\overleftrightarrow{PA}$ is tangent to circle O at A.
  a. If the figure shown is reflected in $\overleftrightarrow{PO}$, what is the image of circle O? of $\overline{PA}$?
  b. Since a reflection is an isometry, what do you know about $\overline{PA}$ and its image?
  c. State the corollary that part (b) proves.

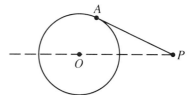

8. Circles P and Q intersect at A and B.
  a. What is the image of A when reflected in $\overleftrightarrow{PQ}$?
  b. What does part (a) tell you about $\overline{AB}$ and $\overleftrightarrow{PQ}$?
  c. Sketch the image of $\overleftrightarrow{XY}$ when reflected in $\overleftrightarrow{PQ}$.
  d. What can you deduce from part (c) about the common external tangents of two circles?

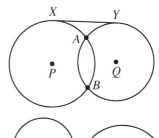

★ **9.** Find an equilateral triangle ABC with vertex B on $\odot P$ and vertex C on $\odot Q$. (*Hint:* Rotate $\odot P$ 60° about A. Its image will intersect $\odot Q$ in two points. Either of these points can be the desired vertex C. How do you find B?)

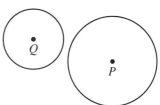

10–18. Work Exercises 19, 20 on page 538; Exercises 36–39 on page 552; Classroom Exercise 14 on page 554 and Exercise 35 on page 556; and Classroom Exercise 13 on page 595.

Constructions **(Chapter 10)**

Objective: Use transformations with concurrency and construction problems. (Requires understanding of Lessons 14-1 through 14-5.)

Concurrency Theorems 10-1 and 10-2 are proved by synthetic methods, while the proofs of concurrency Theorems 10-3 and 10-4 are delayed until coordinate methods can be used. However, all four theorems can be proved with coordinates or without. The following theorem states that three of the four concurrency points are collinear. A proof of this theorem with coordinates is quite involved. (See Exercises 11–14 of Lesson 13-9.) However, the transformational proof below is much shorter and more elegant.

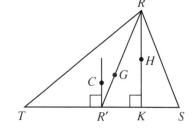

Theorem: The orthocenter, centroid, and circumcenter of a triangle are collinear. (The line on which they lie is called *Euler's line*.)

Given: $\triangle RST$ with orthocenter H, centroid G, and circumcenter C.

Prove: H, G, and C are collinear.

Proof: Consider the dilation $D_{G,\,-\frac{1}{2}}$.

Because G divides each median in a $2:1$ ratio, this dilation maps R to R', the midpoint of $\overline{TS}$. Also this dilation maps altitude $\overline{RK}$ to a parallel line through R'. But since $\overline{RK}$ is perpendicular to $\overline{TS}$, the image of $\overline{RK}$ must be perpendicular to $\overline{TS}$. Thus the image of altitude $\overline{RK}$ is the perpendicular bisector of $\overline{TS}$.

We have just proved that $D_{G,\,-\frac{1}{2}}$ maps an altitude to a perpendicular bisector. Similar reasoning shows that the dilation also maps the other two altitudes to perpendicular bisectors. Therefore the dilation maps the orthocenter H, which is on all three altitudes, to the circumcenter C, which is on all three perpendicular bisectors. And since the dilation $D_{G,\,-\frac{1}{2}}$ also maps G to itself, then H, G, and C must be collinear by the definition of a dilation.

Exercises

Carefully draw a *large* △RST and construct its centroid G, its ortho-center H, and its circumcenter C.

1. G, H, and C should be collinear. Are they?

2. Measure the lengths of $\overline{HG}$ and $\overline{GC}$ and find the ratio $HG:GC$. Does your result agree with what you would expect if you were using the dilation $D_{G, -\frac{1}{2}}$?

3. Draw △R'S'T', the image of △RST by the dilation $D_{G, -\frac{1}{2}}$. What is the ratio of the areas of these two triangles?

4. Locate Q, the midpoint of $\overline{HC}$. Put the point of your compass on Q and draw a circle through R'. This is the famous *nine-point circle*. See page 414.

5. Draw a sketch that shows the locus of points in the coordinate plane whose distance to the x-axis and distance to the y-axis have a sum of 10.

6. Given line l, ⊙P and point M, construct a segment $\overline{XY}$ with X on ⊙P, Y on l, and M as the midpoint of $\overline{XY}$. (*Hint:* Consider a half-turn.) You may find two answers.

7. Construct an isosceles right △ABC with B on ⊙P, C on line l, and right angle at A. (*Hint:* Consider a rotation. What will be the magnitude and center of the rotation?) You may find two answers.

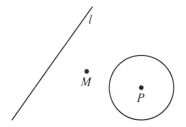
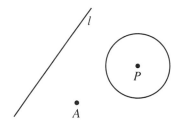

8–16. Work Exercises 21, 22 on page 580; Exercises 18, 20, 21 on page 587; Exercises 31, 38, 39 on pages 591–592; and Exercise 24 on page 597.

Areas (Chapter 11)

Objective: Calculate complex areas by dividing them into portions and using transformations to rearrange the pieces into simpler shapes. (Requires understanding of Lessons 14-1 through 14-5.)

The proof of Theorem 11-2 uses a translation of a triangular portion of a parallelogram to show that a parallelogram and a rectangle have the same area if they have the same base and height. Other transformations, such as rotation, can be used to find areas that would be more difficult without transformations.

An unfamiliar figure can sometimes be divided into pieces that can be rearranged to form a familiar figure whose area is easier to calculate. This method is called *dissection*.

Example   Find the area of the "goblet" that is constructed in a square by drawing one semicircle and two quarter circles, as shown in the diagram at the left below.

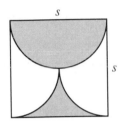

Solution   Divide the bottom of the goblet into halves and rotate them upward as shown in the diagram at the right above. Thus the shaded area is $\frac{1}{2}s^2$.

Exercises

Each figure below is drawn with arcs of radius 4. Use rotations to find the area of each.

1.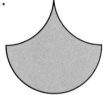

2.

3.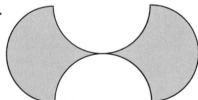

4. The figure shows a series of squares inscribed within each other. Transformation T, which maps region I to region II is achieved by performing a rotation followed by a dilation.

   a. Give the number of degrees in the rotation and the scale factor of the dilation.

   b. What is the image of region I by the transformation T^2 (T performed twice)? by the transformation T^3?

   c. Give the areas of regions I, II, III, IV and V.

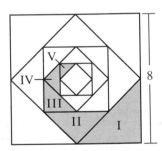

5. The figure shows a series of equilateral triangles and circles inscribed within each other. Transformation T, which maps region I to region II, is the result of performing a rotation followed by a dilation.

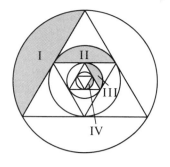

a. Give the number of degrees in the rotation and the scale factor of the dilation.

b. What is the image of region I by the transformation T? T^2? T^3?

c. If region I has area 1, give the areas of regions II, III, and IV.

6. E, F, G, and H are midpoints of the sides of square $ABCD$. If the area of square $PQRS$ is 1, what is the area of square $ABCD$? (*Hint:* Rotate $\triangle APG$ 180° about G. Rotate in a similar manner $\triangle BQF$, CRE, and DSH.)

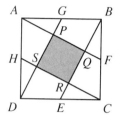

★ 7. Exercise 32 on page 460 is difficult to prove synthetically. Recall that points P, Q, and R divide the sides of $\triangle ABC$ into $2:1$ ratios. Follow the strategy of Exercise 6 above to prove that the area of equilateral $\triangle ABC$ is seven times the area of equilateral $\triangle XYZ$. (*Hint:* Draw an auxiliary line from X to the midpoint, M, of $\overline{CB}$. Rotate $\triangle CXM$ 180° about M. Repeat for the other two sides of $\triangle ABC$.)

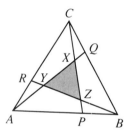

8–10. Work Exercise 31 on page 527, and Exercises 34, 35 on pages 551–552.

Deciding Which Method to Use in a Problem (Chapter 13)

Objective: Learn to recognize clues that indicate whether a coordinate, transformational, or synthetic approach is most suitable for a problem. (Requires completion of Chapters 13 and 14)

When facing a geometry problem, how do you know when to use a coordinate, transformational, or traditional (usually called *synthetic*) approach? Many problems can be solved in more than one way, but often one approach is simpler than another. Here are some tips to help you decide which method may be most suitable for a given problem.

Synthetic Approach

Usually a synthetic approach is best when at least one of the following is true.

1. The lengths of the sides of a figure are given, rather than the coordinates of its vertices.

2. Angle measures other than 90 are given or asked for in the problem. (If lines are parallel or form right angles, however, then a coordinate approach using slopes may be appropriate.)

3. The given information or diagram involves transversals and corresponding angles; congruent lengths, angles, or figures; the areas of similar figures; or the volumes of solids.

Coordinate Approach

1. Usually a coordinate approach is easiest when the problem uses coordinates to name points.
   a. To calculate lengths, use the distance formula.
   b. To locate midpoints, use the midpoint formula.
   c. To show lines are parallel or perpendicular, use the slopes of the lines.
   d. To prove that lines are concurrent, show that their equations have a common solution.

2. Even if a problem does not use coordinates to name points, you can place the coordinate axes on the figure and assign coordinates to the vertices as was shown in Lessons 13-8 and 13-9.
   a. If the figures involved are symmetric, place the axes so that one of them is a line of symmetry. Such a placement reduces the number of variables needed to describe the vertices.
   b. If the figures involved are not symmetric, place the axes so that as many vertices and edges of the figures lie on the x- and y-axes as possible. Distance calculations are simplified whenever zeros appear in the coordinate pairs.

Transformational Approach

1. If the figure has line symmetry, try using a reflection.

2. If the figure has rotational symmetry, try using a rotation.

3. If there are congruent figures placed some distance apart, try using a translation, rotation, glide reflection or a composite of any number of such congruence mappings to map one figure onto the other.

4. If the problem involves similar figures, look for a center of a dilation that would map one figure onto the other. You may first have to rotate one figure so that corresponding sides of similar figures are parallel.

5. If the problem involves calculating the area of an unfamiliar figure, try dissecting the figure and moving the pieces around by transformations until the result is a figure whose area is easy to calculate.

6. Some construction problems can be solved by using reflections (constructing perpendicular bisectors) or by using rotations of angles you can construct, such as 45°, 60°, 90°, or 180°.

Exercises

For Exercises 1–4 complete proofs using the strategies listed, then state which approach is easiest for you.

1. Given: $\overline{AC}$ is the perpendicular bisector of $\overline{BD}$.
   Prove: $AD = AB$ and $CD = CB$.
   a. Use a synthetic proof.
   b. Use a reflection.
   c. Use a coordinate proof.

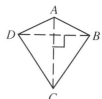

2. Given: O is the midpoint of $\overline{AA'}$, $\overline{BB'}$, and $\overline{CC'}$.
   Prove: $\triangle ABC \cong \triangle A'B'C'$
   a. Use a synthetic proof.
   b. Use a 180° rotation.

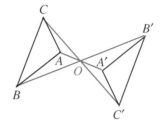

3. Equilateral triangles ABX and BCY are constructed on two sides of $\triangle ABC$ as shown. Prove that $AY = XC$.
   a. Use a synthetic proof.
   b. Find the image of $\overline{AY}$ under the rotation $\mathscr{R}_{B, -60}$.

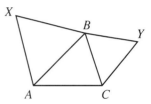

4. A', B', and C' are the midpoints of the sides of $\triangle ABC$. Find the ratio of the areas of $\triangle ABC$ and $\triangle A'B'C'$.
   a. Use a synthetic argument.
   b. Use a dilation to map $\triangle ABC$ to $\triangle A'B'C'$. Where is the center of the dilation? What is the scale factor?

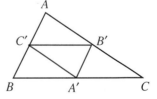

For Exercises 5–20 choose the approach that you feel is best suited to each problem.

5. What kind of figure do you get if you join the midpoints of successive sides of a square? Prove your conjecture.

6. What kind of figure do you get if you join the midpoints of successive sides of a rhombus? Prove your conjecture.

7. The successive midpoints of the sides of quadrilateral $ABCD$ are P, Q, R, and S. Prove that $\overline{PR}$ and $\overline{QS}$ bisect each other.

8. You are given the points $A(-4, 1)$, $B(2, 3)$, $C(4, 9)$, and $D(-2, 7)$.
   a. Show that $ABCD$ is a parallelogram with perpendicular diagonals.
   b. What special name is given to $ABCD$?

9. P is an arbitrary point inside rectangle $ABCD$. Prove the following:
$(PA)^2 + (PC)^2 = (PB)^2 + (PD)^2$

10. Find the area of a quadrilateral with vertices $A(-1, -1)$, $B(9, 4)$, $C(20, 6)$, and $D(10, 1)$.

11. $ABCD$ is a quadrilateral such that if sides $\overline{AB}$ and $\overline{CD}$ are extended they will meet at a right angle. Prove that $(AC)^2 + (BD)^2 = (AD)^2 + (BC)^2$.

12. A ray of light is reflected successively in two perpendicular mirrors. Prove that the final ray is parallel to the initial ray.

★ 13. Find a point X on l and a point Y on m so that $\overline{XY}$ is parallel and congruent to $\overline{AB}$.

★ 14. Find a point X on the triangle and a point Y on line l so that M is the midpoint of $\overline{XY}$.

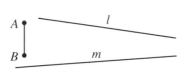

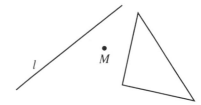

★ 15. Construct three parallel lines so that the middle line is not equidistant from the other two. Then construct an equilateral triangle with one vertex on each of the three lines.

16. Find the length of the tangent line segment from $(9, 13)$ to the circle $x^2 + y^2 = 25$.

17. Given: $\overline{MN}$ is the median of trapezoid $ABCD$;
$\overline{MN}$ intersects the diagonals at X and Y.
Prove: $XY = \frac{1}{2}(AB - DC)$

18. Given: $ABCD$ is a parallelogram whose diagonals meet at O; P and Q are midpoints of $\overline{AB}$ and $\overline{CD}$; R and S are midpoints of $\overline{AO}$ and $\overline{CO}$.
Prove: $PRQS$ is a parallelogram.

★ 19. Graph the circles whose equations are $(x + 9)^2 + (y - 5)^2 = 25$ and $(x - 2)^2 + (y - 4)^2 = 25$. Find a point X on one circle and a point Y on the other such that the y-axis is the perpendicular bisector of $\overline{XY}$.

★ 20. A and B are fixed points 12 units apart. P is an arbitrary point. Once P is chosen, find points X and Y such that $PA = AX$, $\overline{PA} \perp \overline{AX}$, $PB = BY$, and $\overline{PB} \perp \overline{BY}$ as shown. Now locate M, the midpoint of $\overline{XY}$. Show that the position of M does not depend on the position of P.

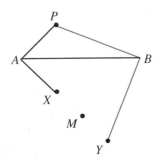

Postulates

Postulate 1   (**Ruler Postulate**)
1. The points on a line can be paired with the real numbers in such a way that any two points can have coordinates 0 and 1.
2. Once a coordinate system has been chosen in this way, the distance between any two points equals the absolute value of the difference of their coordinates.   (p. 12)

Postulate 2   (**Segment Addition Postulate**)   If B is between A and C, then
$$AB + BC = AC.$$ (p. 12)

Postulate 3   (**Protractor Postulate**)   On $\overleftrightarrow{AB}$ in a given plane, choose any point O between A and B. Consider $\overrightarrow{OA}$ and $\overrightarrow{OB}$ and all the rays that can be drawn from O on one side of $\overleftrightarrow{AB}$. These rays can be paired with the real numbers from 0 to 180 in such a way that:
a. $\overrightarrow{OA}$ is paired with 0, and $\overrightarrow{OB}$ with 180.
b. If $\overrightarrow{OP}$ is paired with x, and $\overrightarrow{OQ}$ with y, then $m\angle POQ = |x - y|$.
   (p. 18)

Postulate 4   (**Angle Addition Postulate**)   If point B lies in the interior of $\angle AOC$, then $m\angle AOB + m\angle BOC = m\angle AOC$. If $\angle AOC$ is a straight angle and B is any point not on $\overleftrightarrow{AC}$, then $m\angle AOB + m\angle BOC = 180$.   (p. 18)

Postulate 5   A line contains at least two points; a plane contains at least three points not all in one line; space contains at least four points not all in one plane. (p.23)

Postulate 6   Through any two points there is exactly one line.   (p. 23)

Postulate 7   Through any three points there is at least one plane, and through any three noncollinear points there is exactly one plane.   (p. 23)

Postulate 8   If two points are in a plane, then the line that contains the points is in that plane.   (p. 23)

Postulate 9   If two planes intersect, then their intersection is a line.   (p. 23)

Postulate 10   If two parallel lines are cut by a transversal, then corresponding angles are congruent.   (p. 78)

Postulate 11   If two lines are cut by a transversal and corresponding angles are congruent, then the lines are parallel.   (p. 83)

Postulate 12   (**SSS Postulate**)   If three sides of one triangle are congruent to three sides of another triangle, then the triangles are congruent.   (p. 122)

Postulate 13   (**SAS Postulate**)   If two sides and the included angle of one triangle are congruent to two sides and the included angle of another triangle, then the triangles are congruent.   (p. 122)

Postulate 14   (**ASA Postulate**)   If two angles and the included side of one triangle are congruent to two angles and the included side of another triangle, then the triangles are congruent.   (p. 123)

Postulate 15   (**AA Similarity Postulate**)   If two angles of one triangle are congruent to two angles of another triangle, then the triangles are similar. (p. 255)

Postulate 16   (**Arc Addition Postulate**)   The measure of the arc formed by two adjacent arcs is the sum of the measures of these two arcs. (p. 339)

Postulate 17   The area of a square is the square of the length of a side. ($A = s^2$) (p. 423)

Postulate 18   (**Area Congruence Postulate**)   If two figures are congruent, then they have the same area. (p. 423)

Postulate 19   (**Area Addition Postulate**)   The area of a region is the sum of the areas of its non-overlapping parts. (p. 424)

Theorems

Points, Lines, Planes, and Angles

1-1   If two lines intersect, then they intersect in exactly one point. (p. 23)

1-2   Through a line and a point not in the line there is exactly one plane. (p. 23)

1-3   If two lines intersect, then exactly one plane contains the lines. (p. 23)

Deductive Reasoning

2-1   (**Midpoint Theorem**)   If M is the midpoint of $\overline{AB}$, then
$$AM = \tfrac{1}{2}AB \text{ and } MB = \tfrac{1}{2}AB. \qquad \text{(p. 43)}$$

2-2   (**Angle Bisector Theorem**)   If $\overrightarrow{BX}$ is the bisector of $\angle ABC$, then
$$m\angle ABX = \tfrac{1}{2}m\angle ABC \text{ and } m\angle XBC = \tfrac{1}{2}m\angle ABC. \qquad \text{(p. 44)}$$

2-3   Vertical angles are congruent. (p. 51)

2-4   If two lines are perpendicular, then they form congruent adjacent angles. (p. 56)

2-5   If two lines form congruent adjacent angles, then the lines are perpendicular. (p. 56)

2-6   If the exterior sides of two adjacent acute angles are perpendicular, then the angles are complementary. (p. 56)

2-7   If two angles are supplements of congruent angles (or of the same angle), then the two angles are congruent. (p. 61)

2-8   If two angles are complements of congruent angles (or of the same angle), then the two angles are congruent. (p. 61)

Parallel Lines and Planes

3-1   If two parallel planes are cut by a third plane, then the lines of intersection are parallel. (p. 74)

3-2   If two parallel lines are cut by a transversal, then alternate interior angles are congruent. (p. 78)

3-3   If two parallel lines are cut by a transversal, then same-side interior angles are supplementary. (p. 79)

3-4   If a transversal is perpendicular to one of two parallel lines, then it is perpendicular to the other one also. (p. 79)

3-5   If two lines are cut by a transversal and alternate interior angles are congruent, then the lines are parallel. (p. 83)

3-6   If two lines are cut by a transversal and same-side interior angles are supplementary, then the lines are parallel. (p. 84)

3-7   In a plane two lines perpendicular to the same line are parallel. (p. 84)

3-8   Through a point outside a line, there is exactly one line parallel to the given line. (p. 85)

3-9   Through a point outside a line, there is exactly one line perpendicular to the given line. (p. 85)

3-10   Two lines parallel to a third line are parallel to each other. (p. 85)

3-11   The sum of the measures of the angles of a triangle is 180. (p. 94)

    Corollary 1   If two angles of one triangle are congruent to two angles of another triangle, then the third angles are congruent. (p. 94)

    Corollary 2   Each angle of an equiangular triangle has measure 60. (p. 94)

    Corollary 3   In a triangle, there can be at most one right angle or obtuse angle. (p. 94)

    Corollary 4   The acute angles of a right triangle are complementary. (p. 94)

3-12   The measure of an exterior angle of a triangle equals the sum of the measures of the two remote interior angles. (p. 95)

3-13   The sum of the measures of the angles of a convex polygon with n sides is $(n - 2)180$. (p. 102)

3-14   The sum of the measures of the exterior angles of any convex polygon, one angle at each vertex, is 360. (p. 102)

Congruent Triangles

4-1   **(The Isosceles Triangle Theorem)** If two sides of a triangle are congruent, then the angles opposite those sides are congruent. (p. 135)

    Corollary 1   An equilateral triangle is also equiangular. (p. 135)

    Corollary 2   An equilateral triangle has three 60° angles. (p. 135)

    Corollary 3   The bisector of the vertex angle of an isosceles triangle is perpendicular to the base at its midpoint. (p. 135)

4-2   If two angles of a triangle are congruent, then the sides opposite those angles are congruent. (p. 136)

       Corollary   An equiangular triangle is also equilateral. (p. 136)

4-3   **(AAS Theorem)**   If two angles and a non-included side of one triangle are congruent to the corresponding parts of another triangle, then the triangles are congruent. (p. 140)

4-4   **(HL Theorem)**   If the hypotenuse and a leg of one right triangle are congruent to the corresponding parts of another right triangle, then the triangles are congruent. (p. 141)

4-5   If a point lies on the perpendicular bisector of a segment, then the point is equidistant from the endpoints of the segment. (p. 153)

4-6   If a point is equidistant from the endpoints of a segment, then the point lies on the perpendicular bisector of the segment. (p. 153)

4-7   If a point lies on the bisector of an angle, then the point is equidistant from the sides of the angle. (p. 154)

4-8   If a point is equidistant from the sides of an angle, then the point lies on the bisector of the angle. (p. 154)

Quadrilaterals

5-1   Opposite sides of a parallelogram are congruent. (p. 167)

5-2   Opposite angles of a parallelogram are congruent. (p. 167)

5-3   Diagonals of a parallelogram bisect each other. (p. 167)

5-4   If both pairs of opposite sides of a quadrilateral are congruent, then the quadrilateral is a parallelogram. (p. 172)

5-5   If one pair of opposite sides of a quadrilateral are both congruent and parallel, then the quadrilateral is a parallelogram. (p. 172)

5-6   If both pairs of opposite angles of a quadrilateral are congruent, then the quadrilateral is a parallelogram. (p. 172)

5-7   If the diagonals of a quadrilateral bisect each other, then the quadrilateral is a parallelogram. (p. 172)

5-8   If two lines are parallel, then all points on one line are equidistant from the other line. (p. 177)

5-9   If three parallel lines cut off congruent segments on one transversal, then they cut off congruent segments on every transversal. (p. 177)

5-10   A line that contains the midpoint of one side of a triangle and is parallel to another side passes through the midpoint of the third side. (p. 178)

5-11   The segment that joins the midpoints of two sides of a triangle
(1) is parallel to the third side.
(2) is half as long as the third side. (p. 178)

5-12   The diagonals of a rectangle are congruent. (p. 185)

5-13   The diagonals of a rhombus are perpendicular. (p. 185)

5-14   Each diagonal of a rhombus bisects two angles of the rhombus. (p. 185)

5-15 The midpoint of the hypotenuse of a right triangle is equidistant from the three vertices. (p. 185)

5-16 If an angle of a parallelogram is a right angle, then the parallelogram is a rectangle. (p. 185)

5-17 If two consecutive sides of a parallelogram are congruent, then the parallelogram is a rhombus. (p. 185)

5-18 Base angles of an isosceles trapezoid are congruent. (p. 190)

5-19 The median of a trapezoid
(1) is parallel to the bases.
(2) has a length equal to the average of the base lengths. (p. 191)

Inequalities in Geometry

6-1 **(The Exterior Angle Inequality Theorem)** The measure of an exterior angle of a triangle is greater than the measure of either remote interior angle. (p. 204)

6-2 If one side of a triangle is longer than a second side, then the angle opposite the first side is larger than the angle opposite the second side. (p. 219)

6-3 If one angle of a triangle is larger than a second angle, then the side opposite the first angle is longer than the side opposite the second angle. (p. 220)

    Corollary 1 The perpendicular segment from a point to a line is the shortest segment from the point to the line. (p. 220)

    Corollary 2 The perpendicular segment from a point to a plane is the shortest segment from the point to the plane. (p. 220)

6-4 **(The Triangle Inequality)** The sum of the lengths of any two sides of a triangle is greater than the length of the third side. (p. 220)

6-5 **(SAS Inequality Theorem)** If two sides of one triangle are congruent to two sides of another triangle, but the included angle of the first triangle is larger than the included angle of the second, then the third side of the first triangle is longer than the third side of the second triangle. (p. 228)

6-6 **(SSS Inequality Theorem)** If two sides of one triangle are congruent to two sides of another triangle, but the third side of the first triangle is longer than the third side of the second, then the included angle of the first triangle is larger than the included angle of the second. (p. 229)

Similar Polygons

7-1 **(SAS Similarity Theorem)** If an angle of one triangle is congruent to an angle of another triangle and the sides including those angles are in proportion, then the triangles are similar. (p. 263)

7-2 **(SSS Similarity Theorem)** If the sides of two triangles are in proportion, then the triangles are similar. (p. 263)

7-3   (**Triangle Proportionality Theorem**)   If a line parallel to one side of a triangle intersects the other two sides, then it divides those sides proportionally.   (p. 269)

> **Corollary**   If three parallel lines intersect two transversals, then they divide the transversals proportionally.   (p. 270)

7-4   (**Triangle Angle-Bisector Theorem**)   If a ray bisects an angle of a triangle, then it divides the opposite side into segments proportional to the other two sides.   (p. 270)

Right Triangles

8-1   If the altitude is drawn to the hypotenuse of a right triangle, then the two triangles formed are similar to the original triangle and to each other.   (p. 285)

> **Corollary 1**   When the altitude is drawn to the hypotenuse of a right triangle, the length of the altitude is the geometric mean between the segments of the hypotenuse.   (p. 286)

> **Corollary 2**   When the altitude is drawn to the hypotenuse of a right triangle, each leg is the geometric mean between the hypotenuse and the segment of the hypotenuse that is adjacent to that leg.   (p. 286)

8-2   (**Pythagorean Theorem**)   In a right triangle, the square of the hypotenuse is equal to the sum of the squares of the legs.   (p. 290)

8-3   If the square of one side of a triangle is equal to the sum of the squares of the other two sides, then the triangle is a right triangle.   (p. 295)

8-4   If the square of the longest side of a triangle is less than the sum of the squares of the other two sides, then the triangle is an acute triangle.   (p. 296)

8-5   If the square of the longest side of a triangle is greater than the sum of the squares of the other two sides, then the triangle is an obtuse triangle.   (p. 296)

8-6   (**45°-45°-90° Theorem**)   In a 45°-45°-90° triangle, the hypotenuse is $\sqrt{2}$ times as long as a leg.   (p. 300)

8-7   (**30°-60°-90° Theorem**)   In a 30°-60°-90° triangle, the hypotenuse is twice as long as the shorter leg, and the longer leg is $\sqrt{3}$ times as long as the shorter leg. (p. 300)

Circles

9-1   If a line is tangent to a circle, then the line is perpendicular to the radius drawn to the point of tangency.   (p. 333)

> **Corollary**   Tangents to a circle from a point are congruent.   (p. 333)

9-2   If a line in the plane of a circle is perpendicular to a radius at its outer endpoint, then the line is tangent to the circle.   (p. 333)

9-3   In the same circle or in congruent circles, two minor arcs are congruent if and only if their central angles are congruent.   (p. 340)

9-4   In the same circle or in congruent circles,
(1) congruent arcs have congruent chords.
(2) congruent chords have congruent arcs.   (p. 344)

9-5   A diameter that is perpendicular to a chord bisects the chord and its arc.   (p. 344)

9-6   In the same circle or in congruent circles,
(1) chords equally distant from the center (or centers) are congruent.
(2) congruent chords are equally distant from the center (or centers).   (p. 345)

9-7   The measure of an inscribed angle is equal to half the measure of its intercepted arc.   (p. 350)

> **Corollary 1**   If two inscribed angles intercept the same arc, then the angles are congruent.   (p. 351)

> **Corollary 2**   An angle inscribed in a semicircle is a right angle.   (p. 351)

> **Corollary 3**   If a quadrilateral is inscribed in a circle, then its opposite angles are supplementary.   (p. 351)

9-8   The measure of an angle formed by a chord and a tangent is equal to half the measure of the intercepted arc.   (p. 352)

9-9   The measure of an angle formed by two chords that intersect inside a circle is equal to half the sum of the measures of the intercepted arcs.   (p. 357)

9-10   The measure of an angle formed by two secants, two tangents, or a secant and a tangent drawn from a point outside a circle is equal to half the difference of the measures of the intercepted arcs.   (p. 358)

9-11   When two chords intersect inside a circle, the product of the segments of one chord equals the product of the segments of the other chord.   (p. 362)

9-12   When two secant segments are drawn to a circle from an external point, the product of one secant segment and its external segment equals the product of the other secant segment and its external segment.   (p. 362)

9-13   When a secant segment and a tangent segment are drawn to a circle from an external point, the product of the secant segment and its external segment is equal to the square of the tangent segment.   (p. 363)

Constructions and Loci

10-1   The bisectors of the angles of a triangle intersect in a point that is equidistant from the three sides of the triangle.   (p. 386)

10-2   The perpendicular bisectors of the sides of a triangle intersect in a point that is equidistant from the three vertices of the triangle.   (p. 387)

10-3   The lines that contain the altitudes of a triangle intersect in a point.   (p. 387)

10-4   The medians of a triangle intersect in a point that is two thirds of the distance from each vertex to the midpoint of the opposite side.   (p. 387)

Areas of Plane Figures

11-1   The area of a rectangle equals the product of its base and height.   ($A = bh$) (p. 424)

11-2   The area of a parallelogram equals the product of a base and the height to that base. ($A = bh$)   (p. 429)

11-3   The area of a triangle equals half the product of a base and the height to the base. ($A = \frac{1}{2}bh$)   (p. 429)

11-4 The area of a rhombus equals half the product of its diagonals. $(A = \frac{1}{2}d_1d_2)$
(p. 430)

11-5 The area of a trapezoid equals half the product of the height and the sum of the bases.
$(A = \frac{1}{2}h(b_1 + b_2))$ (p. 435)

11-6 The area of a regular polygon is equal to half the product of the apothem and the
perimeter. $(A = \frac{1}{2}ap)$ (p. 441)

Related Formulas In a circle: $C = 2\pi r = \pi d$ $A = \pi r^2$ (p. 447)

11-7 If the scale factor of two similar figures is $a:b$, then
(1) the ratio of the perimeters is $a:b$.
(2) the ratio of the areas is $a^2:b^2$. (p. 457)

Areas and Volumes of Solids

12-1 The lateral area of a right prism equals the perimeter of a base times the height of
the prism. (L.A. $= ph$) (p. 476)

12-2 The volume of a right prism equals the area of a base times the height of the prism.
$(V = Bh)$ (p. 476)

12-3 The lateral area of a regular pyramid equals half the perimeter of the base times the
slant height. (L.A. $= \frac{1}{2}pl$) (p. 483)

12-4 The volume of a pyramid equals one third the area of the base times the height of
the pyramid. $(V = \frac{1}{3}Bh)$ (p. 483)

12-5 The lateral area of a cylinder equals the circumference of a base times the height
of the cylinder. (L.A. $= 2\pi rh$) (p. 490)

12-6 The volume of a cylinder equals the area of a base times the height of the cylinder.
$(V = \pi r^2h)$ (p. 490)

12-7 The lateral area of a cone equals half the circumference of the base times the slant
height.   (L.A. $= \frac{1}{2} \cdot 2\pi r \cdot l$ or L.A. $= \pi rl$) (p. 491)

12-8 The volume of a cone equals one third the area of the base times the height of the
cone. $(V = \frac{1}{3}\pi r^2h)$ (p. 491)

12-9 The area of a sphere equals 4π times the square of the radius. $(A = 4\pi r^2)$
(p. 497)

12-10 The volume of a sphere equals $\frac{4}{3}\pi$ times the cube of the radius. $(V = \frac{4}{3}\pi r^3)$
(p. 497)

12-11 If the scale factor of two similar solids is $a:b$, then
(1) the ratio of corresponding perimeters is $a:b$.
(2) the ratio of the base areas, of the lateral areas, and of the total areas is $a^2:b^2$.
(3) the ratio of the volumes is $a^3:b^3$. (p. 509)

Coordinate Geometry

13-1 **(The Distance Formula)** The distance d between points (x_1, y_1) and (x_2, y_2) is given
by $d = \sqrt{(x_2 - x_1)^2 + (y_2 - y_1)^2}$. (p. 524)

13-2 An equation of the circle with center (a, b) and radius r is $(x - a)^2 + (y - b)^2 = r^2$.
(p. 525)

13-3 Two nonvertical lines are parallel if and only if their slopes are equal. (p. 535)

13-4 Two nonvertical lines are perpendicular if and only if the product of their slopes is -1. (p. 535)

$$m_1 \cdot m_2 = -1, \text{ or } m_1 = -\frac{1}{m_2}$$

13-5 (**The Midpoint Formula**) The midpoint of the segment that joins points (x_1, y_1) and (x_2, y_2) is the point $\left(\dfrac{x_1 + x_2}{2}, \dfrac{y_1 + y_2}{2}\right)$. (p. 544)

13-6 (**Standard Form**) The graph of any equation that can be written in the form $Ax + By = C$, with A and B not both zero, is a line. (p. 548)

13-7 (**Slope-Intercept Form**) A line with the equation $y = mx + b$ has slope m and y-intercept b. (p. 549)

13-8 (**Point-Slope Form**) An equation of the line that passes through the point (x_1, y_1) and has slope m is $y - y_1 = m(x - x_1)$. (p. 553)

Transformations

14-1 An isometry maps a triangle to a congruent triangle. (p. 573)

    Corollary 1 An isometry maps an angle to congruent angle. (p. 573)

    Corollary 2 An isometry maps a polygon to a polygon with the same area. (p. 573)

14-2 A reflection in a line is an isometry. (p. 577)

14-3 A translation is an isometry. (p. 584)

14-4 A rotation is an isometry. (p. 589)

14-5 A dilation maps any triangle to a similar triangle. (p. 594)

    Corollary 1 A dilation maps an angle to a congruent angle. (p. 594)

    Corollary 2 A dilation $D_{O,k}$ maps any segment to a parallel segment $|k|$ times as long. (p. 594)

    Corollary 3 A dilation $D_{O,k}$ maps any polygon to a similar polygon whose area is k^2 times as large. (p. 351)

14-6 The composite of two isometries is an isometry. (p. 601)

14-7 A composite of reflections in two parallel lines is a translation. The translation glides all points through twice the distance from the first line of reflection to the second. (p. 601)

14-8 A composite of reflections in two intersecting lines is a rotation about the point of intersection of the two lines. The measure of the angle of rotation is twice the measure of the angle from the first line of reflection to the second. (p. 602)

    Corollary A composite of reflections in perpendicular lines is a half-turn about the point where the lines intersect. (p. 602)

Constructions

Glossary

acute angle: An angle with measure between 0 and 90. (p. 17)

acute triangle: A triangle with three acute angles. (p. 93)

adjacent angles: Two angles in a plane that have a common vertex and a common side but no common interior points. (p. 19)

adjacent arcs: Arcs of a circle that have exactly one point in common. (p. 339)

alternate interior angles: Two nonadjacent interior angles on opposite sides of a transversal. Angles 1 and 2 are alternate interior angles. (p. 74)

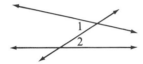

altitude of a parallelogram: Any segment perpendicular to the line containing a base from any point on the opposite side. (p. 424)

altitude of a solid: *See* prism, pyramid, cone, cylinder.

altitude of a trapezoid: Any segment perpendicular to a line containing one base from a point on the opposite base. (p. 435)

altitude of a triangle: The perpendicular segment from a vertex to the line containing the opposite side. In the figure, $\overline{BD}$ and $\overline{AD}$ are altitudes. (p. 152)

 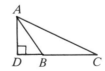

angle: A figure formed by two rays that have the same endpoint. The two rays are called the *sides* of the angle. Their common endpoint is the *vertex*. (p. 17)

angle of depression: When a point B is viewed from a higher point A, as shown by the diagram below, $\angle 1$ is the angle of depression. (p. 317)

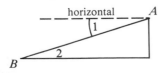

angle of elevation: When a point A is viewed from a lower point B, as shown by the diagram at the left below, $\angle 2$ is the angle of elevation. (p. 317)

apothem: The (perpendicular) distance from the center of a regular polygon to a side. (p. 441)

auxiliary line: A line (or ray or segment) added to a diagram to help in a proof. (p. 94)

axes: Usually, two perpendicular lines used to establish a coordinate system. (p. 523)

axiom: A statement that is accepted without proof. (p. 12)

base of an isosceles triangle: *See* legs of an isosceles triangle.

base of a parallelogram: Any side of a parallelogram can be considered its base. The term *base* may refer to the line segment or its length. (p. 424)

base of a pyramid: *See* pyramid.

bases of a prism: *See* prism.

bases of a trapezoid: *See* trapezoid.

biconditional: A statement that contains the words ''if and only if.'' (p. 34)

bisector of an angle: The ray that divides the angle into two congruent adjacent angles. (p. 19)

bisector of a segment: A line, segment, ray, or plane that intersects the segment at its midpoint. (p. 13)

center of a circle: *See* circle.

center of a regular polygon: The center of the circumscribed circle. (p. 441)

central angle of a circle: An angle with its vertex at the center of the circle. (p. 339)

central angle of a regular polygon: An angle formed by two radii drawn to consecutive vertices. (p. 441)

chord: A segment whose endpoints lie on a circle. (p. 329)

circle: The set of points in a plane that are a given distance from a given point in the plane. The given point is the *center*, and the given distance is the *radius*. (p. 329)

circumference of a circle: The perimeter of a circle given by the limiting number approached by the perimeters of a sequence of regular inscribed polygons. For radius r, $C = 2\pi r$. (p. 446)

circumscribed circle: A circle is circumscribed about a polygon when each vertex of the polygon lies on the circle. The polygon is *inscribed* in the circle. (p. 330)

circumscribed polygon: A polygon is *circumscribed* about a circle when each side of the polygon is tangent to the circle. The circle is *inscribed* in the polygon. (p. 334)

collinear points: Points all in one line. (p. 6)

common tangent: A line that is tangent to each of two coplanar circles. A common *internal* tangent intersects the segment joining the centers. A common *external* tangent does not intersect that segment. (p. 334)

complementary angles: Two angles whose measures have the sum 90. (p. 50)

composite of mappings: A transformation that combines two mappings. The composite of mappings S and T maps P to P'' where $T(P) = P'$ and $S(P') = P''$. Also called a product of mappings. (pp. 599, 605)

concentric circles: Circles that lie in the same plane and have the same center. (p. 330)

concentric spheres: Spheres that have the same center. (p. 330)

conclusion: *See* if-then statement.

concurrent lines: Two or more lines that intersect in one point. (p. 386)

conditional statement: *See* if-then statement.

cone: The diagrams illustrate a *right cone* and an *oblique cone*. Both have circular *bases* and a *vertex V*. In the right cone, h is the length of the *altitude*, l is the *slant height*, and r is the *radius*. (p. 490)

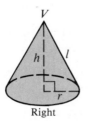
Right

Oblique

congruence mapping: *See* isometry.

congruent angles: Angles that have equal measures. (p. 19)

congruent arcs: Arcs, in the same circle or in congruent circles, that have equal measures. (p. 340)

congruent circles (or spheres): Circles (or spheres) that have congruent radii. (p. 330)

congruent figures: Figures having the same size and shape. (p. 117)

congruent polygons: Polygons whose vertices can be matched up so that the corresponding parts (angles and sides) of the polygons are congruent. (p. 118)

congruent segments: Segments that have equal lengths. (p. 13)

contraction: *See* dilation.

contrapositive of a conditional: The contrapositive of the statement *If p, then q* is the statement *If not q, then not p.* (p. 208)

converse: The converse of the statement *If p, then q* is the statement *If q, then p.* (p. 33)

convex polygon: A polygon such that no line containing a side of the polygon contains a point in the interior of the polygon. (p. 101)

coordinate plane: The plane of the *x*-axis and the *y*-axis. (p. 523)

coplanar points: Points all in one plane. (p. 6)

corollary of a theorem: A statement that can be proved easily by applying the theorem. (p. 94)

corresponding angles: Two angles in corresponding positions relative to two lines. In the figure, $\angle$ 2 and 6 are corresponding angles. (p. 74)

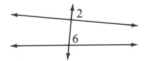

cosine (cos):

$$\text{cosine of } \angle A = \frac{AC}{AB}$$

$$\text{or } \cos A = \frac{\text{adjacent}}{\text{hypotenuse}}$$

(p. 312)

counterexample: An example used to prove that an if-then statement is false. For that counterexample, the hypothesis is true and the conclusion is false. (p. 33)

cube: A rectangular solid with square faces. (p. 476)

cylinder: The diagrams illustrate a *right cylinder* and an *oblique cylinder*. In a right cylinder, the segment joining the centers of the circular *bases* is an *altitude*. The length of an altitude is the *height*, *h*, of the cylinder. A radius of a base is a *radius*, *r*, of the cylinder. (p. 490)

Right   Oblique

decagon: A 10-sided polygon. (p. 101)

deductive reasoning: Proving statements by reasoning from accepted postulates, definitions, theorems, and given information. (p. 45)

diagonal: A segment joining two non-consecutive vertices of a polygon. (p. 102)

diameter: A chord that contains the center of a circle. (p. 329)

dilation: A dilation with center O and nonzero scale factor k maps any point P to a point P' determined as follows:
(1) If $k > 0$, P' lies on $\overrightarrow{OP}$ and $OP' = k \cdot OP$.
(2) If $k < 0$, P' lies on the ray opposite $\overrightarrow{OP}$ and $OP' = |k| \cdot OP$.
(3) The center O is its own image.
If $|k| > 1$, the dilation is an *expansion*.
If $|k| < 1$, the dilation is a *contraction*. (p. 592)

distance from a point to a line (or plane): The length of the perpendicular segment from the point to the line (or plane). (p. 154)

dot product: For vectors (a, b) and (c, d), the number $ac + bd$. The dot product of perpendicular vectors is zero. (p. 543)

equal vectors: Vectors with the same magnitude and the same direction. (p. 540)

equiangular triangle: A triangle with all angles congruent. (p. 93)

equilateral triangle: A triangle with all sides congruent. (p. 93)

expansion: *See* dilation.

exterior angle of a triangle: The angle formed when one side of the triangle is extended. $\angle DAC$ is an exterior angle of $\triangle ABC$, and $\angle B$ and C are *remote interior angles* with respect to $\angle DAC$. *Exterior angle* is also applied to other polygons. (p. 95)

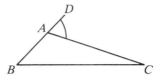

function: A correspondence between sets of numbers in which each number in the first set corresponds to exactly one number in the second set. (p. 571)

geometric mean: If a, b, and x are positive numbers with $\dfrac{a}{x} = \dfrac{x}{b}$, then x is the geometric mean between a and b. (p. 285)

glide: *See* translation.

glide reflection: A transformation in which every point P is mapped to a point P'' by these steps: (1) a glide maps P to P', and (2) a reflection in a line parallel to the glide line maps P' to P''. (p. 584)

glide reflection symmetry: A figure has glide reflection symmetry if there is a glide reflection that maps the figure onto itself. (p. 610)

golden ratio: *See* golden rectangle.

golden rectangle: A rectangle such that its length l and width w satisfy the equation $\dfrac{l}{w} = \dfrac{l + w}{l}$. The ratio $l\!:\!w$ is called the *golden ratio*. (p. 253)

great circle: The intersection of a sphere with any plane passing through the center of the sphere. (p. 331)

half-turn: A rotation through 180°. (p. 589)

height: The length of an altitude of a polygon or solid. (p. 424)

Heron's formula: A formula for finding the area of a triangle when the lengths of its sides are known. (p. 434)

hexagon: A 6-sided polygon. (p. 101)

hypotenuse: In a right triangle the side opposite the right angle. The other two sides are called *legs*. (p. 141)

hypothesis: *See* if-then statement.

identity transformation: The mapping that maps every point to itself. (p. 605)

if-then statement: A statement whose basic form is *If p, then q*. Statement *p* is the *hypothesis* and statement *q* is the *conclusion*. (p. 33)

image: *See* mapping.

indirect proof: A proof in which you assume temporarily that the conclusion is not true, and then deduce a contradiction. (p. 214)

inductive reasoning: A kind of reasoning in which the conclusion is based on several past observations. (p. 106)

inscribed angle: An angle whose vertex is on a circle and whose sides contain chords of the circle. (p. 349)

inscribed circle: *See* circumscribed polygon.

inscribed polygon: *See* circumscribed circle.

intersection of two figures: The set of points that are in both figures. (p. 6)

inverse of a conditional: The inverse of the statement *If p, then q* is the statement *If not p, then not q*. (p. 208)

inverse of a transformation: The inverse of *T* is the transformation *S* such that $S \circ T = I$. (p. 606)

isometry: A transformation that maps every segment to a congruent segment. Also called a *congruence mapping*. (p. 572)

isosceles trapezoid: A trapezoid with congruent legs. (p. 190)

isosceles triangle: A triangle with at least two sides congruent. (p. 93)

kite: A quadrilateral that has two pairs of congruent sides, but opposite sides are not congruent. (p. 193)

lateral area of a prism: The sum of the areas of its lateral faces. (p. 476)

lateral edges of a prism: *See* prism.

lateral edges of a pyramid: *See* pyramid.

lateral faces of a prism: *See* prism.

lateral faces of a pyramid: *See* pyramid.

legs of an isosceles triangle: The two congruent sides. The third side is the *base*. (p. 134)

legs of a right triangle: *See* hypotenuse.

legs of a trapezoid: *See* trapezoid.

length of a segment: The distance between its endpoints. (p. 11)

linear equation: An equation whose graph is a line. (p. 548)

line symmetry: A figure has line symmetry if there is a symmetry line *k* such that the reflection R_k maps the figure onto itself. (p. 609)

locus: The set of all points, and only those points, that satisfy one or more conditions. (p. 401)

logically equivalent statements: Statements that are either both true or both false. (p. 208)

magnitude of a vector $\overrightarrow{AB}$**:** The length *AB*. (p. 539)

major arc: *See* minor and major arcs.

mapping: A correspondence between points. Each point *P* in a given set is *mapped* to exactly one point *P′* in the same or a different set. *P′* is called the *image* of *P*, and *P* is called the *preimage* of *P′*. (p. 571)

measure of a major arc: *See* minor and major arcs.

measure of a minor arc: *See* minor and major arcs.

measure of an angle: A unique positive number, less than or equal to 180, that is paired with the angle. (p. 17)

measure of a semicircle: *See* semicircles.

median of a trapezoid: The segment that joins the midpoints of the legs. (p. 191)

median of a triangle: A segment from a vertex to the midpoint of the opposite side. (p. 152)

midpoint of a segment: The point that divides the segment into two congruent segments. (p. 13)

minor and major arcs: $\overset{\frown}{YZ}$ is a minor arc of $\odot O$. $\overset{\frown}{YXZ}$ is a major arc. The *measure of a minor arc* is the measure of its central angle, here $\angle YOZ$. The *measure of a major arc* is found by subtracting the measure of the minor arc from 360. (p. 339)

n–gon: A polygon of *n* sides. (p. 101)

oblique solid: *See* cone, cylinder, prism.

obtuse angle: An angle with measure between 90 and 180. (p. 17)

obtuse triangle: A triangle with one obtuse angle. (p. 93)

octagon: An 8-sided polygon. (p. 101)

one-to-one mapping (or function): A mapping (or function) from set A to set B in which every member of B has exactly one preimage in A. (p. 571)

opposite rays: Given three collinear points R, S, T: If S is between R and T, then $\overrightarrow{SR}$ and $\overrightarrow{ST}$ are opposite rays. (p. 11)

origin: The intersection point, denoted $O(0, 0)$, of the x-axis and the y-axis in a coordinate plane. (p. 523)

parallel line and plane: A line and a plane that do not intersect. (p. 73)

parallel lines: Coplanar lines that do not intersect. (p. 73)

parallelogram: A quadrilateral with both pairs of opposite sides parallel. (p. 167)

parallel planes: Planes that do not intersect. (p. 73)

pentagon: A 5-sided polygon. (p. 101)

perimeter of a polygon: The sum of the lengths of its sides. (p. 445)

perpendicular bisector of a segment: A line (or ray or segment) that is perpendicular to the segment at its midpoint. (p. 153)

perpendicular line and plane: A line and a plane are perpendicular if and only if they intersect and the line is perpendicular to all lines in the plane that pass through the point of intersection. (p. 128)

perpendicular lines: Two lines that intersect to form right angles. (p. 56)

plane symmetry: A figure in space has plane symmetry if there is a symmetry plane X such that reflection in the plane maps the figure onto itself. (p. 610)

point of tangency: *See* tangent to a circle.

point symmetry: A figure has point symmetry if there is a symmetry point O such that the half-turn H_O maps the figure onto itself. (p. 609)

polygon: A plane figure formed by coplanar segments (*sides*) such that (1) each segment intersects exactly two other segments, one at each endpoint; and (2) no two segments with a common endpoint are collinear. (p. 101)

postulate: A statement that is accepted without proof. (p. 12)

preimage: *See* mapping.

prism: The solids shown are *prisms*. The shaded faces are the *bases* (congruent polygons lying in parallel planes). The other faces are *lateral faces* and all are parallelograms. Adjacent lateral faces intersect in parallel segments called *lateral edges*. An *altitude* of a prism is a segment joining the two base planes and perpendicular to both. The length of an altitude is the *height*, h, of the prism. Figure (A), in which the lateral faces are rectangles, is called a *right prism*. Figure (B) is an *oblique prism*. (p. 475)

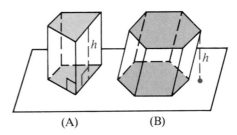

(A)   (B)

product of mappings: *See* composite of mappings.

proportion: An equation stating that two ratios are equal. The first and last terms are the *extremes*; the middle terms are the *means*. (pp. 242, 245)

pyramid: The diagram shows a pyramid. Point V is its *vertex*; the pentagon $ABCDE$ is its *base*. The five triangular faces meeting at V are *lateral faces*; they intersect in segments called *lateral edges*. The segment from the vertex perpendicular to the base is the *altitude*, and its length is the *height*, h, of the pyramid.

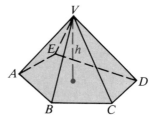

In a *regular pyramid*, the base is a regular polygon, all lateral edges are congruent, all lateral faces are congruent isosceles triangles, and the altitude meets the base at its center. The height of a lateral face is the *slant height* of the pyramid. (p. 482)

Pythagorean triple: Any triple of positive integers a, b, and c, such that $a^2 + b^2 = c^2$. (p. 299)

quadrant: Any one of the four regions into which the plane is divided by the coordinate axes. (p. 523)

quadrilateral: A 4-sided polygon. (p. 101)

radius of a circle: *See* circle.

radius of a regular polygon: The distance from the center to a vertex. (p. 441)

radius of a right cylinder: *See* cylinder.

ratio: The ratio of x to y ($y \neq 0$) is $\dfrac{x}{y}$ and is sometimes written $x:y$. (pp. 241, 242)

ray: The ray AC ($\overrightarrow{AC}$) consists of segment $\overline{AC}$ and all other points P such that C is between A and P. The point named first, here A, is the *endpoint of* $\overrightarrow{AC}$. (p. 11)

rectangle: A quadrilateral with four right angles. (p. 184)

rectangular solid: A right rectangular prism. (p. 475) *See also* prism.

reflection: A transformation in which a *line of reflection* acts like a mirror, reflecting points to their images. A reflection in a line m maps every point P to a point P' such that: (1) if P is not on line m, then m is the perpendicular bisector of $\overline{PP'}$; and (2) if P is on line m, then $P' = P$. (p. 577)

regular polygon: A polygon that is both equiangular and equilateral. (p. 103)

regular pyramid: *See* pyramid.

remote interior angles: *See* exterior angle of a triangle.

rhombus: A quadrilateral with four congruent sides. (p. 184)

right angle: An angle with measure 90. (p. 17)

right solid: *See* cone, cylinder, prism.

right triangle: A triangle with one right angle. (p. 93)

rotation: A rotation about point O through $x°$ is a transformation such that: (1) if point P is different from O, then $OP' = OP$ and $m \angle POP' = x$; and (2) if point P is the same as O, then $P' = P$. (p. 588)

rotational symmetry: A figure has rotational symmetry if there is a rotation that maps the figure onto itself. (p. 609)

same-side interior angles: Two interior angles on the same side of a transversal. (p. 74)

scalar multiple of a vector: The product of the vector (a, b) and the real number k is the scalar multiple (ka, kb). (p. 540)

scale factor: For similar polygons, the ratio of the lengths of two corresponding sides. (p. 249)

scalene triangle: A triangle with no sides congruent. (p. 93)

secant of a circle: A line that contains a chord. (p. 329)

sector of a circle: A region bounded by two radii and an arc of the circle. (p. 452)

segment of a line: Two points on the line and all points between them. The two points are called the *endpoints* of the segment. (p. 11)

segments divided proportionally: $\overline{AB}$ and $\overline{CD}$ are divided proportionally if points L and M lie on $\overline{AB}$ and $\overline{CD}$, respectively, and $\dfrac{AL}{LB} = \dfrac{CM}{MD}$. (p. 269)

semicircles: The two arcs of a circle that are cut off by a diameter. The *measure of a semicircle* is 180. (p. 339)

sides of an angle: *See* angle.

sides of a triangle: *See* triangle.

similarity mapping: A transformation that maps any figure to a similar figure. *See also* dilation. (p. 593)

similar polygons: Two polygons are similar if their vertices can be paired so that corresponding angles are congruent and corresponding sides are in proportion. (p. 249)

similar solids: Solids that have the same shape but not necessarily the same size. (p. 508)

simplest form of a radical: No perfect square factor other than 1 is under the radical sign, no fraction is under the radical sign, and no fraction has a radical in its denominator. (p. 287)

sine (sin):

sine of $\angle A = \dfrac{BC}{AB}$

or $\sin A = \dfrac{\text{opposite}}{\text{hypotenuse}}$

(p. 312)

skew lines: Lines that are not coplanar. (p. 73)

slant height of a regular pyramid: *See* pyramid.

slant height of a right cone: *See* cone.

slope of a line: The steepness of a nonvertical line, defined by $m = \dfrac{y_2 - y_1}{x_2 - x_1}$, $x_1 \neq x_2$, where $P_1(x_1, y_1)$ and $P_2(x_2, y_2)$ are two points on the line. (p. 529)

space: The set of all points. (p. 6)

sphere: The set of all points in space that are a given distance from a given point. (p. 329)

square: A quadrilateral with four right angles and four congruent sides. (p. 184)

straight angle: An angle with measure 180. (p. 17)

sum of two vectors: The sum of the vectors (a, b) and (c, d) is the vector $(a + c, b + d)$. (p. 541)

supplementary angles: Two angles whose measures have the sum 180. (p. 50)

symmetry: A figure in the plane has symmetry if there is an isometry, other than the identity, that maps the figure onto itself. (p. 609)

tangent (tan):

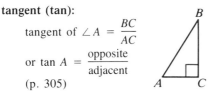

tangent of $\angle A = \dfrac{BC}{AC}$

or $\tan A = \dfrac{\text{opposite}}{\text{adjacent}}$

(p. 305)

tangent circles: Coplanar circles that are tangent to the same line at the same point. (p. 334)

tangent to a circle: A line in the plane of the circle that intersects the circle in exactly one point, called the *point of tangency*. (p. 329)

tessellation: A pattern in which congruent copies of a figure completely fill the plane without overlapping. (p. 610)

theorem: A statement that can be proved. (p. 23)

total area of a prism: The sum of the areas of all its faces. (p. 476)

transformation: A one-to-one mapping from the whole plane to the whole plane. (p. 572)

translation: A transformation that glides all points of the plane the same distance in the same direction, and maps any point (x, y) to the point $(x + a, y + b)$ where a and b are constants. Also called a *glide*. (pp. 583, 584)

translational symmetry: A figure has translational symmetry if there is a translation that maps the figure onto itself. (p. 610)

transversal: A line that intersects two or more coplanar lines in different points. (p. 74)

trapezoid: A quadrilateral with exactly one pair of parallel sides, called *bases*. The other sides are *legs*. (p. 190)

triangle: The figure formed by three segments joining three noncollinear points. Each of the three points is a *vertex* of the triangle and the segments are the *sides*. (p. 93)

vector: Any quantity that has both magnitude and direction. (p. 539)

Venn diagram: A circle diagram that may be used to represent a conditional. (p. 208)

vertex angle of an isosceles triangle: The angle opposite the base. (p. 134)

vertex of an angle: *See* angle.

vertex of a pyramid: *See* pyramid.

vertex of a triangle: *See* triangle.

vertical angles: Two angles whose sides form two pairs of opposite rays. $\angle$ 1 and 2 are vertical angles, as are $\angle$ 3 and 4. (p. 51)

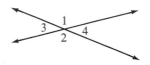

Index

Theorems, 23
  writing proofs of, 43–45
  *See pp. 677–684 for list of
    theorems*
Thinking Skills
  Analysis, *see* Patterns;
    Problem solving; and
    Reasoning, valid and
    flawed
  Applying Concepts, *see*
    Applications, Calculator
    Key-In, Computer Key-
    In, and Mathematical
    modeling
  Classification, *see*, for
    example, 93,
    186 (Cl. Exs. 1–4),
    187 (Exs. 1–10)
  Interpreting, *see* Patterns,
    Problem solving, and
    Reading Geometry
  Reasoning, *see* Deductive
    reasoning, Explorations,
    Hypothesis, Inductive
    reasoning, Logic,
    Probability, Proof, and
    Venn diagrams
  Recall and transfer, *see*
    Applications,
    Mathematical modeling,
    Mixed Review Exercises,
    and Problem solving
  Spatial perception, *see*
    Constructions, Diagrams,
    Drawing geometric
    figures, Model, and
    Visualization
  Synthesis, *see* Conditional,
    Conjunction, Disjunction,
    Problem solving, and Proof
Tiling of the plane, 657
Topology, 275–276
Transformations, 572
  composites or products of,
    599–602, 605–606
  fixed point of, 605
  identity, 605

integrating into Chapters 3–
  13, 657–675
inverse of, 606, 617
Translation, 583–584, 601,
  657
  inverse of, 605–606
Transversal, 74
Trapezoid, 190–191
  altitude of, 435
  area of, 435
  bases of, 190
  height of, 435
  isosceles, 190
  legs of, 190
  median of, 191
Triangle(s), 93, 101
  acute, 93, 296
  altitudes of, 152, 386–387
  angle bisector(s) of, 270,
    386
  angles of, 93–95
  area of, 429, 434, 456,
    465–466
  centroid of, 387, 391
  circumcenter of, 387, 669
  classification of, 93
  concurrent lines in, 386–
    387
  congruent, 117–163
  equiangular, 93, 94, 135,
    136
  equilateral, 93, 135, 136
  exterior angles of, 95, 204
  45°–45°–90°, 300
  incenter of, 387, 669
  isosceles, 93, 134–136
  medians of, 152, 386–387
  obtuse, 93, 296
  orthocenter of, 387, 669
  overlapping, 140
  remote interior angles of,
    95
  right, 93, 94, 141, 285–325
  scalene, 93
  segment joining midpoints
    of two sides, 178,
    267 (Ex. 22)

sides of, 93
similar, 254–256, 263–264
sum of measures of angles,
  21 (Ex. 25), 94, 657
30°–60°–90°, 300–301
vertex of, 93
See also Congruent
  triangles, Isosceles
  triangles, Right triangles
Trigonometry, 305–320
  table of values, 311
Trisection of angle, 99,
  399 (Ex. 18)
Truth table, 644–647

Undefined terms, 5
Uniqueness, 23

Vectors, 539–541
Venn diagrams, 208–209
Vertex
  of angle, 17
  of cone, 493
  of polygon, 101
  of pyramid, 482
  of triangle, 93
Visualization, 4,
    8 (Exs. 19, 20),
    9 (Exs. 29–36),
    77 (Ex. 22),
    194 (Challenge),
    260 (Challenge),
    485 (Exs. 7–16), 565
Volume, 476, 503, 516–517
  approximating, 489 (Ex. 2),
    496, 498–499,
    501 (Ex. 28), 504
  cone, 491
  cylinder, 490–491, 496
  maximum, 481, 502, 515
  prism, 476–477
  pyramid, 483–484
  sphere, 497–498, 504

Selected Answers

The answers in this answer section have been written in condensed form. Key steps are given for certain proofs, and abbreviations, such as CPCT for "corresponding parts of congruent triangles are congruent," are used. Check with your teacher regarding the form in which you should give answers in your own work.

Chapter 1

Written Exercises, Pages 3–4
1. *X* and *F*, 14 m; *X* and *T*, 7 cm, 14 m; *Y* and *F*, 9.5 cm; *F* and *T*, 24 m
3. a, b. See diagram.   **c.** none
5. the distance from *R* to *S*
7. It is twice the area of the inner square.
9. *ab* = *cd*

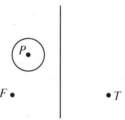

Written Exercises, Pages 7–9
1. True   **3.** True   **5.** True   **7.** True   **9.** False   **11.**

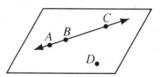

13. *VWT*, *VST*, *VRS*, *VWR*, *WRST*   **15.** $\overleftrightarrow{RW}$, $\overleftrightarrow{RV}$, $\overleftrightarrow{RS}$ **17.** *VRS*, *VST*, *WRST*   **21.** *RSGF* and *FGCB*
23. *ABCD*, *REABF*, *GFBC*   **25. a.** Yes   **b.** No   **27.** No

29.

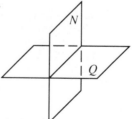

31.

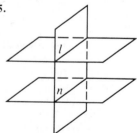

35.

Self-Test 1, Page 10
1. *T*   **2.** *T*   **3.** *V*   **4.** True   **5.** True   **6.** False   **7.**

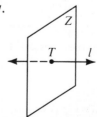

Algebra Review, Page 10

1. $c = 7$ **3.** $c = 17$ **5.** $z = 15$ **7.** $x = 5$ **9.** $x = 1$ **11.** $a = 12$ **13.** $b = -3$
15. $b = -\dfrac{2}{9}$ **17.** $k = \dfrac{1}{2}$ **19.** $e = 24$ **21.** $e = -15$ **23.** $p = 6$ **25.** $t = 8$ **27.** $s = 9$
29. $x = 15$ **31.** $g = 8$ **33.** $w = 8$ **35.** $y = 13$ **37.** $b = 5$ **39.** $h = 20$ **41.** $f = 15$
43. $d = -1$ **45.** $x = 45$

Written Exercises, Pages 15–16

1. 15   **3.** 4.5   **5.** True   **7.** True   **9.** False   **11.** False   **13.** True   **15.** True   **17.** False
19. B **21.** C, G **23.** D **25.** -1 **31. a.** 5   **b.** 10   **c.** 10   **d.** 6   **33.** $x = 6$ **35.** $x = 3$
37. $y = 6$ **39.** $z = 8$; $GE = 10$; $EH = 10$; yes   **41.** $\overline{HN}$ **43.** $\overrightarrow{GT}$ **45.** M **47.** 2 if $AB \geq 3$ cm,
1 if $AB < 3$ cm

Written Exercises, Pages 21–22

1. E; $\overrightarrow{EL}$, $\overrightarrow{EA}$ Answers may vary in Exs. 3–7.   **3.** $\angle DLT$ **5.** $\angle AEL$ **7.** $\angle 7$ **9.** acute   **11.** right
13. straight   **15.** LAS **17.** $\overrightarrow{LE}$, $\angle ALS$ **19.**   **21.**   **23.** 2

25. Yes; the sum of the measures of the angles is 180.   **27.** $180 - t$, t, $180 - t$ **29.** $x = 18$
31. $x = 9$ **33.** $x = 20$ **35. a.** 6; 10   **b.** 15   **c.** $\dfrac{n(n-1)}{2}$

Written Exercises, Pages 25–26

1. If there is a line and a pt. not on the line, then one and only one plane contains them.   **3. a.** a line   **b.** If
two planes intersect, then their intersection is a line.   **5.** Through any 2 pts. there is exactly one line.
7. $ACGE$ **9.** $\overleftrightarrow{AB}$, $\overleftrightarrow{CD}$, $\overleftrightarrow{AD}$, $\overleftrightarrow{BC}$ **11.** $ABCD$, $DCGH$, $ABGH$ **13.** No   **15.** No   **17. a.** Through
any 3 pts. there is at least one plane.   **b.**   **c.** Yes; if 2 pts. are in a plane, then the line
that contains the pts. is in that plane.

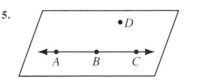

d. Through any 2 pts. there is exactly one line.   **e.** If 2 pts. are in a plane, then the line that contains the pts. is
in that plane.   **19. a.** 3   **b.** 6   **c.** 10   **d.** 15   **e.** 21   **f.** $\dfrac{n(n-1)}{2}$

Self-Test 2, Page 29

1. $\overleftrightarrow{RN}$, $\overleftrightarrow{RC}$, $\overleftrightarrow{NC}$ **2.** $\overrightarrow{NR}$ **3.** No   **4.** $x = 2$ **5.** JOT **6.** $\overrightarrow{OK}$, JOT **7.** 180, straight   **8.** c
9. there is exactly one line   **10.** then $\overleftrightarrow{AB}$ is in Z **11.** their intersection is a line   **12.** there is exactly one
plane that contains j and P

Chapter Review, Page 30

1. infinitely many   **3.** 2   **5.**   **7.**

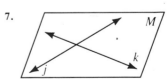

9. U or V **11.** congruent   **13.** $\angle 1$, $\angle 2$, $\angle ADC$; $\angle 1$, $\angle 2$ **15.** obtuse   **17.** True   **19.** False

Chapter 2

Written Exercises, Page 35

1. H: $3x - 7 = 32$, C: $x = 13$ **3.** H: you will, C: I'll try **5.** H: $a + b = a$, C: $b = 0$ **7.** B is between A and C if and only if $AB + BC = AC$. **9.** If points are collinear, then they all lie in 1 line. If points lie in 1 line, then they are collinear. Answers may vary in Exs. 11–15. **11.** $a = 1$, $b = -1$
13.  **15.**

17. True. If $|x| = 6$, then $x = -6$; false. **19.** True. If $5b > 20$, then $b > 4$; true. **21.** True. If Pam lives in Illinois, then she lives in Chicago; false. **23.** True. If $a^2 > 9$, then $a > 3$; false. **25.** False. If $n > 7$, then $n > 5$; true. **27.** False. If $DE + EF = DF$, then points D, E, and F are collinear; true. **29.** Two $\angle$ are $\cong$ if and only if their measures are $=$. **31.** Possible conclusions are: q, not r, s.

Mixed Review Exercises, Page 37

1. $\overline{AM}$; $\overline{MB}$ **2.** $\angle ABX$; $\angle XBC$ **3.** AOB; BOC; AOC **4.** POR; ROQ; 180

Written Exercises, Pages 41–43

1. Given; Add. Prop. of $=$; Div. Prop. of $=$ **3.** Given; Mult. Prop. of $=$; Subtr. Prop. of $=$ **5.** Given; Mult. Prop. of $=$; Dist. Prop.; Add. Prop. of $=$; Div. Prop. of $=$ **7.** 1. $\angle$ Add. Post 2. $\angle$ Add. Post. 3. $m \angle AOD = m \angle 1 + m \angle 2 + m \angle 3$; Substitution Prop. **9.** 1. Given 2. OW, WN; Seg. Add. Post. 3. $DO + OW = OW + WN$ 4. Refl. Prop. 5. $DO = WN$; Subtr. Prop. of $=$ **11.** 1. $m \angle 1 = m \angle 2$; $m \angle 3 = m \angle 4$ (Given) 2. $m \angle 1 + m \angle 3 = m \angle 2 + m \angle 4$ (Add. Prop. of $=$) 3. $m \angle 1 + m \angle 3 = m \angle SRT$; $m \angle 2 + m \angle 4 = m \angle STR$ ($\angle$ Add. Post.) 4. $m \angle SRT = m \angle STR$ (Substitution Prop.)
13. 1. $RQ = TP$; (Given) 2. $RZ + ZQ = RQ$; $TZ + ZP = TP$ (Seg. Add. Post.) 3. $RZ + ZQ = TZ + ZP$ (Substitution Prop.) 4. $ZQ = ZP$ (Given) 5. $RZ = TZ$ (Subtr. Prop. of $=$) **15.** b

Written Exercises, Pages 46–47

1. Def. of midpt. **3.** Def. of $\angle$ bis. **5.** Def. of midpt. **7.** $\angle$ Add. Post. **9.** 60 **11.** 70 **13. a.** 12 **b.** 28 **c.** 6 **d.** 22 **15. a.** $\overline{LM}$ and $\overline{MK}$, $\overline{GN}$ and $\overline{NH}$ **b.** Answers may vary; for example, $\overline{LK} \cong \overline{GH}$ **17.** $AC = BD$ **19.** 1. Given 2. Ruler Post. 3. Given 4. Def. of Midpt. 5. Substitution Prop. 6. $a + b$; Add. Prop. of $=$ 7. Div. Prop. of $=$ **21.** Q: $\dfrac{3a + b}{4}$; T: $\dfrac{5a + 3b}{8}$

Self-Test 1, Page 49

1. H: $\overleftrightarrow{AB}$ and $\overleftrightarrow{CD}$ intersect; C: $\overrightarrow{AB}$ and $\overrightarrow{CD}$ intersect **2.** If $\overrightarrow{AB}$ and $\overrightarrow{CD}$ intersect, then $\overleftrightarrow{AB}$ and $\overleftrightarrow{CD}$ intersect. False **3.** $\overline{AB} \cong \overline{CD}$ if and only if $AB = CD$. **4.** Answers may vary; $m \angle A = 95$ **5.** Substitution Prop. **6.** $x = 3$ **7.** 81 **8.** definitions, postulates

Written Exercises, Pages 52–54

1. 70, 160 **3.** $90 - x$, $180 - x$ **5.** 45, 45 **7.** $\angle AFD$ **9.** $\angle AFD$ and $\angle AFB$ **11.** $\angle BFC$ and $\angle EFD$ **13.** 35 **15.** 25 **17.** 60 **19.** 25 **21.** 25 **23.** 1. Vertical $\angle$ are $\cong$. 2. Given 3. Vert. $\angle$ are $\cong$. 4. $\angle 1 \cong \angle 4$ **25.** $x = 60$, $m \angle A = 76$, $m \angle B = 104$ **27.** $y = 24$, $m \angle C = 16$, $m \angle D = 74$ **29.** $x = \dfrac{1}{2}(90 - x)$; 30 **31.** $180 - x = 6(90 - x)$; 72; 108; 18
33. $x = 33$, $y = 66$

Written Exercises, Pages 58–60

1. a. $90 - x$ **b.** $180 - x$ **3.** Def. of $\perp$ lines **5.** If the ext. sides of 2 adj. $\angle$ are $\perp$, then the $\angle$ are comp. **7.** Def. of $\perp$ lines **9.** 35 **11.** 20 **13.** 1. Given 3. $\angle$ Add. Post. 4. $m \angle AOB + m \angle BOC = 90$ 5. $\angle AOB$ and $\angle BOC$ are comp. $\angle$. **15.** $180 - y$ **17.** $90 - (x + y)$ **19.** No **21.** No **23.** Yes **25.** No **27.** Answers may vary. $m \angle 2 = m \angle 3$; $m \angle DAC = 90$; $m \angle ECA = 90$ **29.** $\overleftrightarrow{XD} \perp \overleftrightarrow{XF}$

Mixed Review Exercises, Page 60

Answers may vary in Exs. 1–7. **1.** $\angle CBF \cong \angle BCG$ **2.** $AC = BD$ **3.** $\angle 3 \cong \angle 4$ **4.** $m \angle 1 = m \angle 2 = 45$ **5.** $CE = BE$ **6.** $\overleftrightarrow{AB} \perp \overleftrightarrow{BF}$ **7.** $m \angle 5 = 90$

Written Exercises, Pages 63–65

1. Seg. Add. Post. **3.** Vertical $\angle\!\!\!\!s$ are $\cong$. **5.** Def. of $\angle$ bis. **7.** Def. of $\perp$ lines **9.** Def. of comp. $\angle\!\!\!\!s$
11. If 2 lines are $\perp$, then they form $\cong$ adj. $\angle\!\!\!\!s$. **13.** Def. of $\perp$ lines **15.** 1. $\angle 5$ are supplementary; Given
2. Def. of supp. $\angle\!\!\!\!s$ 3. Substitution Prop. 5. Subtr. Prop. of = **17. a.** 1. Given 2. If the ext. sides of
2 adj. acute $\angle\!\!\!\!s$ are $\perp$, then the $\angle\!\!\!\!s$ are comp. 3. Given 4. If 2 $\angle\!\!\!\!s$ are comp. of $\cong$ $\angle\!\!\!\!s$, then the 2 $\angle\!\!\!\!s$ are $\cong$.
b. Show that $\angle 3$ and $\angle 6$ are supps. of $\cong$ $\angle\!\!\!\!s$. **19.** 1. $\angle 2 \cong \angle 3$ (Given) 2. $\angle 1 \cong \angle 2$ (Vert. $\angle\!\!\!\!s$ are $\cong$.)
3. $\angle 1 \cong \angle 3$ (Substitution Prop.) 4. $\angle 3 \cong \angle 4$ (Vert. $\angle\!\!\!\!s$ are $\cong$.) 5. $\angle 1 \cong \angle 4$ (Trans. Prop.)
21. 1. $\overline{AC} \perp \overline{BC}$ (Given) 2. $\angle 2$ is comp. to $\angle 1$. (If the ext. sides of 2 adj. acute $\angle\!\!\!\!s$ are $\perp$, then the 2 $\angle\!\!\!\!s$ are
comp.) 3. $\angle 3$ is comp. to $\angle 1$ (Given) 4. $\angle 3 \cong \angle 2$ (If 2 $\angle\!\!\!\!s$ are comp. of the same $\angle$, then the 2 $\angle\!\!\!\!s$ are $\cong$.)
23. $\overrightarrow{OF}$ bisects $\angle COD$. Proof: 1. $\overrightarrow{OE}$ bisects $\angle AOB$. (Given) 2. $\angle 1 \cong \angle 2$ (Def. of $\angle$ bis.) 3. $\angle 2 \cong \angle 3$
(Vert. $\angle\!\!\!\!s$ are $\cong$.) 4. $m\angle 1 = m\angle 3$ (Trans. Prop.) 5. $m\angle 1 = m\angle 4$ (Vert. $\angle\!\!\!\!s$ are $\cong$.) 6. $m\angle 3 = m\angle 4$
(Substitution Prop.) 7. $\overrightarrow{OF}$ bisects $\angle COD$. (Def. of $\angle$ bis.)

Self-Test 2, Page 65

1. Answers may vary; $90 \le m\angle HOK < 180$ **2. a.** 15 **b.** 40 **3.** 53; 37; 53 **4.** $m\angle 4 = 90 - t$,
$m\angle 5 = t$, $m\angle 6 = 90 - t$ **5.** Def. of $\perp$ lines **6.** Vert. $\angle\!\!\!\!s$ are $\cong$. **7.** If the ext. sides of 2 adj. $\angle\!\!\!\!s$ are
$\perp$, then the $\angle\!\!\!\!s$ are comp. **8.** Show that $\angle 1 \cong \angle 2$, so $j \perp k$. **9.** 1. $\angle 1$ is supp. to $\angle 3$; $\angle 2$ is supp. to
$\angle 3$ (Given) 2. $\angle 1 \cong \angle 2$ (If 2 $\angle\!\!\!\!s$ are supp. of the same $\angle$, the 2 $\angle\!\!\!\!s$ are $\cong$.) 3. $j \perp k$ (If 2 lines form $\cong$ adj.
$\angle\!\!\!\!s$, then the lines are $\perp$.)

Extra, Page 66

1. None of it; one **3.** 2; the result is 2 non-Möbius (2-sided) bands linked together. **5.** The result is a
rectangular frame.

Chapter Review, Pages 67–68

1. H: $m\angle 1 = 120$, C: $\angle 1$ is obtuse **3.** Answers may vary; $m\angle 1 = 100$ **5.** Substitution Prop.
7. Div. Prop. of = **9.** Def. of midpt. **11.** $\angle$ Bis. Thm. **13.** $\angle BOA$ or $\angle DOE$ **15.** Def. of $\perp$
lines **17.** If 2 lines are $\perp$, then they form $\cong$ adj. $\angle\!\!\!\!s$. **19.** Show that $\angle 3$ and $\angle 4$ are supps. of $\cong$ $\angle\!\!\!\!s$.

Algebra Review, Page 69

1. $x = 3$, $y = 9$ **3.** $x = -16$, $y = -8$ **5.** $x = 28$, $y = 4$ **7.** $x = 1$, $y = 6$ **9.** $x = 5$, $y = 4$
11. $x = -2$, $y = 3$ **13.** $x = 4$; $y = 1$ **15.** $x = 6$, $y = 2$ **17.** $x = 11$, $y = 16$

Preparing for College Entrance Exams, Page 70

1. C **2.** C **3.** D **4.** E **5.** D **6.** B **7.** E **8.** C

Cumulative Review, Page 71

1. Div. Prop. of = **3.** Def. of $\perp$ lines **5.** Subtr. Prop. of = **7.** If 2 planes int., then their int. is a
line. **9.** Seg. Add. Post. **11.** True **13.** False **15.** False; 3 collinear pts. **17.** True **19.** True
21. $x = 36$ **23.** $x = 10$ **25.** $x = 31$

Chapter 3

Written Exercises, Pages 76–77

1. alt. int. $\angle\!\!\!\!s$ **3.** s-s. int. $\angle\!\!\!\!s$ **5.** corr. $\angle\!\!\!\!s$ **7.** $\overleftrightarrow{PQ}$, $\overleftrightarrow{SR}$; $\overleftrightarrow{SQ}$ **9.** $\overleftrightarrow{PQ}$, $\overleftrightarrow{SR}$; $\overleftrightarrow{PS}$ **11.** $\overleftrightarrow{PQ}$, $\overleftrightarrow{SR}$; $\overleftrightarrow{QR}$
13. corr. $\angle\!\!\!\!s$ **15.** s-s. int. $\angle\!\!\!\!s$ **17.** corr. $\angle\!\!\!\!s$ **19.** Alt. int. $\angle\!\!\!\!s$ are $\cong$. **21. a.** Answers may vary.
b. Same as $m\angle 1 + m\angle 2$ **c.** Same as $m\angle 1 + m\angle 2$ **d.** When 2 nonparallel lines are cut by trans., the sum
of the meas. of s-s. int. $\angle\!\!\!\!s$ is a constant. **23.** $\overleftrightarrow{BH}$, $\overleftrightarrow{CI}$, $\overleftrightarrow{DJ}$, $\overleftrightarrow{EK}$, $\overleftrightarrow{FL}$ **25.** Answers may vary. $\overleftrightarrow{FL}$, $\overleftrightarrow{EK}$, $\overleftrightarrow{DJ}$,
$\overleftrightarrow{CI}$, $\overleftrightarrow{GL}$, $\overleftrightarrow{LK}$, $\overleftrightarrow{JI}$, $\overleftrightarrow{IH}$ **27.** ABHG, BCIH, CDJI, DEKJ **29.** If the top and bottom lie in $\parallel$ planes, then $\overline{CD}$
and $\overline{IJ}$ are the lines of intersection of DCIJ with 2 $\parallel$ planes, and are therefore $\parallel$. **31.** sometimes **33.** always
35. sometimes **37.** always **39.** sometimes

Written Exercises, Pages 80–82

1. $\angle 3$, $\angle 6$, $\angle 8$ **3.** $\angle 4$, $\angle 5$, $\angle 7$, $\angle 10$, $\angle 12$, $\angle 13$, $\angle 15$ **5.** 110, 70 **7.** $x = 60$, $y = 61$
9. $x = 60$, $y = 18$ **11.** $x = 14$, $y = 9$ **13.** 1. Given 2. Def. of $\perp$ lines 3. $l \parallel n$ 4. If 2 $\parallel$ lines are
cut by a trans., then corr. $\angle\!\!\!\!s$ are $\cong$. 5. $m\angle 2 = 90$ 6. Def. of $\perp$ lines **15.** $x = 70$, $y = 12$, $z = 38$

17. a. $m\angle DAB = 64$, $m\angle KAB = 32$, $m\angle DKA = 32$ **b.** More information is needed. **19.** $x = 30$, $y = 5$ **21.** 1. $k \parallel l$ (Given) 2. $\angle 1 \cong \angle 8$ or $m\angle 1 = m\angle 8$ (If 2 $\parallel$ lines are cut by a trans., then alt. int. $\angle$s are $\cong$.) 3. $m\angle 8 + m\angle 7 = 180$ ($\angle$ Add. Post.) 4. $m\angle 1 + m\angle 7 = 180$ (Substitution Prop.) 5. $\angle 1$ is supp. to $\angle 7$ (Def. of supp. $\angle$s) **23. a.** 1. $\overline{AB} \parallel \overline{DC}$; $\overline{AD} \parallel \overline{BC}$ (Given) 2. $\angle A$ is supp. to $\angle B$; $\angle C$ is supp. to $\angle B$ (If 2 $\parallel$ lines are cut by a trans., then s-s. int. $\angle$s are supp.) 3. $\angle A \cong \angle C$ (If 2 $\angle$s are supp. of the same $\angle$, then the 2 $\angle$s are $\cong$.) **b.** Yes, by the same reasoning as in part (a) **25.** 60

Mixed Review Exercises, Page 82

1. a. True **b.** If 2 lines form $\cong$ adj. $\angle$s, then the lines are $\perp$. **c.** True **2. a.** True **b.** If 2 lines are not skew, then they are $\parallel$. **c.** False **3. a.** True **b.** If two $\angle$s are supp., then the sum of their meas. is 180. **c.** True **4. a.** True **b.** If 2 planes do not intersect, then they are $\parallel$. **c.** True

Written Exercises, Pages 87–88

1. $\overline{AB} \parallel \overline{FC}$ **3.** $\overline{AB} \parallel \overline{FC}$ **5.** none **7.** none **9.** $\overline{AE} \parallel \overline{BD}$ **11.** $\overline{AE} \parallel \overline{BD}$ **13.** $\overline{AE} \parallel \overline{BD}$
15. $\overline{FB} \parallel \overline{EC}$; $\overline{AE} \parallel \overline{BD}$ **17.** 1. Given 2. Vert. $\angle$s are $\cong$. 3. Trans. Prop. 4. If 2 lines are cut by a trans. and corr. $\angle$s are $\cong$, then the lines are $\parallel$. **19.** $x = 35$, $y = 20$ **21.** $\angle 1 \cong \angle 4$; $\angle 2 \cong \angle 5$
23. 1. $k \perp t$; $n \perp t$ (Given) 2. $m\angle 1 = 90$; $m\angle 2 = 90$ (Def. of $\perp$ lines) 3. $m\angle 1 = m\angle 2$ or $\angle 1 \cong \angle 2$ (Substitution Prop.) 4. $k \parallel n$ (If 2 lines are cut by a trans. and corr. $\angle$s are $\cong$, then the lines are $\parallel$.)
25. 1. $\overline{BE} \perp \overline{DA}$; $\overline{CD} \perp \overline{DA}$ (Given) 2. $\overline{CD} \parallel \overline{BE}$ (In a plane, 2 lines $\perp$ to the same line are $\parallel$.) 3. $\angle 1 \cong \angle 2$ (If 2 $\parallel$ lines are cut by a trans., then alt. int. $\angle$s are $\cong$.) **27.** $m\angle RST = 110$ **29.** $x = 50$, $y = 20$
31. $x = 12$

Self-Test 1, Page 89

1. sometimes **2.** never **3.** always **4.** sometimes **5.** always **6.** $\angle 3$, $\angle 6$; $\angle 4$, $\angle 5$
7. Answers may vary; $\angle 1$, $\angle 5$; $\angle 2$, $\angle 6$; $\angle 3$, $\angle 7$; $\angle 4$, $\angle 8$ **8.** $\angle 3$, $\angle 5$ or $\angle 4$, $\angle 6$ **9.** $\angle 4$; $\angle 3$
10. $\angle 2$, $\angle 8$; $\angle 4$, $\angle 7$ **11.** $\angle 2$, $\angle 8$ **12.** 65, 115 **13.** $\overline{EB} \parallel \overline{DC}$ **14.** none **15.** $\overline{AE} \parallel \overline{BD}$
16. one, one

Written Exercises, Pages 97–99

1. a. **b.** **c.** 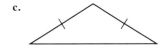 **3.** not possible **5.** 180

7. 95 **9.** 25 **11.** $x = 30$, $y = 80$ **13.** $x = 40$, $y = 50$ **15.** $x = 40$, $y = 50$ **17.** Yes, $n = 5$
19. 30, 60, 90 **21.** $m\angle C > 60$ **23. a.** 22 **b.** 23 **c.** $\angle ABD$ and $\angle C$ are comps. of $\angle CBD$.
25. 1. $\angle ABD \cong \angle AED$ (Given) 2. $\angle A \cong \angle A$ (Refl. Prop.) 3. $\angle C \cong \angle F$ (If 2 $\angle$s of one $\triangle$ are $\cong$ to 2 $\angle$s of another $\triangle$, then the third $\angle$s are $\cong$.) **27.** Given: $\triangle ABC$. Prove: $m\angle 1 + m\angle 2 + m\angle 3 = 180$. Proof:
1. Draw $\overrightarrow{CD}$ through $C \parallel$ to $\overleftrightarrow{AB}$. (Through a pt. outside a line, there is exactly 1 $\parallel$ to a given line.) 2. $\angle 2 \cong \angle 5$ or $m\angle 2 = m\angle 5$ (If 2 $\parallel$ lines are cut by a trans., alt. int. $\angle$s are $\cong$.) 3. $\angle 1 \cong \angle 4$ or $m\angle 1 = m\angle 4$ (If 2 $\parallel$ lines are cut by trans., corr. $\angle$s $\cong$.) 4. $m\angle ACD + m\angle 4 = 180$; $m\angle ACD = m\angle 3 + m\angle 5$ ($\angle$ Add. Post.) 5. $m\angle 1 + m\angle 2 + m\angle 3 = 180$ (Subst.) **29.** $x = 25$, $y = 5$ **31.** $\angle 1 \cong \angle 2 \cong \angle 5$; $\angle 3 \cong \angle 4 \cong \angle 6$

Written Exercises, Pages 104–105

1. 360; 360 **3.** 720; 360 **5.** 1440; 360 **7.** 360; yes **9.** 135 **11.** 120
15. not possible **17.** 24 **21.** $x = 36$; $\overline{AB} \parallel \overline{CD}$ **23.** 108 **25.** 14
27. a. Sketches may vary.
    b. Yes

Written Exercises, Pages 107–109

1. 256, 1024   **3.** $\dfrac{1}{81}, \dfrac{1}{243}$ **5.** 17, 23   **7.** 15, 4   **9.** 500, 250   **11.** none   **13.** none

15. $1234 \times 9 + 5 = 11111$ **17.** $9999^2 = 99980001$

21. True.

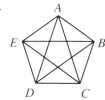

Given: *ABCDE* is a reg. pentagon.
Prove: $\overline{AC} \cong \overline{AD} \cong \overline{BE} \cong \overline{BD} \cong \overline{CE}$

23. False.

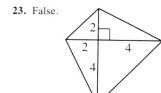

27. a. Opp. ∠ are ≅.   **b.** If both pairs of opp. ∠ of a quad. are ≅, then opp. sides are ∥. Given: *ABCD* is a quad.; $m\angle A = m\angle C$; $m\angle B = m\angle D$ Prove: $\overline{AD} \parallel \overline{BC}$; $\overline{AB} \parallel \overline{CD}$ Proof: 1. $m\angle A + m\angle B + m\angle C + m\angle D = 360$ (The sum of the meas. of the int. ∠ of a quad. is 360.) 2. $m\angle A = m\angle C$; $m\angle D = m\angle B$ (Given) 3. $2m\angle A + 2m\angle B = 360$; $2m\angle B + 2m\angle C = 360$ (Substitution Prop.) 4. $m\angle A + m\angle B = 180$; $m\angle B + m\angle C = 180$ (Div. Prop. of =) 5. $\angle A$ and $\angle B$ are supp.; $\angle B$ and $\angle C$ are supp. (Def. of supp. ∠) 6. $\overline{AD} \parallel \overline{BC}$; $\overline{AB} \parallel \overline{CD}$ (If 2 lines are cut by a trans. and s-s. int. ∠ are supp., then the lines are ∥.)   **c.** Both pairs of opp. ∠ of a quad. are ≅ if and only if opp. sides are ∥.   **29.** 5; 9; 14; 20; $\dfrac{n(n-3)}{2}$

Self-Test 2, Page 110

1. acute   **2.** scalene   **3.** 60   **4.** 105, 35   **5.** 19   **6.** $y = 50, z = 60$ **7.** $x = 1, x = 7$ **8.** 8
9. equilateral, equiangular   **10.** 360, 144   **11.** 6, 60   **12.** 32   **13.** 32   **14.** 36   **15.** 16

Chapter Review, Pages 111–112

1. 2   **3.** alt. int.   **5.** 105, 105   **7.** $y = 20$ **9.** $\overleftrightarrow{DE}$; $\angle A$ is supp. to $\angle ADE$.   **11.** See page 85.
13. 180   **15.** $\angle 3 \cong \angle 6$; If 2 ∠ are supp. of ≅ ∠, then the 2 ∠ are ≅. $\angle 2 \cong \angle 8$; If 2 ∠ of one △ are ≅ to
2 ∠ of another △, then the third ∠ are ≅.   **17.** 160   **19.** 12   **21.** $\dfrac{1}{100}, -\dfrac{1}{1000}$

Algebra Review, Page 113

1. 3   **3.** (0, 0)   **5.** (3, 5)   **7.** (4, 0)   **9.** (−5, 0)   **11.** (−2, 2)   **13.** (−2, −3)   **15.** *K*, *O*, *S*
17. 3   **19.** *M*, *N*, *P*   **21.** *V*, *W*   **23–34.**   **35.** (2, 1)   **37.** (0, 3)
39. (−4, −2)

Cumulative Review, Pages 114–115

1. sometimes   **3.** sometimes   **5.** always   **7.**   **9.** not possible   **11.** $x = 13$;
$m\angle PQR = 156$

13. 15, 75, 90   **15.** 90   **17.** 90   **19.** 60   **21.** 60   **23.** False. If 2 lines are ∥, then they do not
intersect; true.   **25.** True. If an ∠ is not obtuse, then it is acute; false.   **27.** Vert. ∠ are ≅.
29. ∠ Add. Post.   **31.** The meas. of an ext. ∠ of a △ = the sum of the meas. of the 2 remote int. ∠.
33. The sum of the meas. of the ∠ of a △ is 180.   **35.** *X*   **37.** pentagon   **39.** ≅   **41.** biconditional
43. 1080   **45.** 1. $\overline{WX} \perp \overline{XY}$ (Given) 2. $\angle 1$ is comp. to $\angle 2$. (If the ext. sides of adj. ∠ are ⊥, then the ∠
are comp.) 3. $\angle 1$ is comp. to $\angle 3$. (Given) 4. $\angle 2 \cong \angle 3$ (Comps. of same ∠ are ≅.)

Chapter 4

Written Exercises, Pages 120–121

1. $\angle T$ **3.** CA **5.** $\triangle ATC$ **7.** $\angle E, \angle F, \angle S, \angle T$ **9.** $\angle L \cong \angle F, \angle X \cong \angle N, \angle R \cong \angle E, \overline{LX} \cong$ $\overline{FN}, \overline{XR} \cong \overline{NE}, \overline{LR} \cong \overline{FE}$ **11. a.** $\triangle RLA$ **b.** $\overline{RL}$ **c.** $\angle 3$, CPCT; $\overline{LR}$, If 2 lines are cut by a trans. and alt. int. $\underline{\angle}$ are $\cong$, then the lines are $\parallel$.   **d.** $\angle 4$, CPCT; $\overline{PL}, \overline{AR}$, If 2 lines are cut by a trans. and alt. int. $\underline{\angle}$ are $\cong$, then the lines are $\parallel$.   **13.** $C(7, -1)$ **15.** $\triangle ABC \cong \triangle EDF$ **17.** $\triangle ABC \cong \triangle FDE$ **19.** $F(2, 5)$, $F(6, 1)$ **21. a.** Since $NERO \cong MARO$, $\overline{NO} \cong \overline{OM}$. By the def. of midpt., O is the midpt. of $\overline{NM}$.   **b.** $\angle NOR$ and $\angle MOR$ are corr. $\underline{\angle}$ of $\cong$ quads.   **c.** If 2 lines form $\cong$ adj. $\underline{\angle}$, then the lines are $\perp$.   **23.** Yes; yes; yes

Mixed Review Exercises, Page 121

1. 1. $\overline{AD} \perp \overline{BC}; \overline{BA} \perp \overline{AC}$ (Given)   2. $\angle BDA$ and $\angle BAC$ are rt. $\underline{\angle}$. (Def. of $\perp$ lines)   3. $\triangle ABC$ and $\triangle DBA$ are rt. $\underline{\triangle}$. (Def. of rt. $\triangle$)   4. $\angle 1$ and $\angle B$ are comp.; $\angle 2$ and $\angle B$ are comp. (The acute $\underline{\angle}$ of a rt. $\triangle$ are comp.)   5. $\angle 1 \cong \angle 2$ (If 2 $\underline{\angle}$ are comp. of the same $\angle$, then the 2 $\underline{\angle}$ are $\cong$.)   **2.** 1. $\overline{FC}$ and $\overline{SH}$ bis. each other at A. (Given)   2. A is the midpt. of $\overline{FC}$ and $\overline{SH}$. (Def. of bis.)   3. $SA = \frac{1}{2}SH$ and $AC = \frac{1}{2}FC$ (Midpt. Thm.)   4. $FC = SH$ (Given)   5. $\frac{1}{2}FC = \frac{1}{2}SH$ (Mult. Prop. of =.)   6. $SA = AC$ (Substitution Prop.)

Written Exercises, Pages 124–127

1. $\triangle ABC \cong \triangle NPY$; ASA   **3.** $\triangle ABC \cong \triangle CKA$; SSS   **5.** No $\cong$ can be deduced.   **7.** $\triangle ABC \cong \triangle PQC$; SAS   **9.** $\triangle ABC \cong \triangle AGC$; ASA   **11.** $\triangle ABC \cong \triangle BST$; ASA   **13.** No $\cong$ can be deduced. **15.** $\triangle ABC \cong \triangle MNC$; ASA   **17.** 1. Given   2. T; Def. of $\perp$ lines   3. Def. of $\cong$ $\underline{\angle}$ 4. Given   5. $\overline{VT}$; Def. of midpt.   6. UVT; Vert. $\underline{\angle}$ are $\cong$.   7. RSV, UTV; ASA Post.   **19.** 1. E is the midpt. of $\overline{TP}$ and $\overline{MR}$. (Given)   2. $\overline{TE} \cong \overline{PE}; \overline{ME} \cong \overline{RE}$ (Def. of midpt.)   3. $\angle TEM \cong \angle PER$ (Vert. $\underline{\angle}$ are $\cong$.)   4. $\triangle TEM \cong \triangle PER$ (SAS Post.)   **21.** 1. Plane M bis. $\overline{AB}$. (Given)   2. $\overline{AO} \cong \overline{BO}$ (Def. of bis.)   3. $\overline{PO} \perp \overline{AB}$ (Given)   4. $\angle POA \cong \angle POB$ (If 2 lines are $\perp$, then they form $\cong$ adj. $\underline{\angle}$.)   5. $\overline{PO} \cong \overline{PO}$ (Refl. Prop.)   6. $\triangle POA \cong \triangle POB$ (SAS Post.)

23. Given: Isos. $\triangle ABC$ with $\overline{AC} \cong \overline{AB}$;
D is the midpt. of $\overline{CB}$.
Prove: $\triangle ACD \cong \triangle ABD$
Proof: 1. $\overline{AC} \cong \overline{AB}$; D is the midpt. of $\overline{CB}$. (Given)
    2. $\overline{CD} \cong \overline{BD}$ (Def. of midpt.)
    3. $\overline{AD} \cong \overline{AD}$ (Refl. Prop.)
    4. $\triangle ACD \cong \triangle ABD$ (SSS Post.)

27. SSS

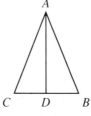

Written Exercises, Pages 130–132

1. 1. Given   2. Given   3. Def. of midpt.   4. Vert. $\underline{\angle}$ are $\cong$.   5. ASA Post.   6. CPCT   7. Def. of midpt. **3.** 1. $\overline{WO} \cong \overline{ZO}; \overline{XO} \cong \overline{YO}$ (Given)   2. $\angle WOX \cong \angle ZOY$ (Vert. $\underline{\angle}$ are $\cong$.)   3. $\triangle WOX \cong \triangle ZOY$ (SAS Post.)   4. $\angle W \cong \angle Z$ (CPCT)   **5.** 1. $\overline{SK} \parallel \overline{NR}; \overline{SN} \parallel \overline{KR}$ (Given)   2. $\angle 1 \cong \angle 3; \angle 2 \cong \angle 4$ (If 2 $\parallel$ lines are cut by a trans., then alt. int. $\underline{\angle}$ are $\cong$.)   3. $\overline{SR} \cong \overline{SR}$ (Refl. Prop.)   4. $\triangle SKR \cong \triangle RNS$ (ASA Post.)   5. $\overline{SK} \cong \overline{NR};$ $\overline{SN} \cong \overline{KR}$ (CPCT)   **7.** 1. $\overline{AD} \parallel \overline{ME}; \overline{MD} \parallel \overline{BE}$ (Given)   2. $\angle A \cong \angle EMB; \angle DMA \cong \angle B$ (If 2 $\parallel$ lines are cut by a trans., then corr. $\underline{\angle}$ are $\cong$.)   3. M is the midpt. of $\overline{AB}$. (Given)   4. $\overline{AM} \cong \overline{MB}$ (Def. of midpt.)   5. $\triangle ADM \cong \triangle MEB$ (ASA Post.)   6. $\overline{MD} \cong \overline{BE}$ (CPCT)   **9.** Either (a) $\angle 1 \cong \angle 2$ or (b) $\overline{QR} \cong \overline{SR}$ can be omitted.   **a.** 1. $\overline{PQ} \cong \overline{PS}; \overline{QR} \cong \overline{SR}$ (Given)   2. $\overline{PR} \cong \overline{PR}$ (Refl. Prop.)   3. $\triangle PQR \cong \triangle PSR$ (SSS Post.)   4. $\angle 3 \cong \angle 4$ (CPCT)   **b.** 1. $\overline{PQ} \cong \overline{PS}; \angle 1 \cong \angle 2$ (Given)   2. $\overline{PR} \cong \overline{PR}$ (Refl. Prop.)   3. $\triangle PQR \cong \triangle PSR$ (SAS Post.)   4. $\angle 3 \cong \angle 4$ (CPCT)   **11.** 1, 2   **13.** 1. $\overline{RS} \perp$ plane Y (Given)   2. $\overline{RS} \perp \overline{ST}; \overline{RS} \perp \overline{SV}$ (Def. of a line $\perp$ to a plane.)   3. $m\angle RST = 90; m\angle RSV = 90$ (Def. of $\perp$ lines)   4. $\angle RST \cong \angle RSV$ (Def. of $\cong$ $\underline{\angle}$)   5. $\angle TRS \cong \angle VRS$ (Given)   6. $\overline{RS} \cong \overline{RS}$ (Refl. Prop.)   7. $\triangle RST \cong \triangle RSV$ (ASA Post.)   8. $\overline{RT} \cong \overline{RV}$ (CPCT)   9. $\triangle RTV$ is isos. (Def. of isos. $\triangle$)   **15.** The wires are of equal length, so $PA = PB = PC$. The stakes are equidistant from the base of the tree, so $TA = TB = TC$. $PT = PT = PT$ by the Refl. Prop. and $\triangle PTA \cong \triangle PTB \cong \triangle PTC$ by SSS. The $\underline{\angle}$ that 3 wires make with the ground are $\cong$ parts of $\cong$ $\underline{\triangle}$.

Self-Test 1, Pages 132–133

1. $\angle P \cong \angle T$; CPCT   **2.** $\overline{KO}$, $\overline{MA}$; $\overline{OP}$, $\overline{AT}$; $\overline{KP}$, $\overline{MT}$ **3.** $\triangle JKX \cong \triangle JKY$; SAS   **4.** No $\cong$ can be deduced.   **5.** $\triangle TRP \cong \triangle TRS$; ASA   **6.** 1. $\angle 1 \cong \angle 2$; $\angle 3 \cong \angle 4$ (Given)   2. $\overline{DB} \cong \overline{DB}$ (Refl. Prop.) 3. $\triangle ADB \cong \triangle CBD$ (ASA Post.)   **7.** 1. $\overline{CD} \cong \overline{AB}$; $\overline{CB} \cong \overline{AD}$ (Given)   2. $\overline{DB} \cong \overline{DB}$ (Refl. Prop.) 3. $\triangle ADB \cong \triangle CBD$ (SSS Post.)   4. $\angle 1 \cong \angle 2$ (CPCT)   **8.** 1. $\overline{AD} \parallel \overline{BC}$ (Given)   2. $\angle 4 \cong \angle 3$ (If 2 $\parallel$ lines are cut by a trans., then alt. int. $\angle\!\!\!\angle$ are $\cong$.)   3. $\overline{AD} \cong \overline{CB}$ (Given)   4. $\overline{DB} \cong \overline{DB}$ (Refl. Prop.) 5. $\triangle ADB \cong \triangle CBD$ (SAS Post.)   6. $\angle 1 \cong \angle 2$ (CPCT)   7. $\overline{DC} \parallel \overline{AB}$ (If 2 lines are cut by a trans. and alt. int. $\angle\!\!\!\angle$ are $\cong$, then the lines are $\parallel$.)

Written Exercises, Pages 137–139

1. 80   **3.** 53   **5.** 5   **7.** 41   **9.** Answers may vary; c, d, b, a   **11.** 1. $\overline{AB} \cong \overline{AC}$ (Given)   2. Let the bis. of $\angle A$ int. $\overline{BC}$ at D. (By the Protractor Post., an $\angle$ has exactly one bis.)   3. $\angle BAD \cong \angle CAD$ (Def. of $\angle$ bis.)   4. $\overline{AD} \cong \overline{AD}$ (Refl. Prop.)   5. $\triangle BAD \cong \triangle CAD$ (SAS Post.)   6. $\angle B \cong \angle C$ (CPCT)   **13.** 1. $\angle 1 \cong \angle 2$ (Given)   2. $\overline{JG} \cong \overline{JM}$ (If 2 $\angle\!\!\!\angle$ of a $\triangle$ are $\cong$, then the sides opp. those $\angle\!\!\!\angle$ are $\cong$.)   3. M is the midpt. of $\overline{JK}$. (Given)   4. $\overline{JM} \cong \overline{MK}$ (Def. of midpt.)   5. $\overline{JG} \cong \overline{MK}$ (Trans. Prop.)   **15.** 1, 3   **17.** 1. $\overline{XY} \cong \overline{XZ}$ (Given) 2. $\angle XYZ \cong \angle XZY$ or $m \angle XYZ = m \angle XZY$ (Isos. $\triangle$ Thm.)   3. $m \angle XYZ = m \angle 1 + m \angle 2$; $m \angle XZY = m \angle 3 + m \angle 4$ ($\angle$ Add. Post.)   4. $m \angle 1 + m \angle 2 = m \angle 3 + m \angle 4$ (Substitution Prop.)   5. $\overline{OY} \cong \overline{OZ}$ (Given)   6. $\angle 2 \cong \angle 3$ or $m \angle 2 = m \angle 3$ (Isos. $\triangle$ Thm.)   7. $m \angle 1 = m \angle 4$ (Subtr. Prop. of =) **19.** 1. $\overline{AB} \cong \overline{AC}$ (Given)   2. $\angle B \cong \angle C$ (Isos. $\triangle$ Thm.)   3. $\overline{AL}$ and $\overline{AM}$ trisect $\angle BAC$, so $\angle 1 \cong \angle 3$. (Given) 4. $\triangle BLA \cong \triangle CMA$ (ASA Post.)   5. $\overline{AL} \cong \overline{AM}$ (CPCT)   **21.** 1. $\overline{OP} \cong \overline{OQ}$; $\angle 3 \cong \angle 4$ (Given) 2. $\angle POS \cong \angle QOR$ (Vert. $\angle\!\!\!\angle$ are $\cong$.)   3. $\triangle POS \cong \triangle QOR$ (ASA Post.)   4. $\overline{OS} \cong \overline{OR}$ (CPCT)   5. $\angle 5 \cong \angle 6$ (Isos. $\triangle$ Thm.)   **23. a.** 40, 40, 60   **b.** $2x, 2x, 3x$ **25. a.** 90   **b.** 90   **27.** $x = 2$, $y = 1$ **29.** $x = 30$, $y = 10$ **31. a.** Key steps of proof: 1. $\triangle JKM \cong \triangle JKN$ and $\triangle LKM \cong \triangle LKN$ (SAS Post.) 2. $\overline{JM} \cong \overline{JN}$ and $\overline{LM} \cong \overline{LN}$ (CPCT)   3. $\triangle JMN$ and $\triangle LMN$ are isos. (Def. of isos. $\triangle$)   **b.** No. They are $\cong$ if and only if $\overline{KJ} \cong \overline{KL}$.   **33.** $m \angle EAF = 9$, $m \angle AFD = 54$, $m \angle DAF = 45$

Written Exercises, Pages 143–145

1. 1. Given   2. Def. of rt. $\triangle$ 3. Given   4. $\overline{XZ} \cong \overline{XZ}$ 5. $\triangle XYZ$; HL   6. $\overline{WZ} \cong \overline{YZ}$; CPCT   **3.** 1. $\overline{EF} \perp \overline{EG}$; $\overline{HG} \perp \overline{EG}$ (Given)   2. $\angle HGE$ and $\angle FEG$ are rt. $\angle\!\!\!\angle$. (Def. of $\perp$ lines)   3. $\triangle HGE$ and $\triangle FEG$ are rt. $\triangle\!\!\!\triangle$. (Def. of rt. $\triangle$)   4. $\overline{EH} \cong \overline{GF}$ (Given)   5. $\overline{EG} \cong \overline{EG}$ (Refl. Prop.)   6. $\triangle HGE \cong \triangle FEG$ (HL)   7. $\angle H \cong \angle F$ (CPCT)   **5.** SAS   **7.** HL   **9. a.** 1. $\overline{PR} \cong \overline{PQ}$ (Given)   2. $\angle PQR \cong \angle PRQ$ (Isos. $\triangle$ Thm.)   3. $\overline{SR} \cong \overline{TQ}$ (Given)   4. $\overline{RQ} \cong \overline{RQ}$ (Refl. Prop.)   5. $\triangle RQS \cong \triangle QRT$ (SAS Post.)   6. $\overline{QS} \cong \overline{RT}$ (CPCT)   **b.** 1. $\overline{PR} \cong \overline{PQ}$ or $PR = PQ$; $\overline{SR} \cong \overline{TQ}$ or $SR = TQ$ (Given)   2. $PR = PS + SR$; $PQ = PT + TQ$ (Seg. Add. Post.) 3. $PS + SR = PT + TQ$ (Substitution Prop.)   4. $PS = PT$ or $\overline{PS} \cong \overline{PT}$ (Subtr. Prop. of =)   5. $\angle P \cong \angle P$ (Refl. Prop.)   6. $\triangle PQS \cong \triangle PRT$ (SAS Post.)   7. $\overline{QS} \cong \overline{RT}$ (CPCT)   **11.** $\overline{PR} \cong \overline{PS}$, $\overline{PQ} \cong \overline{PT}$, $\overline{QR} \cong \overline{TS}$; SSS   **13.** $\angle 3 \cong \angle 4$, $\overline{PQ} \cong \overline{PT}$, $\angle 6 \cong \angle 5$; AAS   **15.** 1. $\angle 1 \cong \angle 2 \cong \angle 3$ (Given)   2. $\overline{ME} \cong \overline{MD}$ (If 2 $\angle\!\!\!\angle$ of a $\triangle$ are $\cong$, then the sides opp. those $\angle\!\!\!\angle$ are $\cong$.)   3. $\overline{EN} \cong \overline{DG}$ (Given)   4. $\triangle MEN \cong \triangle MDG$ (SAS Post.)   5. $\angle 4 \cong \angle 5$ (CPCT)

17. Given: Isos. $\triangle XYZ$ with $\overline{XY} \cong \overline{XZ}$;
  $\overline{ZA} \perp \overline{XY}$; $\overline{YB} \perp \overline{XZ}$
  Prove: $\overline{ZA} \cong \overline{YB}$

Proof: 1. $\overline{ZA} \perp \overline{XY}$; $\overline{YB} \perp \overline{XZ}$ (Given)   2. $m \angle XBY = 90$; $m \angle XAZ = 90$ (Def. of $\perp$ lines)   3. $\angle XBY \cong \angle XAZ$ (Def. of $\cong$ $\angle\!\!\!\angle$)   4. $\angle X \cong \angle X$ (Refl. Prop.) 5. $\overline{XY} \cong \overline{XZ}$ (Given)   6. $\triangle XBY \cong \triangle XAZ$ (AAS Thm.)   7. $\overline{ZA} \cong \overline{YB}$ (CPCT)

Self-Test 2, Page 146

1. 70   **2.** 7   **3.** 30   **4.** $\overline{AB} \cong \overline{AC}$, $\angle A \cong \angle A$, $\angle ANB \cong \angle AMC$, so $\triangle ABN \cong \triangle ACM$ by AAS. **5.** 1. $\overline{BN} \perp \overline{AC}$; $\overline{CM} \perp \overline{AB}$ (Given)   2. $\angle BMC$ and $\angle CNB$ are rt. $\angle\!\!\!\angle$. (Def. of $\perp$ lines)   3. $\triangle BMC$ and $\triangle CNB$ are rt. $\triangle\!\!\!\triangle$. (Def. of rt. $\triangle$)   4. $\overline{MB} \cong \overline{NC}$ (Given)   5. $\overline{BC} \cong \overline{BC}$ (Refl. Prop.)   6. $\triangle BMC \cong \triangle CNB$ (HL) 7. $\overline{CM} \cong \overline{BN}$ (CPCT)

Written Exercises, Pages 148–151

1. a. SSS **b.** CPCT **c.** SAS **d.** CPCT **3. a.** AAS **b.** CPCT **c.** SAS **d.** CPCT **5. a.** SAS
b. CPCT **c.** HL **d.** CPCT **7. a.** 1. $\triangle FLA \cong \triangle FKA$ (SSS) 2. $\angle 1 \cong \angle 2$ (CPCT) 3. $\triangle FLJ \cong \triangle FKJ$
(SAS) 4. $\overline{LJ} \cong \overline{KJ}$ (CPCT) **b.** 1. $\overline{LF} \cong \overline{KF}; \overline{LA} \cong \overline{KA}$ (Given) 2. $\overline{FA} \cong \overline{FA}$ (Refl. Prop.) 3. $\triangle FLA \cong$
$\triangle FKA$ (SSS) 4. $\angle 1 \cong \angle 2$ (CPCT) 5. $\overline{FJ} \cong \overline{FJ}$ (Refl. Prop.) 6. $\triangle FLJ \cong \triangle FKJ$ (SAS) 7. $\overline{LJ} \cong \overline{KJ}$
(CPCT) **9.** Key steps of proof: 1. $\overline{ST} \cong \overline{YZ}; \angle T \cong \angle Z; \angle RST \cong \angle XYZ$ (CPCT) 2. $m \angle KST =$
$\frac{1}{2}m \angle RST; m \angle LYZ = \frac{1}{2}m \angle XYZ$ ($\angle$ Bis. Thm.) 3. $m \angle KST = m \angle LYZ$ (Substitution Prop.) 4. $\triangle KST \cong$
$\triangle LYZ$ (ASA) 5. $\overline{SK} \cong \overline{YL}$ (CPCT) **11.** Key steps of proof: 1. $\triangle GDE \cong \triangle EFG$ (SSS) 2. $\angle DEH \cong$
$\angle FGK$ (CPCT) 3. $\triangle HDE \cong \triangle KFG$ (ASA) 4. $\overline{DH} \cong \overline{FK}$ (CPCT) **17.** isos.; $\overline{AX} \cong \overline{AY}, \overline{AZ} \cong \overline{AZ}$, and
$\angle XAZ \cong \angle YAZ$, so $\triangle XAZ \cong \triangle YAZ$ by SAS. Then $\overline{XZ} \cong \overline{YZ}$ (CPCT) and $\triangle XYZ$ is isos.

Mixed Review Exercises, Page 151

1. Two sides of a $\triangle$ are $\cong$ if and only if the $\angle$ opp. those sides are $\cong$. **2.** sometimes **3.** sometimes
4. always **5. a.** ●——●————● **b.** ●—+——●—+——●
   $\quad A \quad M \quad\quad\quad B$ $\qquad A \quad M \quad\quad B$

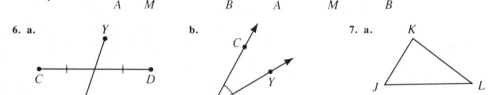
6. a. **b.** **7. a.**

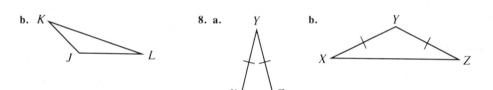
b. **8. a.** **b.**

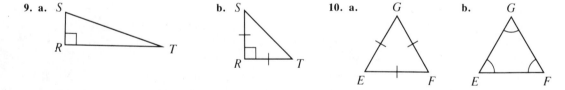
9. a. **b.** **10. a.** **b.**

11. 1. $\overline{BE} \cong \overline{CD}; \overline{BD} \cong \overline{CE}$ (Given) 2. $\overline{BC} \cong \overline{BC}$ (Refl. Prop.) 3. $\triangle EBC \cong \triangle DCB$ (SSS) 4. $\angle EBC \cong$
$\angle DCB$ (CPCT) 5. $\overline{AB} \cong \overline{AC}$ (If 2 $\angle$ of a $\triangle$ are $\cong$, then the sides opp. those $\angle$ are $\cong$.) 6. $\triangle ABC$ is isos.
(Def. of isos. $\triangle$)

Written Exercises, Pages 156–158

Sketches may vary in Exs. 1–5. **1. b.** No **5.** Yes; at the midpt. of the hyp. **7.** $\overrightarrow{KS}, \overrightarrow{KN}$ **9.** bis. of
$\angle S$ **11.** A, F **13.** 1. P is on the $\perp$ bisectors of $\overline{AB}$ and $\overline{BC}$. (Given) 2. $PA = PB; PB = PC$ (If a pt.
lies on the $\perp$ bis. of a seg., then it is equidistant from the end pts. of the seg.) 3. $PA = PC$ (Trans. Prop.)
15. Key steps of proof: 1. Let X be the midpt. of $\overline{BC}$. (Ruler Post.) 2. $\triangle AXB \cong \triangle AXC$ (SSS) 3. $\angle 1 \cong \angle 2$
(CPCT) 4. $\overline{AX} \perp \overline{BC}$ (If 2 lines form $\cong$ adj. $\angle$, then the lines are $\perp$.) 5. $\overline{AX}$ is the $\perp$ bis. of $\overline{BC}$. (Def. of $\perp$
bis.) **17.** Key steps of proof: 1. $\overrightarrow{PX} \perp \overrightarrow{BA}; \overrightarrow{PY} \perp \overrightarrow{BC}; PX = PY$ (Given) 2. $\triangle PXB \cong \triangle PYB$ (HL)
3. $\angle PBX \cong \angle PBY$ (CPCT) 4. $\overrightarrow{BP}$ bis. $\angle ABC$. (Def. of $\angle$ bis.)

21. a. Key steps of proof: 1. $\overline{AB} \cong \overline{AC}$; $\overline{BD} \perp \overline{AC}$; $\overline{CE} \perp \overline{AB}$ (Given)   2. $\triangle ADB \cong \triangle AEC$ (AAS)
3. $\overline{BD} \cong \overline{CE}$ (CPCT)   **b.** The altitudes drawn to the legs of an isos. $\triangle$ are $\cong$.   **23.** Q is on the $\perp$ bis. of $\overline{PS}$,
so $PQ = SQ$. S is on the $\perp$ bis. of $\overline{QT}$, so $QS = TS$. Then $PQ = TS$ by the Trans. Prop.   **25. a.** $\overline{OD}$ is a
$\perp$ bis. of $\overline{AB}$, so $\overline{AD} \cong \overline{BD}$.   **b.** $\overline{OC}$ is a $\perp$ bis. of $\overline{AB}$, so $\overline{AC} \cong \overline{BC}$.   **c.** By parts (a) and (b) above, $\overline{AD} \cong \overline{BD}$
and $\overline{AC} \cong \overline{BC}$. Then since $\overline{CD} \cong \overline{CD}$, $\triangle CAD \cong \triangle CBD$ by SSS and $\angle CAD \cong \angle CBD$ (CPCT).

Self-Test 3, Page 159

1. $\overline{EA} \cong \overline{DB}$ and $\angle AEB \cong \angle BDA$ **2.** 1. $\triangle MPQ \cong \triangle PMN$ (Given)   2. $\overline{MN} \cong \overline{QP}$; $\angle MPQ \cong \angle PMN$
(CPCT)   3. $\overline{MS} \cong \overline{PR}$ (Given)   4. $\triangle MSN \cong \triangle PRQ$ (SAS)   **3. a.** $\overline{LJ}$ or $\overline{KJ}$ **b.** $\overline{KZ}$ **4.** No   **5.** If a
pt. lies on the bis. of an $\angle$, then the pt. is equidistant from the sides of the $\angle$.   **6.** If a pt. is equidistant from
the endpts. of a seg., then the pt. lies on the $\perp$ bis. of the seg.

Chapter Review, Pages 160–161

1. $\triangle QPR$ **3.** $\angle W$ **5.** Yes; SSS   **7.** Yes; ASA   **9.** 1. $\overline{JM} \cong \overline{LM}$; $\overline{JK} \cong \overline{LK}$ (Given)   2. $\overline{MK} \cong \overline{MK}$
(Refl. Prop.)   3. $\triangle MJK \cong \triangle MLK$ (SSS)   4. $\angle MJK \cong \angle MLK$ (CPCT)   **11.** $\overline{ER}, \overline{EV}$ **13.** 25
15. 1. $\overline{GH} \perp \overline{HJ}$; $\overline{KJ} \perp \overline{HJ}$ (Given)   2. $m \angle GHJ = 90$; $m \angle KJH = 90$ (Def. of $\perp$ lines)   3. $\angle GHJ \cong \angle KJH$
(Def. of $\cong$ $\angle$)   4. $\angle G \cong \angle K$ (Given)   5. $\overline{HJ} \cong \overline{HJ}$ (Refl. Prop.)   6. $\triangle GHJ \cong \triangle KJH$ (AAS)
17. 1. ASA   2. CPCT   3. HL   4. CPCT   5. If 2 lines are cut by a trans. and alt. int. $\angle$ are $\cong$, then the lines
are $\parallel$.   **19.** If a pt. lies on the $\perp$ bis. of a seg., then the pt. is equidistant from the endpts. of the seg.

Algebra Review, Page 163

1. $-6, 1$ **3.** $-2, 9$ **5.** $0, 13$ **7.** $-13, 13$ **9.** $-0.2, 0.2$ **11.** 3   **13.** $-6, -2$ **15.** 5
17. $-4, 5$ **19.** $\dfrac{-3 \pm \sqrt{57}}{6}$ **21.** $\dfrac{-5 \pm \sqrt{17}}{2}$ **23.** $\dfrac{5 \pm \sqrt{13}}{2}$ **25.** $1, 9$ **27.** $-7, 2$ **29.** 20
31. 1   **33.** 1.5

Preparing for College Entrance Exams, Page 164
1. A   **2.** C   **3.** D   **4.** C   **5.** B   **6.** C   **7.** E   **8.** D   **9.** B

Cumulative Review, Page 165
1. Seg. Add. Post.   **3.** obtuse   **5.** 16   **7.** 10   **9.** SSS   **11.** $m \angle 5 = 90$, $m \angle 6 = 54$, $m \angle 7 = 36$,
$m \angle 8 = 54$ **13.** No   **15.** Yes; $a \parallel b$ **17.** Key steps of proof: 1. $\overline{MO} \perp \overline{NP}$, $\overline{NO} \cong \overline{PO}$ (Given)
2. $\triangle NQO \cong \triangle PQO$ (HL)   3. $\angle NOQ \cong \angle POQ$ (CPCT)   4. $\triangle MNO \cong \triangle MPO$ (SAS)   5. $\overline{MN} \cong \overline{MP}$
(CPCT)

Chapter 5

Written Exercises, Pages 169–171
1. $\overline{CR}, \overline{CE}$ **3.** $\overline{ER}, \overline{RC}, \overline{CW}$ **5.** $a = 8, b = 10, x = 118, y = 62$ **7.** $a = 5, b = 3, x = 120$,
$y = 22$ **9.** $a = 8, b = 8, x = 56, y = 68$ **11.** 60   **17.** $(3, 2)$ **19.** $x = 3, y = 5$ **21.** $x = 13$,
$y = 5$ **23.** $x = 5, y = 4$ **25.** $5, 2$ **27.** $10, 70$ **29.** 1. $PQRS$ is a $\square$; $\overline{PJ} \cong \overline{RK}$ (Given)
2. $\angle P \cong \angle R$ (Thm. 5-2)   3. $\overline{SP} \cong \overline{QR}$ (Thm. 5-1)   4. $\triangle SPJ \cong \triangle QRK$ (SAS)   5. $\overline{SJ} \cong \overline{QK}$ (CPCT)
31. 1. $ABCD$ is a $\square$; $\overline{CD} \cong \overline{CE}$ (Given)   2. $\overline{AB} \parallel \overline{CD}$ (Def. of $\square$)   3. $\angle CDE \cong \angle A$ (If lines $\parallel$, corr. $\angle \cong$.)
4. $\angle CDE \cong \angle E$ (Isos. $\triangle$ Thm.)   5. $\angle A \cong \angle E$ (Subst.)   **35.** $(6, 0), (0, 8), (12, 8)$

Written Exercises, Pages 174–176
1. Def. of $\square$ **3.** Thm. 5-5   **5.** Thm. 5-6   **7.** Thm. 5-7   **9. a.** Thm. 5-4   **b.** Thm. 5-6
c. Thm. 5-7   **15.** $m \angle DAB = m \angle BCD$, so $m \angle NAM = \dfrac{1}{2} m \angle DAB = \dfrac{1}{2} m \angle BCD = m \angle NCM$. $m \angle DNA =$
$m \angle NAM = m \angle NCM$, so $\overline{AN}$ and $\overline{CM}$ are $\parallel$. $\overline{CN}$ and $\overline{AM}$ are $\parallel$ because $ABCD$ is a $\square$. Then $AMCN$ is a $\square$, by
def. of $\square$.   **17.** Draw $\overline{AC}$ int. $\overline{DB}$ at Z. Since $DZ = ZB$ and $DE = FB$, $EZ = DZ - DE = ZB - FB = ZF$.
Also, $AZ = ZC$. If the diags. of a quad. bis. each other, then the quad. is a $\square$. So $AFCE$ is a $\square$.
19. $x = 18, y = 14$ **21.** $x = 10, y = 2$ **23.** Key steps of proof: 1. $\triangle DAE \cong \triangle BCF$ (AAS)
2. $\overline{DE} \cong \overline{BF}$ (CPCT)   3. $\overline{DE} \parallel \overline{BF}$ (Thm. 3-7)   4. $DEBF$ is a $\square$. (Thm. 5-5)

Written Exercises, Pages 180–182

1. 12, 12   **3.** 4   **5. a.** 40 **b.** 20 **c.** 26 **d.** 34   **7.** D, E **9.** D, F **11.** 6   **13.** 11

15. 3, 2   **17.** $x = 4$, $y = 2$ **19.** 1. $\overline{BE} \parallel \overline{MD}$; M is the midpt. of $\overline{AB}$. (Given)   2. D is the midpt. of $\overline{AE}$. (Thm. 5-10)   3. $\overline{DE} \cong \overline{AD}$ (Def. of midpt.)   4. $ABCD$ is a $\square$. (Given)   5. $\overline{AD} \cong \overline{BC}$ (Thm. 5-1)   6. $DE = BC$ (Trans. Prop.)

23. Given: X, Y, and Z are the midpts. of $\overline{AB}$, $\overline{AC}$, and $\overline{BC}$, resp.; P and Q
   are midpts. of $\overline{BZ}$ and $\overline{CZ}$, resp.
   Prove: $PX = QY$

Key steps of proof: 1. $PX = \frac{1}{2}AZ$; $QY = \frac{1}{2}AZ$ (Thm. 5-11, part (2))

2. $PX = QY$ (Subst.)

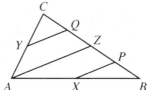

Self-Test 1, Page 182

1. may be   **2.** must be   **3.** must be   **4.** cannot be   **5.** See page 172.   **6. a.** If 3 $\parallel$ lines cut off $\cong$ seg. on one trans., then they cut off $\cong$ seg. on every trans.   **b.** $x = 6$, $y = 19$ **7.** 1. $ABCD$ is a $\square$. (Given) 2. $\overline{AC}$ and $\overline{BD}$ bis. each other. (Diags. $\square$ bis. each other.)   3. O is the midpt. of $\overline{BD}$. (Def. of bis.)   4. M is the midpt. of $\overline{AB}$. (Given)   5. $MO = \frac{1}{2}AD$ (Thm. 5-11)   **8.** 1. $PQRS$ is a $\square$. (Given)   2. $\overline{SR} \parallel \overline{PQ}$; $\overline{SP} \parallel \overline{RQ}$ (Def. of $\square$)   3. $m \angle QPR = m \angle SRP$ (If lines $\parallel$, alt. int. $\angle$ $\cong$.)   4. $\overline{PX}$ bis. $\angle QPR$; $\overline{RY}$ bis. $\angle SRP$. (Given)   5. $m \angle RPX = \frac{1}{2}m \angle QPR$; $m \angle PRY = \frac{1}{2}m \angle SRP$ ($\angle$ Bis. Thm.)   6. $\frac{1}{2}m \angle QPR = \frac{1}{2}m \angle SRP$ (Mult. Prop. of =)   7. $m \angle RPX = m \angle PRY$ (Subst.)   8. $\overline{YR} \parallel \overline{PX}$ (If alt. int. $\angle$ $\cong$, lines $\parallel$.)   9. $RYPX$ is a $\square$. (Def. of $\square$)

Written Exercises, Pages 187–189

1. all   **3.** all   **5.** all   **7.** rhom., sq.   **9.** rect., sq.   **11.** 25, 65, 65, 90   **13.** 10   **15.** $13\frac{1}{2}$

17. 32, 58, 58   **19.** $\frac{1}{2}$ **21.** (2, 5); no   **23.** (4, 6); yes   **25.** 15   **27.** 60   **29.** 1. $ABZY$ is a $\square$. (Given)   2. $\overline{BZ} \cong \overline{AY}$ (Thm. 5-1)   3. $\overline{AY} \cong \overline{BX}$ (Given)   4. $\overline{BZ} \cong \overline{BX}$ (Trans. Prop.)   5. $\angle 1 \cong \angle 2$ (Isos. $\triangle$ Thm.)   6. $\overline{BZ} \parallel \overline{AY}$ (Def. of $\square$)   7. $\angle 2 \cong \angle 3$ (If lines $\parallel$, corr. $\angle$ are $\cong$.)   8. $\angle 1 \cong \angle 3$ (Trans. Prop.)   **31.** 1. $QRST$ is a rect.; $RKST$ and $JQST$ are $\square$. (Given)   2. $\overline{KS} \cong \overline{RT}$; $\overline{JT} \cong \overline{QS}$ (Thm. 5-1)   3. $\overline{RT} \cong \overline{QS}$ (Diags. of rect. $\cong$.)   4. $\overline{JT} \cong \overline{KS}$ (Subst.)   **37.** square   **39.** 6, 8

Mixed Review Exercises, Page 189

1. 13   **2.** 20   **3.** 11   **4.** 9   **5.** 8.2   **6.** -1.5 **7.** 1   **8.** 3.45   **9. a.** 23 **b.** 2 **c.** 4 **d.** -6

Written Exercises, Pages 192–194

1. 12   **3.** 15   **5.** 9   **7.** 4   **9.** 6   **11.** $x = 10$; 40, 40, 140, 140   **13.** $BE = \frac{1}{2}(AD + CF)$

15. 13, 39   **17.** 9, 15   **19.** $CF = 3 \cdot AD$, but $17 \neq 3 \cdot 5$.   **21.** rect.   **23.** rhom.   **25.** $\square$
27. Given: Trap. $ABXY$ with $\overline{BX} \cong \overline{AY}$
   Prove: $\angle 1 \cong \angle 3$; $\angle ABX \cong \angle A$
Key steps of proof: 1. $ABZY$ is a $\square$. (Def. of $\square$)   2. $\angle 1 \cong \angle 2$ (Isos. $\triangle$ Thm.)   3. $\angle 2 \cong \angle 3$ (If lines $\parallel$, corr. $\angle$ $\cong$.)   4. $\angle 1 \cong \angle 3$ (Trans. Prop.)   5. $\angle ABX \cong \angle A$ (Supps. of $\cong$ $\angle$ are $\cong$.)   **29. a.** rect.
b. rect.   **33.** rhom.   **35. a.** Drawings may vary.   **b.** The diags. must be $\cong$.

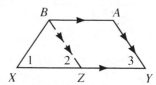

Self-Test 2, Page 195

1. □   **2.** trap.   **3.** rect.   **4.** sq.   **5.** 11   **6.** 17, 67   **7.** 1. $\angle 1 \cong \angle 2 \cong \angle 3 \cong \angle 4$ (Given)
2. $\overline{HG} \parallel \overline{EF}$; $\overline{HE} \parallel \overline{GF}$ (If alt. int. ⩘ ≅, lines ∥.)   3. $EFGH$ is a □. (Def. of □)   4. $\overline{HG} \cong \overline{HE}$ (If 2 ⩘ of a △
are ≅, sides opp. the ⩘ are ≅.) 5. $HGFE$ is a rhom. (Thm. 5-17)   **8. a.** □   **b.** 1. $PQRS$ is a □. (Given)

2. $\overline{PQ} \parallel \overline{SR}$ (Def. of □)   3. X is the midpt. of $\overline{PQ}$; Y is the midpt. of $\overline{SR}$. (Given)   4. $XQ = \frac{1}{2}PQ$; $YR = \frac{1}{2}SR$

(Midpt. Thm.)   5. $PQ = SR$ (Thm. 5-1)   6. $\frac{1}{2}PQ = \frac{1}{2}SR$ (Mult. Prop. of =)   7. $XQ = YR$ (Subst.)

8. $XQRY$ is a □. (Thm. 5-5)   **c.** trap.

Chapter Review, Pages 197–198

1. 110   **3.** 28   **5.** $GS = 5$ or $\overline{SA} \parallel \overline{GN}$ **7.** $\overline{AZ} \cong \overline{GZ}$ **9.** Thm. 5-10   **11.** Thm. 5-11, part (2)
13. □   **15.** rect.   **17.** Key steps of proof: 1. $DO = BO$; $AO = CO$ (Diags. □ bis. each other.)
2. $EO = FO$ (Subtr. Prop. =)   3. $AECF$ is a □. (Thm. 5-7)   4. $\overline{BD} \perp \overline{AC}$ (Diags. of rhom. ⊥.)
5. $\triangle COE \cong \triangle COF$ (SAS)   6. $\overline{CE} \cong \overline{CF}$ (CPCT)   7. $AECF$ is a rhom. (Thm. 5-17)   **19.** $\overline{ZO}, \overline{DI}$ **21.** 4

Cumulative Review, Pages 200–201

1. one   **3.** If you enjoy winter weather, then you are a member of the skiing club.   **5.** Trans. Prop.
7. 180; ∠ Add. Post.   **9.** $\angle 1$; If lines ∥, corr. ⩘ are ≅.   **11.** bis., ⊥   **13. a.** $\triangle RTA$ **b.** $\overline{DB}$
c. $m\angle E$ **15.** 150, 150   **17.** $3r - s$ **19.** bis.   **21.** 72, 36   **23.** ABC, BAC, ACD, CFD
25. $m\angle 1 = m\angle 4 = k$, $m\angle 2 = m\angle 3 = 45 - k$ **27.** $\angle NOM, \angle LMO, \angle NMO$; Thm. 5-14   **29.** $\overline{PQ}$,
$\overline{ON}$; Thm. 5-19   **31.** 1. $\overline{AD} \cong \overline{BC}$; $\overline{AD} \parallel \overline{BC}$ (Given)   2. $ABCD$ is a □. (Thm. 5-5)   3. $\overline{DF} \cong \overline{BF}$ (Diags.
□ bis. each other.)   4. $\angle DFG \cong \angle BFE$ (Vert. ⩘ ≅.)   5. $\overline{DC} \parallel \overline{AB}$ (Def. of □)   6. $\angle CDB \cong \angle ABD$ (If ∥
lines, alt. int. ⩘ ≅.)   7. $\triangle DFG \cong \triangle BFE$ (ASA)   8. $\overline{EF} \cong \overline{FG}$ (CPCT)

Chapter 6

Written Exercises, Pages 206–207

1. a. No   **b.** Yes   **c.** Yes   **d.** No   **e.** Yes   **f.** No   **3. a.** No   **b.** No   **c.** Yes   **d.** Yes   **5.** $j = 2, k = 1$,
$l = 4, m = 3$ **7.** 1. Vert. ⩘ ≅.   2. ∠ Add Post.   3. A Prop. of Ineq.   4. Subst.   **9.** 1. $m\angle ROS >$
$m\angle TOV$ (Given)   2. $m\angle SOT = m\angle SOT$ (Reflex.)   3. $m\angle ROS + m\angle SOT > m\angle TOV + m\angle SOT$
(A Prop. of Ineq.)   4. $m\angle ROS + m\angle SOT = m\angle ROT$; $m\angle TOV + m\angle SOT = m\angle SOV$ (∠ Add. Post.)
5. $m\angle ROT > m\angle SOV$ (Subst.)   **11.** 1. $m\angle 1 > m\angle 2$; $m\angle 2 > m\angle 3$ (Ext. ∠ Ineq. Thm.)   2. $m\angle 1 >$
$m\angle 3$ (A Prop. of Ineq.)   3. $m\angle 3 = m\angle 4$ (Vert. ⩘ ≅.)   4. $m\angle 1 > m\angle 4$ (Subst.)

Written Exercises, Pages 210–212

1. a. If $4n \neq 68$, then $n \neq 17$.   **b.** If $n \neq 17$, then $4n \neq 68$.   **3. a.** If $x + 1$ is odd, then x is even.   **b.** If
x is even, then $x + 1$ is odd.   **5.** True. If I don't live in Calif., then I don't live in L.A.; true. If I live in
Calif., then I live in L.A.; false. If I don't live in L.A., then I don't live in Calif.; false.   **7.** False. If M is not
the midpt. of $\overline{AB}$, then $AM \neq MB$; false. If M is the midpt. of $\overline{AB}$, then $AM = MB$; true. If $AM \neq MB$, then M is
not the midpt. of $\overline{AB}$; true.   **9.** True. If $n \leq -3$, then $-2n \geq 6$; true. If $n > -3$, then $-2n < 6$; true. If
$-2n \geq 6$, then $n \leq -3$; true.   **11.** If you are a senator, then you are at least 30 years old.   **a.** No concl.
b. She is at least 30 years old.   **c.** No concl.   **d.** He is not a senator.   **13. a.** It is raining.   **b.** I am happy.
c. No concl.   **d.** No concl.   **15. a.** No concl.   **b.** $\angle ABC$ and $\angle DBF$ are not vert. ⩘.   **c.** No concl.
d. $\angle RVU \cong \angle SVT$, $\angle RVT \cong \angle SVU$ **17. a.** Diags. are ≅.   **b.** No concl.   **c.** No concl.   **d.** $STAR$ is not
a rect.   **19.** contrapositive   **21.** Statement: If $m\angle A + m\angle B \neq 180$, then $m\angle D + m\angle C \neq 180$.
Contrapositive: If $m\angle D + m\angle C = 180$, then $m\angle A + m\angle B = 180$.   Given: $m\angle D + m\angle C = 180$ Prove:
$m\angle A + m\angle B = 180$ Proof: 1. $m\angle D + m\angle C = 180$ (Given)   2. $\overleftrightarrow{AD} \parallel \overleftrightarrow{BC}$ (If s-s. int. ⩘ supp., lines ∥.)
3. $m\angle A + m\angle B = 180$ (If ∥ lines, s-s. int. ⩘ supp.)

Mixed Review Exercises, Page 212

1. sometimes   **2.** sometimes   **3.** always   **4.** never   **5.** always   **6.** always   **7.** sometimes
8. $m\angle 1 = 60$, $m\angle 2 = 75$, $m\angle 3 = 45$, $m\angle 4 = 60$ **9.** 95

Written Exercises, Pages 216–217

1. Assume temp. that $m \angle B \neq 40$.   **3.** Assume temp. that $a - b = 0$.   **5.** Assume temp. that $\overleftrightarrow{EF} \parallel \overleftrightarrow{GH}$.
7. Assume temp. that $\angle Y$ is a rt. $\angle$. Since $m \angle X = 100$, this contradicts Thm. 3-11 Cor. 3. The temp. assumption must be false. It follows that $\angle Y$ is not a rt. $\angle$.   **11.** Assume temp. that planes P and Q do not intersect, that is, they are $\parallel$. The lines in which plane N intersects planes P and Q, $\overleftrightarrow{AB}$ and $\overleftrightarrow{CD}$, must be $\parallel$. This contradicts the given info. that $\overleftrightarrow{AB} \nparallel \overleftrightarrow{CD}$. The temp. assumption must be false. It follows that planes P and Q intersect.   **15.** Assume temp. that n does not int. k. Since n and k are coplanar, n and k must be $\parallel$. Then P is on n and l, and n and l are both $\parallel$ to k. This contradicts the thm. which states that through a pt. outside a line there is exactly 1 line $\parallel$ to the given line. The temp. assumption must be false. It follows that n does int. k.
17. Assume temp. that there is an n-sided reg. polygon with an interior $\angle$ of meas. 155. Then the meas. of each ext. $\angle$ is 25 and $25n = 360$. This contradicts the fact that there is no whole number n such that $25n = 360$. The temp. assumption must be false. It follows that there is no reg. polygon with an interior $\angle$ of meas. 155.

Self-Test 1, Page 218

1. True   **2.** True   **3.** False   **4.** False   **5.** If $\triangle ABC$ is not acute, then $m \angle C = 90$. False
6. If $m \angle C = 90$, then $\triangle ABC$ is not acute. True   **7.** C   **8. a.** $ABCD$ is not a rhom.   **b.** No concl.
c. No concl.   **d.** $GHIJ$ is a $\square$.   **9.** Assume temp. that $AC \neq 14$.   **10.** d, b, a, c

Written Exercises, Pages 222–223

1. 3, 15   **3.** 0, 200   **5.** $a - b, a + b$ **7.** $\angle 2$ **9.** $\angle 3$ **11.** $\overline{WT}$ **13.** $\overline{WY}$
15. $c > d > e > b > a$ **17.** $m \angle 2 > m \angle X > m \angle XZY > m \angle Y > m \angle 1$ **19.** 1. $EFGH$ is a $\square$; $EF > FG$ (Given)   2. $HG > EH$ (Thm. 5-1 and Subst.)   3. $m \angle 1 > m \angle 2$ (Thm. 6-2)

Written Exercises, Pages 231–232

1. $m \angle 1 > m \angle 2$; SSS Ineq.   **3.** $>; >$ **5.** $<; >$ **7.** $<$ **9.** $>$ **11.** 1. $m \angle SUV > m \angle STU$ (Ext. $\angle$ Ineq. Thm.)   2. $\overline{TU} \cong \overline{US} \cong \overline{SV}$ (Given)   3. $m \angle SVU = m \angle SUV$ (Isos. $\triangle$ Thm.)   4. $m \angle SVU > m \angle STU$ (Subst.)   5. $ST > SV$ (Thm. 6-3)   **13.** Key steps of proof: 1. $m \angle P > m \angle Q$ (SSS Ineq. Thm.)
2. $m \angle PCA + m \angle A + m \angle P = 180$; $m \angle QCB + m \angle QBC + m \angle Q = 180$ (Thm. 3-11)   3. $m \angle PCA = m \angle A$; $m \angle QCB = m \angle QBC$ (Isos. $\triangle$ Thm.)   4. $m \angle PCA < m \angle QCB$ (Subst.)

Self-Test 2, Page 233

1. $\overline{XY}$ **2.** $\overline{OD}$ **3.** $<$ **4.** $=$ **5.** $>$ **6.** 1, 11   **7.** cannot be   **8.** must be   **9.** may be

Chapter Review, Pages 235–236

1. $>$ **3.** $=$ **5.** $>$ **7.** No concl.   **9.** Barbara is at least 18 years old.   **11.** $m \angle T$ **13.** $<$
15. $>$ **17.** $=$

Algebra Review, Page 237

1. $\dfrac{1}{5}$ **3.** $\dfrac{a}{2}$ **5.** $\dfrac{1}{3}$ **7.** $-4y^2$ **9.** $\dfrac{ab}{2c}$ **11.** $3x - 2y$ **13.** $\dfrac{1}{3}$ **15.** $t + 1$ **17.** $\dfrac{b + 5}{b - 7}$
19. $\dfrac{3(x - 4)}{3x - 4}$

Preparing for College Entrance Exams, Page 238

1. A   **2.** A   **3.** B   **4.** B   **5.** B   **6.** E   **7.** E   **8.** C

Cumulative Review, Page 239

1. 57   **3. a.** Yes; SAS   **b.** Yes; ASA   **c.** No   **d.** Yes; AAS   **5. a.** $\overline{YZ}$ **b.** $\overline{XZ}$ **7.** 109, 71
9. Assume temp. that $\angle Q$, $\angle R$, and $\angle S$ are all $120°$ angles. Then $m \angle P > 0$ and $m \angle Q + m \angle R + m \angle S + m \angle P > 360$. This contradicts the thm. that states the sum of the int. $\angle$s of a quad. $= 360$. Therefore, the temp. assumption must be false. It follows that $\angle Q$, $\angle R$, and $\angle S$ are not all $120°$ angles.

Chapter 7

Written Exercises, Pages 243–244

1. 5:3 **3.** 1:5 **5.** 3:16 **7.** 2 to 1 **9.** $\frac{1}{3}$ **11.** $\frac{17}{1}$ **13.** 12:6:5 **15.** 1:9 **17.** 3:4

19. 8:5 **21.** $\frac{3}{4b}$ **23.** $\frac{3}{a}$ **25.** 132, 48 **27.** 37.5, 52.5 **29.** 72, 90, 90, 144, 162, 162 **31.** 50, 70, 110, 130; 2 s-s. int. $\angle$ are supp. **33. a.** 104 **b.** 0.310 **35.** 52.5

Written Exercises, Pages 247–248

1. 6 **3.** 21 **5.** $\frac{4}{7}$ **7.** $\frac{y+3}{3}$ **9.** $2\frac{2}{5}$ **11.** $\frac{14}{15}$ **13.** -3 **15.** 2 **17.** 11 **19.** 3

21. 21; 12; 28 **23.** 8; 24; 20 **25.** 8; 4; 15 **27.** 27; 36; 12 **29.** By the means-ext. prop., $\frac{a+b}{b} = \frac{c+d}{d}$ is equiv. to $ad + bd = bc + bd$, or $ad = bc$. **33.** 20 **35.** $\frac{1}{2}$ **37.** 4 or $-\frac{9}{5}$ **39.** $x = 16$, $y = 4$ **43.** 3:2

Written Exercises, Pages 250–252

1. always **3.** sometimes **5.** always **7.** sometimes **9.** always **11.** never **13.** sometimes

15. 4:5 **17.** 135 **19.** 12 **21.** 4k **23.** Prop. 2 **25.** $x = 8$, $y = 18$, $z = 12$ **27.** $x = 6\frac{1}{4}$, $y = 6\frac{2}{3}$, $z = 5$ **29.** **31.** $RS = RS$, but $ZR > XR$, so $\frac{RS}{RS} = 1 \neq \frac{ZR}{XR}$.

33. $C'(9, 1)$, $D'(8, 2)$, or $C'(5, 1)$, $D'(6, 2)$ **35.** 90; sq. **37. a.** $-3 + 3\sqrt{5}$ **b.** $\frac{1 + \sqrt{5}}{2}$; 1.62

Self-Test 1, Page 252

1. 3:5 **2.** 3 to 10 **3.** $\frac{2a}{3b}$ **4.** 6 **5.** 10 **6.** 3 **7.** No **8.** Yes **9.** Yes **10.** 45, 60, 75 **11.** 2:3 **12.** 12 **13.** 15 **14.** 12 **15.** 100, 100, 100, 120, 140, 160

Written Exercises, Pages 257–260

1. $\sim$ **3.** $\sim$ **5.** No concl. **7.** $\sim$ **9.** No concl. **11.** $x = 6$, $y = 4$ **13.** $x = 9$, $y = 5$ **15.** 27 m **17.** 0.55 cm **19.** $x = 2$, $y = 6$ **21. a.** 1. $\overline{EF} \parallel \overline{RS}$ (Given) 2. $\angle XFE \cong \angle XSR$; $\angle XEF \cong \angle XRS$ (If lines $\parallel$, corr. $\angle \cong$.) 3. $\triangle FXE \sim \triangle SXR$ (AA $\sim$) **b.** 1. $\triangle FXE \sim \triangle SXR$ (Part (a), above) 2. $\frac{FX}{SX} = \frac{EF}{RS}$ (Corr. sides of $\sim$ $\triangle$ are in prop.) **23.** 1. $\angle B \cong \angle C$ (Given) 2. $\angle 1 \cong \angle 2$ (Vert. $\angle \cong$.)

3. $\triangle MLC \sim \triangle MNB$ (AA $\sim$) 4. $\frac{NM}{LM} = \frac{BM}{CM}$ (Corr. sides of $\sim$ $\triangle$ are in prop.) 5. $NM \cdot CM = LM \cdot BM$ (means-ext. prop.) **25.** Key steps of proof: 1. $\angle B \cong \angle Y$ (Corr. $\angle$ of $\sim$ $\triangle$ are $\cong$.) 2. $\triangle ADB \sim \triangle XWY$ (AA $\sim$) 3. $\frac{AD}{XW} = \frac{AB}{XY}$ (Corr. sides of $\sim$ $\triangle$ are in prop.) **27.** Key steps of proof: 1. $\triangle AHE \sim \triangle ADG$ (AA $\sim$) 2. $\frac{AE}{AG} = \frac{HE}{DG}$ (Corr. sides of $\sim$ $\triangle$ are in prop.) 3. $AE \cdot DG = AG \cdot HE$ (means-ext. prop.)

29. Key steps of proof: 1. $\triangle ABC \sim \triangle ADB$ (AA $\sim$) 2. $\frac{AB}{AD} = \frac{AC}{AB}$ (Corr. sides of $\sim$ $\triangle$ are in prop.) 3. $(AB)^2 = AD \cdot AC$ (means-ext. prop.) **31.** 20

Written Exercises, Pages 266–267

1. $\triangle BAC \sim \triangle EDC$; SAS $\sim$ **3.** $\triangle LKM \sim \triangle NPO$; SAS $\sim$ **5.** $\triangle ABC \sim \triangle AEF$; AA $\sim$ **7.** $\triangle ABC \sim \triangle TRI$; 2:3 **9.** $\triangle ABC \sim \triangle ITR$; 2:5 **11.** 1. $\frac{DE}{GH} = \frac{DF}{GI} = \frac{EF}{HI}$ (Given)

14 / *Selected Answers*

2. $\triangle DEF \sim \triangle GHI$ (SSS $\sim$)   3. $\angle E \cong \angle H$ (Corr. $\angle$ of $\sim \triangle$ are $\cong$.)   **13.** 1. $\dfrac{VW}{VX} = \dfrac{VZ}{VY}$ (Given)

2. $\angle V \cong \angle V$ (Reflex.)   3. $\triangle VWZ \sim \triangle VXY$ (SAS $\sim$)   4. $\angle 1 \cong \angle 2$ (Corr. $\angle$ of $\sim \triangle$ are $\cong$.)   5. $\overline{WZ} \parallel \overline{XY}$

(If corr. $\angle \cong$, lines $\parallel$.)   **15.** 1. $\dfrac{JL}{NL} = \dfrac{KL}{ML}$ (Given)   2. $\angle MLN \cong \angle KLJ$ (Vert. $\angle \cong$.)   3. $\triangle MLN \sim \triangle KLJ$

(SAS $\sim$)   4. $\angle J \cong \angle N$ (Corr. $\angle$ of $\sim \triangle$ are $\cong$.)

17. Given: $\triangle ABC \sim \triangle DEF$; $\overline{AM}$ and $\overline{DN}$ are
       medians.

    Prove: $\dfrac{AM}{DN} = \dfrac{AB}{DE}$

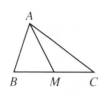 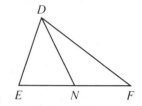

Key steps of proof: 1. $BM = \dfrac{1}{2}BC$; $EN = \dfrac{1}{2}EF$

(Midpt. Thm.)   2. $\dfrac{BC}{EF} = \dfrac{AB}{DE}$ (Corr. sides of $\sim \triangle$ are

in prop.)   3. $\dfrac{BM}{EN} = \dfrac{AB}{DE}$ (Subst.)   4. $\triangle ABM \sim \triangle DEN$ (SAS $\sim$)   5. $\dfrac{AM}{DN} = \dfrac{AB}{DE}$ (Corr. sides of $\sim \triangle$

are in prop.)

Mixed Review Exercises, Page 268

1. a. $\overline{GC}$; $\overline{EF}$ **b.** 18   **c.** 3   **d.** 90   **2. a.** Midpt., $\overline{RV}$ **b.** 2   **c.** 4

Written Exercises, Pages 272–273

1. a. No   **b.** Yes   **c.** Yes   **d.** No   **e.** Yes   **f.** Yes   **3.** 7.5   **5.** 26   **7.** 18   **9.** 14.5   **11.** 4
13. $AN = 10$ **15.** $RT = 8$; $AN = 18$; $NP = 12$; $TP = 25$ **17.** $AR = 8$; $NP = 6$; $AP = 12$ **21.** 22.5
23. 78   **25.** 0.5

Self-Test 2, Page 274

1. SSS $\sim$ **2.** AA $\sim$ **3.** SAS $\sim$ **4. a.** $\triangle EDC$ **b.** ED; EC; DC **c.** 10; x; 14   **d.** 10; y; 18   **5.** r

6. p **7.** h **8.** a **9.** 12   **10.** 14   **11.** $6\dfrac{2}{3}$

Extra, Page 276

1. c   **3.** b   **5.** d

Chapter Review, Pages 277–278

1. 3 : 5   **3.** $\dfrac{2y}{3x}$ **5.** No   **7.** Yes   **9.** $\angle J$ **11. a.** 12   **b.** 50   **13. a.** $\triangle UVH$ **b.** AA $\sim$

15. UH; $\dfrac{RT}{UV}$ **17.** $\triangle NCD \sim \triangle NAB$; AA $\sim$ **19.** No   **21.** 2   **23.** 14.4

Algebra Review, Page 280

1. 6   **3.** $2\sqrt{6}$ **5.** $10\sqrt{3}$ **7.** $\dfrac{\sqrt{15}}{3}$ **9.** 1   **11.** 13   **13.** 12   **15.** 162   **17.** $12\sqrt{3}$ **19.** $10\sqrt{2}$
21. 5   **23.** 12   **25.** $\sqrt{65}$ **27.** $2\sqrt{2}$ **29.** 7

Cumulative Review, Pages 281–283

True-False Exercises   **1.** F   **3.** F   **5.** T   **7.** F   **9.** F   **11.** F   **Multiple-Choice Exercises**   **1.** d
3. d   **5.** c   **Always-Sometimes-Never Exercises**   **1.** S   **3.** S   **5.** N   **7.** A   **9.** S   **11.** A
13. A   **15.** S   **Completion Exercises**   **1.** 120   **3.** obtuse   **5.** 108   **7.** rect.   **9.** 36
Algebraic Exercises   **1.** 6   **3.** 84   **5.** 20   **7.** 7   **9.** 6   **11.** 15   **13.** 16 cm, 20 cm, 28 cm
15. $x = 6$, $y = 3.5$ **Proof Exercises**   **1.** 1. $\overline{SU} \cong \overline{SV}$; $\angle 1 \cong \angle 2$ (Given)   2. $\overline{QS} \cong \overline{QS}$ (Reflex.)
3. $\triangle QUS \cong \triangle QVS$ (SAS)   4. $\overline{UQ} \cong \overline{VQ}$ (CPCT)   **3.** Key steps of proof: 1. $\overline{QR} \cong \overline{QT}$; $\angle R \cong \angle T$;
$\overline{RU} \cong \overline{TV}$ (CPCT)   2. $\overline{RS} \cong \overline{TS}$ (Seg. Add. Post and Add. Prop. $=$)   3. $\triangle QRS \cong \triangle QTS$ (SAS)
5. 1. $\overline{EF} \parallel \overline{JK}$; $\overline{JK} \parallel \overline{HI}$ (Given)   2. $\overline{EF} \parallel \overline{HI}$ (Thm. 3-10)   3. $\angle 2 \cong \angle 3$; $\angle F \cong \angle H$ (If lines $\parallel$, alt. int. $\angle \cong$.)
4. $\triangle EFG \sim \triangle IHG$ (AA $\sim$)

Chapter 8

Written Exercises, Pages 288–290

1. $2\sqrt{3}$ **3.** $3\sqrt{5}$ **5.** $20\sqrt{2}$ **7.** $18\sqrt{10}$ **9.** $6\sqrt{5}$ **11.** $\dfrac{\sqrt{21}}{7}$ **13.** $6\sqrt{3}$ **15.** $\dfrac{\sqrt{3}}{9}$ **17.** 9

19. $10\sqrt{10}$ **21.** $11\sqrt{10}$ **23.** 9   **25.** $3\sqrt{5}$ **27.** 9   **29.** 3   **31.** $x = 10$, $y = 2\sqrt{29}$, $z = 5\sqrt{29}$

33. $x = \dfrac{\sqrt{2}}{6}$, $y = \dfrac{\sqrt{3}}{6}$, $z = \dfrac{\sqrt{6}}{6}$ **35.** $x = 5.4$, $y = 9.6$, $z = 7.2$ **37.** $x = \sqrt{2}$, $y = 2$, $z = \sqrt{2}$

39. $x = 4$, $y = 2\sqrt{5}$, $z = 3\sqrt{5}$ **41. a.** cd, ce **b.** $a^2 + b^2 = cd + ce = c(d + e) = c^2$ **43.** Key steps of proof: 1. $\triangle PST \sim \triangle TRQ$ (SAS$\sim$) 2. $m\angle PTS = m\angle TQR$ (Corr. $\angle$ of $\sim$ $\triangle$ are $\cong$.) 3. $m\angle QTR + m\angle TQR = 90$ (Thm. 3-11 Cor. 4) 4. $m\angle PTS + m\angle QTR = 90$ (Subst.) 5. $m\angle PTQ + m\angle PTS + m\angle QTR = 180$ and $m\angle PTQ = 90$ ($\angle$ Add. Post.)

Written Exercises, Pages 292–294

1. 5   **3.** 8   **5.** $10\sqrt{3}$ **7.** 8   **9.** 25   **11.** $8\sqrt{2}$ **13.** 3   **15.** $4\sqrt{2}$ **17.** 68   **19.** 3
21. $3\sqrt{5}$ **23.** 12   **25.** 10   **27.** 17   **29.** 20   **31. a.** 5 **b.** 4.8   **33.** 13   **35.** $e\sqrt{3}$ **37.** 12
39. 12

Mixed Review Exercises, Page 294

1. AC **2.** $>$, A **3.** $>$ **4.** $\overline{AB}$ **5.** B, C **6.** AB **7.** BX, CX

Written Exercises, Pages 297–298

1. acute   **3.** rt.   **5.** obt.   **7. a.** rt. **b.** rt.   **9.** $(ST)^2 = 13^2 - 12^2 = 25$; $(ST)^2 = (RS)^2 + (RT)^2 = 25$. By the conv. of the Pythag. Thm., $\triangle RST$ is a rt. $\triangle$.   **11.** acute   **13.** obt.   **15.** $12 < x \le 16$
17. $\overline{RM}$; $\angle RST$ is obt. and $\angle STU$ is acute, so $RT > SU$.   **19. a.** is greater than the sum of the squares of the other 2 sides, then the $\triangle$ is an obt. $\triangle$.   **b.** 1. $n^2 = j^2 + k^2$ (Pythag. Thm.) 2. $l^2 > j^2 + k^2$ (Given) 3. $l^2 > n^2$ and $l > n$ (Subst.) 4. $m\angle S > m\angle V = 90$ (SSS Ineq. Thm.) 5. $\triangle RST$ is obt. $\triangle$. (Def. of obt. $\triangle$)

Written Exercises, Pages 302–303

1. 4; $4\sqrt{2}$ **3.** $\sqrt{5}$; $\sqrt{10}$ **5.** $3\sqrt{2}$; $3\sqrt{2}$ **7.** $4\sqrt{2}$; 8   **9.** $7\sqrt{3}$; 14   **11.** 5; 10   **13.** 5; $5\sqrt{3}$

15. $\sqrt{3}$; $2\sqrt{3}$ **17.** $12\sqrt{2}$ **19.** 36   **21.** $x = 4$, $y = \dfrac{4\sqrt{3}}{3}$ **23.** $x = 6\sqrt{2}$, $y = 12$ **25.** $x = 8\sqrt{2}$,

$y = 4\sqrt{6}$ **27.** $OB = \sqrt{2}$, $OC = 2$, $OD = 2\sqrt{2}$, $OE = 4$ **29.** 16, $16\sqrt{3}$ **31.** A $30°-60°-90°$ $\triangle$ with hyp. 2 has legs 1 and $\sqrt{3}$. Any $\triangle$ with sides in ratio $1:\sqrt{3}:2$ is $\sim$ to this $30°-60°-90°$ $\triangle$ and thus is a $30°-60°-90°$ $\triangle$.   **33.** $GH = GI = 6$, $JG = 6\sqrt{3}$, $HI = 6\sqrt{2}$, $JH = 12$ **35. a.** $4\sqrt{2} + 4\sqrt{6}$

b. $(1 + \sqrt{3}):2$ **37.** $\dfrac{3j + j\sqrt{3}}{4}$ **39.** $4\sqrt{2}$

Self-Test 1, Page 304

1. $3\sqrt{5}$ **2. a.** 4 **b.** $2\sqrt{5}$ **c.** $4\sqrt{5}$ **3. a.** rt. **b.** acute **c.** obtuse   **4.** $4\sqrt{5}$ **5.** $20\sqrt{2}$ cm
6. $6\sqrt{3}$ cm   **7.** 12

Written Exercises, Pages 308–310

1. 13.7   **3.** 48.3   **5.** 55.4   **7.** 57°   **9.** 27°   **11.** 31°   **13.** $w = 60$, $z \approx 54$ **15.** $w = 75$, $z \approx 89$ **17.** $w = 160$, $z \approx 117$ **19.** about 4°   **21.** 65°   **23.** 174 cm   **25. a.** 0.7002, 0.4663, 1.1665 **b.** 60°, 1.7321 **c.** No **d.** No   **27. a.** 5 **b.** 22°   **29.** about 136 ft

Written Exercises, Pages 314–316

1. $x \approx 21$, $y \approx 28$ **3.** $x \approx 89$, $y \approx 117$ **5.** $x \approx 28$, $y \approx 10$ **7.** $v° \approx 26°$ **9.** $x \approx 9$, $v° \approx 63°$
11. $v° \approx 37°$, $w° \approx 106°$ **13. a.** $\sqrt{115}$ **b.** $y \approx 40$, $x \approx 10.7$ **c.** Yes; $\sqrt{115} \approx 10.7$ **15.** about 149 m
17. 0.4 m   **19. a.** $AB = AC \approx 16$ **b.** ≈ 15 **21.** length ≈ 17 cm, width ≈ 5 cm   **23.** about 12 cm

Written Exercises, Pages 318–320

1. about 32 m   **3.** about 50 m   **5.** about 2.3 km   **7.** Heidi; ≈ 63 cm longer   **9.** about 440 m
11. $\approx 14°$ **13. a.** $\angle A$ **b.** A; a player at A has a wider $\angle$ over which to aim at the goal.

Self-Test 2, Page 320

1. $\frac{7}{24}$ **2.** $\frac{24}{25}$ **3.** $\frac{7}{25}$ **4.** $\frac{24}{7}$ **5.** 74   **6.** 74   **7.** 109   **8.** 113   **9.** about 45 m

Chapter Review, Pages 323–324

1. 6   **3.** $5\sqrt{6}$ **5.** $3\sqrt{5}$ **7.** $7\sqrt{2}$ **9.** acute   **11.** rt.   **13.** $5\sqrt{3}$ **15.** 16   **17. a.** 1.5   **b.** $\frac{2}{3}$

c. 34   **19. a.** $\frac{12}{13}$ **b.** $\frac{12}{13}$ **c.** 67   **21.** 57   **23.** 23

Preparing for College Entrance Exams, Page 326

1. A   **2.** C   **3.** B   **4.** C   **5.** E   **6.** A   **7.** C   **8.** A   **9.** B   **10.** C

Cumulative Review, Page 327

1. Seg. Add. Post.   **3.** corollary   **5.** contrapositive   **7.** $1:\sqrt{2}$ **9. a.** If a $\triangle$ is equiangular, then it is isos.   **b.** If a $\triangle$ is isos., then it is equiangular.   **11.** 36   **13.** 20   **15.** Since $\overline{AX}$ is a median, $\overline{BX} \cong \overline{CX}$. Since $\overline{AX}$ is an altitude, $\angle AXB \cong \angle AXC$. Thus, $\triangle AXB \cong \triangle AXC$ (SAS) and $\overline{AB} \cong \overline{AC}$ (CPCT). By def., $\triangle ABC$ is isos.   **17.** 1. $\angle WXY \cong \angle XZY$ (Given)   2. $\angle Y \cong \angle Y$ (Reflex.)   3. $\triangle XYW \sim \triangle ZYX$ (AA~)
4. $\frac{XY}{ZY} = \frac{WY}{XY}$ or $(XY)^2 = WY \cdot ZY$ (Corr. sides of $\sim$ $\triangle$ are in prop.)

Chapter 9

Written Exercises, Pages 330–331

1. The midpts. lie on a diam. $\perp$ to the given chords. of the hyp.   **d.** 5   **5.** 8, 22   **9.**

3. b. It is equidist. from the vertices.   **c.** at the midpt.   **13.** 24   **15.** $12\sqrt{3}$ **17. a.** rhom.; $\odot Q \cong \odot R$ so $\overline{QC}$, $\overline{QD}$, $\overline{RC}$, and $\overline{RD}$ are $\cong$.   **b.** Diags. of rhom. are $\perp$ bis. of each other.   **c.** 16   **19.** $4\sqrt{6}$

Extra, Page 332

1. 4 odd, 1 even; cannot be traced   **3.** 2 odd, 6 even; can be traced   **5.** There are more than 2 odd vertices.

Written Exercises, Pages 335–337

1. 8   **3.** 12   **5.** 8.2   **7. a.** $\overline{AB} \cong \overline{CD}$ Proof: 1. Draw $\overleftrightarrow{AB}$ and $\overleftrightarrow{CD}$ int. at Z. (Through any 2 pts. there is ex. 1 line.)   2. $ZA + AB = ZB$; $ZC + CD = ZD$ (Seg. Add. Post.)   3. $ZB = ZD$ (Thm. 9-1 Cor.)
4. $ZA + AB = ZC + CD$ (Subst.)   5. $ZA = ZC$ (Thm. 9-1 Cor.)   6. $\overline{AB} \cong \overline{CD}$ (Subtr. Prop. =)   **b.** Yes
9. a. square; $\overline{XZ} \perp \overline{OX}$, so $\overline{XZ} \parallel \overline{OY}$. Similarly, $\overline{ZY} \perp \overline{OX}$, so $OXZY$ is a rect. Since $OX = OY$, $OXZY$ is a square.
b. $5\sqrt{2}$ **11.** $\overline{AR} \perp \overline{RS}$ and $\overline{BS} \perp \overline{RS}$ (Thm. 9-1) so $\overline{AR} \parallel \overline{BS}$. Then $\angle A \cong \angle B$ and $\triangle ARC \sim \triangle BSC$ (AA~), so $\frac{AC}{BC} = \frac{RC}{SC}$. (Corr. sides of $\sim$ $\triangle$ are in prop.)   **13.** Two planes tan. to a sphere at the endpts. of a diam. are $\parallel$.
15. $RA = RC$ and $SB = SC$ (Thm. 9-1 Cor.), so $PR + RS + SP = PA + PB$ (Subst.)   **17.** 15 (trapezoid)
19. a. G is the midpt. of $\overline{EF}$. Key steps of proof: 1. $GE = GH$; $GH = GF$ (Thm. 9-1 Cor.)   2. $GE = GF$ (Trans. Prop.)   **b.** $m \angle EHF = 90$. Key steps of proof: 1. $m \angle E = m \angle GHE$; $m \angle F = m \angle GHF$ (Isos. $\triangle$ Thm.)   2. $m \angle E + m \angle GHE + m \angle F + m \angle GHF = 180$ (Thm. 3-11)   3. $2m \angle GHE + 2m \angle GHF = 180$ (Subst.)   4. $m \angle EHF = 90$ (Div. Prop. =)   **21. a.** 8   **b.** infinitely many   **23.** $2\sqrt{2}$

Mixed Review Exercises, Page 337

1. 15   **2.** $9\sqrt{2}$ **3.** $2\sqrt{7}$

Written Exercises, Pages 341–343

1. 85 **3.** 150 **5.** 52 **7.** 30 **9. a.** 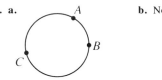 **b.** No

11. $m\overgroup{BD}$: 34, 44; $m\angle COD$: 100, 88, 104, $p + q$;

$m\angle CAD$: 50, 44, 50, $\frac{1}{2}(p + q)$

13. d. The opp. $\angle$s of inscr. quad. are supp.

15. Key steps of proof: 1. Draw $\overline{OY}$. (Through any 2 pts. there is ex. 1 line.) 2. $m\angle WOY = m\overgroup{WY} = 2n$ (Def. meas. of arc, Arc Add. Post.) 3. $m\angle WOY = m\angle Z + m\angle OYZ$ (Thm. 3-12) 4. $m\angle Z = m\angle OYZ$ (Isos. $\triangle$ Thm.) 5. $m\angle Z = n$ (Subst. and Div. Prop. $=$) **17.** $r \approx 4700$ km **19.** $r \approx 5300$ km **23.** ≈ 3800 km

Written Exercises, Pages 347–348

1. 8 **3.** $9\sqrt{2}$ **5.** 80 **7.** 24 **9.** $10\sqrt{5}$ **11.** $2\sqrt{21}$ cm **13.** $2\sqrt{21}$ cm **15.** 1. $\angle J \cong \angle K$ (Given) 2. $\overline{JZ} \cong \overline{KZ}$ (If 2 $\angle$s of $\triangle \cong$, sides opp. the $\angle$s are $\cong$.) 3. $\overgroup{JZ} \cong \overgroup{KZ}$ (In same $\odot$, $\cong$ chords have $\cong$ arcs.) **17.** $10\sqrt{3}$ **19.** 26 cm **21.** ≈ 74 **23.** If 2 $\odot$s are concentric and a chord of the outer $\odot$ is tan. to the inner circle, then the pt. of tan. is the midpt. of the chord. **25.** $18\sqrt{3}$ **27.** 2.8 cm

Self-Test 1, Page 349

1. a. $\overline{QB}, \overline{QC}$ **b.** $\overline{BC}$ **c.** $\overline{AC}$ or $\overline{BC}, \overleftrightarrow{AC}$ **2. a.** **b.**
3. 15 **4.** two concentric $\odot$s
5. $4\sqrt{10}$ cm **6. a.** 50, 310
b. In same $\odot$, $\cong$ chords have $\cong$ arcs.

Written Exercises, Pages 354–356

1. $x = 30, y = 25, z = 15$ **3.** $x = 110, y = 100, z = 100$ **5.** $x = 50, y = 130, z = 65$ **7.** $x = 104, y = 104, z = 52$ **9.** $x = 50, y = 100, z = 35$ **11. a.** If the arcs between 2 chords are $\cong$, then the chords are $\parallel$. **b.** False, the chords may int. **13.** 1. Thm. 9-1; def. of $\perp$ lines 2. Def. of semicircle 3. Subst. **17.** $\triangle ADE \sim \triangle BCE, \triangle EDC \sim \triangle EAB$ **19.** $x = 80, m\angle D = 20$ **21.** $x = 10, m\angle A = 55$ **23.** rect.; $m\overgroup{AB} = 120$ and $m\overgroup{AQ} = 60$, so $\overgroup{BAQ}$ is semicir. and $\angle BAQ$ is rt. $\angle$. Similarly, $\angle AQP$, $\angle QPB$, and $\angle PBA$ are rt. $\angle$s, so $AQPB$ is rect. **29.** $\frac{ab}{c}$

Mixed Review Exercises, Page 357

1. $\overline{LM}$ **2.** $\overleftrightarrow{LM}$ **3.** $\overline{NP}$ **4.** 14 **5.** $360 - x$ **6.** 6

Written Exercises, Pages 359–361

1. 90 **3.** 25 **5.** 55 **7.** 35 **9.** 90 **11.** 60 **13.** 30 **15.** 30 **17.** 40 **19.** 90

21. 115 **23.** 100, 90, 86, 84 **27.** $b - a = c$ **29.** Key steps of proof: 1. $m\angle ABP = \frac{1}{2}m\overgroup{AP}$ (Thm. 9-7) 2. $m\angle Q = \frac{1}{2}(m\overgroup{AB} - m\overgroup{PC})$ (Thm. 9-10) 3. $m\angle Q = \frac{1}{2}(m\overgroup{AC} - m\overgroup{PC})$ (Subst.) 4. $m\angle Q = \frac{1}{2}m\overgroup{AP}$ (Arc Add. Post.) **31.** $m\overgroup{CE} = 3m\overgroup{BD}$

Written Exercises, Pages 364–366

1. 10 **3.** $\sqrt{21}$ **5.** 6 **7.** 8 **9.** 5 **11.** 1. $\overline{UT}$ is tan. to $\odot O$ and $\odot P$. (Given) 2. $UV \cdot UW = (UT)^2, UX \cdot UY = (UT)^2$ (Thm. 9-13) 3. $UV \cdot UW = UX \cdot UY$ (Subst.) **13.** 4 or 12 **15.** 6 **17.** 9 **19.** 4 **21. a.** Pythag. Thm. **c.** Thm. 9-13 **23.** 20 m **25.** 1. $AX \cdot XB = PX \cdot XQ, CX \cdot XD = PX \cdot XQ$ (Thm. 9-11) 2. $AX \cdot XB = CX \cdot XD$ (Trans. Prop.) **27.** $2\sqrt{10}$

Self-Test 2, Page 367

1. 40 **2.** 150 **3.** 82 **4.** 9 **5.** $x = 70, y = 50$ **6.** 35 **7.** 10 **8.** 12

Chapter Review, Pages 369–370

1. chord, secant   **3.** diam.   **5.** tan.   **7.** 10   **9.** 100   **11.** ∠YPW   **13.** 120   **15.** In same ⊙, ≅ chords are equally distant from the center.   **17.** 50, 50   **19.** 105   **21.** 40   **23.** 9

Cumulative Review, Pages 372–373

1. 3, 2x + 3   **3.** 1. $\overline{MN}$ is median of trap. (Given)   2. $\overline{MN} \parallel \overline{ZY} \parallel \overline{WX}$ (Thm. 5-19)   3. V is the midpt. of $\overline{WY}$. (Thm. 5-10)   4. $\overline{MN}$ bis. $\overline{WY}$. (Def. of bis.)   **5.** 7, 7√3   **7.** 9, 81, 90   **9.** 2√2   **11.** 1. ∠1 ≅ ∠2, ∠2 ≅ ∠3 (Given)   2. $\overline{AB} \parallel \overline{DC}$ (If alt. int. ⦟ ≅, lines ∥.)   3. $\overline{AD} \parallel \overline{BC}$ (If corr. ⦟ ≅, lines ∥.)   4. ABCD is a ▱. (Def. of ▱)   5. $\overline{AB} \cong \overline{DC}$ (Opp. sides of a ▱ are ≅.)   **13. a.** If ∠A ≇ ∠C, then quad. ABCD is not a ▱.   **b.** If quad. ABCD is not a ▱, then ∠A ≇ ∠C.   **15.** 20   **17. a.** inside   **b.** on   **c.** on   **19. a.** Janice likes to dance.   **b.** no concl.   **c.** no concl.   **d.** Kim is not Bill's sister.   **21.** always   **23.** always   **25.** sometimes   **27.** 0, 1   **29.** 1. AB > AC (Given)   2. m∠ACB > m∠ABC (Thm. 6-2)   3. BD = EC (Given)   4. BC = BC (Reflex.)   5. BE > CD (SAS Ineq. Thm.)

Chapter 10

Written Exercises, Pages 378–379

9.

13. 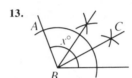 $m\angle ABC = \frac{3}{4}x$

15. c. They are the same pt., which is equidistant from the sides of the △.
19. Methods may vary; for example, see the figure at the right.

165°

Mixed Review Exercises, Page 380

1. midpt.   **2.** ▱   **3.** rect.   **4.** rhom.   **5.** 5√2   **6.** 108

Written Exercises, Pages 383–385

1. Const. 5   **3.** Const. 4   **5.** Const. 7   **7.** Extend $\overrightarrow{HJ}$; use Const. 6.
11. Methods may vary; for example, see the figure at the right.
15. b, c. Yes; yes   **19.** Const. $\overline{AB}$ so that AB = a. Const. the ⊥ bis. of

$\overline{AB}$ int. $\overline{AB}$ at M, so that AM = $\frac{1}{2}a$. Const. $\overline{MC} \perp \overline{AB}$ so that MC = $\frac{1}{2}a$.

With ctrs. A and C and radius AM, draw arcs int. at D. Draw $\overline{AD}$ and $\overline{CD}$.
23. Const. a square with sides of length b.

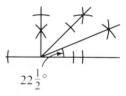

$22\frac{1}{2}°$

Written Exercises, Pages 388–389

1. a. any acute △   **b.** any obt. △   **c.** any rt. △   **3.** 2, 4   **5.** 3.8, 5.7   **7.** Const. 3   **9.** The pt. of int. of the ⊥ bis. of $\overline{XY}$, $\overline{XZ}$, and $\overline{YZ}$ is equidistant from all 3 towns. It would be wiser to build it equidistant from X and Z, near Y.   **11. a.** GD = $\frac{1}{3} \cdot AD = \frac{1}{3} \cdot BE = GE$ **b.** GB   **c.** ∠GBA, ∠GED, ∠GDE   **13.** 3, −1

15. Key steps of proof: 1. Draw $\overline{BD}$ int. $\overline{AC}$ at Y. (Through any 2 pts. there is ex. 1 line.)   2. $\overline{BM}$ and $\overline{CY}$ are medians of △BDC. (Def. of median)   3. CX = $\frac{2}{3}$CY (The medians of a △ int. in a pt. that is $\frac{2}{3}$ of the dist. from each vertex to the opp. side.)   4. CX = $\frac{2}{3} \cdot \frac{1}{2} \cdot AC = \frac{1}{3}AC$ (Subst.)   **17. a.** pts. in the interior of ∠XPY   **b.** pts. in the interior of the ∠ vert. to ∠XPY

Self-Test 1, Page 390

1. Const. 4   **2.** Draw $\overline{ST}$. With ctrs. S and T, and radius ST, draw arcs int. at R. Draw $\overrightarrow{SR}$; $m \angle RST = 60$. Use Const. 3.   **3.** Const. 6   **4.** Const. 7   **5.** Methods may vary. Const. $\overline{JK}$ such that $JK = 2AB$ (Const. 1). Const. lines $\perp$ to $\overline{JK}$ at J and K. Const. $\overline{JM} \cong \overline{KL} \cong \overline{AB}$. Draw $\overline{ML}$.   **6.** lines that contain the altitudes, medians, $\angle$ bis., $\perp$ bis. of the sides   **7.** midpt. of the hyp.   **8.** $4\sqrt{3}$, $2\sqrt{3}$

Mixed Review Exercises, Page 391

1. 12   **2.** tan., 10   **3.** 3   **4.** 141

Written Exercises, Pages 395–396

1. Const. 8   **3.** Const. 10   **5.** Const. 10   **7.** Const. 11   **9.** Draw $\odot O$ with radius r. Choose pt. A on $\odot O$, and with ctr. A and radius r, mark off $\overset{\frown}{AB}$. With ctr. B and radius r, mark off $\overset{\frown}{BC}$. Similarly, mark off $\overset{\frown}{CD}$, $\overset{\frown}{DE}$, and $\overset{\frown}{EF}$. Draw $\overline{AC}$, $\overline{EC}$, and $\overline{AE}$.   **11. a.** Draw $\odot O$. Draw diam. $\overline{AE}$. Const. $\perp$ diam. $\overline{CG}$. Bis. 2 adj. rt. $\angle$s to form 8 $\cong$ arcs. Connect consec. pts. to form octagon $ABCDEFGH$.   **b.** Draw overlapping squares $ACEG$ and $BDFH$.   **13.** Draw the diags. of the square int. at O. Draw the $\odot$ with ctr. O and radius = half the length of a diag.   **15.** Divide the $\odot$ into 6 $\cong$ arcs as in Ex. 9. At every other pt., const. a tan. to the $\odot$.   **17.** Const. a $\parallel$ to l through O, int. $\odot O$ at P. Const. a tan. to $\odot O$ at P.

Written Exercises, Page 399

1. Const. 12   **3. b.** No   **c.** Let the 5 $\cong$ seg. from Ex. 3(a) be $\overline{AW}$, $\overline{WX}$, $\overline{XY}$, $\overline{YZ}$, and $\overline{ZB}$. $AX:XB = 2:3$ **5.** Const. 13   **7.** Const. 14   **9.** Use $\dfrac{z}{w} = \dfrac{y}{x}$ or $\dfrac{z}{y} = \dfrac{w}{x}$ with Const. 13.   **11.** Const. 14, 12   **13.** Const. 1, 14   **15.** Draw a line and const. $\overline{AB}$ so that $AB = 3$ and $\overline{BC}$ so that $BC = 5$. Use Const. 14.   **17.** Divide $\overline{CD}$ into 7 $\cong$ parts: $\overline{CU}$, $\overline{UV}$, $\overline{VW}$, $\overline{WX}$, $\overline{XY}$, $\overline{YZ}$, and $\overline{ZD}$. Then $CV:VX:XD = 2:2:3$. Use SSS to const. a $\triangle$.

Self-Test 2, Page 401

1. Const. 9   **2.** Const. 11   **3.** Use Const. 12 to divide a seg. $\overline{AB}$ into 3 $\cong$ parts, $\overline{AX}$, $\overline{XY}$, and $\overline{YB}$. Then $AY:YB = 2:1$.   **4.** Const. 13   **5.** Const. 14   **6.** $\perp$, F, $\perp$, G **7.** Const. the $\perp$ bis. of 2 sides of $\triangle TRI$, int. at O. Draw a $\odot$ with ctr. O and radius OT.

Written Exercises, Pages 404–405

1. the $\perp$ bis. of $\overline{AB}$ **3.** 2 $\parallel$ lines 4 cm apart with h halfway between them   **5.** the seg. joining the midpts. of $\overline{AD}$ and $\overline{BC}$ **7.** diag. $\overline{BD}$ **9.** a plane $\parallel$ to both planes and halfway between them   **11.** a sphere with ctr. E and radius 3 cm   **13. a.** Use Const. 3 to bis. $\angle HEX$.   **b.** Use Const. 3 to bis. the $\angle$ formed by j and k; locus is 2 $\perp$ lines.   **15.** Const. the $\odot$ with diameter AB, and exclude pts. A and B.   **17.** Const. 2 $\odot$s with radius EF, one with ctr. E and one with ctr. F, and exclude pts. E and F and the other 2 pts. of the int. of the $\odot$s with $\overleftrightarrow{EF}$.   **19.** a line $\perp$ to the plane of the square at the int. of the diags.

Written Exercises, Pages 407–410

1. a. $\perp$ bis. of $\overline{AB}$ int. $\odot O$ in 2 pts.   **b.** $\perp$ bis. of $\overline{AB}$ doesn't int. $\odot O$.   **c.** $\perp$ bis. of $\overline{AB}$ is tan. to $\odot O$.
3. a. a $\odot$ with ctr. D and radius 1 cm   **b.** a $\odot$ with ctr. E and radius 2 cm
c.

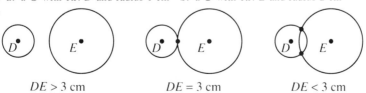

$DE > 3$ cm $DE = 3$ cm $DE < 3$ cm

d. The locus is 0, 1, or 2 pts., depending on the int. of $\odot D$ and $\odot E$.

5. the int. of $\odot P$, with radius 3 cm, and l (2 pts.)   **7.** the int. of $\odot A$, with radius 2 cm, and $\odot B$, with radius 2 cm (2 pts.)   **9.** the int. of $\odot A$, with radius 2 cm, and the bis. of $\angle A$ (1 pt.)   **11.** 0, 1, or 2 pts.   **13.** 0, 1, or 2 pts.   **15.** 0 pts., 1 pt., or a $\odot$ **17.** 2 $\odot$s   **19.** 0 pts. $(d > 5)$, 2 pts. $(d = 5)$, 2 $\odot$s $(d < 5)$
21. a. the $\perp$ bis. plane of $\overline{RS}$ **b.** the $\perp$ bis. plane of $\overline{RT}$ **c.** line, line   **d.** the $\perp$ bis. plane of $\overline{RW}$ **e.** pt., pt.
23. a. infinitely many   **b.** 2   **c.** none

Written Exercises, Pages 412–413

1. The locus is the 2 ∥ lines.   **3.** The ⊙, of radius a, has ctr. at the int. pt. of the bis. of ∠XYZ and a line that is ∥ to $\overrightarrow{YZ}$ and a units from $\overrightarrow{YZ}$.   **5.** The locus is a pair of lines, both ∥ to $\overline{AB}$ and r units from $\overline{AB}$. Const. may vary in Exs. 7-17.   **7.** Const. $j \perp k$ at M. Const. $\overline{MA}$ on j so that $MA = s$ and then ≅ segs. $\overline{AB}$ and $\overline{AC}$, B and C on k, so that $AB = AC = t$.   **9.** Const. $\overline{AB}$ so that $AB = t$. Const. the ⊥ bis. of $\overline{AB}$ to locate midpt. M of $\overline{AB}$. Draw an arc with ctr. M and radius r, and an arc with ctr. A and radius s intersecting at C. Draw $\overline{AC}$ and $\overline{BC}$.

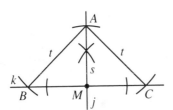

Ex. 7

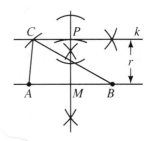

Ex. 9

11. Const. ∠A with meas. n. Const. a line ∥ to and s units from $\overrightarrow{AX}$ in order to locate point C. Const. $\overline{BC} \perp \overline{AC}$.
13. Const. $\overline{AB}$ such that $AB = t$. Const. the ⊥ bis. of $\overline{AB}$ to locate midpt. M of $\overline{AB}$. Const. line k, ∥ to and r units from, $\overleftrightarrow{AB}$. With ctr. M and radius s, draw an arc int. k at pt. C. Draw $\overline{AC}$ and $\overline{BC}$.

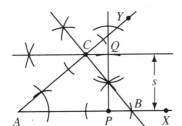

Ex. 11

Ex. 13

Self-Test 3, Page 414

1. the bis. of the vert. ∡ formed by j and k (2 lines)   **2.** the sphere with ctr. P and radius t **3.** the ⊥ bis. plane of $\overline{WX}$ **4.** the int. of the bis. of ∠DEF and a pair of rays ∥ to $\overrightarrow{EF}$ and each 4 cm from $\overrightarrow{EF}$ (1 pt.)
5. the int. of ⊙A with radius 4 cm and a line ∥ to s and t halfway between them. (0, 1, or 2 pts.)   **6.** Use Const. 4 to const. the ⊥ bis. of 2 sides of △RST. The locus is the pt. of int. of the ⊥ bis.   **7.** Const. may vary; for example, const. ∠X ≅ ∠1. Const. $\overline{XY}$ ≅ $\overline{BC}$ on one side of ∠X. Const. a line from Y ⊥ to the other side of ∠X.

Extra, Pages 414–415

3. Some of the pts. are the same: L and R, M and S, N and T. The ⊙ has ctr. H.   **5.** Key steps of proof:

$\overline{NM}$ ∥ $\overline{AB}$; $\overline{XY}$ ∥ $\overline{AB}$; $NM = \frac{1}{2}AB$; $XY = \frac{1}{2}AB$ (Thm. 5-11) 2. $XYMN$ is a ▱. (Thm. 5-5) 3. $\overline{NX}$ ∥ $\overline{CH}$ (Thm. 5-11)   4. $\overline{NM} \perp \overline{NX}$ (Thm. 3-4) 5. $XYMN$ is a rect. (Thm. 5-16)

Chapter Review, Pages 416–417

1. Const. 1   **3.** Const. 3   **5.** Const. 5   **7.** Const. 7   **9.** ∠ bis.   **11.** 1:2   **13.** Const. 9
15. Const. 10   **17.** Const. 13   **19.** a line ∥ to l and m and halfway between them   **21.** a plane ∥ to both planes and halfway between them   **23.** the int. of the ⊥ bis. of $\overline{PQ}$ and ⊙P with radius 8 cm (2 pts.)   **25.** 0 pts., 1 pt., a ⊙, a ⊙ and a pt., or 2 ⊙s, dep. on the int. of 2 planes ∥ to Q and 1 m from Q and a sphere with ctr. Z and radius 2 m.

27. Const. $\overline{RS}$ so that $RS = a$. Const. the $\perp$ bis. of $\overline{RS}$ to locate midpt. M. Draw an arc with ctr. M and radius b and an arc with ctr. R and radius c, int. at T. Draw $\overline{TR}$ and $\overline{TS}$.

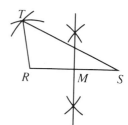

Algebra Review, Page 419

1. 1.69　**3.** $\dfrac{19}{3}$　**5.** $18\sqrt{2}$　**7.** 2826　**9.** 42　**11.** $-\dfrac{1}{2}$　**13.** 54

15. $15\sqrt{2}$　**17.** 96　**19.** cd　**21.** πrl　**23.** πd^2　**25.** $x = \dfrac{c - by}{a}, a \neq 0$

27. $n = \dfrac{S}{180} + 2$　**29.** $h = \pm\sqrt{xy}$　**31.** $h = \dfrac{2A}{b}, b \neq 0$

Preparing for College Entrance Exams, Page 420
1. B　**2.** C　**3.** E　**4.** A　**5.** B　**6.** C　**7.** A　**8.** C　**9.** E

Cumulative Review, Page 421
1. never　**3.** sometimes　**5.** always　**7.** never　**9.** 107　**11.** 15　**13.** Methods may vary: SSS,

SAS, ASA　**15. a.** 3.6　**b.** $4\dfrac{2}{7}$　**17.** 1. $m\angle 1 = 45$ (Given)　2. $m\overset{\frown}{PQ} = 90$ (Thm. 9-7)　3. $m\angle O = 90$

(Def. meas. of arc)　4. $\overline{OP} \cong \overline{OQ}$ (All radii of a $\odot$ are $\cong$.)　5. $m\angle OQP = m\angle OPQ$ (Isos. $\triangle$ Thm.)
6. $m\angle OQP + m\angle OPQ = 90; 2m\angle OPQ = 90; m\angle OPQ = m\angle OQP = 45$ (Thm. 3-11 Cor. 4, algebra)
7. $\triangle OPQ$ is a $45° - 45° - 90°$ $\triangle$. (Def. of $45° - 45° - 90°$ $\triangle$)　**19.** Methods may vary. Draw line k and pts. P
and Q on k so that $PQ < AB$. Const. line $l \perp$ to k at P and line $m \perp$ to k at Q. Draw an arc with ctr. Q and
radius AB int. l at S. Draw an arc with ctr. P and radius AB int. m at R. Draw $\overline{RS}$.

Chapter 11

Written Exercises, Pages 426–427
1. 60 cm²　**3.** 5 cm　**5.** 24　**7.** $2x^2 - 6x$　**9.** 36 cm²; 26 cm
11. 5 cm; 80 cm²　**13.** $a^2 - 9; 4a$　**15.** $x - 3; 4x - 6$　**17.** 130

19. 48　**21.** 39.4　**23.** $40xy$　**25.** $\dfrac{d^2}{2}$　**27.** 144 m²　**29. a.** 768 ft²

b. 3 cans　**31.** 14 m × 28 m　**35. a.** length $= \dfrac{1}{2}(40 - 2x) = 20 - x$

b. $20x - x^2$　**c.** See figure at right.　**d.** 10 m × 10 m

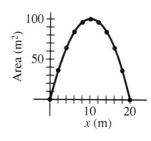

Written Exercises, Pages 431–433

1. 29.9 m²　**3.** 12　**5.** $16\sqrt{3}$　**7.** 40　**9.** 84　**11.** 16　**13.** $30\sqrt{2}$　**15.** $\dfrac{25\sqrt{3}}{2}$　**17.** 240

19. $2r^2$　**21.** 18.2　**23.** 73.5　**25.** $\triangle DFE \sim \triangle DGF \sim \triangle FGE$; 20, 4, 16　**27.** 40; 20　**29. a.** 2:3

b. 20　**31. a.** $A = \dfrac{1}{2}ab$　**b.** $A = \dfrac{1}{2}ch$　**c.** $h = \dfrac{ab}{c}$　**d.** 4.8; 5　**33. a.** $b = s, h = \dfrac{s\sqrt{3}}{2}$;

$A = \dfrac{1}{2} \cdot s \cdot \dfrac{s\sqrt{3}}{2} = \dfrac{s^2\sqrt{3}}{4}$　**b.** $\dfrac{49\sqrt{3}}{4}$　**35.** 10; 20　**37.** 41.5　**39.** 936 cm²; 504 cm²

Written Exercises, Pages 436–438

1. 70; 10　**3.** 6; $3\dfrac{3}{4}$　**5.** 5; 18　**7.** 1; 4　**9.** 9　**11.** 108　**13.** $\dfrac{27\sqrt{3}}{4}$　**15.** 24　**17.** 128

Answers may vary in Exs. 19–21.　**19.** 42.0　**21.** 87.8　**23.** 15; 74　**25.** $\triangle ABC$: $36\sqrt{3}$; $\triangle ACD$: $72\sqrt{3}$;

$ADEF$: $108\sqrt{3}$　**27.** 12.5 cm²; 112.5 cm²　**29.** $\dfrac{175 - 25\sqrt{3}}{2}$　**31.** 156

Mixed Review Exercises, Page 440

1. 52   **2.** 146   **3.** 135   **4.** 18   **5.** $10\sqrt{2}$ cm   **6.** 15 cm, $15\sqrt{3}$ cm   **7.** 10 m   **8.** $\dfrac{15}{17}$

Written Exercises, Pages 443–444

1. 8; 256   **3.** $\dfrac{7\sqrt{2}}{2}$; $\dfrac{7}{2}$ **5.** 3; $18\sqrt{3}$; $27\sqrt{3}$ **7.** $\dfrac{4\sqrt{3}}{3}$; $\dfrac{2\sqrt{3}}{3}$; $4\sqrt{3}$ **9.** $2\sqrt{3}$; 24; $24\sqrt{3}$ **11.** $4\sqrt{3}$; $24\sqrt{3}$;

$72\sqrt{3}$ **13.** $36\sqrt{3}$ **15.** $216\sqrt{3}$ **17. a.** $m\angle AOX = \dfrac{1}{2}m\angle AOB = \dfrac{1}{2}\left(\dfrac{360}{10}\right) = \dfrac{1}{2}(36) = 18$ **b.** 0.3090;

$\dfrac{OX}{1}$, 0.9511   **c.** 6.18   **d.** 0.2939   **e.** 2.939   **19.** $a \approx 0.707$; $p \approx 5.656$; $A = 2$ **21.** $p \approx 6.2112$; $A \approx 3$

Self-Test 1, Page 444

1. 81   **2.** 60   **3.** $40\sqrt{3}$ **4.** $4\sqrt{3}$ cm^2 **5.** $6\sqrt{13}$ cm^2 **6.** 40   **7.** 39   **8.** $150\sqrt{3}$ **9.** $5xy$
10. 49

Written Exercises, Pages 448–450

1. 14π; 49π **3.** 5π; $\dfrac{25}{4}\pi$ **5.** 10; 100π **7.** 5; 10π **9. a.** 132; 1386   **b.** $44k$; $154k^2$

11. ≈ 57 in.; ≈ 254 in.^{2} **13.** 24 oz   **15.** 984 ft^2 **17.** 15-in. pizza   **19.** Area I $= \dfrac{a^2\pi}{8}$,

Area II $= \dfrac{b^2\pi}{8}$, Area III $= \dfrac{c^2\pi}{8}$. Since $a^2 + b^2 = c^2$, Area I + Area II = Area III.   **21. a.** 198,000 cm (or
1.98 km)   **b.** 10,000   **23. a.** π; 3π; 5π; 7π **b.** $(2n + 1)\pi$ **25.** 3π; 3π; 3π **27.** 32π **29.** 1:2
31. $0.14r^2$ **33. a.** r **b.** $A = \pi r^2$ **35.** Radius is hypot. of rt. $\triangle$ in which radii of $\odot O$ and $\odot P$ are legs.

Algebra Review, Page 451

1. $\dfrac{4}{3}\pi$ **3.** 32π **5.** $5\pi\sqrt{2}$ **7.** 12π

Written Exercises, Pages 453–455

1. 2π; 12π **3.** 2π; 3π **5.** $\dfrac{3}{2}\pi$; $\dfrac{9}{8}\pi$ **7.** π; $\dfrac{9}{4}\pi$ **9.** $3\pi\sqrt{2}$; 15π **11.** 6   **13.** $4\pi - 8$

15. $12\pi + 8$ **17.** $\dfrac{4\pi + 6\sqrt{3}}{3}$ **19.** $(100\pi - 192)$ cm^2 **21. a.** 49; 98   **b.** 52 cm^2 **23.** 1343 m^2

25. $(24\pi - 18\sqrt{3})$ cm^2 **27. b.** $72\pi - 108\sqrt{3}$ **29.** $\dfrac{48\sqrt{3} - 22\pi}{3}$; trapezoid; 60

Written Exercises, Pages 458–460

1. 1:4; 1:16   **3.** $r:2s$; $r^2:4s^2$ **5.** 3:13; 9:169   **7.** 3:8; 3:8   **9.** 1:25,000,000,000,000

11. 1:2; 1:4   **13.** $\triangle ABE \sim \triangle DCE$; 36:25; $6\dfrac{2}{3}$ **15.** 125 cm^2 **17. a.** 9:7   **b.** 5:4

19. a. 3:4   **b.** 3:7   **21.** Answers may vary. $\triangle ABC \sim \triangle CDA$, 1:1; $\triangle ABG \sim \triangle CEG$, 9:25;
$\triangle ABF \sim \triangle DEF$, 9:4; $\triangle AGF \sim \triangle CGB$, 9:25; $\triangle EFD \sim \triangle EBC$, 4:25; $\triangle ABF \sim \triangle CEB$, 9:25
23. a. 1:9   **b.** 1:4   **c.** 1:8   **25. a.** 16:81   **b.** 4:9   **c.** 4:9   **d.** 1:1   **e.** 16:169   **27.** 9:40
29. 4:5

31. Each of the small $\triangle$s has area $= \dfrac{1}{6} \cdot$ area of the orig. $\triangle$.

Written Exercises, Pages 463–464

1. $\dfrac{1}{4}$ **3.** $\dfrac{1}{4}$ **5. a.** $\dfrac{1}{25}$ **b.** 3   **7.** $\dfrac{\pi}{200} \approx 0.016$ **9.** 0.04   **11.** $\dfrac{2}{3}$ **13.** 0.125 m^2 **15. b.** 7.5 mm

17. a. $\dfrac{13}{50}$ **b.** $\dfrac{1}{10}$

Self-Test 2, Page 465

1. 88; 616   **2.** 81π **3. a.** 6π **b.** 36π **c.** $36\pi - 72$ **4.** 16:49   **5.** 2:3   **6. a.** 4:9 **b.** 3:2

7. $64 - 16\pi$ **8.** $36\pi - 27\sqrt{3}$ **9.** $\dfrac{3}{5}$ **10.** $\dfrac{\pi}{4}$

Extra, Page 466

Answers may vary in Exs. 1–8.   **1.** 55.2   **3.** 39.7   **5.** 75.3   **7.** 178.2

Chapter Review, Page 470

1. 64   **3.** 18 cm^2 **5.** 9   **7.** 7   **9.** $30 + 4\sqrt{2}$; 52   **11.** $9\sqrt{3}$ **13.** 188.4; 2826   **15.** $8\pi\sqrt{2}$; 32π

17. $24\pi + 9\sqrt{3}$ **19.** 16π **21.** 1:4   **23.** $\dfrac{9}{25}$

Cumulative Review, Pages 472–473

1. False   **3.** False   **5.** False   **7.** True   **9.** True   **11.** False   **13.** ∥, skew   **15.** -45 **17.** 5
19. a sphere with ctr. P and radius 4 cm, along with its interior   **21.** Key steps of proof: 1. $\triangle ABC \cong$
$\triangle DCB$ (HL)   2. $\angle 1 \cong \angle 2$ (CPCT)   3. $\overline{CE} \cong \overline{BE}$ (Thm. 4-2)   4. $\triangle BCE$ is isos. (Def. isos. $\triangle$)
23. Assume temp. that there is a $\triangle$ whose sides have lengths x, y, and $x + y$, then the length of the longest side equals the sum of the lengths of the other two sides. This contradicts the $\triangle$ Ineq. Thm., if 2 sides of a triangle have lengths x and y, then the third side must be greater than $x + y$. Therefore, the temp. assumption must be false. It follows that no $\triangle$ has sides of length x, y, and $x + y$.   **25.** 5   **27.** 4.5   **29.** 17   **31.** 61
33. Const. a seg. of length $2x$. Use Const. 13 with $a = y$, $b = 2x$, and $c = x$ to find a seg. with length t;
$\dfrac{y}{2x} = \dfrac{x}{t}$; $ty = 2x^2$; $t = \dfrac{2x^2}{y}$.   **35.** $32\sqrt{6}$ **37. a.** 46   **b.** $\dfrac{1}{4}$

Chapter 12

Written Exercises, Pages 478–480

1. 40; 88; 48   **3.** 3; 54; 90   **5.** 6, 168, 108   **7.** 54; 27   **9.** 10; 600   **11.** 5; 125   **13.** 390
15. 4; 8   **17.** 240; $240 + 32\sqrt{3}$; $160\sqrt{3}$ **19.** 252; 372; 420   **21.** 180; 228; 216   **23.** 675 cm^3

25. 1.8 kg   **27.** 19 kg   **29.** $50x^3$; $120x^2$ **31.** 198 cm^2 **33.** ≈ 336 **35.** $V = Bh = \dfrac{1}{2}aph =$

$\dfrac{1}{2} \cdot \dfrac{x\sqrt{3}}{6} \cdot 3x \cdot x = \dfrac{1}{4}x^3\sqrt{3}$ **39.** 6 cm

Written Exercises, Pages 485–487

1. 6; $\sqrt{34}$ **3.** 25; $\sqrt{674}$ **5.** 3; $\sqrt{41}$ **7.** 36   **9.** 192   **11.** 60; 96; 48   **13.** 260; 360; 400

15. 6 cm   **17. a.** 15 cm; 13 cm   **b.** 384 cm^2 (V-ABCD is not reg.)   **19.** Vol. pyr. $= \dfrac{1}{6} \cdot$ vol. rect. solid

21. 8; $\sqrt{73}$ **23. a.** 3; 6; $6\sqrt{3}$ **b.** $45\sqrt{3}$; $36\sqrt{3}$ **25.** 144; $24\sqrt{39}$ **27.** ≈ 66 cubic units   **29.** $\dfrac{x^3\sqrt{2}}{12}$

31. 246

Mixed Review Exercises, Page 487

1. 12π; 36π **2.** 22π; 121π **3.** π; $\dfrac{\pi}{4}$ **4.** $6\pi\sqrt{3}$; 27π **5.** 5; 25π **6.** 9; 81π **7.** 7; 14π

8. $\sqrt{15}$; $2\pi\sqrt{15}$ **9. a.** 144π mm^2 **b.** 576 mm^2 **10. a.** 32   **b.** $8\pi\sqrt{2}$

Written Exercises, Pages 492–495

1. 40π; 72π; 80π **3.** 24π; 56π; 48π **5.** 4   **7.** 48π **9.** 5; 20π; 36π; 16π
11. 5; 156π; 300π; 240π **13.** 12; 9; 324π; 432π **15.** 8; 17; 255π; 480π **17. a.** 1:4 **b.** 1:4
c. 1:8   **19.** 1:3   **21.** 24 cm   **23.** 25 min   **25.** 2   **27. a.** cyl. with $r = 6$, $h = 10$; $V = 360\pi$
b. cyl. with $r = 10$, $h = 6$; $V = 600\pi$ **29. a.** $270\sqrt{3}$; $180\sqrt{3}$ **b.** 720; $240\sqrt{2}$ **c.** $540\sqrt{3}$; 360
31. cyl. with $r = s$, $h = s$; $V = \pi s^3$ **33.** 16π cm^3 **35.** $18\pi\sqrt{2}$ cm^3 **37.** 60π; $18\sqrt{91}$ **39.** 1200π

Self-Test 1, Page 496

1. 162; 322; 360   **2.** 624; 1200; 960   **3.** 140π in.^{2}; 340π in.^{2}; 700π in.^{3} **4.** 180 cm^2; $(180 + 108\sqrt{3})$ cm^2; $270\sqrt{3}$ cm^3 **5.** 135π; 216π; 324π **6.** 8000 m^3 **7.** 6   **8.** 6:5

Written Exercises, Pages 500–502

1. 196π; $\dfrac{1372\pi}{3}$ **3.** π; $\dfrac{\pi}{6}$ **5.** 4; $\dfrac{256\pi}{3}$ **7.** 8π; $\dfrac{8\pi\sqrt{2}}{3}$ **9.** 4, 8   **11.** 1 cm   **13.** 21π cm^2

15. Vol. of hemisphere = 4 · Vol. of sphere   **17.** 358 million km^2 **19.** $\dfrac{1750\pi}{3}$ m^3 **21.** 6 cans

23. a. 32 cm   **b.** cone: $8\pi\sqrt{1088} \approx 829$; sphere: $4\pi \cdot 8^2 \approx 804$ **25.** $2\pi r^3$ **27.** $\dfrac{4}{3}\pi r^3$

29. 81π in.^{2}; 121.5π in.^{3} **31. a.** $h = 2x$; $r^2 = 10^2 - x^2 = 100 - x^2$; $V = \pi r^2 h = \pi(100 - x^2)(2x) = 2\pi x(100 - x^2)$ **b.** $\dfrac{4000\pi\sqrt{3}}{9}$ **33.** 144π cm^2

Mixed Review Exercises, Page 507

1. $\dfrac{2}{3}$ **2.** 10   **3.** $x = 15$; $y = 6$; $z = 9$ **4. a.** 32; 48   **b.** $\dfrac{2}{3}$ **c.** They are both $\dfrac{2}{3}$.   **5. a.** 48; 108

b. $\dfrac{4}{9}$ **c.** $\dfrac{4}{9} = \left(\dfrac{2}{3}\right)^2$ **6. a.** True   **b.** False   **c.** True   **d.** False   **e.** False   **f.** True   **g.** True   **h.** False

Written Exercises, Pages 511–513

1. Yes   **3. a.** 3:4   **b.** 3:4   **c.** 9:16   **d.** 27:64   **5. a.** 4:1   **b.** 16:1   **c.** 64:1   **7. a.** 2:3   **b.** 2:3
c. 4:9   **9.** Paint for actual airplane = 40,000 times paint for model   **11.** 81π cm^2 **13.** 18.5 kg
15. the larger ball   **17.** 108 ft^3 **19. a.** 9:16   **b.** 9:16   **c.** 9:7   **d.** 27:64   **e.** 27:37

21. 54 cm^3; 196 cm^3 **23.** $\dfrac{4}{3}\pi a^3 : \dfrac{4}{3}\pi b^3 = a^3 : b^3$ **25.** $r_1 : r_2 = l_1 : l_2$;

L.A.$_1$: L.A.$_2$ = $\pi r_1 l_1 : \pi r_2 l_2 = r_1^2 : r_2^2$ **27.** $B_1 : B_2 = e_1^2 : e_2^2$; $h_1 : h_2 = e_1 : e_2$;
$V_1 : V_2 = B_1 h_1 : B_2 h_2 = e_1^2 e_1 : e_2^2 e_2 = e_1^3 : e_2^3$ **29.** $6\sqrt[3]{4}$

Self-Test 2, Page 513

1. 36π cm^2; 36π cm^3 **2.** 16π m^2 **3.** $\dfrac{22{,}000\pi}{3}$ cm^3 **4.** 25π cm^2 **5. a.** $\dfrac{16}{3}$ **b.** 9:4   **6. a.** 2:5

b. 8:125

Extra, Page 517

1. 22   **3.** $\dfrac{56\pi}{3}$ **5.** 8 cm^2

Chapter Review, Pages 518–519

1. lateral edge   **3.** 236; 240   **5.** $\dfrac{160\sqrt{3}}{3}$ **7.** 900; 1020; 17   **9.** 24π; 56π **11.** $2\sqrt{10}$ cm

13. 616 m^2 **15.** $\dfrac{5324\pi}{3}$ cm^3 **17.** 1:9   **19.** 64:27

Preparing for College Entrance Exams, Page 520
1. A   **2.** C   **3.** E   **4.** D   **5.** B   **6.** C   **7.** B

Cumulative Review, Page 521
1. True   **3.** False   **5.** False   **7.** False   **9.** True   **11.** Key steps of proof: 1. $\triangle WXY \cong \triangle YZW$ (HL)   2. $\angle XYW \cong \angle ZWY$ (CPCT)   3. $\overline{WZ} \parallel \overline{XY}$ (If alt. int. $\angle$s $\cong$, lines $\parallel$.)   **13. a.** AA$\sim$
b. If $\triangle JKL \sim \triangle XYZ$, then $\angle J \cong \angle X$ and $\angle K \cong \angle Y$. True   **15.** 8   **17.** 8   **19.** 125π cm^3

Chapter 13

Written Exercises, Pages 526–528

1. 7   **3.** 4   **5.** $\sqrt{10}$ **7.** 10   **9.** $2\sqrt{13}$ **11.** $5\sqrt{2}$ **13.** Yes; A **15.** No   **17.** $(-3, 0); 7$
19. $(j, -14); \sqrt{17}$ **21.** $(x - 3)^2 + y^2 = 64$ **23.** $(x + 4)^2 + (y + 7)^2 = 25$ **27.** $AM = AY = 3\sqrt{5}$
29. $\dfrac{JA}{RF} = \dfrac{6}{3} = \dfrac{2}{1}, \dfrac{AN}{FK} = \dfrac{2\sqrt{5}}{\sqrt{5}} = \dfrac{2}{1}, \dfrac{JN}{RK} = \dfrac{4\sqrt{2}}{2\sqrt{2}} = \dfrac{2}{1}$. The $\triangle$ are $\sim$ by SSS $\sim$ Thm.   **31.** 39   **33.** $(10, 0)$,
$(6, 8), (8, 6), (0, 10), (-6, 8), (-8, 6), (-10, 0), (6, -8), (8, -6), (0, -10), (-6, -8), (-8, -6)$
35. $x^2 + (y - 6)^2 = 100$ **37.** $x^2 + (y - 2)^2 = 9$ **39. a.** 5; 10   **b.** 15   **c.** Dist. between ctrs. = sum
of radii.   **41.** Quad. $RAYJ$ is a $\square$.   **43.** 15, 5, 2, -1, -11 **45.** $(-2, 4); 6$

Written Exercises, Pages 532–534

1. a. k **b.** n, r **c.** l, x-axis   **d.** s, y-axis   **3.** 1   **5.** -1 **7.** -1 **9.** 0   **11.** $\dfrac{3}{2}$

13. $-\dfrac{2}{5}; 2\sqrt{29}$ **15.** $\dfrac{3}{4}; 10$ Answers will vary in Exs. 17–19. Examples are given.

17. $(-8, -2), (2, 2)$ **19.** $(-4, -4), (4, -6)$ **21.** Slope of $\overline{PQ}$ = slope of $\overline{QR}$ = $-\dfrac{1}{3}$ **23.** 9

25. $m(r - p) + q$ **27. a.** SAS   **b.** $m\angle BOS = m\angle BOR + m\angle ROS = m\angle BOR + m\angle AOB = 90$
c. -1 **29. a.** $RS = \sqrt{58}; RT = 2\sqrt{2}; ST = 5\sqrt{2}$ **b.** $(RT)^2 + (ST)^2 = 8 + 50 = 58 = (RS)^2$ **c.** -1
31. 1   **33.** $a = 2\sqrt{3} + 1$

Algebra Review, Page 534

1. -216 **3.** $\dfrac{1}{9}$ **5.** $-\dfrac{1}{64}$ **7.** $\dfrac{27}{125}$ **9.** 1   **11.** 2   **13.** r^{13} **15.** r^5 **17.** 1   **19.** b^8
21. $6y^6$ **23.** $5b^4$

Written Exercises, Pages 537–538

1. a. $\dfrac{2}{3}$ **b.** $\dfrac{2}{3}$ **c.** $-\dfrac{3}{2}$ **3.** $\dfrac{7}{2}, \dfrac{7}{2}; 0; 0$ **5. a.** $-3; -3$ **b.** Slope of $\overline{LM}$ = slope of $\overline{PN}$ **c.** $\dfrac{1}{3}; -\dfrac{1}{7}$

d. Slope of $\overline{MN} \neq$ slope of $\overline{LP}$ **e.** trap.   **7.** Slope of $\overline{AC} = -\dfrac{5}{2}$, slope of $\overline{AB} = \dfrac{3}{7}$, slope of $\overline{BC} = \dfrac{8}{5}$; slope

of alt. to $\overline{AC} = \dfrac{2}{5}$, slope of alt. to $\overline{AB} = -\dfrac{7}{3}$, slope of alt. to $\overline{BC} = -\dfrac{5}{8}$ **9.** Slope of $\overline{RS} = \dfrac{6}{5}$, slope of $\overline{ST} =$

$-\dfrac{5}{6}; \dfrac{6}{5}\left(-\dfrac{5}{6}\right) = -1$ **11. a.** Slope of $\overline{AB} = \dfrac{2 - (-4)}{4 - (-6)} = \dfrac{3}{5}$, and slope of $\overline{DC} = \dfrac{8 - 2}{6 - (-4)} = \dfrac{3}{5}; \overline{AB} \parallel \overline{DC}$.

Slope of $\overline{AD} = \dfrac{2 - (-4)}{-4 - (-6)} = 3$, slope of $\overline{BC} = \dfrac{8 - 2}{6 - 4} = 3; \overline{AD} \parallel \overline{BC}$.   **b.** $AB = 2\sqrt{34} = DC, AD = 2\sqrt{10} =$

BC **13. a.** Slope of $\overline{RS}$ = slope of $\overline{UT} = \dfrac{4}{3}$; slope of $\overline{RU}$ = slope of $\overline{ST} = -\dfrac{3}{4}; RSTU$ is a $\square$. $\overline{RS} \perp \overline{RU}$, so

$RSTU$ is a rect.   **b.** $RT = US = 5\sqrt{5}$ **15.** trap.   **17.** rect.   **19.** $\dfrac{3}{4}$ **21. a.** True   **b.** True   **c.** True

Written Exercises, Pages 541–543

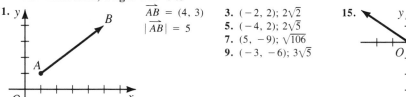

1. $\overrightarrow{AB} = (4, 3)$ **3.** $(-2, 2); 2\sqrt{2}$ **15.**
$|\overrightarrow{AB}| = 5$ **5.** $(-4, 2); 2\sqrt{5}$
7. $(5, -9); \sqrt{106}$
9. $(-3, -6); 3\sqrt{5}$

17. 9   **19.** -12 **21.** $(7, 0)$ **23.** $(4, 4)$ **25.** $(4, 2)$ **27.** $(6, 8)$; 10   **29. a.** $(18, 18)$, $(9, 9)$, $(6, 6)$ **b.** $(11, 12)$, $(8, 9)$ **31.** $|(ka, kb)| = \sqrt{(ka)^2 + (kb)^2} = \sqrt{k^2(a^2 + b^2)} = |k|\sqrt{a^2 + b^2} = |k| \cdot |(a, b)|$ **33. a.** 1. Def. of vector sum   2. Subst.   3. $k[(a, b) + (c, d)] = k(a, b) + k(c, d)$ 4. Def. of vector sum   **b.** Thm. 5-11

Mixed Review, Page 543

1. -2 **2.** 6; 6   **3.** 16   **4.** $a\sqrt{3}$ **5.** 120   **6.** $2x$; $x\sqrt{3}$ **7.** 45   **8.** $(-3, 5)$ **9.** 25   **10.** 18   **11. a.** $(DE)^2 + (EF)^2 = 25 + 100 = 125$; $(DF)^2 = 121 + 4 = 125$ **b.** Slope of $\overline{DE}$ · slope of $\overline{EF} = -\dfrac{4}{3} \cdot \dfrac{3}{4} = -1$ **12. a.** $\dfrac{2}{3}$ **b.** $\dfrac{1}{4}$

Written Exercises, Pages 545–547

1. $(3, 3)$ **3.** $(0, -2)$ **5.** $(1.9, 0.4)$ **7.** $2\sqrt{41}$; $\dfrac{-5}{4}$; $(-1, -3)$ **9.** 17; $-\dfrac{15}{8}$; $\left(-3, \dfrac{7}{2}\right)$ **11.** $(9, 5)$

13. 1. The midpt. of $\overline{AB}$ is $M(4, 2)$. Slope of $\overline{AB} = \dfrac{4 - 0}{8 - 0} = \dfrac{1}{2}$, slope of $\overline{PM} = \dfrac{2 - 6}{4 - 2} = -2$; $\dfrac{1}{2}(-2) = -1$, so $\overline{PM} \perp \overline{AB}$.   2. $PA = 2\sqrt{10} = PB$, so P is on the $\perp$ bis. of $\overline{AB}$.   **15.** $\left(-\dfrac{5}{2}, \dfrac{1}{2}\right)$, $\left(\dfrac{11}{2}, \dfrac{1}{2}\right)$; 8

17. a. $\left(\dfrac{9}{2}, \dfrac{9}{2}\right)$ **b.** $\square$ **c.** slope of $\overline{PQ} = $ slope of $\overline{OR} = \dfrac{3}{7}$; slope of $\overline{PO} = $ slope of $\overline{QR} = 3$ **d.** $PQ = OR = \sqrt{58}$; $PO = QR = 2\sqrt{10}$ **19. a.** $(-3, 4)$ **b.** 5; 5; 5   **c.** Thm. 5-15   **d.** $(x + 3)^2 + (y - 4)^2 = 25$

21. a. $J\left(-\dfrac{1}{2}, \dfrac{3}{2}\right)$, $K(3, 6)$, $L\left(\dfrac{17}{2}, \dfrac{9}{2}\right)$; $M(5, 0)$ **b.** rhom.; $JK = KL = LM = JM = \dfrac{\sqrt{130}}{2}$

23. $\left(\dfrac{5}{8}x_1 + \dfrac{3}{8}x_2, \dfrac{5}{8}y_1 + \dfrac{3}{8}y_2\right)$

Self-Test 1, Page 547

1. a. 2   **b.** $(4, 1)$ **2. a.** 10   **b.** $(4, -3)$ **3. a.** $10\sqrt{2}$ **b.** $(3, 2)$ **4. a.** $\sqrt{29}$ **b.** $\left(-4, \dfrac{9}{2}\right)$

5. $x^2 + y^2 = 81$ **6.** $(x + 1)^2 + (y - 2)^2 = 25$ **7.** $(-2, 3)$; 6   **8.** $\dfrac{4}{7}$ **9.** $-\dfrac{3}{5}$ **10.** vertical

11. a. 2   **b.** $-\dfrac{1}{2}$ **12. a.** $(6, -2)$ **b.** $(-3, -3)$ **c.** $(0, 4)$ **13. a.** $2\sqrt{10}$ **b.** $3\sqrt{2}$ **c.** 4

14. a. $(4, -9)$ **b.** $(22, -9)$ **15.** $(-15, 18)$

Written Exercises, Pages 550–552

1.

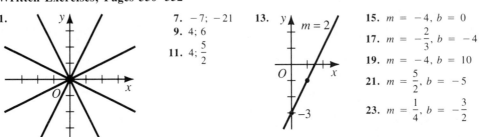

7. -7; -21 **9.** 4; 6   **11.** 4; $\dfrac{5}{2}$

13.

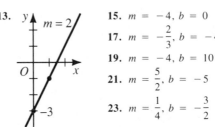

15. $m = -4$, $b = 0$ **17.** $m = -\dfrac{2}{3}$, $b = -4$ **19.** $m = -4$, $b = 10$ **21.** $m = \dfrac{5}{2}$, $b = -5$ **23.** $m = \dfrac{1}{4}$, $b = -\dfrac{3}{2}$

25. $(1, 2)$ **27.** $(4, 3)$ **29.** $(2, -3)$ **31. a.** Both have slope -2.   **b.** No   **c.** There is no sol.

33. a. 2; $-\dfrac{1}{2}$ **b.** They are $\perp$; 2 nonvert. lines are $\perp$ iff the prod. of their slopes is -1.   **35. b.** $(2, -1)$, $(-1, 5)$, $(-4, -4)$ **c.** $22\dfrac{1}{2}$ **37.** $(3, 4)$, $(-5, 0)$

Written Exercises, Pages 555–556

1. $y = 2x + 5$ **3.** $y = \dfrac{1}{2}x - 8$ **5.** $y = -\dfrac{7}{5}x + 8$ **7.** $y = -\dfrac{1}{4}x + 2$ **9.** $y = \dfrac{1}{2}x + 4$

11. $y - 2 = 5(x - 1)$ **13.** $y - 5 = \dfrac{1}{3}(x + 3)$ **15.** $y = -\dfrac{1}{2}(x + 4)$ **17.** $y = 2x - 1$

19. $y = \dfrac{1}{3}x + 2$ **21.** $x = 2$ **23.** $x = 5$ **25.** $y - 7 = 3(x - 5)$ **27.** $5x + 8y = -31$

29. $y = -\dfrac{5}{3}x + \dfrac{34}{3}$ **31.** $y = x$ **33.** $\dfrac{2}{3}$ or $-\dfrac{2}{3}$ **35.** (2, 0) **37. a.** $y = x,\ x = 3,\ x + 2y = 9$
b. $C(3,\ 3)$ **c.** $CQ = CR = CS = 3\sqrt{10}$ **d.** $(x - 3)^2 + (y - 3)^2 = 90$ **39. a.** slope of $\overline{CG} = -1 =$
slope of $\overline{GH}$ **b.** $GH = 2\sqrt{2},\ GC = \sqrt{2}$

Written Exercises, Pages 558–559

1. $(0, b),\ (a, 0)$ **3.** $(-f, 2f),\ (f, 2f)$ **5.** $(h + m, n)$ **7.** $\left(\dfrac{s}{2},\ \dfrac{s\sqrt{3}}{2}\right)$ **9.** $(\sqrt{a^2 - b^2},\ b),$
$(\sqrt{a^2 - b^2} + a,\ b)$

Written Exercises, Pages 562–563

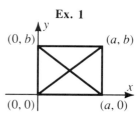

Ex. 1

1. Plan for proof: Use the distance formula twice to show that the length of each diag. is $\sqrt{a^2 + b^2}$. **5.** Plan for proof: Find the coords. of the midpts. G and H of $\overline{NP}$ and $\overline{MO}$, resp. Use slopes to show that $\overline{NM} \parallel \overline{GH} \parallel \overline{OP}$. Use the Dist. Formula to show that $GH = \dfrac{1}{2}(OP - NM)$. **9.** Plan for proof: Use the eq. of the $\odot$ to show that $b^2 - a^2 = -c^2$. Then use slopes to show that $\overline{CA} \perp \overline{CB}$.

Self-Test 2, Page 563

1. $\dfrac{2}{5};\ -4$ **2.**

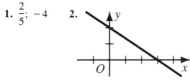

3. $y = -\dfrac{1}{2}x + \dfrac{5}{2}$ **4.** $y = 5$ **5.** $(1,\ -1)$ **6.** $(2e,\ 0)$
7. $(c + g,\ h)$ **8.** $(c - g,\ h)$ **9.** slope of $\overline{GO} =$
slope of $\overline{LD} = -\dfrac{1}{4};$ slope of $\overline{OL} =$ slope of $\overline{DG} = 3$

Extra, Page 565
1. y-axis **3.** x-axis **5.** xy-plane **7.** yz-plane

Chapter Review, Page 567

1. $4\sqrt{5};\ 10;\ 2\sqrt{5}$ **3.** $(-3, 0);\ 10$ **5.** $(x + 6)^2 + (y + 1)^2 = 9$ **7.** -19 **9.** 0 **11.** $\dfrac{3}{4},\ -\dfrac{4}{3}$

13. a. $(4, 3)$ **b.** 5 **c.** $(-8, -6)$ **15.** $\left(4, -\dfrac{3}{2}\right)$ **17.** $(0, b)$ **19.**

21. $(2, 1)$ **23.** $y = 2x + 4$ **25.** $M\left(\dfrac{a}{2} + b,\ c\right),\ N\left(\dfrac{a}{2},\ 0\right);$
slope of $\overline{ON} = 0 =$ slope of $\overline{MQ},\ \overline{ON} \parallel \overline{MQ};$ slope of $\overline{OM} =$
$\dfrac{2c}{a + 2b} =$ slope of $\overline{NQ},\ \overline{OM} \parallel \overline{NQ}$

Cumulative Review, Page 569
1. obtuse **3.** No; no; draw a fig. in which the diags. do not bis. each other. **5. a.** Since $\angle AEB \cong \angle CED$
(Vert. $\angle$ $\cong$.) and $\dfrac{AE}{CE} = \dfrac{BE}{DE},\ \triangle AEB \sim \triangle CED$ (SAS~); therefore $\angle B \cong \angle D$ (Corr. $\angle$ of $\sim$ $\triangle$ are $\cong$.).
b. $x = 18$ **c.** $4{:}9$ **7. a.** $\dfrac{1}{3}$ **b.** $\dfrac{1}{3}$ **c.** $2\sqrt{2}$ **d.** $\dfrac{2\sqrt{2}}{3}$ **9.** $364\pi;\ 820\pi$ **11.** $6\dfrac{2}{3}$ **13.** 89

15. Given: $\overrightarrow{BD}$ bis. $\angle ABC$; $\overrightarrow{BD} \perp \overline{AC}$
  Prove: $\triangle ABC$ is isos.
Key steps of proof: 1. $\triangle ABD \cong \triangle CBD$ (ASA)
  2. $\overline{AB} \cong \overline{CB}$ (CPCT)
17. Plan for proof: Show that the slopes of the bases $= 0$, and that the slope of the median $= 0$.

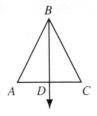

Chapter 14

Written Exercises, Pages 574–576
1. 33, 4 **3.** 10, 10; no **5. a.** $A'(4, 2)$, $B'(8, 4)$, $C'(6, -2)$ **b.** Yes **c.** $(8, 8)$ **7. a.** $A'(0, 12)$, $B'(12, 18)$, $C'(6, 0)$ **b.** No **c.** $(4, 2)$ **9. a.** $A'(12, 4)$, $B'(8, 6)$, $C'(10, 0)$ **b.** Yes **c.** $(0, 6)$
11. a. Yes **b.** preserves **c.** No **13.** Let X be int. of $\overleftrightarrow{AC}$ and $\overleftrightarrow{DB}$. P on $\overline{DC}$ maps to pt. P', where $\overrightarrow{PX}$ int. $\overline{AB}$. Not an isom. **15.** $A'(5, 3)$, $B'(7, 3)$, $C'(9, 5)$, $D'(7, 5)$. Area $ABCD = 4$, perimeter $ABCD = 8$; Area $A'B'C'D' = 4$, perimeter $A'B'C'D' = 4 + 4\sqrt{2}$ **17.** Yes **19.** No **21.** If A and B are diff. pts., $AB > 0$. But $A'B' = 0$. **23. a.** $A'(-6, 1)$, $B'(-3, 4)$, $C'(-1, -3)$

Written Exercises, Pages 580–582
1. 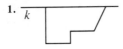 **3.** 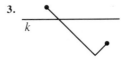 **5.**   **7. a.** $(2, -4)$ **b.** $(-2, 4)$ **c.** $(4, 2)$ **9. a.** $(0, 2)$ **b.** $(0, -2)$ **c.** $(-2, 0)$ **11. a.** $(-3, 2)$ **b.** $(3, -2)$ **c.** $(-2, -3)$

13. Examples: WOW, AHA. **15.** An isos. $\triangle$ with m the $\perp$ bis. of the base. **17.** If P is not on plane X, then X is $\perp$ to and bis. $\overline{PP'}$. If P is on plane X, $P' = P$. **19.** Let X and Y be the pts. where $\overline{PP'}$ and $\overline{QQ'}$ int. m, resp. $\triangle XYQ \cong \triangle XYQ'$, so $XQ = XQ'$ and $\angle QXY \cong \angle Q'XY$. Then $\angle PXQ \cong \angle P'XQ'$ and since $XP = XP'$, $\triangle XPQ \cong \triangle XP'Q'$ by SAS. Then $PQ = P'Q'$. **21.** Const. k, the $\perp$ to t through A, int. t at P; const. $\overline{PA'}$ on k so that $AP = PA'$; $A' = R_t(A)$. **23.** Yes **25.** Path from B to H hits walls first at X, then at Y. Because reflect. is isom., $YH = YH'$, $XH = XH''$. Thus $BX + XY + YH = BX + XY + YH' = BX + XH' = BX + XH'' = BH''$. **27.** Aim for image of hole under reflect. in two walls, as in Ex. 25. **29.** Yes

31. $y = -x - 5$ **33.** $x = 2$ **35.** $y = x$ **37.** $y = -\dfrac{5}{3}x + 6$ **39. a.** $(6, 3)$ **b.** $(10, -2)$

c. $(13, 1)$ **d.** $(10 - x, y)$

Written Exercises, Pages 586–587
1. a. $A'(-4, 6)$, $B'(-2, 10)$, $C'(1, 5)$; yes **c.** Yes; yes **3.** $(8, 4)$ **5.** $(-4, -3)$ **7.** $A'(1, 4)$, $B'(4, 6)$, $C'(5, 10)$; $A''(-1, 4)$, $B''(-4, 6)$, $C''(-5, 10)$ **9.** $(-x, y + 4)$ **11.** a, b, c, d
13. $(x - 4, y + 9)$ **15. a.** $A'(2, -2)$, $B'(3, 2)$ **b.** $\square$; $10 + 2\sqrt{17}$ **17.** The midpts. of $\overline{AA'}$, $\overline{BB'}$, and $\overline{CC'}$ lie on the reflecting line. **19.** Let translation T map P to P' and Q to Q'. Let reflection R_k map P' to P'' and Q' to Q''. Since T and R_k are isom., $PQ = P'Q'$ and $P'Q' = P''Q''$. By trans. $PQ = P''Q''$, so the glide reflection is also an isom.

Written Exercises, Pages 590–592
Answers may vary in Exs. 1–5. **1.** $\mathscr{R}_{O, 440}$ **3.** $\mathscr{R}_{A, 90}$ **5.** $\mathscr{R}_{O, 180}$ **7.** C **9.** E **11.** D **13.** D
15. rotation **17.** half-turn **19.** rotation **21.** reflection **23.** 6 **25.** a, b, c, d
27.   **29.**   **31.** Const. the $\perp$ bis. of $\overline{AA'}$ and $\overline{BB'}$. They int. at O.

33. b. $A'(3, 0)$, $B'(1, -4)$ **c.** slope of $\overleftrightarrow{AB} = -\dfrac{1}{2}$, slope of $\overleftrightarrow{A'B'} = 2$; the lines are $\perp$ **d.** A rotation is an isom. **e.** An isom. maps any $\triangle$ to a $\cong \triangle$. **f.** $(y, -x)$

35. Extend $\overrightarrow{OF}$ to int. l' at G and let H be the int. of l and l'. $m \angle F'GO = 90 - x$ so $m \angle GHF = 90 - (90 - x) = x$.   **37. a.** $\mathcal{R}_{C,\,90}$ **b.** $\overline{AD}$ is the image of $\overline{BE}$ under an isom.   **c.** If a rotation of 90° maps $\overline{BE}$ to $\overline{AD}$, then one of the $\angle$ between $\overline{BE}$ and $\overline{AD}$ has meas. 90. (Result from Ex. 35.)   **39.** Locate X and Z as you did B and C in Ex. 38, using $\mathcal{R}_{A,\,90}$ instead of $\mathcal{R}_{A,\,60}$. With ctrs. X and Z and radius AX, draw arcs int. at Y.

Mixed Review Exercises, Page 592

1. $ODE,\ OFG$ **2.** $2{:}3$ **3.** $x = \dfrac{9}{2},\ y = \dfrac{10}{3},\ z = 5,\ w = \dfrac{9}{2}$ **4.** $3{:}5$ **5.** $4{:}9$ **6.** $9{:}25$ **7.** $4{:}25$

Written Exercises, Pages 596–597

1. $A'(12, 0),\ B'(8, 4),\ C'(4, -4)$ **3.** $A'(3, 0),\ B'(2, 1),\ C'(1, -1)$ **5.** $A'(-12, 0),\ B'(-8, -4),$ $C'(-4, 4)$ **7.** $A'(6, 0),\ B'(7, -1),\ C'(8, 1)$ **9.** 4; expansion   **11.** $\dfrac{1}{3}$; contraction   **13.** 4; expansion   **15.** b, d   **17.** a, b, c   **19.** $3{:}2,\ 9{:}4$ **21.** $2{:}1,\ 4{:}1$ **23. a.** $16{:}9$ **b.** $64{:}27$ **25. a.** Slope of $\overline{PQ} = \dfrac{y_2 - y_1}{x_2 - x_1} =$ slope of $\overline{P'Q'}$.   **b.** ∥   **27.** $(4, 2),\ k = 3$

Self-Test 1, Page 597

1. An isom. is a one-to-one mapping from the whole plane onto the whole plane that maps every seg. to a ≅ seg.   **2.** $-1, 3$ **3.** $(1, -2),\ (-1, 2)$ **4. a.** $(3, -5)$ **b.** $(-3, 5)$ **c.** $(5, 3)$ **5.** a, b   **6.** Answers may vary; for example, $\mathcal{R}_{O,\,330}$ **7.** B **8.** C **9.** $\overline{AB}$ **10.** $\overline{OB}$ **11.** M **12.** $\overline{AO}$ **13.** L **14.** NDO **15.** C **16.** Q **17.** C **18.** L

Written Exercises, Pages 603–605

1. a. 1   **b.** $2x^2 - 7$ **c.** 9   **d.** $(2x - 7)^2$ **3. a.** 8   **b.** 27   **c.** $\left(\dfrac{x + 1}{2}\right)^3$ **d.** 14   **e.** 63   **f.** $\dfrac{x^3 + 1}{2}$ **11. a.** Q **b.** S **c.** M **d.** Q **e.** Q **13.** b   **15.** a, b, c, d   **17.** $(-3, -1)$ **19.** $(9, 2)$ **21.** $(4, -8)$ **23.** $(3, -3)$ **25. a.** $Q(-2, 5)$ **b.** 90   **c.** slope of $\overline{OP} = \dfrac{2}{5}$, slope of $\overline{OQ} = -\dfrac{5}{2}$; $\dfrac{2}{5}\left(-\dfrac{5}{2}\right) = -1$ **d.** $(-y, x),\ (y, -x)$ **27.** Construct B' so that k is the ⊥ bisector of $\overline{BB'}$. Construct line j, the ⊥ bis. of $\overline{AB'}$.   **29.** translation

Written Exercises, Pages 607–608

1. $\dfrac{1}{4}$ **3.** $\dfrac{3}{2}$ **5.** C **7.** A **9.** C **11.** A **13.** C **15.** I **17.** H_O **19.** $(x + 6, y - 8)$ **21.** $S^{-1}{:}(x, y) \to (x - 5, y - 2)$ **23.** $S^{-1}{:}(x, y) \to \left(\dfrac{1}{3}x, -2y\right)$ **25.** $S^{-1}{:}(x, y) \to \left(x + 4, \dfrac{1}{4}y\right)$ **27.** $T{:}(x, y) \to \left(x + 2, y - \dfrac{1}{2}\right)$ **29. a.** 2 units rt., 2 units left

Written Exercises, Pages 612–614

1. a. 5   **b.** No   **c.** $\mathcal{R}_{O,\,72},\ \mathcal{R}_{O,\,144},\ \mathcal{R}_{O,\,216},\ \mathcal{R}_{O,\,288}$ **3. a.** 4   **b.** Yes   **c.** $\mathcal{R}_{O,\,90},\ \mathcal{R}_{O,\,180},\ \mathcal{R}_{O,\,270}$
5. A, B, C, D, E, K, M, T, U, V, W, Y   **7.** H, I, N, O, S, X, Z

9.   **11.**   **13.**   **15.**   **17.**

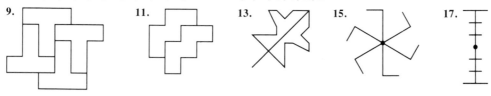

19. a. The ellipse has line symm. about two axes and pt. symm. about the int. of the axes.   **b.** If $a = b$, the ellipse becomes a ⊙ and the solid formed is a sphere with vol. $\dfrac{4}{3}\pi a^3$.   **c.** The ellipsoid has plane symm. about the

inf. many planes that contain b; plane symm. about the single plane that contains a and is $\perp$ to b; inf. many rot. symmetries about b; 180° rot. symm. about the inf. many lines $\perp$ to b and containing the int. of the axes.
21. a. non-isos. trap. **b.** isos. trap. **c.** not poss. **23. a.** reg. octagon **b.** **c.**
25. a. at the midpts. of the sides **b.** translational

Self-Test 2, Page 615
1. A **2.** A **3.** B **4.** P **5.** y-axis **6.** No **7.** P **8.** T **9.** translation **10. a.** $D_{O, \frac{1}{8}}$
b. $\mathcal{R}_{O, 70}$ **c.** R_y **d.** $S^{-1}:(x, y) \to (x - 2, y + 3)$

Extra, Pages 616–617
1. Let t be the $\perp$ bis. of the base. **3.**

$\circ$	I	R_t
I	I	R_t
R_t	R_t	I

$\circ$	I	$\mathcal{R}_{O, 120}$	$\mathcal{R}_{O, 240}$
I	I	$\mathcal{R}_{O, 120}$	$\mathcal{R}_{O, 240}$
$\mathcal{R}_{O, 120}$	$\mathcal{R}_{O, 120}$	$\mathcal{R}_{O, 240}$	I
$\mathcal{R}_{O, 240}$	$\mathcal{R}_{O, 240}$	I	$\mathcal{R}_{O, 120}$

5. a. 2 **b.** 4
9. a. No; there is no identity.
b. Yes; the rot. symmetries
c. I, H_O

Chapter Review, Page 619
1. $\cong$ **3. a.** $(6, 1)$, $\left(\frac{3}{2}, 5\right)$ **b.** No **5.** $y = -2x + 1$ **7.** $(12, 2)$ **9.** a, c **11.** $(1, 3)$
13. $(x + 1, y - 6)$ **15.** I **17.** No **19.** Yes

Preparing for College Entrance Exams, Page 621
1. B **2.** A **3.** E **4.** C **5.** C **6.** E **7.** A **8.** E **9.** B **10.** D

Cumulative Review, Pages 622–625
True-False Exercises **1.** T **3.** T **5.** F **7.** F **9.** F **11.** T **13.** F **15.** T **17.** T
19. F **Multiple-Choice Exercises** **1.** d **3.** b **5.** d **7.** c **9.** c **11.** b

Completion Exercises **1.** Add. Prop. $=$ **3.** 124 **5.** 33 **7.** $-\dfrac{3}{2}$ **9.** $(-5, 2)$ **11.** 22.5

13. $\dfrac{15}{17}$ **15.** $27\sqrt{7}$ **17.** $324\pi, 135\pi$ **19.** $(4, -2)$ **21.** $1:7$ **Always-Sometimes-Never**

Exercises **1.** N **3.** S **5.** A **7.** N **9.** S **11.** A **13.** N **15.** A **17.** S
Construction Exercises **1.** Const. 3 **3.** Const. 11 **5.** Const. $l \perp m$ at A; on m mark off $AB = y$; from B, locate D on l such that $BD = x$; const. $n \perp l$ at D; on n mark off $DC = y$; draw $\overline{BC}$. **7.** On a line, mark off $AB = x$, $BC = x$, $CD = x$. Use Const. 14 to const. the geom. mean of AD and y. **Proof Exercises** **1.** Key steps of proof: 1. $\triangle OQP \sim \triangle OSR$ (AA$\sim$) 2. $\dfrac{PO}{RO} = \dfrac{PQ}{RS}$ (Corr. sides of $\sim \triangle$ are in prop.) **3.** Key steps of proof: 1. $OS = OR$; $OP = OQ$ (If 2 $\angle$ of a $\triangle$ are $\cong$, sides opp. those $\angle$ are $\cong$.) 2. $PR = QS$ (Add. Prop. $=$, Seg. Add. Post.) 3. $\triangle PSR \cong \triangle QRS$ (SAS) **5.** Plan for proof: Let the coords. be $R(-2a, 0)$, $S(2a, 0)$, and $T(0, 2b)$. Midpts. of the segs. are $M(-a, b)$, $N(0, 0)$, and $P(a, b)$. Use the Dist. Form. to show that $NM = NP = \sqrt{a^2 + b^2}$.

Logic

Exercises, Pages 645–646
1. I like the city and you like the country. **3.** You don't like the country.
5. I like the city or you don't like the country. **7.** I don't like the city or you don't like the country. **9.** It is not true that "I like the city or you like the country." **11.** $p \vee q$ **13.** $\sim(p \vee q)$ **15.** $\sim(p \wedge q)$
17. Yes

19.

p	q	$\sim q$	$p \vee \sim q$
T	T	F	T
T	F	T	T
F	T	F	F
F	F	T	T

Exercises, Page 647

1. If you like to paint, then you are an artist. **3.** If you are not an artist, then you do not draw landscapes. **5.** If you like to paint and you are an artist, then you draw landscapes. **7.** If you draw landscapes or you are an artist, then you like to paint. **9.** $b \to k$ **11.** $(\sim b \vee \sim k) \to s$ **13.** $\sim(b \to s)$ **15. a.** Yes; no **b.** Yes; yes **17.**

p	q	$p \to q$	$\sim(p \to q)$
T	T	T	F
T	F	F	T
F	T	T	F
F	F	T	F

19. $\sim(p \to q)$ and $p \wedge \sim q$

Exercises, Page 649

1. 1. Given 2. Step 1, Simplification 3. Given 4. Steps 2, 3, Modus Ponens **5.** 1. $a \wedge b$ (Given) 2. a (Step 1, Simplification) 3. $a \to \sim c$ (Given) 4. $\sim c$ (Steps 2, 3, Modus Ponens) 5. $c \vee d$ (Given) 6. d (Steps 4, 5, Disj. Syllogism) **7.** Given: $w \to g$; $g \to p$; $w \wedge y$. Prove: p.

Exercises, Pages 650–651

1. a. T, T, F, T **b.** T, T, F, F **c.** T, T, T, T (tautology) **d.** T, T, T, F **5. a.** The sandwich costs $3.50. **b.** Perhaps it's not true that if I have enough money I'll buy milk. Maybe I'm allergic to milk. Or maybe the statement that milk costs a dollar is wrong.

Exercises, Page 652

1. 1. Given 2. Step 1, Double Neg. 3. Given 4. Steps 2, 3, Modus Tollens **5.** 1. $p \vee \sim q$ (Given) 2. $\sim q \vee p$ (Step 1, Comm. Rule) 3. q (Given) 4. $\sim(\sim q)$ (Step 3, Double Neg.) 5. p (Steps 2, 4, Disj. Syllogism) **9.** Given: $c \to t$; $\sim c \to \sim s$; s. Prove: t

Exercises, Page 654

1. $p \wedge r$ **3.** $s \wedge (t \vee p)$ **5.** $(t \vee s) \wedge (\sim t \vee s)$ **7.** Electricity passes through $p \vee \sim p$ but never through $p \wedge \sim p$.

Handbook

Exercises, Pages 658–659

1. The sum is 360. **3. b.** Vert. $\angle$s are $\cong$. **c.** Opp. $\angle$s of a $\square$ are $\cong$. **5. b.** 360 **7.** 360; Thm. 3-14

Exercises, Pages 659–660

5. c. $\angle A \cong \angle B$ because the 2 $\angle$s overlap exactly. **7. a.** A trans. maps fig. I to fig. III. **b.** The distance bet. corr. pts. in figures I and III is twice the distance bet. l and m.

Exercises, Pages 661–662

3. No **5.** $S(3, 2)$ **7. b.** $M(1, 3)$, $N(3, 4)$ **c.** slope of $\overline{MN}$ = slope of $\overline{OI} = \frac{1}{2}$; $\overline{MN} \parallel \overline{OI}$; $OMNI$ is a trap.

9. a. slope of $\overline{AB}$ = slope of $\overline{DC} = -\frac{3}{4}$, slope of $\overline{AD}$ = slope of $\overline{BC} = \frac{4}{3}$; Thm. 13-4 **b.** rect.

11. a. $DE = EF = FG = GD = \sqrt{40} = 2\sqrt{10}$ **b.** slope of $\overline{DF} = 1$, slope of $\overline{EG} = -1$ **13–33.** Refer to Sel. Ans. of specified pages.

Exercises, Pages 663–664

1. **3.** Construct the bridge between $(3, 3)$ and $(4, 3)$. **7.** $\sqrt{5}$

Exercises, Pages 664–665

1. about $\frac{2}{3}$; rotation 90° **3.** $\overrightarrow{RS} \parallel \overrightarrow{R'S'}$ **5.** 2; 3 **7.** $y = 2x + 3$ **9.** Refer to Sel. Ans. for p. 527.

Exercises, Pages 666–667

1. rt. **3.** obt. **5. a.** slope of $\overline{AB} = \frac{3}{2}$, slope of $\overline{BC} = -\frac{2}{3}$; $\overline{AB} \perp \overline{BC}$ **b.** $(AB)^2 = 13$, $(BC)^2 = 52$, $(AC)^2 = 65$ **7.** $\mathcal{R}_{N, -90}$ followed by $D_{N, 2}$; 2 **9.** The left fig. has area c^2; the right fig. has area $b^2 + a^2$. This suggests the Pyth. Thm. **11–21.** Refer to Sel. Ans. of specified pages.

Exercises, Pages 668–669

1. a. $(x - 6)^2 + y^2 = 25$ **b.** Yes **c.** No **3. b.** $M(-1, 7)$ **c.** slope of $\overline{OM} = -7$; slope of $\overline{AB} = \frac{1}{7}$; Thm. 13-4 **5.** radius of outer $\odot = \sqrt{225} = 15$, radius of inner $\odot = \sqrt{25} = 5$, dist. bet. ctrs. of $\odot$s is 10; $10 + 5 = 15$ **7. a.** $\odot O$; the other tan. to $\odot O$ from P **b.** $\overline{PA}$ and its image are $\cong$. **c.** Thm. 9-1 Cor. **9.** To find B, rotate C $-60°$ about A. **11–17.** Refer to Sel. Ans. of specified pages.

Exercises, Page 670

3. 1:4 **5.** $|x| + |y| = 10$

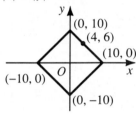

7. Rotate $\odot P$ 90° about A to locate two points, C_1 and C_2, then rotate each $-90°$ about A to get B_1 and B_2.

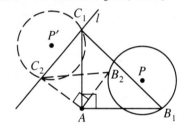

9–15. Refer to Sel. Ans. of specified pages.

Exercises, Pages 671–672

1. 32 **3.** 64 **5. a.** $-60°$, $\frac{1}{2}$ **b.** reg. II; reg. III; reg. IV **c.** $\frac{1}{4}$, $\frac{1}{16}$, $\frac{1}{64}$

Exercises, Pages 674–675

1. Plans for proofs: **a.** Use Thm. 4-5 **b.** Reflect $\triangle ADC$ to $\triangle ABC$. The isom. preserves dist. **c.** Assign coords. $A(0, a)$, $B(b, 0)$, $C(0, -c)$, $D(-b, 0)$ and use the dist. form. **3. a.** Plan for proof: Use SAS to prove $\triangle XBC \cong \triangle ABY$. **b.** $\overline{XC}$ Answers may vary in Exs. 5–17. **5.** square; use a coord. approach. **7.** Use a syn. or coord. approach. **9.** Use a syn. approach. Draw $\overline{XY} \parallel \overline{BC}$ through P with X on $\overline{AB}$ and Y on $\overline{DC}$. **11.** Use a syn. approach. Extend $\overrightarrow{AB}$ and $\overrightarrow{DC}$ to int. at rt. $\angle X$. Then use the Pyth. Thm. **13.** Use a transf. approach. Trans. l toward m a dist. AB. **15.** Use a transf. approach. See the figure at the right. Const. $\parallel$ lines j, k, l. Choose A on k. Rotate l 60° about A to get B on j. Rotate B $-60°$ about A to get C on l.

17. Use a syn. approach. Note that $XN = \frac{1}{2}AB$ and $YN = \frac{1}{2}DC$.

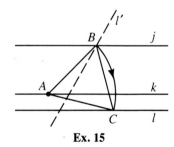

Ex. 15

Acknowledgments

Book designed by Ligature, Inc.
Cover concept and design by Ligature, Inc.
Technical art: Precision Graphics, ANCO/Boston

Cover photographs: John Payne Photo, Ltd., (background) Jim Brandenburg/Westlight

Photographs

xvi *(background)* Astromedia, division of Kalmbach Publishing Co. and The Press Syndicate of the University of Cambridge, *(left)* Jim Richardson/Westlight, *(right)* Chuck O'Rear/Westlight **4** Tad Goodale **5** Paul Von Stroheim **9** Lou Jones **11** Tad Goodale **24** Chuck O'Rear/Westlight **27** Landslides **32** *(left)* Rene Sheret/Marilyn Gartman Agency, *(right)* Hank Morgan/Rainbow **36** *(top)* David Muench, *(bottom)* Yoram Kahana/Peter Arnold, Inc. **43** Seth Resnick/Picture Group **49** *(left)* Margaret Berg/Berg and Associates, *(right)* University of California, Berkeley **54** Tad Goodale **56** Landslides **72** *(background)* courtesy of Spieth-Anderson International Inc. and American Athletic, Inc., *(left)* David Madison, *(right)* Roland Weber/Masterfile **75** Cary Wolinsky/Stock Boston **100** *(top left)* Dan McCoy/Rainbow, *(top right)* Larry Lee/Westlight, *(bottom)* Index Stock **101** The Flag Research Center **104** Shostal **116** *(left)* Robert Carr/Stock Boston, *(right)* America Hurrah **128** Thomas Russell **133** *(left)* FPG, *(middle)* Steve Elmore/Stock Market, *(right)* Swanke, Hayden, Connell Architects **134** Tom and Michelle Grimm/After Image **140** Barbara Burten **166** *(top)* Hans Hofmann, 1961, oil on canvas, 151 × 182 cm, Mr. and Mrs. Frank G. Logan Prize Fund, 1962.775, The Art Institute of Chicago. All Rights Reserved. *(bottom)* Richard Hirneisen/The Stock Shop **171** *(left)* Stephen R. Brown/Stock Market **179** Owen Franken/Stock Boston **190** Rob Outlaw **196** *(top)* Rob Outlaw, *(bottom)* Peter Chapman **202** *(left)* Breck Kent, *(right)* Jim Olive/Uniphoto Picture Agency **209, 210** FPG **213** *(left)* NASA, *(right)* Ken Lax/Stock Shop, *(bottom)* NASA **214** Peter Chapman **224** FPG **240** *(background)* U. S. Geological Survey, *(left)* Kathleen Norris Cook, *(right)* Jim Richardson/Westlight **242** Lee Boltin **244** Peter Chapman **253** *(left)* Gail Page, *(right)* The Dallas Museum of Art, Foundation for the Arts Collection, gift of the James H. and Lillian Clark Foundation **262** Eric Kroll/Taurus **284** *(background)* Bruce Roberts-Goodson, A.M.S.N.A.M.E., *(left)* John Terence Turner/FPG, *(right)* Sharon Green/Sportschrome, Inc. **303** Ed Brunette **304** *(left)* Department of Mathematics, Massachusetts Institute of Technology, *(right)* Sovfoto

315 Dominique Berretty/Black Star **317** Karl Hentz/Image Bank **320** Cameramann International **321** Charles F. Norton/Deck House, Inc. **328** *(left)* Lou Jones, *(right)* Jeffrey L. Cook/The Stock Broker **334** Jonathan Barkin/Picture Cube **338** *(left)* Alfred Borcover, *(right)* Culver Pictures **343** NASA **368** Rothwell/FPG **374** *(background)* Chilton Book Company/Chrysler Plymouth Company, *(left)* Takeshi Takahara/Photo Researchers, *(right)* Jay Freis/Image Bank **379** *(bottom)* courtesy of Cruft Photo Lab, Harvard University/Paul Donaldson, *(left)* Department of the Navy **390** Collection of Whitney Museum of American Art, New York, gift of the Howard and Jean Lipman Foundation, Inc., photo by Geoffrey Clements **400** *(top)* Ellis Herwig/Picture Cube, *(right)* Tom Tracey/Stock Shop, *(left)* Sepp Seitz/Woodfin Camp and Associates **401** Seth Goltzer/Stock Market **403, 409** Richard Haynes, Jr. **410** Brian Milne/Animals, Animals **422** *(left)* Gerrit Rietvald, *Red-Blue Chair*, 1918, Acc #L1884.139, Milwaukee Art Museum, *(right)* Cecile Brunswick/Peter Arnold **440** Wayne Sorce **447** Landslides **452** FPG **467** NASA **474** *(background)* courtesy of Anthony Belluschi Architects, Ltd., *(left)* Tom Grill/Comstock, Inc., *(right)* © Greg Murphy **488** *(top)* M. Timothy O'Keefe/Tom Stack and Associates, *(bottom)* Christopher Crowley/Tom Stack and Associates **497** Van Dusen Corporation **501** Judy Gibbs **505** *(top)* Owen Franken/Stock Boston, *(bottom)* City of Montreal, Archives Division **507** *(left)* Buckminster Fuller Institute, *(right)* DeClan Haun/Black Star **508** FPG **511** Barbara Burten **522** *(background)* U. S. Department of Commerce, NOAA, *(left)* Milton and Joan Mann/Cameramann International, *(right)* James Warren/Westlight **529** Story Litchfield/Stock Boston **533** Richard Haynes, Jr. **535** Grant Haller/Leo deWys, Inc. **539** Dick Luria/Folio Inc. **570** *(background, left)* © Wayne Eastep, *(right)* Tom Campbell/FPG **577** Larry Sutton/FPG **581** Kay Chernush/Image Bank **583** Focus on Sports **588** Braniff/FPG **592** Eric Neurath/Stock Boston **598** IBM **611** *(left)* Philip A. Savoie/Bruce Coleman, Inc., *(right)* Robert P. Carr/Bruce Coleman, Inc. **612** *(top left)* Carl Roessler/Animals, Animals, *(top right)* Bruce Coleman, Inc., *(bottom left)* Steve Solum/Bruce Coleman, Inc., *(bottom right)* David Stone

Rest easy baby, Rest easy!